PLASMONICS – PRINCIPLES AND APPLICATIONS

Edited by **Ki Young Kim**

Plasmonics – Principles and Applications
http://dx.doi.org/10.5772/2633
Edited by Ki Young Kim

Contributors

Paolo Di Sia, Guy A. E. Vandenbosch, Xingyu Gao, Brian Ashall, Dominic Zerulla, Valeria Lotito, Christian Hafner, Urs Sennhauser, Gian-Luca Bona, Henrique T. M. C. M. Baltar, Krystyna Drozdowicz-Tomsia, Ewa M. Goldys, V. A. G. Rivera, F. A. Ferri, O. B. Silva, F. W. A. Sobreira, E. Marega Jr., Yongqi Fu, Jun Wang, Daohua Zhang, Joong Wook Lee, DaiSik Kim, Amir Reza Sadrolhosseini, A. S. M. Noor, Mohd. Maarof Moksin, Taikei Suyama, Akira Matsushima, Yoichi Okuno, Toyonori Matsuda, Yu-Hang Yang, Ta-Jen Yen, K. Sathiyamoorthy, C. Vijayan, V.M. Murukeshan, Young Chul Jun, Mahnaz M. Abdi, Wan Mahmood Mat Yunus, Majid Reayi, Afarin Bahrami, G. Ruffato, G. Zacco, F. Romanato, Witold Jacak, Fuyi Chen, Jian Liu, Negash Alemu, Qin Chen, Xiaohua Shi, Yong Ma, Jin He, Sotirios Papaioannou, Konstantinos Vyrsokinos, Dimitrios Kalavrouziotis, Giannis Giannoulis, Dimitrios Apostolopoulos, Hercules Avramopoulos, Filimon Zacharatos, Karim Hassan, Jean-Claude Weeber, Laurent Markey, Alain Dereux, Ashwani Kumar, Sergey I. Bozhevolnyi, Alpaslan Suna, Oriol Gili de Villasante, Tolga Tekin, Michael Waldow, Odysseas Tsilipakos, Alexandros Pitilakis, Emmanouil E. Kriezis, Nikos Pleros

Published by InTech

Janeza Trdine 9, 51000 Rijeka, Croatia

Copyright © 2012 InTech

Notice

Publishing Process Manager Dragana Manestar
Technical Editor InTech Prepress, Novi Sad
Cover InTech Design team

Additional hard copies can be obtained from orders@intechopen.com

Plasmonics – Principles and Applications, Edited by Ki Young Kim
p. cm.
ISBN 978-953-51-0797-2

For EU product safety concerns:
IN TECH d.o.o., Prolaz Marije Krucifikse Kozulić 3, 51000 Rijeka, Croatia,
info@intechopen.com or visit our website at intechopen.com.

Contents

Preface

Although plasmonic phenomena have been identified more than half century ago, recent accomplishments in this technological area have witnessed remarkable developments in both fundamental principles and quite various practical applications due to the current advancements in computational resources and nanoscale fabrication techniques, elevated needs for novel applications including photovoltaics, subwavelength imaging, optical interconnects, spectroscopy, microscopy, lithography, etc., and recent efforts of combination with conventionally and newly developed physics and techniques such as extraordinary light transmission, metamaterials, graphenes, nanoparticles, terahertz radiation, and many more.

The book attempts to encompass state-of-the-art research in modern plasmonic technologies, resulting in 21 excellent chapters, which has been sub-categorized into 4 sections; (1) Modeling and computational methods for plasmonics, (2) Plasmonic structures for light transmission, focusing, and guiding, (3) Emerging concepts with plasmonics, and (4) Plasmonics applications. Some chapters straddle the border of two or more sections due to the inherent nature of plasmonics research.

This book concerns itself with advanced recent research results on plasmonic technologies covering the aforementioned topics. Although it is a collection of specific technological issues, I strongly believe that the readers can obtain generous and overall ideas and knowledge of the state-of-the-art technologies in plasmonics. I hope the readers will be inspired to start or to improve further their own research and technologies and to expand potential applications, hopefully triggering new interdisciplinary research work. Lastly, special words of thanks should go to all the scientists and engineers who have devoted a great deal of time to writing excellent chapters in this book.

Ki Young Kim

Section 1

Modeling and Computational Methods for Plasmonics

Chapter 1

Modelling at Nanoscale

Paolo Di Sia

Additional information is available at the end of the chapter

http://dx.doi.org/10.5772/50755

1. Introduction

One of the most important aspects at nanoscale concerns the charge transport, which can be influenced by particles dimensions and assumes different characteristics with respect to those of bulk. In particular, if the mean free path of charges, due to scattering phenomena, is larger than the particle dimensions, we have a mesoscopic system, in which the transport depends on dimensions and in principle it is possible to correct the transport bulk theories by considering this phenomenon. A similar situation occurs also in a thin film, in which the smallest nanostructure dimension can be less than the free displacement and therefore requires variations to existing theoretical transport bulk models. Therefore a rigorous knowledge of transport properties is to be acquired. From a theoretical viewpoint, various techniques can be used for the comprehension of transport phenomena, in particular analytical descriptions based on transport equations. Many existing and used theories at today regard numerical approaches, not offering analytical results, which would be of great mathematical interest and in every case suitable to be implemented through the experimental data existing in literature and continuously found by the experimentalists.

To establish the applicability limit of a bulk model and to investigate the time response of systems at nanoscale, recently it has appeared a novel theoretical approach, based on correlation functions obtained by a complete Fourier transform of the frequency-dependent conductivity of the studied system [1]. With this model it is possible to calculate exactly the expressions of the most important connected functions, i.e. the velocities correlation function, the mean free displacement and the diffusion coefficient.

At nanoscale we are in the middle between classical and quantum effects, macroscopic and microscopic properties of the matter [2]. Actually one of the most important experimental technique for the study of the frequency-dependent complex-valued far-infrared photoconductivity $\sigma(\omega)$ is the Time-resolved THz Spectroscopy (TRTS), an ultrafast non-contact optical probe; data are normally fitted via Drude-Lorentz, Drude-Smith or effective medium models [3]. The quantum effects in the nanoworld have opened new ways in a lot of old and new technological sectors.

In the following it is illustrated an overview of the fundamental models that, starting from the Drude model, have attempted to analyze and explain in increasing accuracy the transport phenomena of the matter at solid state level and in particular at nanoscale, until the recent developments concerning the variations of Drude-Lorentz-like models, which involve in particular the concept of plasmon.

2. The Drude model

The most meaningful characteristics of metals are their elevated properties of electric and thermal conduction. Over the years this fact has brought to think in terms of a model in which the electrons are relatively free and can move under the influence of electric fields. Historically two models of the elementary theory of metals are born:

The Drude model, published in 1900 and based on the kinetic theory of an electron gas in a solid. It is assumed that all the electrons have the same average kinetic energy E_m;

The previous model integrated with the foundations of quantum mechanics, called Sommerfeld model.

In the Drude model the valence electrons of atoms are considered free inside the metal; all the electrons move as an electronic gas. It is assumed that all the electrons have the same energy E_m and that it exists a mechanism of collisions among ions and electrons, allowing the thermal equilibrium for the electrons; this fact implies the application of the kinetic theory of gases to such electronic gas. Drude published the theory three years after the discovery of the electron from J. J. Thomson. The free electrons have only kinetic energy, not potential energy; the average energy E_m is therefore $(3/2)k_BT$. It can be correlated to an average quadratic speed v_m from the relation $E_m = (3/2)k_BT = mv_m^2/2$, denoting m the mass of the free electron. At environment temperature it is $v_m \cong 10^7$ cm/s, representing the average thermal speed of the electrons. It was assumed also that the electrons have collisions as instant events, i.e. the time of the diffraction is very smaller with respect to every other considered time. Through such collisions, the electrons acquire a thermal equilibrium corresponding to the temperature T of the metal. If the electrons don´t collide, they move in linear way following the Newton laws. The presence of a constant electric field determines an extra average velocity (the drift velocity) given by $v_d = -(eE/m)t$. The relaxation time τ is defined as the average time between two collisions; it is so possible to get a mean free path, defined by $l_{mfp} = v_m\,\tau$. The current density is $\vec{J}=\sigma_{cond}\vec{E}$, with σ_{cond} electric conductivity. This result, obtained by Drude, is an important goal of the classic theory for the conduction of metals (said "Drude theory").

3. The Drude-Lorentz model

The Lorentz model (1905) is a refining of the Drude model, in which the statistical aspects are specified. The electrons are considered as free charges, with charge "-e"; they are described by a maxwellian velocity distribution. Considering an electron gas in a spatial region with a constant electric field, the drift velocity of the electrons is constant; this corresponds to a current density $\vec{J}$ proportional to the applied field $\vec{J}=\sigma_0\vec{E}$, with $\sigma_0=ne^2\tau/m$ (n is the electron density). Estimating the relaxation time τ, Drude and Lorentz have obtained values of conductivity in good accordance with the experiments. In presence of an electric field of the form $E(t)=E_0 e^{-i\omega t}$, the complex conductivity assumes the form $\sigma_\omega=\sigma_0/(1-i\omega\tau)$. Such model, said "Drude-Lorentz model", has received great success, but has also underlined series difficulties.

4. The most utilized Drude-Lorentz-like models

One of the most utilized models for describing experimental transport data is the Drude-Lorentz model [4,5]; with such model the main transport parameters are obtainable. Starting from the Drude-Lorentz model, it is possible to obtain the velocities correlation function, from this the quadratic average distance crossed by the charges as a function of time and examine directly the possible compatibility with the Einstein relation.

Considerable variations of this model were made in this years; the most used are the following:

1. the "Maxwell-Garnett (MG) model": in this model the dielectric function is given by a Drude term with an additional "vibrational" contribution at a finite frequency ω_o [6], leading to a dielectric function of the form:

$$\varepsilon_{//}(\omega)=1-\frac{\omega_p^2}{\omega(\omega+i/\tau)}+\frac{\omega_s^2}{\omega_0^2-\omega^2-i\gamma\omega} \tag{1}$$

   where the amplitude ω_s, the resonant frequency ω_0 and the damping constant γ are material dependent constants. The MG model usually describes an isotropic matrix containing spherical inclusions isolated from each other, such as the metal particles dispersed in a surrounding host matrix.

2. In the "effective medium theories (EMTs)" the electromagnetic interactions between pure materials and host matrices are approximately taken into account [7]. The commonly used EMTs include the Maxwell-Garnett (MG) model and the "Bruggeman (BR) model", which is a variation of the MG model.

   Generally, in the THz regime, the dielectric function $\varepsilon_m(\omega)$ consists of contributions of the high-frequency dielectric constant, conduction free electrons, and lattice vibration [8]:

$$\varepsilon_m(\omega)=\varepsilon_\infty-\frac{\omega_p{}^2}{\omega^2+i\gamma\omega}+\sum_j\frac{\varepsilon_{st_j}\,\omega_{TO_j}{}^2}{\omega_{TO_j}{}^2-\omega^2-i\Gamma_j\omega} \tag{2}$$

   in which ε_∞ is the high-frequency dielectric constant, the second term describes the contribution of free electrons or plasmons and the last term stands for the optical phonons.

When the response originates mainly from the contribution of free electrons or plasmons, it is usually adopted the Drude model:

$$\varepsilon_m(\omega)=\varepsilon_\infty-\frac{\omega_p^{\;2}}{\omega^2+i\gamma\omega} \tag{3}$$

which described well the dielectric properties of metals and semiconductors [7].

If instead the interaction of a radiation field with the fundamental lattice vibration plays a dominant role and results in absorption of electromagnetic wave, due to the creation or annihilation of lattice vibration, the dielectric function $\varepsilon_m(\omega)$ mainly consists of the contributions of the lattice vibrations, expressed by the classical pseudo-harmonic phonon model in the first approximation [7]:

$$\varepsilon_m(\omega)=\varepsilon_\infty+\frac{\varepsilon_{st}\,\omega_{TO}^{\;2}}{\omega_{TO}^{\;2}-\omega^2-i\gamma\omega} \tag{4}$$

5. The Smith model

Smith has started from the response theory for the optical conductivity, considering an electric field impulse applied to a system, in order to examine the answer with respect to the current. He has utilized the following Fourier transform for the frequency-dependent complex conductivity:

$$\sigma(\omega)=\int_0^\infty j(t)\exp(i\omega t)dt \tag{5}$$

A field impulse, which exceeds every other acting force on the system, permits to assume the hypothesis that the electrons initially can be considered totally free; therefore it holds:

$$j(0)=n^*e^2/m \tag{6}$$

with n^* effective electron density. After calculation, the real part of $\sigma(\omega)$ results:

$$\int_0^\infty \mathrm{Re}\,\sigma(\omega)d\omega=\frac{\pi}{2}j(0)=\frac{\omega_p^2}{8} \tag{7}$$

If the initial current decays exponentially to its initial value with relaxation time τ, it is possible to write:

$$j(t)/j(0)=\exp(-t/\tau) \tag{8}$$

from which the standard Drude formula is obtainable:

$$\sigma(\omega)=(n^*e^2\tau/m)/(1-i\omega\tau) \tag{9}$$

Eq. (9) can be considered the first term of a series of the form:

$$j(t)/j(0)=\exp(-t/\tau)\left[1+\sum_{n=1}^{\infty}c_n(t/\tau)^n/n!\right] \tag{10}$$

The c_n factors hold into account of the original electrons speed remained after the n-th collision. The analytical form of the complex conductivity is therefore [9]:

$$\sigma(\omega)=\frac{n^*e^2\tau}{m(1-i\omega\tau)}\left[1+\sum_{n=1}^{\infty}\frac{c_n}{(1-i\omega\tau)^n}\right] \tag{11}$$

6. Plasmonics

6.1. Introduction

A lot of nanostructures, in particular the nanostructured metals, show very complex and interesting optical properties. One of the most interesting and promising phenomena encountered in these structures are electromagnetic resonances, due to collective oscillations of the conduction electrons, said "plasmons". Plasmon modes exist in different geometries and in various metals, in particular in noble metals such as gold and silver. Under determined circumstances, plasmons are excited by light; this leads to strong light scattering and absorption and to an increase of the local electromagnetic field. The interest in plasmon modes has started to the beginning of the 20th century, with Zenneck (1907), Mie (1908), Sommerfeld (1909) and other scientists; recent advances regarding the structure, the manipulation and the observation at nanoscale, have increased the study of this scientific topic. The theoretical efforts have encountered also the growing demand at technological level, in particular for semiconductor based integrated electronic components, optical applications and new nano-components. Actually it remains an important challenge for research and development processes, like the guide of light in integrated optical systems and the interface with electronic components. A lot of nanostructures are believed to be one of the key ingredients of the future optoelectronic devices [10].

6.2. Plasmons

The concept of plasma is very useful in the description of some aspects of the interaction radiation/conductive matter; the free electrons of the conductive material (for example a metal) are considered as an electron fluid with high density, of order of $10^{23}\,cm^{-3}$. Such concept is the base of the classic Drude model, which assumes that the material contains stopped positive ions and a gas of classical not interacting electrons, whose motion results slowed by a force of viscous friction, due to the collisions with the ions, and characterized from a relaxation time τ. The motion of electrons results so casual, due to the continuous collisions with the lattice. The plasma frequency is defined as the proper frequency of the collective motion of electrons in the following way:

$$\omega_p^2 = \frac{\sigma_0}{\varepsilon_0 \tau} = \frac{ne^2}{\varepsilon_0 m} \tag{12}$$

with σ_0 conductivity of material and common meaning of the other mathematical symbols. The dielectric constant of metal can be written as a function to the plasma frequency; in first approximation it results:

$$\varepsilon(\omega) = 1 - \frac{\omega_p^2}{\omega^2} \tag{13}$$

From the comparison between ω and ω_p, we can deduce the behaviour of the electromagnetic waves arriving to the metal. If it holds $\omega < \omega_p$, the dielectric constant is negative, therefore its square root is imaginary pure; this involves the reflection of the incident wave. In the contrary case, the square root of the dielectric constant is real and the incident wave can propagate in the medium with a small attenuation.

The plasma oscillations are the fluctuations of charge density, which happen to the frequency ω_p and propagate in the metal. The quanta of such fluctuations inside the volume are the "volume plasmons".

There is also a mechanism on the metal surface, characterized by the "surface plasmons". Normally it appears in the wavelength range between the visible and the infrared for the interface air/metal. Localized surface electromagnetic excitations can exist at the surface of nanoparticles and metallic nanostructures: the "localized surface plasmons" (LSP) [11]. The frequency and the intensity of the radiation are very sensitive with respect to the dimension, form and morphology of the nanostructures [12]. Nanoparticles and nanostructures with smaller dimensions with respect to the wavelength of the exciting light are characterized by a wide absorption band, normally in the range of the visible and near infrared spectrum.

6.3. Related theoretical models

The models concerning plasmonics simulate the extinction of the "localized surface plasmons resonance" (LSPR) from nanoparticles and nanoarrays; the application regards the calculation of the light absorption and the scattering. The most used models are:

- the Mie theory [13];
- the Gans theory [14];
- the discrete-dipole approximation method (DDA) [15];
- the finite-difference time-domain method (FDTD) [16].

6.3.1. The Mie theory

The Mie theory is a theoretical approach concerning the optical properties of the nanoparticles. When the nanoparticle dimension is smaller than the wavelength of the incident light, such theory predicts that the extinction caused by a metallic nanosphere is estimated in the quasi-static and dipole limit. The relative cross section of the process results:

$$\sigma_{ext} = \frac{24\pi N_A a^3 \varepsilon_m^{3/2}}{\lambda \ln(10)} \left[\frac{\varepsilon_i}{(\varepsilon_r + \chi \varepsilon_m)^2 + \varepsilon_i^2} \right] \tag{14}$$

where N_A is the surface density, *a* the radius of the metallic nanosphere, ε_m the dielectric constant of the medium surrounding the nanosphere, λ the wavelength, ε_i and ε_r imaginary and real part of the dielectric function of the metallic nanosphere respectively and χ form factor concerning the nanoparticle. The localized resonance depends also on the interparticle space and on the dielectric constant of the substrate.

6.3.2. The Gans theory

The Gans theory extends the Mie theory to spheroidal particles case. In the case of polarization of incident light parallel to the symmetry axis of the spheroid, the cross section of the process is given by:

$$\sigma_{ext} = \frac{2\pi V \varepsilon_m^{3/2}}{3\lambda} \sum_j \left[\frac{(1/P_j)^2 \varepsilon_j}{(\varepsilon_r + \frac{1-P_j}{P_j}\varepsilon_m)^2 + \varepsilon_j^2} \right] \tag{15}$$

with *V* volume of nanoparticle and P_j depolarization factors along the three cartesian aces, which hold into account of the anisotropic form of particles.

6.3.3. The discrete-dipole approximation method (DDA)

With this method the nanoparticles are divided in a cubic array of *N* polarizable dipoles, with polarizabilities α_i determined by the dielectric function of nanoparticles. The induced dipole P_i of every element results $P_i = \alpha_i E_{loc,i}$ in an applied plane wave field. The cross section is determined by:

$$\sigma_{ext} = \frac{4\pi k}{\left|\vec{E}_{inc}\right|^2} \sum_{j=1}^{N} \mathrm{Im}\left(\vec{E}_{inc,j}^{\,*} \cdot \vec{P}_j\right) \tag{16}$$

with $k = \omega/c$ (in vacuum). Such method can be applied for the calculation of the absorption, scattering, extinction and other optical properties of nanoparticles of various forms and dimensions. It is considered one of the most important methods for the understanding of the structural characteristics and optical properties of nanomaterials and nanostructures.

6.3.4. The finite-difference time-domain method (FDTD)

This is a method of numerical calculation in order to resolve the Maxwell equations directly in the time domain. Being a time-domain method, the solutions can cover a wide frequency range with a single process of simulation. It is a versatile and useful technique in applications where the resonance frequencies are not exactly known. A great variety of magnetic and dielectric materials can be modelled in relatively simple way with such method [17].

7. Linear response theory: a new interesting idea

We start considering a system with an hamiltonian of the form:

$$H = H_0 + H_1 \tag{17}$$

with H_1 having small effects with respect to H_0, and of the form:

$$H_1 = \lambda A e^{-i\omega t} e^{\eta t} \tag{18}$$

being λ a real quantity and η positive, so that in remote past the perturbation is negligible (adiabatic representation: $\lim_{t \to -\infty} H_1 = 0$). In the case of an electric field of frequency ω we have:

$$H_1 = e\vec{E} \cdot \vec{r} \tag{19}$$

If the electric field is constant in space and its evolution depends on time, it is writable as follows:

$$\vec{E} = \vec{E}_0 e^{i\omega t} \tag{20}$$

The time dependent corresponding current is:

$$\vec{J}(t) = \sigma(\omega)\vec{E}(t) \tag{21}$$

The conductivity $\sigma(\omega)$ is in general a complex function of the frequency ω and can be deduced from linear response theory. Following the standard time-dependent approach, it is possible to find a general formula for the linear response of a dipole moment density $\vec{B} = e\vec{r}/V$ in the β direction with the electric field $\vec{E}$ in the α direction, where V is the volume of the system. This permits to deduce the susceptivity $\chi(\omega)$, which is correlated to $\sigma(\omega)$ through the relation:

$$1 + 4\pi\chi(\omega) = 1 + 4\pi i \frac{\sigma(\omega)}{\omega} \tag{22}$$

Analytical calculations permit to write the real part of the complex conductivity $\sigma(\omega)$ as:

$$\mathrm{Re}\,\sigma_{\beta\alpha}(\omega)=\frac{e^2\,\omega\pi}{V\hbar}S_{\beta\alpha}(\omega)\left(1-e^{-\hbar\omega/KT}\right) \tag{23}$$

with:

$$S_{\beta\alpha}(\omega)=\int_{-\infty}^{+\infty}dt\left\langle \vec{r}^{\alpha}(0)\,\vec{r}^{\beta}(t)\right\rangle_T e^{-i\omega t} \tag{24}$$

The part $\langle .....\rangle_T$ in the integral (24) is the thermal average, and the exponential factor arises from equilibrium thermal weights for Fermi particles. By considering the identity

$$v=\frac{d}{dt}r=\frac{i}{\hbar}\left[H,r\right],$$

Eq. (23) can be written in a form containing the velocities correlation function instead of the position correlation function. Assuming the high temperature limit $\hbar\omega<<KT$ (valid in such contests), we obtain:

$$\mathrm{Re}\,\sigma_{\beta\alpha}(\omega)=\frac{e^2}{2VKT}\int_{-\infty}^{+\infty}dt\left\langle \vec{v}^{\alpha}(0)\vec{v}^{\beta}(t)\right\rangle_T e^{-i\omega t} \tag{25}$$

The integral in Eq. (25) spans the entire t axis, so we can perform the complete inverse Fourier transform of this equation. It gives:

$$<\vec{v}^{\alpha}(0)\,\vec{v}^{\beta}(t)>_T=\frac{KTV}{\pi e^2}\int_{-\infty}^{+\infty}d\omega\,\mathrm{Re}\,\sigma_{\beta\alpha}(\omega)e^{i\omega t} \tag{26}$$

The new introduced key idea is the possibility to perform a complete inversion of Eq. (26) on temporal scale, i.e. considering the entire time axis (-∞, +∞), not the half time axis (0, +∞), as usually considered in literature [9,18]. This idea is viable if we consider the real part of the complex conductivity $\sigma(\omega)$. Via contour integration by the residue theorem in Eq. (26), the integral is determined by the poles of $\mathrm{Re}\,\sigma(\omega)$. This leads to an exact formulation and gives a powerful method to describe the velocities correlation function (and consequently the mean square displacement and the diffusion coefficient) in analytical way.

8. The other important functions in the nano-bio-context

Another interesting quantity at nanolevel is the mean squared displacement of particles at equilibrium, defined as:

$$R^2(t)=\left\langle [\vec{R}(t)-\vec{R}(0)]^2\right\rangle \tag{27}$$

where $R(t)$ is the position vector at time t. Through a coordinate transformation relative to the integration region it is possible to rewrite the relation (27) as follows:

$$R^2(t) = 2\int_0^t dt'(t-t')\langle\vec{v}(t')\cdot\vec{v}(0)\rangle \tag{28}$$

The mean squared displacement can therefore be evaluated through the velocities correlation function. Through Eq. (26) it is possible to deduce also the diffusion coefficient D:

$$D(t) = \frac{dR^2(t)}{2dt} = \int_0^t dt'\langle\vec{v}(t')\cdot\vec{v}(0)\rangle \tag{29}$$

The three relations (26), (28), (29) are fundamental in deducing the most important characteristics concerning the transport phenomena.

9. About the complex conductivity $\sigma(\omega)$

Let us consider a particle in a region in which there is an electric time-oscillating field, direct along z-axis, with an elastic-type and a friction-type force acting on system; the dynamic equation of the particle can be written as:

$$m\ddot{z} = -kz - \lambda\dot{z} + eE_o e^{-i\omega t} \tag{30}$$

where m is the mass of the particle, k the strenght constant of the oscillators, $\lambda = m/\tau$ and τ is the relaxation time. We consider solutions of the form:

$$z(t) = z_0 e^{-i\omega t} \tag{31}$$

The current density in the field direction is:

$$j_z = ne\dot{z} = nez_0(-i\omega)e^{i\omega t} = \sigma E \tag{32}$$

From Eq. (32), via analytical calculations, it is possible to obtain the frequency dependent complex conductivity:

$$\sigma = \frac{i\omega ne^2}{m(\omega^2 - \omega_0^2 + i\lambda\omega)} \tag{33}$$

The real part of the complex conducivity (33) results [1,19]:

$$\mathrm{Re}\,\sigma(\omega) = \frac{\tau ne^2}{m}\frac{\omega^2/\tau^2}{\left(\omega_0^2 - \omega^2\right)^2 + \omega^2/\tau^2} \tag{34}$$

10. About the Drude-Lorentz-like models: a new promising "plasmon model"

Recently it has been published a new formulation of the Drude-Lorentz model [1,20], based on linear response theory and resonant plasmonic mode; it is able to accommodate some of the observed departures and gives a detailed description of the dynamic response of the carriers at nano-level. One of the peculiarities of this new model is the inversion of relation (26), extending the integration on time to the entire time axis (-∞, +∞); this procedure is not trivial and it was introduced for the first time in the nano-bio-context. One of the main advantages is the possibility to obtain analytical relations for the description of the dynamic behaviour of nanosystems. Such relations are functions of parameters, experimentally obtainable through Time-resolved techniques. An important consequence of the application of such formulation turns out to be the mathematical justification of the unexpected experimental results of high initial mobility of charge carriers in devices based on semiconductor nanomaterials, as the dye sensitized solar cells (DSSCs). The increase of the diffusion coefficient implies a raise of the efficiency of such electro-chemical cells. The new formulation is able to describe very adequately the properties of transport of nano-bio-materials, but it holds also for other objects, like ions [19].

Many experimental data have indicated that plasmon models describe nanostructured systems in particularly effective way. At quantum level, the key factors incorporating the quantum behaviour are the weights f_i, related to plasma frequencies by:

$$\omega_{p_i}{}^2 = \frac{4\pi N e^2}{m} f_i \tag{35}$$

where N is the carrier density, m and e respectively the mass and the charge of the electron.

The velocities correlation function in quantum case assumes the form:

$$\langle \vec{v}(0)\cdot\vec{v}(t)\rangle = \left(\frac{KT}{m}\right)\sum_i\left(f_i \exp\left(-\frac{t}{2\tau_i}\right)\left[\cos\left(\frac{\alpha_{iR}}{2}\frac{t}{\tau_i}\right) - \frac{1}{\alpha_{iR}}\sin\left(\frac{\alpha_{iR}}{2}\frac{t}{\tau_i}\right)\right]\right) \tag{36}$$

with:

$$\alpha_{iR} = \sqrt{4\tau_i^2\,\omega_i^2 - 1}\,; \tag{37}$$

$$\langle \vec{v}(0)\cdot\vec{v}(t)\rangle = \frac{1}{2}\left(\frac{KT}{m}\right)\sum_i\left(\left(\frac{f_i}{\alpha_{iI}}\right)\left[\left(1+\alpha_{iI}\right)\exp\left(-\frac{(1+\alpha_{iI})}{2}\frac{t}{\tau_i}\right) - \left(1-\alpha_{iI}\right)\exp\left(-\frac{(1-\alpha_{iI})}{2}\frac{t}{\tau_i}\right)\right]\right) \tag{38}$$

with:

$$\alpha_{iI} = \sqrt{1 - 4\tau_i^2\,\omega_i^2}\,. \tag{39}$$

(K is the Boltzmann's constant, T the temperature of the system, ω_i and τ_i frequencies and decaying times of each mode respectively).

Integrating these expressions through Eqs. (28) and (29), we obtain the mean squared displacement R^2 and the diffusion coefficient D [20,21].

11. Some interesting results of the new model

The model has demonstrated at today high generality and good accordance with experiments. It has been applied in relation to the most commonly used and studied materials at nano-level, i.e. zinc oxide (ZnO), titanium dioxide (TiO_2), gallium arsenide (GaAs), silicon (Si) and carbon nanotubes (CN), also at nano-bio-sensoristic level [19,22].

In Figure 1 it is showed R^2 for doped Silicon. For this systems the conductivity is the contribution of a Drude-Lorentz term and a Drude term. At large times the Drude-Lorentz term leads to an R^2 approaching a constant value (Figure 2), while the Drude term alone is the dominant one. Therefore, for sufficiently large times, only the Drude term survives.

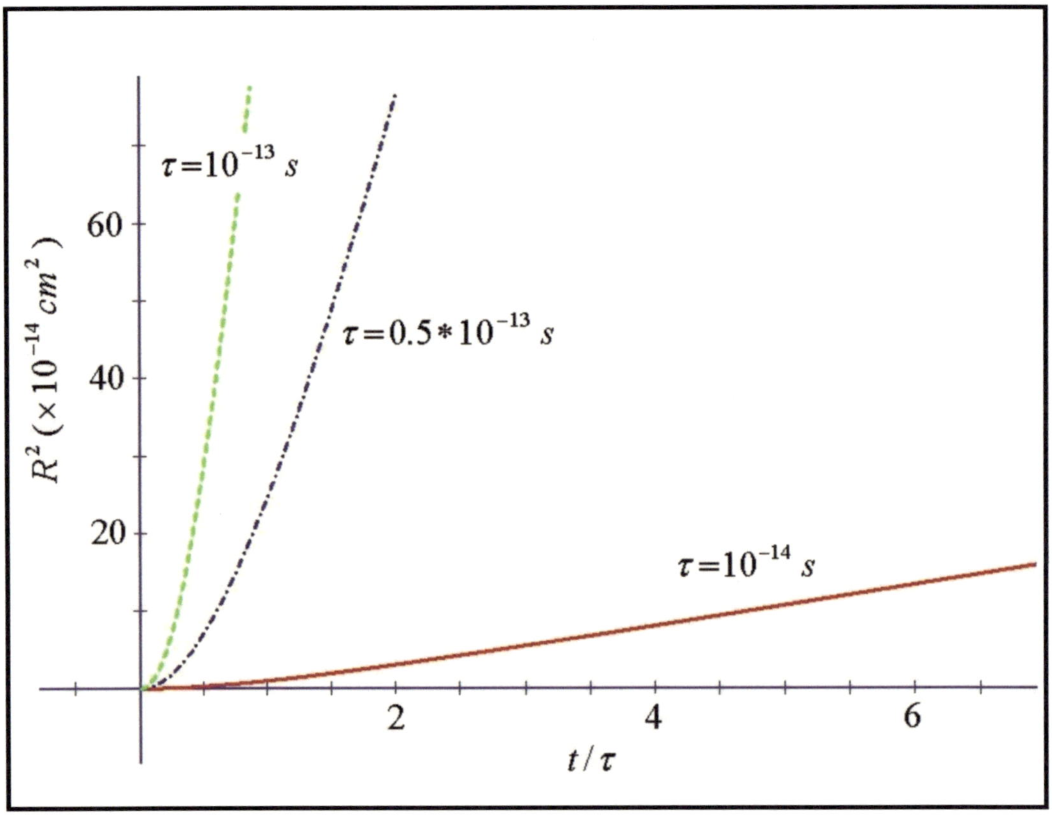

Figure 1. R^2 vs $x = t/\tau$ for some representative values of τ, typical of doped Silicon [23] ($\omega_0 = 0$, T=300K) (Drude model). A complete description of R^2 for Silicon requires the evaluation of the contribution of the Drude-Lorentz part.

We can observe that the linear relation at large times becomes quadratic at smaller times. The cross-over between the two regimes occurs at times comparable to the scattering time.

This means that diffusion occurs after sufficient time has elapsed, so that scattering events become significant, while at smaller times the motion is essentially ballistic.

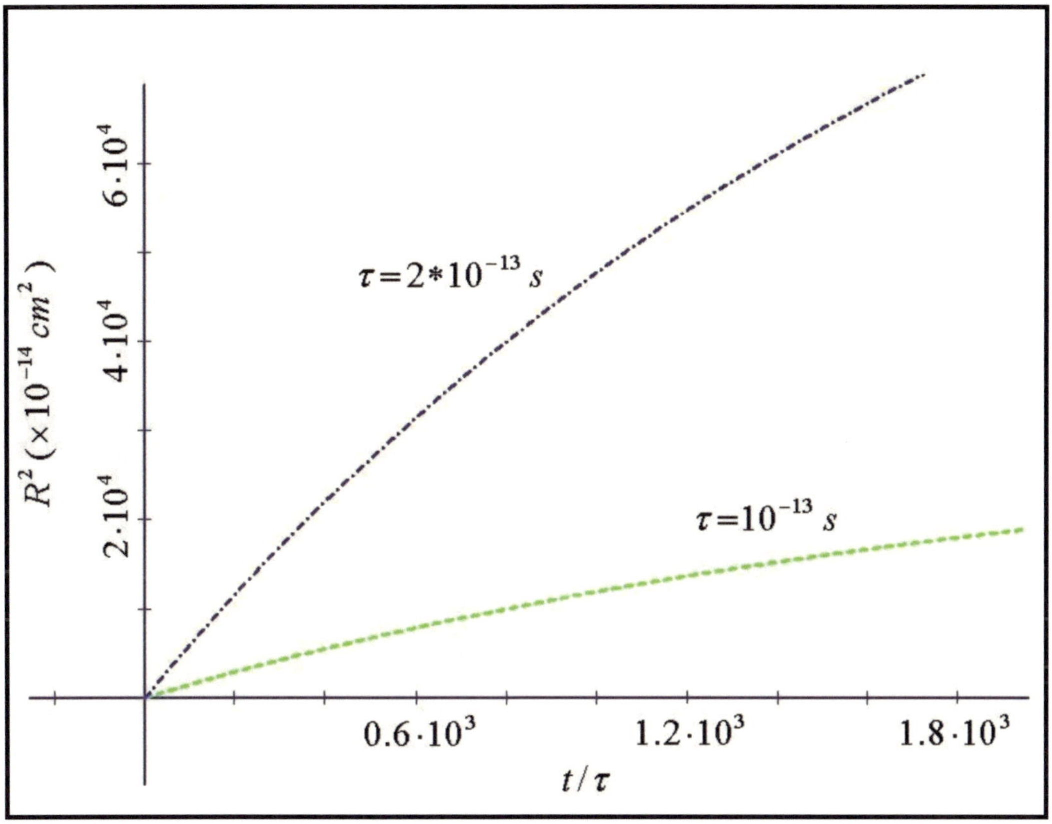

Figure 2. R^2 vs $x=t/\tau$ for 2 values of τ ($\omega_0=1.12\cdot 10^{11}\,Hz$ dot-dashed line; $\omega_0=2.24\cdot 10^{11}\,Hz$ dashed line) for TiO_2 (m=6m_e, T=300K). Saturation values occur at sufficiently large t.

The Drude behaviour is contrasted by the plasmon behaviour. It has reported representative cases at large times $x=t/\tau>>1$ (Figure 2) for TiO_2 nanoparticle films with small frequencies ω_0, i.e. close to the Drude case. It is observable that all of the curves reach a plateau value at sufficiently long times and that the slope within a given time interval increases with τ. As consequences of this behaviour, the plateau of R^2 may become larger than the size of the nanoparticles composing the films, depending on the parameters ω_0 and τ, indicating that carriers created at time $t=0$ will have enhanced mobility in the nanoporous films at small times, on the order of few τ, in contrast with a commonly expected low mobility in the disordered TiO_2 network. Secondly, the time derivative of R^2 decreases to zero at very large times, indicating localization by scattering and therefore leading to decreasing mobility and to the existence of a strictly insulating state for $t\rightarrow\infty$

Few cases corresponding to short time behaviour and large frequencies are showed in Figure 3, with the scattering time in the range 10^{-13}-10^{-14}s.

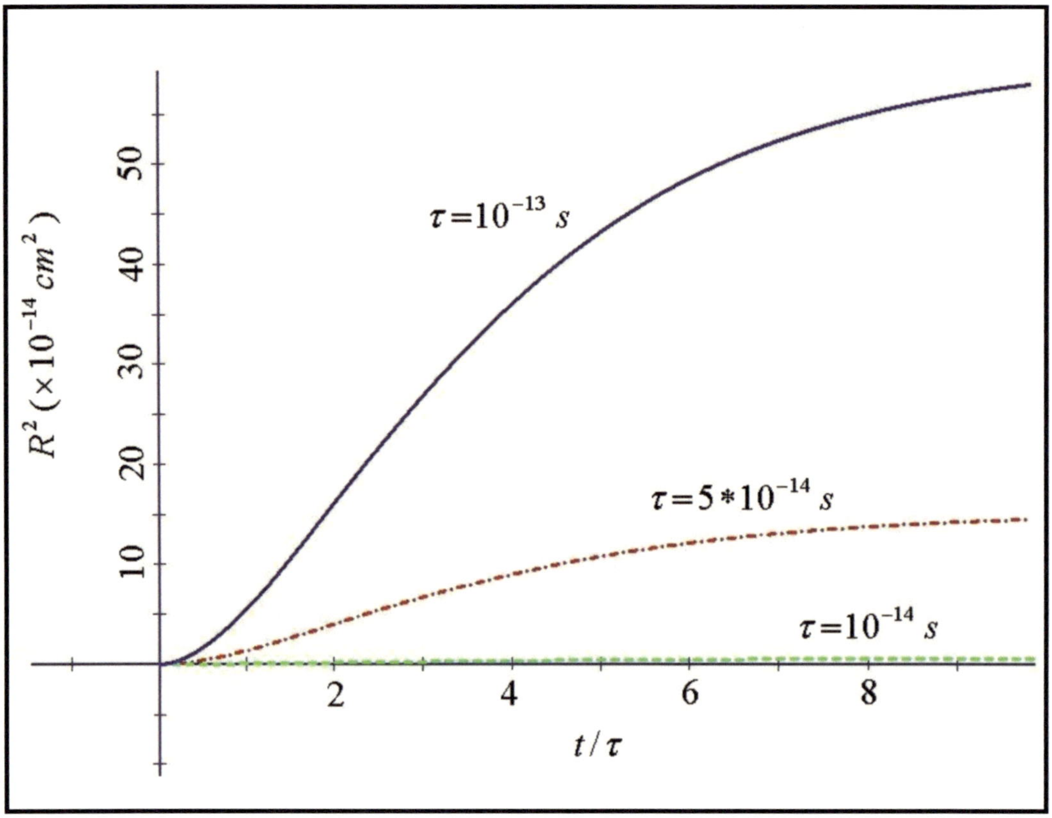

Figure 3. R^2 vs $x = t/\tau$ for 3 values of τ ($\omega_0 = 0.5 \cdot 10^{13}\ Hz$ solid line; $\omega_0 = 10^{13}\ Hz$ dot-dashed line; $\omega_0 = 0.5 \cdot 10^{14}\ Hz$ dashed line) for TiO_2 (m=6m_e, T=300K).

In relation to the behaviour of the velocities correlation function, excluding the Drude model ($\omega_0 = 0$), characterized by simple exponential decreasing functions with time, the correlation function of velocities corresponds to either a damped oscillatory behaviour (Eq. (36)), or to a superposition of two exponentials (Eq. (38)), with quite different decay times depending on the value of α_I (Figures 4, 5).

At $\alpha \cong 1$, i.e. close to Drude behaviour, proper case of some systems [3], the current results the superposition of a short $\approx\tau$ and long $\tau/(1-\alpha_I)$ time decay modes. The current is a damped oscillating function of time (Eq. (36)) when $\omega_0\tau >> 1$, and will have double exponential behaviour for $\omega_0\tau << 1$ (Eq. (38)).

From this analysis, it is viable the possibility that the previous results can give an explanation of the ultra-short times and high mobilities, with which the charges spread in mesoporous systems, of large interest in photocatalitic and photovoltaic systems [24]; the relative short times (few τ), with which charges can reach much larger distances than typical dimensions of nanoparticles, indicate easy charges diffusion inside the nanoparticles. The unexplained fact, experimentally found, of ultrashort injection of charge carriers (particularly in Grätzel's cells) can be related to this phenomenon.

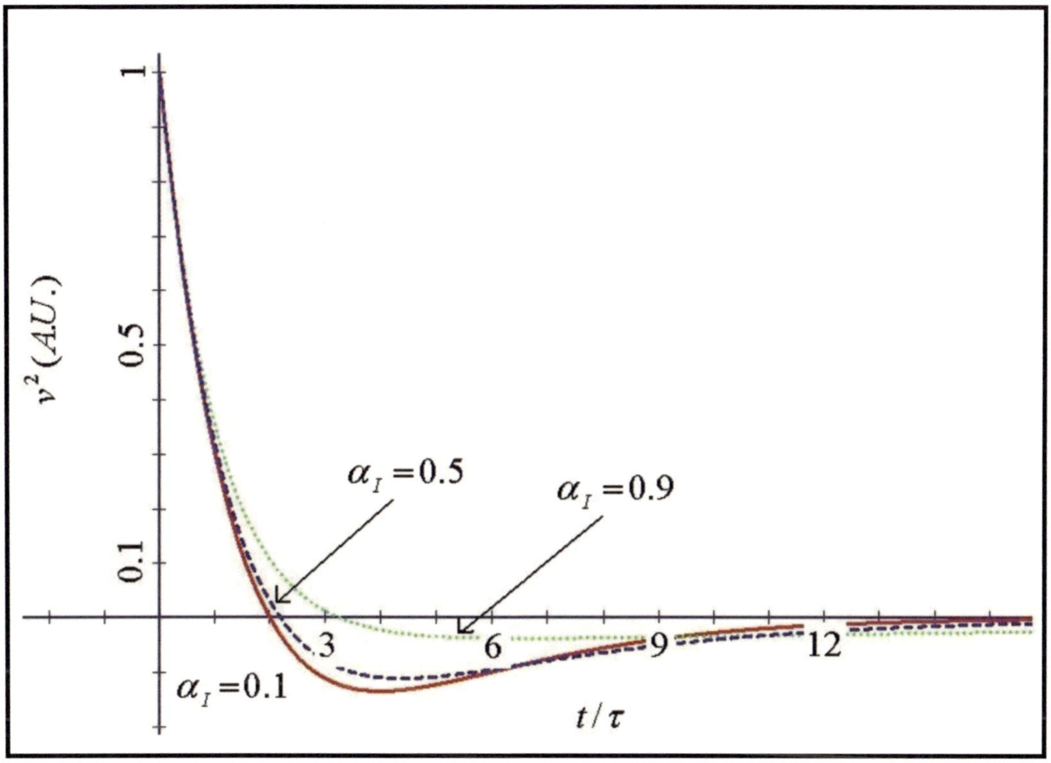

Figure 4. Velocities correlation function vs $x = t/\tau$ for some values of α_I (m=6m_e, T=300K).

In systems with charge localization, where the carrier mean free path is comparable to the characteristic dimension of the nanoparticles, the conductivity response becomes more complicated. The response at low frequencies is characterized by an increasing real part of the conductivity and by a negative imaginary part. Deviations from the Drude model become strong in nanostructured materials, such as photoexcited TiO_2 nanoparticles, ZnO films, InP nanoparticles [25], semiconducting polymer molecules [26], and carbon nanotubes [27].

In isolated GaAs nanowires the electronic response exhibits a pronounced surface plasmon mode, that forms within 300 fs, before decaying within 10 ps as a result of charge trapping at the nanowire surface. The conductivity in this case was fitted by using the Drude model for a plasmon and the mobility found to be remarkably high, being roughly one-third of that typical for bulk GaAs at room temperature.

The Smith model with $c_1 = -1$ is obtained as a limit of the new introduced plasmon model when $\alpha_I \to 0$. On performing this limit in Eq. (38) one finds the expression obtained by Smith with a scattering time twice the plasmon scattering time τ. The situation $\alpha_I \to 0$ corresponds to $\tau \approx 1/2\omega_0$. From the other hand, both Smith's and the new model reduce to the Drude model in the limit $\omega_0 \to 0$. So, although the two models are analytically different, their predictions are expected to be quite similar. From Figure 4 it is possible to note that, for

$\alpha_I \to 0$, the lower curve is of the form of the Smith curve. The backscattering mechanism invoked by Smith arises in a natural way in this new model, without further assumptions on successive scattering events.

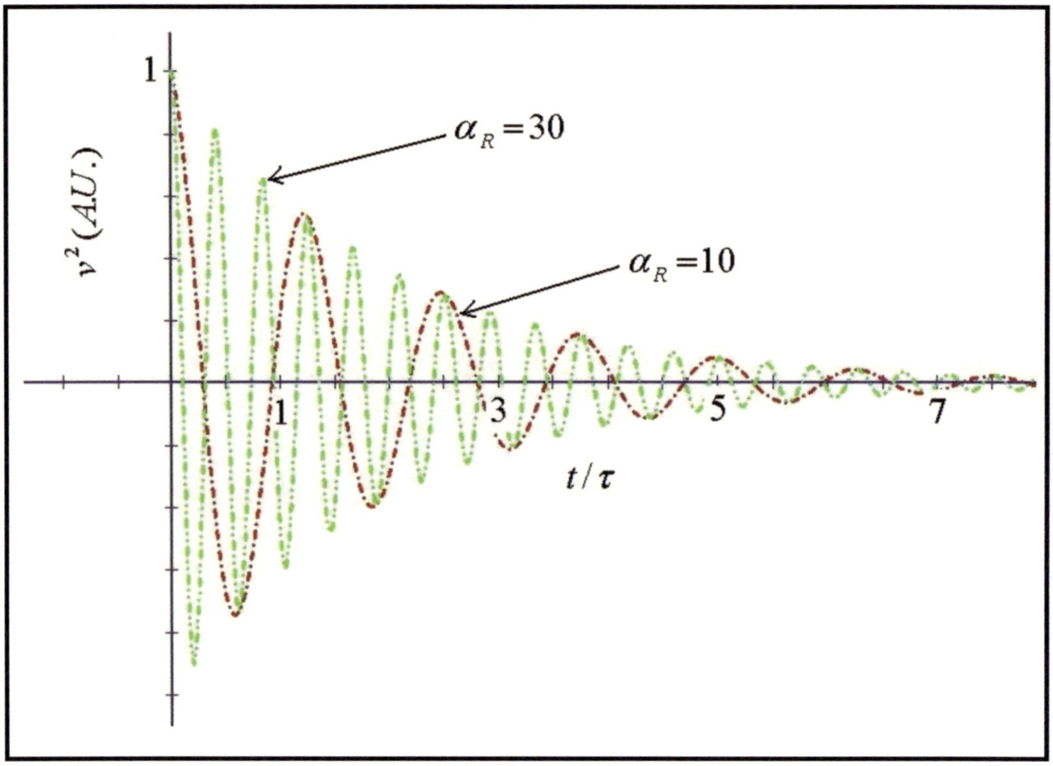

Figure 5. Velocities correlation function vs $x = t/\tau$ for two values of α_R (m=6m_e, T=300K). Evident exponentially damped oscillations are displayed in this case.

Time evolution of charges can be systematically studied by means of time resolved techniques. The THz technique allows a detailed investigation of very short time behaviour, such as the photoinjection, as well as longer time behaviour such as thermalized motion. The studies have indicated a common mechanism of the short time domain in nanostructures, where carriers are close to Drude behaviour with a rather large diffusion coefficient, followed by a range with time decreasing mobility. This latter stage is characterized by decay of the response as the superposition of a short time and a longer time exponentials.

The time response of ZnO films, nanowires, and nanoparticles to near-UV photoexcitation has been investigated in THz experiments. Films and nanoparticles show ultrafast injection, but with the addition of a second slower component. For ZnO nanoparticles, double exponential decay indicates characteristic times $\tau_1 \approx 94\,ps$ and $\tau_2 \approx 2.4\,ns$. From Eq. (38) it is possible to deduce a ratio of two different relaxation times τ_1 and τ_2; it is $\tau_1 / \tau_2 = (1-\alpha_I)/(1+\alpha_I)$. Therefore, on using the experimental values, we find $\alpha_I = 0.92$, from which we deduce $\omega_0 \tau \cong 0.2$. For τ of the order of 10^{-13}s, we obtain $\omega_0 \cong 1.5 \cdot 10^{12}\, Hz$, which is of the correct order of magnitude for the resonance in the infrared. The same procedure for injection times of films (500 nm grains) leads to $\omega_0 \cong 2.5 \cdot 10^{12}\, Hz$ (for $\tau \cong 10^{-13}\, s$).

In GaAs nanowires [27,28] a resonance at $\omega_o = 0.3 - 0.5\, THz$ has been suggested to explain the frequency-dependent conductivity obtained from THz experiments. A long characteristic obtained time $\tau_c \approx 1.1 \div 5 \cdot 10^{-12} s$ is in accordance with the new model, corresponding to $\tau_c = \tau / (1 - \alpha_I) = 5.10^{-12} s$ and leading to $\alpha_I = 0.1 \div 0.8$ for $\tau = 10^{-12} s$ and $\omega_0 \tau = 0.3 \div 0.49$.

Similar considerations can be applied to single walled carbon nanotubes (SWCN), where the resonance state is at $\omega_o = 0.5\, THz$ with Drude-Lorentz scattering time $\tau \approx 10^{-13}$ s and to sensitized TiO_2 nanostructured films. In the latter the conductivity of electrons injected from the excited state of the dye molecule into the conduction band of TiO_2 is initially more Drude-like, and then evolves into a conduction dominated by strong backscattering as the carriers equilibrate with the lattice. These results conform to the backscattering mechanism of the new plasmon model.

In Figure 6 it is presented the fitting of data of GaAs photoconductivity [27,28] with the new model.

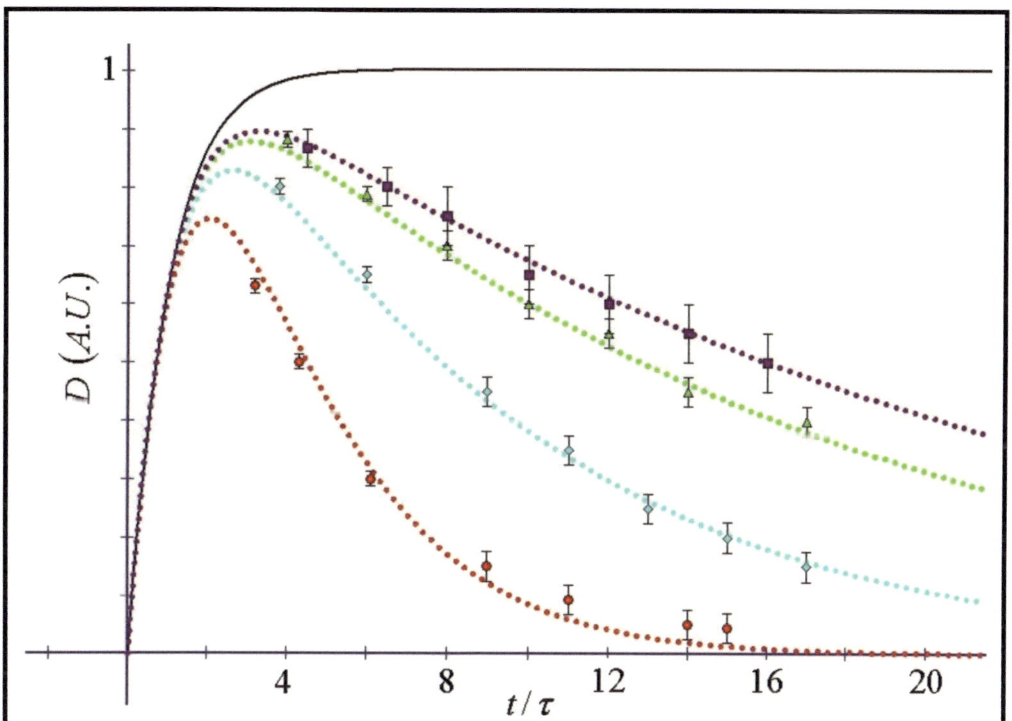

Figure 6. Behaviour of the diffusion coefficient D for α_I (Eq. (39) at classical level [1]) varying in the interval [0.1-1] and $\tau = 10^{-12}$ s. Note that $D \to 0$ for $t \to \infty$, indicating absence of diffusion at very long times.

In this figure the upper curve is the Drude result corresponding to $\omega_0 = 0$. It is the highest value of the diffusion coefficient, as ω_0 varies from zero to its maximum value. The dots in Figure 6 are experimental data derived from THz spectroscopy, the error bars on dots represent the distribution of the experimental data and the lines the predictions of the new plasmon model. D tends to a vanishing value as $t \to \infty$, the larger ω_0 the faster this vanishing occurs. The results can be interpreted as follows: at early time, of the order of τ, the system behaves as Drude-like, irrespective of ω_0, with carriers assuming large mobility values (the numerical evaluation gives $D \approx 7.5$ cm²/s for the case of Figure 6); at increased times, charges become progressively localized as a result of scattering, significantly in agreement with the conclusions drawn by THz spectroscopy [29,30].

12. Conclusions

It is hoped that this new plasmon model will be useful to describe experimental data as an alternative to other generalizations of the Drude model. The principal findings of it are the reversal of the current in small systems like nanoparticles and the time dependence of the transport parameter D, which describes the dynamics of the system at short and long time, with increased mobilities at very short times followed by localization at longer times. This unusual behaviour, with respect to the predicted Drude-like, converges as a whole in the peculiar frequency dependence of the optical conductivity. The strength of the new model consist of its ability to accommodate this behaviour and include previous models, like the Smith model [19].

Author details

Paolo Di Sia
Free University of Bozen-Bolzano, Bruneck-Brunico (BK), Italy

13. References

[1] Di Sia P. An Analytical Transport Model for Nanomaterials. Journal of Computational and Theoretical Nanoscience 2011; 8: 84-89.
[2] Wolf EL. Nanophysics and Nanotechnology: An Introduction to Modern Concepts in Nanoscience. John Wiley & Sons; 2006.
[3] Schmuttenmaer CA. Using Terahertz Spectroscopy to Study Nanomaterials. Terahertz Science and Technology 2008; 1(1): 1-8.
[4] Ziman M. Principles of the Theory of Solids. Cambridge University Press; 1979.
[5] Kittel C. Introduction to Solid State Physics. Wiley New York; 1995.
[6] Levy O, Stroud D. Maxwell Garnett theory for mixtures of anisotropic inclusions: Application to conducting polymers. Physical Review B 1997; 56(13): 8035-8046.
[7] Han J, Zhang W, Chen W, Ray S, Zhang J, He M, Azad AK, Zhu Z. Terahertz Dielectric Properties and Low-Frequency Phonon Resonances of ZnO Nanostructures. The Journal of Physical Chemistry C 2007; 111(35): 13000-13006.

[8] Han JG, Wan F, Zhu ZY, Liao Y, Ji T, Ge M, Zhang ZY. Shift in low-frequency vibrational spectra of transition-metal zirconium compounds. Applied Physics Letters 2005; 87(17): 172107-172109.
[9] Smith NV. Classical generalization of the Drude formula for the optical conductivity. Physical Review B 2001; 64(15): 155106-155111.
[10] Shipway AN, Katz E, Willner I. Nanoparticle arrays on surfaces for electronic, optical and sensoric applications. ChemPhysChem 2000; 1: 18-52.
[11] Hutter E, Fendler JH. Exploitation of Localized Surface Resonance. Advanced Materials 2004; 16(19): 1685-1706.
[12] Willets KA, Van Duyne RP. Localized Surface Plasmon Resonance Spectroscopy and Sensing. Annual Review of Physical Chemistry 2007; 58: 267-297.
[13] Stuart DA, Haes AJ, Yonzon CR, Hicks EM, Van Duyne RP. Biological applications of localised surface plasmonic Phenomenae. IEE Proceedings-Nanobiotechnology 2005; 152(1): 13-32.
[14] Gulati A, Liao H, Hafner JH. Monitoring Gold Nanorod Synthesis by Localized Surface Plasmon Resonance. The Journal of Physical Chemistry B 2006; 110(45): 22323–22327.
[15] Zhang ZY, Zhao Y-P. Optical properties of helical Ag nanostructures calculated by discrete dipole approximation method. Applied Physics Letters 2007; 90: 221501-221503.
[16] Sherry LJ, Chang S-H, Schatz GC, Van Duyne RP. Localized Surface Plasmon Resonance Spectroscopy of Single Silver Nanocubes. Nano Letters. 2005; 5(10): 2034–2038.
[17] Zhang W, Yue Z, Wang C, Yang S, Niu W, Liu G. Theoretical models, Fabrications and Applications of Localized Surface Plasmon Resonance sensors. 978-1-4244-4964-4/10/$25.00©2010 IEEE 2010; pp. 4; Project supported by the National Natural Science Foundation of China (Grant N. 60574091, 60871028).
[18] Sassella A, Borghesi A, Pivac B, Pavesi L. Characterization of porous silicon by microscopic Fourier transform infrared spectroscopy; 266-267. 9th International Conference on Fourier transform spectroscopy; 1994.
[19] Di Sia P. Classical and quantum transport processes in nano-bio-structures: a new theoretical model and applications. PhD Thesis. Verona University – Italy; 2011.
[20] Di Sia P. An Analytical Transport Model for Nanomaterials: The Quantum Version. Journal of Computational and Theoretical Nanoscience 2012; 9: 31-34.
[21] Di Sia P. Oscillating velocity and enhanced diffusivity of nanosystems from a new quantum transport model. Journal of Nano Research 2011; 16: 49-54.
[22] Di Sia P. New theoretical results for high diffusive nanosensors based on ZnO oxides. Sensors and Transducers Journal 2010; 122(1): 1-8.
[23] Pirozhenko I, Lambrecht A. Influence of slab thickness on the Casimir force. Physical Review A 2008; 77: 013811-013818.
[24] Grätzel M. Solar Energy Conversion by Dye-Sensitized Photovoltaic Cells. Inorganic Chemistry 2005; 44(20): 6841-6851.
[25] Nienhuys H-K, Sundström V. Influence of plasmons on terahertz conductivity measurements. Applied Physics Letters 2005; 87: 012101-012103.

[26] Hendry E, Koeberg M, Schins JM, Nienhuys HK, Sundström V, Siebbeles LDA, Bonn M. Interchain effects in the ultrafast photophysics of a semiconducting polymer: THz time-domain spectroscopy of thin films and isolated chains in solution. Physical Review B 2005; 71: 125201-125210.

[27] Parkinson P, Lloyd-Hughes J, Gao Q, Tan HH, Jagadish C, Johnston MB, Herz LM. Transient Terahertz Conductivity of GaAs Nanowires. Nano Letters 2007; 7(7): 2162–2165.

[28] Parkinson P, Joyce HJ, Gao Q, Tan HH, Zhang X, Zou J, Jagadish C, Herz LM, Johnston MB. Carrier Lifetime and Mobility Enhancement in Nearly Defect-Free Core–Shell Nanowires Measured Using Time-Resolved Terahertz Spectroscopy. Nano Letters 2009; 9(9): 3349–3353.

[29] Di Sia P, Dallacasa V, Dallacasa F. Transient conductivity in nanostructured films. Journal of Nanoscience and Nanotechnology 2011; 11: 1-6.

[30] Di Sia P, Dallacasa V. Anomalous charge transport: a new "time domain" generalization of the Drude model. Plasmonics 2011; 6(1): 99-104.

Chapter 2

Computational Electromagnetics in Plasmonics

Guy A. E. Vandenbosch

Additional information is available at the end of the chapter

http://dx.doi.org/10.5772/51166

1. Introduction

In electromagnetics, numerical techniques have been essential in the development of new technology in the last two decades. The rapidly growing computer capacity and calculation speeds make accurate solutions of very complex problems feasible. This has been especially true in the design of microwave and millimeter wave components and antennas. Whereas 30 years ago, the design of an antenna was based on simple analytical models, or trial and error strategies, nowadays, simulations seem to be as crucial to the design as real measurements.

The situation is quite different in plasmonics. Plasmonics is a quite novel research field and the application of computational electromagnetics in plasmonics can be categorized as "very recent". There are many challenges that still need to be faced and "missing links" that have to be solved.

The plasmonic structures targeted are structures in the order of magnitude of a wavelength at plasmonic frequencies, i.e. at near IR and optical frequencies, and beyond. Although this frequency range is totally different from the traditional range where computational tools have been developed, i.e. the microwave range, in most cases no special modeling techniques have to be used. By far most plasmonic topologies reported in literature have been analyzed / designed with the well-known numerical techniques implemented within in-house developed or commercial software packages. This means that in this chapter the major numerical techniques can be overviewed in a general sense, referring to standard literature. These techniques will not be derived or explained here in full detail. Instead, this chapter focuses on those aspects that come into the picture when the structure is plasmonic.

After the section on techniques and tools, this chapter will focus on the performance of these techniques and tools for plasmonic structures. This is done through an overview of benchmarks available in literature and by considering a few thoroughly analyzed structures. Missing links will be pointed out and suggestions will be given for the future.

2. Fundamental physical modeling differences

This section discusses the differences between classical topologies handled with computational electromagnetics and plasmonic topologies. Classical means structures at much lower frequencies, for example at microwave frequencies, where computational electromagnetics is a well-developed mature field.

2.1. Frequency

First of all, it is essential to point out that the interaction between light and plasmonic structures in the frequency bands considered can still be analyzed with a high degree of accuracy using classical electromagnetic theory. Although the frequency is orders of magnitude higher in plasmonics compared to microwaves, Mawell's laws are the same. They are linear, and thus scalable. As long as quantum effects do not have to be taken into account, which is still the case for the plasmonic applications considered, since the structures are not that small [1], there is no fundamental problem. The underlying formulation of Maxwell's equations can remain unaffected.

The fact that at this small scale, no quantum effects have to be taken into account is really a crucial observation. It means that the concept of a "scatterer", a device able to be excited by electromagnetic waves rather than particles, still works. Basically, the coupling between an electromagnetic (light-) wave and a plasmonic scatterer is thus the same as it is at microwave frequencies, and can be studied in the same way.

2.2. Volumetric currents

A consequence of the high frequencies is the small scale: the elementary building blocks of the topologies considered are at nanoscale. There are two important issues related to this. First, nanoscale fabrication technology of today is only able to generate 3D type structures (i.e. volumes). Thin 2D sheets, where the thickness of the metal is orders of magnitude smaller than the transversal dimensions of the pattern, as commonly used at microwave frequencies (for example a conducting strip or patch), are not possible. Second, for nanostructures operating at plasmonic frequencies the skin depth may be comparable with the structural dimensions, so that currents do flow over the complete volume and the device has to be described with volumetric currents. A surface current description is not sufficient.

2.3. Material characteristics

At microwave frequencies, in most cases the material properties are constant over the frequency bands considered. Also, apart from the losses, most metals behave more or less in the same way, i.e. as good conductors. At IR and optical frequencies however, most metals have very dispersive properties. Permittivities and conductivities may vary orders of magnitude over these frequency bands. The real part of the permittivity may even be negative. In Fig. 1, the permittivity of Cu and Ag are depicted, both real and imaginary part, which corresponds to conductivity. These data were obtained through experimental ellipsometry. Note that the difference between the two metals is enormous. It is evident that this strong variation has a serious impact on the behavior of a device over the frequency band. The variability of material characteristics thus has to be taken into account fully into the modeling.

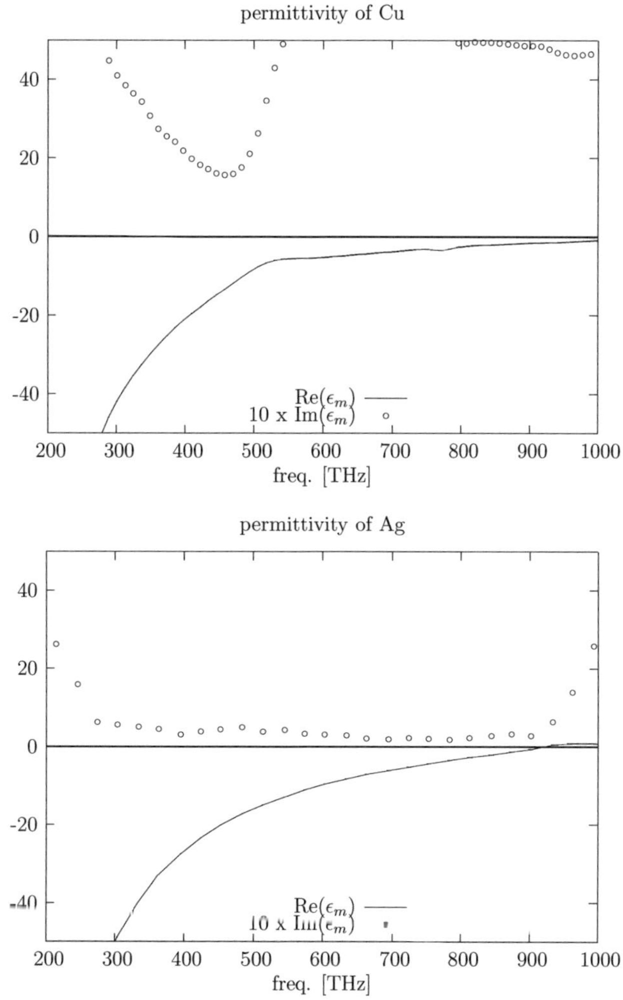

Figure 1. Top: permittivity of Cu; bottom: permittivity of Ag.

3. Modeling techniques

In this section the full wave modeling techniques are introduced and categorized on the basis of their solution method: Finite Elements (FE), Finite Differences in the Time Domain (FDTD), Finite Integration Technique (FIT), and Integral Equations (IE) solved by the Method of Moments (MoM). Based on their theoretical specificities, the application of each method in the case of plasmonics is discussed.

The cradle of computational electromagnetics can be found in the microwave research community. Since this community traditionally is dealing with structures in the order of wavelengths, right from its beginning days, it had no choice than to try to rigorously solve Maxwell's equations. Most standard reference works on full-wave computational electromagnetics by consequence can be found within this community. A history and a comprehensive overview of the different numerical techniques and of their application in computational electromagnetics (CEM) may be found in [2]. Recent work on the last developments in CEM [3] concentrates on the two main approaches of differential and integral methods. A Good review and perspectives concerning the relationship between differential and integral equations (IE) modeling is recommended in a review paper by Miller [4]. A thorough discussion on the different techniques with clarifying examples is also given in [5].

3.1. Differential equation techniques

In time domain, Maxwell's equations are

$$\nabla \times \mathbf{E} = -\frac{d\mathbf{B}}{dt} \tag{1}$$

$$\nabla \times \mathbf{H} = \frac{d\mathbf{D}}{dt} + \mathbf{J} \tag{2}$$

with $\mathbf{E}$ and $\mathbf{B}$ the electric field and the magnetic induction, respectively, $\mathbf{H}$ and $\mathbf{D}$ the magnetic field and the electric induction, respectively, and $\mathbf{J}$ the electric current flowing. In free space, and in a homogeneous, isotropic, time-invariant, linear medium (most materials behave like this in the microwave frequency range), the following relations hold

$$\mathbf{D} = \varepsilon \mathbf{E} \tag{3}$$

$$\mathbf{B} = \mu \mathbf{H} \tag{4}$$

They are called the constitutive relations, with ε and μ the permittivity and permeability of the medium surrounding the observation point considered. It is crucial to point out that in plasmonics in general (3) and (4) cannot be used. For example, the very dispersive properties of metals at optical frequencies (and beyond) prohibit this for these materials.

In time domain, the time variation has to be determined. In frequency domain, it is assumed that all field quantities are varying in a sinusoidal way. This means that the variation with time is known, and only the variation of the fields in space has to be determined. Using complex notation for the frequency domain with a $\exp(j\omega t)$ time dependency, Maxwell's equations become

$$\nabla \times \mathbf{E} = -j\omega \mathbf{B} \tag{5}$$

$$\nabla \times (\mu^{-1}\mathbf{B}) = j\omega\varepsilon \mathbf{E} + \mathbf{J} \tag{6}$$

with $\omega = 2\pi \cdot$ frequency the pulsation. Note that in (5) and (6) ε and μ are used. In general, here they are depending on frequency, and in this way dispersion is fully taken into account (see also section 2.3). In the field of plasmonics, this is an important difference between computational tools in time and in frequency domain. Computational tools in frequency domain may use directly the well-known concepts of permittivity and permeability, albeit that they become frequency dependent. Computational tools in time domain cannot use these concepts directly. As will be shown later, this complicates things.

From a mathematical perspective, Maxwell's equations are differential equations relating vector fields to each other. Differential equation methods in Computational Electromagnetics are methods that directly consider Maxwell's equations (or the Helmholtz wave equations derived from them), with little analytical preprocessing. Basically, these differential equations, which are valid in any point of space and time, are solved by approximating them by difference equations, which are valid in a discrete set of points in space and time. This is done by chopping up space (and time if time domain is considered) in little pieces in which the field variation has a pre-described profile. This reduces the problem of fields varying over space (and time) to a discrete (matrix) problem that can be handled on a computer. The differences between the several differential equation methods are related to the different ways in which space (and time) can be chopped up.

Since the number of unknowns is proportional to the volume and the resolution considered, differential equation methods are particularly suitable for modeling small full three-dimensional volumes that have complex geometrical details, for example smaller closed-region problems involving inhomogeneous media [6]. Intrinsically, differential equations are less suited for open problems. The reason is that in principle they require a discretization of the entire space under consideration. This space is limited in case of closed problems, but corresponds to infinite space in case of open problems. In practice, this problem is solved by the introduction of techniques like Absorbing Boundary Conditions, and Perfectly Matched Layers (PML) [7]. They mimic the wave arriving at the boundary as travelling further to infinity. The quality of these truncating techniques nowadays is very high so that, in practice, the intrinsic problem with open structures has been overcome, albeit it in an approximate numerical way.

The most popular differential equation-based methods are the Finite Element Method (FEM), for example utilized in Ansoft's HFSS software package, and the Finite-Difference Time Domain method (FDTD), which is employed for example by CST's Time Domain transient solver (in the particular case of Cartesian grids), and by Lumerical.

3.1.1. *The Finite Element Method (FEM) [7], [8]*

FEM is a method based on solving partial differential equations. It is most commonly formulated based on a variational expression. It subdivides space in elements, for example tetrahedra. Fields inside these elements are expressed in terms of a number of basic functions, for example polynomials. These expressions are inserted into the functional of the equations, and the variation of the functional is made zero. This yields a matrix eigenvalue equation whose solution yields the fields at the nodes. FEM gives rise to a very sparse matrix equation, which can be solved using dedicated matrix algebra technology, leading to very fast solution times, considering the huge number of unknowns.

Its first formulations were developed as matrix methods for structural mechanics. This lead to the idea to approximate solids and Courant (1942) introduced an assembly of triangular elements and the minimum of potential energy to torsion problems [9]. The first paper on the application of FEM to electrical problems appeared in 1968 [10]. An extensive review on the history of FEM in electromagnetics was published in an issue of the Antennas and Propagation Magazine [11]. FEM normally is formulated in the frequency domain, i.e. for time-harmonic problems. This means that, as for IE-MoM, the solution has to be calculated for every frequency of interest.

Numerous references can be given developing, explaining, and using FEM. A good book to start with is [8]. A software tool using FEM and very widely spread is Ansoft HFSS.

3.1.2. The Finite-Difference Time-Domain technique (FDTD) [12], [13], [14]

The nature of Maxwell's differential equations is that the time derivative of the H-field is dependent on the curl of the E-field, and the time derivative of the H-field is dependent on the curl of the E-field. These basic properties result in the core FDTD time-stepping relation that, at any point in space, an updated value of an E/H-field in time is dependent on the stored value of the E/H-field and the numerical curl of the local distribution of the H/E-field in space. The numerical translation into a time-stepping algorithm was introduced by Yee in 1966. Indeed, swapping between E-field and H-field updates allows to define a marching-on-in-time process wherein sampled fields of the continuous electromagnetic waves under consideration are used. These waves can be seen to propagate in the Yee lattice, a numerical three-dimensional space lattice comprised of a multiplicity of Yee cells, see Fig. 2. More specifically, Yee proposed a leapfrog scheme for marching-on in time wherein the E-field and H-field updates are staggered so that E-field updates are observed midway during each time-step between successive H-field updates, and vice versa. A huge advantage of FDTD is that this explicit time-stepping scheme avoids the need to solve simultaneous equations, so no matrix inversions are necessary. It also yields dissipation-free numerical wave propagation. Negative is that this scheme results in an upper bound on the time-step to ensure numerical stability. This means that simulations may require many thousands of time-steps for completion. The use of the Yee lattice has proven to be very robust in numerical calculations.

FDTD is extremely versatile since the interaction of an electromagnetic wave with matter can be mapped into the space lattice by assigning appropriate values of permittivity to each electric field component, and permeability to each magnetic field component. This can be done without seriously compromising the speed of the method.

The fact that time is observed directly, and the versatility of the method, make it probably the most efficient technique for complex 3D transient problems. An implementation of the

FDTD technique dedicated to plasmonics and photonics can be found in the commercial tool Lumerical [15].

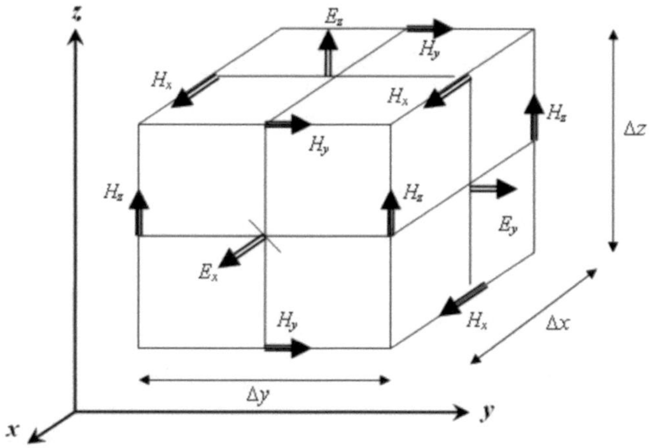

Figure 2. Standard Cartesian Yee cell used in the FDTD technique.

3.1.3. The Finite Integration Technique (FIT) [16]

The Finite Integration Technique was introduced by Weiland in 1977. The word integration does not imply any relation with integral equations. FIT first describes Maxwell's equations on a grid space. The matrix equations for the electromagnetic integral quantities obtained by FIT possess some inherent properties of Maxwell's equations, for example with respect to charge and energy conservation. This makes them very attractive from a theoretical point of view. FIT can be formulated on different kinds of grids, e.g. Cartesian or general non-orthogonal ones, which is a clear advantage. In the time-domain, the resulting discrete grid equations of FIT are, at least in "some" cases, identical to the discrete equations derived with the classical Finite-Difference Time-Domain (FDTD) method. In contrast to FIT, which is applied to the integral form of the field equations, FDTD (as a subset of the finite integration method) is applied to the differential form of the governing Maxwell curl equations. Theoretical links between FDTD and the FIT approach realized in CST may be found in [17]. A comparative study on Time Domain methods was presented in [18]. In some sense, FIT can thus be considered as a powerful generalization of the FDTD technique. A software tool using FIT and very widely spread is CST Microwave Studio.

3.2. Integral Equation techniques (IE)

Integral equation methods make use of Maxwell's equations in integral equation form to formulate the electromagnetic problem in terms of unknown currents flowing on the object to be described. These currents are induced by a field incident on the object. This incident field can be a real incident field, travelling in space, or a bounded wave feeding the object for example via a transmission line. An integral equation solution is fundamentally based on the combination of two equations.

In the case of Electric Field Integral Equations (the case of prime importance to the plasmonics community), the solution is based on considering the electric field. The first equation is

$$\mathbf{E}^{sca}(\mathbf{r}) = \int_{V'} \mathbf{G}(\mathbf{r},\mathbf{r}') \cdot \mathbf{J}^{ind}(\mathbf{r}')dV' \tag{7}$$

which gives the scattered electric field $\mathbf{E}^{sca}$ generated in an arbitrary observation point $\mathbf{r}$ in space in terms of an induced volumetric electric current distribution $\mathbf{J}^{ind}$ flowing within the volume V' of the object considered. The kernel $\mathbf{G}(\mathbf{r},\mathbf{r}')$ is a so-called dyadic Green's function, which is a tensor, and which relates a Dirac impulse type current flowing in $\mathbf{r}'$ to the electric field it generates. The main strength of the integral equation technique is that in many cases the dyadic Green's function can be calculated either directly analytically, such as in homogeneous media, or as some kind of inverse Fourier transform of an analytic function, as is the case for example in layered media. It is crucial to understand that the dyadic Green's function is actually a solution of Maxwell's equations and thus rigorously takes into account the background medium, for example a stack of dielectric layers. This means that unknown currents only have to be assumed on the objects embedded within this background medium. This results in an enormous reduction of unknowns compared to differential equation techniques, which have to model these dielectric layers in just the same way as the objects embedded within them. Basically, the problem formulation automatically covers the entire surrounding space without making any fundamental approximations. As a consequence, the corresponding solution is automatically valid in every point of the background medium. Far field radiation phenomena, surface waves in layered structures, etc., that are vital for efficient and accurate analysis, are analytically included in the solution.

In the case of plasmonics, the second equation is

$$\begin{aligned} \mathbf{J}^{ind}(\mathbf{r}') &= j\omega\,(\varepsilon(\mathbf{r}') - \varepsilon_0)\,\mathbf{E}^{ind}(\mathbf{r}') \\ &= j\omega\,(\varepsilon(\mathbf{r}') - \varepsilon_0)\,(\mathbf{E}^{sca}(\mathbf{r}') + \mathbf{E}^{sca}(\mathbf{r}')) \end{aligned} \tag{8}$$

It expresses the boundary condition which has to be enforced within the object under consideration. It is a relation between the total field, which is the sum of the incident and the scattered field, and the volumetric electric current, which can be considered as partially being a conduction current (due to the imaginary part of the permittivity), and a polarization current (due to the real part of the permittivity).

Combining (7) and (8) yields the equation from which the currents can be solved. This can be done by chopping up the object considered (only the object, and thus not entire space !!!) in little pieces in which the current variation has a pre-described profile (a so-called basis function). This reduces the problem of currents varying over the object to a discrete (matrix) problem that can be handled on a computer. IE-MoM gives rise to a dense matrix equation, which can be solved using standard matrix algebra technology.

Next to the reduced number of unknowns, there are other theoretical advantages linked to the IE-MoM method. The second advantage is that, if properly formulated, IE-MoM *is variationally stable* since most of the output parameters are expressed in integral form over the equivalent currents. This means that even if the calculated currents differ considerably from the exact solution, integral parameters over both currents may remain very similar. Further, this will be illustrated by showing that even with a rather rough mesh high quality physical results may be obtained. A third advantage is that MoM does not heavily suffer from field singularities for example near sharp edges, since they are analytically incorporated inside the Green's functions. For differential equation methods special care (= a fine mesh) should be taken to describe correctly these field singularities.

The main disadvantage of IE-MoM is that, although there are many efforts in that direction, there is still a lack of matrix solvers operating on dense matrices which are comparable in efficiency to the solvers used in the differential equation techniques (mainly FEM), which yield sparse matrices. This has precluded the use of the very flexible volumetric IE-MoM technique for modeling complex structures on widespread computer systems.

IE-MoM is normally applied in the frequency domain, i.e. for time-harmonic problems. This means that the solution has to be determined at each frequency of interest.

Much more details can be found in the classic and basic Method of Moments (MoM) book [19]. There are many variants of the method. In general, boundary equations can be enforced also at the boundaries of volumes (utilizing Surface Integral Equations (SIE) [20]), next to inside the entire volumes themselves (applying Volume Integral Equations (VIE) [21] at the inside of the components, as described above). Also, the integral equations can be written down in different forms (dyadic form, mixed-potential form, hybrid forms, etc. [22]), which give rise to specific implementations. Further, the first theoretical developments in the field of computational plasmonics are appearing in literature [23].

In the following sections, several aspects linked with the use of integral equation techniques for plasmonic structures are discussed.

3.2.1. 3D volumetric current

The fact that a 3D volumetric current has to be described, as mentioned already, invokes the need for either a real volumetric MoM implementation or the more common surface approach, where the plasmonic component is described by applying the equivalence principle at its surface. The surface approach has already been used, for example in [24]. The volumetric approach as described in [21], has very recently been introduced in the plasmonics community [25]. For realistic structures, where volumes can be embedded within a layer structure, it offers a very flexible solution technique.

3.2.2. Material characteristics

Although strongly varying material characteristics may seem trivial to implement in an integral equation scheme, in some cases more severe consequences occur. For example, some Green's function calculation schemes are based on the fact that the material properties do not change over the frequency band of interest. Any particular implementation of IE-MoM needs this essential relaxation into its formalism, which possibly needs to be adapted.

An advantage of a frequency domain technique like IE-MoM compared to a time domain technique like FDTD is that the ellipsometric measurement data can be used directly in the tool, without any further fitting. FDTD techniques developed for the optical range tend to fit the dielectric response of the metal to the experimentally determined dielectric permittivity using a Drude model [26] or more sophisticated models (for example in Lumerical). However, this may create problems (see further).

3.2.3. Occurrence of layer structures

As explained above, IE technique solvers are formulated making use of Green's functions. These Green's functions can be formulated for multi-layered structures, such that the background medium of the structure may consist of an arbitrary number of horizontal, infinitely stretched, dielectric and metallic layers, which are taken into account analytically. This environment is particularly interesting for plasmonic structures, because they are traditionally deposited on a flat layered dielectric substrate, for example glass. This glass can be contained in the background environment. The only remaining components are local scattering objects with medium or small dimensions compared to the wavelength, such as the dipole in section 5.2.1, but also dots, rods, monomers, dimmers, rings, discs, etc.. With volumetric integral equations these components are replaced by equivalent volume currents, which appear as the primary unknowns in the resulting integral equations. Since the substrates used in normal circumstances are huge compared to the wavelength, edge effects are extremely small and can be neglected. It may thus be concluded that plasmonic structures have specific features which make a IE-MoM solution quite attractive.

4. Software tools

In the following sections, several solvers are briefly described. Several commercial solvers and one academic solver are considered. It has to be emphasized that the author does not claim that this overview is complete. Since the widespread use of computational electromagnetics in plasmonics is quite recent, it is highly probably that there are more solvers that the author is not aware of.

4.1. Commercial software tools

4.1.1. HFSS [27]: FEM

Since it was one of the first tools in the market, and also due to its generality and flexibility, HFFS is one of the tools heavily used in industrial microwave and millimeter wave design environments. The purpose of HFSS is to extract parasitic parameters (S, Y, Z), visualize 3D electromagnetic fields (near- and far-field), and generate SPICE models, all based on a 3D

FEM solution of the electromagnetic topology under consideration. This software is extremely popular and is used for all kinds of purposes. Numerous results for plasmonic topologies can be easily found by googling the words HFSS and plasmonics.

The first step in a new simulation project is to define the geometry of the structure. The geometry is modeled in the GUI or can be imported from another program (AutoCAD, STEP, ...). Then the user continues by defining material properties, boundary conditions, and excitations to the different domains and surfaces of the geometry. If desired one can control the meshing process manually. Finally, the solution parameters are defined. The most important are the solution frequency, the order of the base functions and the matrix solver type. The direct matrix solver is default, but an iterative solver can be used as well. Very useful is the automatic adaptive mesh generation and refinement, which in many cases frees the designer of worrying about which mesh/grid to choose. First a solution for an initial mesh is determined and the solution is assessed. If the solution does not qualify, the mesh is refined and a new solution is computed. This procedure is repeated until one of the exit criteria is fulfilled. It is important to note that HFSS offers no curved elements for a better approximation of curved objects. This means that the mesh along a curved surface has to be chosen very fine. This can be achieved with the 'surface approximation' parameter. Finally the solution, i.e. electric and magnetic fields, currents, and S-parameters can be visualized in 1D, 2D and 3D. All solutions can be exported as files. An 'Optimetrics' toolbox is offered for optimizations, parameter, sensitivity and statistical analysis.

4.1.2. COMSOL multiphysics [28]: FEM

Comsol Multiphysics uses the finite element method to solve partial differential equations in 2D and 3D. It is a multiphysics code, which means that it can handle not only electromagnetics, but also acoustics, mechanics, fluid dynamics, heat transfer, etc.. The geometry, material parameters and boundary conditions have to be set up in the graphical user interface (GUI). Then, after defining boundaries and domains the mesh is generated. This includes the selection of the order of the basis functions and the curved mesh elements (order higher than one means curved element). They are used for a better approximation of curved boundaries. The meshing can be steered by specifying a number of nodes and their distribution on each edge of the model. Comsol Multiphysics offers a wide variety of matrix solvers to solve the system of equations. There are direct solvers, more suitable for smaller problems, and iterative solvers suitable for larger problems. Comsol Multiphysics can generate 2D and 3D plots. Quantities derived from the electromagnetic field, like Poynting vector, energy, and energy loss, are available as predefined variables. Comsol Multiphysics does not allow parameter sweeps or optimization. COMSOL is very easily combined with MATLAB. A simulation set up with the GUI can be saved as a Matlab script, which can be easily modified. The Matlab optimization toolbox can be used to optimize a topology.

4.1.3. JCMsuite [29]: FEM

JCMsuite is a software package based on FEM. It contains dedicated tools for certain simulation problems appearing in nano-optics and plasmonics. It incorporates scattering tools, propagation mode tools, and resonance mode tools. The time-harmonic problems can be formulated in 1D, 2D and 3D. The topologies can be isolated or may occur in periodic patterns, or a mixture of both. Interesting is that dedicated tools for problems posed on cylindrically symmetric geometries are available. The electromagnetic fields are discretized with higher order edge elements. JCMsuite contains an automatic mesh generator (with adaptive features), goal-oriented error estimators for adaptive grid refinement, domain-decomposition techniques and fast solvers.

4.1.4. CST [30]: FIT

CST Microwave Studio (CST MWS) is based on the finite integration technique (FIT). It allows to choose the time domain as well as the frequency domain approach. Despite the presence of transient, eigenmode, and frequency domain solvers within CST MWS, the transient solver is considered as the flag ship module. The Time Domain Solver calculates the broadband behavior of electromagnetic devices in one simulation run with an arbitrarily fine frequency resolution. The modeling of curved structures using the Perfect Boundary Approximation® technique and the modeling of thin perfectly electric conducting sheets with the Thin Sheet Technique® tries to cope with the typical difficulties inherent to FDTD methods for classical structures. Several mesh types can be chosen. The automatic mesh generator detects the important points inside the structure (fixpoints) and locates mesh nodes there. The user can manually add fixpoints on a structure, as well as fully control the number of mesh lines in each coordinate with regards to the specified wavelength. Energy based adaptation of the mesh allows to refine it in a predefined number of passes, providing a mesh refinement of sophisticated design features for the price of longer overall simulation times. Although the FIT in principle can handle material parameters changing over the dielectric volumes defined, this is not implemented yet.

CST, as a general purpose software package being a real competitor for HFSS in the traditional microwave field, has gained a lot of popularity in the last few years. Also for the analysis and design of plasmonic structures, more and more results obtained with CST can be found in literature (just google the words CST and plasmonics). A problem sometimes observed with CST is a ripple in the frequency response in case the tool settings are not appropriate. This is due to the fact that the flagship of CST is inherently a time domain solver.

4.1.5. Lumerical [15]: FDTD

Lumerical is the leading software tool in the plasmonics and photonics community. It is based on the FDTD algorithm, and can be used for 2D and 3D topologies. The topology has to be generated in the GUI or can be imported with GDSII/SEM files. Basic shapes include triangles, rectangular blocks, cylinders, conic surfaces, polygons, rings, user-defined (parametric) surfaces, spheres and pyramids. Several boundary conditions can be used: absorbing (PML), periodic, Bloch, symmetric, asymmetric, and metal boundaries. A non-uniform mesh can be used and automesh algorithms are provided. Since it is a dedicated time domain solver, it has a sophisticated library of Lorentz, Drude, Debye models for the material parameters. Excitation of the structure is possible with waveguide sources, dipoles, plane waves, focused beams and diffraction-limited spots. A scripting language is available to customize simulation. Data can be exported to Matlab or in ASCII.

4.2. Non-commercial software tools

4.2.1. MAGMAS 3D [31]: IE-MoM

MAGMAS 3D is the IE-MoM code developed at the Katholieke Universiteit Leuven, Belgium. It was originally developed in cooperation with the European Space Agency for planar and quasi-planar antenna and scattering structures embedded in a multilayered dielectric background medium and operating in the microwave frequency range. Starting from the topology considered, frequency band needed, and type of excitation, it calculates the network, radiation, and scattering characteristics of the structure under consideration. It is based on a full-wave Mixed-Potential formulation of Electric Field Integral Equations (MP-EFIE), originally applied only to 2D surface currents [22], [32]. New theoretical techniques were developed and implemented within the framework: the Expansion Wave Concept (EWC) [33], [34], the Dipole Modeling Technique (DMT) [35], special de-embedding procedures [36], etc.. In 2007, it was extended with the capability to handle volumetric 3D currents with the VIE technique introduced in [21]. This was crucial in view of its application in plasmonics. In 2009, this led to the first verified simulated results in this field [25]. Now, it is extensively used for plasmonics. The MAGMAS mesh is based on a combination of rectangular and triangular mesh cells. Full mesh control is available in manual meshing mode. A Graphical User Interface is available.

To the best knowledge of the author, to date (June 2012), it is the only IE-MoM framework able to handle arbitrary plasmonic structures embedded in a multilayered environment, with the very general and flexible VIE technique. No similar commercial solvers exist, in contrast to the situation at microwave frequencies, where a multitude of IE solvers can be found.

4.2.2. Other non-commercial tools found in literature

At this moment plasmonics is a mainly experimentally driven research field. Whereas for example in the field of traditional antenna research, numerous papers can be found on modeling as such, this is much less the case in plasmonics, especially in the high impact journals. Most papers are concerned with the description of the physical phenomena occurring and occasionally just mention the numerical tool used.

Most researchers in the plasmonics field use the commercial tools available, and as far as I can see, the main one is Lumerical. In the publications that do treat the numerical modeling of plasmonic structures as such, mainly the FDTD or the FEM technique is applied [37], [38]. Very few papers consider the IE-MoM technique, and if they do, it is the Surface IE technique [39], [40], [41]. In [42] Chremmos uses a magnetic type scalar integral equation to describe surface plasmon scattering by rectangular dielectric channel discontinuities. Even the use of the magnetic current formalism to describe holes, classical at microwave frequencies, already has been used in plasmonics [43]. However, these dedicated developments cannot be categorized under the title „analysis framework" in the sense that they do not allow to handle a wide range of different topologies.

5. Benchmarking

Nowadays, physicists and engineers rely heavily on highly specialized full wave electromagnetic field solvers to analyze, develop, and optimize their designs. Computer-aided analysis and optimization have replaced the design process of iterative experimental modifications of the initial design. It is evident that the underlying solution method for a software tool may significantly influence the efficiency and accuracy by which certain structure types are analyzed. Nevertheless, the commercial focus increasingly switches from such key theoretical considerations to improvements in the area of layout tools and system-level design tools. Therefore, users may get the wrong impression that a given solver is automatically suited to solve any kind of problem with arbitrary precision. This is of course not true.

This section verifies the plausibility of such expectations by presenting a benchmark study for a few plasmonic structures. The study focuses on the capabilities and limitations of the applied EM modeling techniques that usually remain hidden for the user.

5.1. Benchmarking in literature

In literature, not many comparisons between solvers in the field of plasmonics can be found. In [44] Hoffmann et al. consider a single plasmonic topology, a pair of Au spheres, and analyze this structure with several electromagnetic field solvers: COMSOL Multiphysics (FEM), JCMsuite (FEM), HFSS (FEM), and CST Microwave Studio (the FEM solver available in the Studio is used, not the flagship FDTD solver). The output parameter considered is the electric field strength in between the two spheres. Note that all solvers tested are based on the Finite Elements technique in the frequency domain, so it can be expected that the main issues with these solvers for this topology are the same. Very interesting is the fact that in the paper itself CST is categorized as "inaccurate". However, afterwards this was corrected. It seems that wrong material settings were used, and after correction accurate results were obtained. This issue led to the fact that this benchmark example is now presented as a reference example on the CST website.

In [45], several numerical methods are tested for 2D plasmonic nanowire structures: not only the Finite Element Method (FEM) and the Finite Difference Time-Domain (FDTD) technique, but also less „commercial" methods like the Multiple Multipole Program (MMP), the Method of Auxiliary Sources (MAS), and the Mesh-less Boundary Integral Equation (BIE) method are tested. By comparing the results, several conclusions can be drawn about their applicability and accuracy for plasmonic topologies. Differential techniques like FEM and FDTD can reach a high level of accuracy only with a high discretization. In 2D, this is readily affordable on present-day computer systems. There, these techniques have a clear advantage in terms of speed, matrix size, and accuracy. In 3D, due to the rocketing size of the problem, this may not always be that obvious. The advantages may thus disappear in a full 3D analysis of geometrically complicated structures. This happens because matrices become denser or more ill-conditioned. The main conclusion of the paper is that the most efficient method depends on the problem dimension and complexity. This is a similar conclusion as was reached in [46] within the context of the analysis and design of planar antennas.

5.2. Comparison between differential and integral equation techniques

In this section, the differential and integral equation techniques are compared. This is done by choosing a representative solver from each category and using it in the analysis of a basic plasmonic topology. In the category differential techniques, Lumerical is chosen, as it is the most widespread commercial solver in the plasmonic community. Since, as far as we know, there are no commercial integral equation solvers in the plasmonics area, MAGMAS 3D is chosen in this category, the in-house developed solver at Katholieke Universiteit Leuven. The basic topology selected is the plasmonic dipole. It is a structure that can function as a scatterer, but also as a real nano-antenna. The main result of this section is that it draws conclusions on different computational aspects involved in the modeling of nanostructures with the two solvers.

5.2.1. Plasmonic dipole

The plasmonic dipole is depicted in Fig. 3. It is a structure consisting of two nanorods with a gap in between. When the gap is "shortcircuited", it functions as a single rod, when the gap is open, it functions as a device generating an enhanced electric field there. When it is connected to other nanocircuits, it may function as a nano-antenna, both in transmit (Fig. 3b), and in receive (Fig. 3c). The two rods may be fabricated from metals like Au, Ag, Cu, Al, Cr, etc.. In this section the width W and height H of the dipole are kept constant at 40 nm. This value is well-chosen, since it is a value which can be fabricated with sufficient accuracy using present day nanofabrication technology. The gap width can have different values, depending on the case considered. Note however that 10 nm is about the minimum that can be fabricated with reasonable accuracy nowadays.

5.2.1.1. Quality of the input

It cannot be over-emphasized that Computational Electromagnetics solvers in general produce correct results consistently for a multitude of different structures only if two conditions are met:

1. The user has sufficient general background in the field, and a more specific knowledge about the solution technique implemented in the solver that he is using. If not, there is a huge danger that the solver is not used in a proper way. For example, although this is not necessary to "operate" the solver, if the user has no basic knowledge and does not

realize how the discretization / meshing scheme works, it is impossible to understand and assess the effect of a proper meshing. In the majority of the cases, there will be no problem with that, and automatic meshing schemes will be able to produce good results. However, in unusual and/or challenging cases, often met in scientific research, this issue will be a crucial factor.

2. The user needs to have sufficient knowledge about the actual tool that he is using and its peculiarities, more specific about the way a certain problem has to be imported into the solver. A very good illustration of this has already been addressed in section 5.1 where CST was actually not properly used and produced incorrect results that were published in open literature. Afterwards, this was corrected and the problem disappeared. However, following the proper line of reasoning is not always straightforward. Even experienced researchers sometimes easily can make mistakes. It is the opinion of the author that this is happening too much, even in a lot of peer-reviewed scientific papers.

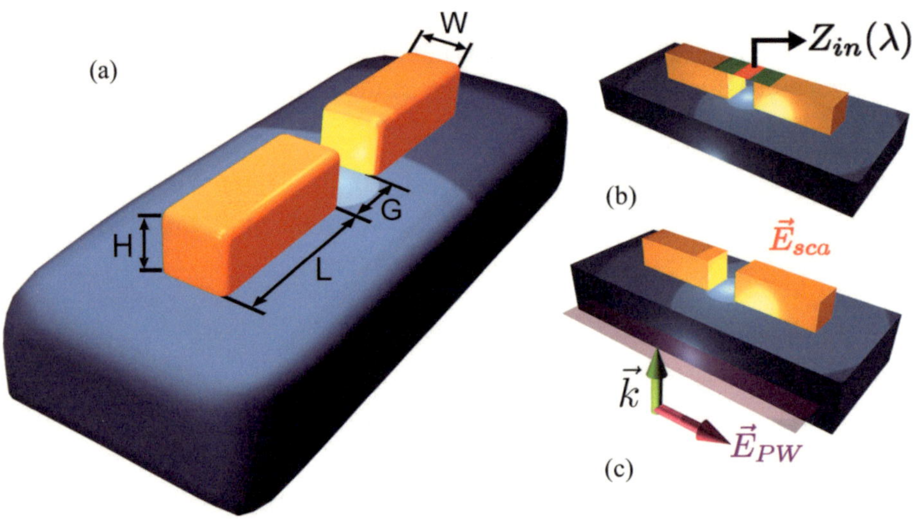

Figure 3. The dipole model studied. a) W and H are set equal to 40 nm and the gap G can vary. b) The dipole as transmitting antenna with a model for the feeding structure located in the gap. c) The dipole as receiving antenna excited by a plane wave.

5.2.1.2. Modeling of the feed

The first parameter studied is the impedance that is seen when the nanodevice operates as a transmitting or receiving antenna. In Fig. 4, the impedance simulated with both MAGMAS and Lumerical is given for an Al dipole of 200 nm length and with a gap of 10 nm. The input impedance is calculated as $Z = V/I$, where Z, V and I are the input impedance, the voltage over the gap G, and the source current, respectively. The exciting source is located in the middle of the gap. MAGMAS uses a physical current filament of finite width w. Lumerical does not have a built-in physically feasible current source model. It uses a (less realistic) pure electrical dipole source, which means that the current I has to be evaluated "manually" by the user based on the magnetic field distribution and Ampere's law. This also means that the impedance will be somewhat depending on the resolution chosen during meshing. It is seen that there is an excellent agreement in the trend of the curves, i.e. the change of input impedance with frequency for different widths of the source current filament in MAGMAS (5 nm and 13.33 nm) is the same as the one in Lumerical. However, in general there is an offset. Only for a specific width of the current filament in MAGMAS, nl. 5 nm, the two excitation mechanisms become almost completely equivalent and even the offset between the two solvers disappears. A similar agreement was also observed for both gold and silver dipoles.

The main conclusion of this section is that there is still an issue with the modeling of (localized) feeds, especially in Lumerical. The reason is that the plasmonics community is used to consider objects as scatterers, excited by an incident wave. In this case, the models implemented are satisfactory. However, in view of designing nanocircuits, localized feeds and impedances derived from them become important. The basic rule is that the feeding model has to correspond as closely as possible to the actual feed topology that is going to be used later in practice. In this respect, the localized feeding models available in Lumerical today are unsatisfactory.

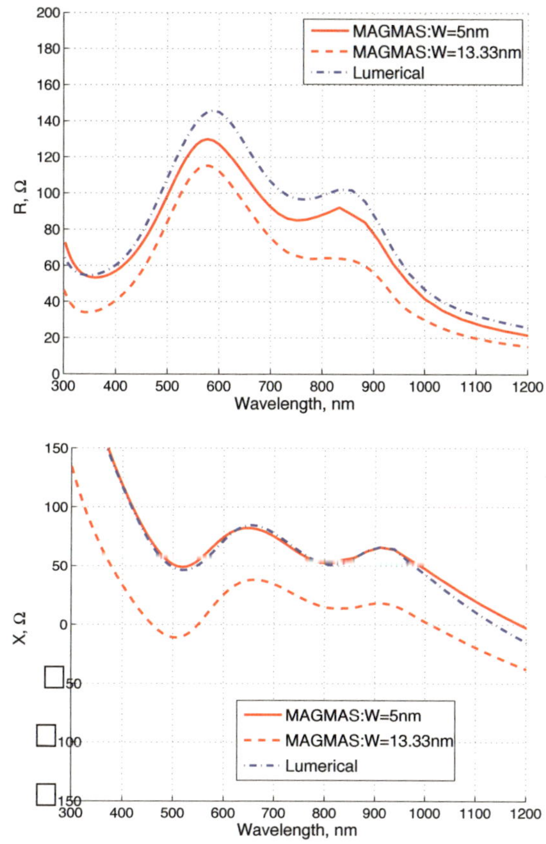

Figure 4. Comparison of impedances obtained with MAGMAS and Lumerical for an aluminum dipole on a glass substrate, L = 200 nm and gap = 10 nm. Two widths w are used in MAGMAS for the source current filament. R and X stand for real and imaginary part.

5.2.1.3. Modeling of losses

The importance of taking into account losses is obvious. Metals at plasmonic frequencies may be very lossy and neglecting this would result in totally erroneous results. However, there is a difference between time domain and frequency domain solvers. Since Lumerical is an FDTD based tool, it works in time domain. This means that it cannot work directly with (complex) permittivities and permeabilities. The material data have to be transformed from functions of frequency into functions of time. This is a very difficult process, involving a convolution, and is subject to various constraints. Local curve fitting of permittivities is very difficult to implement in a time domain solver. Lumerical fits the sampled measurement data with its own multi-coefficient material model, which "provides a superior fit compared to the standard material models (such as Lorentz, Debye, Drude, etc)". However, this produces artefacts, as is illustrated in Fig. 5. There, the relative error is given between the measured imaginary part of the permittivity (which represents losses) and the value obtained for this imaginary part after fitting, for three metals. It is seen that for gold and aluminum, the error is in the order of 20 %, which is already quite elevated. However, in the case of silver, the error reaches a full 100 %, which is unacceptable. In some cases, it will result in a totally wrong prediction for example of the radiation efficiency of nano-antennas [47]. If one forces the material characteristics to be exactly the same in the two solvers, for example by using the interpolated values also in MAGMAS, instead of the real measured ones, the two solvers produce very similar results, see Fig. 6.

The main conclusion is that a highly accurate prediction of losses in Lumerical is still an issue. The material models have to be further refined, especially for silver.

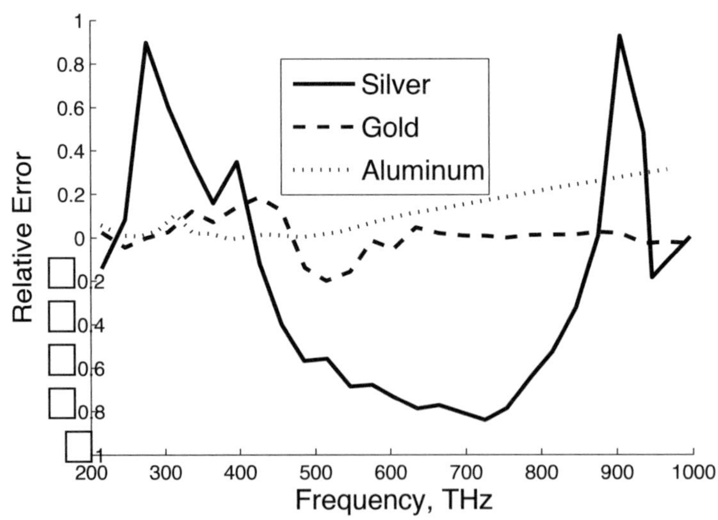

Figure 5. Relative error between interpolated and measured imaginary part of permittivity for silver, gold and aluminum.

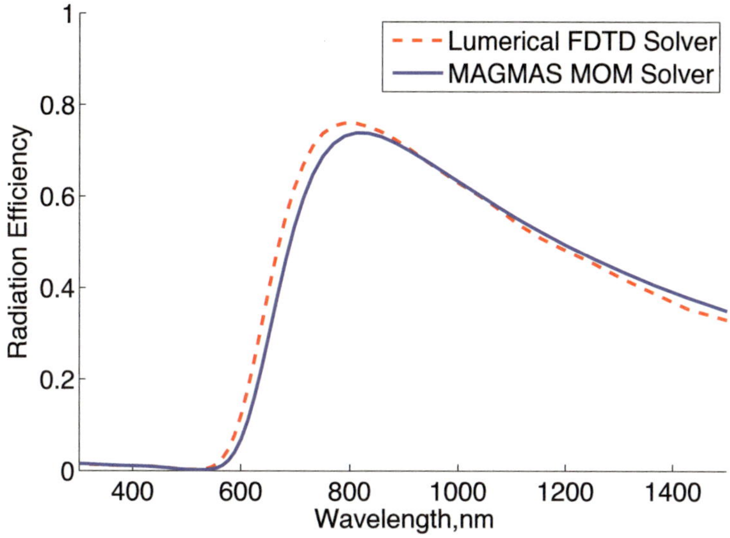

Figure 6. Comparison between FDTD and IE-MoM for a gold dipole in free space, L = 250 nm and the gap is 10 nm. Identical material parameters are used in both solvers.

5.2.1.4. Meshing

It is well known that the mesh quality and resolution are key factors in the accuracy of any solver. The electromagnetic coupling between nearby segments may differ considerably due to the specific meshing used, especially in parts of the structure where rapid topological changes occur. The question is whether the meshes used in the solvers are adequate. This issue was investigated by performing a convergence study in terms of the resolution of the meshes [48].

The extinction cross section of a gold monomer (no gap, or in other words G = 0 nm) calculated using MAGMAS 3D with different meshes is plotted in Fig. 7(a). The analysis of these data demonstrates clearly the stability of IE-MoM. The results obtained even with a very rough mesh 6x1x1 (41.7 *nm* x 40 *nm* x 40 *nm*) provide already a very good estimation of the antenna resonance properties. The differences in extinction cross section calculated with the rough and fine meshes are almost negligibly small. This result is obtained thanks to the variational stability of IE-MoM. On the adjacent figure Fig. 7(b) extinction cross sections are calculated using Lumerical. As expected, in general the results obtained with the FDTD method depend considerably on the chosen mesh. The calculated wavelengths as a function of the mesh cell size are plotted in Fig. 7(c). It should be also noted that in contrast to Fig. 7(a) (IE-MoM) in Fig. 7(b) (FDTD) not only peak positions but also their levels depend clearly on the mesh.

The conclusion is that in Lumerical a fine mesh is mandatory for reliable calculations of the monomer resonant wavelength. Thanks to its variational approach, in IE-MoM, a highly dense mesh is not always needed, especially when scattering problems are studied.

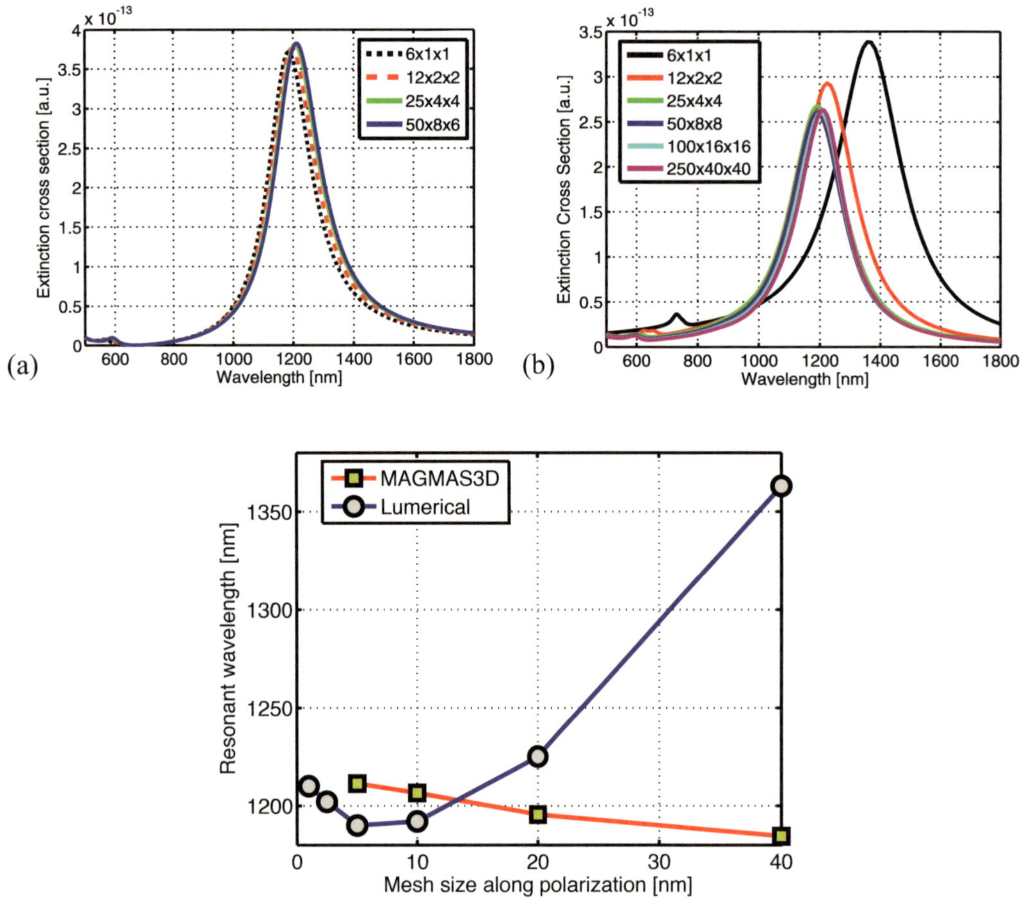

Figure 7. a and b: extinction cross section of a 250 nm x 40 nm x 40 nm gold monomer on a substrate with n=1.5 (a. MAGMAS 3D, b. Lumerical), c. convergence performance of both solvers.

5.2.1.5. Calculation speed

Information on the calculation times for the convergence study of the previous section is given in Table 1. It has to be emphasized that calculation times cannot be directly compared. The computers used were different, and the FDTD tool used 10 processors in parallel. Further, since it is a time domain technique, a calculation time per frequency point cannot be given for FDTD. Nevertheless, combining the data in this table with the data concerning the convergence, given in Fig. 7, it is easily seen that IE-MoM outperforms FDTD for this particular case.

5.2.2. Plasmonic nanojets

In this section it is illustrated what can be reached with computational electromagnetics in the field of plasmonics. The results in this section are extracted from the paper [49], where full details can be found.

IE-MoM		FDTD	
Mesh	Calculation time per frequency point (s)	Mesh	Total calculation time (s) (10 processors in parallel)
6 x 6 x 1	5.4	6 x 1 x 1	38
12 x 2 x 2	10.4	12 x 2 x 2	55
25 x 4 x 4	33.6	25 x 4 x 4	225
50 x 8 x 6	238	50 x 8 x 8	672
		100 x 16 x 16	5893
		250 x 40 x 40	20100

MoM: Intel(R) Xeon(R) CPU E5335 @ 2 GHz, 32 Gb memory
FDTD: d1585g6 4x hc CPU @ 2.8 GHz, 128 Gb memory

Table 1. Calculation times as a function of the mesh for the gold monomer.

Similarly to a pebble hitting a water surface, the energy deposition by femtosecond laser pulses on a gold surface can produce local melting and back-jet. Contrary to water though, a gold surface can be nanopatterned and the excitation of surface plasmon resonances leads to the appearance of hotspots, literally. This was explicitly proven for the first time in a nanopatterned gold surface, composed of a G shaped periodic structure. It was observed there that laser-induced melting and back-jet occur precisely in the plasmonic hotspots. The aid of computational electromagnetics to rigorously analyze the structure by predicting the hotspots was crucial in explaining this phenomenon. Much more details can be found in [49]. It is clear that this type of insight into the basic interaction mechanisms of light with matter at the nanoscale opens up new possibilities for applications.

Another example of what can be reached by applying computational electromagnetics in plasmonics can be found in [50].

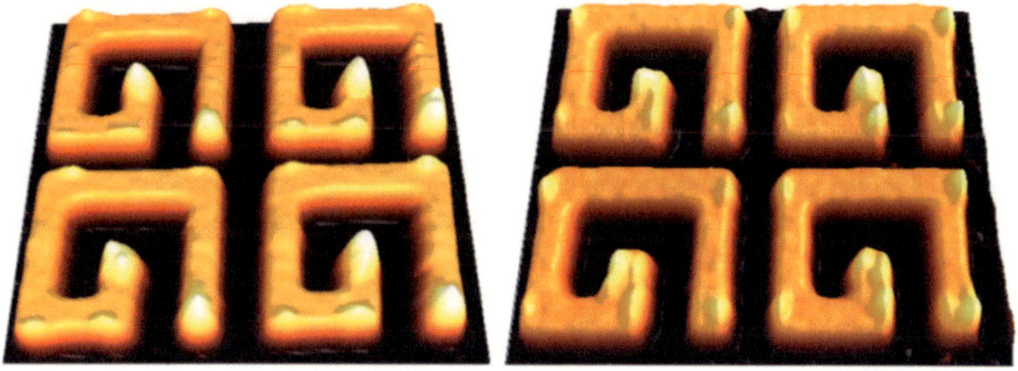

Figure 8. Illuminating the sample with femtosecond laser pulses produces a polarization dependent pattern of nanobumps with sharp tips on the sample surface [49]. Left: horizontally polarized light, right: vertically polarized light.

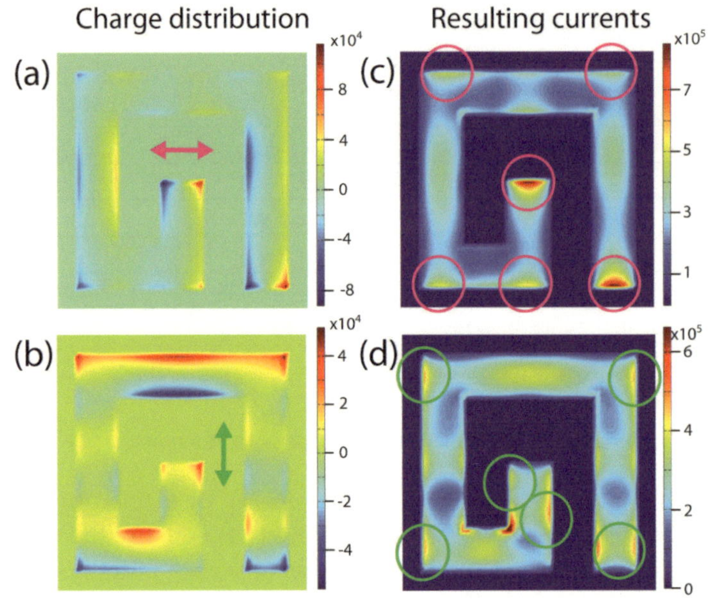

Figure 9. The pattern of melted nanobumps matches that of the plasmonic currents, where Ohmic losses locally increase the temperature [49]. The electric field of light incident on the nanostructures causes the surface charges to oscillate in response to the direction of light polarization: horizontal, in (a), and vertical, in (b). The locations of the resulting electric currents are shown in (c) and (d) respectively. The current maxima are indicated with colored circles matching the pattern of nanobumps in Fig. 8.

6. Conclusions

In this chapter, a brief introductory overview has been given on the use of existing computational techniques and software tools for the analysis and design of plasmonic structures. This field is booming, and many modeling techniques developed at lower frequencies, i.e. in the microwave range, are now being transferred to the plasmonics community, of course with the necessary adaptations. Although the differential equation techniques are the most widespread, both in scientific literature, and in the commercial scene, it is proven that integral equation techniques are a valid alternative. In some cases they are even superior. The author sees a lot of opportunities in this area.

In most cases, the tools are used to analyze structures. Few papers really make dedicated designs. However, this will change in the future. Inevitably, the plasmonics community will follow a similar patch as the traditional antenna community. Whereas 30 years ago, the design of an antenna was based on the accumulated expertise and know-how of many years, nowadays most antennas are designed almost in a single pass by experienced designers using commercial tools. This is possible since the tools in this field have sufficient accuracy and matureness in order to be able to do that. It is my belief that as soon as plasmonics will make the unavoidable shift from the physics to the engineering community, this last community will aim at conceiving applications which will involve designs heavily based on computational electromagnetics.

I would like to end this chapter with the following guideline. The use of two different solvers, based on different theoretical methods (integral and differential) may provide an excellent means to characterize the quality of simulation results. If the two results are in good agreement, it is highly likely that the results are correct. If the two results are in disagreement, a deeper investigation of the structure and its modeling is absolutely necessary.

Author details

Guy A. E. Vandenbosch
Katholieke Universiteit Leuven, Belgium

Acknowledgement

The author gratefully acknowledges the following persons: Zhongkun Ma for the impedance and radiation efficiency comparison between MAGMAS and Lumerical, Dr. Vladimir Volski for the convergence study with MAGMAS and Lumerical, and V. K. Valev for making available the plasmonic nanojet figures.

We also would like to express our gratitude to the KU Leuven Methusalem project and its beneficiary Prof. V. V. Moshchalkov, whose activities created a real stimulus to become personally active in this field, and the fund for scientific research Flanders (FWO-V) for the financial support.

7. References

[1] I. Ahmed, E. H. Khoo, E. Li, and R. Mittra, "A hybrid approach for solving coupled Maxwell and Schrödinger equations arising in the simulation of nano-devices", IEEE Antennas and Wireless Propagation Letters, Vol. 9, pp. 914- 917, 2010.

[2] D. B. Davidson,"A review of important recent developments in full-wave CEM for RF and microwave engineering," IEEE 3rd Int. Conf. Comp. Electromagnetics and Its Applications, pp. PS/1-PS/4, Nov. 2004.

[3] C. W. Townbridge and J. K. Sykulski,"Some Key Developments in Computational Electromagnetics and Their Attribution," IEEE Antennas and Propag. Magazine, vol. 42, no. 6, pp. 503 – 508, Apr. 2006

[4] E. K. Miller,"A Selective Survey of Computational Electromagnetics," IEEE Trans Antennas Propag, vol. 36, no. 9, Sept. 1988, pp. 1281 – 1305

[5] F. Peterson, S. L. Ray, and R. Mittra, "Computational methods for electromagnetics", IEEE Press – Oxford University Press, 1998.

[6] Awadhiya, P. Barba, and L. Kempel, "Finite-element method programming made easy???", IEEE Antennas and Propagation Magazine, vol. 45, pp. 73 – 79, Aug. 2003.

[7] J. M. Jin, The Finite Element Method in Electromagnetics, second edition, John Willey & Sons, Inc., New York, 2002.

[8] J. L. Volakis, A. Chatterjee, and L. C. Kempel, Finite element method for electromagnetics, IEEE Press, Oxford University Press, 1997.
[9] R. L. Courant, "Variational methods for the solution of problems of equilibrium and vibration," Bulletin of the American Mathematical Society, 5, pp. 1-23, 1943.
[10] S. Ahmed, "Finite-element method for waveguide problems," Electronics Letters, vol. 4, Issue 18, pp.387 – 389, Sept. 1968.
[11] R. Coccioli, T. Itoh, G. Pelosi, P. P. Silvester, "Finite-Element Methods in Microwaves: A Selected Bibliography," IEEE Antennas and Propag. Magazine, vol. 38, Issue 6, pp.34 – 48, Dec. 1996.
[12] K. S. Yee, "Numerical Solution of Initial Boundary Value Problems Involving Maxwell's Equations in Isotropic Media," IEEE Trans on Antennas Propag, vol 14, pp. 302-307, 1966.
[13] Taflove, Computational electrodynamics: the finite difference time domain method, Artech House, 1997.
[14] D. M. Sullivan, "Electromagnetic simulation using the FDTD method", Wiley – IEEE Press., 2000, ISBN 978-0-7803-4747-2.
[15] www.lumerical.com
[16] T. Weiland, "A discretization method for the solution of Maxwell's equations for six-component fields", Electronics and Communications AEÜ, vol 31, No. 3, 116–120, 1977.
[17] Bossavit and L. Kettunen,"Yee-like schemes on a tetrahedral mesh, with diagonal lumping," Int. J. Numer. Model., vol. 42, pp. 129 – 142, 1999.
[18] M. Celuch-Marcysiak and W. K. Gwarek, "Comparative study of the time-domain methods for the computer aided analysis of microwave circuits," Int. Conf. on Comp. in Electromagnetics., pp. 30-34, Nov. 1991.
[19] R. F. Harrington, "Field Computation by Moment Methods", New York: Macmillan, 1968.
[20] T. K. Sarkar, E. Arvan, and S. Ponnapalli, "Electromagnetic scattering from dielectric bodies", IEEE Trans. Antennas Propag., vol. 37, pp. 673-676, May 1989.
[21] Y. Schols and G. A. E. Vandenbosch, "Separation of horizontal and vertical dependencies in a surface/volume integral equation approach to model quasi 3-D structures in multilayered media ", IEEE Trans. Antennas Propagat., vol. 55, no. 4, pp. 1086-1094, April 2007.
[22] M. Vrancken and G.A.E. Vandenbosch, Hybrid dyadic-mixed-potential and combined spectral-space domain integral-equation analysis of quasi-3-D structures in stratified media, IEEE Trans Microwave Theory Tech, vol 51, pp. 216-225, 2003.
[23] X. Zheng, V. K. Valev, N. Verellen, Y. Jeyaram, A. V. Silhanek, V. Metlushko, M. Ameloot, G. A. E. Vandenbosch, and V. V. Moshchalkov, "Volumetric Method of Moments and conceptual multi-level building blocks for nanotopologies", Photonics Journal, Vol. 4, No. 1, pp. 267-282 , Jan. 2012.
[24] J. Smajic, C. Hafner, L. Raguin, K. Tavzarashvili, and M. Mishrickey, "Comparison of Numerical Methods for the Analysis of Plasmonic Structures", *Journal of Computational and Theoretical Nanoscience*, vol. 6, no. 3, pp. 763-774, 2009.

[25] G. A. E. Vandenbosch, V. Volski, N. Verellen, and V. V. Moshchalkov, "On the use of the method of moments in plasmonic applications", Radio Science, 46, RS0E02, doi:10.1029/2010RS004582, 21 May 2011.

[26] R. Qiang, R. L. Chen, and J. Chen, "Modeling electrical properties of gold films at infrared frequency using FDTD method", *International Journal of Infrared and Millimeter Waves*, vol. 25, no. 8. pp. 1263-1270, 2004.

[27] www.ansoft.com

[28] www.comsol.com

[29] www.jcmwave.com

[30] CST GmbH – Computer Simulation Technology, CST Microwave Studio 2006 user manual, 2006, http://www.cst.com

[31] G. A. E. Vandenbosch, MAGMAS 3D, http://www.esat.kuleuven.be/telemic/antennas/magmas/

[32] G. A. E. Vandenbosch and A. R. Van de Capelle, Mixed-potential integral expression formulation of the electric field in a stratified dielectric medium - application to the case of a probe current source, *IEEE Trans. Antennas Propagat.*, vol. 40, pp. 806-817, July 1992.

[33] F. J. Demuynck, G. A. E. Vandenbosch and A. R. Van de Capelle, The expansion wave concept, part I: efficient calculation of spatial Green's functions in a stratified dielectric medium, *IEEE Trans. Antennas Propagat.*, vol. 46, pp. 397-406, Mar. 1998.

[34] G. A. E. Vandenbosch and F. J. Demuynck, The expansion wave concept, part II: a new way to model mutual coupling in microstrip antennas, *IEEE Trans. Antennas Propagat.*, vol. 46, pp. 407-413, March 1998.

[35] B. L. A. Van Thielen and G. A. E. Vandenbosch, Method for the calculation of mutual coupling between discontinuities in Planar Circuits, *IEEE Trans. Microwave Theory and Techniques*, vol. 50, pp. 155-164, January 2002.

[36] E. A. Soliman, G. A. E. Vandenbosch, E. Beyne and R.P. Mertens, Multimodal Characterization of Planar Microwave Structures, *IEEE Trans. Microwave Theory and Techniques*, vol. 52, no. 1, pp. 175-182, January 2004.

[37] R. Kappeler, D. Erni, X.-D. Cui, L, Novotny, "Field computations of optical antennas", *Journal of Computational and Theoretical Nanoscience*, vol. 4, no. 3, pp. 686-691, 2007.

[38] R. Qiang, R. L. Chen, and J. Chen, "Modeling electrical properties of gold films at infrared frequency using FDTD method", *International Journal of Infrared and Millimeter Waves*, vol. 25, no. 8. pp. 1263-1270, 2004.

[39] B. Gallinet and O. J. Martin, "Scattering on plasmonic nanostructures arrays modeled with a surface integral formulation", *Photonics and Nanostructures – Fundamentals and Applications*, vol. 8, no. 4, pp. 278-284, 2009.

[40] A. M. Kern and O. J. F. Martin, "Surface integral formulation for 3D simulations of plasmonic and high permittivity nanostructures", J Opt. Soc. Am. A, vol. 26, no. 4, pp. 85-94, 2010.

[41] B. Gallinet, A. M. Kern, and O. J. F. Martin, "Accurate and versatile modeling of electromagnetic scattering on periodic nanostructures with a surface integral approach", J Opt. Soc. Am. A, vol. 27, no. 10, pp. 2261-2271, 2010.

[42] I. Chremmos, "Magnetic field integral equation analysis of surface plasmon scattering by rectangular dielectric channel discontinuities", J Opt. Soc. Am. A Opt. Image Sci. Vis., vol. 27, no. 1, pp. 85-94, 2010.

[43] J. Alegret, P. Johansson, and M. Kall, "Green's tensor calculations of Plasmon resonances of single holes and hole pairs in thin gold films", *New Journal of Physics*, vol. 10, no. 10, 2008.

[44] J. Hoffmann, C. Hafner, P. Leidenberger, J. Hesselbarth, S. Burger, Comparison of electromagnetic field solvers for the 3D analysis of plasmonic nano antennas, Proc. SPIE Vol. 7390, pp. 73900J-73900J-11 (2009).

[45] J. Smajic, C. Hafner, L. Raguin, K. Tavzarashvili, and M. Mishrickey, "Comparison of Numerical Methods for the Analysis of Plasmonic Structures", *Journal of Computational and Theoretical Nanoscience*, vol. 6, no. 3, pp. 763-774, 2009.

[46] Vasylchenko, Y. Schols, W. De Raedt, and G. A. E. Vandenbosch, "Quality assessment of computational techniques and software tools for planar antenna analysis", IEEE Antennas Propagat. Magazine, Feb. 2009.

[47] G. A. E. Vandenbosch and Z. Ma, "Upper bounds for the solar energy harvesting efficiency of nano-antennas", Nano Energy, Vol. 1, No. 3, pp. 494-502, May 2012.

[48] F. Pelayo G. De Arquer, V. Volski, N. Verellen, G. A. E. Vandenbosch, and V. V. Moshchalkov, "Engineering the input impedance of optical nano dipole antennas: materials, geometry and excitation effect", IEEE Trans. Antennas Propagat., Vol. 59, No. 9, pp. 3144-3153, Sep. 2011.

[49] V. K. Valev, D. Denitza, X. Zheng, A. Kuznetsov, B. Chichkov, G. Tsutsumanova, E. Osley, V. Petkov, B. De Clercq, A. V. Silhanek, Y. Jeyaram, V. Volskiy, P. A. Warburton, G. A. E. Vandenbosch, S. Russev, O. A. Aktsipetrov, M. Ameloot, V. V. Moshchalkov, T. Verbiest, "Plasmon-enhanced sub-wavelength laser ablation: plasmonic nanojets", Advanced Materials, Vol. 24, No. 10, pp. OP29-OP35, March 2012.

[50] V. K. Valev, A. V. Silhanek, B. De Clercq, W. Gillijns, Y. Jeyaram, X. Zheng, V. Volskiy, O. A. Aktsipetrov, G. A. E. Vandenbosch, M. Ameloot, V. V. Moshchalkov, T. Verbiest, "U-shaped switches for optical information processing at the nanoscale", Small, Vo. 7, No. 18, pp. 2573-2576, Sep. 2011.

Chapter 3

Numerical Simulations of Surface Plasmons Super-Resolution Focusing and Nano-Waveguiding

Xingyu Gao

Additional information is available at the end of the chapter

http://dx.doi.org/10.5772/50751

1. Introduction

Surface plasmons(SPs) are evanescent waves that propagate along the surface of dispersive media [1]. When a light beam incidents onto the surface of metal, the electromagnetic field interacts with the free electrons inside the metal, which leads to the oscillation of free electrons to excite the SPs in the exit medium. Noble metals with nanostructure geometry have special optical properties because they can excite localized surface plasmons (LSPs) under the illumination of light field. Since the pioneer study of Ritchie [2], the special optical properties of SPs in nano-scale have been tightly investigated. Various metallic nano-structures are reported for nano-plasmonic devices, including thin film[3,4], nanowires[5,6], nanorods[7,8], nano-hole array[9,10], nano-slits[11] et. al. Due to the sub-wavelength excitation range and strong field enhancement properties of SPs, they have been widely applied in super-resolution optical microscopy[12], nano-photonic trapping technology[13], biology and medical sciences[14] and nano-photonic waveguide[15].

In this chapter we firstly introduce a numerical algorithm for implementing the relative permittivity model of dispersive media in finite difference time domain(FDTD) method, i.e. piecewise linear recursive revolution(PLRC) method. Next the super-resolution phenomenon derived from the surface plasmons excited at the focal region is presented and analyzed. In the third part, we explore two kinds of plasmonic nano-waveguide: parallel nanorods and metal-dielectric-metal structure. Their optical properties, such as long distance waveguiding from the focal region, turning waveguiding effect and optical switch effect, will be demonstrated in detail. Finally, we conclude our research results in recent years and look forward the applications of surface plasmons in nano-technologies.

2. FDTD method for plasmonics simulation

The FDTD method was firstly proposed by Yee in 1966[16]. Because of its strong and precise power for simulating the propagation of the electromagnetic field, it was quickly applied in many research fields associate with the electromagnetics and optics. The detailed discrete differential equations of Maxwell equations can be found in Ref[17]. In this section, we mainly present the FDTD algorithm for dispersive media which is referred to as piecewise linear recursive revolution(PLRC) method and the verification of the program code.

2.1. 3D FDTD algorithm for dispersive media

The recursive revolution(RC) method for implementing the relative permittivity models of dispersive media in finite difference time domain(FDTD) algorithm was first proposed by Luebbers et al. [18] in 1990. It had been testified to perform faster calculation speed and fewer memory space requirement than the auxiliary differential equation(ADE) method [19], Z-transform method [20] and shift operator(SO) method [21]. In 1996, Kelley and Luebbers proposed the improved PLRC method [22] which remained the speed and low memory of RC method, and provided the accuracy existing in ADE method. The PLRC method for Drude model and Lorentz model has been presented respectively[22, 23]. In this section, we combine these two sets of formulations and propose the PLRC method for Drude-Lorentz model.

The curl equations of Maxwell's equations are presented as:

$$\nabla \times \mathbf{H} = \varepsilon \frac{\partial \mathbf{E}}{\partial t} \tag{1}$$

$$\nabla \times \mathbf{E} = -\mu \frac{\partial \mathbf{H}}{\partial t} \tag{2}$$

where $\mathbf{E}$ is the electric field vector and $\mathbf{H}$ is the magnetic field vector, respectively. ε and μ are the permittivity and permeability of the medium, respectively.

In dispersive medium, the displacement vector $\mathbf{D}$ has the linear relation with the electric field vector $\mathbf{E}$ in frequency domain as:

$$\mathbf{D}(\omega) = \varepsilon(\omega)\mathbf{E}(\omega) = \varepsilon_0\varepsilon_r(\omega)\mathbf{E}(\omega) \tag{3}$$

where $\varepsilon_r(\omega)$ is the relative permittivity function of dispersive medium. It can be expressed by Drude-Lorentz model[24]as:

$$\varepsilon_r(\omega) = \varepsilon_\infty + \frac{\omega_d^2}{j\omega\gamma_d - \omega^2} + \frac{\Delta\varepsilon\omega_L^2}{\omega_L^2 + 2j\omega\delta_L - \omega^2} = \varepsilon_\infty + \chi_d(\omega) + \chi_L(\omega) \tag{4}$$

where ε_∞ is the relative permittivity at the infinite frequency, ω_d is the plasma frequency, γ_d is the oscillation rate of electrons, $\Delta\varepsilon$ is the difference of the relative permittivity between infinite frequency and zero frequency, ω_L is the oscillation rate of Lorentzian dipoles, δ_L is the decreasing coefficient, $\chi_d(\omega)$ and $\chi_L(\omega)$ are Drude susceptibility term and Lorentz susceptibility term respectively. The update equation for the electric field in FDTD format is:

$$\mathbf{E}^{n+1} = \left(\frac{\varepsilon_\infty - \xi^0}{\varepsilon_\infty - \xi^0 + \chi^0}\right)\mathbf{E}^n + \left(\frac{\Delta t/\varepsilon_\infty}{\varepsilon_\infty - \xi^0 + \chi^0}\right)\nabla \times \mathbf{H} + \left(\frac{1}{\varepsilon_\infty - \xi^0 + \chi^0}\right)\boldsymbol{\Psi}^n \tag{5}$$

where $\boldsymbol{\Psi}^n$ is an intermediary variable which includes the contributions from Drude term and Lorentz term, respectively. The parameters in Equ.(5) are combination of Drude and Lorentz terms. The contribution from Drude term for variable $\boldsymbol{\Psi}^n$ is updated as:

$$\boldsymbol{\Psi}_d^n = \left(\Delta\chi_d^n - \Delta\xi_d^n\right)\mathbf{E}^n + \Delta\xi_d^n\mathbf{E}^{n-1} + \boldsymbol{\Psi}^{n-1}\exp\left(-\gamma_d\Delta t\right) \tag{6}$$

where

$$\chi_d^n = \frac{\omega_d^2}{\gamma_d}\left[\Delta t - \frac{1}{\gamma_d}\left(1 - e^{-\gamma_d\Delta t}\right)\right] \tag{7}$$

$$\Delta\chi_d^n = -\frac{\omega_d^2}{\gamma_d^2}\left(1 - e^{-\gamma_d\Delta t}\right)^2 \tag{8}$$

$$\xi_d^n = \frac{\omega_d^2}{\gamma_d}\left[\frac{\Delta t}{2} - \frac{1}{\gamma_d^2}\left(1 - e^{-\gamma_d\Delta t}\right) + \frac{1}{\gamma_d}e^{-\gamma_d\Delta t}\right] \tag{9}$$

$$\Delta\xi_d^n = -\frac{\omega_d^2}{\gamma_d^2}\left[\frac{1}{\gamma_d\Delta t}\left(1 - e^{-\gamma_d\Delta t}\right)^2 - \left(1 - e^{-\gamma_d\Delta t}\right)e^{-\gamma_d\Delta t}\right] \tag{10}$$

The contribution from Lorentz term for variable $\boldsymbol{\Psi}^n$ is updated as:

$$\widehat{\boldsymbol{\Psi}}_L^n = \left(\Delta\widehat{\chi}_L^n - \Delta\widehat{\xi}_L^n\right)\mathbf{E}^n + \Delta\widehat{\xi}_L^n\mathbf{E}^{n-1} + \widehat{\boldsymbol{\Psi}}_L^{n-1}\exp\left((-\alpha + j\beta)\Delta t\right) \tag{11}$$

where "^" means it is a complex variable. $\alpha=\delta_L$, $\beta=(\omega_L^2-\delta_L^2)$, $\gamma = \Delta\varepsilon\omega_L^2$, and

$$\widehat{\chi}_L^n = -\frac{j\gamma}{-\alpha+j\beta}\{1 - \exp[(-\alpha + j\beta)\Delta t]\} \tag{12}$$

$$\Delta\widehat{\chi}_L^n = \frac{j\gamma}{-\alpha+j\beta}\{1 - \exp[(-\alpha + j\beta)\Delta t]\} \tag{13}$$

$$\widehat{\xi}_L^n = \frac{j\gamma}{\Delta t(\alpha-j\beta)^2}\{[(\alpha - j\beta)\Delta t + 1]\exp[(-\alpha + j\beta)\Delta t] - 1\} \tag{14}$$

$$\Delta\widehat{\xi}_L^n = \frac{j\gamma}{\Delta t(\alpha-j\beta)^2}\{[(\alpha - j\beta)\Delta t + 1]\exp[(-\alpha + j\beta)\Delta t] - 1\}\{1 - \exp[(-\alpha + j\beta)\Delta t]\} \tag{15}$$

Finally the variable $\boldsymbol{\Psi}^n$ is the sum of Drude term and the real part of Lorentz term:

$$\boldsymbol{\Psi}^{\mathbf{n}} = \boldsymbol{\Psi}_d^n + \mathrm{Re}\left(\widehat{\boldsymbol{\Psi}}_L^n\right) \tag{16}$$

The parameters in Equ.(5) are also the sum of Drude term and the real part of Lorentz term:

$$\chi^{\mathbf{n}} = \chi_d^n + \mathrm{Re}\left(\widehat{\chi}_L^n\right) \tag{17}$$

$$\xi^{\mathbf{n}} = \xi_d^n + \mathrm{Re}\left(\widehat{\xi}_L^n\right) \tag{18}$$

From above equations we can see that the Drude and Lorentz terms are deduced respectively, so that the PLRC method can be applied for both Drude and Drude-Lorentz models. For describing the relative permittivity of silver[25], we adopt the values of the parameters of Drude-Lorentz model as $\varepsilon_\infty=4.6$, $\omega_d=1.401\times10^{16}$Hz, $\gamma_d=4.5371\times10^{13}$Hz, $\Delta\varepsilon=3.428$, $\omega_L=2.144\times10^{16}$Hz, and $\delta_L=1.824\times10^{18}$Hz.

2.2. Verification of the program code

We adopt the classical Krestschmann-Type SPR device shown in Fig.1(a) to verify our FDTD code. The two-dimensional(2D) simulation conditions are: n_1=1.78, n_2=1.132, θ=42º, thickness of silver film d=50nm, and sampling interval $\Delta x=\Delta y$=5nm. Under these configurations, the SPR wavelength calculated by SPs dispersion relation is λ_{SP}=500nm. A modulated 2D TE Gaussian pulse with the central wavelength 400nm is incident from the medium n_1 onto the metal film. The transmission enhancement coefficients in the visible range in the medium n_2 calculated by ADE, SO and PLRC methods are shown in Fig.1(b). These three methods demonstrate the same SPR peak and enhancement strength under the same condition. However, in the shorter wavelength region the ADE method shows larger error, and in the longer wavelength region the SO method shows larger error. So in the total wavelength region, the PRLC method shows better precision than those two methods. We will use the PLRC method and the parameters of Drude-Lorentz model mentioned above throughout this chapter.

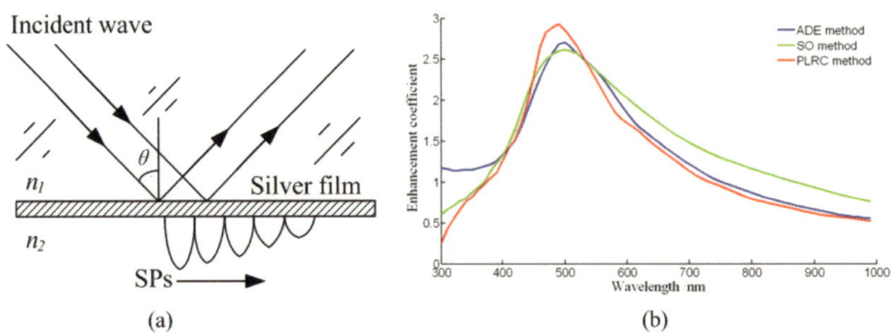

Figure 1. Verification of our FDTD code for dispersive media. (a) Configuration of Krestschmann-Type SPR device; (b) Compare of the enhancement coefficients of the three methods.

3. Application of surface plasmons in super-resolution focusing

A tightly focused evanescent field can be generated by a centrally obstructed high numerical aperture objective lens [26], and a super-resolved evanescent focal spot of λ/3 has been obtained [27]. The enhancement of the electromagnetic (EM) field by tight focusing enables nano-lithography using evanescent field [28]. It has been demonstrated that a tightly focused beam can be further modulated by a negative-refraction layer together with a nonlinear layer [29] or a saturable absorber [30] to approximately λ/4~λ/5 close vicinity of the focus. Considering the difficulty in realizing a negative-refraction layer in practice, here we introduce another mechanism of light modulation in the tightly focused region. Due to the use of high numerical aperture objective lens, the focused evanescent field is highly depolarized, which offers strong transverse and longitudinal polarization components.

Therefore the deployment of nano-plasmonic structure, which is polarization sensitive, offers new mechanism to modulate the focused evanescent field. In this section, we will give two research results for the application of surface plasmons in super-resolution focusing.

3.1. Simulation of radially polarized focusing through metallic thin film

As shown in Fig.2, the configuration of tightly focused evanescent field is based on the scanning total internal reflection microscopy[26]. The refractive indices of immersion oil and the coverslip glass are 1.78. The thin silver film is coated on the surface of the coverslip glass. An annular radially polarized incident beam is focused on the upper surface of coverslip glass by an objective. We define the annular coefficient ε=d/D which produces a ring beam illumination onto the silver film. In the simulation we set ε=0.606 to make sure that the incident angle is larger than the critical angle for the total internal reflection of the ring beam.

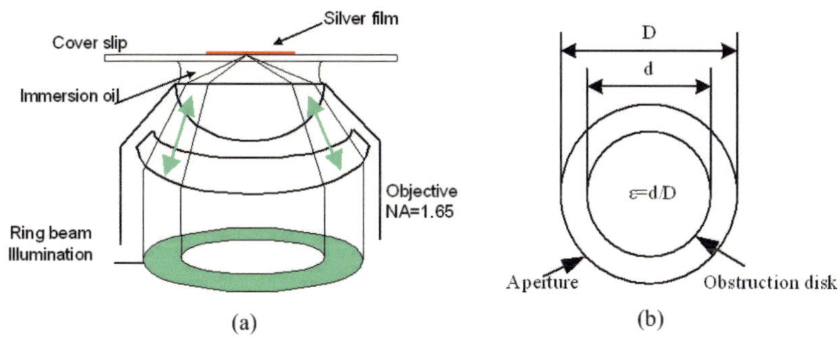

Figure 2. The configuration of focal ring beam illumination for the excitation of SPs. (a) Scheme of total internal reflection focusing; (b) Illustration of annular coefficient.

The electromagnetic field of radially polarized focal beam is expressed as:

$$\mathbf{E}(\mathrm{r},\boldsymbol{\Psi},\mathrm{z}) = ik(cos\Psi \boldsymbol{I}_1\mathbf{i} + sin\Psi \boldsymbol{I}_1\mathbf{i} + \boldsymbol{I}_0\mathbf{K}) \tag{19}$$

where

$$\boldsymbol{I}_0 = \int_0^{\alpha} \sqrt{\cos\theta}\,\sin^2\theta J_0(krsin\theta)exp(-ikzcos\theta)d\theta \tag{20}$$

$$\boldsymbol{I}_1 = \int_0^{\alpha} \sqrt{\cos\theta}\,\sin\theta cos\theta J_1(-krsin\theta)exp(-ikzcos\theta)d\theta \tag{21}$$

where **i**, **j** and **k** are unit vectors in the x, y and z directions, respectively. $\mathrm{r} = \sqrt{\mathrm{x}^2 + \mathrm{y}^2}$ is transverse radial coordinate. Ψ denotes the angle between r and $+x$ axis. θ denotes the incident angle of light. $J_i(x)$ is the ith order Bessel function of the first kind. It should be pointed out that the electric fields under the radially polarized focal beam is circular symmetrical. The simulation conditions are set as: incident wavelength λ=532nm, NA=1.65, sampling interval Δx=Δy=Δz=5nm, thickness of silver film d=60nm. In this simulation, we calculate the electric field in the transverse plane with the defocal distance z=-600nm as the input source.

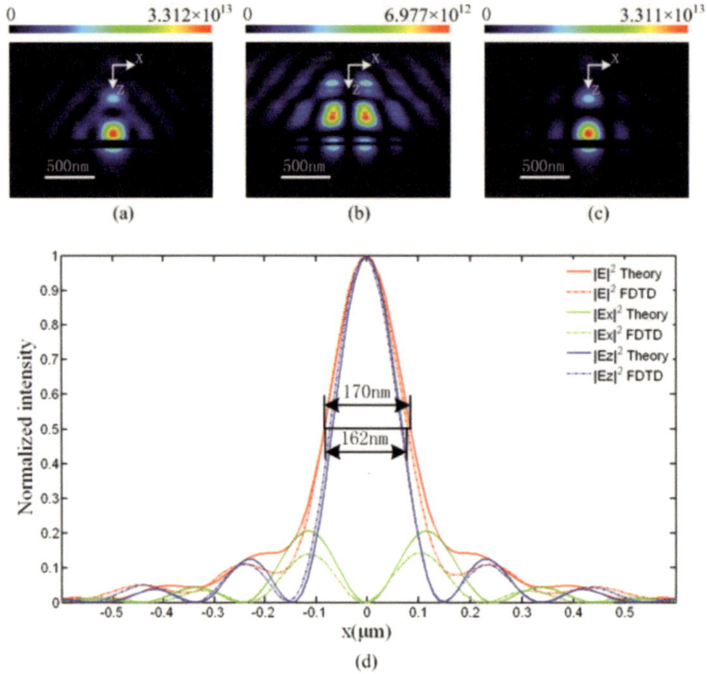

Figure 3. The intensity distribution of radially polarized beam focusing through silver film. Colorbar unit: V^2/m^2. (a) $|E_{total}|^2$; (b) $|E_x|^2$; (c)$|E_z|^2$; (d)Comparison of the sizes of the SPs enhanced focus with that of theoretical normal focus.

Under the conditions mentioned above, after 2000 time steps simulation, the time averaged intensity distributions of E_{total}, E_x and E_z in *xz* plane are shown in Fig.3. The intensity of total electric field is the combination of E_x and E_z components. The E_z field shows a single evanescent focus which is formed by the SPs excited by the silver film and the total electric field is dominated by E_x component. In order to illustrate the super-resolved SPs focus phenomenon, we plot the SPs focus simulated by FDTD and the theoretical radially polarized focus calculated by Debye theory along *x* axis in Fig.3(d). The intensity distributions of the SPs focus and the theoretical focus are normalized by the maximum values of their total electric fields. It is demonstrated that the proportion of E_x intensity and Ex intensity of SPs focus is 0.145, while that of the theoretical focus is 0.207, which indicates that the E_z component is enhanced by the SPs. So that the focal size of SPs enhanced focus(FWHM=162*nm*) is smaller than that of the theoretical focus(FWHM=170*nm*).

It is obviously that in the total electric field distribution, the reflected field is much stronger than the transmitted field. Under the annular coefficient $0.606<\varepsilon<1.0$ condition, the total ring beam includes a wide range of incident angle($34.18^{o}<\theta<67.97^{o}$). However, only the narrow angle ranges around $\theta=36.72°$ satisfy the SPs dispersion relation. We simulate the thin ring beam illumination ($0.645<\varepsilon<0.65$) which contains the SPR incident angle and the thin ring beam illumination ($0.795<\varepsilon<0.8$) which is far from the SPR incident angle in order to demonstrate the differences between the SPR focus and non-SPR focus.

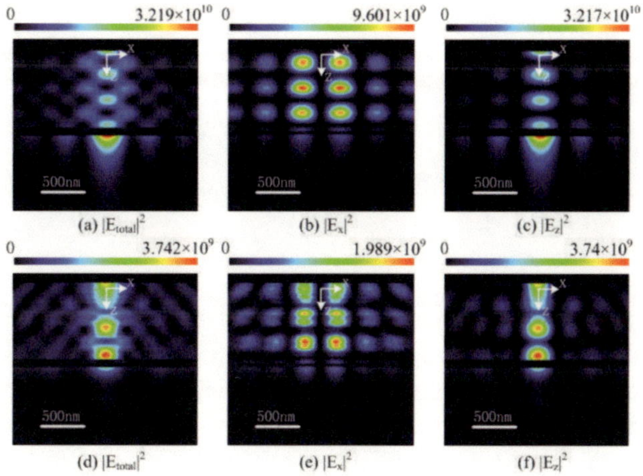

Figure 4. The intensity distributions of SPR focus (a), (b), (c) and non-SPR focus (d), (e), (f) in xz plane. Colorbar unit: V^2/m^2.

The simulation results are shown in Fig.4. Comparing Fig.4(a) and (d), it can be seen that under the SPs dispersion relation, the SPR significantly enhances the focus below the silver film, which leads to a much high transmission rate of the focal field. While for the non-SPR focus, the reflection is much stronger than the transmission, so most part of the focus energy is reflected back. Comparing the E_x components of the two conditions in Fig.4(b) and (e), they demonstrate almost the same distribution with no SPR phenomenon. That is because the E_x component is parallel to the silver film, i.e. s-polarization, it can't excite the SPR. Finally, comparing the E_z components that are perpendicular to the silver film, i.e. p-polarization, of the two conditions in Fig.4(c) and (f), it is obviously demonstrated that the E_z component contributes to the SPR focus dominantly. For the SPR focus, the E_z component is significantly enhanced by the SPR, while for the non-SPR focus, the reflection of E_z component is much larger than its transmission. The SPs dispersion relation satisfied by the ring beam with $0.645<\varepsilon<0.65$ is the main mechanism for the super-resolved SPs focus. This phenomenon gives rise to a simple approach for achieving a super-resolution focus beyond diffraction limit, which has potential applications in nano-lithography and nano-trapping.

3.2. Simulation of focusing through two parallel nanorods

3.2.1. Numerical simulation model

Our simulation configuration is shown in Fig.5. Two silver nanorods are lying on the interface of two dielectric media with the separation (D) of 120nm centre-to-centre. The refractive index of lower medium (n_1) is 1.78 and the refractive index of upper medium (n_2) is 1.0. The size of the nanorods is 150nm in length (l) and 50nm in diameter (d). The numerical aperture of the objective is 1.65, and the pure focused evanescent field is generated by inserting a centrally placed obstruction with normalized radius ε_c=0.606, corresponding to the total internal refraction condition.

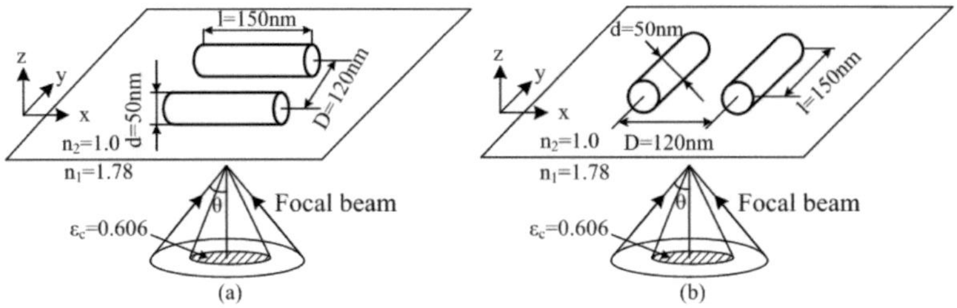

Figure 5. Configuration of nano-plasmonic waveguide. (a) Nanorods lie along x direction. (b) Nanorods lie along y direction.

If the illuminating beam polarizes along the *x* direction, the electric field at the focal region, calculated with high angle vectorial Debye theory, is highly depolarized, and each polarization component can be expressed as:

$$\mathbf{E}(\mathrm{r},\boldsymbol{\Psi},\mathrm{z}) = \frac{\pi \mathrm{i}}{\lambda}\{[\boldsymbol{I}_0 + \cos(2\Psi)\boldsymbol{I}_2]\mathbf{i} + \sin(2\Psi)\boldsymbol{I}_2\mathbf{J} + 2\mathrm{i}\cos(2\boldsymbol{\Psi})\boldsymbol{I}_1\mathbf{K}\} \tag{22}$$

where

$$\boldsymbol{I}_0 = \int_0^{\alpha}\sqrt{\cos\theta}\,\sin\theta(1+cos\theta)J_0(krsin\theta)exp(-ikzcos\theta)d\theta \tag{23}$$

$$\boldsymbol{I}_1 = \int_0^{\alpha}\sqrt{\cos\theta}\,\sin^2\theta J_1(-krsin\theta)exp(-ikzcos\theta)d\theta \tag{24}$$

$$\boldsymbol{I}_2 = \int_0^{\alpha}\sqrt{\cos\theta}\,\sin\theta(1-cos\theta)J_2(krsin\theta)exp(-ikzcos\theta)d\theta \tag{25}$$

where **i**, **j** and **k** are unit vectors in the *x*, *y* and *z* directions, respectively. $\mathrm{r} = \sqrt{\mathrm{x}^2+\mathrm{y}^2}$ is transverse radial coordinate. Ψ denotes the angle between *r* and +*x* axis. θ denotes the incident angle of light. $J_i(\mathrm{x})$ is the *i*th order Bessel function of the first kind. The electric and magnetic fields are calculated at a plane one wavelength before the interface as the input source for the FDTD simulation. The total field/scatter field technique is used to eliminate the light propagating to –z direction, so that the incident focal beam only propagates in forward direction.

In the simulation, the wavelength of incident focal beam is 532*nm*. It should be noted that the annular beam illumination includes a wide range of incident angle ($34.18°<\theta<69.97°$), which corresponds to a wide range of surface plasmon resonance (SPR) wavelengths (340*nm*<λ<1120*nm*). Our simulations show that different wavelengths selected in this spectrum range make little difference for the excitation of LSPs by the nano-plasmonic waveguide. The grid sizes Δx, Δy and Δz for each dimension are set to 2.34*nm*. According to the discretization in space domain, the discretization in time domain that satisfies the Courant stability condition is adopted as:

$$\Delta \mathrm{t} = \frac{0.985}{\mathrm{c}\sqrt{1/\Delta \mathrm{x}^2+1/\Delta \mathrm{y}^2+1/\Delta \mathrm{z}^2}} \tag{26}$$

where *c* is the velocity of light in vacuum. Under these configurations, the iteration of program runs 400 time steps for a period. In order to obtain the stable state of the optical field, 10 periods are adopted, i.e. 4000 time steps are calculated in the simulation.

3.2.2. *Simulation results and discussion*

The intensity distribution of a focused evanescent field under the linearly polarized illumination is shown in Fig.6 and agrees well with previous theory[26, 27]. Under the conditions described above, the intensity of E_y component is one order smaller than either E_x or E_z, so the overall impact of E_y on the intensity distribution is less significant. Due to the depolarization effect, a strong longitudinal E_z component appears, with its strength comparable to the illuminating polarization component E_x, i.e., $|E_z|^2/|E_x|^2 \approx 0.85$. As a result, the intensity distribution of the total field is splitted to two lobes shown in Fig.6 (a).

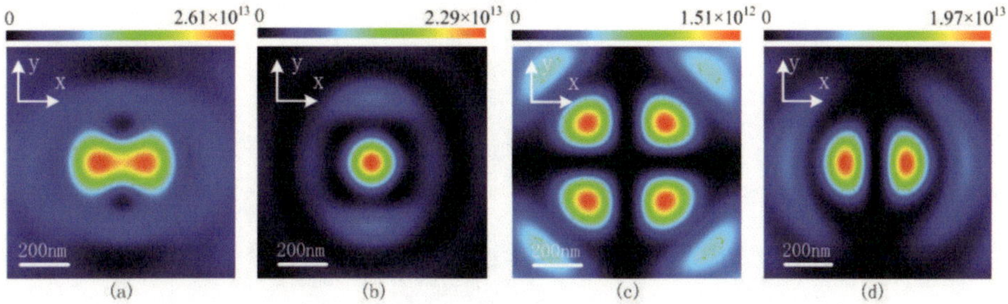

Figure 6. The intensity distribution of evanescent electric field for linearly polarized focal beam at the interface. (a) $I=|E_x|^2+|E_y|^2+|E_z|^2$ (b) $|E_x|^2$ (c) $|E_y|^2$ (d) $|E_z|^2$ (Unit: V^2/m^2)

The modulation of the focused evanescent field by a pair of silver nanorods is demonstrated in Fig. 7, where the intensity distributions in planes of different distances from the interface are illustrated. In the left column, when there is no nanorod on the interface, the focal spot splits into two lobes at different distances above the interface. In the middle column, the nanorods are lying along *x* direction, parallel to the dominant transverse polarization component E_x. In the right column, the nanorods are lying along *y* axis, perpendicular to the dominant transverse polarization component E_x. In Fig.7 (a), it is noted that at the *xy* plane 25*nm* above the interface, the electric field is significantly enhanced by the LSPs between the nanorods, if the nanorods are lying in the *y* direction, resulting in a strong localized field between the two nanorods. While the nanorods are lying in the *x* direction, the electric field shows less enhancement and localization. At the plane 60*nm* above the interface (10*nm* above the nanorods, Fig.7 (b)), it is observed that the longitudinal electric field component E_z displays a strong enhancement at both ends of each nanorod, forming four strong intensity lobes. When the nanorods are lying along *x* direction, the dominant transverse electric field component E_x is not significantly enhanced, so the electric field is dominated by the longitudinal component E_z, which shows four strong intensity lobes. However when the nanorods lie in the *y* direction, due to the significant enhancement of E_x component between the nanorods, the four strong lobes become less evident. At the horizontal planes that further away from the nanorods, i.e. at the planes 75*nm* (Fig.7 (c)) and 100*nm* (Fig.7 (d)) above the interface, super-resolved focal spots are demonstrated. In particular, with the significant enhancement of dominant transverse electric field component E_x, the focal spots show narrower distribution and stronger strength, when the nanorods lie in the direction perpendicular to the E_x component.

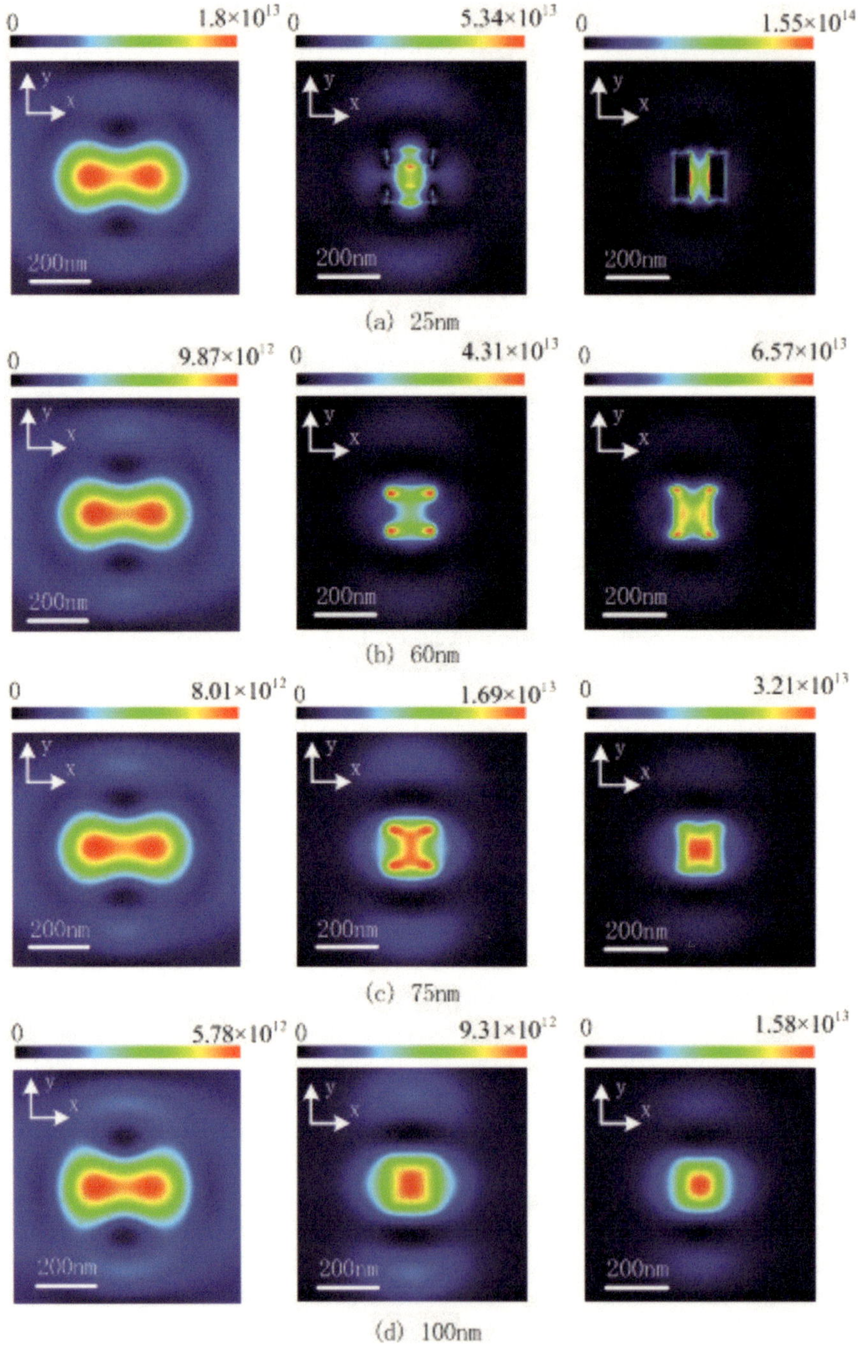

Figure 7. Intensity distributions for linearly polarized focal beam in xy plane of different distances from the interface. In the left column there is no nanorod. In the middle column the nanorods are lying along x direction. In the right column the nanorods are lying along y direction. (Unit: V^2/m^2)

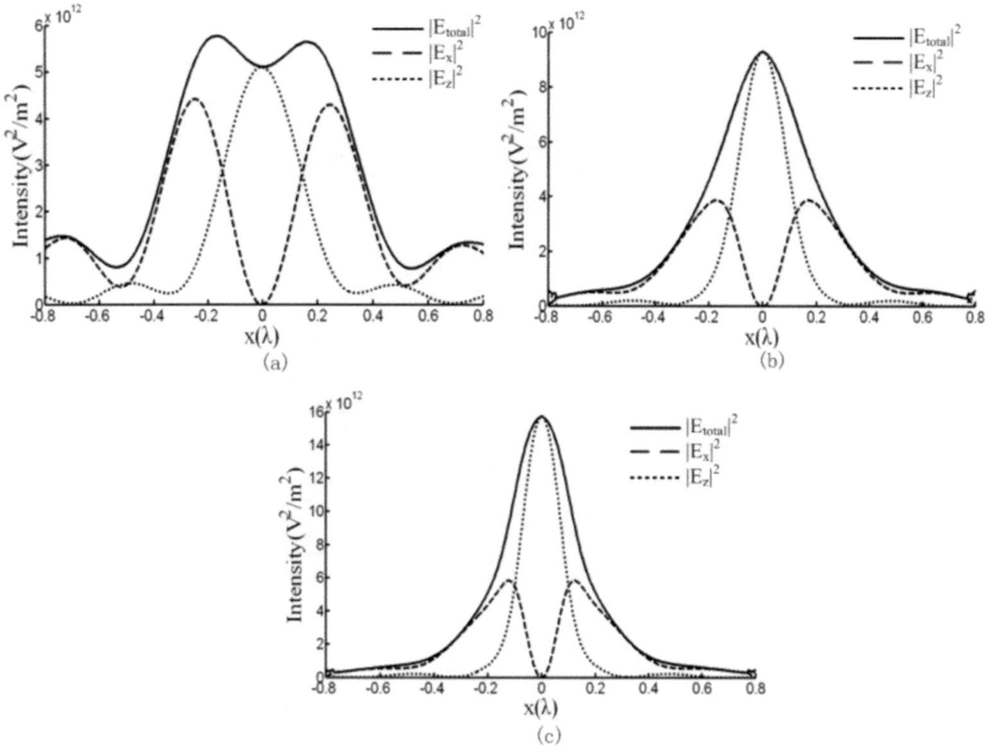

Figure 8. Cross sections for polarization components of linearly polarized light along x axis at the xy plane 100nm above the interface. (a) Without nanorods; (b) Nanorods lie along x direction; (c) Nanorods lie along y direction.

The detailed analysis of the LSPs effect on each polarization component at the plane 100*nm* above the interface is illustrated in Fig.8. When the nanorods are lying along x direction (Fig. 8(b)), the enhancement of E_z component is insignificant compared with the case that the nanorods lie in the *y* direction. Nevertheless, it still produces a narrower focal spot with 46.6% reduction in FWHM and 61% increase in strength. When the nanorods are lying in *y* direction (Fig. 1(c)), the dominant transverse polarization E_z is not only significantly enhanced in strength, e.g. by a factor of 2.71, but also becomes more localized, e.g. the full width at half maximum (FWHM) is reduced from 388*nm* to 149*nm*. This phenomenon indicates the LSPs excited by the E_z component couple between the nanorods and re-distribute the energy of electromagnetic field. The longitudinal component E_z is also enhanced by a factor of 1.4, and becomes narrower, e.g. the distances between two intensity peaks reduced from 266*nm* to 127*nm*. As a result, a super-resolution focal spot can be formed outside the waveguide. The above analysis shows that the nano-plasmonic waveguide provides strongest enhancement to the transverse polarization component perpendicular to the nanorods, followed by the longitudinal polarization component, and the enhancement for the transverse polarization component parallel to the nanorods is least significant.

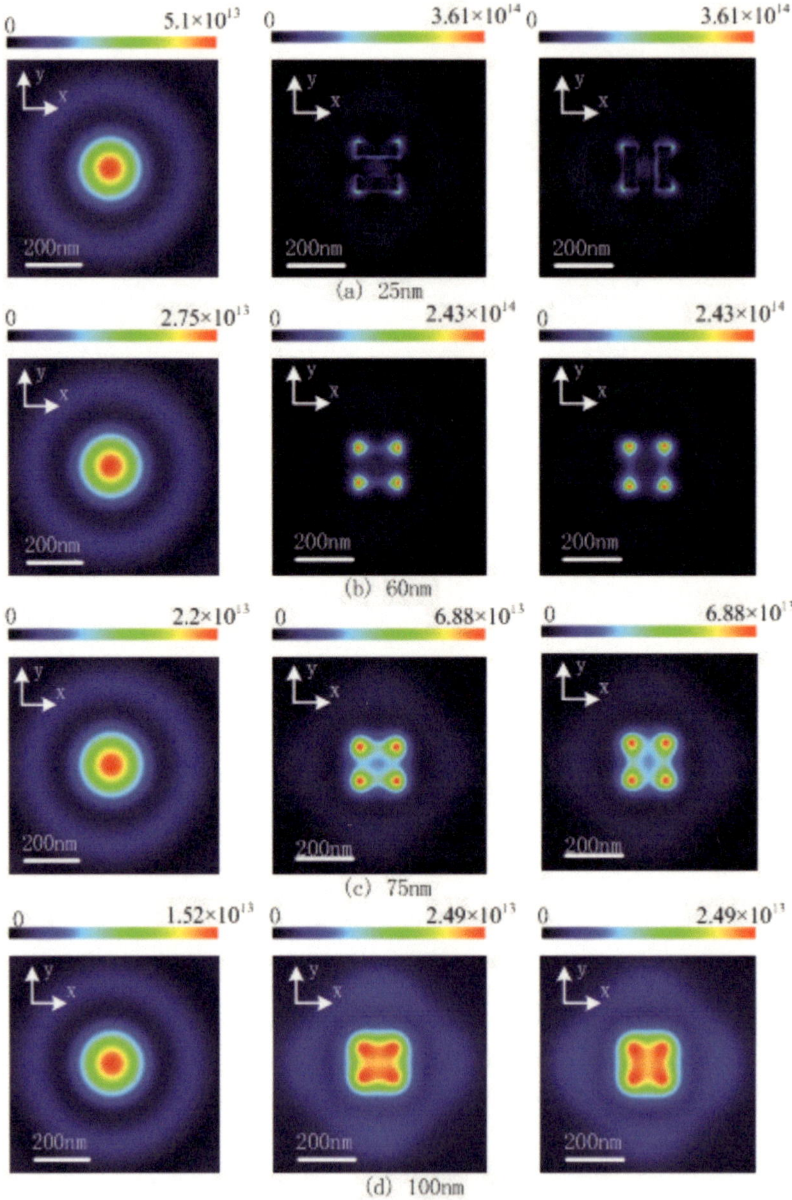

Figure 9. Intensity distributions for radially polarized focal beam in xy plane of different distances from the interface. In the left column there is no nanorod. In the middle column the nanorods are lying along x direction. In the right column the nanorods are lying along y direction. (Unit: V^2/m^2)

It is well known that the focal spot for radially polarized beam is circularly symmetrical. According to the analysis demonstrated in the previous section, it is expected that the circular symmetry would be broken due to the LSPs effect which is polarization sensitive. Fig.9 shows the intensity distributions of evanescent radially polarized focal beam at xy

planes of different distances above the interface for three cases, including without nanorods, nanorods lying in the x and y directions, respectively. The focal spot is circularly symmetrical and decay exponentially further from the interface without the presence of the nanorods, as shown in the left column of Fig.9. With strong LSPs effect from the nano-plasmonic waveguide, the circular symmetry is broken, and the focal spot shows a strong intensity lobe at each of the four ends of the two nanorods (Figs. 9 (b-d)). It is noted that the intensity of the longitudinal component E_z is approximately one order of magnitude stronger than that of the transverse components. As a result, the intensity distribution at the focal region shows four strong intensity lobes produced by LSPs effect excited by the longitudinal polarization component E_z.

In this section we demonstrate a new method to modulate highly focused evanescent field with a nano-plasmonic waveguide. The modulation of focus is based on the mechanism that the LSPs are polarization sensitive and the focus is strongly depolarized by a high numerical aperture objective. For a simple nano-plasmonic waveguide that consists of two silver nanorods lying on the interface between two dielectrics, LSPs effect is strongest for the polarization component perpendicular to the nano-plasmonic waveguide. A super-resolved focal spot with significantly enhanced strength can be achieved, when the nanorods are lying perpendicular to the dominant polarization component. The design of the nano-plasmonic waveguide structure gives rise to a new approach to further improve the tightly focused evanescent field to achieve the resolution beyond diffraction limit, and thus facilitates potential applications in nano-trapping and nano-lithography.

4. Plasmonic nano-waveguiding

In this section we will propose the waveguiding properties of two types of plasmonic nano-structures: parallel long nano-wires and the metal-dielectric-metal waveguide. They all present interesting optical nano-waveguiding properties and can be applied in some special research areas.

4.1. Focusing through parallel nanorods that perpendicular to the interface

In recent years, metallic nanorods are intensely investigated and widely applied in the field of super-long waveguiding [31]. It has been reported that the optical properties of single nanorod is sensitive to the polarization of incident wave and the rod aspect ratio [32]. While for the two parallel nanorods and U-shaped nanorods, their optical properties are more complicated and interesting [33]. The SPPs excited along the surface of single nanorod couple between the nanorods, which leads to the extinction spectrum whose resonance peaks are dependent on the geometry of the nanorods. The U-shaped nanorods show stronger resonance strength and more longitudinal couple modes than the two parallel nanorods.

In some cases, such as super-resolution imaging or focusing, the more complicated focusing beam with three polarization components should be considered as the incident source.

However, the interaction between the focal beam and the two nanorods has not been investigated thoroughly yet. Here we fist present the SPR excitation spectrums of two parallel silver nanorods and the π-shaped silver nanorods using the FDTD method. Second we simulate the focusing process through these two structures using 3D FDTD method, respectively. The waveguide effect and electromagnetic field transfer efficiency of these two structures are compared and analyzed. Finally, the focusing process through the angular π-shaped nanorods structure is simulated which presents the ability of guiding the focal field to different directions.

4.1.1. Simulation modeling and resonance spectrums

The schemes of two silver nanorods structures are shown in Fig.10. In Fig.10(a), the length of nanorods is l, the diameter of nanorods is a and the distance between the two nanorods is d. In Fig.10(b), the length of longitudinal nanorods is l_1, the length of transverse nanorod is l_2. In our simulations, we set a=50*nm*, l=l_1=500*nm*, and l_2=260nm constantly, and change the distance d from 30*nm* to 100*nm* gradually to calculate their SPR spectrums.

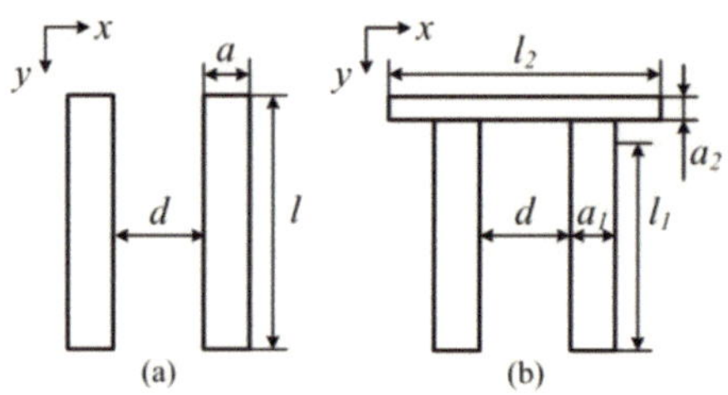

Figure 10. Structures of silver nanorods waveguide. (a) Two parallel nanorods structure; (b) π-shaped nanorods structure

An x-polarized Gaussian temporal profile pulse with the central wavelength of 400*nm* is injected from the upper side and propagates through the nano-structures. The spectrum width of the pulse is wide enough to cover the wavelength range from 300*nm* to 1000*nm*. The temporal electromagnetic field coupling between the nanorods at the central cross section of the nanorods is recorded, and transformed to frequency domain using Fourier transform, and divided by the incident wave magnitude at each wavelength component to calculate the enhancement coefficient spectrum.

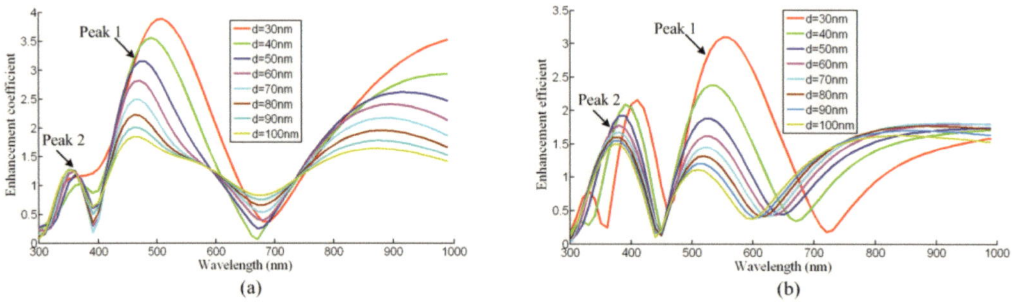

Figure 11. Electromagnetic field enhancement coefficient of the SPR coupling between the two silver nanorods. (a) Two parallel nanorods structure; (b) π-shaped nanorods structure.

As shown in Fig.11, the spectrums perform two peaks for both structures, but the positions of the peaks of the two structures under the same distance *d* are different. In Fig.11(a), for the two parallel nanorods structure, the main resonance peak 1 shows blue shift with the increase of the distance *d*, while the sub-resonance peak 2 remains at the wavelength about 350*nm*. In Fig. 11(b), the main resonance peak 1 also shows blue shift with the increase of the distance *d*. The peak 2 remains at about 390*nm* but becomes stronger. Due to the reflection by the transverse nanorod in the π-shaped nanorods structure, their enhancement coefficients are a little smaller than those of the two parallel nanorods structure.

4.1.2. *Focusing through nanorods structures*

Based on the results obtained above, we will investigate the focusing processes through the nanorods structures in this section. Due to the depolarization effect of high numerical aperture objective, the incident linearly polarized beam would change its polarization state after propagating through the objective, i.e. the strong longitudinal polarization component occurs. The interaction effect between the incident focal beam and the two nanorods structures are more complicated than that of the single polarized incident beam case. Here the focal beam is calculated by the vectorial Debye theory and induced into the 3D FDTD simulation region with total-field/scatter-field method. The refractive index of upper medium is 1.78 and the refractive index of lower medium is 1.0. The numerical aperture of the objective is 1.65, and the pure focused evanescent field is generated by inserting a centrally placed obstruction disk with the normalized radius ε_c=0.606, corresponding to the total internal refraction condition at the interface between the two media. We fix the distance *d*=70*nm* throughout the following simulations, so that the tops of the two parallel nanorods are all covered in the focal region. The wavelength of focal beam is 532*nm* which corresponding to the resonance peak of π-shaped nanorods structure at *d*=70*nm* condition. We place the nanorods structures normally onto the interface, and set the length of the longitudinal parallel nanorods to 1250*nm* to observe the long waveguide effect for transferring the electromagnetic field from the focus region to the far field. The 3D discretized cell sizes for three directions are $\Delta x=\Delta y=\Delta z=4.68nm$ and the time step Δt=8.27288e-18*s* which satisfies the Courant stability condition. After 4000 time steps simulation, the total electromagnetic field is stable and the time averaged intensity distributions are obtained as shown in Fig.12.

The long waveguide effect produced by the nanorods structures are clearly shown in Fig.12. Without the nano-structures, the pure evanescent focal field would decay exponentially away from the interface and only can propagate no more than 100*nm*. However, when the nanorods structures are implemented onto the interface, the SPPs are excited by the evanescent focal field along the surfaces of the nanorods. The electromagnetic fields along the inner surface of the longitudinal nanorods are stronger than those along the outer surface of the longitudinal nanorods, which means that the SPPs between the two nanorods couple with each other and form a long waveguide. The SPPs propagate along the nanorods with a resonance mode, with each space resonance period about 150*nm* in length. The electromagnetic field can be transferred to the lower end of the nanorods where is the far field more than 1μm.

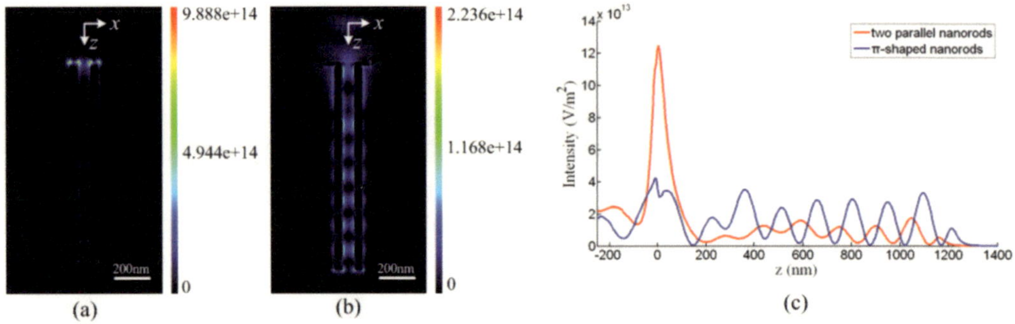

Figure 12. Focusing through long nanostructure waveguide. (a) Two parallel nanorods structure; (b) π-shaped nanorods structure; (c) Plot of intensity along the z axis. Color bar unit: V^2/m^2.

The two parallel nanorods structure and the π-shaped nanorods structure show almost the same SPPs resonance mode, except their excitation strength. In Fig.12(a), the electromagnetic field is strongly enhanced at the top of the two nanorods and decay exponentially very fast, which leads to four strong field points at the top of the nanorods. In Fig.12(b), the focal field is evenly enhanced by the transverse nanorod, so there is no strong field point shown at the top side. The comparison of the electromagnetic field intensities along z axis of the two situations is shown in Fig.12(c). At the interface of the two media(0 position along z axis), the two parallel nanorods show stronger intensity than the π-shaped nanorods. However, at the distances larger than 200*nm* along z axis, the π-shaped nanorods show stronger intensity than the two parallel nanorods. There are two reasons for this phenomenon. First, under d=70*nm* condition, the incident wavelength 532*nm* is almost the SPR wavelength for the π-shaped nanorods structure as shown in Fig.11(b), but not the SPR wavelength for the two parallel nanorods structure as shown in Fig.11(a). Second, the transverse nanorod of π-shaped nanorods interacts with the longitudinal polarization component of the focal beam, so that the longitudinal SPPs along the surfaces of the nanorods are excited, which gives contribution to the total SPPs intensity in the nanorods waveguide. The π-shaped nanorods structure shows stronger electromagnetic field enhancement and better energy transfer efficiency in the far field than the two parallel nanorods structure. This result gives a method to transfer the focal field from the focus region to the far field.

Finally, we present the simulation of focusing through the angular π-shaped nanorods structure as shown in Fig.13(a). The two legs of the π-shaped nanorods are bent by the angle of 90°, so that the ends of the nanorods shift from the optical axis with the distance of l_2. In our simulation, we set l_1=500*nm*, l_2=500*nm*, l_3=260*nm*, d=70*nm* and a=50*nm*. The incident wavelength keeps 532*nm*. The simulation results for different cross section planes are shown in Fig.13(b), (c) and (d). In Fig.13(b), the intensity distribution in the xz plane shows the same pattern with that in Fig.12(b). In Fig.13(c), the intensity distribution in the yz plane shows an "L" shaped optical path which formed by the SPPs coupled between the two nanorods. This phenomenon testifies that the SPPs can propagate not only along the straight path, but also along the angular path with the angular structure of the nanorods. The focal electromagnetic field can be transferred not only along the optical axis, but also to the directions that perpendicular to the optical axis. In Fig.13(d), the xy plane is the cross section through the centers of the lower two transverse nanorods. It shows the same resonance length with the longitudinal part of the waveguide. It can be predicted that if we change the bend angle to other degrees, the SPPs also can be transferred to other directions. Due to the limit memory of computer, we can't simulate the waveguiding effect of longer nanorods structures, but it can be seen that the electromagnetic field waveguiding ability of the two nanorods structures investigated in this paper must be far more than $1\mu m$.

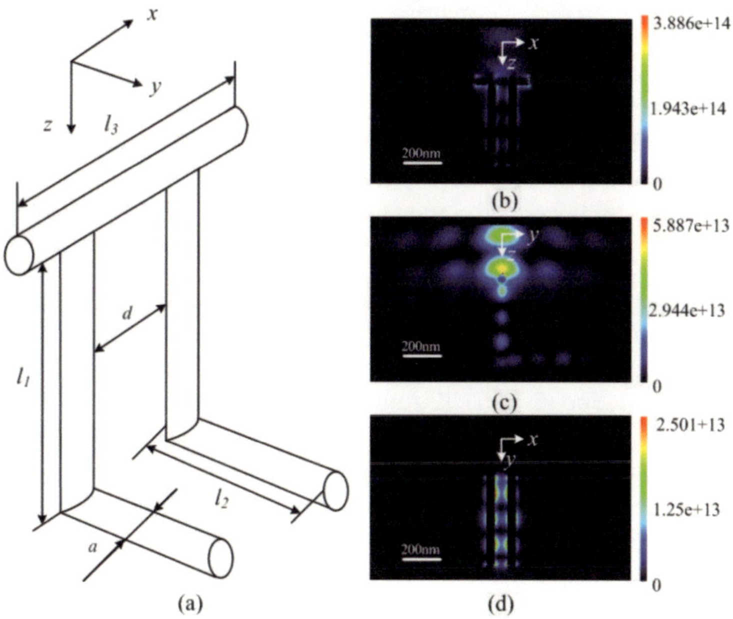

Figure 13. Focusing through angular π-shaped nanorods structure. (a) The configuration of the angular π-shaped nanorods structure; (b) The intensity distribution in xz plane; (c) The intensity distribution in yz plane.

In this section, the optical properties of two silver nanorods plasmonic waveguide structures are simulated with the FDTD method. The SPR spectrums of the two parallel nanorods and the π-shaped nanorods structures are calculated, which show different SPR resonance peaks at the same distance condition for both structures. The focusing processes through these two types of structures show almost the same SPPs resonance mode with each resonance period about $150nm$. However, at the far field, the electromagnetic field enhancement of the π-shaped nanorods structure is stronger than that of the two parallel nanorods, due to the contribution of longitudinal SPPs component excited by the transverse nanorod. At last, the focusing process through the angular π-shaped nanorods structure indicates that the focus field can be transferred not only to the far field along the optical axis, but also to any other directions that the ends of the two nanorods direct to. The two nanorods plasmonic waveguide structures studied in this paper have potential applications in nano-biosensing, nano-trapping or nano-waveguiding.

4.2. MDM nano-waveguides

Plasmonic waveguides based on the principles of SPPs have gained great attentions in recent years due to their ability of confining and guiding optical field in sub-wavelength scale. Among various types of plasmonic waveguides, the metal-dielectric-metal (MDM) waveguide is considered to be a key element in the fields of waveguide couplers[34, 35], sub-wavelength scale light confinement[36, 37], wavelength filters[38] and integrated optical devices[39, 40]. The remarkable advantages of the MDM waveguide, including the strong confinement of optical field in nano-scale gaps, the high sensitivity of its transmission characteristics to the waveguide structures, and the facility of its fabrication, attracted a great deal of effort to be devoted to develop the MDM based nano-plasmonic devices. It has been reported that the transmission of MDM waveguide coupled with stub structure could be changed with the length of the stub[38]. Further more, the stub filled with absorptive medium was considered to be a resonance cavity that acts as an optical switch controlled by the pumping field[39]. An improved transmission model[40] was also developed to describe the transmittance of multi-stubs MDM structure.

The investigation approaches for the transmission characteristics of MDM waveguide include theoretical transmission line theory(TLT) and FDTD method [40]. In this section, we first provide the formulas of the optical transmission characteristics of the MDM waveguide with a stub structure deduced by TLT. And then the FDTD method will be employed to numerically study the optical switch effect of the stub structure in terms of changing the length and the refractive index of the stub, respectively. The simulation results coincide with the calculation results of TLT and the physical mechanisms of the optical switch effect are analyzed and discussed.

4.2.1. Scheme modeling and transmission line theory

The scheme of the MDM waveguide coupled with a single stub structure is shown in Fig.14(a). Firstly we assume that the medium in the stub is air, i.e. n_1=n_2=1.0. The silver is employed as the dispersive medium for the MDM waveguide. The width of the waveguide and the stub is d=100nm. A transverse electric (TE) plane wave with a wavelength λ=600nm is incident from the left side into the waveguide. The transmission line modeling of the schematic is shown in Fig.14(b).

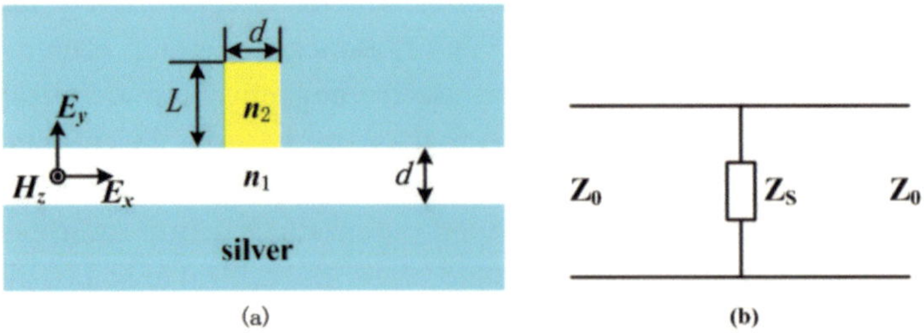

Figure 14. (a) The scheme of MDM waveguide coupled with a stub structure with the same width d and the length L. (b) The transmission line modeling of the scheme.

According to the transmission line theory, the stub can be considered as admittance. Assume that the phase shift only occurs when the SPPs are reflected by the end of the stub. Z_0 and Z_s are the characteristic impedances of the loss-free transmission lines corresponding to the MDM waveguide and the stub[40], respectively. Their relation is expressed as:

$$\frac{1}{Z_s} = -j\frac{1}{Z_0}\tan\left(\frac{2\pi L}{\lambda_{SP}}\right) \tag{27}$$

where L is the length of the stub and λ_{SP} is the propagating wavelength of the SPPs. From Fig.14(b), the amplitude transmission of the electric field can be expressed as:

$$t = \frac{2Y_0}{2Y_0+Y_s} \tag{28}$$

where $Y_0=1/Z_0$ and $Y_S=1/Z_S$. Therefore, the energy transmission of the MDM waveguide coupled with a single stub is finally expressed as[38]:

$$T = \frac{4}{4+\tan^2\left(\frac{2\pi L}{\lambda_{SP}}\right)} \tag{29}$$

From Equ.(29) we can see that the energy transmission is the function of L and λ_{SP}. In the stub, the SPPs wavelength λSPS can be changed by the refractive index of the medium in the stub. As a result, the transmission would be modulated periodically by changing the length L and the refractive index n_2 of the stub linearly.

4.2.2. Numerical simulation modeling and results

We simulated the transmission of the MDM waveguide with L ranging from 0 to 1μm by 2D FDTD algorithm. The relative permittivity of silver is described by Drude-Lorentz model to fit the experimental data of relative permittivity of silver and implemented into FDTD program with the PLRC method. The discretized cell size is Δx=Δy=5nm. After 3000 time steps simulation, the results are shown in Fig.15.

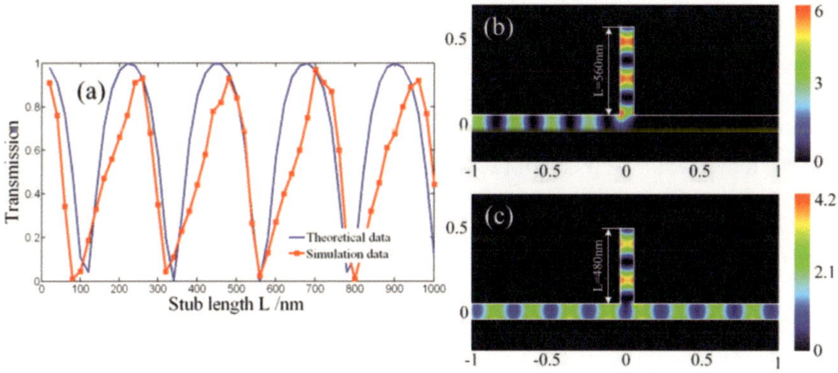

Figure 15. The transmission of MDM waveguide coupled with a stub structure as a function of the stub length. (a) Compare of the transmission calculated by transmission line theory and FDTD simulation. (b) The intensity distribution of the MDM waveguide at the "off" state with L=560nm. (c) The intensity distribution of the MDM waveguide at the "on" state with L=480nm.

From Fig.15(a) we can see that the simulation results agree well with the theoretical data calculated by Equ.(29). The SPPs excited along the surfaces of the metal layers propagate in the waveguide with a resonance mode. When passing through the stub, a part of the SPPs propagate into the side-coupled stub. The SPPs reflected from the end of the stub interfere with the passing SPPs, which leads to a modulation of the superposition wave. If we change the length of the stub, the phase change of the interference would cause the transmission vary from 0 to 1, so that the MDM waveguide coupled with a single stub structure presents the optical switch effect on the incident wave. As the incident wavelength λ=600*nm*, the λ_{SP} is approximately to be 480*nm*. The transmission shows about 4.5 periods for L linearly ranging from 0 to 1μm in length. The lowest transmissions are less than 1%, and the highest transmissions are more than 92%, which means the transmission of this structure switches between "on" and "off" states with the variation of the length of the stub. As shown in Fig.15(b), when the transmission is at the "off" state with L=560*nm*, the reflected light from the end of the stub modulates the interference intensity to be 0, which causes the propagating SPPs in the waveguide stopped by the stub significantly. As shown in Fig. 15(c), when the transmission is at the "on" state with L= 480*nm*, the SPPs can be transferred to the end of the waveguide without any loss. For both states, the superposition of the waves going into the stub and reflected by the end of the stub makes the field intensity in the stub stronger than that in the waveguide. Moreover, the field intensity of "off" state is stronger than that of "on" state due to the storage of the stopped field energy in the stub by the interference.

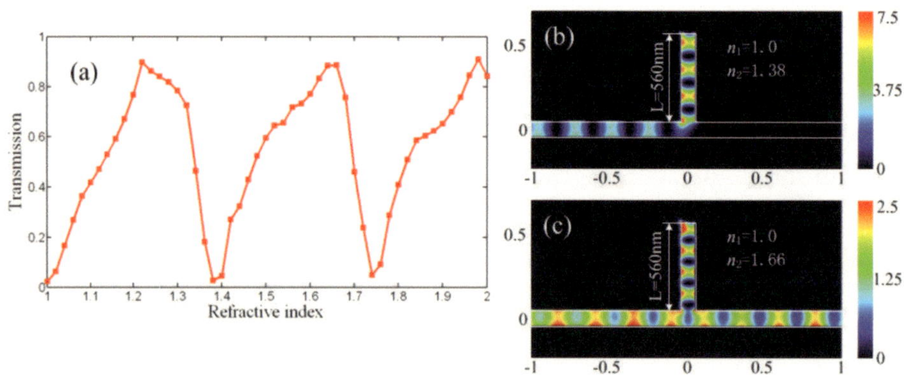

Figure 16. (a)The transmission of the MDM waveguide coupled with a stub structure as a function of the refractive index n_2 of the stub. (b) The "off" state when n_2=1.38. (c) The "on" state when n_2=1.66.

Another approach for modulating the phase of the reflected wave from the end of the stub is changing the refractive index of the stub. We carried out a series of simulations with the refractive index n_2 ranging from 1.0 to 2.0, and keeping n_1=1.0. The simulation results are shown in Fig.16. In Fig.16(a), it is obviously that the transmission demonstrate a periodic distribution as a function of the refractive index n_2. This phenomenon is similar with the distribution as shown in Fig.15(a). The mismatch of n_1 and n_2 leads to the phase difference between the reflected wave in the stub and the passing wave in the waveguide which also processes a phase modulation effect for the interference wave. Fig.16(b) and (c) show the phenomena of "off" and "on" states of the transmission when n_2=1.38 and n_2=1.66, respectively. It is clearly shown that the effective wavelength of the propagating SPPs in the stub is changed according to the refractive index n_2. The field intensity at "off" state is also much higher than that at "on" state because of the storage of the SPPs energy in the stub.

We have numerically studied the transmission characteristics of the MDM waveguide coupled with stub structure as functions of the length and the refractive index of the stub, respectively. The 2D FDTD simulation results show that the transmission rates obtained by both approaches change as periodical distributions, which implies that the MDM waveguide can be treated as an optical switch device controlled by the length and the refractive index of the stub. The physical mechanism of this phenomenon is the phase modulation of the interference of the reflected SPPs wave from the end of the stub and the passing SPPs wave in the waveguide. The results help us to further apply the MDM waveguide as an optical switch element in nano-scale optical chips and optical integrated devices.

5. Conclusions

The importance of surface plasmons in the applications of nano-photonics has been proved in many examples. In this chapter we focused on two fields: super-resolution focusing and nano-waveguiding. Super-resolution focusing is the key element for nano-lithography, high density optical data storage, and super-resolution imaging. We have presented that the high density focal field can be re-distributed by some specified metallic nano-structures such as nano-film and parallel nanorods, so that a super-resolution focusing can be generated in some particular space areas. This method can be further applied in optical nano-trapping due to the small and enhanced plasmonic focus. While for nano-waveguiding there have been many nano-structures and methods reported previously. Here we have presented the parallel nanorods and MDM structures, respectively. When the parallel nanorods are perpendicularly put on the interface of two materials, the incident focal field generates surface plasmons along their surfaces which form a nano-waveguide. The fields of focus can propagate along the waveguide for a long distance. The MDM with a stub structure showed a optical switch effect with the altering of the length and the refractive index of the stub, which has a potential application in optical communications and optical sensing.

Author details

Xingyu Gao
Institute of Opto-mechatronics, Guangxi Key Laboratory of Manufacturing System & Advanced Manufacturing Technology, School of Mechanical & Electrical Engineering, Guilin University of Electronic Technology, Guilin Guangxi, China

6. References

[1] William L.Barnes, Alain Dereux and Thomas W. Ebbesen. Surface plasmon subwavelength optics. Nature 2003; 424(6950) 824-830.

[2] R.H.Ritchie. Plasma Losses by Fast Electrons in Thin Film. Phys. Rev. 1957; 106(5) 874-881.

[3] Hiroshi Kano, Seiji Mizuguchi, and Satoshi Kawata, Excitation of surface-plasmon polaritons by a focused laser beam, J. Opt. Soc. Am. B 1998; 15(4) 1381-1386.

[4] Lihong Shi and Lei Gao. Subwavelength imaging from a multilayered structure containing interleaved nonspherical metal-dielectric composites. Phys. Rev. B 2008; 77(19) 195121.

[5] Thierry Laroche and Christian Girard. Near-field optical properties of single plasmonic nanowire. Appl. Phy. Lett. 2006; 89(23) 233119.

[6] Stephen K. Gray and Teobald Kupka. Propagation of light in metallic nanowire arrays: Finite-difference time-domain studies of silver cylinders. Phy. Rev. B 2003; 68(4) 045415.

[7] Yuan-Fong Chao, Min Wei Chen and Din Ping Tsai. Three-dimensional analysis of surface plasmon resonance models on a gold nanorod. Appl. Opt. 2009; 48(3) 617-622.

[8] Xingyu Gao and Xiaosong Gan. Modulation of evanescent focus by localized surface plasmons waveguide. Opt. Express 2009; 17(25) 22726-22734.

[9] Fu Min Huang, Nikolay Zheludev, Yifang Chen, et al.. Focusing of light by a nanohole array. Appl. Phy. Lett. 2007; 90(9) 091119.

[10] Yakov M. Strelniker. Theory of optical transmission through elliptical nanohole arrays. Phy. Rev. B 2007; 76(8) 085409.

[11] Haofei Shi, Changtao Wang, Chunlei Du, et al. Beam manipulating by metallic nano-slits with variant widths. Opt. Express 2005; 13(18) 6815-6820.

[12] Michael G. Somekh, Shugang Liu, Tzvetan S. Velinov, et al. High-resolution scanning surface-plasmon microscopy. Appl. Opt. 2000; 39(34) 6279-6287.

[13] Maurizio Righini, Giovanni Volpe, Christian Girard, et al. Surface Plasmon Optical Tweezers: Tunable Optical Manipulation in the Femtonewton Range. Phy. Rew. Lett. 2008; 100(18) 186804.

[14] Patrick Englebienne, Anne Van Hoonacker and Michel Verhas. Surface plasmon resonance: principles, methods and applications in biomedical sciences. Spectroscopy, 2003; 17: 255~273.

[15] Jason M. Montgomery and Stephen K. Gray. Enhancing surface plasmon polariton propagation lengths via coupling to asymmetric waveguide structures. Phys. Rev. B 2008; 77(12) 125407.

[16] K.S.Yee. Numerical Solution of Initial Boundary Value Problems Involving Maxiwell's Equations in Isotropic Media. IEEE Trans. Antennas Propagat. 1966; 14(3) 802-807.

[17] A. Taflove and S. C. Hagness. Computational Electrodynamics: The Finite-Difference Time-Domain Method. 3rd ed, Norwood, MA: Artech House; 2005.

[18] Raymond J. Luebbers, Forrest P. Hunsberger, Karl S. Kunz, et al. A Frequency-Dependent Finite-Difference Time-Domain Formulation for Dispersive Materials. IEEE Trans. Electromag. Compat. 1990; 32(3) 222-227.

[19] Rose M. Joseph, Susan C. Hagness, and Allen Taflove. Direct time integration of Maxwell's equations in linear dispersive media with absorption for scattering and propagation of femtosecond electromagnetic pulses. Opt. Lett. 1991; 16(8) 1412-1414.

[20] Dennis M. Sullivan. Frequency-Dependent FDTD Methods Using Z Transforms. IEEE Trans. Antennas Propag. 1992; 40(10) 1223-1230.
[21] Ge Debiao, Wu Yueli, Zhu Xiangqin. Shift operator method applied for dispersive medium. Chinese J. Radio Sci. 2003; 18(4) 359-362.
[22] David F. Kelley and Raymond J. Luebbers. Piecewise Linear Recursive Convolution for Dispersive Media Using FDTD. IEEE Trans. Antennas Propag. 1996; 44(6) 792-797.
[23] Jun SHibayama, Taichi Takeuchi, Naoki Goto, et al. Numerical Investigation of a Kretschmann-Type Surface Plasmon Resonance Waveguide Sensor. J. Lightw. Technol. 2007; 25(9) 2605-2611.
[24] Xingyu Gao, Zexin Xiao, Lihua Ning. Surface plasmon enhanced super-resolution focusing of radially polarized beam. OSA-IEEE Topical Conference: Advances in Optoelectronics & Micro/nano-Optics. 2010:5713551.
[25] P.B. Johnson and R. W. Christy. Optical Constants of the Noble Metals. Phys. Rew. B 1972; 6(12) 4370-4397.
[26] James W. M. Chon and Min Gu. Scanning total internal reflection fluorescence microscopy under one-photon and two-photon excitation: image formation. Appl. Opt. 2004; 43(5) 1063-1071.
[27] Baohua Jia, Xiaosong Gan, and Min Gu. "Direct measurement of a radially polarized focused evanescent field facilitated by a single LCD," Opt. Express 13, 6821-6827.
[28] Baohua Jia, Xiaosong Gan, and Min Gu. Direct measurement of a radially polarized focused evanescent field facilitated by a single LCD. Opt. Express 2005; 13(18) 6821-6827.
[29] A. Husakou and J. Herrmann. Subdiffraction focusing of scanning beams by a negative-refraction layer combined with a nonlinear layer. Opt. Express 2006; 14(23) 11194-11203.
[30] A. Husakou and J. Herrmann. Focusing of Scanning Light Beams below the Diffraction Limit without Near-Field Spatial Control Using a Saturable Absorber and a Negative-Refraction Material. Phy. Rev. Lett. 2006; 96(1) 013902.
[31] R. F. Oulton, V. J. Sorger, D. A. Genov, et al. A hybrid plasmonc waveguide for subwavelength confinement and long-range propagation. Nature Photonics 2008; 2(8) 496-500.
[32] Yuan-Fong Chau, Din Ping Tsai, Guang-Wei Hu, et al. Subwavelength optical imaging throught a silver nanorod. Opt. Eng., 2007; 46(3) 039701.
[33] Zhongyue Zhang and Yiping Zhao. Optical properties of U-shaped Ag nanostructures. J. Phy: Condens. Matter 2008; 20(34) 345223.
[34] G. Veronis and S. Fan. Subwavelength light bending by metal slit strctures. Opt. Express 2005; 13(24) 9652-9659.
[35] R. A. Wahsheh, Z. Lu and M. A. G. Abushagur. Nanoplasmonic couplers and splitters. Opt. Express 2009; 17(21) 19033-19040.
[36] Ki Young Kim, Young Ki Cho, and Heung-Sik Tae. Light transmission along dispersive plasmonic gap and its subwavelength guidance characteristics. Opt. Express 2006; 14(1) 320-330.

[37] Y. C. Jun, R. D. Kekatpure, J. S. White, and M. L. Brongersma. Nonresonant enhancement of spontaneous emission in metal-dielectric-metal plasmon waveguide structures. Phy. Rev. B 2008; 78(17) 153111.

[38] Changjun Min and Georgios Veronis. Absorption switches in metal-dielectric-metal plasmonic waveguides. Opt. Express 2009; 17(13) 10757-10766.

[39] A. Pannipitiya, I. D. Rukhlenko, M. Premaratne, H. T. Hattori, and G. P. Agrawal. Improved transmission model for metal-dielectric-metal plasmonic waveguides with stub structure. Opt. Express 2010; 18(6) 6191-6204.

[40] Georgios Veronis, Zongfu Yu, Şükrü Ekin Kocabaş, David A. B. Miller, Mark L. Brongersma, and Shanhui Fan. Metal-dielectric-metal plasmonic waveguide devices for manipulating light at the nanoscale. Chin. Opt. Lett. 2009; 7(4) 302-308.

Section 2

Plasmonic Structures for Light Transmission, Focusing and Guiding

Chapter 4

Surface Plasmons on Complex Symmetry Nanostructured Arrays

Brian Ashall and Dominic Zerulla

Additional information is available at the end of the chapter

http://dx.doi.org/10.5772/51336

1. Introduction

In recent years it has become accepted that the direction of the plasmonics community is becoming increasingly applied. This is a natural progression, whereby scientific advances are inevitably applied to appropriate technologies. Indeed, in order for the community of plasmonics to continue growing, or at least to maintain the current status, real world technological applications are required. However, this is not to say that the level of fundamental SP research will decrease, as there are still many questions to be answered or clarified on a fundamental level. The dramatic growth of plasmonics in the modern era can be predominantly contributed to four components: nanoscale fabrication techniques, computation power, SPP applications, and "the promise of plasmonics" [1].

The focus of this chapter draws inspiration from all four of the above points. In particular, following an introduction and description of nanostructure fabrication techniques and design considerations in the first section, the second section will detail farfield analysis techniques used for the examination of the light diffracted from structured arrays, and the subsequent identification of plasmons based on their farfield signatures. Following this, the excitation of SPPs on tailor designed 3 fold symmetric structures will be discussed, with advantages resulting from this symmetry breaking explored. Unlike rotationally symmetric structures, such 3-fold symmetric structures are inherently capable of symmetry breaking as a result of their orientational dependencies. In particular, this section (Sec. 3) will focus on the engineering of the SPP nearfield distributions on complex nanostructures. For this, the use of a PEEM (Photo Emission Electron Microscope) to map the plasmon nearfields on the surface of an array of the structures will be presented. It will be shown that the location and intensity of the focused nearfields can be controlled by changing the polarisation of the excitation light, enabling the switching of the plasmon energy localisation [2]. In section 4, specific symmetry and geometric properties of nanostructures will be shown to have an impact on the propagation of SPPs. In particular, it will be demonstrated and justified how in certain orientations, arrays of rotor shaped nanostructures have interesting wave-guiding interactions with propagating SPPs [3]. One result of this is a shift from P polarised illumination at which the classical farfield SPP related minimum reflectivity occurs. Following this, the first instance of plasmon mediated polarisation reorientation observed in the farfield, with no associated directional change of the farfield light, will be described and accompanied by supporting simulations [4, 5]. Finally, section 5 will deal with aspects of ultrafast dynamics of propagating SPPs. In particular, a tailor designed architecture will be examined for the possibility of generating broadband, ultrashort plasmon pulses [6]. Furthermore, the temporal modification of the illumination pulse resulting from SPP excitation will be investigated.

2. Plasmon active nanostructures and their fabrication

The field of plasmonics has taken a big leap in recent years, with one of the major attributors being the development of techniques to fabricate the micro- and nano-structures needed to control the flow and storage of electromagnetic energy on a very small scale. This local excitation and control of SPPs requires structuring techniques with nanoscale precision, of which electron-beam lithography and focused ion beam irradiation have proven to be the most important because of their ability to make diverse structures with high resolution. In this section, these structuring techniques will be briefly discussed, along with the implications of sample quality.

2.1. Nanostructuring

Electron-Beam Lithography (EBL) is a process that uses a focused beam of electrons to form patterns for material deposition on (or removal from) the sample substrate. In comparison to optical lithography, which uses light for the same purpose, EBL offers higher patterning resolution because of the shorter wavelength possessed by the 10-50 keV electrons that it employs. This small diameter focused beam of electrons is scanned over a surface, negating the need for masks required in optical lithography for the projection of patterns. An EBL system simply draws the pattern over the resist wafer using the electron beam as its drawing pen. Thus, EBL systems produce the resist pattern in a serial manner. This makes it slow compared to optical systems, but gives a user more control of the structure shape and allows for different rate of lithography at different locations.

The *Focused Ion Beam* (FIB) technique was developed during the late 1970s and the early 1980s, with the first commercial devices available in the late 1980's [7]. The technology enables localised milling and deposition of conductors and insulators with high precision, hence its success in device modification, mask repair, process control and nanopatterning [8–10]. When energetic ions hit the surface of a solid sample, they lose energy to the electrons of the solid as well as to its atoms. The most important physical effects of incident ions on the substrate are: sputtering of neutral and ionized substrate atoms (this effect enables substrate milling), electron emission (this effect enables imaging), displacement of atoms in the solid (induced damage) and emission of phonons (heating). Chemical interactions include the breaking of chemical bonds, thereby dissociating molecules (this effect is exploited during deposition). The best resolution of FIB imaging and milling is comparable to the minimum ion beam spot size, typically below 10 nm. In crystalline materials, such as aluminium and silver, the ion penetration depth varies due to channeling along open columns in the lattice structure. The removal of sample material is achieved using a high ion current beam, resulting in a physical sputtering of sample material. By scanning the beam over the substrate, an arbitrary shape can be etched.

Thin Metal Films: A basic requirement for experimental research on SPPs is the ability to make high purity, smooth, and often thin, metallic films. A number of techniques are available for this, the most typical being resistive thermal evaporation and e-beam evaporation.

The principle of vacuum evaporation is simple: the substrate and the coating material are both placed in an evacuated enclosure, some distance apart. The coating material is then heated to its vaporisation pressure point, so that it evaporates. Sufficient thermal energy is supplied to enable individual atoms to escape from the surface of the molten material. These atoms travel in a straight line through the vacuum towards the substrate where they adhere to the surface. Several methods may be used to melt the evaporant, such as resistive heating in which the evaporant is loaded in a boat shaped crucible made of a metal with a considerably higher melting point than the evaporant, through which a large DC current is passed. Many metals may be evaporated very successfully using resistive heating, and of particular importance here, this list includes Silver, Gold, Aluminium, Nickel, Platinum and Chromium; some of which are important as plasmon active substrates, and others important for adhesion layers [11].

An alternative deposition technique is electron beam heating. Here, a hot wire filament is used as a thermionic electron emitter, where the electrons are accelerated and guided toward the evaporant (either directly, or contained in a crucible). The evaporant is heated by the kinetic energy of the electrons and subsequently is coated on the sample. This method can typically generate smoother films in comparison to the resistive heating method, as the material is evaporated from its surface in a much more controlled manner than resistive heating, where the sample is typically completely melted and prone to sputtering.

In either system, the rapid condensation typically produces grained films which are relatively rough on an atomic scale, but on the scale of the wavelength of the radiation they are typically very smooth. Therefore, roughness induced re-radiation of an excited plasmon will be at an acceptably low level for the majority of plasmon experiments.

2.2. Three fold symmetric structures

Experimental examinations presented in this chapter primarily deal with arrays of structures of 120° symmetry properties, along with a reference sample of a more typical geometry (ring or doughnut shape). The geometry of four example structures are depicted in Fig. 1. They are designed such that they relinquish the widely investigated circular symmetry in favor of a 120°, or 3 fold, symmetry. One of the primary and original concepts behind this 3 fold symmetry design is that it permits the reduction of the footprint of the structure in comparison to a ring design, while maintaining surface plasmon resonance conditions. For a ring shape structure, an optimum nearfield resonance will occur where the ring diameter is an (low) integer multiple of the plasmon wavelength. If this is the case, a plasmon propagating around the structure will not destructively interfere with itself, but instead each circulation plasmon wave will constructively add to the other propagating waves. The same is true for the other structures displayed in Fig. 1, but as a result of the more complex shape, the plasmon propagation will be more complex and exhibit a smaller footprint. Additionally, new nearfield focuses will be introduced, for example at the structure centers. The structure geometries were analytically designed according to Lissajous type functions, or more specifically Epitrochoide geometries [12]. Using these functions, the geometries of the structures are defined according to the following equations:

Figure 1. Sample designs of structures (top) and extended structure design for production (bottom) [12].

$$x = (R + r)cos\phi - (r + \rho)cos\left(\frac{R + r}{r}\phi\right) \quad (1)$$

$$y = (R + r)sin\phi - (r + \rho)sin\left(\frac{R + r}{r}\phi\right) \quad (2)$$

Where ϕ is an angle between 0 and 2π, r is the radius of a circle that is rotated about the boundary of a larger circle of radius R, and ρ is the distance from the outer boundary of the smaller circle to its center. Numerous variations of the above described structures were manufactured in arrays using EBL [13], and were prepared as follows:

Following a 200°C, 1 hour bake, ZEP 520 photo-resist was spun on to silicon plates to a thickness of 100 nm. The desired structures designs were then written onto the photo-resist using a Joel JBX-6000FS/E EBL system and subsequently etched (with SF_6 and C_4FH) to reveal the desired surface profile. Following this, a 5 nm thick adhesion layer of Platinum was deposited, and finally an 80 nm thick silver film deposition, to facilitate SPP excitation, was made. The overall array size was typically 200 μm^2 to 400 μm^2 (varying with sample scale and grating constant), meaning the overall arrays were typically visible by eye; their visibility aided by their natural diffraction. In the majority of the samples, these geometries represent the shape of the raised topography. However, in one manufacturing phase, the geometries represent lowered (trough) locations resulting in "rotor" shaped nanostructures; the importance of which will be discussed in section 4 of this chapter.

2.3. Farfield optical analysis

The characterisation of the structured arrays begins with a farfield diffraction analysis, using a semi-spherical scanner [Fig. 2]. Here, the sample was mounted on a sample goniometer, and was illuminated with a polarised laser (HeNe or Ti:Sa). Prior to incidence on the sample, the laser beam was also passed through an iris to reduce the beam diameter to approximately 0.5 mm (slightly larger than the typical size of the structure arrays). Additionally, for CW laser operation, a Fresnel Rhombus was placed in the laserline between the polariser and the iris, allowing for variation of the polarisation with no intensity dependence related to a fundamental polarisation preference of the laser. The detector used (photomultiplier tube) was positioned behind an aperture which had an azimuthal angular acceptance of 0.15° [Fig. 2].

Performing farfield diffraction characterisation allows for the quantification of the quality of the sample from both a plasmon perspective, and sample quality. An example of a complete θ, ϕ farfield scan is presented in Fig. 3a. The sample examined is an array of the 3-fold symmetric shamrock shaped design (grating constant of 1.5 μm.). Here, the sample was examined under P polarisation illumination, and with the sample goniometer at angles of $\theta = 45°$, $\phi = 0°$ and $\alpha = 45°$ (diffraction pattern rotated by 45° to its normal). Series of scans such as this allow for the identification of SPP resonances (see section 4), checking the array pitch by measuring the diffraction angles, and making some sample quality checks, related to the sharpness of the diffraction and reflection channels. Using the scanner, high resolution mapping of individual diffraction orders is possible, enabling detailed characterisation of the spatial intensity distributions of the diffraction orders. Similar analysis (called spot analysis) is widely used in low energy electron diffraction (LEED) and related surface science techniques, where it permits the characterisation of the reciprocal space structure of a surface at atomic resolutions. However, the use of spot analysis is rarely used in visible laser spectroscopy techniques, as employed here. One such spatial intensity scan is presented in Fig. 3b, for a shamrock sample under P polarisation illumination, and with the sample goniometer at angles of $\theta = 55°$, $\phi = 0°$ and $\alpha = 0°$. Immediately visible from the farfield intensity map is that the spot does not have a uniform (Gaussian) spatial distribution. This spatial distribution is not generated by the laser beam profile, which is close to (as a result of the small aperture) a Gaussian profile which is spatially cut approximately half way down its wings. This was confirmed by examining the laser profile of the reflection and diffraction from a symmetric ring shaped structure. This comparative check of the farfield spatial intensity profile of the ring structure and the shamrock structure confirms that the uneven spatial profile in Fig. 3b is as a result of the nanostructures symmetry properties. This is further confirmed by performing a Fast Fourier Transform (FFT) on the image, which reveals 6 preferential symmetry directions, directly related to the 3 fold symmetry of the nanostructures [Fig. 3c].

3. PEEM as a tool for imaging plasmonic fields

In the this section, a nearfield examination of the plasmon enhanced electromagnetic fields on the above discussed reduced symmetry structures will be presented. For this, a PEEM (photo emission electron microscope) is used to map the plasmon nearfields on the surface

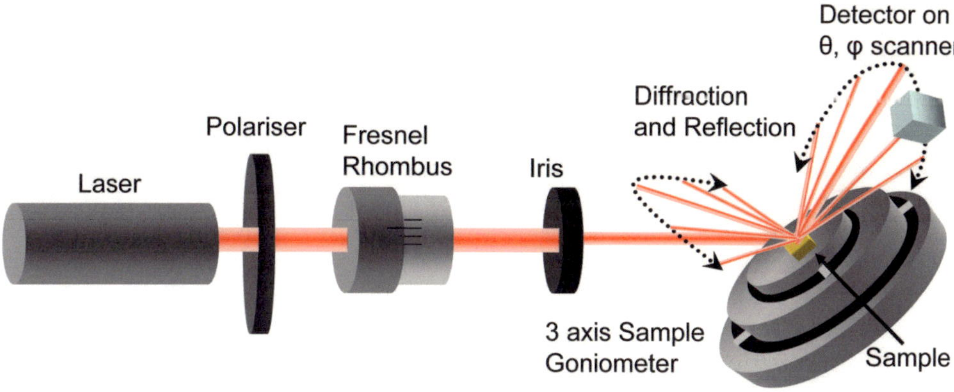

Figure 2. Set-up for farfield diffraction pattern (and individual diffraction spot) analysis [14].

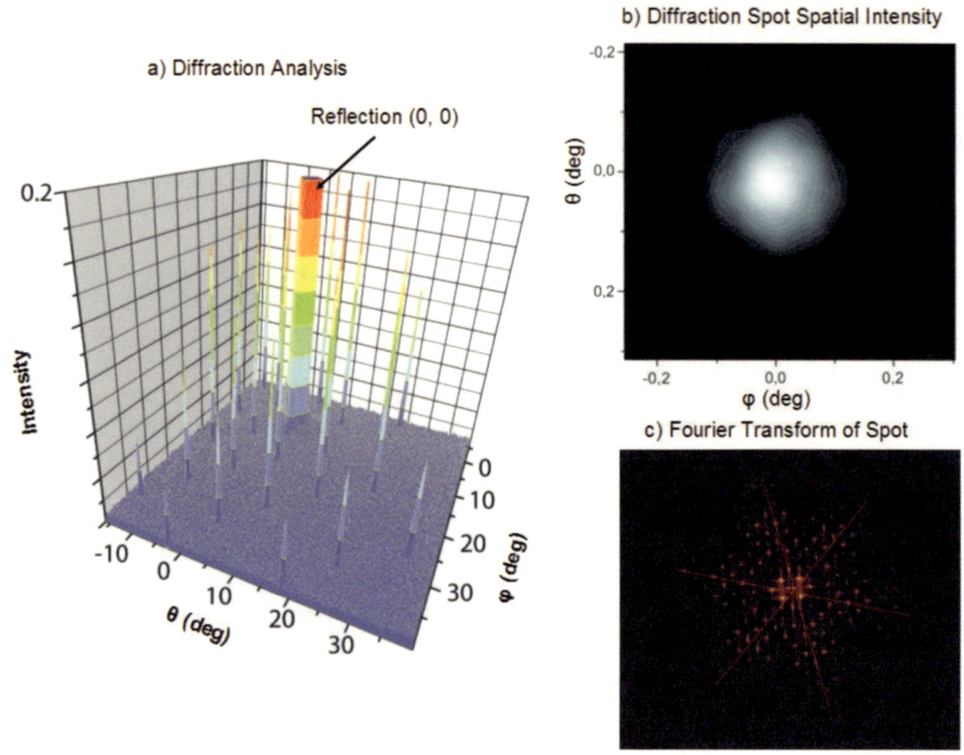

Figure 3. FIX C a) Sample diffraction characterisation. Reflection taken as the $(\theta, \phi) = (0, 0)$ point, θ and ϕ are the scan angles. b) High resolution farfield angular intensity map of single diffraction order (0, -2) from a 3-fold symmetric shamrock structure. c) Fast Fourier Transformation (FFT) of (b) demonstrates 3 preferential symmetry directions (indicated by dotted red lines) of the intensity spot, as a result of three fold symmetry of the structure [14].

of an array of threefold symmetric structures. In addition to the experimental observation of plasmon energy localization control [2], the powerful use of a PEEM for the non-perturbative subdiffraction limited imaging of plasmonic fields will be discussed.

3.1. Nearfield imaging

At the heart of much plasmon research are tools that allow researchers to examine plasmon effects in the nearfield. This was initially made possible with the development of scanning probe techniques, and their modification for the purpose of examining nearfield SPP properties directly at the surface at which the SPP is confined. The first scanning probe technique applied to the investigation of SPPs was scanning tunneling microscopy (STM), relying on the detection of changes to the tunnel current by SP induced variations in the local density of states [15–18] or the farfield scattered light due to the local SPP interaction with a STM tip [19].

Now, a wide range of techniques for the imaging of sub diffraction scale plasmon processes are available, with Scanning Nearfield Optical Microscopy (SNOM, or sometimes NSOM) being the most popular. This is a technique for optical investigation below the farfield resolution limit (diffraction limit), which is achieved by exploiting the properties of evanescent waves. The technique involves placing the detector (typically a very small probe) very close (distance smaller than the wavelength) to the specimen surface, allowing for surface inspection with high spatial resolving power. With this technique, the resolution of the image is not limited by the wavelength of the illuminating light, but rather by the size of the detector probe, along with other considerations.

Irish scientist E. H. Synge is given credit for conceiving and developing the idea for an imaging instrument that would image by exciting and collecting diffraction in the nearfield. His original idea, proposed in 1928 [20, 21], was based on the usage of intense planar light from an arc under pressure behind a thin, opaque metal film with a small aperture of about 100 nm. The aperture was to remain within 100 nm of the surface, and information was to be collected by point by point scanning. He foresaw the illumination and the detector movement being the biggest technical difficulties [22].

Current generation SNOM techniques are very powerful tools for studying SPPs; however the perturbation and alteration of the SPP field associated with the introduction of a tip is one of the major concerns and drawbacks of SNOM techniques. The use of probes (coated or uncoated fibers, or SPM tips etc) in the nearfield proximity of metal surface results in a perturbation of the electromagnetic field due to the tip / surface interaction. However, until recently, there was no known way to investigate the nearfield plasmon information without influencing the plasmon itself to some extent with the measurement device.

However, in 2005, two independent groups demonstrated the use of a PEEM for the observation of plasmonic nearfields [23, 24]. In PEEM [Fig. 4c], photo-electrons emitted from the surface of a metal are imaged with electron optics, and these electrons are collected at a distance on the millimeter scale, meaning there is no influence of the collection optics on the plasmon before or during measurements.

PEEM is closely related to the more recently developed Low Energy Electron Microscopy (LEEM), with the predominant different being that the principle of PEEM is the photoelectric effect. PEEM has already proven to be a powerful tool in material science, surface physics and chemistry, thin film magnetism, polymer science, and biology [25]. Historically, the invention of PEEM dates to the early 1930's, shortly after the introduction of electron lenses. The first working PEEM was built by Bruche in 1932, and the principal design of his PEEM is still used. In Bruche s PEEM, UV light from a mercury lamp was focused onto a sample, and the emitted photoelectrons were accelerated by a potential difference of 10 to 30 kV between the sample (cathode) and the anode of the PEEM, and subsequently focused onto a phosphor screen. A PEEM forms an image of a surface based on the spatial distribution of photoelectrons emitted. Importantly for plasmonics, photoelectron emission has been shown to be enhanced by the increase of the local electrical field upon excitation of SPPs [26].

The PEEM system we have in our lab is a SPECS PEEM P90, and depending on the photo-emission flux, this system is capable of imaging with a lateral resolution of 5 nm. The illumination frequency at which one can operate is dependent on the sample work function, which the illumination energy must be above for direct photoelectron mapping. Therefore, PEEM imaging of direct (or single) photo-emission processes presents a high resolution map of the surface workfunction threshold. For most plasmonic experiments, the surface under investigation is largely a smooth single material surface, with specifically designed sharp nanoscale structures for controlling the plasmons. On such a surface, the dominant variations in workfunction arise from the sharp topographic variations, and so these can be mapped with high resolution using single photoelectron PEEM imaging. In order to map the plasmon nearfields, multiphoton photoelectron emission is used (typically 2 Photon Photo-Emission - 2PPE). This imaging technique is particularly useful for the imaging of plasmon effects, as the local multiphoton photo-emission is extremely sensitive to local field intensities (it varies with the square of the field strength), which are dramatically stronger at plasmon localised points [23, 24, 27].

A comparison of imaging plasmons using the PEEM technique to SNOM techniques reveals a number of advantages and disadvantages:

Advantages:

First and foremost, PEEM has the ability to image the plasmon without perturbing the plasmon field. By comparison, SNOM techniques require placing a probe within the plasmon field, and so inherently altering the field.

Unlike SNOM, PEEM is not a scanning imaging technique. Like an optical microscope, it captures all field information simultaneously, but with nanometer resolution. This means that the technique is extremely quick, allowing for real time, very fast monitoring of plasmons. In fact, the imaging rate is solely dependent on the required camera integration time, which is in turn dependent on the photoelectron emission rate, and ultimately on the illumination power.

As the 2PPE process is dependent on the square of the power of the local field strength, the contrast of the PEEM image is also a squared contrast. Therefore, to a first approximation, in a 2PPE PEEM image, an area that is twice as bright as another area, actually indicates that there is only a field strength difference of the square-root of the difference between the intensities. This squared dependence on the 2PPE results in a high contrast level when compared to SNOM techniques.

Disadvantages:

The main problem with PEEM is that the work function of typical plasmon active materials restricts 2PPE to the blue end of the visible spectrum, meaning that blue excited plasmons are most appropriate. However, with higher power systems (e.g. amplified Ti:Sa systems) observation of plasmons excited at 800 nm would be possible with 3 photon photo-emission. In comparison, SNOM keeps a fundamental advantage here, where its performance increases as the excitation frequency is reduced, and significantly, it works very well at telecom frequencies.

Another disadvantage of PEEM is the associated cost of the equipment. Imaging photo-electrons requires extremely good vacuum conditions, which has a high associated cost, and also a relatively large space requirement. Additionally, the imaging optics, similar to those in a Transmission Electron Microscope (TEM) or Scanning Electron Microscope (SEM), are costly. Furthermore, for successful plasmon imaging with a PEEM, femtosecond laser systems are required for excitation; again a significant cost. A typical SNOM would cost considerably less, requiring a relatively low quality SPM as its basis (atomic scale resolution is not necessary or usable), simple CW laser light sources as illumination, and no vacuum requirements.

Despite these disadvantages of a PEEM system, the advantage and power of direct real time imaging without perturbing the plasmon field cannot be ignored.

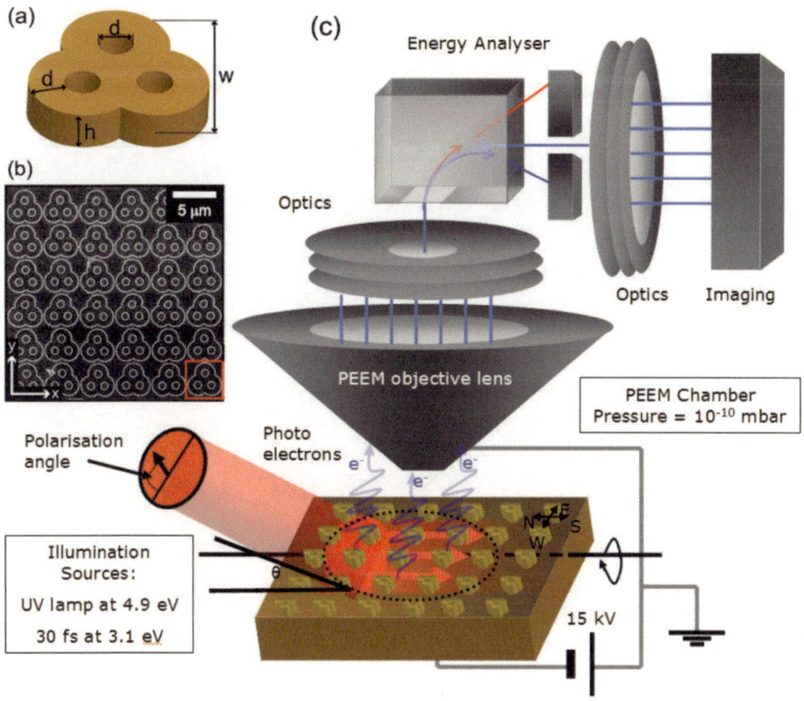

Figure 4. (a) Geometry of the structures. Structures in scale *A* have dimensions: w = 2700 nm, d = 600 nm, h = 100 nm; structures in scale *B* have dimensions w = 3600 nm, d = 800 nm, h = 100 nm. (b) SEM image of array of structures in scale *A* [2]. (c) Schematic of the technique for laser excited optical nearfield imaging with a PEEM [14]. For the presented results, the illumination angle (θ) is fixed at 25°. Note: Not all PEEMs are equipped with an energy analyser as depicted here for a Specs 90 PEEM.

3.2. Nearfield analysis of 3-fold symmetric structures

SPPs are intrinsically accompanied by strong electromagnetic nearfields, and surfaces can be actively modified to influence the excitation conditions of SPPs and hence nearfields. In particular, the selective addressability of nearfields on a surface is of interest, as shown in a demonstration of adaptive nearfield shaping [27]. In the following section, an investigation of the SPP electromagnetic nearfields excited on some of the structures described above is presented; the geometry and dimensions of which are depicted in Fig. 4. To investigate the nearfield distributions of an array of these structures, a Focus IS PEEM was used, described in detail in [28]. To record the PEEM images, two different light sources were used: a mercury-discharge lamp (UV illumination) with high-energy cutoff at 4.9 eV and a frequency-doubled Ti:Sa laser system, delivering 400 nm (3.1 eV) pulses of 100 fs duration and 20 mW at 80 MHz repetition rate. While the energy of the UV illumination is sufficient for the electrons to overcome the work function of the structured silver surface of 4.64 to 4.74 eV [29], the energy of the laser photons is too low. Hence at least two photon processes are required to generate photoelectrons; i.e. two photon photo-emission (2PPE). The laser power bandwidth product coupled to the plasmon nearfield enhancement resulted in readily observable 2PPE processes. As discussed above, as these 2PPE processes are very sensitive to the intensity of nearfields, nonlinear PEEM becomes a highly suitable tool to investigate and map plasmonic nearfield processes.

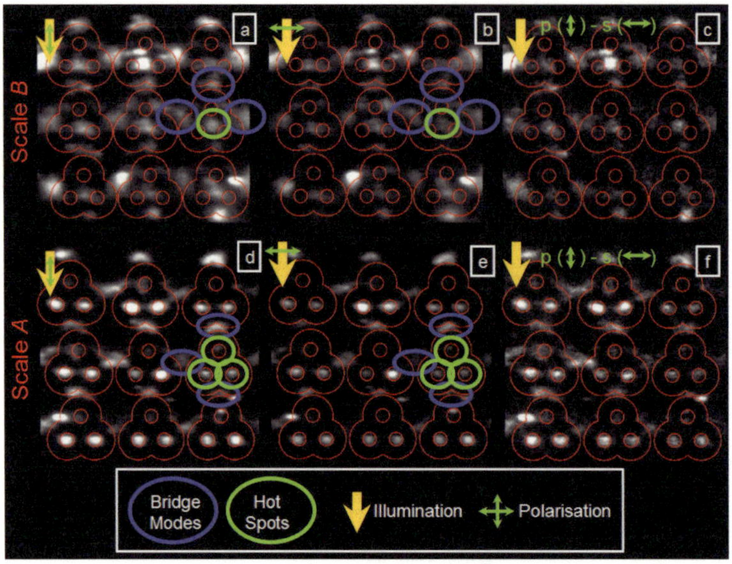

Figure 5. Contrast enhanced 2PPE PEEM images of a 3 by 3 cut out of the array of structures (a, b, d, and e). Images (c) and (f) show the differential image of S polarisation subtracted from P polarisation. 400 nm, ∼ 100 fs laser illumination incident as indicated by yellow arrows (at an angle of $\theta = 25°$. Polarisation is indicated by green arrows, P and S corresponding to vertical and horizontal respectively. For better recognition, the contour of the structures is highlighted (red). Deviations from the regular array pattern are due to spherical aberrations of the PEEM electron optics. Green circles represent hot spots as discussed in text and blue circles indicate bridge modes [2].

3.3. Plasmon energy localisation control observed by PEEM

For our presented investigation, firstly, the structures have been imaged using CW excitation at 4.9 eV to map the work function of the structure surface. These maps can be used as a reference for the 2PPE PEEM images to distinguish photo-emission effects due to the local electronic and morphologic structure of the sample surface from optical nearfields. Additionally, these work function images are used as a basis for aligning the actual structure geometries to the 2PPE PEEM images, whose size is independently confirmed by SEM (e.g. Fig. 4b) and AFM. Following this sample workfunction mapping, the structures have been investigated using the pulsed laser excitation for different illumination polarisation conditions. These 2PPE PEEM images were background subtracted and normalised via a sample independent reference beam on the microchannel plate in the imaging system of the PEEM, and subsequently contrast enhanced. In these images it is important to remember that intensity differences do not correlate linearly to the nearfield intensities, as discussed above. From these 2PPE PEEM images [Fig. 5], we can clearly identify nearfield plasmon effects. These excitations can be assigned to two different locations, at which they occur periodically across the entire array of structures:

Firstly, areas between the individual structures are excited, indicated by the bright intensities, which connect the structures of the array with each other vertically and horizontally. This emission occurs identically for structures of scales *A* and *B*, as highlighted by blue indicators in Fig. 5. These gap modes are due to field enhancement effects, which occur between the exposed edges of one structure towards its neighbors in the array.

Secondly, the other location where noteworthy enhancement takes place lies within the contour of the individual structures. These hot spots appear in each structure of the array. For scale A structures, these excitations are up to 280% brighter than the bridge excitations between the structures, depending predominantly on the intensity of the individual spot itself. This indicates an increased localisation of nearfield intensity in hot spots on the surface. Comparing Figs. 5a and 5d shows, additionally, that the strength and location of these excitations is dependent on structure size: for the larger structures the hot spots are located within the holes of the structure of the circles pointing along the x axis (there are no comparable excitations in the holes of the circles oriented in the y direction). Whereas, for the smaller structures the excitations are centered between the holes of the two circles along the x axis. Their intensity is much less pronounced and similar to the bridge excitations on the same array. Their excitation is promoted by geometric conditions resulting in the excitation of localized SPPs. The change of the location and intensity of the excitations within the contours of the individual structures is dependent on structure size and can hence be attributed to different interference and resonance conditions for the SPPs, determined purely by the geometric considerations.

To examine any polarisation dependence of these effects, P and S incident polarisations were examined for structures of both sizes. Images in Figs. 5a and 5d are acquired with P polarised light, images in Figs. 5b and 5e with S polarised illumination light. For comparison, S polarisation images are subtracted from P polarisation images [Figs. 5c and 5f]. The subtraction images demonstrate directly that the excitations within the contours of the individual structures generated with P polarisation have a stronger intensity than the ones generated with S polarisation for structures of both structure sizes. Surprisingly, the locations of the excitations are independent of the polarisation used. A quantitative analysis of the hot spots in the larger structures shows that the P intensity is about 50% higher than the S intensity. For the smaller structures the enhancement factor for the excitations within the contours of the structures varies in the range from 20% to 80%. The deviations in nearfield intensities from one structure to another can be attributed to the roughness peaks [30], but the location of the hot spots is predominantly determined by the geometric arrangement and shape of the structures.

3.4. Conclusion

A focusing of the SPP nearfields by threefold symmetrical to well defined locations is demonstrated. Observation of this effect is achieved using PEEM as a tool to map electromagnetic nearfields. The location of the focused nearfields varies with, and can be chosen by, the original structure size and design. More importantly, the intensity of locally fixed nearfields (hot spots) can be influenced by changing the polarisation of the excitation light by rotating the polarisation from S to P orientation. This enables one to switch the energy localisation on and off. Hence this approach presents a step towards a predictable design of structured surfaces, which focus energy in a spatially selective and switchable manner [2].

4. SPPs on 3-fold symmetry nanostructures

Of primary interest in this section are two arrays featuring nanostructures based on two different symmetry classes. The first array (the ring array) is designed with nanostructured rings invariant under C_∞ transformations and will be presented here as a reference array. In the second array (the rotor array), the structures resemble the shape of a triquetra rotor [Figs.

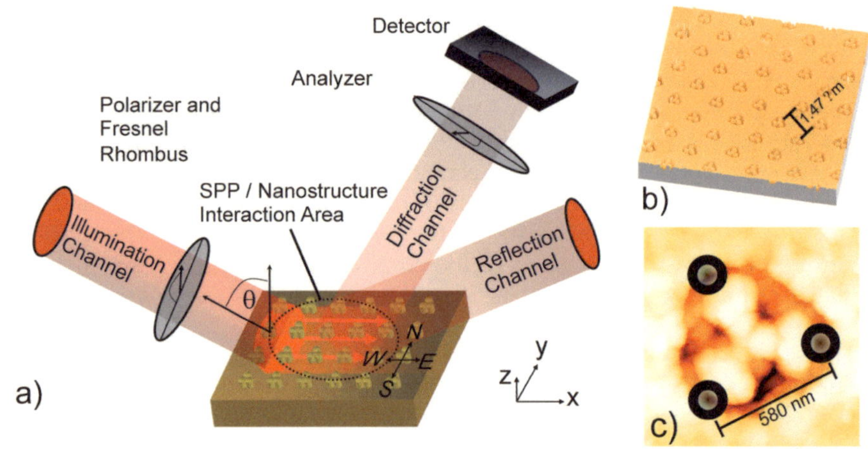

Figure 6. a) Visualisation of SPP excitation, propagation and re-emission processes, on a nanostructure arrayed surface [4]. Compass notation indicates the 4 examined interaction orientations - the displayed orientation in the schematic is E. b) AFM image of nanostructures array. c) Individual rotor structure including indication of scattering points used in the simulations.

6b and 6c]. They have a threefold symmetry and are invariant under C_3 transformations. The nanostructures are arranged to form a squared array, as indicated in Figs. 6a and 6b. As the rotor array is examined in four 90° separated orientations (α), compass notation (N, S, E, and W) will be used as identification labels [Fig. 6a]. Both arrays are housed on a 1 mm^2 silver coated section of silicon wafer and were prepared using e-beam lithography followed by etching to reveal the desired surface profile, as described above.

The experimental setup [Figs. 2 and 6a] for the farfield polarisation examination is as follows: A laser source ($\lambda = 632.8$ nm) is collimated, polarised (extinction ratio of 10,000:1), and made incident on the sample which is housed on a rotation table on a fine adjust goniometer [Fig. 2]. A Fresnel Rhombus is positioned in the beam-line between the polariser and the sample. This allows for polarisation angle (β) variation with uniform beam intensity, independent of any fundamental polarisation of the laser. The detector (photodiode) is mounted on a computer controlled, highly resolving, angular scanner which has the sample goniometer at its fulcrum. A polarising analyser (extinction ratio of 10 000:1, analyser angle = γ) can be positioned on the semi-spherical scanner in front of the detector, depending on the experimental requirements.

4.1. Plasmon excitation on nanostructured arrays

Certain anomalies in the intensity of light diffracted from a grating are known to correspond to the excitation of SPPs [30]. They are apparent from sharp changes in the reflected intensity of P polarised light when the grating vector is parallel to the plane of incidence. In this case, the grating changes the in-plane wave vector of the incident photon field by the addition or subtraction of integer multiples of the grating wave vector [31, 32].

In order to locate the grating induced SPPs, angular scans for S and P polarisation were recorded in a direction about an axis parallel to the S oscillation, while maintaining the sample perpendicular to the laser in the plane of P oscillation. Such angular scans were carried out on a number of available diffraction orders for both samples in all four structure orientations. Here, for consistency and clarity, we predominantly limit our presentation to the

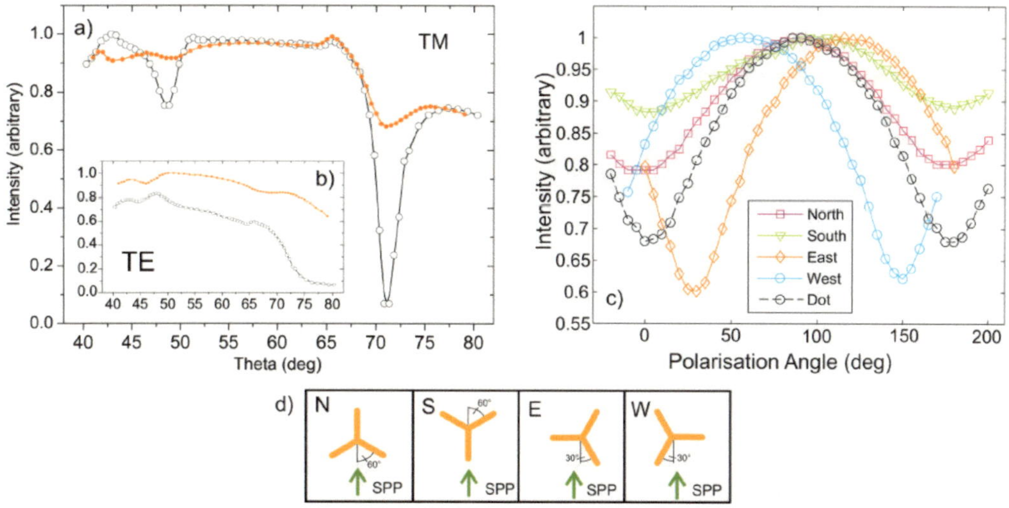

Figure 7. Angular scans of the intensity of the (+2, 0) diffraction order for the reference ring array (unfilled) and the rotor array in the North orientation (red, filled) for P (TM) polarised (a) and S (TE) polarised (b) light. c) Fixed angle polarisation scan for the ring array and for N,S,E and W orientations of the rotor array (P = 0°, 180°, S = 90°). Note: The analyser indicated in Fig. 6a is absent for the results displayed in here. d) Four orientations of the rotors primary axis with respect to the excited propagating SPP [3].

(2, 0) diffraction order (however, the effect is not limited to this particular diffraction order [14]). For the reference (ring) array, the P polarisation curve reveals two pronounced minima at 71.3° and 48.5°, where the SPP extracts energy [Fig. 7a]. From this figure, it is clear that for the ring array the coupling efficiency for P polarised light is very strong; for the SPP at 71°, $I_{SPP}/I_{max} > 10\%$. For the rotor array, the excitation efficiencies at P polarised illumination are considerably lower than for the ring array. In terms of the angular resonance scan, the only effect the nanostructure design can have is on the efficiency at which light can be coupled to the surface. Generally, for a grating with single periodicity, the further one deviates from a sinusoidal cross-section profile, the lower coupling efficiency one gets, as a Fourier analysis of the profile reveals smaller (but broader) peaks [33].

4.2. SPP illumination polarisation dependence

As the ring array is completely invariant under 0°, 90°, 180°, and 270° rotations (α), angular scans at these orientations result in an identical plot to the corresponding plot in Fig. 7a. However, as a result of the 120° symmetry of the rotor structures, no overall 90° or 180° rotational symmetry is conserved. Therefore, if the structures themselves are to have an effect on the SPP resonance conditions, examinations with the structures in N, S, E, and W orientations should present individual differences. To this extent, it is observed that the rotor structures have a definite impact on the polarisation angle at which SPP related minimum reflectivities occur [Fig. 7c]. In order to investigate this phenomenon, the arrays were mounted at an angle such that the examined diffraction order was in resonance for P polarised light. The polarisation direction was subsequently rotated using the Fresnel rhombus in 5° increments through a full polarisation rotation. In contrast to such a scan using the ring array, which shows minimum reflectivity for purely P polarised light, for the rotor array it

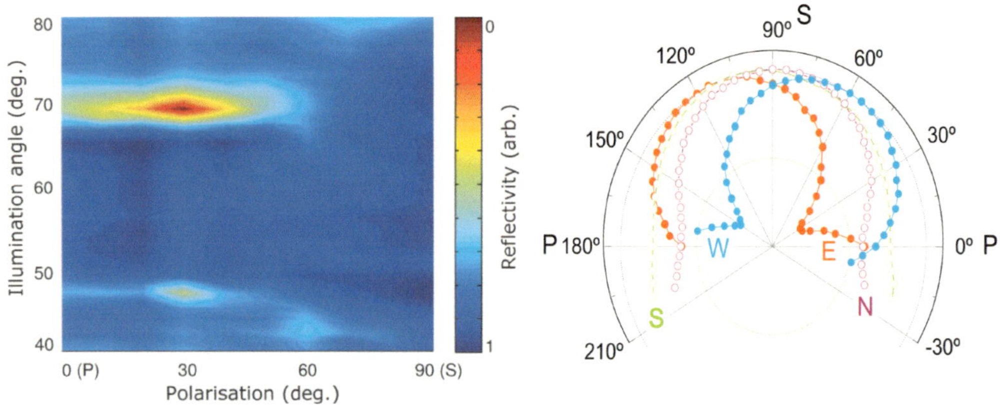

Figure 8. Left: Measured reflectivity for an illumination angular range, $\theta = 40°$ to $80°$, and illumination polarisation of P to S. Peaks at $\theta = 70°$ and $47°$ indicate plasmon modes. Right: Intensity of the ($+2^{nd}$) diffraction order as a function of polarisation angle (β) for rotor nanostructures in the 4 orientations (N, S, E, W) at the SPP excitation illumination angle ($\theta = 70°$) [4]. Intensities are individually normalised to 1. Note: The analyser indicated in Fig. 6a is absent for the results displayed in these plots.

is observed that the minimum reflectivity does not necessarily occur for incoming P polarised light [Fig. 7c]. While the minimum reflectivities for N and S orientations are found at P polarisation, this is not the case for E and W orientations, where the minimum is shifted from P polarisation by +30° and -30°, respectively [3].

Before this illumination polarisation shift of the reflectivity minima can be confirmed to be solely a result of an interaction between propagating SPPs with the rotor structures, it must be confirmed that it is not purely a grating artifact; as it is well known that a complex grating topography can present changes in farfield intensity, independent of plasmonic effects [33, 35]. Therefore, we carried out a complete angular and illumination polarisation characterisation of our gratings for all 4 illumination orientations. An example of such an angular/polarisation scan for the SPP/rotor interaction in the E orientation is presented in Fig. 8; demonstrating that:

a) The SPP excitations at illumination angles (θ) of 70° and 47° are the only pronounced intensity variations.

b) For this SPP/nanostructure orientation (E) both plasmon reflectivity minima are shifted to a polarisation angle (β) of TM + 30°.

From a qualitative point of view, an explanation of this shift of illumination polarisation corresponding to SPP related minimum reflectivity can be found in considering the symmetry of the structures with respect to the incoming illumination, and hence initial SPP propagation direction [Fig. 7d]. Turning our attention to processes occurring after excitation; consider the rotors in the E orientation. A grating induced SPP would propagate along the silver surface, where upon reaching the boundary of a rotor structure, the SPP wave would undergo a number of different processes with different probabilities [36]. As the SPP wave impinges on the boundary, the portion of the wave that is not transmitted or reradiated interacts with the boundary in two manners:
The first is through reflection [37, 38] and the second, and more interesting here, is through a guiding effect which can occur when a propagating SPP is made incident on a guiding surface feature [39–42]. It is anticipated that our triangular trough boundary acts much like

a waveguide; behaving as a gutter collecting the SPPs, and guiding them into propagation within the trough. This behaviour can only occur with relatively high efficiencies if there is an acute angle between the original SPP direction and the new guided direction, e.g. in analogy to skimming a stone on water. Following this, the SPP can be reradiated where its polarisation would be determined by the polarisation of the originally guided wave [43, 44]. As a result of the phase shift between the two optical channels contribution to this diffracted mode [45], these two light components will predominantly destructively interfere. Naturally, this destructive interference will be at a maximum where the polarisation states of the two interfering components are matching. It is important to remember that changing the incoming illumination polarisation angle does not change the associated polarisation orientation of the excited SPP; it only alters its relative excitation strength. For the rotor structure in the E orientation, for this polarisation matching to occur, the polarisation of the incoming light would be set to P +30°; matching the twist in the SPPs associated polarisation as described above. By a similar argument, with the rotors in the W orientation, this polarisation matching would occur at P - 30°.

4.3. Plasmon mediated polarisation twisting

In order to investigate the origin and processes involved in this polarisation minimum shift, we have used a polariser/analyser set-up as in Fig. 6a. With the illumination angle set at the SPP excitation angle (θ = 70°), and the illumination polarisation set to P, we have recorded the intensity monitored by the detector, as a function of analyser polarisation. Such an examination would typically present a $\cos^2$ function of the angle between the polariser and the analyser. With the exception of an intensity offset, this is exactly what is observed for illumination angles off SPP resonance, and also for the symmetric SPP/nanostructure interaction orientations (N and S). However, for such a scan in the E and W orientations at the SPP excitation angle, we observe a deviation from a $\cos^2$ function [Fig. 9 left panel]. Most notably, we observe a 5° shift in analyser angle at which we observe a maximum. This deviation from a $\cos^2$ function indicates that our plasmon-rotor interaction is causing an additional polarising function. More specifically, this is proof that this interaction is twisting the polarisation of the light involved in the SPP excitation and re-emission process.

Although in this setup we observe a shift of only 5°, instead of the 30° shift observed in Fig. 8, both results are in fact consistent with each other. The apparent difference originates from that fact that one measurement takes into account SPP and regular diffraction channel processes, while the other isolates the SPP excitation processes [4].

This plasmon mediated polarisation twisting process is further confirmed by actively altering the distribution of light following the two paths. The variation in this ratio can be accurately controlled by tuning the illumination angle (θ), as presented in the right panel of Fig. 9. This plot of observed maxima in the polariser analyser experiment as a function of illumination angle, not only confirms the process of polarisation twisting, but also demonstrates the external active control over it that can be readily achieved (steps of 0.25° twisting readily realisable). Furthermore, the range of polarisation twisting could be greatly increased by improving both structure design and surface quality, currently limiting the SPP excitation efficiency to below 20%. If this excitation efficiency is increased, for example to 50%, the farfield measurable polarisation reorientation effect would be increased to 15°, and modifying the design of the structures to re-orientate the plasmon more efficiently would also increase the effect.

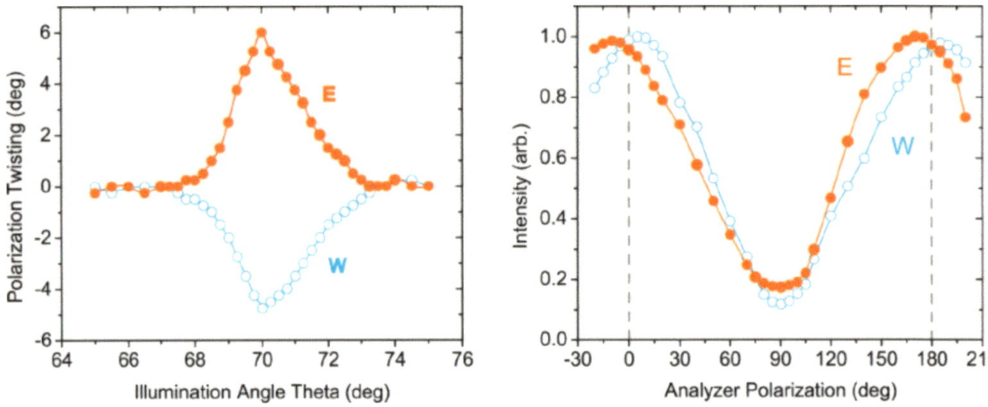

Figure 9. Left: Plot of normalised intensity as a function of analyser polarisation (γ), for illumination polarisation (β) of P. Right: Plot of polarisation twisting degree as a function of illumination angle (θ), for illumination polarisation of P [4].

4.4. Plasmon / rotor interaction simulations

To further understand the origin of the observed polarisation twisting, numerical simulations based on elastic SPP scattering [46] have been performed. In these simulations, we examine the interaction between the SPP associated electromagnetic field with the 120° symmetric structures. Initially, to allow us to focus solely on the origin of the polarisation twisting some simplifications of the processes contributing to the experimental observations have been deliberately made. This approach allows us to easily define and focus on the details we are interested in (namely the plasmon **E** field scattering) and so understand the real fundamental roots of polarisation twisting process. In these initial simulations performed by Dr. Vohnsen, a scalar effective polarisability representative of the scattering strength of each nanostructure has been used. Additionally, an ideal planar incident SPP has been assumed, absorption losses have been neglected (as these are negligible on the scale of an individual rotor structure), and for simplicity multiple SPP scattering between the structures has been omitted. By implementing these simplifications, a focus can be made on the origin of the plasmon **E** field re-orientation, which ultimately determines the polarisation re-orientation. Figs. 10a and 10d show results obtained with an individual rotor nanostructure in the N and W orientations and illuminated by an incident SPP from below. For the SPP incident in the W orientation, a change in the main **E** field direction of approximately 3° with respect to the direction of incidence is observed. This is also observed where a array of the structures is considered, as shown in Fig. 10e. This is caused by the interference of the incident SPP and the asymmetric configuration of the three-particle rotor model. In comparison to Figs. 10a and 10b where no resultant redirection of the **E** field occurs, the redirection of the plasmon visible in Figs. 10d and 10e implies that the propagation direction of the plasmon on the surface is redirected. However, the grating conditions place strict restrictions on the direction (or channels) at which light can leave the grating. Indeed, it is the additional conditions imposed by the grating that enables a polarisation twisting that, importantly, is not accompanied by a farfield relocation (spatial shift) of the light. This is typically not the case, where the polarisation *and* spatial conditions of the light are defined by the plasmon **E** field prior to reradiation. Therefore, for our structures, the polarisation of the reradiated light is defined by the plasmon **E** field immediately prior to the reradiating process, but the spatial direction of the light is defined by the grating conditions. For this reason, we can label the effect we observe as a true polarisation twisting.

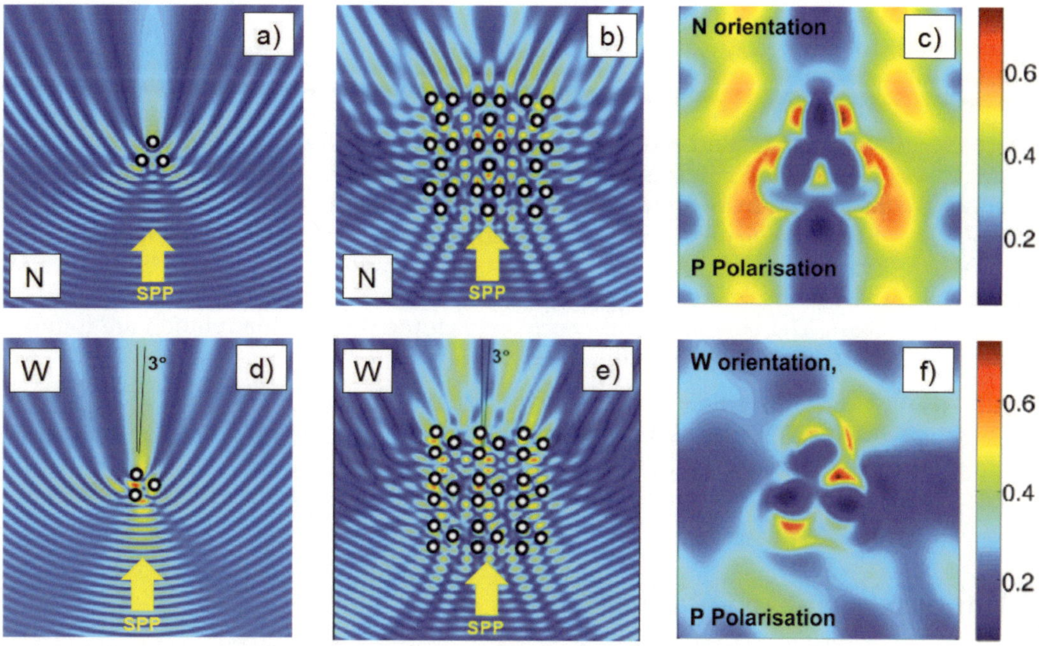

Figure 10. a&d) 10 x 10 micron field amplitude images of the incident and elastic scattered lossless SPPs, interacting with individual rotor representative structures in the N (a) and W (d) orientations [4]. b&e) Extension to interaction of SPPs with 3 by 3 arrays of the structures in the N (b) and W (e) orientations [14]. c&f) FDTD simulated nearfield energy distribution for the rotor structures in N (c) and W (f) orientations for P polarisation [5].

The difference in the degree of polarisation twisting between the experimental observations and the simulations is accounted for in the deliberate simplifications made in the simulations; especially in the substitution of the complex structures with just 3 scattering points. Regardless of these simplifications, the primary function of these simulations here is to identify and understand the dominating mechanisms contributing to the experimentally observed polarisation twisting. This has been confirmed to be as a result of an asymmetrical in-plane SPP scattering while the SPP/rotor interaction is in specific orientations (E and W). However, full FDTD calculations (by the group of Prof. Runge) demonstrate the experimentally observed 30° reorientation [5]. In these simulations, the near- and far-field properties of the rotor structure in the time-domain is calculated using the program FDTD solutions of Lumerical Solutions Inc. From these simulations, the electromagnetic nearfields, and the farfield reflectivities are calculated, as displayed in Figs. 10c and 10f. Importantly, as with the elastic SPP scattering simulations discussed above, in Fig. 10f a reorientation of the plasmon **E** field is again observed, but for these more complete simulations, the true degree of reorientation of ($\sim 30°$) is apparent, corroborating the experimental findings presented above.

4.5. Conclusion

In this section, it has been shown that symmetry properties nanostructures can be designed to control the propagation of SPPs on the surface. In particular, it has been demonstrated and justified how in certain orientations, rotor shaped nanostructures have interesting wave-guiding interactions with propagating SPPs, resulting in a shift from P polarised illumination at which the farfield SPP related minimum reflectivity occurs [3]. Building on this, the first instance of plasmon mediated polarisation reorientation observed in the farfield with no associated directional change of the farfield light was described. For this, an experimental demonstration of how tailor designed topographic structures of threefold symmetry can be used to alter the polarisation of an EM wave by a selective amount was made. It was isolated and confirmed that the primary process involved in this polarisation twist is from the interaction of a propagating SPP wave with the nanostructures. Specifically, the polarisation orientation of the light is determined by the **E** field orientation of the plasmon directly before its re-emission, and the farfield spatial location is determined by the grating conditions. This results in the observed polarisation twisting with no associated farfield directional change. The only apparent restrictions on the polarisation rotation are found to be the initial plasmon excitation conditions [4].

Finally, using Green's function based simulations, the interaction between a propagating plasmon wave and 120° three-fold symmetric structures was examined, confirming that the origin of the farfield polarisation twisting is an asymmetrical in-plane SPP scattering occurring in the nearfield. This computational observation is further confirmed by FDTD calculations of the same structures [5].

5. Ultrafast broadband plasmonics

Time scales associated with SPPs vary from 100's of attoseconds to 100's of nanoseconds. The lower limit is a theoretical limit defined by the inverse spectral width of a broadband plasmonic resonance [47], and is one of the fastest time scales in optics. However, as of yet, little experimental output has come from examining freely propagating plasmons at metal / dielectric interfaces on these ultrashort time scales. The reason for this is as a result of the combination of the difficulty in making accurate measurements on a suitable time scale, but more importantly, the difficulty in accessing plasmon modes of suitable bandwidth to support these ultrafast processes. Despite these difficulties, understanding these ultraquick processes is of key importance to the field of nanoplasmonics, and could have potential applications in, for example, ultrafast computations, and data control and storage on the nanoscale. Recently, a system that provides access to the efficient excitation of broadband, propagating plasmon modes, capable of supporting SPP pulses with temporal lengths on the 20 fs scale has been designed [6]. To achieve this feat, a surface array of tailor designed, reduced symmetry nanostructures has been specifically architected to enable the appropriate control of the plasmon dispersion relation.

5.1. Ultrafast plasmonics

The vast majority of the experimental work on ultrafast SPP dynamics have either dealt with temporal dephasing of particle plasmons [48–53], or have been aimed at understanding the processes through which unexpected levels of optical transmission in subwavelength perforated thin metal films occurs. Indeed, for the ultrafast dynamics of SPPs, the major focus of experimental research has been geared toward the understanding of extraordinary optical transmission (EOT); first observed in the visible regime by Ebbesen et al. [31]. The complete underlying processes of EOT is still somewhat of a debate [32, 54–57]; and as a result, in order to understand the process of EOT in more detail, researches have examined the temporal characteristics of EOT using ultrashort pulse illumination.

The first of these experimental examinations was carried out by the group of Ebbesen [58] where the transit time of a $\sim$100 fs pulse passing through a metal film perforated with an array of subwavelength apertures was considered. Light transmitted through a subwavelength aperture was coherent with the incident pulse, and showed a 7 fs total transient time over the 0.3 mm layer of silver film. The authors report that these delay times support the general picture in which the resonant coupling of light with metallic surface modes is responsible for the relatively large transmission and the slow group velocity of light inside these subwavelength apertures. In the following years, further theoretical and experimental examinations were made based on similar designs, but instead of assigning this delay to the finite transit time for light propagation through the nanoholes, it was assigned to the SPP lifetime for such a structure. In a theoretical work studying the propagation of 10 fs pulses (shorter than the damping time of SP excitations at the interfaces of the metal film), pronounced temporal oscillations in the transmitted light were predicted [59]. It was concluded that these oscillations reflect the temporal character of the coupling of SPPs at both interfaces via photon tunneling through the nanohole channel. More recent experiments have also confirmed a modification of the dephasing rates due to interactions between localised particle plasmons and optical waveguide modes, and subsequent modification of the photonic density of states [60]. In [61] it was demonstrated that SP transmission peaks through the nanohole arrays were homogeneously broadened by the SP radiative lifetime. From the same group an experimental study of ultrafast light propagation through plasmonic nanocrystals using light pulses much shorter in duration than the SPP damping time, was made [62]. Here, phase-resolved measurements of the time structure of the transmitted light allowed for the identification of two different contributions to the EOT effect to be nonresonant tunneling and SPP re-radiation.

Other recent reports of specific significance on the topic of ultrafast SPP dynamics have used pump probe experiments, combined with SP induced photoelectron imaging (PEEM) to achieve nanoscale spatial, and femtosecond (even sub fs) temporal imaging. In [63], ultrafast laser spectroscopy and PEEM were combined to image the quantum interference of localised SP waves. This technique permitted imaging of the spatio-temporal evolution of SP fields with a 50 nm spatial resolution taken at a 330 as frame rate. Using a similar technique, SP dynamics in silver nanoparticles have been studied [26], and an investigation of the optical nearfield was demonstrated by mapping photoemitted electrons from specially tailored Ag nanoparticles deposited on a Si substrate [23]. While on the topic of active plasmon control, femtosecond optical switching of a propagating SPP signal was reported in [64]. Here, experimental examination and theoretical analysis show that femtosecond plasmon pulses can be generated, transmitted, modulated and decoupled for detection in a single device.

Regarding the future of ultrafast plasmonics; a number of publications have recently been made indicating some of the potential directions of nanoscale spatial and fs temporal plasmonics research [47, 65, 66]. One of the key requirements to achieve the potential that ultrafast plasmonics can offer is the ability to access plasmon modes of suitable bandwidth; this will be the focus of the next section.

5.2. Broadband ultrashort propagating plasmon pulses

For a typical Ti:Sa laser system, following group delay dispersion compensation, near transform limited (sub 20 fs) illumination pulses can be generated. In order to permit the excitation of SPP pulses of comparable temporal duration to such illumination pulses, the first requirement is that the SPP excitation mechanism simultaneously envelopes the

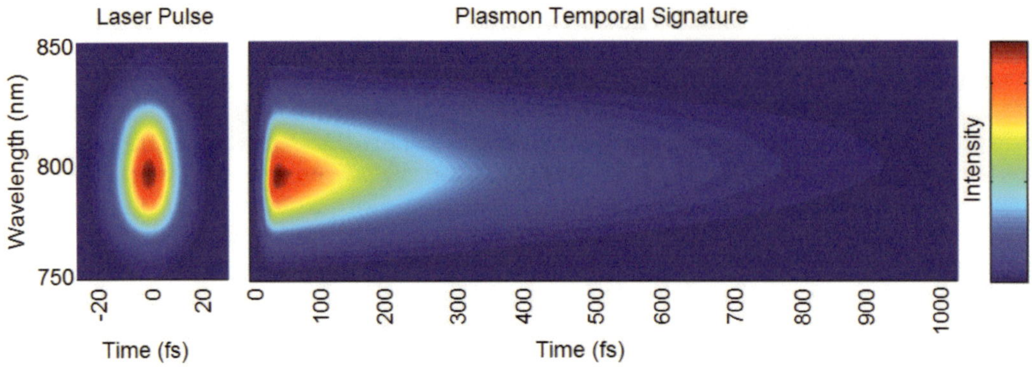

Figure 11. Simulation of the temporal signature of a broadband propagating SPP (right) driven by a sub 20 fs broadband laser pulse from a Ti:Sa laser (left).

complete spectrum of the illumination pulse. In order to spectrally envelope a sub 20 fs pulse bandwidth, an SPP coupling acceptance bandwidth exceeding 80 nm is required. Furthermore, for propagating plasmons, the illumination pulse must remain spatially and temporally optimized prior to interaction with the surface. This implies that the required broadband excitation mechanism must couple this complete illumination bandwidth at a single angle of incidence. However, for high efficiency SPP excitation architectures (i.e. typical grating or attenuated total reflection coupling) the SPP dispersion relation varies rapidly with illumination frequency and angle [30]. For example, in an ATR configuration, the full bandwidth of a spatially and temporally unchirped ultrashort pulse cannot be coupled to an SPP simultaneously [69], thus prohibiting the generation of SPP pulses of comparable duration to the driving ultrashort laser pulse. For some nanofeature based excitation mechanisms (e.g. nanoparticle, rough surface [30], slit [70], etc.) a suitable broad range of momenta can be inherently provided, allowing for broadband plasmon excitation. However, such excitation schemes are limited to comparably weak SPP generation [70], with only a very small percentage of the illumination light coupled to the desired SPP mode. Furthermore, such coupling techniques are not suitable for propagating plasmon pulse generation, as the excited plasmons will have a range of group and phase velocities, will not co-propagate, and so will be strongly spatially and temporally chirped. With these restrictions in mind, a primary research objective is to overcome these obstacles and realize a highly efficient broadband SPP excitation mechanism.

For the laser excitation wavelength of a Ti:Sa laser centered at 800 nm [67], the unperturbed SPP lifetime is $\sim$ 230 fs for a silver / air interface. A theoretical prediction of the temporal evolution of a broadband SPP excited by a 20 fs broadband pulse is presented in Fig. 11. This simulation shows how a plasmon propagating on a silver surface would evolve and decay in time. It also demonstrates that on this planar silver/dielectric interface, the SPP dispersion relation implies that as the plasmon pulse propagates, its wavefront will become distorted due to the fact that the 1/e decay time ranges from approximately 155 fs at the blue end of the Ti:Sa spectrum (750 nm) to 260 fs at the red end (850 nm). This introduces an additional restriction on accessing non-distorting ultrashort plasmon pulses; as it means that a broadband plasmon propagating at a planar metal interface will always become chirped as it propagates.

Therefore, one of the primary challenges in ultrafast plasmonics is the design of a plasmon excitation mechanism that allows for the efficient excitation of broadband, non-distorting plasmon modes. As justified above, this feat will not be possible for planar interface

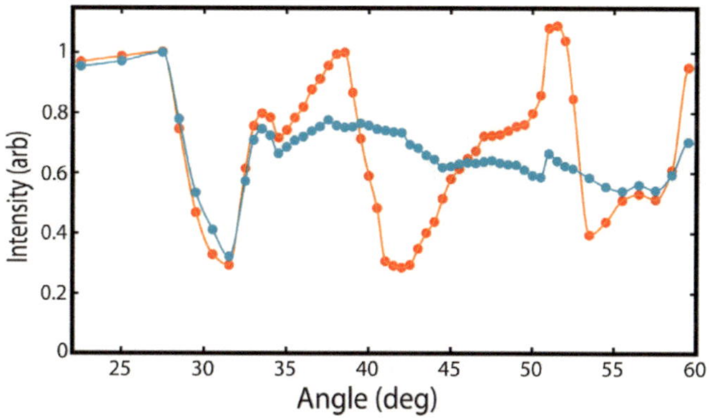

Figure 12. Intensity reflectivity scans for a 800 nm CW laser (red) and a broadband (centered at 800 nm) sub 20 fs laser (blue). Three pronounced plasmonic modes are apparent for CW light located at 31°, 42° and 55°, but only one pronounced mode is visible for the broadband pulse at 31°.

arrangements. Thus, in order to achieve this goal, our experiments deal with SPP excitation on an array of reduced symmetry nanostructures. The initial design for the structures was conceived and optimized using reciprocal space analysis. The geometry was chosen to present a range of inter-structure distances to suit the near Gaussian spectral distribution of the laser, and the physical dimensions of the surface features were optimized for SPP excitation in the near IR [6].

Confirmation of SPP excitation in these tailor designed nanostructured arrays for CW light (at 800 nm) and sub 20 fs pulses of broadband light (at 760-840 nm) is presented in Fig. 12. Three SPP modes are clearly identifiable (as reflectivity minima) for the CW laser excitation at 31°, 42° and 55°. However, for the broadband source, only one sharp SPP mode at 31° is apparent. Importantly, because the plasmon mode at 31° appears identical for both broadband and CW light sources, this indicates that this mode has at least a comparable bandwidth to the broadband source. For such grating excited plasmons, it is typically not possible to excite SPPs of comparable bandwidth to a sub 20 fs Ti:Sa laser. This is because of a strong momentum variation typically exhibited by plasmon modes over such a broad spectrum; as is the case for the modes at 42° and 55° observed only for the CW scan in Fig. 12. However, for SPP excitation on the array of reduced symmetry nanostructures it is found that a plasmon mode whose angular variation over the spectral range is very low is accessible [6]. Indeed, this mode has been found to be suitable for broadband plasmon excitation, coupling the complete illumination spectrum of a 17.5 fs, Ti:Sa system. Further angular and spectral examinations of the excitation of SPPs confirm the presence of multiple plasmonic modes for the CW source; but importantly they also reveal multiple modes for the broadband source. These additional modes are found to have the typical strong spectral variation [30], and so are not suitable for broadband SPP excitation. However, for the mode at 31° plasmons of very high bandwidth (and so very short temporal characteristics) can be excited [6].

5.3. Conclusion

In final section, a method for the excitation of broadband plasmonic modes in the near-IR regime was presented [6]. This has been achieved using a tailor designed reduced symmetry periodic surface that grants access to an SPP mode which has a fixed momentum value over the entire bandwidth of an ultrafast Ti:Sa laser. For this sample, as a result of the well ordered array basis, a high SPP coupling efficiency is achieved, and as a result of the reduced symmetry nanostructures, the range of momenta provided by the grating in a fixed direction is increased, presenting the possibility of accessing efficient, co-propagating, broadband SPPs. This ability to generate broadband, ultrashort SPP pulses that exhibit no spatial or temporal chirping in their excitation is an important step toward accessing the previously predicted ultraquick optical processes associated with SPPs.

Acknowledgements

The majority of the work presented here is based on sections of the PhD thesis of Dr Brian Ashall, carried out under the supervision of Dr Dominic Zerulla.

The authors would like to thank Dr Michael Berndt, Prof. Martin Aeschlimann, Dr Brian Vohnsen, Prof. Erich Runge, Dr Jose Francisco Lopez-Barbera and Dr Stephen Crosbie for their valuable contributions. The authors acknowledge Science Foundation Ireland and Enterprise Ireland for ongoing funding of our research in the field of plasmonics.

Author details

Brian Ashall
School of Science, Technology, Engineering and Mathematics, Institute of Technology Tralee, Ireland

Dominic Zerulla
School of Physics, College of Science, University College Dublin, Belfield, Dublin 4, Ireland

6. References

[1] H. A. Atwater, The Promise of Plasmonics, Scientific American, April Issue (2007).
[2] M. Berndt, M. Rohmer, B. Ashall, C. Schneider, M. Aeschlimann, and D. Zerulla, Opt. Lett., 34:959 (2009).
[3] B. Ashall, M. Berndt, and D. Zerulla, Appl. Phys. Lett., 91: 203109 (2007).
[4] B. Ashall, B. Vohnsen, M. Berndt, and D. Zerulla, Phys. Rev. B, 80:245413 (2009).
[5] D. Leipold, S. Schwieger, B. Ashall, D. Zerulla, and E. Runge, Photonics and Nanostructures 8 297 (2010).
[6] B. Ashall, J. F. Lopez-Barbera, and D. Zerulla, Manuscript submitted to New Journal of Physics (2012).
[7] J. Melngailis, J. Vac. Sci. Technol. B, 5, 469 (1987).
[8] D. K. Stewart, A. F. Doyle, and J.D. Jr Casey, Proc. SPIE, 276, 4337 (1995).
[9] B. W. Ward, N. P. Economou, D. C. Shaver, J. E. Ivory, M. L. Ward, and L. A. Stern, Proc. SPIE, 92, 923 (1988).
[10] T. H. Taminiau, R. J. Moerland, F. B. Segerink, L. Kuipers, and N. F. van Hulst, Nano Lett., 7, 28 (2007).
[11] X. Jiao, J. Goeckeritz, S. Blair, and M. Oldham, Plasmonics, 4, 37 (2009).
[12] M. Berndt, Anregung von Oberflächenplasmonen auf mesoskopischen Strukturen, Diploma Thesis, Heinrich-Heine-Universität, Düsseldorf, (2007).
[13] EBL system available via the National Access Program, Tyndall National Institute in Cork.
[14] B. Ashall, Surface Plasmon Polaritons on Nanostructures Surfaces, Doctor of Philosophy Thesis, University College Dublin (2009).

[15] R. Möller, U. Albrecht, J. Boneberg, B. Koslowski, P. Leiderer, and K. Dransfeld, J. Vac. Sci. Technol. B, 9, 506 (1991).

[16] N. Kroo, J.P. Thost, M. Völcker, W. Krieger, and H. Walther, Europhys. Lett., 15, 289 (1991).

[17] D.W. Pohl and D. Courjon, Near Field Optics, Kluwer, (1993).

[18] I. I. Smolyaninov, A. V. Zayats, and O. Keller, Phys. Lett. A, 200, 438 (1995).

[19] M. Specht, J.D. Pedarnig, W.M. Heckl, and T.W. Hansch, Phys. Rev. Lett., 68, 476 (1992).

[20] E. H. Synge, Phil. Mag., 6, 356 (1928).

[21] E. H. Synge, Phil. Mag., 13, 297 (1932).

[22] B. Hecht, B. Sick, U. P. Wild, V. Deckert, R. Zenobi, O. J. F. Martin, and D. W. Dieter, J. Chem. Phys., 18, 112 (2000).

[23] M. Cinchetti, A. Gloskovskii, S. A. Nepjiko, G. Schönhense, H. Rochholz, and M. Kreiter, Phys. Rev. Lett., 95, 047601 (2005).

[24] A. Kubo, K. Onda, H. Petek, Z. Sun, Y. S. Jung, and H. K. Kim, Nano Letters, 5, 1123 (2005).

[25] J. Feng and A. Scholl, Science of Microscopy - Chapter 9: Photoemission Electron Microscopy, Springer New York (2007).

[26] J. Lehmann, M. Merschdorf, W. Pfeiffer, A. Thon, S. Voll, and G. Gerber, Phys. Rev. Lett., 85, 2921 (2000).

[27] M. Aeschlimann, M. Bauer, D. Bayer, T. Brixner, F. J. García de Abajo, W. Pfeiffer, M. Rohmer, C. Spindler, and F. Steeb, Nature, 446, 301 (2007).

[28] M. Munzinger, C. Wiemann, M. Rohmer, L. Guo, M. Aeschlimann, and M. Bauer, New J. Phys., 7, 68 (2005).

[29] H. B. Michaelson, J. Appl. Phys, 48, 4729 (1977).

[30] H. Raether, Surface Plasmons on Smooth and Rough Surfaces and on Gratings, Springer Verlag, Berlin (1988).

[31] T. W. Ebbesen, H. J. Lezec, H. F. Ghaemi, T. Thio, and P. A. Wolff, Nature, 391, 667 (1998).

[32] H. F. Ghami, T. Thio, D. E. Grupp, T. W. Ebbesen, and H. J. Lezec, Phys. Rev. B, 58, 6779 (1998).

[33] R. Petit, Electromagnetic Theory of Gratings, Springer Berlin, Springer Topics in Current Physics Vol 22 (1980).

[34] S. Rehwald, M. Berndt, F. Katzenberg, S. Schwieger, E. Runge, K. Schierbaum, and D. Zerulla, Phys. Rev.B, 76, 085420 (2007).

[35] C.H. Wilcox, Scattering Theory for Diffraction Gratings, Springer Berlin, Springer Applied Mathematical Sciences Vol 46 (1984).

[36] G. I. Stegeman, A. A. Maradudin, and T. S. Rahman, Phys. Rev. B, 23, 2576 (1981).

[37] R. F. Wallis, A. A. Maradudin, and G. I. Stegeman, Appl. Phys. Lett., 42, 764 (1983).

[38] P. Dawson, F. de Fornel, and J-P. Goudonnet, Phys. Rev. Lett., 72, 2927 (1994).

[39] Y. Satuby and M. Orenstein, Appl. Phys. Lett., 90, 251104 (2007).

[40] S. I. Bozhevolnyi, J. Erland, K. Leosson, P. M. W. Skovgaard, and J. M. Hvam, Phys. Rev. Lett., 86, 3008 (2001).

[41] S. I. Bozhevolnyi, V. S. Volkov, K. Leosson, and A. Boltasseva, Appl. Phys. Lett., 79, 1076 (2001).

[42] B. Steinberger, A. Hohenau, H. Ditlbacher, A. L. Stepanov, A. Drezet, F. R. Aussenegg, A. Leitner, and J. R. Krenn, Appl. Phys. Lett., 88, 094104 (2006).

[43] G. Isfort, K. Schierbaum, and D. Zerulla, Phys. Rev. B, 73, 033408 (2006).

[44] G. Isfort, K. Schierbaum, and D. Zerulla, Phys. Rev. B, 74, 033404 (2006).

[45] J. M. Pitarke, V. M. Silkins, E. V. Chulkov, and P. M. Echenique, Rep. Prog. Phys., 70 (2007).

[46] S. I. Bozhevolnyi and V. Coello, Phys. Rev. B, 58, 10899 (1998).
[47] M. I. Stockman, M. F. Kling, U. Kleineberg, and F. Krausz, Nat. Photon., 1, 539 (2007).
[48] C. Sönnichsen, T. Franzl, T. Wilk, G. von Plessen, J. Feldmann, O. Wilson, and P. Mulvaney, Phys. Rev. Lett., 88, 077402 (2002).
[49] B. Lamprecht, J. R. Krenn, A. Leitner, and F. R. Aussenegg, Phys. Rev. Lett., 83, 4421 (1999).
[50] B. Lamprecht, A. Leitner, and F. R. Aussenegg, Appl. Phys. B, 68, 419 (1999).
[51] A. Wokaun, J. P. Gordon, and P. F. Liao, Phys. Rev. Lett., 48, 957 (1982).
[52] T. Klar, M. Perner, S. Grosse, G. von Plessen, W. Spirkl, and J. Feldman, Phys. Rev. Lett., 80, 4249 (1998).
[53] J. Bosbach, C. Hendrich, F. Stietz, T. Vartanyan, and F. Trager, Phys. Rev. Lett., 89, 257404 (2002).
[54] U. Schroter and D. Heitmann, Phys. Rev. B, 58, 15419 (1998).
[55] J. A. Porto, F. J. Garcia-Vidal, and J. B. Pendry, Phys. Rev. Lett., 83, 2845 (1999).
[56] T. Thio, H. F. Ghaemi, H. J. Lezec, P. A. Wolff, and T. W. Ebbesen, J. Opt. Soc. Am. B, 16, 1743 (1999).
[57] M. M. J. Treacy, Appl. Phys. Lett., 75, 606 (1999).
[58] A. Dogariu, T. Thio, L. J. Wang, T. W. Ebbesen, and H. J. Lezec, Opt. Lett., 26, 450 (2001).
[59] R. Müller, V. Malyarchuk, and C. Lienau, Phys. Rev. B, 68, 205415 (2003).
[60] T. Zentgraf, A. Christ, J. Kuhl, and H. Giessen, Phys. Rev. Lett., 93, 243901 (2004).
[61] D. S. Kim, S. C. Hohng, V. Malyarchuk, Y. C. Yoon, Y. H. Ahn, K. J. Yee, J.W. Park, J. Kim, Q. H. Park, and C. Lienau, Phys. Rev. Lett., 91, 143901 (2003).
[62] C. Ropers, Müller, C. Lienau, G. Stibenz, G. Steinmeyer, D-J. Park, Y-C. Yoon, and D-S. Kim, Ultrafast dynamics of light transmission through plasmonic crystals, International Conference on Ultrafast Phenomena, Niigata, Japan (2004).
[63] A. Kubo, K. Onda, H. Petek, Z. Sun, Y. S. Jung, and H. K. Kim, Nano Lett., 5, 1123 (2005).
[64] K. F. MacDonald, Z. L. Sámson, M. I. Stockman, and N. I. Zheludev, Nature Photonics, 3, 55 (2008).
[65] M. I. Stockman, N. J. Phys., 10, 025031 (2008).
[66] J. Lin, N. Weber, A. Wirth, S. H. Chew, M. Escher, M. Merkel, M. F. Kling, M. I. Stockman, F. Krausz, and U. Kleineberg, J. Phys.: Condens. Matter, 21, 314005 (2009).
[67] Griffin C Ti:Sa Laser, Kapteyn-Murnane Laboratories Inc.
[68] R. Trebino, Frequency-Resolved Optical Gating: The Measurement of Ultrashort Pulses, Kluwer Academic Publishers (2000).
[69] S. E. Yalcin, Y. Wang, and M. Achermann, Appl. Phys. Lett., 93, 101103 (2008).
[70] J. Chen, Z. Li, M. Lei, S. Yue. J. Xiao, and Q. Gong, Opt. Exp., 19, 26463 (2011).

Chapter 5

Novel SNOM Probes Based on Nanofocusing in Asymmetric Structures

Valeria Lotito, Christian Hafner, Urs Sennhauser and Gian-Luca Bona

Additional information is available at the end of the chapter

http://dx.doi.org/10.5772/50806

1. Introduction

Scanning near-field optical microscopy (SNOM) has become a microscopy tool of paramount importance for nanostructure investigation in different fields ranging from material to life science. In fact, such technology combines the advantages offered by classical optical microscopy with high resolution typical of scanning probe microscopy (SPM) [1,2]. More specifically, it allows overcoming the diffraction limit by exploitation of the non-propagating evanescent waves in the near-field zone, which are not retrieved in conventional optical microscopy. Still it retains some important features of this technique, e.g. the possibility to carry out real time observation of the sample in its native environment in a non-invasive way and with little sample preparation and, more importantly, the availability of a wide variety of contrast mechanisms, like polarization, absorption and fluorescence. Such advantageous characteristics are usually not affordable with non-optical scanning probe techniques, which require more awkward sample preparation and provide little information on non-morphological properties of the sample. On the other hand, similarly to SPM, SNOM enables to get also high resolution information on sample topography by scanning a sharp probe in the region of interest of the sample and using a non-optical short range interaction between the probe and the sample to control probe-sample distance. Hence, SNOM allows getting nano-scale insight into a specimen, casting light on both topographic and optical properties of the sample.

As suggested by the name itself, optical properties of SNOM rely on the near-field interaction between the probe and the sample: the probe can be used for near-field excitation of the sample, whose response is collected in the far-field or to detect the near-field response of the sample broadly illuminated in the far-field or for both. Independently of the specific excitation/detection scheme, a crucial role in the ultimate attainable optical resolution is played by the probe itself, whose design, modelling and optimization has been encouraged by a vast wealth of numerical methods, shedding light into the optical behaviour of the probe at very small scales.

The best known probe configuration, used by the pioneers of SNOM, is the so-called aperture probe, which is still commonly used today [1,3,4]. Such a structure is based on a metal-coated tapered dielectric (typically an optical fiber) with an aperture left at the very apex. In this case, resolution is mostly dictated by the size of the aperture, which cannot be decreased at will not only because of technological limits, but also due to the dramatic slump in signal throughput with decreasing aperture size. The latter cannot be simply improved by increasing the input power because of the potential risk for thermal damage. Significant heating can occur as a consequence of multiple back-reflections from the taper and metal absorption, resulting in aperture expansion, tip contraction and elongation, possible partial detachment or even breakdown of the metal layer due to mechanical stresses arising from the different thermal expansion coefficients of the fiber and the metal coating [1,5-7]. All these phenomena inherently cause degradation of the probe behaviour and impose severe limits on the injected power.

Another major factor influencing the performance of aperture probes is the taper cone angle. In fact, such probes can be viewed as a tapered hollow metal waveguide filled with a dielectric, in which the mode structure changes as a function of the characteristic dimension of the dielectric core [8]. Guided modes run into cut-off one after the other as the diameter decreases, until, at a well-defined diameter, even the last guided mode, the fundamental HE_{11} mode, runs into cut-off; for smaller diameters of the dielectric core, the energy in the core decays exponentially because of the purely imaginary propagation constant of all the modes [1]. The amount of light that reaches the probe aperture depends on the distance separating the aperture and the HE_{11} cut-off diameter: the larger the opening angle of the tapered structure, the better the light transmission of the probe is, as the final cut-off diameter approaches the probe apex [1].

Taper profile depends of course on the adopted fabrication method: for fiber-based probes, two main approaches are used for taper formation, either fiber heating and pulling (consisting in locally heating the fiber using a CO_2 laser and subsequently pulling apart the two parts) and chemical etching (which basically consists in dipping the fiber in a HF solution with an organic overlayer) [1,2]. The second approach usually results in shorter tapers, i.e. larger opening angles, beneficial for higher signal throughput in spite of a higher surface roughness, which can be reduced by resorting to some variants of the etching process like the so-called tube etching process. Even shorter tapers can be achieved by using the selective etching methods in which the different etching rates of the core and the cladding of the fiber are exploited to get probes in which only the core of the fiber is tapered, resulting in tapered length of few micrometers against the hundreds of micrometers of fibers produced by different techniques. The following steps consist in metal deposition (by thermal or e-beam evaporation or sputtering) and aperture formation, which can be done either by exploiting shadowing effects during metal evaporation or by punching or with focused ion beam (FIB) milling [1].

Several attempts have been carried out to overcome the fundamental limitations of aperture probes, i.e. poor throughput and poor resolution. Two main routes have been followed to improve their performance, i.e. optimization of the taper profile and of the aperture shape.

As pointed out earlier, chemical etching leads to an advantageous reduction of the overall taper length and a better control of the overall taper shape. Using selective chemical etching, probes based on multiple tapers like the double taper, the triple taper and the steeple-on-mesa taper profiles have been realized: the goal of these structures was to benefit from the advantages of large taper angles in signal throughput at least for part of the probe because the use of a single taper with a large taper angle would not be convenient during approach to the sample [9-13]. An asymmetric edged probe with a sharp edge at the foot of the conical taper has shown to improve the transmission of HE_{11} mode with proper linearly polarized excitation along the direction of the asymmetry [14]. Also the taper profile of probes produced by heating and pulling has been optimized: for this fabrication process the overall probe profile can be varied by controlling some process parameters; parabolic profiles have shown to give rise to higher throughput compared to conical profiles [15]. Theoretical studies have been carried out to assess the influence of taper profile on probe performance to determine optimal shapes in terms of throughput: for example, structures based on the alternation of conical and cylindrical sections or analytical expressions for improved taper profiles have been proposed as well as corrugations at the interface between the dielectric core and the metal coating or on the external metal coating; in the latter two cases, the increase in the throughput is due to the exploitation of surface plasmon polaritons (SPPs) [16-19].

Another route followed for the improvement of the performance of aperture probes has been the implementation of other aperture shapes different from the typical circular design: rectangular, square, slit, elliptical, C-shaped, I-shaped or dumbbell, H-shaped, bowtie, connected and separated double aperture, triangular, rod hole and tooth hole, gap apertures have been analysed and/or fabricated in fiber- and cantilever-based probes [20-28]. The improved throughput and field localization for some preferential input polarizations is due to the strong asymmetry of such aperture shapes and to the excitation of SPPs.

Despite the efforts in the optimization of the aperture probe to overcome its fundamental limits in throughput and resolution, the current design of aperture probes is not able to sustain routinely ultrahigh resolution imaging because of the limit of the skin depth of the metal coating that increases the effective aperture size, thus making the field distribution at the aperture significantly larger than its physical size [29]. Better throughput and resolution are expected if an apertureless metal or fully metal-coated probe is used, due to the combination of electrostatic lightning rod effect and surface plasmon excitation for proper illumination conditions. For example, due to the lightning rod effect, strong enhancement is expected at the tip apex, even though a real probe apex does not represent a real singularity (i.e. not defined first and second order derivatives) because of the finite conductivity of real metals and the finite tip radius [29]. In the next paragraph, we will see how plasmonic effects in apertureless probes can be used to get high resolution.

2. Nanofocusing in SNOM probes: The axisymmetric structure

Apertureless probes represent a promising alternative probe configuration. The fundamental problem with such probes is given by the fact that they are usually illuminated externally either by a focusing lens or using a prism-based total internal reflection configuration, resulting in a strong background due to far-field illumination, which could be detrimental for some sensitive samples and can be reduced only partially by on-axis illumination [30]. In order to reduce such background some groups have envisaged different solutions like far-field excitation of SPPs at the wide end of a tapered metal nanowire by using grating geometry: such SPPs propagate towards the apex of the nanowire leading to nanofocusing, that is delivery and concentration/focusing of the optical energy at the nanoscale, which means a region much smaller than the dimensions allowed by the diffraction limits [30,31].

Nanofocusing in metal tapered nanorods has been theoretically studied starting from the analysis of the SPP modes of a plasmonic waveguide consisting of a metal nanowire whose axis coincides with the *z* axis and whose dielectric function $\varepsilon_m(\omega)$ is uniform in space with ω being the excitation angular frequency; the wire is surrounded by a dielectric with dielectric function ε_d [31]. The three fundamental SPP modes supported by a metal nanowire are the transverse magnetic TM mode and two hybrid modes HE_1 and HE_2 (Figure 1). Note that we adopted this notation to indicate the modes, although the classification in TM_{0n} (or TM_n), TE_{0n} (or TE_n), HE_{mn}, EH_{mn} (where m is related to the angular symmetry and n to the radial variation of the mode) commonly used for circular dielectric wires might still be extended to circular metallic wires. Moreover, we considered the modes with lower losses.

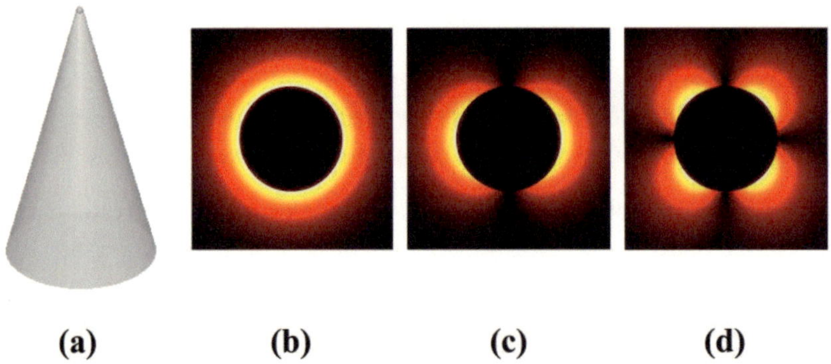

Figure 1. Sketch of: (a) fully metal probe; normalized electric field distributions of (b) TM plasmon mode, (c) HE_1 plasmon mode, (d) HE_2 plasmon mode.

The TM SPP mode can exist for arbitrarily small diameters of the nanowire and, hence, has been shown to be suitable for nanofocusing. This mode is axially uniform and is characterized by a magnetic field having only ϕ component; both electric and magnetic fields are independent of ϕ. Assuming that the radius *R(z)* of the nanowire is a smooth function of *z* (which decreases from microscale to nanoscale for *z* going from large negative values to zero), the so-called eikonal or Wentzel-Kramers-Brillouin (WKB) or quasi-classical or geometric optics approximation can be applied to study the SPP modes and back-reflections can be neglected [31]. In order to study their propagation in a tapered metal rod, one can use a staircase approximation and interpret the taper as a series of cylindrical nanowires with smoothly decreasing diameters. The effective index $n_{eff}(R) = n_{eff}(R(z))$ for the TM mode of the plasmonic waveguide at a point z can be determined from:

$$\left[\frac{\varepsilon_d}{\kappa_d}\frac{K_1(\kappa_d R)}{K_0(\kappa_d R)} + \frac{\varepsilon_m}{\kappa_m}\frac{I_1(\kappa_m R)}{I_0(\kappa_m R)}\right] = 0 \tag{1}$$

where I_n and K_n are the modified Bessel functions of the first and second kind, $\kappa_d = k_0\sqrt{n_{eff}^2 - \varepsilon_d}$ and $\kappa_m = k_0\sqrt{n_{eff}^2 - \varepsilon_m}$. Under the plasmonic condition $\mathrm{Re}\{\varepsilon_m\} < -\varepsilon_d$, one gets the solutions for the SPP modes. For a thick wire ($k_0 R >> 1$), the solution is the same as for a flat surface:

$$n_{eff} = \sqrt{\frac{\varepsilon_m \varepsilon_d}{\varepsilon_m + \varepsilon_d}} \tag{2}$$

For a thin, nano-scale radius wire ($k_0 R << 1$) one gets approximately with logarithmic precision [31]:

$$n_{eff}(R) \approx \frac{1}{k_0 R}\sqrt{-\frac{2\varepsilon_d}{\varepsilon_m}}\left[\ln\sqrt{-\frac{4\varepsilon_m}{\varepsilon_d}} - \gamma\right]^{-1} \tag{3}$$

where $\gamma \approx 0.57721$ is the Euler constant. From the previous expression it can be inferred that at the tip ($k_0 R \to 0$), $n_{eff} \to \infty$; therefore, the wave number increases and the SPPs asymptotically stop as both the phase and group velocity tend to zero because they are both proportional to $k_0 R$ [31]. Besides, the study of the propagation of the SPPs through the tapered metal rod carried out with the staircase approximation has revealed an anomalous increase in the SPP field amplitudes, as SPPs approach the tip, because the electric field varies as $\propto R^{-\frac{3}{2}}$ [32]. As a consequence, the simultaneous wavelength decrease and amplitude increase lead to a concurrent energy localization and, hence, to nanofocusing. Several theoretical studies have shown such a phenomenon [32-35]. The TM SPP mode can be excited using an axially symmetric grating under radially polarized excitation [36,37] or double-sided E-symmetric excitation [38] or asymmetric excitation via grating coupling on just one side of the tip [39].

An even more promising solution compared to fully metal probes in terms of simplification of the experimental set-up is the excitation of SPP modes on the metal coating of a fully metal-coated dielectric probe used under internal back excitation [29,40].

As for fully metal probes, the adiabatic approximation has been used to evaluate analytically the wave propagation in the fully metal-coated tapered dielectric probe, considering the cone as a succession of cylinders made up of a dielectric core of radius r_1 surrounded by a metal coating of thickness δ located in its turn in a dielectric medium [41]: in a first approximation, waveguide modes (WGM) have been calculated as those supported by a dielectric core surrounded by an infinitely thick metallic coating neglecting the external dielectric, while SPP modes have been calculated as those supported by a metal wire surrounded by a dielectric neglecting the internal dielectric core. Dispersion relations for waveguide modes and for SPP modes have thus been determined (the latter coincides with the dispersion equation for the metal wire). At a certain value of r_1, the wavevector of the WGM can match the one of the SPP mode: in these conditions, energy transfer from the WGM into the SPP mode is possible and, as a result, surface plasmons are excited bringing about nanofocusing. The transfer of energy and the field profiles can be determined with this approach using the coupled mode theory [41,42]. In this way, the SPPs can be excited using the WGM, thereby overcoming the problem of background inherent in the use of fully metal probes under external illumination.

The energy transfer from WGM to SPP modes has been thoroughly investigated in several studies, taking into account not only the "outer" SPP modes at the outer metal surface, but also the "inner" SPP modes at the inner metal surface; in some studies, non-conical taper profiles have been scrutinized as well [40,43-45]. The WGM has to present proper characteristics in order for the excited surface plasmons to exhibit axial symmetry, thereby interfering constructively at the tip apex and therefore generating nanofocusing. The three fundamental WGMs are a pair of orthogonal linearly polarized modes followed by a radially polarized one. If the fully metal-coated structure is excited by a linearly polarized HE_{11} mode, then the electric fields of the excited surface plasmons have opposite polarities on the opposite sides of the probe, giving rise to destructive interference at the very end of the probe; on the contrary, a radially polarized WGM excites SPPs with axial symmetry, which interfere constructively at the tip apex. These mechanisms are responsible for the high field confinement in case of radially polarized excitation compared to the linearly polarized case. The eventual outcome of this process is the creation of an ultrasmall hot spot in the region close to the tip apex in the former situation, as opposed to broader and weaker two-lobed electric field intensity distributions for the latter one. In particular, the size of the achievable hot spot in case of radially polarized excitation (and, hence, the ultimate attainable resolution) is mostly limited by the diameter of the metal apex, which can be decreased at will. Such behaviour has been confirmed in both theoretical and experimental studies [29,40,46-50] and is sketched in Figure 2.

Because of its attractive characteristics, the fully metal-coated dielectric probe has been chosen as the starting point for structural optimization; as explained in more detail in the next paragraph, the goal of the work will be the achievement of field localization (a feature of utmost importance for SNOM applications) under linearly polarized excitation, bringing about a further substantial simplification in experimental set-ups.

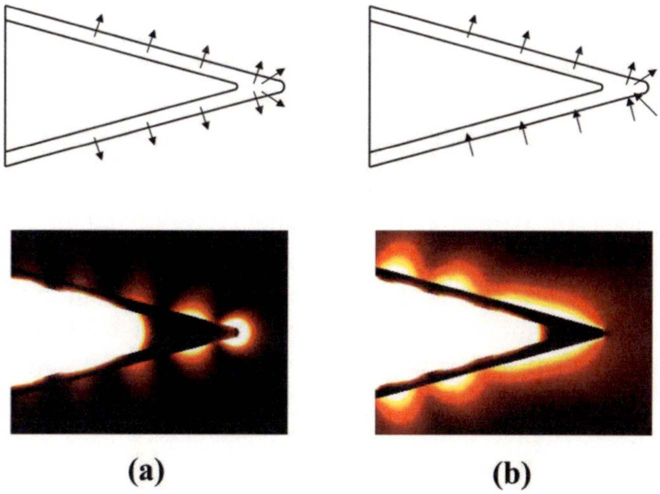

Figure 2. Sketch of the behaviour of a fully metal-coated dielectric probe under: (a) radially; (b) linearly polarized excitation. Arrows indicate the electric fields associated with the surface plasmons (adapted from [46]).

3. Nanofocusing in SNOM probes: The asymmetric structure

As previously explained, axisymmetric fully metal-coated probes under internal back excitation hold the promise for high-resolution applications, as they can allow the achievement of a strongly localized hot spot, whose size is mostly limited by the diameter of the metal apex, which can be decreased at will. However, as pointed out, the resolution of fully metal-coated tips is highly sensitive to the polarization state of the input field: such desirable field localization properties are affordable only under radially polarized excitation.

The reason for the strong polarization-dependent behaviour is due to the different characteristics of the surface plasmons excited by the different waveguide modes on the external metal coating: those excited by a linearly polarized mode interfere destructively at the tip apex, resulting in a weak and broad near-field distribution contrarily to those excited by a radially polarized mode which interfere constructively bringing about field localization.

The fundamental drawback of radially polarized excitation, despite its potential attractiveness, resides in its cumbersome injection procedure, which is extremely sensitive to misalignments that could impair the potential benefits stemming from its use [50,51].

In order to get similar superfocusing effects under a more easily excitable linearly polarized mode, one could break the axial symmetry of the fully metal-coated probe so as to avoid the destructive interference between the SPPs excited by a linearly polarized mode on the opposite sides of the axisymmetric structure. If z is the direction of the probe axis and an asymmetry is introduced in the tip structure along x, field confinement under x linearly polarized mode could be expected. This idea stems from the numerical investigation of the effects of unintentional asymmetries like single and multiple air spherical bubbles, which have been shown to have a weak field localization effect due to the coupling between the linearly polarized mode along the direction of the asymmetry and the radially polarized one [52].

Of course, superfocusing based on random defects could not be easily forecast, but stronger and more easily predictable focusing effects can be achieved by the introduction of intentional modifications. The asymmetry can be either present in the probe structure itself or in the probe illumination scheme. Unilateral and bilateral slits in the metal coating have been numerically studied as a form of structural modification. Tip on aperture probes and probes based on a monopole antenna can also be included in this category [53,54]. An asymmetric illumination scheme can be used as an alternative to a structural asymmetry [55]: surface plasmons are excited only on one side of an apertureless probe via an opening close to the probe base in the offset aperture probe [56], while asymmetric single-sided SPP excitation results in field localization in an axisymmetric apertureless probe [38].

We have to underline how all these asymmetric structures are based on the introduction of an asymmetry along one specific direction and, for this reason, we can indicate them as "directional" asymmetries [55,57]. Hence, field localization occurs only for the linearly polarized mode that is oriented in the direction of the asymmetry: for example, if the asymmetry in the structure or in the illumination is along x, field localization is expected under x linearly polarized excitation, but not under y linearly polarized excitation, because no asymmetry is present along the y direction. This is due to the fact that all these forms of asymmetries exhibit a plane of symmetry and, hence, cannot guarantee field localization for arbitrarily oriented linearly polarized modes. Such asymmetries still require alignment between the linearly polarized mode and the asymmetry itself, even though injection procedures are considerably simplified because radially polarized excitation is no longer necessary.

The practical implication of such a behaviour in experimental applications is apparent if considering the difference in imaging properties using asymmetric tips based on a tip-on-aperture and on a monopole antenna grown on the rim of an aperture tip with different orientation of the input linear polarization, with dramatic variations in resolution and signal intensity upon rotation of the input excitation from the direction of the asymmetry to the orthogonal one [53,54]. However, the control of the direction of the input polarization is not an easy task and often imposes a determination *a posteriori* of the effective polarization direction close to the asymmetry by comparison of the field distribution observed close to the tip apex with simulation results [14].

Therefore, it would be desirable to attain superfocusing effects for arbitrarily oriented linearly polarized excitation, by using an "adirectional" asymmetry, which means a suitable modification likely to create an asymmetry along all spatial directions [55,57].

In the coming paragraphs, we will report the numerical analysis of novel forms of asymmetry based both on new types of structural directional asymmetries and on the pioneering concept of adirectional design. In particular, we will illustrate the properties of:

1. directional asymmetries based on: (i) an oblique cut; (ii) asymmetric corrugations in the metal coating;
2. adirectional asymmetries based on: (i) a spiral corrugation; (ii) azimuthal corrugations arranged in a spiral-like fashion.

Before tackling the design and the characteristics of these specific structures, in the next paragraph a short overview of the issues and difficulties encountered in probe modelling is presented.

4. Optimization of probe structures: Challenges of tip modelling

The theoretical study of the behaviour of SNOM probes is essential not only to get insight into the characteristics of commonly used probe structures and identify potential problems in imaging, but also to detect possible routes for optimization and predict the implications of the use of novel probe configurations. With such an approach, experimental efforts can be devoted to those probes that theoretically exhibit the most promising features.

Analytical solutions have been determined by Bethe and Bouwkamp for an aperture in an infinitely thin perfectly conducting screen, which laid down the foundation for further theoretical treatments. In the search for a model that could describe more faithfully the aperture probe behaviour, analytical studies that took into account the finite thickness of the metal screen have been carried out [58-59]. All of these treatments, however, suffer from neglecting the finite conductivity of the metal cladding used in real SNOM probes and provide little resemblance to the actual tip geometries used in experiments [60].

Other semi-analytical approaches have been based on a staircase approximation in which the longitudinally non-uniform waveguide (the tapered part of the probe) is considered as a succession of cylindrical sections of decreasing radius and the eigenmodes of the uniform waveguides obtained by infinitely stretching along the axis of the probe at each cross section are computed, as illustrated earlier. An analytical evaluation of the power transmitted by an aperture probe based on such an approximation using a mode matching theory has been done [1]. Starting also from this staircase approximation, different probe profiles like parabolic, exponential and mixed shapes based on the alternation of conical and cylindrical sections along the taper have been examined using the cross-section method [16, 17].

As discussed in paragraph 2, the staircase approximation has also been used for the analysis of nanofocusing in fully metal and fully metal-coated dielectric probes using local mode theory in a weakly non-uniform optical waveguide and the eikonal approximation [31, 34, 40, 43, 44].

Although analytical expressions are useful to get an understanding of probe behaviour, their range of applicability is limited. For example, the eikonal approximation can be applied as long as the adiabatic criterion is satisfied, i.e. for small local taper angles

[31,34,40,43]. On the other hand, the study of probes that do not satisfy such requirement can be of great interest: for example, optimal conditions for nanofocusing on tapered metal rods have been found in nonadiabatic conditions, which fail to be treated analytically [61]. Moreover, analytical methods turn out to be not adequate to model abrupt variations in the tapered profile as they often suppose a weak longitudinal non-uniformity.

Whenever more challenging probe geometries need to be studied, a numerical approach becomes mandatory. For this reason, the development of optimized probe structures has been accompanied and favoured by the flourishing of a vast range of numerical methods giving a glimpse at nanoscale mechanisms [9,62-64].

Such methods have been extensively used to investigate novel aperture shapes using, for example, the finite difference time domain method (FDTD) or the field susceptibility technique and in most of the cases considering apertures in thin metal films [21,24,26-28]. Furthermore, the analysis of wave propagation in tapered structures (either traditional conical tapers or structures modified with corrugations or multiple tapers) with an aperture at the end has been carried out using the finite difference beam propagation method [65], the FDTD method [18,19], the body of revolution FDTD method (BOR-FDTD) [48], the multiple multipole method (MMP) [66]. Fully metal probes and fully metal-coated pyramidal probes with different shapes have also been analysed both in case of direct illumination at the metal apex or with far-field excitation further away from the metal apex as described in paragraph 2 [36,37,61,67] using FDTD, BOR-FDTD, finite integration technique (FIT), and the finite element method (FEM). The fully metal-coated dielectric structure which, as anticipated, is the one of interest in this chapter, has been intensively numerically investigated especially under internal back excitation and its polarization-dependent properties have been carefully examined: the need for a radially polarized excitation has been pointed out as essential to get field localization [40,46-50].

Even numerical treatment is challenging because near-field probes involve different length scales: while phenomena of major interest for near-field interactions occur in the mesoscopic (sizes of the order of the incident wavelength) and nanoscopic (structures smaller than 100 nm) regimes [63], the overall probe structure, especially for fiber-based probes, can include sections much bigger than the incident wavelength. Computational cost for modelling a probe in its entirety would become prohibitive. Therefore, the overall computational domain is typically restricted to the very end of the probe, often using two-dimensional approximations or exploiting symmetry properties of the structure.

In the light of the previous considerations, we have investigated our novel probes based on appropriate modifications introduced in fully metal-coated dielectric probes adopting a numerical approach and restricting the analysis to the very end of the probe. An overview of the numerical approach adopted for our simulations is provided in the next paragraph, together with the description of the developed computational model. The model has been tested first to study the behaviour of the fully metal-coated probe, which, as reported in paragraph 2, has been already numerically and experimentally investigated and represents the reference for comparison for the new probe configurations.

5. Description of the computational model

Among the different numerical methods used in the past to simulate fully metal-coated SNOM probes, ranging from the MMP [46,68], to FIT [40,49,50,52,69], or the FDTD method [48], we have preferred FEM, chosen also in [47].

FEM is a tool used for the solution of differential equations in many disciplines, ranging from electromagnetics to solid and structural mechanics, from fluid dynamics to acoustics and thermal conduction [70]. A point of strength of FEM is its ability to deal with a complex geometry. Unstructured grids can accommodate for complex geometries in a much more straightforward way than other methods using Cartesian grids like finite difference methods.

The irregular domain is discretized into smaller and regular subdomains, known as finite elements, thereby replacing a domain having an infinite number of degrees of freedom by a system with a finite number of degrees of freedom. The essential principle behind FEM is a piecewise approximation: the solution of a complex problem is obtained by splitting the region of interest into smaller regions and approximating the solution over each subregion by a simple function [71].

In particular, our three-dimensional (3D) computational model for the simulation of the electromagnetic modes in the investigated probe configurations has been developed with the help of a commercial software (Comsol Multiphysics) based on FEM. The computational process is articulated in a two-dimensional (2D) analysis to calculate the eigenmodes at the input port, followed by the 3D simulation of the propagation of the first three eigenmodes, i.e., the two lowest order linearly polarized modes and the radially polarized one. Second order elements with minimum size of about 0.8 nm have been used. Simulations have been run on a 64 bit workstation with 32 GB of RAM.

The first examined structure is the fully metal-coated dielectric probe under radially and linearly polarized excitation. Although in this case one could benefit from the symmetry properties of the structure, the axisymmetric probe represents only the starting point for the search of optimized configurations, illustrated in the following paragraphs. As we will see, such structures are characterized by strong asymmetries, which impose the need for full 3D analysis. Therefore, for the sake of a better comparison, a 3D modelling has been adopted also for the reference axisymmetric structure, without resorting to any of the simplifications used in previous works to handle axially symmetric structures. Only the very end of the tip is examined due to the high computational burden of the simulations because of the different scales of the metal layer and the dielectrics. In previous works, simulations involving larger portions could be carried out only when the structure was less computationally challenging due the rotational symmetry of the probe allowing reduction of the problem complexity either with the BOR-FDTD method or even with approximate 2D simulations [18,48].

Figure 3 reports the sketch of the simulated axisymmetric probe, consisting of a silica core (n=1.5) surrounded by an aluminium coating (n=0.645+5.029i at the operating wavelength

λ=532 nm). The radii of the inner silica cone and of the outer metallic hollow cone are 225 nm and 275 nm, respectively. Both cones (having an apex angle of 30°) are rounded, with the radius of curvature of the inner cone being 10 nm, the one of the outer cone amounting to 20 nm. The overall modelling domain is a 1.6 μm cylinder with radius 1 μm. The probe axis lies on the z axis.

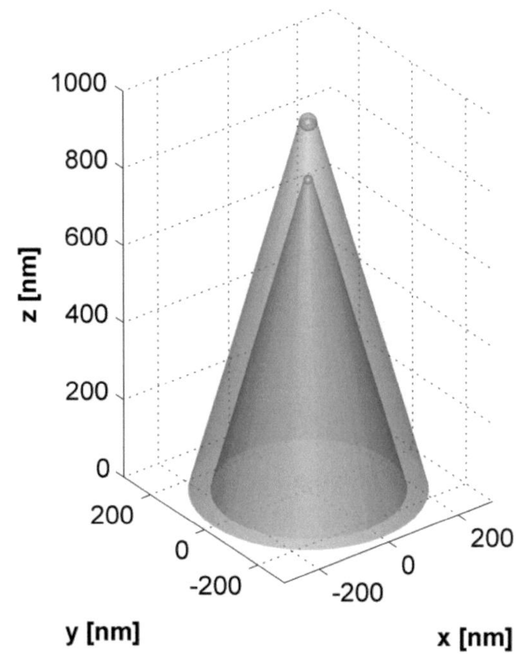

Figure 3. Sketch of the simulated axisymmetric fully metal-coated dielectric probe.

In Figure 4, the first three eigenmodes, i.e. the two lowest order x and y linearly polarized modes (H and V) and the radially polarized one (R), are reported together with the corresponding near-field distributions (square of the norm of the electric field) taken over a square area 600 nm by 600 nm centred around the probe apex at 10 nm from the apex; normalization to the maximum value of the electric field intensity distribution for the radial polarization has been done in order to emphasize the relative field strengths [51].

A highly localized hot spot with a full width at half maximum (FWHM) of 38 nm is observed for the R mode due to the constructive interference of the surface waves along the taper. As shown in [49], the size of the hot spot is influenced by the final rounding in the metal coating that, in our simulations, was chosen to be 20 nm in radius just for convenience as a reference for comparison and can be decreased at will. On the contrary, destructive interference of surface waves at the tip apex gives rise to two-lobed distributions, polarized mainly along the x and y axis, under H and V linearly polarized excitation, respectively; the average size (measured as the distance over which the field is more than or equal to half of its peak value) is approximately 400 nm and the peak value is about 50 times smaller than the peak of the R spot.

These results, in agreement with those previously obtained for such a probe, represent the reference for comparison for the novel probe configurations scrutinized in the following paragraphs, in which the characteristics of the near-field distributions at 10 nm from the apex of the modified probes are normalized to the peak value obtained in the axisymmetric probe under R excitation for an easier comparison.

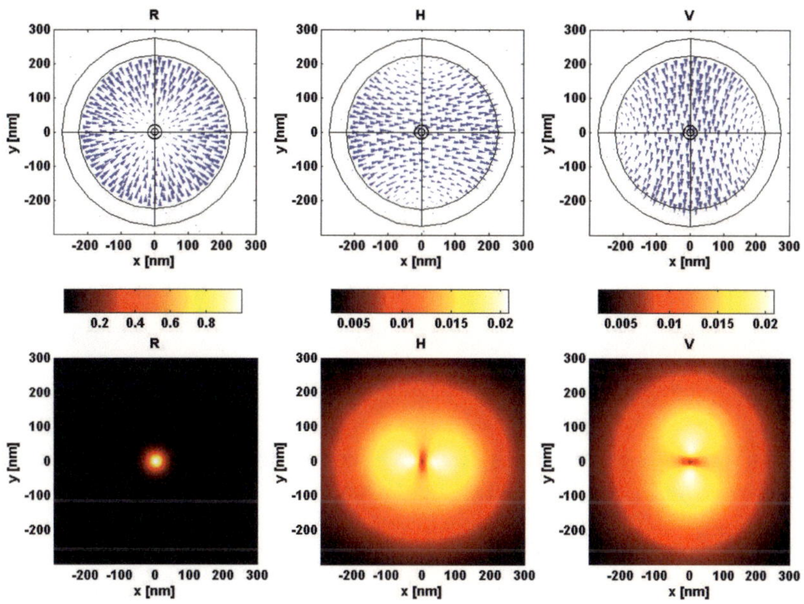

Figure 4. Illustration of input modes (upper row) and corresponding normalized near-field distributions at 10 nm from the apex of a standard axisymmetric fully metal-coated probe (lower row) (adapted from [51]).

6. Numerical investigation of probes with directional asymmetries under linearly polarized excitation

As anticipated, we have designed and investigated the behaviour of two different structural modifications that can be classified as directional asymmetries, one based on an oblique cut close to the tip apex stripping off both the metal coating and the inner core, and the other consisting of asymmetric corrugations in the metal coating. The sketches of the two structures are reported in Figure 5.

The first tip (Figure 5(a)) is cut along a plane which is neither orthogonal nor parallel to the tip axis: the cut angle (defined as the angle between a plane orthogonal to the axis of the tip and the plane of the cut itself) and the cut height (meant as the height of the new tip apex after the cut, measured from the bottom of the computational domain) can be varied [72].

The structure with asymmetric corrugations (Figure 5(b)) is based on the introduction on the outer metal surface of semicircular corrugations (either bumps or grooves), modelled by joining five truncated toroids of radius 20 nm with hemispherical terminations having the same radius: the toroidal sections are filled with air in case of a groove or with metal in case of a bump [51]. The case of metal oxide filling has been considered as well. Corrugations are limited to just one half of the tip, hence their angular extension is less than 180°.

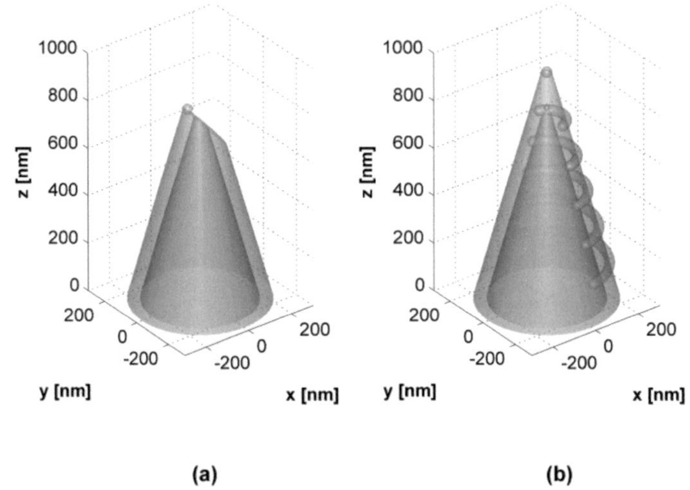

Figure 5. Sketch of the simulated structures based on an adirectional asymmetry: (a) probe with oblique cut; (b) asymmetrically corrugated fully metal-coated probe.

In both the cases, the asymmetry is present along the *x* direction.

6.1. Probe with oblique cut

For this probe structure, the behaviour of the probe under variable cut angle at constant cut height and for variable cut height at constant cut angle has been analysed [72].

In the first case, the angle has been varied from 20° to 60° (with a step of 10°) at a height of 816 nm. Normalized near-field patterns under H and V polarized excitation are shown in Figure 6.

As expected, field localization is achieved only under H polarized excitation. The two initially separated lobes of the V mode tend to merge as the angle becomes steeper until getting completely intermingled, while the initially asymmetric spots obtained under H excitation become progressively more symmetric about their centre along the *x* axis as the cut angle increases. The FWHM and the peak value normalized to the one obtained in an axisymmetric probe under radially polarized excitation are reported in Figure 7 as a function of the cut angle. No dramatic changes occur in the size of the spots (always between 37 nm and 41 nm), while an increase in the H peak occurs with steeper cut angles.

A similar analysis carried out for a cut height varying from 741 nm to 841 nm (with a step of 25 nm) at a cut angle of 30° revealed once again field localization under H polarized excitation. The results of the simulations with variable cut height suggest the use of cuts involving a larger fraction of the originally axisymmetric probe because both the FWHM and the peak value of the achieved hot spot undergo deterioration with increasing cut height. However, the approach to samples with steeper topographic variations would be hampered by cuts at lower heights unless a large cut angle is used at the same time, imposing a trade-off with the quality of the achievable H hot spot in the choice of the cut height.

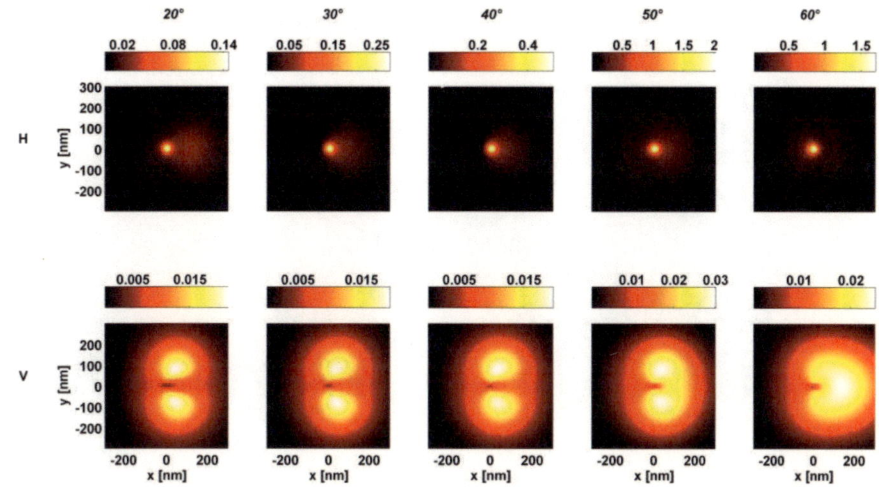

Figure 6. Normalized near-field intensity distributions at 10 nm from the apex of a cut probe with cut height of 816 nm and cut angle variable from 20° to 60° under H (upper row) and V (lower row) excitation (adapted from [72]).

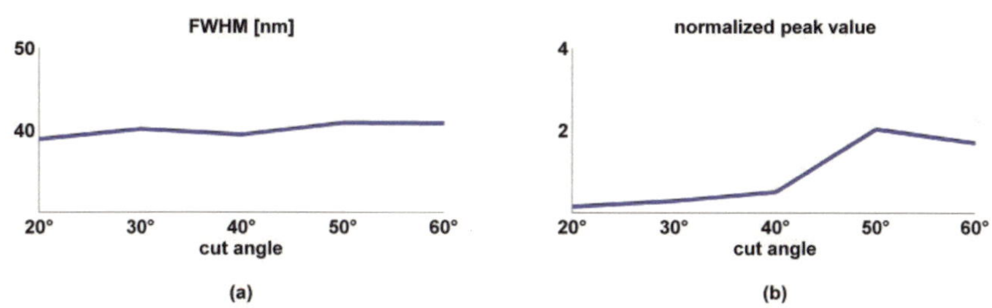

Figure 7. Characteristics of the near-field intensity distributions at 10 nm from the apex of the cut probe under H polarized excitation for variable cut angle: (a) FWHM; (b) peak value normalized to the one achieved in an axisymmetric probe under radially polarized excitation (adapted from [72]).

6.2. Probe with asymmetric corrugations

As we said, the five corrugations could consist in either grooves or bumps on the outer metal coating and are equally spaced, with the z-spacing amounting to 150 nm and the first bottom one centred at 150 nm from the input port [51].

First, the effect of a variation in the sequence of grooves and bumps has been considered and all the possible permutations of air indentations and metal bumps for the same structure have been analysed (at constant azimuthal extension of 160°). The different configurations have been named after the initial of the filling material (*a* for air groove and *m* for metal bump) starting from the bottom corrugation. As an example, the normalized near-field distributions for the configuration *amama* is reported in Figure 8. As expected, field localization is achieved under H excitation, while the V distribution maintains an almost two-lobed pattern.

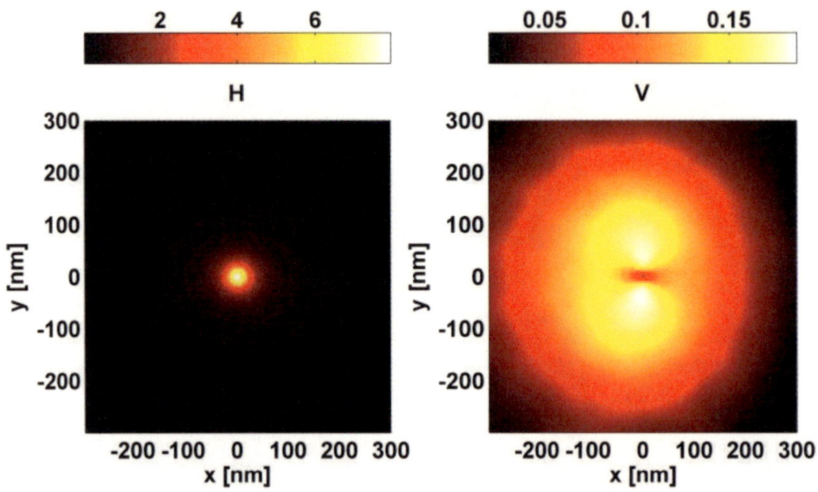

Figure 8. Normalized near-field intensity distributions at 10 nm from the apex of a probe with asymmetric corrugations (*amama* configuration) (adapted from [51]).

A systematic analysis of the FWHM and the peak value of the hot spot achieved under H polarized excitation for all the possible permutations of air indentations and metal bumps for the same structure has revealed that, except for only one configuration, the size of the near-field distribution undergoes a significant shrinkage in all the cases with the creation of a real ultrasmall spot in most of the cases with a FWHM comparable to that observed for the R mode excitation in the axisymmetric probe (Figure 9(a)). Although the H peak value generally increases compared to the case of the standard axisymmetric probe, only few material combinations give rise to values comparable or, in two cases, even much superior to the radial peak for the axisymmetric probe, with the best results given by the *amama* configuration (Figure 9(b)). Similar trends in terms of alternation of metal and dielectric have been observed upon replacement of the air grooves with metal oxide bumps (in this case aluminium oxide) and considering all the possible permutations of metal and metal oxide in the five semirings. However, the substitution of air with metal oxide gives rise to higher peak values due to better coupling between inner and outer SPPs when air is substituted with metal oxide, because the coupling of surface modes at two adjacent metal-dielectric interfaces becomes more efficient when the indices of refraction of the two dielectrics are closer [73].

Next, the influence of a variation of a geometric parameter, i.e. the azimuthal extension of the corrugations, has been studied. The *amama* configuration has been chosen and the angular extension of each corrugation has been changed between 110° to 160° with a step of 10°. No significant variations in the FWHM for the H hot spots have been observed, while the peak value increases with increasing azimuthal extension of the corrugations.

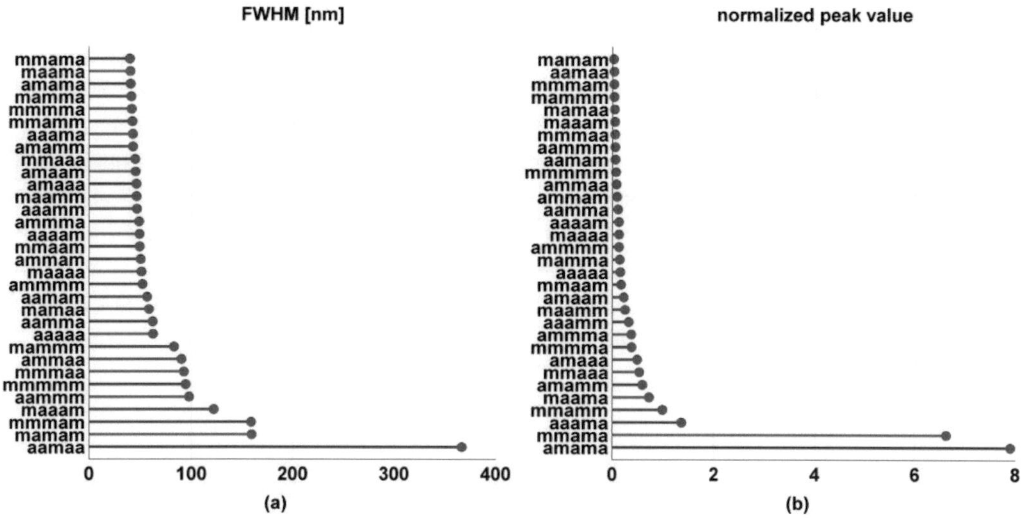

Figure 9. Characteristics of the near-field intensity distributions at 10 nm from the apex of the probe with asymmetric corrugations under H polarized excitation for each of the material permutations in the five semirings: (a) FWHM; (b) peak value normalized to the one achieved in an axisymmetric probe under radially polarized excitation (adapted from [51]).

6.3. Analysis of the behaviour of the directional asymmetries for arbitrary orientation of the asymmetry with respect to the input polarization

Our finite element based simulations have shown that carefully designed asymmetries introduced in an originally axisymmetric fully metal-coated tip can produce field localization under an excitation linearly polarized along a proper polarization direction. The presence of the asymmetry causes the electric fields associated with SPPs on the opposite sides of the tip not to have opposite phases any longer, a phenomenon that leads to destructive interference under linearly polarized injection in an axisymmetric structure [51,57]. A global analysis of the results obtained for both a cut probe and a tip with asymmetric corrugations has shown that this effect is enhanced when the asymmetry is extended over a broader region (as is the case for steeper cut angles or lower cut heights in case of the probe with oblique cut or a larger azimuthal extension of the corrugation for the tip with asymmetric corrugations). In this way, superfocusing can be achieved with a linearly polarized injection, which is much easier than a radially polarized one, with an enormous simplification in experimental applications.

However, so far we have considered the behaviour of the probe for two specific orientations of the input linearly polarized excitation with respect to the asymmetry (located along x), i.e. alignment along the direction of the asymmetry (H polarization) or alignment along the direction orthogonal to the asymmetry (V polarization). Field localization occurs for input polarization aligned along the preferential direction of the asymmetry, while no significant variation compared to the axisymmetric probe is shown for input polarization orthogonal to such direction.

In this section, to better understand how the misalignment from the preferential spatial direction can affect the performance of a probe with a directional asymmetry, we show the behaviour of a probe based on asymmetric corrugations for variable mutual orientation of the direction of the input linearly polarized excitation with respect to the one of the asymmetry [55,57,74]. Such mutual alignment has been defined as the angle α between the direction of the input linear polarization and the angle bisector of the corrugations (Figure 10(a)); the mutual orientation specified by α was varied from -85° to 90° to encompass all the possible mutual positions. Note that α=0° and α=90° correspond to x and y linearly polarized excitations so far labelled as H and V, with α=0° representing the position of best alignment of the input linear polarization with respect to the directional asymmetry and α=90° representing the maximum misalignment.

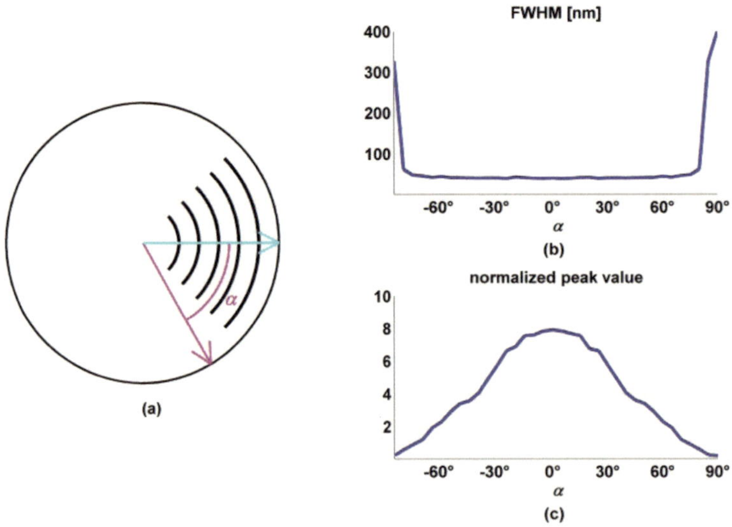

Figure 10. Behaviour of the probe with asymmetric corrugations for variable orientation of the input linearly polarized excitation with respect to the asymmetry: (a) schematic of the xy projection of the probe: the angle α is the one between the direction of the input linear polarization (magenta line) and the angle bisector of the corrugations (cyan line); (b) FWHM of the near-field distribution for the *amama* configuration under variable α; (c) peak value of the near-field distribution normalized to the one of the axisymmetric probe under radially polarized excitation for the *amama* configuration under variable α (adapted from [55]).

Figure 10 (b) and (c) report the FWHM and the normalized peak value as a function of α for the *amama* configuration and azimuthal extension of the corrugations equal to 160°. As the misalignment of the input polarization from the preferential direction of the asymmetry increases, both the peak value and the shape of the near-field intensity pattern change. In particular, the peak value decreases and the distribution becomes gradually broader. This is due to the fact that, as α increases, the asymmetry perceived by the input linearly polarized excitation progressively disappears. In retrospect, the different behaviour can be explained if recalling that a structure asymmetric along x appears symmetric for a y linearly polarized excitation, which brings about destructive interference of the excited SPPs, similarly to what happens for an axisymmetric structure under linearly polarized excitation. If we consider the average of both the FWHM and the peak value over all the possible mutual positions, from the graphs it can be inferred that maximum deviations of the peak value and of the spot size from the corresponding averages are about 100% and 450%, respectively.

Similar analyses have been run for various forms of directional asymmetries with different structural parameters. The maximum deviation of the peak ratio and of the spot size from the average value over all the mutual orientations of the linearly polarized excitation with respect to the asymmetry is reported in order to highlight the sensitivity to mode orientation (Figure 11).

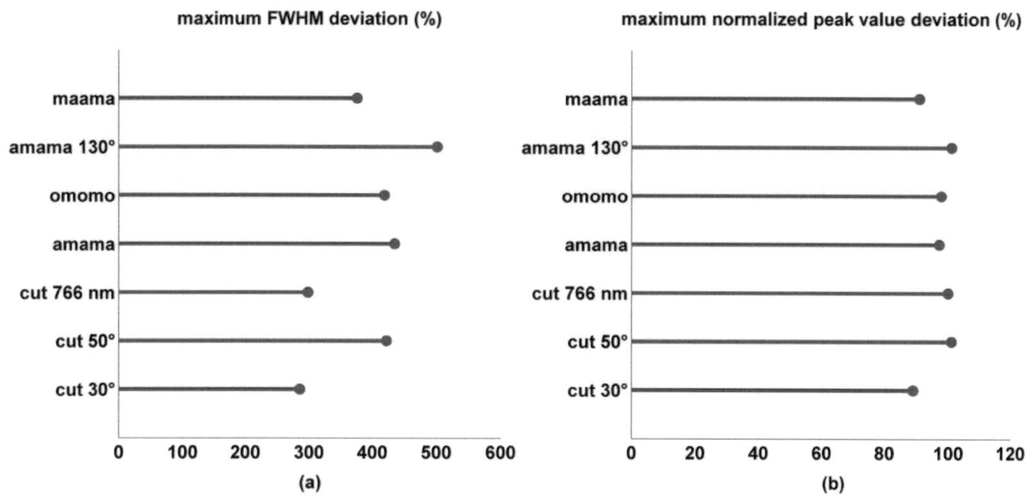

Figure 11. Characteristics of the near-field intensity distributions for directional asymmetries under linearly polarized excitation for variable orientation α of the asymmetry with respect to the input linear polarization: (a) maximum deviation from the average FWHM; (b) maximum deviation from the average normalized peak value; *omomo* is a probe with asymmetric corrugations and alternation of oxide (*o*) and metal (*m*) bumps (adapted from [57]).

Variations in the peak value are above 90% and, more importantly, those in spot size exceed 280%. Such strong variations are due to the intrinsic directional nature of the asymmetries, with degradation in peak intensity and resolution as a consequence of misalignments from the preferential direction of the asymmetry.

7. Numerical investigation of probes with adirectional asymmetries under linearly polarized excitation

In order to reduce the sensitivity of the probe behaviour to the direction of the input polarization, we have introduced the concept of adirectional asymmetry. Two different implementations of this structural modification have been considered, both based on a spiral design: in fact, the spiral intrinsically fits the specification of lack of rotational and reflection symmetry and offers an interesting case study to investigate the feasibility of the concept of orientation-insensitive field localization [55,57]. First, the effects of a spiral corrugation on the outer metal surface of a fully metal-coated probe have been investigated. Then, another implementation based on azimuthal corrugations arranged in a spiral-like fashion will be discussed. A sketch of the two structures is shown in Figure 12.

The spiral corrugation is formed by joining a tapered helix-shaped 3D object (with circular cross-section of radius *r*) with two hemispherical terminations (having the same radius *r*). The spiral winding appears as a semicircular spiral corrugation and is placed between 150 nm and 750 nm along the *z* direction; the pitch along *z* is 300 nm. The spiral corrugation can take on the form of either a groove, i.e. it is filled with air, or a bump, corresponding to metal filling; the effect of metal oxide filling has been analysed as well [57].

The azimuthal corrugations are formed by joining truncated toroids with hemispherical terminations as in paragraph 6. However, in order to create a spiral arrangement (and hence an adirectional asymmetry distributed over all spatial directions), the corrugations are shifted one with respect to the other [55].

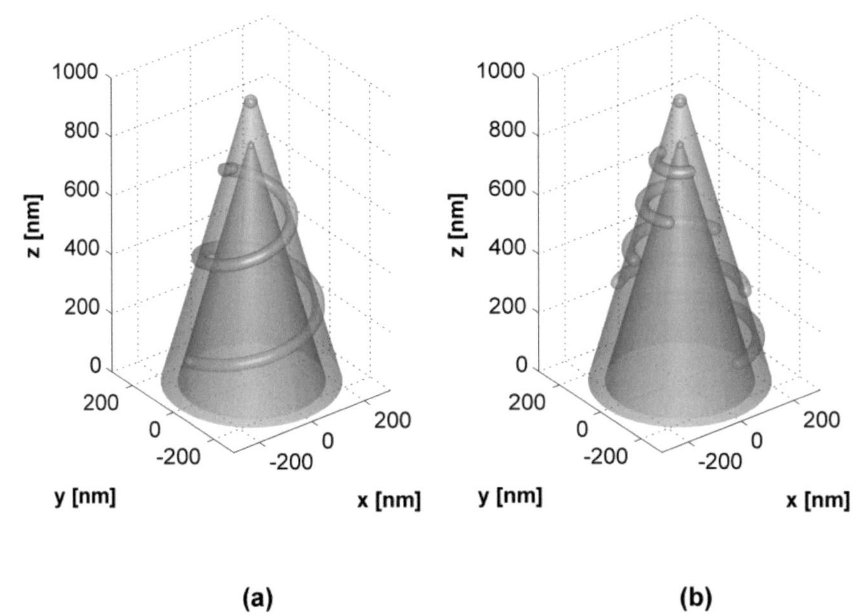

Figure 12. Sketch of the structures based on an adirectional asymmetry: (a) probe with spiral corrugation; (b) probe with spiral-arranged azimuthal corrugations.

7.1. Probe with spiral corrugation

The near field distributions obtained for a spiral metal corrugation with radius $r = 25$ nm under H and V polarized excitation are illustrated in Figure 13.

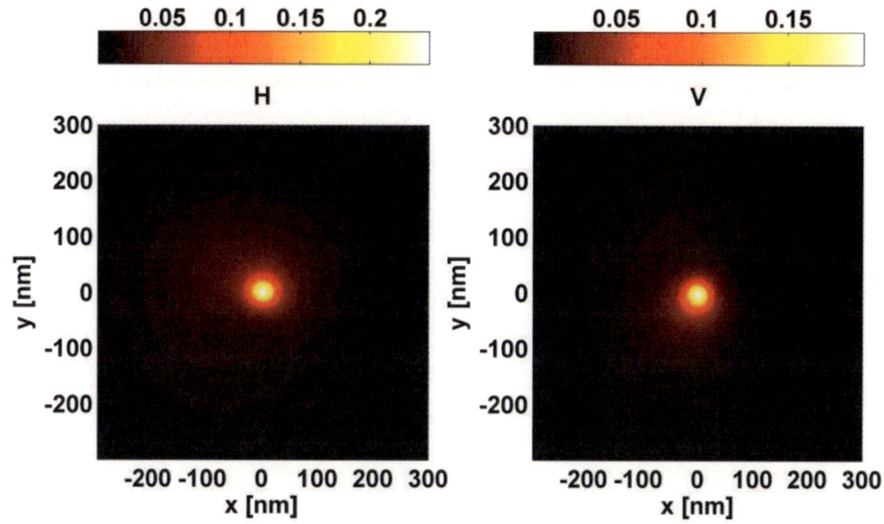

Figure 13. Normalized near-field intensity distributions at 10 nm from the apex of a probe with a spiral metal corrugation of radius $r = 25$ nm (adapted from [57]).

Strong field localization is observed for both the orthogonal input polarizations, with the creation of ultrasmall spots very similar in terms of both FWHM and peak value. As illustrated below, the intensity (that, in this case, is still 5 times lower than the one of the radial hot spot of the axisymmetric probe) can be adjusted by changing either the filling material or the radius of the corrugation. Similar near-field distributions have been observed upon variation in the chirality of the spiral winding (that is whether it wraps the tip in clockwise or counter-clockwise direction).

An analysis of the FWHM and the peak value (normalized to the peak achieved in the reference axisymmetric probe under radially polarized excitation) under different mutual orientations α of the direction of the linearly polarized excitation with respect to the asymmetry has been carried out to assess the properties of the adirectional asymmetry: according to Figure 14, α=0° corresponds to x linearly polarized excitation, with the two extremes of the spiral winding located along the x axis.

The most noticeable feature is that only minor fluctuations occur in the FWHM, which means that the spot size is almost insensitive to variations in the mutual orientation. We should remind that the FWHM is related to the eventual achievable resolution and, hence, its robustness with respect to variations in the orientation of the input polarization implies a substantial simplification in experimental applications. Although the peak value still depends on the mutual orientation, the maximum deviation from its average value calculated over all the mutual positions is below 20%, still tolerable if the average value were sufficiently high for detection. Note that, contrarily to the case of directional asymmetry where α=0° and α=90° represented the best and worst alignment, for the adirectional asymmetry different orientations are almost equivalent.

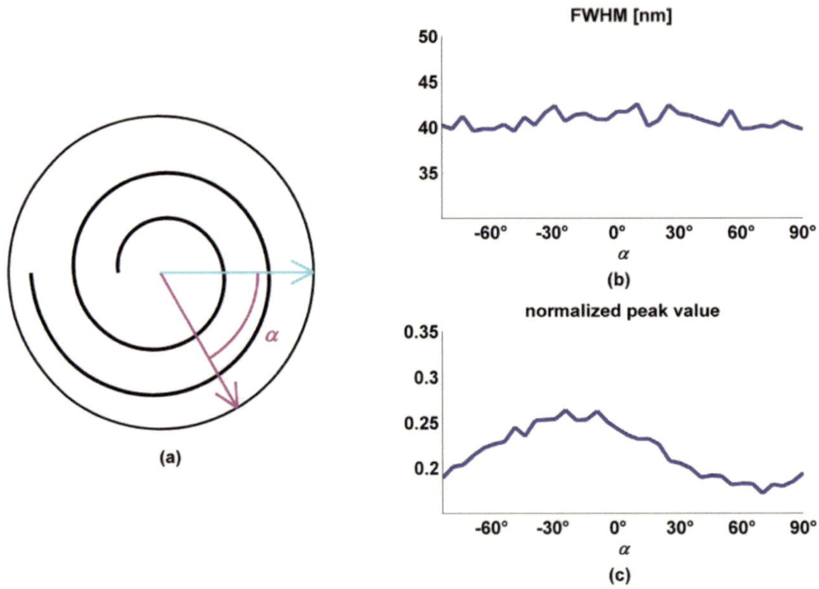

Figure 14. Behaviour of the probe with spiral corrugation for variable orientation of the input linearly polarized excitation with respect to the asymmetry: (a) schematic of the xy projection of the probe: the angle α is the one between the direction of the input linear polarization (magenta line) and the line along which the two extremes of the spiral are aligned (cyan line); (b) FWHM of the near-field distribution for the spiral metal corrugation under variable α; (c) peak value of the near-field distribution normalized to the one of the axisymmetric probe under radially polarized excitation for the spiral metal corrugation under variable α (adapted from [57]).

Similarly to the case of directional asymmetries, the FWHM and the peak value can be optimized by varying the characteristics of the spiral asymmetry. In particular, we changed the radius r of the spiral from 15 nm to 30 nm with a step of 5 nm. Moreover, we considered the case of replacement of the metal spiral bump with a spiral groove (air filling) or with an aluminium oxide spiral bump. In particular, for each structure based on a particular combination of filling material and radius r, the behaviour of the probe under all the possible mutual orientations of the linearly polarized excitation with respect to the spiral winding was examined, by calculating the average value of the peak ratio and the FWHM over all the mutual orientations and considering these values together with the maximum deviation from the corresponding average as figures of merit for comparison.

As for the directional asymmetries, overall better results (i.e. smaller spot size and higher peak values) have been obtained for stronger asymmetries due to larger r values. However, differently from the case of directional asymmetries, the data about the maximum deviation confirm that field localization is achieved irrespective of the mutual orientation between the linearly polarized mode and the asymmetry introduced in the structure. The deviation in FWHM dips as the radius increases, while the peak value shows an opposite trend with a change of the spiral radius: however, the maximum deviations are still acceptable, especially if combined with a rise in the average value occurring when the metal bump is substituted with an air groove or, even better, an aluminium oxide bump. The behaviour of the probe as a function of the filling material can be explained as a result of better coupling mechanisms between inner and outer SPPs, as also pointed out in paragraph 6.2.

7.2. Probe with corrugations arranged in a spiral-like fashion

The spiral corrugation can be replaced by a series of azimuthal corrugations shifted one with respect to the other to create a spiral arrangement, as visible in the 2D projection of Figure 15; the angle α between the direction of the input linearly polarized excitation and the angle bisector of the bottom corrugation is also defined. The parameters of this new structure can be varied to improve the field localization and enhancement for any orientation of the input linearly polarized excitation.

The effect of variations in the shift angle β (identical for any two consecutive corrugations) has been examined as well as the impact of variable radius and variable azimuthal extension γ of the single corrugations (which means the arc of circumference over which corrugations are spanned). According to the sketches in Figure 12 and Figure 15, the angle bisector of the bottom corrugation is along the x axis. The z spacing between consecutive corrugations, which can be either grooves or bumps, is 150 nm. We focused our attention on the structure based on alternation of grooves and bumps starting from the bottom corrugation, previously labelled as *amama*, because the studies reported in paragraph 6.2 for null shift angle β between consecutive corrugations showed that it resulted in the best performance.

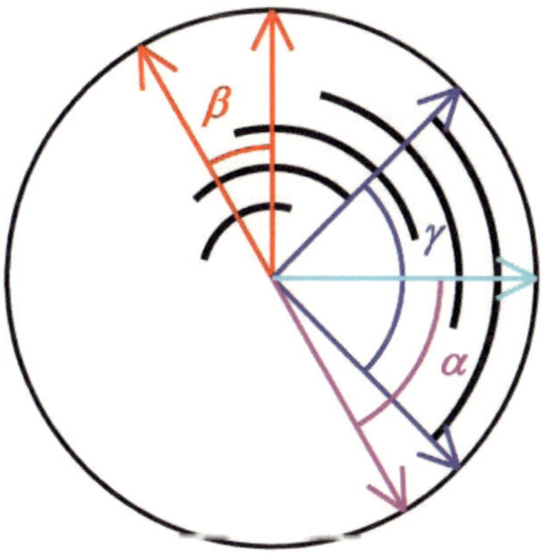

Figure 15. Schematic of the 2D projection of the probe with spiral-arranged corrugations and relevant parameters; the black arcs represent the corrugations; the angle α is the one between the direction of the input linear polarization (magenta line) and the angle bisector of the bottom corrugation (cyan line); the angle β is the shift angle measured between the angle bisectors of two consecutive corrugations, while the angle γ is the angular extension of each corrugation. The angle bisector of the bottom corrugation is along the x axis (adapted from [55]).

Figure 16 reports the square of the normalized near field distributions under H and V polarized excitation for a shift angle β of 45° (adirectional asymmetry), total azimuthal extension γ for each of the five corrugations equal to 160° and radius of 20 nm. The two distributions show close resemblance in both peak value and FWHM.

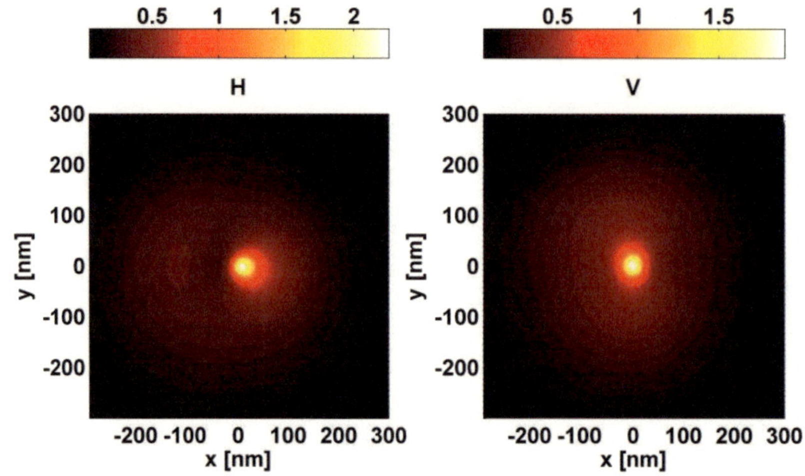

Figure 16. Normalized near-field intensity distributions at 10 nm from the apex of a probe with azimuthal corrugations arranged in a spiral-like fashion (β=45°; γ=160°; radius 20 nm) (adapted from [55]).

In order to further confirm that field localization is insensitive to the alignment of the input polarization, also for this asymmetry we studied the behaviour of the probe under variable orientation of the corrugations with respect to the input linear polarization, and, in particular, the FWHM and the peak value normalized to the one of the axisymmetric probe under radially polarized excitation.

Figure 17 shows deviations of the peak value from the average below 20% and those of the spot size less than 10% for the probe with adirectional asymmetry, which demonstrates the possibility to get superfocusing under arbitrarily oriented linearly polarized excitation. In particular, the ultimate achievable resolution is almost insensitive to the input polarization direction. Similar trends were observed by changing the chirality, i.e. the handedness, of the spiral-like arrangement.

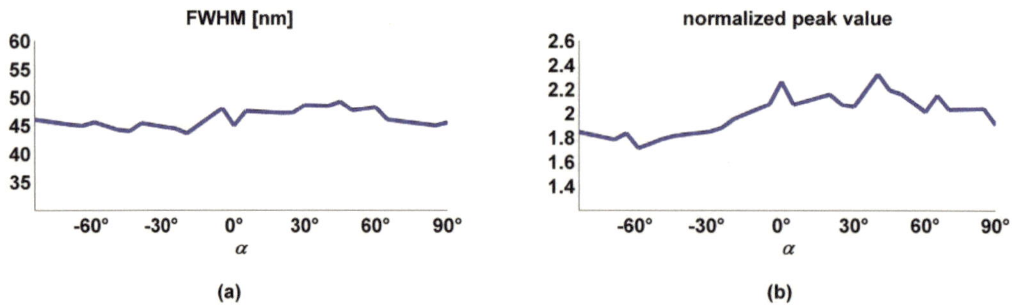

Figure 17. Behaviour of the probe with spiral-arranged corrugations for variable orientation α of the input linearly polarized excitation with respect to the asymmetry (β=45°; γ=160°; radius 20 nm): (a) FWHM of the near-field distribution for variable α; (b) peak value of the near-field distribution normalized to the one of the axisymmetric probe under radially polarized excitation for variable α (adapted from [55]).

The effect of some geometric parameters (more specifically the shift angle β, the angular extension γ and the radius of the corrugation) on the probe behaviour has been analysed as well. By increasing the shift angle from 0° to 45° (with a step of 7.5°), a better confinement and an increase in the peak value has been observed, together with smaller variations in these parameters under variable orientation of the corrugations with respect to the input linearly polarized excitation. An increase in the angular extension and in the radius has also been shown to have an overall beneficial effect on probe behaviour.

8. Interaction of asymmetric probes with fluorescent molecules

Beyond the overall intensity distribution, also the analysis of the single vectorial components of the near-field distribution is of major interest and provides insight into the field localization mechanisms. A full vectorial picture of the electric field distribution close to the tip apex can be gained by scanning single fluorescent molecules with a SNOM probe. In the seminal work by Betzig and Chichester, the near-field distribution of aperture probes was mapped by fluorescent molecules, which act as point detectors for the components of the electric field aligned along their absorption dipole direction [75]. The interaction of single fluorescent molecules with aperture probes has been thoroughly investigated both experimentally and theoretically, not only as a tool for probe characterization, but also for its importance in biological as well as in material science [60,76-78].

For all these reasons, we have investigated the interaction of a single fluorescent molecule with an axisymmetric probe under radially polarized excitation and with an asymmetric probe under linearly polarized excitation in order to identify the similarities between the vectorial components of the near-field distributions close to the apex of the probes [79]. In most of the calculations, we considered the cut probe as an example of asymmetric probe due to the simplicity to fabricate this probe using FIB milling, as will be discussed in the next paragraph.

Figure 18 shows the quantities necessary to describe the dipole moment of the fluorescent molecules.

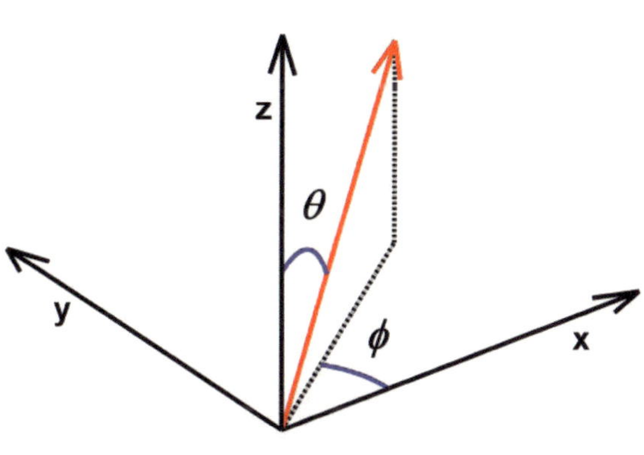

Figure 18. Representation of the orientation of the dipole (red vector) through the polar angle θ and the azimuthal angle ϕ.

The fluorescence intensity I emitted by a molecule is related to the square of the dot product of the local electric field E and the normalized absorption dipole moment p far from saturation [75]:

$$I \propto \left| E \cdot p \right|^2 \quad (4)$$

Therefore, the fluorescence intensity I depends on both the absolute value and the direction of the electric field, which shows the ability of a fluorescent molecule to provide a vectorial picture of the local electric field. The normalized absorption dipole moment describing the 3D dipole orientation is given by:

$$p = \begin{pmatrix} \sin\theta\cos\phi \\ \sin\theta\sin\phi \\ \cos\theta \end{pmatrix} \quad (5)$$

where θ and ϕ represent the polar and azimuthal angle, respectively (Figure 18). In our case the local field $E=(E_x,E_y,E_z)$ coincides with the electric field distribution on the xy plane at a distance z of 10 nm above the probe apex.

The interaction of an axisymmetric probe under radially polarized excitation with a fluorescent molecule with polar and azimuthal orientation varying from 0° to 90° (with a step of 30°) was considered. The resulting fluorescence intensity distributions calculated according to equation (4) on a 400 nm by 400 nm square area centered on the probe apex are shown in Figure 19, with an indication of the peak intensity for a molecule with a specific dipole orientation normalized to the peak of the total local electric field intensity $\left| E \right|^2$ (number reported in white). As visible, the radially polarized excitation gives rise to a strong longitudinal local field at the tip apex (aligned along the probe axis z and corresponding to θ=0°), made up of a single hot spot, and weaker orthogonal transverse components (corresponding to θ=90° and ϕ=0° and θ=90° and ϕ=90°, respectively) of almost equal magnitude appearing as two-lobed patterns.

A variation in the azimuthal angle of the fluorophore corresponds to almost identical intensity distributions, simply rotated according to ϕ, because of the axial symmetry of both the probe and the input excitation. On the contrary, a decrease in the peak intensity and a gradual transition from a single spot to a two-lobed distribution occurs upon a change of the polar angle θ from 0° to 90°, because of the overlap of the dipole absorption moment dominantly with the weaker transverse components of the local electric field rather than with the stronger longitudinal component.

The most remarkable feature is the dominance of a strong longitudinal field component, which can be beneficial for the imaging of molecules with a mainly longitudinal dipole moment. Moreover, molecules with identical polar angle and different azimuthal orientation give rise to similar fluorescent patterns apart from the rotation by ϕ, which allows the determination of the azimuthal component of the fluorophore dipole moment. On the contrary, using aperture probes, molecules with variable azimuthal angle result in strongly different patterns due to the significant differences in orthogonal transverse components close to the aperture, with variations in peak intensities by even two orders of magnitude [77].

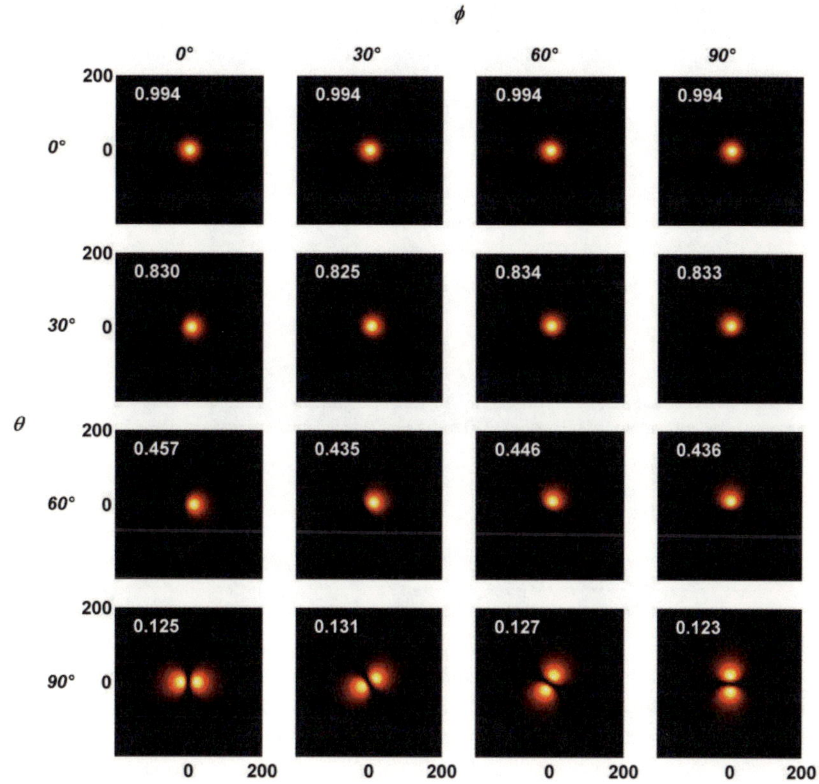

Figure 19. Simulated fluorescence intensity maps for single molecules with different orientations (as specified by the polar angle θ and the azimuthal angle ϕ) excited by an axisymmetric fully metal-coated probe under radially polarized excitation and located at 10 nm from the tip apex. All the maps are reported on the same colour scale, with each plot normalized to its peak intensity value (adapted from [79]).

A similar analysis has been run for a cut probe. Figure 20 shows the fluorescence intensity maps for a structure with a cut angle of 30° and a cut height of 766 nm under x linearly polarized excitation (i.e. oriented along the direction of the asymmetry; α=0°). Also in this case, the dominance of the longitudinal component is clearly visible, with double-lobed orthogonal transverse components slightly different in magnitude. The close resemblance with the fluorescence distributions obtained for an axisymmetric probe under radially polarized excitation is significant, which makes the asymmetric structure eligible to replace axisymmetric probes in single fluorescent molecule studies, because the advantages pointed out for the axisymmetric probe can be obtained under an easier linearly polarized excitation.

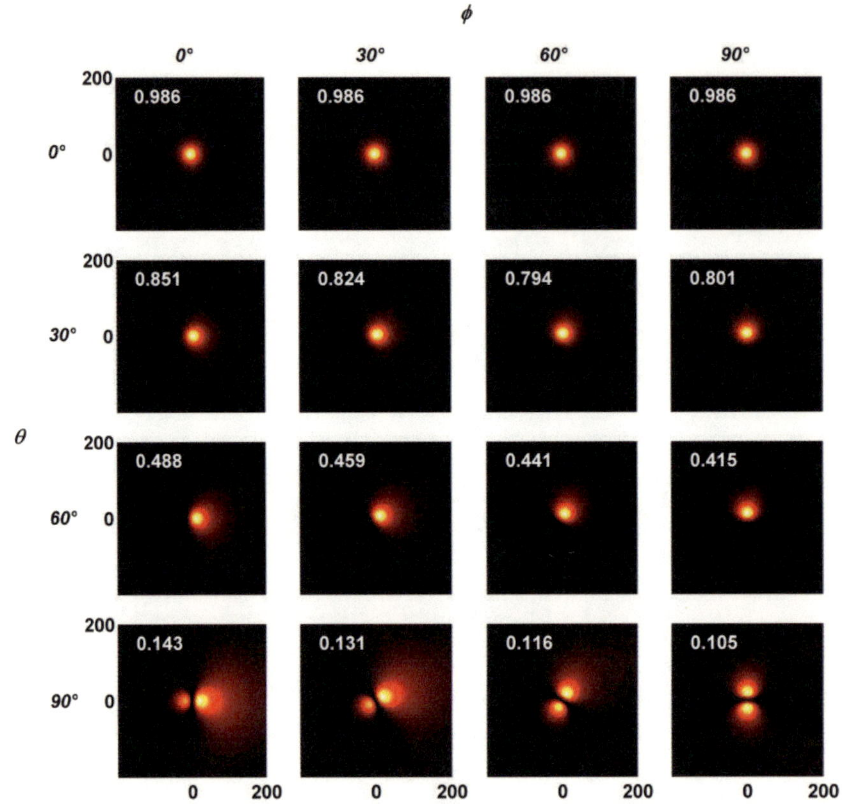

Figure 20. Simulated fluorescence intensity maps for single molecules with different orientations (as specified by the polar angle θ and the azimuthal angle ϕ) excited by a cut probe under linearly polarized excitation along the direction of the asymmetry (x) and located at 10 nm from the tip apex. All the maps are reported on the same colour scale, with each plot normalized to its peak intensity value (adapted from [79]).

The study of the interaction of different asymmetric probe structures with fluorescent molecules has revealed that all the asymmetric structures exhibit a highly confined strong longitudinal electric field component and weaker transverse components, like the axisymmetric probe under radially polarized excitation.

Moreover, the analysis of the intensity maps obtained for a fluorescent molecule with longitudinal dipole moment under variable α using either the previously considered cut probe or a probe with two-turn spiral corrugation in form of an aluminium bump with 25 nm radius (as in paragraph 7.1) has confirmed once again how adirectional asymmetries are more robust to variations in the direction of the input polarization compared to the directional ones. If the asymmetry lies along one specific spatial direction, alignment of the input linearly polarized excitation with respect to the asymmetry is still necessary. On the contrary, light funnelling to a highly confined mainly longitudinally polarized distribution independent of the orientation of the input linear polarization can be achieved by extending the asymmetry over all the spatial directions.

9. Hints on probe fabrication

In the introduction, we have briefly summarized the fundamental steps to create an aperture fiber probe. Our modified asymmetric probes can still benefit from this well developed technology. The overall fabrication process consists in the creation of the taper and subsequent metallization. Next, the desired modification is introduced by FIB milling. In our case, for first tests on the fabrication of our novel asymmetric probes, fiber-based SNOM tips have been bought and properly modified using a dual beam FIB/SEM (scanning electron microscope) system (FEI Strata DB 235). The system includes an ion beam and an electron beam column tilted one with respect to the other by 52°, allowing simultaneous nanostructuring and imaging of the probe.

FIB milling has been applied in the past to the fabrication of apertures in aperture SNOM probes, either to improve the quality of apertures created by some other method or to create an aperture in a fully metal-coated probe. Two different approaches have been used for this purpose, i.e. head-on drilling and slicing [80,81]. In the first approach the Ga^+ beam is scanned from the top across the tip surface using a previously generated pattern (the ion beam is aligned along the tip axis); although in this way a better control over shape and size of the aperture can be achieved by choosing an appropriate milling pattern, deep drilling into the tip structure can occur and the ion beam needs to be guided to the center of the tip apex [80-82]. In the second approach, the metal-coated fiber probe is sliced off transversally, i.e. the ion beam is incident at 90° with respect to the probe axis; the size of the aperture can be controlled only indirectly by estimating the correct location for slicing and the size of the milled area; small beam sizes are used to remove thin slices of material corresponding to low ion currents [80,81,83-85].

The slicing approach, properly adapted, is suitable for the fabrication of our probes based on an oblique cut. However, differently from the case of conventional aperture probes, the angle of incidence of the ion beam with respect to the probe axis is no longer 90°, but needs to be chosen according to the desired cut angle. The definition of cut angle corresponds to the one given in paragraph 6: hence, a simple aperture probe would correspond to a cut angle of 0°.

The fiber probes used to fabricate and test the probe based on an oblique cut, bought from Lovalite, have been produced by heating and pulling and subsequent deposition of aluminium. For the fabrication, a step-by-step procedure was adopted in which a milling step consisting in scanning the ion beam over a small rectangular area to remove as little material as possible was followed by an imaging step in order to check the results of the previous nanostructuring step. To get a first knowledge of the aperture size and metal thickness, first simple aperture probes with diameters between about 70 nm and 200 nm were realized. Then cut probes were produced. Figure 21 shows an example of an aperture probe and of a probe with an oblique cut, with a cut angle of 50°.

FIB milling could also be applied for the creation of corrugations, even a spiral one: in fact, spiral lenses have been fabricated on planar structures [86,87]. In the wake of these encouraging results, we are confident that a spiral conical corrugation could be produced as well.

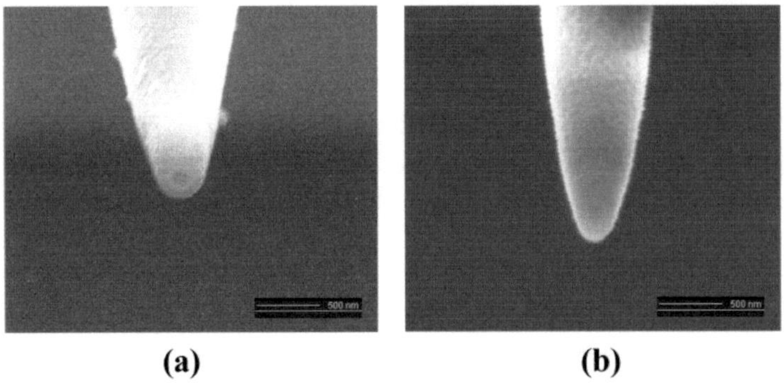

Figure 21. SEM image of: (a) aperture probe; (b) cut probe.

10. Conclusions

An extensive analysis of nanofocusing in asymmetric SNOM probe structures has been presented. The introduction of an asymmetry in an originally axisymmetric fully metal-coated probe structure has been shown to be effective in the achievement of field localization under linearly polarized excitation. In fact, in an apertureless axisymmetric structure, field localization, essential for high-resolution applications, can be attained only under radially polarized excitation. In this case, the surface plasmons excited by the radially polarized waveguide mode interfere constructively at the tip apex leading to the creation of an ultrasmall hot spot. On the contrary, those excited by a linearly polarized mode interfere destructively producing a broad, weak and asymmetric near-field distribution. However, the promising radially polarized excitation requires complicated injection procedures, extremely sensitive to misalignments.

Breaking the symmetry of the probe can help to avoid the destructive interference occurring under linearly polarized excitation allowing field localization at the tip apex under a more straightforward input excitation. Depending on the characteristics of the asymmetry, we have distinguished two different categories of asymmetry, which we called directional and adirectional asymmetries. Both the asymmetries break the axial symmetry of the original probe. However, the directional asymmetry consists in an asymmetry along one specific spatial direction; more specifically the structural modification exhibits a plane of symmetry. The adirectional asymmetry is devoid of reflection symmetry. In the first case, field localization is expected under linearly polarized excitation along the preferential spatial direction; in the second case, the field localization effect is almost insensitive to the direction of the input linearly polarized excitation. Hence, although both the categories of asymmetry obviate the need for a radially polarized excitation, adirectional asymmetries waive any requirement on the input polarization, with a further simplification of experimental procedures.

Probe design and modelling has been carried out adopting a numerical approach based on the finite element method. Two different implementations of directional asymmetries (a probe based on an oblique cut and one on asymmetric corrugations on the external metal coating) have been considered as well as two different configurations of the pioneering adirectional probe design (a probe with a spiral corrugation and one with azimuthal corrugations arranged in a spiral-like fashion). A 3D analysis of mode propagation has confirmed that properly tailored asymmetries can bring about field localization effects under linearly polarized excitation comparable to those achieved in an axisymmetric probe under radially polarized excitation. Moreover, the adirectional design guarantees more robustness against variations in the input polarization direction. Structural parameters can be optimized to get better near-field distributions in terms of FWHM and peak value. In general, probes with more extended asymmetries have shown to result in better overall performance.

The fabrication of the modified structures can benefit from well-established technologies for probe fabrication. In particular, the asymmetry itself can be introduced using FIB nanostructuring, as shown, for example, for the probe with an oblique cut. In this case, the slicing approach, already used for the creation of aperture probes, has been properly adapted for the creation of the novel probe.

In conclusion, these new promising probe concepts could result in significant headways towards high-resolution SNOM applications.

Author details

Valeria Lotito*
Electronics/Metrology Laboratory, Empa,
Swiss Federal Laboratories for Materials Science and Technology, Dübendorf, Switzerland
Department of Information Technology and Electrical Engineering, ETH Zurich, Zurich, Switzerland

Christian Hafner
Department of Information Technology and Electrical Engineering, ETH Zurich, Zurich, Switzerland

Urs Sennhauser
Electronics/Metrology Laboratory, Empa,
Swiss Federal Laboratories for Materials Science and Technology, Dübendorf, Switzerland

Gian-Luca Bona
Direction, Empa,
Swiss Federal Laboratories for Materials Science and Technology, Dübendorf, Switzerland
Department of Information Technology and Electrical Engineering, ETH Zurich, Zurich, Switzerland

Acknowledgement

The authors gratefully acknowledge the support of the Swiss National Science Foundation (project number 200021-115895) and the contribution of Dr. Konstantins Jefimovs (Empa) for probe nanostructuring.

---

*Corresponding Author

11. References

[1] Novotny L, Hecht B. Principles of Nano-Optics. New York: Cambridge University Press; 2006.

[2] Hecht B, Sick B, Wild U, Deckert V, Zenobi R, Martin O, Pohl D. Scanning near-field optical microscopy with aperture probes: fundamentals and applications. Journal of Chemical Physics 2000;112 7761-7774.

[3] Lewis A, Isaacson M, Harootunian A, Muray A. Development of a 500 Å spatial resolution light microscope. Ultramicroscopy 1984;13 227-231.

[4] Pohl DW, Denk W, Lanz M. Optical stethoscopy: Image recording with resolution $\lambda/20$. Applied Physics Letters 1984;44 651-653.

[5] Ambrosio A, Fenwick O, Cacialli F, Micheletto R, Kawakami Y, Gucciardi PG, Kang DJ, Allegrini M. Shape dependent thermal effects in apertured fiber probes for scanning near-field optical microscopy. Journal of Applied Physics 2006;99 084303.

[6] La Rosa AH, Yakobson BI, Hallen HD. Origins and effects of thermal processes on near-field optical probes. Applied Physics Letters 1995;67(18) 2597-2599.

[7] Gucciardi PG, Colocci M, Labardi M, Allegrini M. Thermal-expansion effects in near-field optical microscopy fiber probes induced by laser light absorption. Applied Physics Letters 1999;75(21) 3408-3410.

[8] Novotny L, Hafner C. Light propagation in a cylindrical waveguide with a complex metallic dielectric function. Physical Review E 1994;50 4094-4106.

[9] Antosiewicz TJ, Marciniak M, Szoplik T. On SNOM resolution improvement. In: Sibilia C, Benson TM, Marciniak M, Szoplik T (eds.) Photonic crystals: physics and technology. Milan: Springer; 2008 p. 217–238.

[10] Saiki T, Matsuda K. Near-field optical fiber probe optimized for illumination-collection hybrid mode operation. Applied Physics Letters 1999;74(19) 2773-2775.

[11] Yatsui T, Kourogi M, Ohtsu M. Increasing throughput of a near-field optical fiber probe over 1000 times by the use of a triple-tapered structure. Applied Physics Letters 1998;73(15) 2090-2092.

[12] Mononobe S, Saiki T, Suzuki T, Koshihara S, Ohtsu M. Fabrication of a triple-tapered probe for near-field optical microscopy in UV region based on selective chemical etching of a multistep index fiber. Optics Communications 1998;146 45-48.

[13] Nakamura H, Sato T, Kambe H, Sawada K, Saiki T. Design and optimization of tapered structure of near-field fibre probe based on finite-difference time-domain simulation. Journal of Microscopy 2001;202(1) 50-52.

[14] Yatsui T, Kourogi M, Ohtsu M. Highly efficient excitation of optical near-field on an apertured fiber probe with an asymmetric structure. Applied Physics Letters 1997;71(13) 1756-1758.

[15] Garcia-Parajo M, Tate T, Chen Y. Gold-coated parabolic tapers for scanning near-field optical microscopy: fabrication and optimisation. Ultramicroscopy 1995;61 155-163.

[16] Arslanov NM. The optimal form of the scanning near-field optical microscopy probe with subwavelength aperture. Journal of Optics A: Pure and Applied Optics 2006;8(3) 338-344.

[17] Bakunov MI, Bodrov SB, Hangyo M. Intermode conversion in a near-field optical fiber probe. Journal of Applied Physics 2004;96(4) 1775-1780.

[18] Antosiewicz TJ, Szoplik T. Corrugated metal-coated tapered tip for scanning near-field optical microscope. Optics Express 2007;15(17) 10920–10928.

[19] Wang Y, Wang Y-Y, Zhang X. Plasmonic nanograting tip design for high power throughput near-field scanning aperture probe. Optics Express 2010;18(13) 14004-14011.

[20] Matteo JA, Fromm DP, Yuen Y, Schuck PJ, Moerner WE, Hesselink L. Spectral analysis of strongly enhanced visible light transmission through single C-shaped nanoapertures. Applied Physics Letters 2004;85(4) 648-650.

[21] Jin EX, Xu X. Finite-difference time-domain studies on optical transmission through planar nano-apertures in a metal film. Japanese Journal of Applied Physics 2004;43(1) 407-417.

[22] Danzebrink HU, Dziomba T, Sulzbach T, Ohlsson O, Lehrer C, Frey L. Nano-slit probes for near-field optical microscopy fabricated by focused ion beams. Journal of Microscopy 1999;194(2/3) 335-339.

[23] Butter JYP, Hecht B. Aperture scanning near-field optical microscopy and spectroscopy of single terrylene molecules at 1.8 K. Nanotechnology 2006;17 1547-1550.

[24] Tanaka K, Tanaka M. Optimized computer-aided design of I-shaped subwavelength aperture for high intensity and small spot size. Optics Communications 2004;233 231-244.

[25] Minh PN, Ono T, Tanaka S, Esashi M. Spatial distribution and polarization dependence of the optical near-field in a silicon microfabricated probe. Journal of Microscopy 2001;202(1) 28-33.

[26] Bortchagovsky E, Colas des Francs G, Naber A, Fischer UC. On the optimum form of an aperture for a confinement of the optically excited electric near field. Journal of Microscopy 2008;229(2) 223-227.

[27] Li Z-B, Zhou W-Y, Kong X-T, Tian J-G. Polarization dependence and independence of near-field enhancement through a subwavelength circle hole. Optics Express 2010;18(6) 5854-5860.

[28] Li Z-B, Zhou W-Y, Kong X-T, Tian J-G. Near-field enhancement through a single subwavelength aperture with gaps inside. Plasmonics 2011;6(1) 149-154.

[29] Bouhelier A, Novotny L., Near field optical excitation and detection of surface plasmons. In Brongersma ML, Kik PG (eds.) Surface Plasmon Nanophotonics, Dordrecht The Netherlands: Springer; 2007 p. 139-153.

[30] Hartschuh A. Tip-enhanced near-field optical microscopy. Angewandte Chemie International Edition 2008;47 8178-8191.

[31] Stockman MI. Nanofocusing of optical energy in tapered plasmonic waveguides. Physical Review Letters 2004;93(13) 137404.

[32] Babadjanyan AJ, Margaryan NL, Nerkararyan KV. Superfocusing of surface polaritons in the conical structure. Journal of Applied Physics 2000;87(8) 3785-3788.

[33] Kurihara K, Otomo A, Syouji A, Takahara J, Suzuki K, Yokoyama S. Superfocusing modes of surface plasmon polaritons in conical geometry based on the quasi-separation of variables approach. Journal of Physics A – Mathematical and theoretical. 2007;40(41) 12479–12503.

[34] Vogel MW, Gramotnev DK. Optimization of plasmon nano-focusing in tapered metal rods. Journal of Nanophotonics 2008;2 1-17.

[35] Ruppin R. Effect of non-locality on nanofocusing of surface plasmon field intensity in a conical tip. Physics Letters A 2005;340(1-4) 299-302.

[36] Baida FI, Belkhir A. Superfocusing and light confinement by surface plasmon excitation through radially polarized beam. Plasmonics 2009;4(1) 51-59.
[37] Maier SA, Andrews SR, Martìn-Moreno L, Garcìa-Vidal FJ. Terahertz surface plasmon-polariton propagation and focusing on periodically corrugated metal wires. Physical Review Letters 2006;97 176805.
[38] Lee JS, Han S, Shirdel J, Koo S, Sadiq D, Lienau C, Park N. Superfocusing of electric or magnetic fields using conical metal tips: effect of mode symmetry on the plasmon excitation method. Optics Express 2011;19(13) 12342-12347.
[39] Ropers C, Neacsu CC, Elsaesser T, Albrecht M, Raschke MB, Lienau C. Grating-coupling of surface plasmons onto metallic tips: a nanoconfined light source. Nano Letters 2007;7(9) 2784-2788.
[40] Ding W, Andrews SR, Maier SA. Internal excitation and superfocusing of surface plasmon polaritons on a silver-coated optical fiber tip. Physical Review A 2007;75 063822.
[41] Janunts NA, Baghdasaryan KS, Nerkararyan KV, Hecht B. Excitation and superfocusing of surface plasmon polaritons on a silver-coated optical fiber tip. Optical Communications 2005;253(1-3) 118–124.
[42] Nerkararyan K, Abrahamyan T, Janunts E, Khachatryan R, Harutyunyan S. Excitation and propagation of surface plasmon polaritons on the gold covered conical tip. Physics Letters A 2006;350(1-2) 147-149.
[43] Babayan AE, Nerkararyan KV. The strong localization of surface plasmon polariton on a metal-coated tip of optical fiber. Ultramicroscopy 2007;107(12) 1136-1140.
[44] Nerkararyan KV, Hakhoumian AA, Babayan AE. Terahertz surface plasmon-polariton superfocusing in coaxial cone semiconductor structures. Plasmonics 2008;3 27-31.
[45] Abrahamyan T, Nerkararyan K. Surface plasmon resonance on vicinity of gold-coated fiber tip. Physics Letters A 2007;364(6) 494–496.
[46] Bouhelier A, Renger J, Beversluis MR, Novotny L. Plasmon-coupled tip-enhanced near-field optical microscopy. Journal of Microscopy 2003;210(3) 220-224.
[47] Chen W, Zhan Q. Numerical study of an apertureless near field scanning optical microscope probe under radial polarization illumination. Optics Express 2007;15(7) 4106-4111.
[48] Liu L, He S. Design of metal-cladded near-field fiber probes with a dispersive body-of-revolution finite-difference time-domain method. Applied Optics 2005;44(17) 3429-3437.
[49] Vaccaro L, Aeschimann L, Staufer U, Herzig HP, Dändliker R. Propagation of the electromagnetic field in fully coated near-field optical probes. Applied Physics Letters 2003;83(3) 584-586.
[50] Tortora P, Descrovi E, Aeschimann L, Vaccaro L, Herzig HP, Dändliker R. Selective coupling of HE_{11} and TM_{01} modes into microfabricated fully metal-coated quartz probes. Ultramicroscopy 2007;107(2-3) 158-165.
[51] Lotito V, Sennhauser U, Hafner C. Effects of asymmetric surface corrugations on fully metal-coated scanning near field optical microscopy tips Optics Express 2010;18(8) 8722-8734.
[52] Nakagawa W, Vaccaro L, Herzig HP. Analysis of mode coupling due to spherical defects in ideal fully metal-coated scanning near-field optical microscopy probes. Journal of the Optical Society of America A 2006;23(5) 1096-1105.

[53] Frey HG, Keilmann F, Kriele A, Guckenberger R. Enhancing the resolution of scanning near-field optical microscopy by a metal tip grown on an aperture probe. Applied Physics Letters 2002;81(26) 5030-5032.

[54] Taminiau TH, Segerink FB, Moerland RJ, Kuipers L(K), van Hulst NF. Near field driving of a optical monopole antenna. Journal of Optics A: Pure and Applied Optics 2007;9 S315-S321.

[55] Quong MC, Elezzabi AY. Offset-apertured near-field scanning optical microscope probes. Optics Express 2007;15(16) 10163-10174.

[56] Lotito V, Sennhauser U, Hafner C, Bona G-L. Fully metal-coated scanning near-field optical microscopy probes with spiral corrugations for superfocusing under arbitrarily oriented linearly polarized excitation. Plasmonics 2011;6 327-336.

[57] Lotito V, Sennhauser U, Hafner C, Bona G-L. A novel nanostructured scanning near field optical microscopy probe based on an adirectional asymmetry. Journal of Computational and Theoretical Nanoscience 2012;9 486-494.

[58] Roberts A. Electromagnetic theory of diffraction by a circular aperture in a thick, perfectly conducting screen. Journal of the Optical Society of America A 1987;4 1970-1983.

[59] Roberts A. Near zone fields behind circular apertures in thick, perfectly conducting screens. Journal of Applied Physics 1989;65 2896-2899.

[60] Dunn RC. Near-field scanning optical microscopy. Chemical Reviews 1999;99 2891-2927.

[61] Issa NA, Guckenberger R. Optical nanofocusing on tapered metallic waveguides. Plasmonics 2007;2 31-37.

[62] Ohtsu M. Progress in Nano-electro-optics III Industrial applications and dynamics of the nano-optical system. Berlin Heidelberg: Springer; 2005.

[63] Girard C, Dereux A. Near field optics theories Report on Progress in Physics 1996;59 657-699.

[64] Girard C. Near fields in nanostructures. Report on Progress in Physics 2005;68 1883-1933.

[65] Moar PN, Love JD, Laudoceur F, Cahill LW. Waveguide analysis of heat-drawn and chemically etched probe tips for scanning near-field optical microscopy. Applied Optics 2006;45(25) 6442-6456.

[66] Novotny L, Pohl DW, Hecht B. Scanning near field optical probe with ultrasmall size. Optics Letters 1995;20 970-972.

[67] Wang XQ, Wu S-F, Jian G-S, Pan S. The advantages of a pyramidal probe tip entirely coated with a thin metal film for SNOM. Physics Letters A 2003;319(5-6) 514-517.

[68] Frey HG, Bolwien C, Brandenburg A, Ros R, Anselmetti D. Optimized apertureless optical near-field probes with 15 nm optical resolution. Nanotechnology 2006;17(13) 3105-3110.

[69] Nakagawa W, Vaccaro L, Herzig HP, Hafner C. Polarization mode coupling due to metal-layer modifications in apertureless near-field scanning optical microscopy probes. Journal of Computational and Theoretical Nanoscience 2007;4 692–703.

[70] Bondeson A, Rylander T, Ingelström P. Computational Electromagnetics. New York: Springer; 2005.

[71] Rao SS. The finite element method in engineering. Burlington MA USA: Butterworth-Heinemann; 1999.

[72] Lotito V, Sennhauser U, Hafner C. Finite element analysis of asymmetric scanning near field optical microscopy probes. Journal of Computational and Theoretical Nanoscience 2010;7 1596–1609.

[73] Janunts NA, Nerkararyan KV. Modulation of light radiation during input into a waveguide by resonance excitation of surface plasmons. Applied Physics Letters 2001;79(3) 299–301.

[74] Lotito V, Sennhauser U, Hafner C. Numerical analysis of novel asymmetric SNOM tips. PIERS Online 2011;7(4) 394-400.

[75] Betzig E, Chichester RJ. Single molecules observed by near-field scanning optical microscopy Science 1993;262(5138) 1422-1425.

[76] Hollars W, Dunn RC. Probing single molecule orientations in model lipid membranes with near-field scanning optical microscopy. Journal of Chemical Physics 2000;112(18) 7822-7830.

[77] Veerman JA, M. Garcia-Parajo F, Kuipers L, van Hulst NF. Single molecule mapping of the optical field distribution of probes for near field microscopy. Journal of Microscopy 1999;194(2-3), 477-482.

[78] Moerland RJ, van Hulst NF, Gersen H, Kuipers L. Probing the negative permittivity perfect lens using near-field optics and single molecule detection. Optics Express 200513(5) 1604-1614.

[79] Lotito V, Sennhauser U, Hafner C, Bona G-L. Interaction of an asymmetric scanning near field optical microscopy probe with fluorescent molecules. Progress In Electromagnetics Research 2011;121 281-299.

[80] Heinzelmann H, Freyland JM, Eckert R, Huser T, Schürmann G, Noell W, Staufer U, de Rooij NF. Towards better scanning near-field optical microscopy probes – progress and new developments. Journal of Microscopy 1999;194(2-3) 365-368.

[81] Muranishi M, Sato K, Osaka S, Kikukawa A, Shintani T, Ito K. Control of aperture size of optical probes for scanning near-field optical microscopy using focused ion beam technology. Japanese Journal of Applied Physics 1997;36(7B), L942-L944

[82] Lacoste T, Huser T, Prioll R, Heinzelmann H. Contrast enhancement using polarization-modulation scanning near-field optical microscopy (PM-SNOM). Ultramicroscopy 1998;71 333–340.

[83] Veerman JA, Otter AM, Kuipers L, van Hulst NF. High definition aperture probes for near-field optical microscopy fabricated by focused ion beam milling. Applied Physics Letters 1998;72 115–3117.

[84] Pilevar S, Edinger K, Atia A, Smolyaninov I, Davis C. Focused ion-beam fabrication of fiber probes with well-defined apertures for use in near-field scanning optical microscopy. Applied Physics Letters 1998;72 3133–3135.

[85] Zhang Y, Dhawan A, Vo-Dinh T. Design and fabrication of fiber-optic nanoprobes for optical sensing. Nanoscale Research Letters 2011;6(18) 6 pp.

[86] Yang S, Chen W, Nelson RL, Zhan Q. Miniature circular polarization analyzer with spiral plasmonic lens. Opics Letters 2009;34(20) 3047-3049.

[87] Chen W, Abeysinghe DC, Nelson RL, Zhan Q. Experimental confirmation of miniature spiral plasmonic lens as a circular polarization analyzer. Nano Letters 2010;10 2075–2079.

Chapter 6

Propagating Surface Plasmons and Dispersion Relations for Nanoscale Multilayer Metallic-Dielectric Films

Henrique T. M. C. M. Baltar, Krystyna Drozdowicz-Tomsia and Ewa M. Goldys

Additional information is available at the end of the chapter

http://dx.doi.org/10.5772/51218

1. Introduction

The study of propagating surface plasmons (PSPs) is an important aspect of the understanding of the interaction between light and metallic surfaces. One of the key concepts related to PSPs, is the dispersion relation. This relation is the basis for understanding of coupling of light to PSPs, by using special approaches to match the wavevector. Moreover, it can be used to predict the matching of localised surface plasmons (LSPs) and PSPs to achieve highly enhanced electromagnetic field and/or tailored transmission.

This chapter is concerned with a detailed analysis of propagating surface plasmons (PSPs), and the calculations of the dispersion relations in nanoscale multilayer metallic-dielectric films, starting from the fundamental Maxwell's equations. Furthermore, we discuss the symmetric IMI (insulator-metal-insulator) and MIM (metal-insulator-metal) geometries, as well as their asymmetric variants. We will also describe the PSP in the IIMI (insulator-insulator-metal-insulator) geometry.

There is a vast literature on the subject, however, numerous authors assume the materials as lossless by using the Drude's model without damping. We initially model the dielectric function of the metals in our multilayer structures by using the lossless Drude's model. Then, we model the metals by applying complex values of permittivities acquired experimentally. The Drude's model without damping, with its simplicity, provides a basic understanding, as all the calculated quantities are purely real or imaginary numbers. Nevertheless, real structures behave differently from the predictions of this simple model. The dispersion relation changes markedly, when complex permittivities are used, and the changes are more pronounced around the surface plasmon frequency. This leads to wavevector limitations and to the existence of a region of anomalous dispersion, called quasi-bound mode [4]. Therefore, it is necessary to extend the analysis to include complex permittivity, which has been tabulated for the most common materials.

There are some previous studies on the behaviour of the IMI and MIM structures using complex permittivity, but the discussions are restrained to symmetric structures and/or do not show the entire dispersion curve [4, 5, 32]. In many cases, such symmetries are not applicable to the real experiments. For instance, assays with analytes deposited over a metallic layer on a substrate are, in this sense, an asymmetric problem.

Other experimentally important structure is the IIMI. Such structure is used for surface-enhanced fluorescence (SEF). A dielectric separation layer between metal and fluorophore can be introduced to maximise the fluorescence by reducing the possibility of quenching of fluorophores in direct contact with metal surfaces [6]. To the best of our knowledge, there are no published studies of the dispersion relation for such type of asymmetric structures. These new calculations are presented in this work.

2. Permittivity of metals

The Drude's model, taking into account only free electrons, is applicable to metals and leads to the following expression for the relative permittivity [16, 23]:

$$\varepsilon_r = 1 - \frac{\omega_p^2}{\omega^2 + i\Gamma\omega} \tag{1}$$

Here, ω_p is called the plasma frequency, and the damping constant Γ is the electron scattering rate (the inverse of the collision time for the conduction electrons). When a metal is assumed lossless (very long time between collisions, $\Gamma \to 0$), the permittivity becomes real:

$$\text{Drude's permittivity without damping:} \quad \varepsilon_r = 1 - \frac{\omega_p^2}{\omega^2} \tag{2}$$

In this study, we also model the metal permittivity by the tabulated values of silver from Lynch and Hunter [20]. In the case of asymmetric MIM geometry, we will also use the tabulated permittivity of gold from Johnson and Christy [13]. In Figure 1, we plotted the real and imaginary parts of the permittivities of silver and gold. For the Drude's model without damping, we used the plasma frequency of 13.1 Prad (2.08 PHz, $\lambda_p = 144$ nm) for silver and 13.4 Prad (2.13 PHz, $\lambda_p = 141$ nm) for gold.

3. Theoretical framework

3.1. Source-free Maxwell's equations in time-harmonic regime

The Maxwell's equations hold for any arbitrary time-dependence of the electric field. For materials in a linear regime we can consider these equations for each frequency component separately [1]. The time-dependent electric or magnetic fields can be written as the real part of a complex field, denoted by F:

$$\vec{F}(\vec{r}, t) = \vec{F}_0(\vec{r}) \cos\left(\vec{k} \cdot \vec{r} - \omega t + phase\right) = Re\left[\vec{F}(\vec{r})\, e^{-i\omega t}\right] \tag{3}$$

[1] The non-linear regime is when the material properties such as permittivity and permeability depend on frequency and intensity of the field, and will not be considered here.

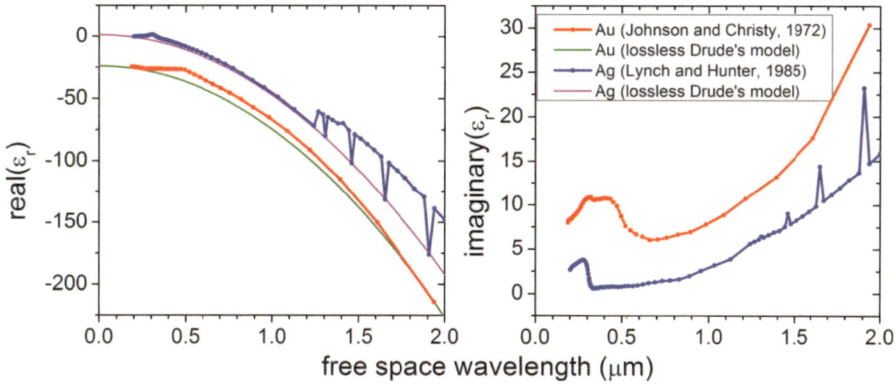

Figure 1. Permittivities of metals. Silver is modelled by the Drude's model without damping ($\omega_p = 13.1\,\text{Prad}$, $\lambda_p = 144\,\text{nm}$) and the tabulated values of Lynch and Hunter (1985) [20]. Gold is modelled by the Drude's model without damping ($\omega_p = 13.4\,\text{Prad}$, $\lambda_p = 141\,\text{nm}$) and the tabulated values of Johnson and Christy (1972) [13]. For easiness of visualisation, the real part of gold permittivity was shifted -25 units, and the imaginary part, $+5$ units.

Thus a field can be described by a vector phasor $\vec{F}(\vec{r})$, which contains information on the magnitude, direction and phase [3].

In plasmonics, it is common to deal with structures without any external current sources and/or charges. In this situation, we apply the source-free macroscopic Maxwell's equations, in which the displacement charges and currents are fully incorporated in the permittivities and permeabilities of the materials.

In the time-harmonic regime, the source-free macroscopic Maxwell's equations can be written as [3]

$$\begin{aligned} \text{Gauss' law of electricity:} &\quad \vec{\nabla} \cdot \vec{D}(\vec{r}) = 0 \\ \text{Gauss' law of magnetism:} &\quad \vec{\nabla} \cdot \vec{B}(\vec{r}) = 0 \\ \text{Faraday's law of induction:} &\quad \vec{\nabla} \times \vec{E}(\vec{r}) = i\omega\vec{B}(\vec{r}) \\ \text{Ampère's circuital law:} &\quad \vec{\nabla} \times \vec{H}(\vec{r}) = -i\omega\vec{D}(\vec{r}) \end{aligned} \tag{4}$$

3.2. Boundary conditions

The source-free Maxwell's equations in their integral form applied to the interface between two different media (1 and 2) lead to four boundary conditions [10, 29]:

$$\begin{aligned} \hat{n} \cdot \left(\vec{D}_2 - \vec{D}_1\right) &= 0 \\ \hat{n} \cdot \left(\vec{B}_2 - \vec{B}_1\right) &= 0 \\ \hat{n} \times \left(\vec{E}_2 - \vec{E}_1\right) &= 0 \\ \hat{n} \times \left(\vec{H}_2 - \vec{H}_1\right) &= 0 \end{aligned} \tag{5}$$

These boundary conditions describe the continuity of the components of the fields perpendicular and parallel to the interface. Moreover, the momentum parallel to the interface is also continuous [12]. In such case, one can write the boundary conditions as

$$\vec{D}_{2\perp} = \vec{D}_{1\perp} \tag{6a}$$
$$\vec{B}_{2\perp} = \vec{B}_{1\perp} \tag{6b}$$
$$\vec{E}_{2\|} = \vec{E}_{1\|} \tag{6c}$$
$$\vec{H}_{2\|} = \vec{H}_{1\|} \tag{6d}$$
$$\vec{k}_{2\|} = \vec{k}_{1\|} \tag{6e}$$

3.3. Electromagnetic waves

The Maxwell's equations represent a coupled system of differential equations. With some algebraic manipulation, it is possible to transform these four equations (4) into two uncoupled ones, representing the homogeneous vector wave equations for the electric and magnetic fields. In the time-harmonic regime, these wave equations become the homogeneous vector Helmholtz' equations [3]:

$$\begin{aligned} \nabla^2\vec{E} + k^2\vec{E} = 0 \\ \nabla^2\vec{H} + k^2\vec{H} = 0 \end{aligned} \tag{7}$$

In which

$$k = \frac{\omega}{v} = nk_0 = \sqrt{\mu\varepsilon}k_0 \tag{8}$$

is the wavenumber in a medium with index of refraction $n = \sqrt{\mu\varepsilon}$, and $k_0 = \omega/c$ is the free-space wavenumber.

3.4. Light polarisation

Due to the linearity of the homogeneous Helmholtz' equations, we can divide the problem of a wave incident to a planar surface into two parts: for an s- and for a p-polarised wave[2]. These waves are depicted in Figure 2, in which we define the plane of incidence as the x-z plane.

In these conditions, for a plane wave propagating in the x-z plane, the magnitudes of the phasor fields are dependent on the coordinates x and z, but constant along the y direction. In order to find the surface plasmons — which are waves bound to the interfaces —, we can write the magnitudes of the electric and magnetic phasor fields as

$$\begin{aligned} \vec{E}(\vec{r}) = \vec{E}(x,z) = \vec{E}(z)\,e^{i\beta x} \\ \vec{H}(\vec{r}) = \vec{H}(x,z) = \vec{H}(z)\,e^{i\beta x} \end{aligned} \tag{9}$$

Where β is the propagation constant, in the x direction. As the parallel wavevector to the interface must be constant (Eq. 6e), we have

$$\beta = k_{x,1} = k_{x,2} \tag{10}$$

[2] The terms s- and p-polarised are the initials of the German words: *senkrecht* (perpendicular) and *parallel*. In this nomenclature, s or p means that the electric field is perpendicular (s) or parallel (p) to the plane of incidence [23]. The s-polarisation is also called TE (transverse electric); and the p-polarisation, TM (transverse magnetic) [3, 29].

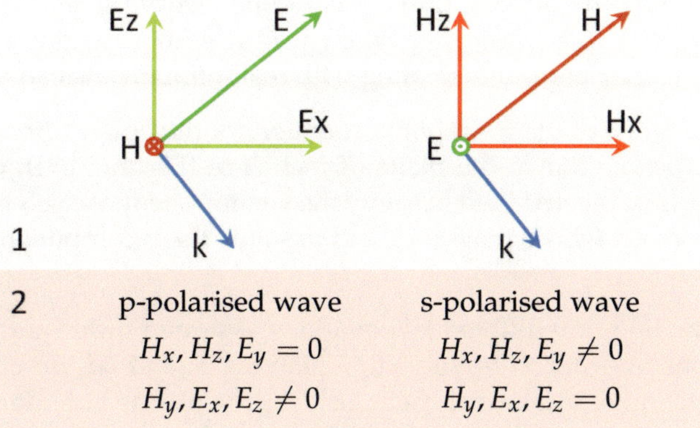

Figure 2. Interface between two media (1 and 2), showing the polarisation vectors of a p-polarised (left) and an s-polarised waves (right). The plane of incidence coincides with the page. A p-polarised wave (TM) is characterised by the magnetic field perpendicular to the plane of incidence; while an s-polarised (TE) wave is characterised by the electric field perpendicular to the plane of incidence.

In this case, the differential operators with respect to the coordinates x and y can be written as

$$\frac{\partial}{\partial x} = i\beta \qquad \frac{\partial}{\partial y} = 0 \tag{11}$$

This makes it possible to express the divergence and Laplacian of any of the fields as

$$\vec{\nabla} \times \vec{F}(x,z) = \begin{vmatrix} e_x & e_y & e_z \\ i\beta & 0 & \frac{\partial}{\partial z} \\ F_x & F_y & F_z \end{vmatrix} = \left(-\frac{\partial F_y}{\partial z}\right)\hat{e}_x + \left(\frac{\partial F_x}{\partial z} - i\beta F_z\right)\hat{e}_y + \left(i\beta F_y\right)\hat{e}_z \tag{12a}$$

$$\nabla^2 \vec{F}(x,z) = \frac{\partial^2 \vec{F}}{\partial z^2}(x,z) - \beta^2 \vec{F}(x,z) \tag{12b}$$

3.4.1. Revisiting Helmholtz' equations and boundary conditions

The Laplacian in Equation 12b and the fields in Equation 9 can be used to simplify the Helmholtz' equations (7) to one dimension [21]:

$$\frac{\partial^2 \vec{E}}{\partial z^2}(z) - \gamma_z^2 \vec{E}(z) = 0 \tag{13a}$$

$$\frac{\partial^2 \vec{H}}{\partial z^2}(z) - \gamma_z^2 \vec{H}(z) = 0 \tag{13b}$$

With

$$\gamma_j^2 = -k_{z,j}^2 = \beta^2 - \varepsilon_j k_0^2 \tag{14}$$

$k_{z,j}$ is the wavevector along the z axis in the medium j $(= 1,2)$. We define the propagation constant γ as

$$\gamma = ik_z \tag{15}$$

This definition is commonly done in transmission-line theory [3], and it is also used in most of the literature about PSPs. We note that, when dealing with a theoretical situation of lossless materials, working with γ implies in dealing with real numbers instead of imaginary.

It is sufficient to solve one of the Helmholtz' equation for the polarisation whose field (electric or magnetic) is perpendicular to the plane of incidence (Figure 2). In other words, for the s-polarised wave, whose electric field has only the y-component, we solve Equation 13a; while for the p-polarised wave, whose magnetic field has only the y-component, we solve Equation 13b.

The Helmholtz' equations are differential equations of second order, and each one needs two boundary conditions in order to be solved. Equations 6c and 6d, describing the continuity of the parallel electric and parallel magnetic fields, provide one boundary condition for each of the Helmholtz' equations. Below, we provide one additional boundary condition for each field (electric and magnetic).

3.4.2. *s-polarisation*

Expanding the divergence (Equation 12a) in the Faraday's law (Equation 4) for an s-polarised wave (Figure 2), we get

$$\left(-\frac{\partial E_y}{\partial z}\right)\hat{e}_x + \left(i\beta E_y\right)\hat{e}_z = \left(i\omega B_x\right)\hat{e}_x + \left(i\omega B_z\right)\hat{e}_z \tag{16}$$

- **z component**: The continuity of the perpendicular $\vec{B}$ field (Equation 6b), B_z, implies the continuity of βE_y. As β is continuous (Equation 10), this condition is equivalent to the previous condition of continuity of E_y (Equation 6c).
- **x component**: For homogeneous isotropic materials, $\vec{B} = \mu\vec{H}$. Therefore,

$$i\omega H_x = -\frac{1}{\mu}\frac{\partial E_y}{\partial z} \tag{17}$$

This equation, together with the continuity of the parallel magnetic field (Equation 6d), leads to the continuity of $\frac{1}{\mu}\frac{\partial E_y}{\partial z}$.

In summary, the conditions for the s-polarised wave are:

$$\text{Continuity conditions for s-polarised wave:} \quad \begin{aligned} E_{y,1} &= E_{y,2} \\ \frac{1}{\mu_1}\frac{\partial E_{y,1}}{\partial z} &= \frac{1}{\mu_2}\frac{\partial E_{y,2}}{\partial z} \end{aligned} \tag{18}$$

3.4.3. *p-polarisation*

For the p-polarised wave, we use the same approach as for the s-polarised one. Expanding the divergence (Equation 12a), in the Ampère's law (Equation 4), we obtain

$$\left(-\frac{\partial H_y}{\partial z}\right)\hat{e}_x + \left(i\beta H_y\right)\hat{e}_z = -\left(i\omega D_x\right)\hat{e}_x - \left(i\omega D_z\right)\hat{e}_z \tag{19}$$

- **z component**: The continuity of the perpendicular $\vec{D}$ field (Equation 6a), D_z, implies the continuity of βH_y. As β is continuous (Equation 10), this condition is equivalent to the previous condition of continuity of H_y (Equation 6d).
- **x component**: For homogeneous isotropic materials, $\vec{D} = \varepsilon \vec{E}$. Therefore,

$$i\omega E_x = \frac{1}{\varepsilon} \frac{\partial H_y}{\partial z}. \tag{20}$$

This equation together with the continuity of the parallel electric field (Equation 6c) leads to the continuity of $\frac{1}{\varepsilon}\frac{\partial H_y}{\partial z}$ [2].

In summary, the conditions for the p-polarised wave are:

$$\text{Continuity conditions for p-polarised wave:} \quad \begin{aligned} H_{y,1} &= H_{y,2} \\ \frac{1}{\varepsilon_1}\frac{\partial H_{y,1}}{\partial z} &= \frac{1}{\varepsilon_2}\frac{\partial H_{y,2}}{\partial z} \end{aligned} \tag{21}$$

4. PSP at a planar interface

Now we discuss the conditions satisfied by a wave propagating at a planar interface between two half-spaces. So as to solve the problem of an s-polarised wave propagating at an interface, we will use Equation 13a, and the two boundary conditions stated in Equations 18. For the p-polarised wave, we will use Equation 13b and the boundary conditions in Equations 21.

4.1. s-polarised wave

In order to solve Equation 13a, we postulate a solution of a certain form, and verify that it indeed satisfies our equation and boundary conditions. We are looking for a wave bound to the interface, that vanishes away from it. With this in mind, and inspecting Equation 13a, we postulate our solution to be

$$\vec{E}\,(z) = \begin{cases} E_1 e^{-\gamma_1 z}\hat{e}_y, & z > 0 \\ E_2 e^{\gamma_2 z}\hat{e}_y, & z < 0 \end{cases} \tag{22}$$

with the real parts of γ_j positive in order to have a field decaying away from the interface. Applying the boundary conditions to this ansatz, we find

$$\begin{aligned} E_1 &= E_2 \\ \frac{\gamma_1}{\mu_1}E_1 &= -\frac{\gamma_2}{\mu_2}E_2 \end{aligned} \tag{23}$$

These conditions are equivalent to [27]

$$\frac{\gamma_1}{\mu_1} = -\frac{\gamma_2}{\mu_2} \tag{24}$$

For the materials found in nature, the permeability at optical frequencies is close to unity[3] [27]. This leads to $\gamma_1 = -\gamma_2$, what is impossible as the real components of both γ_1 and γ_2 are

[3] The relative permeability of unity in the optical regime for most of the materials implies that the index of refraction is $n = \sqrt{\varepsilon\mu} = \sqrt{\varepsilon}$.

positive. However, magnetic surface plasmons can be achieved by the use of metamaterials [24], for which the magnetic permeability can be engineered to be negative at frequencies higher than in natural materials. For example, ref. [19] shows magnetic plasmon propagation at infra-red along a chain of split-ring resonators.

4.2. p-polarised wave

Similarly to the case of an s-polarised wave, we propose the following solution for the Equation 13b:

$$\vec{H}(z) = \begin{cases} H_1 e^{-\gamma_1 z} \hat{e}_y, & z > 0 \\ H_2 e^{\gamma_2 z} \hat{e}_y, & z < 0 \end{cases} \tag{25}$$

Applying the boundary conditions:

$$\begin{aligned} H_1 &= H_2 \, (= H_0) \\ \frac{\gamma_1}{\varepsilon_1} H_1 &= -\frac{\gamma_2}{\varepsilon_2} H_2 \end{aligned} \tag{26}$$

These conditions lead to [7, 21, 23, 27, 28, 34]

$$\frac{\gamma_1}{\varepsilon_1} = -\frac{\gamma_2}{\varepsilon_2} \tag{27}$$

This condition combined with Equation 14 and the index of refraction[3] $n = \sqrt{\varepsilon}$ leads to the dispersion relation[4] [4, 7, 21, 23, 34]:

$$\text{Dispersion relation for half-spaces:} \quad \beta^2 = k_0^2 \frac{\varepsilon_1 \varepsilon_2}{\varepsilon_1 + \varepsilon_2} \tag{28}$$

Substituting this β^2 in Equation 14, one obtains the normal component of the wavevector [4, 23]:

$$k_{z,j}^2 = -\gamma_j^2 = k_0^2 \frac{\varepsilon_j^2}{\varepsilon_1 + \varepsilon_2} \tag{29}$$

4.2.1. Conditions for PSPs

In this subsection, we limit our discussion on the conditions for the existence of PSPs to lossless materials. Subsequently, we will discuss the effect of complex permittivities on β and γ. In lossless materials, all permittivities as well as the propagation constants (Equations 28 and 29) are real. Without loss of generality, one can define the surface wave propagating at the direction $+x$. Hence, β is positive (Equation 28). 1) This requires that $\varepsilon_1 \varepsilon_2$ and $\varepsilon_1 + \varepsilon_2$ are both positive or both negative. Furthermore, as we are looking for the modes bound to the interface, we want the field propagating at the boundary, but not along the z axis. Therefore, γ must be positive real (Equation 25) (or $k_{z,1}$ and $k_{z,2}$ must be positive imaginary, Equation 15), leading to evanescent, exponentially decaying fields away from the interface. 2) This requirement is achievable only if $\varepsilon_1 + \varepsilon_2$ is negative (Equation 29). As a result, the conditions for the existence of PSPs are [23]

[4] The dispersion relation is an equation that relates the wavevector along the direction of propagation and the frequency.

$$\text{Conditions for PSPs:} \quad \begin{array}{c} \varepsilon_1 \varepsilon_2 < 0 \\ \varepsilon_1 + \varepsilon_2 < 0 \end{array} \tag{30}$$

The first condition states that one of the permittivities must be positive, and the other one, negative. The second condition, indicates that the absolute value of the negative permittivity must be higher than the positive one [23].

As already seen in Figure 1 and Equation 2, lossless Drude's metals have negative permittivities for $\omega < \omega_p$ (or $\lambda > \lambda_p$). The absolute values of these permittivities increase with the wavelength, exceeding the values found in dielectrics. Semiconductors [26] and graphene [9, 25] have also been used for their plasmonic properties in the terahertz range. Although in graphene there is no strict surface plasmon, due to its two-dimensional geometry, which practically prevents transverse oscillations of electrons [11]. As metals are the most common materials used for PSPs, a planar interface that supports bound waves will henceforth be called insulator-metal (IM) geometry.

Other important consideration to take into account, when calculating the wavevectors, is the position of β in relation to the light line. If β is at right of the light line ($\beta^2 > \varepsilon_j k_0^2$) (Equation 14) in the medium j, γ_j is real ($k_{z,j}$ is imaginary). If β is at left ($\beta^2 < \varepsilon_j k_0^2$), γ is imaginary ($k_{z,j}$ is real), leading to the propagation of the wave away from the interface [2].

4.2.2. *Complex β and γ in lossy materials*

When the permittivity of any medium in the structure is complex, the normal and perpendicular wavevectors components are complex. The magnetic field (Equations 9 and 25) can be re-written in the following form:

$$\vec{H}(\vec{r}) = \vec{H}(z)\, e^{i\beta x} = \begin{cases} H_0 e^{i\beta x} e^{-\gamma_1 z} \hat{e}_y, & z > 0 \\ H_0 e^{i\beta x} e^{\gamma_2 z} \hat{e}_y, & z < 0 \end{cases} \tag{31}$$

At the interface ($z = 0$), this converges to

$$\vec{H}(x, y, 0) = H_0 e^{i\beta x} \hat{e}_y = H_0 e^{iRe(\beta)x} e^{-Im(\beta)x} \hat{e}_y \tag{32}$$

Looking at this equation, we remind that the meaning of the (positive) real part of β is that the wave propagates on the interface towards $+x$. If the imaginary part of β is negative, the field increases exponentially (Equation 32), which is un-physical in our situation. Consequently, the imaginary part of β must be positive, leading to an exponential decay with x. Lossless materials ($Im\,(\beta = 0)$) imply an endless propagation of the PSP along the interface. One can define the propagation length as the distance the wave travels along the interface until its energy decays to e^{-1} (≈ 0.368) of its original value. As the energy is proportional to the square of the field ($energy \propto |H|^2 \propto e^{-2Im(\beta)x}$), the propagation length is [7, 21, 34]

$$propagation\ length \equiv L_{sp} = \frac{1}{2Im\,(\beta)} \tag{33}$$

Now we analyse the component of the wavevector that is perpendicular to the interface, γ_j. For a fixed x, Equation 31 can be re-written as

$$\vec{H}(constant, y, z) = H_0 e^{i\beta x} \times \begin{cases} e^{-Re(\gamma_1)z} e^{-iIm(\gamma_1)z} \hat{e}_y, & z > 0 \\ e^{Re(\gamma_2)z} e^{iIm(\gamma_2)z} \hat{e}_y, & z < 0 \end{cases} \tag{34}$$

For bound modes, we remind that $Re(\gamma_j)$ must be positive (or $Im(k_{z,j})$ must be negative). Equation 14 tells γ has two solutions in anti-phase in the complex plane. One must choose, therefore, the solutions in the first and fourth quadrants.

4.2.3. Dispersion curve

Applying the lossless Drude's model (Equation 2) to the dispersion relation for the IM geometry (Equation 28), one obtains the result presented in Figure 3A. In this figure, we plotted the solution for two different dielectrics: free-space and a material with $n = 1.5$ throughout the entire spectral range. The dispersion is divided into two branches with a region between them where β is imaginary. In this region, called plasmon bandgap [4], the wave is evanescent and it reflects back to the dielectric. For the Drude's model without damping (Equation 2), the index of refraction of a metal is real when the angular frequency is larger than the plasma frequency ($\omega > \omega_p$). Therefore, the material is transparent and the wave propagates freely [12, 28]. This radiation mode is called Brewster mode [28]. Under the plasmon bandgap, there is a region limited by what is called the surface plasmon frequency, the frequency in which β diverges. It occurs when the permittivity of the metal (ε_m) cancels that one of the dielectric (ε_d) (Equation 28) [4, 21, 34]:

$$\varepsilon_m = -\varepsilon_d \Rightarrow \omega_{sp} = \frac{\omega_p}{\sqrt{1+\varepsilon_d}} \tag{35}$$

The higher the index of refraction of the dielectric is, the lower the surface plasmon frequency is. From this frequency down, light propagates in a bound mode. For a very high wavevector[5], the electric field has an electrostatic character, as the wavelength along the interface ($= 2\pi/\beta$) tends to zero. Likewise, the group velocity ($d\omega/dk$) [21] and the decay lengths into the metal and dielectric ($= 1/Re(\gamma_j)$) tend to zero. This high confinement of light at the interface indicates a true surface mode, called Fano mode [28]. For very low frequencies, the dispersion converges to the light line, in a regime called Sommerfeld-Zenneck waves [21, 28].

Using the complex experimental values of permittivity, the dispersion relation (Equation 28) is shown is Figure 3B. As the metal is no longer modelled as lossless, β is complex, leading to a finite propagation length (Equation 33), drawn in Figure 3C. Here, the wavenumber does not diverge as previously. It bends backwards filling the region previously called plasmon bandgap and connects to the Brewster mode. This region of anomalous dispersion (negative phase velocities $= d\omega/dk$) is called quasi-bound mode [4].

In the infra-red, the imaginary part of the permittivity of silver is very small compared to the real part. As a consequence, the propagation length is around 10–1000 µm. In the region of the quasi-bound mode, the imaginary component reaches values comparable to the real component (up to 0.83 of the real part for n=1, up to 1.9 for n=1.5), leading to a decrease in the propagation length down to 10–100 nm.

5. Physical interpretation

An electromagnetic wave inside a material induces polarisation in the medium. The coupled excitation resulting from the interaction between an incident electromagnetic wave and the

[5] In surface plasmons, the wavevector is higher than it would be in the dielectric, leading to a wavelength shorter than the one in the dielectric. This has applications in super-resolution light microscopy [15, 33].

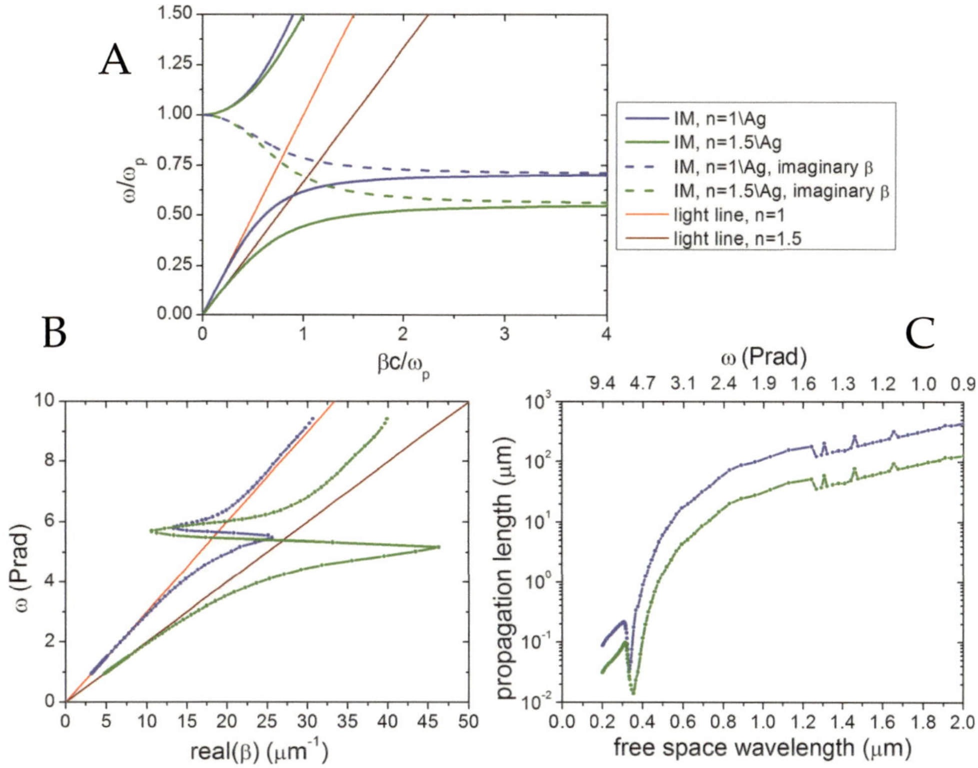

Figure 3. Dispersion curve and propagation length (Equation 28) for PSPs on an interface between half-spaces of a metal and insulators with refractive indices of 1 or 1.5. A) Drude's metal without damping. The imaginary β in the bandgap is also shown. B) Dispersion curve and C) propagation length for tabulated values of the metal (silver) permittivity.

material polarisation is called polariton [22, 31]. The way the coupling occurs is responsible for permittivity, permeability and light velocity being different from those ones in vacuum [22]. When the polariton is restricted to a small region around an interface, it is called surface polariton [31]. There are diverse types of polaritons according to the induced excitation, such as phonon-polariton [8], exciton-polariton [30], plasmon-polariton etc. The latter is the interest in the context of PSPs.

Plasmon is technically a quantum of charge density oscillation in a plasma [18]. The plasmons can be divided into two main types: bulk plasmons — arising from fluctuations of the free charge density inside the material which propagate as a longitudinally-polarised charge-density wave [18] — and surface plasmons (SPs) — which propagate along the interfaces of materials under specific circumstances. The SPs can be subdivided in localised surface plasmons (LSPs) and propagating surface plasmons (PSPs). In the former, the charge oscillations are confined to a nanostructure; while in the latter, the charge oscillation propagates relatively large distances, from tens of nanometers to hundreds of micrometers (Figure 3C).

We showed in the previous section that electromagnetic waves can propagate along the interfaces between dielectrics and metals. In response to an incidence electric field, the electrons of the metal induce an opposing field determined by the dielectric function of the metal. The fact that in metals the charge density is confined to a small region close to the surface, leads to high enhancement of the electric field, which can be as high as about 100 times[6] [27].

6. Exciting PSPs

As demonstrated previously, the wavenumber of a PSP is higher than the wavenumber in the dielectric (Figure 3A and B). Hence, light incident at an IM interface typically cannot excite PSPs. In order to do so, it is necessary to artificially increase the wavenumber of the light, so that the wavevectors in both materials can be matched. Some of the techniques to excite PSPs include the Otto and Kretschmann configurations, and the application of a grating or other periodic structure.

The Otto and Kretschmann configurations [7, 23, 27, 34] relies on the proximity of a dielectric with a higher index of refraction than the dielectric of the interface where the PSP is being excited. As the wavenumber is directly proportional to the index of refraction (Equation 8), one can use the evanescent waves that arise from a total internal reflection on a high-refractive index material to excite PSPs. In the Otto configuration, the reflectance comes from the side of the interface dielectric; while in the Kretschmann one, the evanescent waves come across a thin metal layer to excite the interface on the other side.

Diffraction gratings provide light with an additional momentum on the grating axis due to the spatial periodicity, allowing the coupling of light to PSPs [7, 27]. For a one-dimensional lattice, the additional momentum is given by

$$\Delta k = m\frac{2\pi}{a} \tag{36}$$

Where a is the period of the grating, and m is an integer.

7. PSP at a thin layer

The actual geometry of a thin layer (medium 2) between two media (1 and 3) is presented in Figure 4A. As previously discussed for an IM geometry, only p-polarised wave can excite PSP in naturally-occurring materials. This statement is also valid for other geometries. In a p-polarised wave, there is an electric field perpendicular to the surface (Figure 2). As D_z is continuous (Equation 6a), E_z changes, resulting in the creation of charge density at the interface. In an s-polarised wave, the electric field has no component parallel to the surface, therefore it is continuous and it does not produce charge density at the surface [14]. For modes bound to the interface in p-polarisation, we postulate the magnetic field — stronger at the interface and exponentially decaying away from it — as

$$H_y(z) = \begin{cases} H_1 e^{-\gamma_1 z}, & z > \frac{a}{2} \\ H_2^- e^{-\gamma_2 z} + H_2^+ e^{\gamma_2 z}, & \frac{a}{2} > z > -\frac{a}{2} \\ H_3 e^{\gamma_3 z}, & -\frac{a}{2} > z \end{cases} \tag{37}$$

Applying the continuity conditions (Equations 21) to the ansatz given in Equation 37, we

[6] This enhancement of the electric field may be applied for surface-enhanced spectroscopies, such as surface-enhanced Raman spectroscopy (SERS), surface-enhanced fluorescence (SEF) [17] and surface-enhanced infra-red absorption (SEIRA) [1].

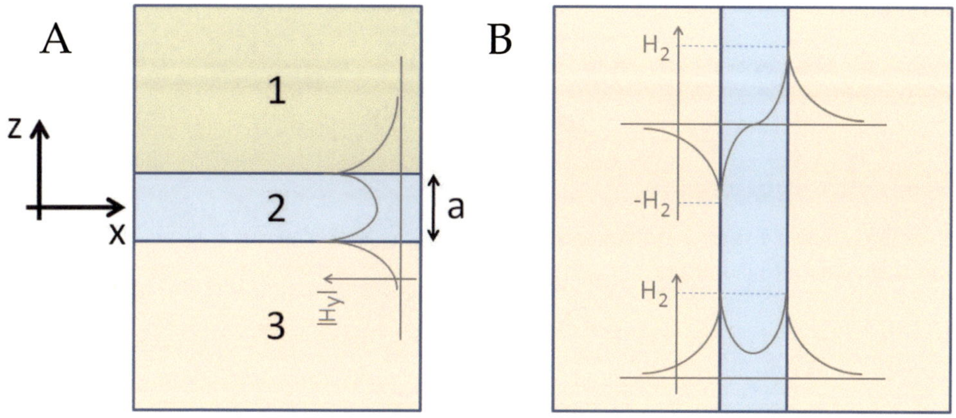

Figure 4. Planar thin layer geometry. A) Geometry of a planar thin layer surrounded by two half-spaces. It also shows the absolute value of the magnetic field expected for modes bound to the interfaces. B) The field magnitudes for the symmetric and anti-symmetric PSPs in a symmetric thin layer.

obtain four equations, two for each boundary.

- Boundary 1–2 ($z = a/2$)

$$H_1 e^{-\gamma_1 a/2} = H_2^- e^{-\gamma_2 a/2} + H_2^+ e^{\gamma_2 a/2} \tag{38}$$

$$\frac{\gamma_1}{\varepsilon_1} H_1 e^{-\gamma_1 a/2} = \frac{\gamma_2}{\varepsilon_2} H_2^- e^{-\gamma_2 a/2} - \frac{\gamma_2}{\varepsilon_2} H_2^+ e^{\gamma_2 a/2} \tag{39}$$

- Boundary 2–3 ($z = -a/2$)

$$H_3 e^{-\gamma_3 a/2} = H_2^- e^{\gamma_2 a/2} + H_2^+ e^{-\gamma_2 a/2} \tag{40}$$

$$\frac{\gamma_3}{\varepsilon_3} H_3 e^{-\gamma_3 a/2} = -\frac{\gamma_2}{\varepsilon_2} H_2^- e^{\gamma_2 a/2} + \frac{\gamma_2}{\varepsilon_2} H_2^+ e^{-\gamma_2 a/2} \tag{41}$$

To simplify the notation, we define

$$R_i = \frac{\gamma_i}{\varepsilon_i} \tag{42}$$

Substituting $H_1 e^{-\gamma_1 a/2}$ in Equation 39 by 38, $H_3 e^{-\gamma_3 a/2}$ in 41 by 40, and multiplying by $e^{\gamma_2 a/2}$, we obtain

$$R_1 \left(H_2^- + H_2^+ e^{\gamma_2 a}\right) = R_2 \left(H_2^- - H_2^+ e^{\gamma_2 a}\right) \tag{43}$$

$$R_3 \left(H_2^+ + H_2^- e^{\gamma_2 a}\right) = R_2 \left(H_2^+ - H_2^- e^{\gamma_2 a}\right) \tag{44}$$

The original system of four variables is now reduced to two: H_2^+ and H_2^-. Re-arranging the terms, we can obtain the following relationship in a matrix form:

$$\begin{pmatrix} e^{\gamma_2 a}(R_1 + R_2) & R_1 - R_2 \\ R_3 - R_2 & e^{\gamma_2 a}(R_2 + R_3) \end{pmatrix} \begin{pmatrix} H_2^+ \\ H_2^- \end{pmatrix} = \begin{pmatrix} 0 \\ 0 \end{pmatrix} \tag{45}$$

This is a homogeneous system of linear equations. Its non-zero solutions occur when the determinant of the matrix of coefficients is zero, resulting in [21]

$$\text{Dispersion relation for thin layer:} \quad e^{2\gamma_2 a} = \frac{R_1 - R_2}{R_1 + R_2}\frac{R_3 - R_2}{R_3 + R_2} \tag{46}$$

γ_2 and R_i are given by Equations 14 and 42, respectively.

7.1. Symmetric structures

Suppose that media 1 and 3 are composed of the same material. In this case, the dispersion relation (Equation 46) becomes

$$e^{2\gamma_2 a} = \left(\frac{R_1 - R_2}{R_1 + R_2}\right)^2 \tag{47}$$

This equation has two roots [28]:

$$e^{\gamma_2 a} = \pm\frac{R_1 - R_2}{R_1 + R_2} \tag{48}$$

Which can be re-arranged to

$$\frac{R_1}{R_2} = -\frac{e^{\gamma_2 a} \pm 1}{e^{\gamma_2 a} \mp 1} = -\frac{e^{\gamma_2 a/2} \pm e^{-\gamma_2 a/2}}{e^{\gamma_2 a/2} \mp e^{-\gamma_2 a/2}} \tag{49}$$

Or [21]

$$\frac{R_1}{R_2} = \begin{cases} -coth\left(\frac{\gamma_2 a}{2}\right) \\ -tanh\left(\frac{\gamma_2 a}{2}\right) \end{cases} \tag{50}$$

Returning to Equation 43 and re-arranging, we obtain

$$\frac{R_1}{R_2} = \frac{H_2^- - H_2^+ e^{\gamma_2 a}}{H_2^- + H_2^+ e^{\gamma_2 a}} = -\frac{e^{\gamma_2 a/2} - e^{-\gamma_2 a/2}\frac{H_2^-}{H_2^+}}{e^{\gamma_2 a/2} + e^{-\gamma_2 a/2}\frac{H_2^-}{H_2^+}} \tag{51}$$

Comparing Equations 50 and 51 leads to [4]

$$\text{Dispersion relation for symmetric thin layer:} \quad \begin{matrix} \frac{R1}{R2} = -coth\left(\frac{\gamma_2 a}{2}\right) \Rightarrow H_2^+ = -H_2^- \\ \frac{R1}{R2} = -tanh\left(\frac{\gamma_2 a}{2}\right) \Rightarrow H_2^+ = H_2^- \end{matrix} \tag{52}$$

Therefore, in symmetric structures, the PSPs has two branches: a symmetric (or even) and an anti-symmetric (or odd). A sketch of the field magnitude is shown in Figure 4B.

There are two possibilities for a symmetric structure: IMI (insulator-metal-insulator) and MIM (metal-insulator-metal). We will show in sequence the dispersion curves for the symmetric IMI, symmetric MIM and their respective asymmetric variants.

7.2. Dispersion curves for a thin layer

Applying lossless Drude's metal for the dispersion relation for symmetric IMI (Equations 52), we obtain the curves depicted in Figure 5A and B. In 5A, the metal is immersed in free space, while in 5B, it is immersed in a dielectric with $n = 1.5$. The dispersion curves of the light lines in the dielectrics and of the respective IM geometries are also plotted for comparison. In the thin layer case, the dispersion relationship splits in two modes. The even mode appears below the dispersion of the IM, while the odd mode appears above it. At very low frequencies, both of the modes converge to the light line, as does the IM dispersion. For frequencies close to ω_{sp}, the wavevectors of both modes tend to infinity. The bulk mode appears only for the even mode, very similar to the one of the IM dispersion, but with higher wavevectors. The thinner is the layer, the farther the modes are from the IM dispersion. For thicker layers, the two modes converge to the IM dispersion [32], as expected due to lower interactions between the two interfaces. We note that more than one solution is possible for the odd mode, as shown in the inset in Figure 5A. Indeed, for the 25 nm layer, there were up to three solutions for each frequency.

For the symmetric MIM structures, the dispersion relation for a lossless Drude's metal is plotted in Figure 5C. The shape changes markedly from the IM dispersion. Now, the odd mode appears below the dispersion of the IM. This is the opposite of the result for the IMI geometry. Furthermore, the solution for the bulk mode appears only from the equation (52) of the odd mode. The even mode partially coincides with the plasmon bandgap, producing a quasi-bound mode. Close to ω_{sp}, the wavevectors of both modes tend to infinity. For very low frequencies, the odd mode does not tend to the light line as in the IMI structure. In addition, as expected, the thinner the layer is, the farther its dispersion is from the IM dispersion. The inset shows that, for the 50 nm layer, the even mode has two solutions for a range of frequencies.

The dispersion curve (Equation 46) for a 25 nm asymmetric IMI, is presented in Figure 5D. The mode that would be equivalent to the even mode in the symmetric variant appears under the IM dispersion with the high refractive-index dielectric. For very low frequencies, this mode converges to the light line of the same dielectric. The mode equivalent to the odd mode in the symmetric variant appears above the IM dispersion with the low refractive-index dielectric. Close to the respective ω_{sp}, the wavevector of this mode tends to infinity. For lower frequencies, there is a cut off. This mode disappears when it reaches the light line of the high refractive-index dielectric (see arrow in Figure 5D).

For a 25 nm asymmetric MIM (Figure 5E), the mode under the IM dispersions tends to infinity in the lower ω_{sp} (silver); while the mode above the IM dispersions tend to infinity in the higher ω_{sp} (gold). Silver has a lower plasma frequency than gold. Therefore, the IM dispersion of silver and vacuum remains under the IM dispersion of gold and vacuum. Between the ω_{sp} of the two half-spaces, there is a gap. In addition, the bulk mode also presents a cut-off frequency above the plasma frequencies of the two metals, when it reaches the IM dispersion of silver and vacuum (see arrow in Figure 5E).

The dispersion curves of the symmetric variants of IMI and MIM modelling the metals by the experimental values of permittivity are presented in Figure 6. As metals are modelled as lossy, the propagation lengths are also plotted. For the symmetric IMI with 25 and 50 nm lossy metal layer (Figure 6A and B), the odd modes are present in the entire frequency range. However, the even mode wavevectors of the IMI enters the bandgap region and decreases down to zero and re-appear in the bulk region. In the region of real surface plasmons, its wavevector is greater than the one of the odd mode. The odd mode, with smaller wavevectors, has propagation lengths longer than the lengths of the IM (Figure 6B), which — in its turn — are typically longer than the propagation length of the even mode. The even mode may have a longer propagation length only in the bulk mode, region not important for PSPs.

For the symmetric MIM structure with complex permittivities (Figure 6C and D), the even mode does not present a solution. At ω_{sp}, the odd mode wavevector is greater than the one in the IM dispersion. Its propagation length is shorter than the one of the IM case.

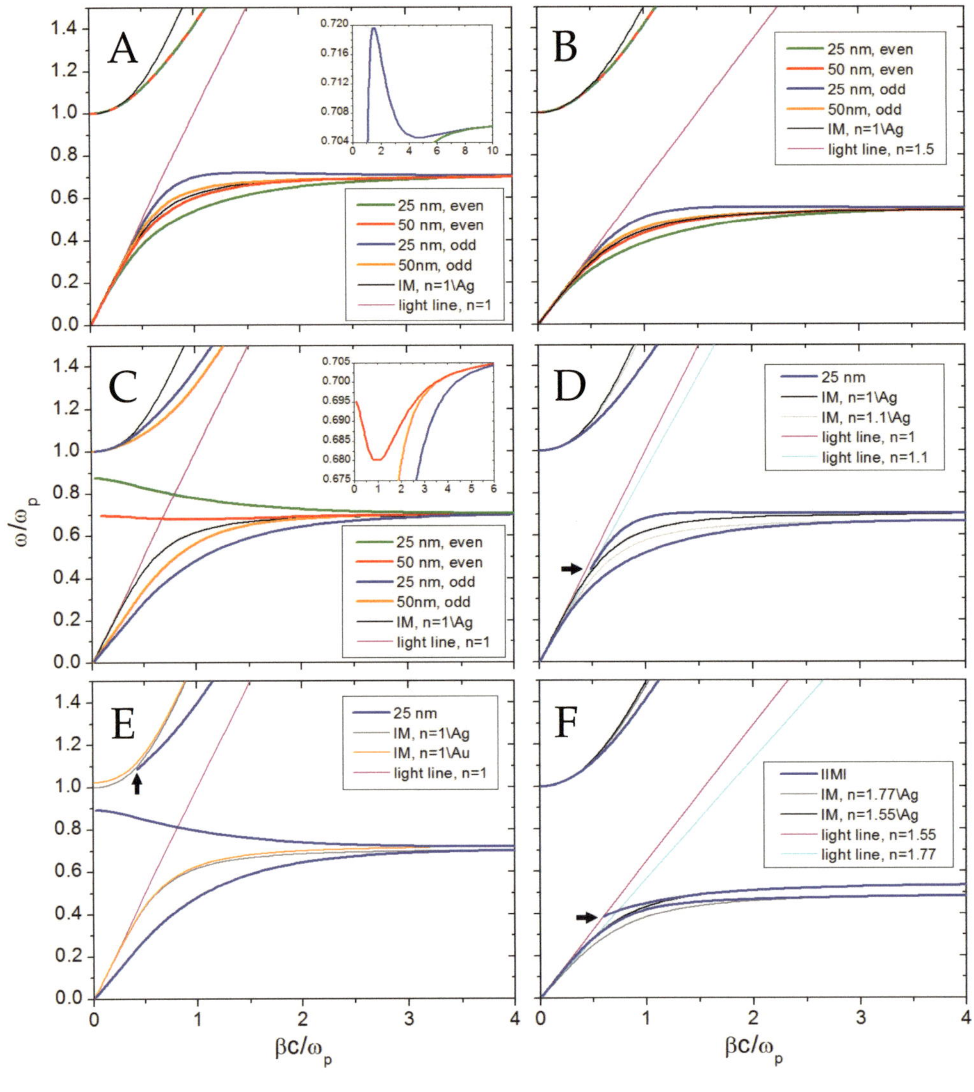

Figure 5. Dispersion relations for diverse geometries and lossless Drude's metals. A) Symmetric IMI: 25 and 50 nm metal layer immersed in free space. B) Similar to the previous geometry, but the metal is immersed in a dielectric with n=1.5. C) Symmetric MIM: 25 and 50 nm free-space layer immersed in metal. D) Asymmetric IMI: 25 nm metal layer surrounded by free space and other dielectric with n=1.1. E) Asymmetric MIM: 25 nm free space surrounded by silver and gold. F) IIMI: free space, 15 nm layer of a dielectric with n=1.77, 45 nm metal layer and a dielectric with n=1.55. The ω_p on the axes is the one of the silver. The insets in A and C show selected curves with different axes limits.

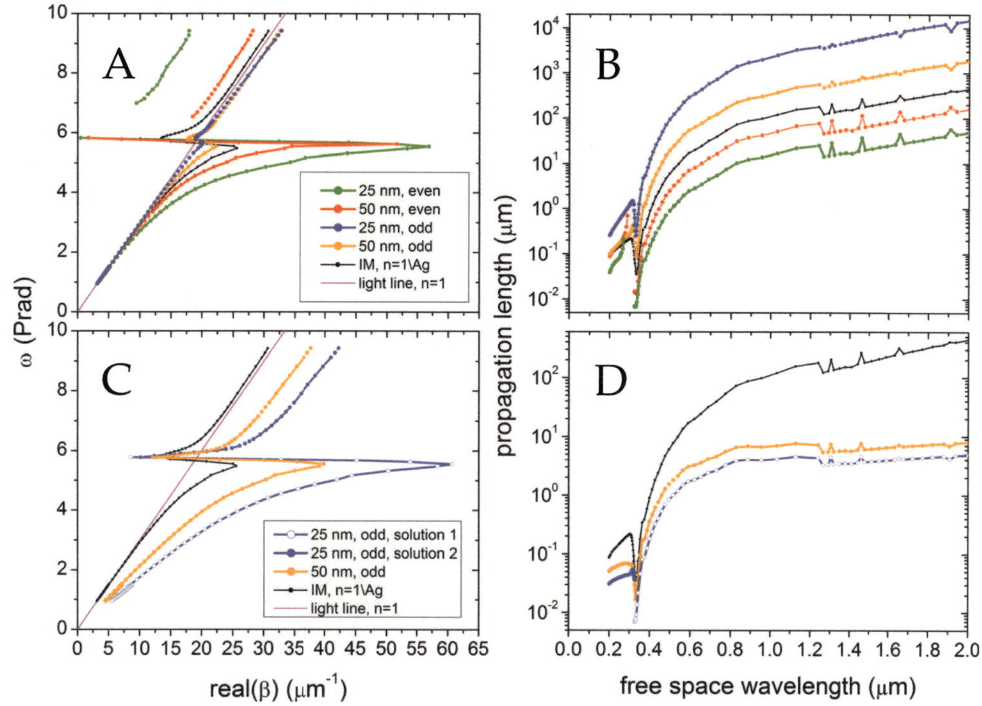

Figure 6. Dispersion curves and propagation lengths for symmetric geometries modelling the metals by the tabulated permittivities. A, B) IMI: 25 and 50 nm silver layer immersed in free space. C, D) MIM: 25 and 50 nm free-space layer immersed in silver.

The dispersion curves and propagation lengths of the asymmetric variants of IMI and MIM for tabulated permittivities are presented in Figure 7. For the IMI structure (Figure 7 A and B), instead of two solutions, three solutions appear. One of them (solution 3) has very short propagation length throughout the spectrum, and very low wavevector, except in the bulk region. Therefore, this mode does not produce PSP. The mode with great wavevector (solution 1), as the one in the symmetric case, enters the region of quasi-bond mode and decreases down to zero, with a bandgap, appearing again in the bulk region. Another mode (solution 2), with wavevectors close to the light line ($n = 1.1$), presented regions of cut-off in low frequencies as well as in the bulk region.

The MIM structure (Figure 7C and D) shows only one solution in the whole frequency spectrum. It is interesting to note that this solution presents two regions of PSPs, and consequently two quasi-bond mode regions. These regions corresponds to the ω_{sp} of the IM dispersion with silver and the ω_{sp} of the IM dispersion with gold. The propagation lengths are shorter than in any of the IM cases, with silver or with gold.

8. PSP at an IIMI structure

The geometry of an IIMI structure is depicted in Figure 8A. The structure consists of a thin metallic layer (m) deposited on a dielectric substrate (4), and covered by a thin dielectric layer (2) with free space on the top of the whole structure (1). The solution for the magnetic field of

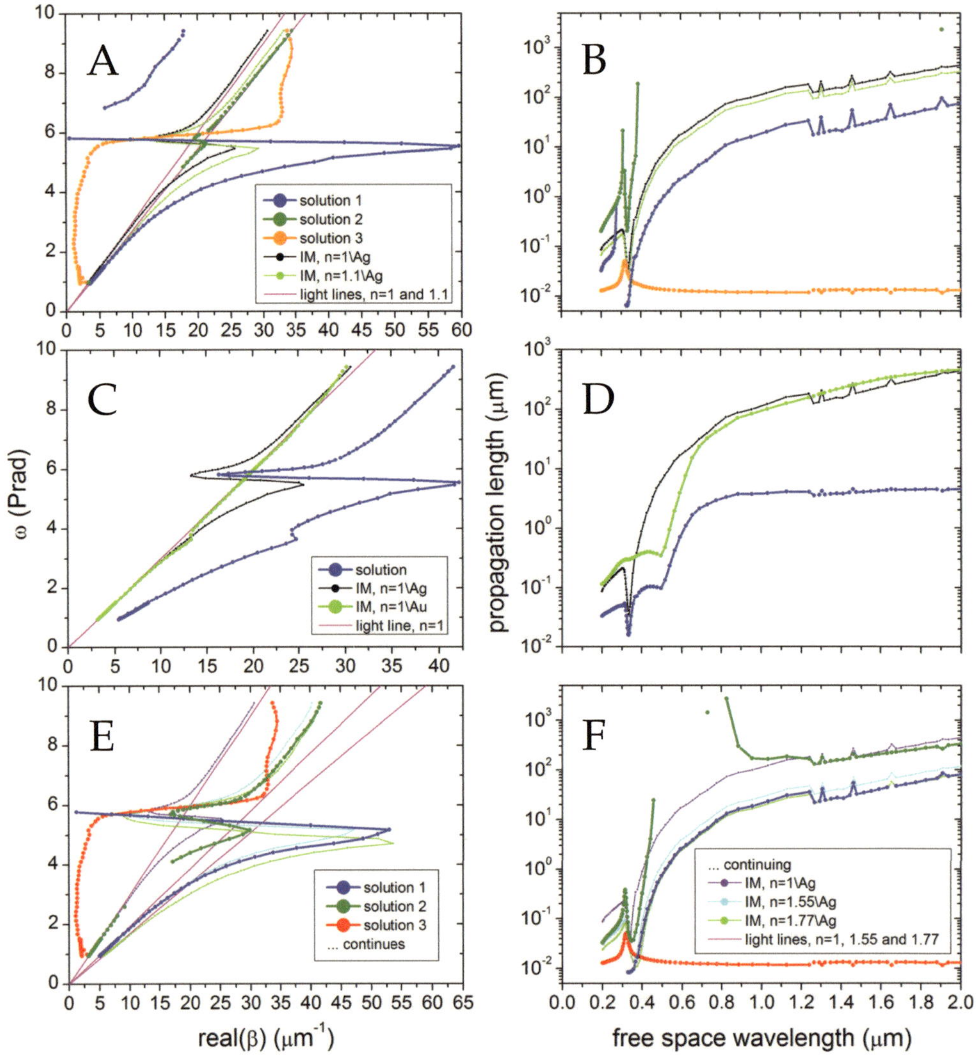

Figure 7. Dispersion relations and propagation lengths for asymmetric geometries modelling the metals by the tabulated permittivities. A, B) IMI: 25 nm silver layer surrounded by free space and other dielectric with n=1.1. C, D) MIM: 25 nm free-space layer surrounded by silver and gold. E, F) IIMI: free space, 15 nm layer of a dielectric with n=1.77, 45 nm silver layer and a dielectric with n=1.55.

the p-polarised wave decaying away from the interfaces is postulated as

$$H_y = e^{i\beta x} \times \begin{cases} A_1 e^{-\gamma_1 z}, & a+b<z \\ A_2^- e^{-\gamma_2 z} + A_2^+ e^{\gamma_2 z}, & a<z<a+b \\ A_m^- e^{-\gamma_m z} + A_m^+ e^{\gamma_m z}, & 0<z<a \\ A_4 e^{\gamma_4 z}, & z<0 \end{cases} \tag{53}$$

One should not expect to have a PSP at the 1–2 interface, as both materials are dielectrics. Nevertheless, it is necessary to take into account evanescent waves decaying away from this interface. Otherwise, the IIMI structure would simply behave like an IMI.

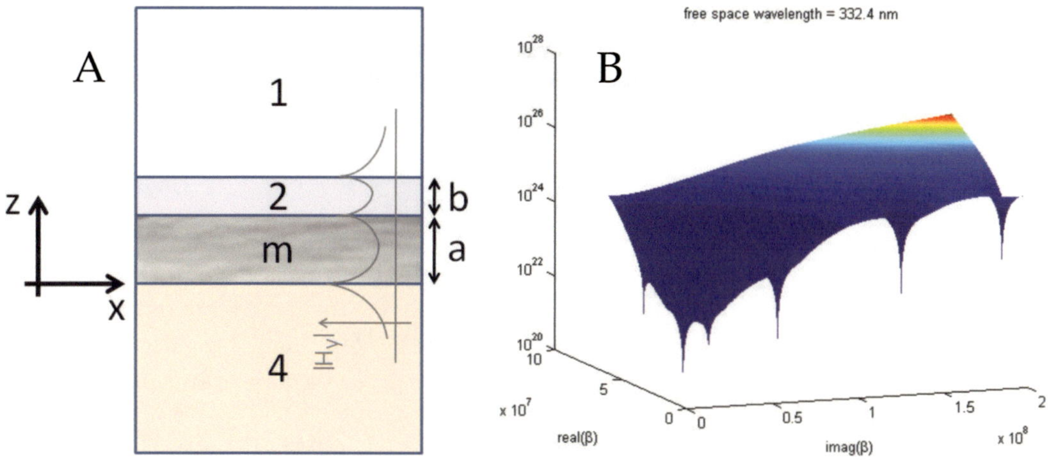

Figure 8. A) Geometry of the IIMI structure. The metal layer is located on the top of a dielectric half-space, and it is covered by a layer of other dielectric with free space above. A sketch of the absolute value of the magnetic field, as expected by modes bound to the interfaces, is also presented. B) Example of absolute values surface of the determinant of the matrix **C** (Equation 63) in the space of complex β for $\lambda = 332.4\,nm$. The minima which converge to zero represent solutions for the dispersion equation of the IIMI structure (Equation 64).

Applying the continuity conditions (Equation 21), we obtain six equations, two for each boundary.

- Boundary 1–2 ($z = a + b$)

$$A_1 e^{-\gamma_1(a+b)} = A_2^- e^{-\gamma_2(a+b)} + A_2^+ e^{\gamma_2(a+b)} \tag{54}$$

$$\frac{\gamma_1}{\varepsilon_1} A_1 e^{-\gamma_1(a+b)} = \frac{\gamma_2}{\varepsilon_2} A_2^- e^{-\gamma_2(a+b)} - \frac{\gamma_2}{\varepsilon_2} A_2^+ e^{\gamma_2(a+b)} \tag{55}$$

- Boundary 2–m ($z = a$)

$$A_2^- e^{-\gamma_2 a} + A_2^+ e^{\gamma_2 a} = A_m^- e^{-\gamma_m a} + A_m^+ e^{\gamma_m a} \tag{56}$$

$$\frac{\gamma_2}{\varepsilon_2} A_2^- e^{-\gamma_2 a} - \frac{\gamma_2}{\varepsilon_2} A_2^+ e^{\gamma_2 a} = \frac{\gamma_m}{\varepsilon_m} A_m^- e^{-\gamma_m a} - \frac{\gamma_m}{\varepsilon_m} A_m^+ e^{\gamma_m a} \tag{57}$$

- Boundary m–4 ($z = 0$)

$$A_m^- + A_m^+ = A_4 \tag{58}$$

$$-\frac{\gamma_m}{\varepsilon_m} A_m^- + \frac{\gamma_m}{\varepsilon_m} A_m^+ = \frac{\gamma_4}{\varepsilon_4} A_4 \tag{59}$$

Substituting $A_1 e^{-\gamma_1(a+b)}$ in Equation 55 by Equation 54 yields

$$\frac{\gamma_1}{\varepsilon_1}\left(A_2^- e^{-\gamma_2(a+b)} + A_2^+ e^{\gamma_2(a+b)}\right) = \frac{\gamma_2}{\varepsilon_2}\left(A_2^- e^{-\gamma_2(a+b)} - A_2^+ e^{\gamma_2(a+b)}\right) \tag{60}$$

Substituting A_4 in 59 by 58,

$$\frac{\gamma_m}{\varepsilon_m}\left(-A_m^- + A_m^+\right) = \frac{\gamma_4}{\varepsilon_4}\left(A_m^- + A_m^+\right) \tag{61}$$

With Equations 60 and 61, we reduced the system to 4 variables: A_2^-, A_2^+, A_m^- and A_m^+. These two equations in addition to Equations 56 and 57 can be written in a matrix form:

$$\mathbf{C}\begin{pmatrix} A_2^- \\ A_2^+ \\ A_m^- \\ A_m^+ \end{pmatrix} = \begin{pmatrix} 0 \\ 0 \\ 0 \\ 0 \end{pmatrix} \tag{62}$$

In which $\mathbf{C}$ is a 4×4 matrix of coefficients:

$$\mathbf{C} = \begin{pmatrix} R_1 e^{-\gamma_2(a+b)} - R_2 e^{-\gamma_2(a+b)} & R_1 e^{\gamma_2(a+b)} + R_2 e^{\gamma_2(a+b)} & 0 & 0 \\ 0 & 0 & R_4 + R_m & R_4 - R_m \\ e^{-\gamma_2 a} & e^{\gamma_2 a} & -e^{-\gamma_m a} & -e^{\gamma_m a} \\ R_2 e^{-\gamma_2 a} & -R_2 e^{\gamma_2 a} & -R_m e^{-\gamma_m a} & R_m e^{\gamma_m a} \end{pmatrix} \tag{63}$$

γ_i and R_i are given by Equations 14 and 42, respectively. For a homogeneous system of linear equations, the non-zero solutions occur when the determinant of the matrix of coefficients is zero:

$$\text{Dispersion relation for IIMI:} \quad |\mathbf{C}| = 0 \tag{64}$$

8.1. Dispersion curves for an IIMI structure

The dispersion curve of this last structure for a Drude's model without damping is presented in Figure 5F. For comparison, the light lines of the dielectrics adjacent to the metal are drawn, as well as the half-spaces of the same dielectrics with the metal. This structure behaves close to an asymmetric IMI. It presents one mode, under the lower half-spaces dispersion. The wavevector of this mode tends to infinity at ω_{sp} of the lower half-spaces dispersion. The mode above the two half-spaces dispersion tends to infinity at ω_{sp} of the other half-spaces dispersion. For lower frequencies, the same mode reaches the low refractive-index light line and it is cut-off (see arrow in Figure 5F).

The dispersion curve and propagation length of the same geometry, but with the silver modelled by the experimental permittivity is drawn in Figure 7E and F. As in the case of the asymmetric IMI structure, the actual one presented three modes. One of the modes that has no PSP character (solution 3). Other mode presents small wavevectors and a cut-off, but longer propagation lengths. Finally, a third mode with great wavevectors and PSP, that bends in the quasi-bound mode, reaching zero. This mode does not appear in the bulk region.

9. Conclusion

We showed, mathematically, the principal properties of propagating surface plasmons. We presented the analysis of plasmons that appear at planar interfaces, such as IM, symmetric and assymetric variants of IMI and MIM, and IIMI. Only for the IM geometry, there is an explicit function $\beta = f(\omega)$ (Equation 28). For the other geometries, the dispersions are implicit functions[7] (Equations 46, 52 and 64). For these geometries, one can re-arrange the dispersion relations as $f(\beta, \omega) = 0$, and numerically find the zeros for β in the first quadrant, as we can define the propagation towards $+x$, and lossy materials lead to exponential decays of the fields [32].

When modelling the metals as lossless by using the lossless Drude's permittivities, the propagation constants are real. In this case, the wavevectors parallel to the interface diverge at the surface plasmon frequencies. When the metals are modelled as lossy by using complex experimental values of permittivities, the wavevectors do not diverge any more, and the dispersion curves bend forming quasi-bond modes, regions with anomalous dispersion ($d\omega/dk < 0$).

Figure 8B shows an example the surface $|f(\beta, \omega)|$ for the IIMI structure at one specific frequency and using tabulated permittivity of silver. This surface presents many minima. The minima that converge to zero are all solutions of the dispersion relation. Infinite solutions are possible for one single frequency, but the most interesting ones are the solutions with relatively long propagation lengths — from tens of nanometers up. Nevertheless, the solutions with very short propagation lengths are also important for matching the boundary conditions and launching the propagating surface plasmons [32].

Author details

Baltar Henrique T. M. C. M., Drozdowicz-Tomsia Krystyna and Goldys Ewa M.
Faculty of Science of Macquarie University, Department of Physics and Astronomy, Sydney, Australia

10. References

[1] Aroca, R. F., Ross, D. J. & Domingo, C. [2004]. Surface-enhanced infrared spectroscopy, *Applied Spectroscopy* 58(11): 324A–338A.
[2] Bozhevolnyi, S. I. [2009]. *Introduction to surface plasmon-polariton waveguides*, Pan Stanford, chapter 1.
[3] Cheng, D. K. [1989]. *Field and wave electromagnetics*, 2 edn, Addison-Wesley.
[4] Dionne, J. A., Sweatlock, L. A., Atwater, H. A. & Polman, A. [2005]. Planar metal plasmon waveguides: frequency-dependent dispersion, propagation, localization, and loss beyond the free electron model, *Physical Review B* 72(7).
[5] Dionne, J. A., Verhagen, E., Polman, A. & Atwater, H. A. [2008]. Are negative index materials achievable with surface plasmon waveguides? a case study of three plasmonic geometries, *Optics Express* 16(23): 19001–19017.
[6] Fort, E. & Gresillon, S. [2008]. Surface enhanced fluorescence, *Journal of Physics D: Applied Physics* 41(1): 013001.
[7] Homola, J. [2006]. *Electromagnetic theory of surface plasmons*, Springer series on chemical sensors and biosensors, Springer, chapter 1.
[8] Huber, A., Ocelic, N., Kazantsev, D. & Hillenbrand, R. [2005]. Near-field imaging of mid-infrared surface phonon polariton propagation, *Applied Physics Letters* 87(8): 081103.
[9] Hwang, E. H. & Das Sarma, S. [2007]. Dielectric function, screening, and plasmons in two-dimensional graphene, *Physical Review B* 75(20): 205418.
[10] Iskander, M. F. [1992]. *Electromagnetic fields and waves*, Prentice-Hall.

[7] Functions not with the strict meaning, as one input may produce many outputs.

[11] Jablan, M., Buljan, H. & Soljačić, M. [2009]. Plasmonics in graphene at infrared frequencies, *Physical Review B* 80(24): 245435.
[12] Jackson, J. D. [1962]. *Classical electrodynamics*, 1 edn, John Wiley and Sons.
[13] Johnson, P. B. & Christy, R. W. [1972]. Optical constants of the noble metals, *Physical Review B* 6(12): 4370–4379.
[14] Junxi, Z., Lide, Z. & Wei, X. [2012]. Surface plasmon polaritons: physics and applications, *Journal of Physics D: Applied Physics* 45(11): 113001.
[15] Kawata, S. [2001]. *Near-field optics and surface plasmon polaritons*, Springer.
[16] Lavoie, B. R., Leung, P. M. & Sander, B. C. [2011]. Metamaterial waveguides, *ArXiv e-prints* .
[17] Le Ru, E. C. & Etchegoin, P. G. [2008]. Surface-enhanced raman scattering (sers) and surface-enhanced fluorescence (sef) in the context of modified spontaneous emission, *ArXiv e-prints* .
[18] Le Ru, E. C. & Etchegoin, P. G. [2009]. *Principles of surface-enhanced Raman spectroscopy and related plasmonic effects*, Elsevier.
[19] Liu, H., Genov, D. A., Wu, D. M., Liu, Y. M., Steele, J. M., Sun, C., Zhu, S. N. & Zhang, X. [2006]. Magnetic plasmon propagation along a chain of connected subwavelength resonators at infrared frequencies, *Physical Review Letters* 97(24): 243902.
[20] Lynch, D. W. & Hunter, W. R. [1985]. *Comments on the optical constants of metals and an introduction to the data for several metals*, Academic Press, Orlando.
[21] Maier, S. A. [2007]. *Plasmonics: fundamentals and applications*, Springer.
[22] Mills, D. L. & Burstein, E. [1974]. Polaritons: the electromagnetic modes of media, *Reports on Progress in Physics* 37(7): 817–926.
[23] Novotny, L. & Hecht, B. [2006]. *Principles of nano-optics*, Cambridge University, Cambridge.
[24] Pendry, J. B., Holden, A. J., Robbins, D. J. & Stewart, W. J. [1999]. Magnetism from conductors and enhanced nonlinear phenomena, *IEEE Transactions on Microwave Theory and Techniques* 47(11): 2075–2084.
[25] Rana, F. [2008]. Graphene terahertz plasmon oscillators, *IEEE Transactions on Nanotechnology* 7(1): 91–99.
[26] Rivas, J. G., Kuttge, M., Bolivar, P. H., Kurz, H. & Sánchez-Gil, J. A. [2004]. Propagation of surface plasmon polaritons on semiconductor gratings, *Physical Review Letters* 93(25): 256804.
[27] Sarid, D. & Challener, W. A. [2010]. *Modern introduction to surface plasmons: theory, mathematica modeling, and applications*, Cambridge.
[28] Sernelius, B. E. [2001]. *Surface modes in physics*, Wiley-VCH.
[29] Staelin, D. H., Morgenthaler, A. W. & Kong, J. A. [1994]. *Electromagnetic waves*, Prentice-Hall.
[30] van Vugt, L. K. [2007]. *Optical properties of semiconducting nanowires*, PhD thesis.
[31] Welford, K. [1991]. Surface plasmon-polaritons and their uses, *Optical and Quantum Electronics* 23(1): 1–27.
[32] Zakharian, A. R., Moloney, J. V. & Mansuripur, M. [2007]. Surface plasmon polaritons on metallic surfaces, *IEEE Transactions on Magnetics* 43(2): 845–850.
[33] Zayats, A. V. & Richards, D. [2008]. *Near-optics and near-field optical microscopy*, Artech House.
[34] Zayats, A. V., Smolyaninov, I. I. & Maradudin, A. A. [2005]. Nano-optics of surface plasmon polaritons, *Physics Reports* 408(3-4): 131–314.

Chapter 7

Light Transmission via Subwavelength Apertures in Metallic Thin Films

V. A. G. Rivera, F. A. Ferri, O. B. Silva, F. W. A. Sobreira and E. Marega Jr.

Additional information is available at the end of the chapter

http://dx.doi.org/10.5772/50807

1. Introduction

The optical properties of subwavelength apertures in metallic films have been the focus of much research activity around the world since the extraordinary optical transmission (EOT) phenomenon was reported over a decade ago (Ebbesen et al., 1998).

EOT is an optical phenomenon in which a structure containing subwavelength apertures in an opaque screen transmits more light than might naively be expected on the basis of either ray optics or even knowledge of the transmission through individual apertures. The phenomenon was discovered serendipitously for two-dimensional (2D) periodic arrays of subwavelength holes in metals (Garcia-Vidal et al., 2010). Surprisingly, such arrays may, for certain wavelengths, exhibit transmission efficiencies normalized to the total area of the holes that exceed unity. In other words, for these wavelengths a periodic array of subwavelength holes transmits more light than a large macroscopic hole with the same area as the sum of all the small holes. The surprise is compounded by the fact that a single subwavelength aperture generally transmits light with an efficiency that is substantially below unity.

This remarkable transmission enhancement has potential applications in photolithography, near-field microscopy, and photonic devices (Lal et al., 2007). Although the detailed picture of the transmission enhancement is still being investigated, the excitation of surface plasmon-polaritons (SPPs) is proposed to be involved in the process (Weiner, 2009). Ever since the first experimental report on EOT through subwavelength apertures, considerable theoretical effort has been devoted to interpreting the essential physics of the process in slit arrays (Porto et al., 1999; Takahura, 2001; Xie et al., 2005). Experimental studies subsequent to the initial report were also performed, which demonstrated a number of surprising features. For instance, spectral transmission measurements (Lezec et al., 2004) revealed that suppression, as well as enhancement, was a characteristic property of slit arrays. Additionally, interferometric studies (Gay et al., 2006a, 2006b; Kalkum et al., 2007) showed that the contribution of transient diffracted surface modes is as important as the SPP guided mode in the immediate vicinity of the subwavelength object. A more recent investigation (Pacifici et al., 2008) with the aim to confront the question of how transmission minima and maxima depend on array periodicity showed a minimum in transmission at slit separations equal to the wavelength of the SPP mode, and maxima occurring approximately at half-integer multiples.

These previous studies focused attention on the properties of arrays fabricated in metallic films. It was already shown that a single slit is an interesting structure, since it combines the compactness of a single defect with the directionality of an array launcher. The extended dimension of the long axis imposes directionality on the transmitted light beam, the divergence of which can easily be controlled (Laluet et al., 2008). In a very recent investigation dealing with single subwavelength slits, it was possible to observe that the slit transmission is notably affected by the film thickness, and increases linearly with increasing slit width for a fixed film thickness (Ferri et al., 2011).

In this chapter, we selected some fundamental subjects of high and general interest involved in the phenomenon of light transmission via subwavelength apertures in metallic thin films. The manuscript will cover issues on both theory and experiment, such as (1) fabrication and measurement setup characteristics, (2) different materials and metallic structures, and (3) phenomena from the metallic film/dielectric interface. In addition, numerical simulations were performed in order to investigate the optical transmission through subwavelength apertures.

The paper is organized as follows. In Section 2, we comment on the classical theories of diffraction by subwavelength apertures. In Section 3, a rapid discussion about preceding experiments and interpretations are presented. In Section 4, we give a simple introduction to the plasmon-polariton. In Section 5, some characteristics of focused-ion beam (FIB) nanofabrication are presented. In this manner, the topics covered by the above mentioned sections are basically a compilation of information found in literature. In Section 6, we in fact demonstrate some applications made by our research group investigating the (1) influence of the metallic film thickness, as well as the (2) use of multilayered metallic thin films on the optical transmission through subwavelength slits. A short summary is provided in Section 7.

2. Diffraction by subwavelength apertures

The wave nature of light implies modifications in the transmission through apertures like the phenomenon of diffraction. This common process, that even in simple geometries is very complex, have been extensively studied and many different models and approximations were developed based on the classical theory of diffraction (Jackson, 1999; Bouwkamp, 1954). Probably the simplest geometry, and maybe for this reason one that received the most attention, is that of a circular aperture with radius r in an infinitely thin and perfect conducting screen (Fig. 1).

If the radius r of the aperture is some orders of magnitude larger than the wavelength λ_0 of the impinging radiation, i.e., $r \gg \lambda_0$, the problem can be treated with the Huygens-Fresnel principle and its mathematical formulation can be given as a good approaching by the

Kirchhoff scalar theory of light diffraction (Jackson, 1999). This theory is based on a scalar wave theory, and, thus, it does not take into account effects due to the polarization of light. In the case of normal incidence of light through a circular aperture, it is easy to show that the transmitted intensity (dI) per unit solid angle (dΩ) in the far-field region, also known as Fraunhofer diffraction limit, is given by $I(\theta) \cong I_0\left(k^2r^2/4\pi\right)\left|2J_1(krsin\theta)/krsin\theta\right|^2$, where the incident intensity I_0 is equally distributed on the aperture area πr^2. The wavenumber of the incident light is $k=2\pi/\lambda_0$, θ is the angle between the normal to the aperture and the direction of the emitted radiation, and $J_1(krsin\theta)$ is the Bessel function of the first kind. The pattern described by this formula is the well known Airy pattern, composed by a central bright spot surrounded by concentric bright rings of decreasing intensity, caused by the interference of light rays originated inside the aperture (see Fig. 1). The ratio of the total transmitted intensity to I_0, given by $T = \int I(\theta)d\Omega/I_0$, is called the transmission coeficient. For large apertures, with $r \gg \lambda_0$, in which case the treatment is outlined here is valid, $T \approx 1$.

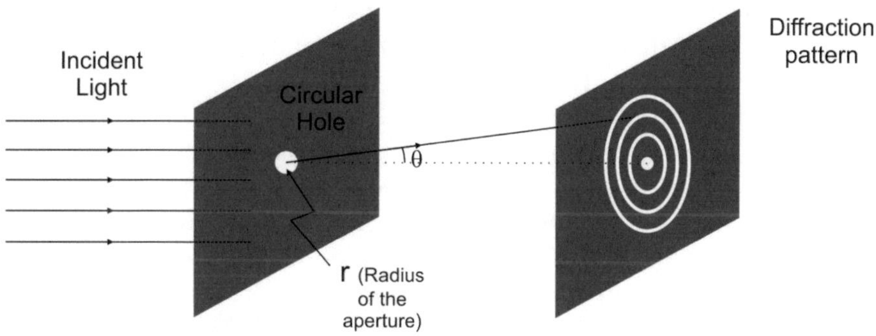

Figure 1. Transmission of light through a circular aperture or radius r in an infinitely thin opaque screen

It has been proposed (Weiner, 2009) that the existence of surface waves such as SPPs are involved in the transmission process. For this reason, the regime of subwavelength apertures $r \ll \lambda_0$ is much more interesting, because near-field effects are expected to contribute dominantly in the transmission process. The problem arising here is that even in an approximate analysis of a perfectly conducting screen in the limit of zero width, we must use a vector description via Maxwell's equations. In Kirchhoff's method, the basic assumption is that the electromagnetic field in the aperture is the same as if the opaque screen is not present, a case which does not fulfil the boundary condition of zero tangential electric filed on the screen. For large holes, in which $r \gg \lambda_0$, this basic failure is less severe, because the diffracted fields are small when compared to the directly transmitted ones. Nevertheless, for subwavelength apertures this approximation is inadequate even in a first order treatment of the problem.

Assuming that the incident light intensity I_0 is constant over the area of the aperture, Bethe and Bouwkamp arrived at an exact analytical solution for light transmission through a sub-wavelength circular hole in an ideal perfectly conducting and infinitely thin screen (Bethe, 1944; Bouwkamp, 1950a; Bouwkamp, 1950b). For normal incidence, the aperture can be described as a magnetic dipole located in the plane of the hole. The transmission coefficient for an incident plane wave is then given by (Maier, 2007)

$$T = \frac{64}{27\pi^2}(kr)^4 \propto \left(\frac{r}{\lambda_0}\right)^4 \quad (1)$$

There is a weak total transmission for a subwavelength aperture due to the scaling with $(r/\lambda_0)^4$, smaller by an amount of $(r/\lambda_0)^2$ compared to Kirchhoff's scalar theory. This scaling of $T \propto \lambda_0^{-4}$ is in agreement with the theory of light scattering by small objects due to Rayleigh. The case described in Eq. (1) is that of normal light incidence with both transverse electric (TE) and transverse magnetic (TM) polarization. This is not the case when radiation is incident in the screen at another angle, in this case an additional electric dipole in the normal direction is needed to describe the different behaviour of the process. More radiation is transmitted for TM than for TE polarization in this case (Bethe, 1944).

There are two major approximations in the Bethe-Bouwkamp theory of light transmission through a circular aperture in a screen. The screen is said to be made of an ideal perfectly conducting screen, and so perfectly opaque to the transmission of radiation, and its thickness is taken to be infinitely small. One of these assumptions could be omitted by taking numerical simulations for the problem of screens with finite thickness (Maier, 2007). However, when discussing the transmission properties through real apertures, i.e., in real metals, the finite conductivity, and so the transmission, should be taken into account. The thin films used in optical experiments cannot be taken as perfectly opaque screens, and we could not employ the Bethe-Bouwkamp theory. On the other hand, if the film thickness is higher than some skin depths, that is, if we are dealing with a "thick" film, it could be taken as an opaque screen. It has been shown that for apertures fulfilling these conditions, localized surface plasmons have a significant influence in the transmission process (Degiron et al., 2004).

3. Early experiments and implications

In the paper published by Ebbesen et al., it was affirmed the measurement of "transmission efficiency...orders of magnitude greater than predicted by standard aperture theory". A typical result obtained in the original experiments is shown in the right panel of Fig. 2. The figure shows the transmitted power as a function of the incident light wavelength λ through a circular aperture of radius r. The transmission intensity is normalized to the cross-sectional area of the hole $A=\pi r^2$. Using the experimental parameters of the caption of Fig. 2, the efficiency predicted by Bethe theory, given by Eq. (1), is 0.34%, while the results obtained by Ebbesen et al. reported peak efficiencies of a factor more than two. In this case, the enhancement of light transmission over that expected by Bethe theory is about 600 times. In a following paper (Ghaemi et al., 1998) the same group reported peak transmission efficiencies "that are about 1000 times higher than that expected for subwavelength holes". To obtain an EOT it is plausible to say that would be necessary at least a three-order of magnitude increase over the predictions of Bethe formula.

It was proposed in these early reports that transmission enhancements were caused by a new phenomenon not taken into account in electromagnetic (E-M) vector field Bethe theory nor in the scalar diffraction Kirchhoff model: the resonant excitation of surface plasmons

(SP) waves supported by the periodic array structure in the metal film. The physics of SP waves shall be discussed in the next Section. About a year later Treacy suggested (Treacy, 1999) "dynamic diffraction" as another model to investigate the problem of light transmission through a periodic array of holes or slits. It was pointed out that the oscillation at frequency ω of the optical field would induce currents within the skin depth of the metal. The periodic structuring of holes in the metal gives rise to Bloch modes of the E-M field induced in the metal within the skin depth and consistent with its periodicity. There will be an oscillating current associated to each of these Bloch modes. Then, Treacy invoked "interband scattering" as a way to distribute these energy among these Bloch modes and from there to the propagating modes and surface waves at the aperture exit. It was not exactly clarified how does this redistribution occurs, but Treacy suspected that the success of "dynamic diffraction" for interpreting X-ray scattering in crystals might be useful for the understanding of light transmission through these new structures as well.

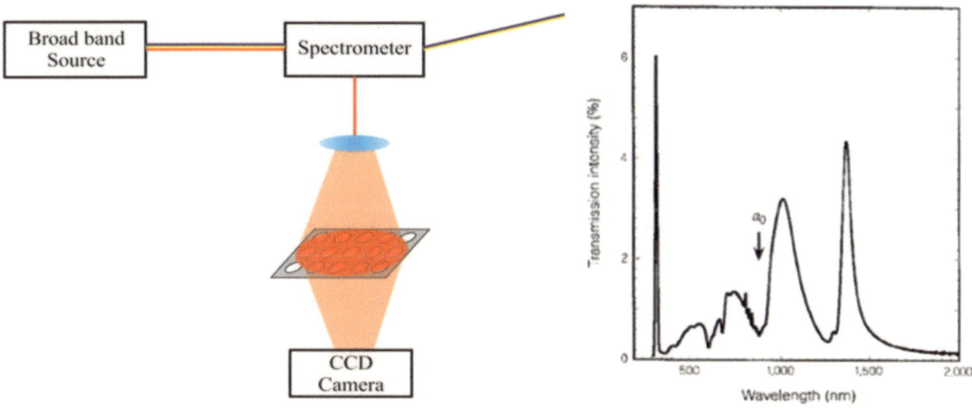

Figure 2. Left panel: schematic of the original spectral transmission experiments. A broadband, incoherent light source is spectrally filtered by a scanning spectrophotometer and focused onto an array of subwavelength structures (holes or slits). A charge-coupled device (CCD) camera records the transmission intensity through the structure as a function of wavelength of the input light (adapted from Weiner, 2009). Right panel: transmission spectrum from an array of holes in an Ag metal film evaporated onto a transparent substrate as reported by Ebbesen et al. Array periodicity: 900 nm, hole diameter: 150 nm, metal film thickness: 200 nm. The point indicated a_0 marks the array periodicity

A much more complete presentation of dynamic diffraction was published two years later by Treacy (Treacy, 2002). In this work it was clarified the relation between this approach and earlier interpretations of transmission in terms of "resonant" excitation of SP waves. This pioneering paper pointed out the way forward by emphasizing critical factors in the proper analysis of the problem. The most important thing to observe is that Bloch modes are determined by the E-M field present on and below the metal surface where are placed the periodic structures. These modes obey the periodic boundary conditions of the structure independently of the wavelength of the incident light. The simplest case is that of a one dimensional (1D) structure, with arrays and/or grooves placed periodically. In this case the modes are parallel to the surface in the metal and in the dielectric media, and evanescently vanishing in the normal direction. In parallel to this there are propagating modes. These propagating modes are responsible for the transmission of light through the structures. The transmission of the E-M filed is defined by both, propagating and evanescent fields, as a result of linear combinations of Bloch modes.

4. Plasmon-polaritons

It is important to stress here that metals play a more important role in plasmonics than dielectric media (Huang et al., 2007). In a metal, the optical as well as the electric properties are very different from dielectrics because of the existence of huge free electrons. These electrons have a fast response to varying fields leading to a different response than that in a dielectric media.

In the Drude model for free electrons, the dielectric constant is given by

$$\varepsilon_m = 1 - \frac{\omega_P{}^2}{\omega(\omega + i\gamma)} \tag{2}$$

The constant ω_P is called the bulk plasma frequency and is a constant that depends on the metal. The constant γ is associated to the scattering of electrons in the Drude model. In "good" metals, where the scattering process is reduced we can neglect the damping ($\gamma = 0$). For high frequencies the dielectric constant is positive and there are modes whose dispersion relation is given by $\omega^2 = \omega_P{}^2 + c^2k^2$. In this expression, c is the light velocity in vacuum. These modes are known as bulk plasmon-polaritons (BPPs) and are a result of the coupling between light and the free electrons in metals. For light at low frequencies, which is the case of visible light for metals, then $\omega < \omega_P$, and light propagation is forbidden by the negative permittivity.

For the case of a metal, whose real part of permittivity is negative, light incident normally to the surface gives rise to evanescent modes. But even in metals, there are propagating modes in the surface of the metal, provided that the surface of the metal is interfaced with a dielectric (or vacuum). These modes are the so-called SPP waves mentioned before. A typical geometry for this kind of problem is a metal and a dielectric separated by an infinite plane surface, as shown in the left panel of Fig. 3. This surface wave is based on the coupling between the surface free charges along the metal and light. In this case, the dispersion relation is given by (Zayats et al., 2005):

$$k_{SPP} = \frac{\omega}{c}\sqrt{\frac{\varepsilon_m \varepsilon_d}{\varepsilon_m + \varepsilon_d}} \tag{3}$$

Here, ε_d is the permittivity of dielectric. The condition for the propagation of the SPP is that k_{SPP} is real. As the permittivity of metal is negative it is necessary that $\varepsilon_m + \varepsilon_d < 0$, in this case $\omega < \omega_{SP} \equiv \omega_P / \sqrt{1 + \varepsilon_d}$, where ω_{SP} is the surface plasmon frequency. The dispersion relation for the SPPs and BPPs are plotted schematically in the right panel of Fig. 3. The left panel of Fig. 3 shows the mechanism of how the surface plasmon propagates along the metal/dielectric surface.

Some features of the SPP propagation along these flat interfaces between a metal and a dielectric, as summarized by Huang et al., are addressed in the following.

It was discussed that SPP modes are the result of a coupling between the free electrons in the metal side and light (E-M field). The modes given by the dispersion relation in Eq. (3) are called TM modes. These are the only allowed modes, in the case of an interface between two non-magnetic media, and are characterized by a magnetic field normal to the propagation direction (in the left panel of Fig. 3 it is normal to the plane of the paper) and an electric filed that has components parallel ($E_{//}$) and normal ($E_{\perp}$) to the direction of propagation (the plane of the paper in Fig. 3). The ratio of the normal and parallel components of electric field inside the dielectric $(E_{\perp}/E_{//})_d = \sqrt{\varepsilon_m/\varepsilon_d}$ and inside the metal $(E_{\perp}/E_{//})_m = -\sqrt{\varepsilon_d/\varepsilon_m}$ shows that, inside the metal, as the electric field is almost completely concentrated in the direction of propagation, the free electrons present a movement of back and forth in the direction of propagation. These are the electron density waves shown in Fig. 3.

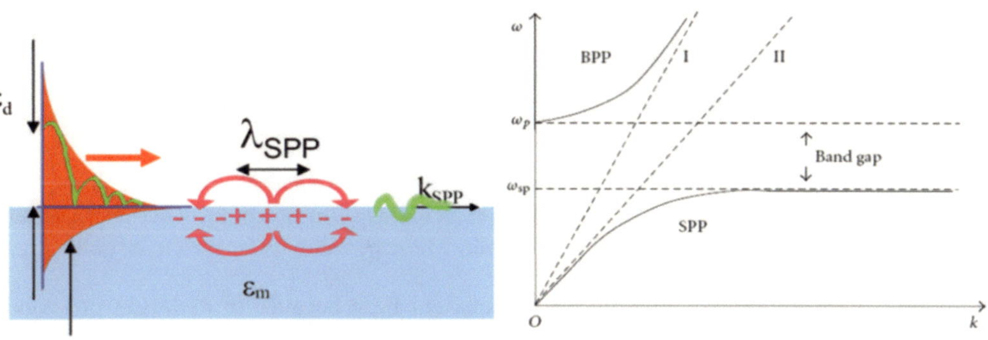

Figure 3. Plasmon-polariton modes associated with the metals. Left panel: SPP mode in a schematic view of infinite metal/dielectric interface. Right panel: dispersion relation of BPP and SPP modes. The dashed lines I and II denote the light dispersion $\omega = ck$ and $\omega = ck/\sqrt{\varepsilon_d}$, respectively. The frequency range between ω_{SP} and ω_P corresponds to a gap where the electromagnetic wave cannot propagate via either bulk or surface modes (Huang et al., 2007)

The SPP modes propagate along the interface between media with a larger propagation constant ($k_{SPP} > k_0\sqrt{\varepsilon_d}$), so the wavelength and propagation velocity are smaller than in vacuum. Taking into account that the metal permittivity still has a negative part, that is responsible for losses, the propagation of SPP is reduced to a finite value given by the propagation length $L_{SPP} \approx {\varepsilon'_m}^2/k_0\varepsilon''_m\varepsilon_d^{3/2}$, where ε'_m and ε''_m are the real and imaginary parts of metal permittivity, respectively. In visible and near-infrared region, L_{SPP} can take values of several micrometers, as can be seen in Fig. 4, that shows the real and imaginary parts of permittivity for Au and Ag [Fig. 4(a)], and the propagation length for each of these media with an interface with air [Fig. 4(b)].

The confinement of SPP modes along the surface is characterized by its evanescent behaviour in either side of the interface, which happens because of the larger propagation constant. On the dielectric side there is a higher penetration of the E-M field, given by the decaying length $\delta_d \approx \sqrt{|\varepsilon_m'|}/k_0\varepsilon_d$. On the metal side, where the decaying length is $\delta_m \approx 1/k_0\sqrt{|\varepsilon_m'|}$, there is a higher confinement than in the dielectric side. The relation between these lengths δ_d/δ_m is about some tens. This is what is desired in practice. Moreover, there is a strong enhancement of the fields near the interface of the media.

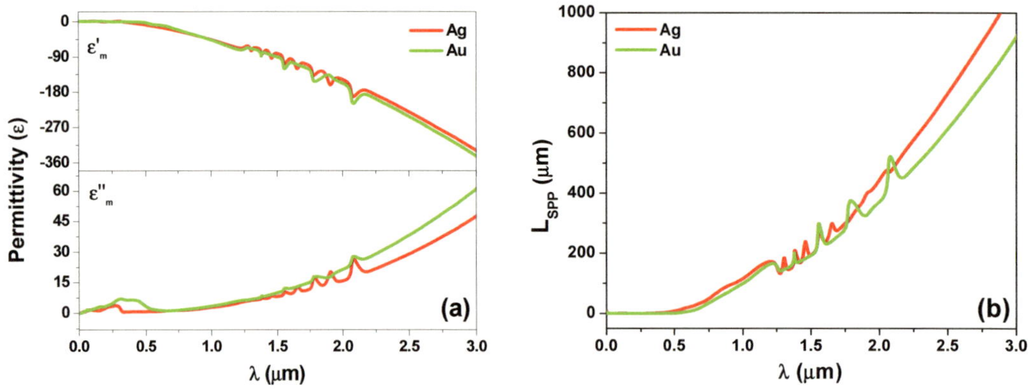

Figure 4. (a) Real ε_m' and imaginary ε_m'' parts of permittivity for Au and Ag as obtained using data from Palik, and the corresponding (b) Propagation length L_{SPP} as a function of the wavelength. In this case it is considered an interface between metal and air, whose permittivity is taken to be unity

SPP modes cannot be excited directly by the incident light because of its larger propagation constant. Special techniques have been used (Zayats et al., 2005) to compensate the phase mismatch as well as the difference in the wavevectors. Some of these techniques employ, for example, prism coupling by attenuated total reflection and diffraction gratings. It is also possible to couple plasmons by near-field excitation with a near-field optical microscope. An efficient mode to excite SPP is a subwavelength hole or slit (Zayats et al., 2005; Yin et al., 2004; Lalanne et al., 2005), where the diffraction components can ensure momentum conservation. Because of its localized feature, a direct observation of SPP is very difficult. Some methods, like the observation by using near-field microscopy (Hecht et al., 1996) can be used for this purpose. It is also possible to map SPP modes by recording the scattered light from a metal surface (Bouhelier et al., 2001) prepared with structured nanoscale corrugations. Another method, called fluorescence imaging (Ditlbacher et al., 2002a), has been proposed and consists in a metal surface covered with fluorescent molecules which emit radiation with intensity proportional to the electric field. This method was successfully used to observe interference, beam splitting and reflection of SPP modes (Ditlbacher et al., 2002b).

It is also important to note that even in more complex cases of metal/dielectric interfaces, SPP modes can exist with some parallel characteristics. A good example is that of a thin metal film where the thickness is of the same order of the skin depth. In this case, the SPP modes excited in each of its surfaces couple together and give rise to the so-called long-range SPP modes (Sarid, 1981). There is also a current research interest in particle plasmon-polariton modes that are excited in particles with have dimensions much smaller than the E-M wavelength. A detailed discussion on this topic can be found in (Hutter et al., 2004).

5. Focused-ion beam nanofabrication

Plasmonic structures can be obtained by many standard techniques, some of them are: optical lithography, electron-beam lithography, focused-ion beam (FIB) lithography, atomic layer deposition, soft lithography and template stripping (Lindquist et al., 2012). Taking into account that in the present work FIB lithography was extensively used, we will briefly discuss here only this fabrication technique. A comprehensive discussion of the other mentioned techniques (as well as FIB lithography) is very well presented in the work by Lindquist et al.

FIB lithography has been extensively used for direct fabrication of metallic nanostructures, by making patterns on substrates (Melngailis, 1987; Orloff et al., 1993; Langford et al., 2007). It can also be used for the deposition of various metals by using ion-beam induced deposition (Tao et al., 1990), for doping semiconductors (Melngailis, 1987; Moberlychan et al. 2007), and for preparing transmission electron microscope (TEM) samples (Reyntjens et al., 2001; Mayer, 2007). These methods proved to be very useful to make tests of device designs and geometries, fix masks or electrical traces, or to produce high-resolution ion-beam images, and have been an essential tool for the development of the field of plasmonics.

FIBs impinging on a surface offer a very different form of nanopatterning compared with other conventional methods that use resist, exposure and development. In general, the accelerating potential of the ions is of tens of kilovolts, the current beams range from many orders of magnitude, from picoamps to several nanoamps. Depending on the column optics, ion source and beam current, the beam spot sizes can range from ~ 5 nm up to a micrometer. Numerous ion species can be used in the setup, such as Al, Au, B, Be, Cu, Ga, Ge, Fe, In, Li, P, Pb and Si, the most commonly used being the semiconductor dopants (Melngailis, 1987; Orloff et al., 1993). In particular, Ga is widely used due to its low melting temperature (30 °C), low volatility and low vapour pressure (Volkert et al., 2007).

The FIB system is based on a liquid-metal ion source which is used to produce a smaller and brighter ion beam (Volkert et al., 2007). A metal source is heated up, such that it flows down and wets a sharp tungsten needle. An extraction voltage applied between the metal source and an extraction aperture forces the liquid metal to be pulled into an extremely sharp "Taylor–Gilbert" cone (Volkert et al., 2007; Forbes et al., 1996). The balance between the electrostatic force produced by the extraction voltage and the surface tension forces in the liquid (Orloff et al., 1993), the liquid source can have a tip size on the order of several nanometers (Melngailis, 1987). Then, the ions are extracted from the tip of this cone by field emission. As opposed to e-beam imaging systems, the "lenses" used in FIB are electrostatic and not magnetic, this happens because the Lorentz force in heavy ions, as those used, is much smaller than in electrons with the same kinetic energy (Volkert et al., 2007). Some other similarities and differences in these two kind of systems is presented in Fig. 5(a). To make the patterns a fast beam blanker is used. In commercial systems many patterns can be drawn by using pre-fabricated CAD proprietary files or using options within the control software. The final size and the resolution of the focused beam can, in general, be affected by chromatic aberration, i.e., the energy dispersion of ions in the beam, but with a good approximation it can be regarded as a Gaussian profile. The minimum beam size is on the order of ~ 5 nm. A schematic view of the ion source is shown in Fig. 5(b).

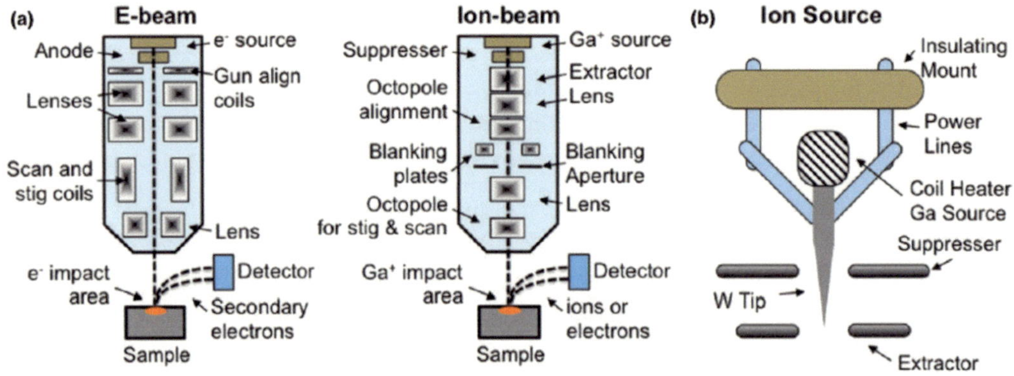

Figure 5. (a) Schematic view showing the similarities and differences between typical e-beam and FIB systems. (b) Scheme of the gallium liquid-metal source (Lindquist et al., 2012)

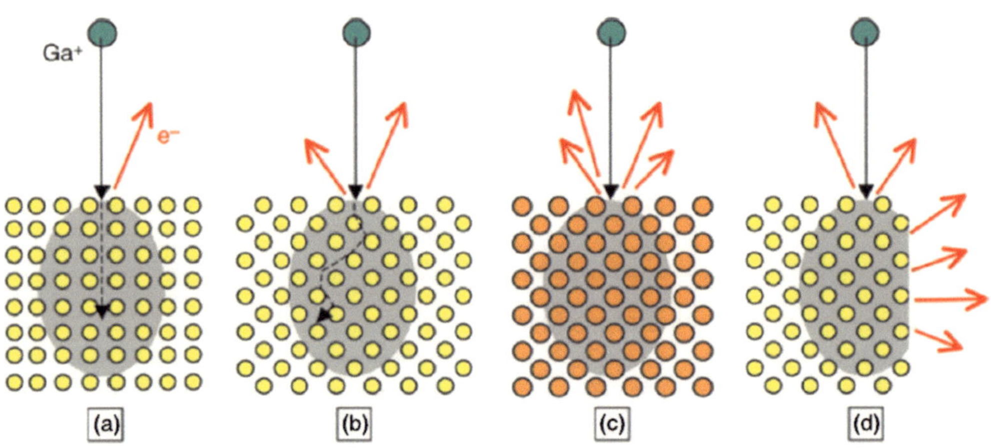

Figure 6. The crystal orientation of a sample can affect the FIB sputtering rates, shown in (a) and (b). (c) The sputtering rate is also affected by the mass of the atoms (orange atoms are more massive) and (d) by the local geometry of the sample (Lindquist et al., 2012). Original figure from Volkert et al.

In the process of milling a substrate by using FIB, many effects can be produced, with approximately one to five atoms removed per incident ion depending on the ion energy or substrate. It is even possible to displace atoms from their equilibrium positions, to induce chemical reactions and use the emerging electrons for imaging. The Gaussian profile of the beam is not the only factor that the shape of the milled groove depends (Melngailis, 1987). Redeposition and self-focusing effects can lead to large geometric differences depending on whether the patterns designed were done in single of multiple steps, even with the same overall dose. Also, milling a trench with large total ion doses deviates from the Gaussian profile, giving an unexpectedly deep, V-shaped groove (Melngailis, 1987). The FIB milling process depends on many aspects such as the material to be patterned, ion-beam incident angle, redeposition of sputtered materials and even the crystal orientation, as outlined in Fig. 6. Grain orientation-dependent FIB sputtering can also lead to severe surface roughness on polycrystalline samples, however, care can be taken to produce well-defined structures (Lindquist et al., 2012).

FIB can also be used for the deposition of various metals (such as W, Pt, C and Au) via site-specific chemical vapour deposition (CVD), this can be done by using a gas injection system (GIS) (Volkert et al., 2007). A high-efficiency deposition can be achieved, with about 1 μm per minute accumulation rates, by adjusting the gas precursor flow rates and the ion-beam current density. The reaction of the ion beams with the precursor materials offers the ability to weld micromanipulators to specific parts of a substrate in situ. With subsequent FIB milling and thinning, those parts are cut free, and are often mounted to a TEM imaging grid (Mayer, 2007). A deficiency of this kind of system is that metals deposited via FIB have a high contamination by carbon. It can also be used other GIS systems, instead of depositing metals, to enhance the inherent FIB milling etch rate.

FIB instruments offer many significant advantages, like a direct write, maskless, high-resolution nanofabrication with the ability to sputter, image, analyse and deposit. It is possible to design 2D and 3D patterns (Langford et al., 2007). However, it has some limitations as a patterning tool, particularly for metals. FIB milling is a serial lithography technique, such as e-beam lithography, patterning only one spot or device at a time, unlike optical lithography that patterns the whole wafer with one short exposure. Large area patterning is not feasible. For high-resolution (< 100 nm) features, FIB milling can also be slow since very low currents (~ pA) must be used. Along with FIB-induced sample damage (Mayer, 2007), Ga ions are implanted at atomic fractions of 1–50% near the sample surface (Volkert et al., 2007). The plasmonic properties of the patterned metal films can be degraded is this process. For high surface roughness of metals the SPP propagation length can be strongly reduced (Lindquist et al., 2012). As such, the advantages and disadvantages of FIB milling need to be taken into account when fabricating new optical or electronic devices. When combined with a template stripping technique, many of these roughness and contamination issues are minimized, since FIB is then used to only pattern a reusable template, leaving the resulting metal films smooth and contamination-free (Lindquist et al., 2012).

6. Applications: Optical transmission through subwavelength single slits

6.1. Influence of metallic film thickness

6.1.1. Motivation

By focusing attention on the properties of arrays fabricated in metallic films with fixed thickness, some previous studies mentioned in Section 1 apparently missed the important role played by the film thickness. For example, in the designing of surface-plasmon-based sensors, a proper choice of the thickness of the metallic film for the optimization of the device sensitivity is very important (Fontana, 2006). There are only a few theoretical (Xie et al., 2005; Fontana, 2006; Janssen et al., 2006) and experimental (Shou et al., 2005; Kim et al., 2006; Pang et al., 2007) investigations taking into account the influence of film thickness on the considered process. Therefore, the present studies are motivated by the necessity to understand the physics of this phenomenon and to develop optimum designs practices for subwavelength structure fabrication. We present here a systematic study of the optical transmission through subwavelength slits fabricated in Ag and Au film samples possessing different thicknesses. The influence of slit width was also considered. The present work deals with single slits. The extended dimension of the long axis imposes directionality on the transmitted light beam, the divergence of which can easily be controlled (Laluet et al., 2008). A slit is thus an interesting structure since it combines the compactness of a single defect with the directionality of an array launcher. Moreover, in order to remove measurement ambiguities existing in setups employing an incoherent and broadband light source dispersed through a scanning spectrophotometer, as shown in the left panel of Fig. 2, we have measured the transmission intensity through the slits using coherent and monochromatic spectral sources (Pacifici et al., 2008; Ferri et al., 2011).

6.1.2. Experimental considerations

A series of Ag films with thicknesses of 120, 160, 200, 270, and 330 nm and a set of Au samples with thicknesses of 120, 180, 260, 360, and 450 nm, as measured by a Talystep profilometer, were thermally evaporated onto BK7 glass substrates. Slits with widths in the range of approximately 70–150 nm in the Ag films, and 120–270 nm in the Au films, were milled with an FEI focused ion beam QUANTA 3D 200i (Ga^+ ions, 30 keV). In order to verify the depth of the slits, the gallium ions' source was calibrated using atomic force microscopy. For example, the right panel of Fig. 7 shows a scanning electron micrograph of a slit with 150 nm of width fabricated in the 200 nm thick Ag film.

We have undertaken a series of high-resolution measurements of the optical transmission through the slits. The transmission measurement setup consists of 488.0 nm (for Ag) and 632.8 nm (for Au) wavelength light beams from Ar ion and HeNe lasers, respectively, with a power of about 1 μW, aligned to the optical axis of a microscope. The beam is focused at normal incidence onto the sample surface by a 20× microscope objective (with an NA of 0.4) in TM polarization (magnetic field component parallel to the long axis of the slits). Light intensity transmitted through each slit is then gathered by an optical fibre and detected with a CCD array detector. It was used on a multimode fibre with an NA of 0.22 and a core diameter of 200 μm. Light intensity is obtained by integrating the signal over the entire region of interest in the CCD image and subtracting the background originating from electronic noise. The transmitted intensity of every slit was recorded in the far-field by the CCD as the sample was stepped using an x-y translation stage. The left panel of Fig. 7 shows the schematic of the measurement setup.

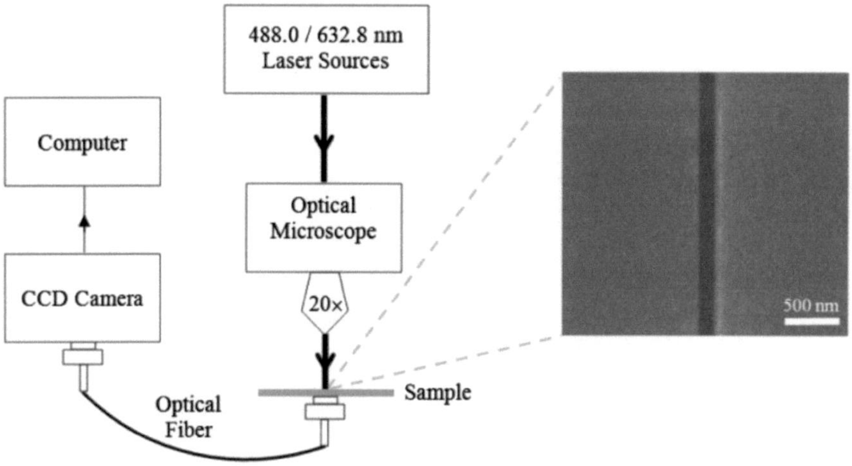

Figure 7. Left panel shows a schematic of the optical transmission experiment. 488.0 nm (for Ag) and 632.8nm (for Au) Ar ion and HeNe laser light sources, respectively, are normally focused onto the sample surface by a 20× microscope objective lens. A CCD camera records the transmission intensity through the slits as the sample surface was stepped. Right panel shows a scanning electron micrograph (taken with 40000× magnification) of a typical structure. The considered slit has approximately 150 nm of width and was focused-ion-beam milled through a 200 nm thick Ag layer. In the experiments, the thicknesses of the Ag and Au films were varied in the range of 100–450 nm. The width is varied from 70 to 270 nm (Ferri et al., 2011)

6.1.3. Results and discussion

Fig. 8 shows the physical picture adopted in this work to investigate the light transmission through the subwavelength slits. The essential elements of the model, which describes a plasmonic damped wave with amplitude decreasing as the inverse of the film thickness (Gay et al., 2006b), are represented in the sketch of Fig. 8. Basically, an incident monochromatic light beam with wave vector k_0 in air is linearly polarized perpendicular to the slit of subwavelength width w, milled in a metallic film with thickness t, and deposited on a dielectric substrate (BK7 glass).

The far-field intensity enhancement for the single slits involves multiple coupling processes (see Fig. 8). Initially, the incident laser light generates SPs on the metal film. Because of vertical plasmon coupling, which depends on the film depth (t), surface charges are induced on the top metal film and simultaneously a strong electric field is generated inside the slit. Subsequently, an SPP mode (Maier, 2007), i.e., an electromagnetic excitation propagating at the interface between the dielectric and the metallic conductor, evanescently confined in the perpendicular direction, is generated on the metal film/BK7 interface. The SPP evanescent mode travels along the interface toward the slit, where it reconverts to a propagating wave and interferes with the travelling field directly transmitted through the slit. Additionally, penetration of the incident field inside the film enables the excitation of localized SP resonances (Maier, 2007) on the rim of the aperture, which contribute to the superposed output field. In this way, induced dipole moments at each rim form an "antenna coupling", which radiatively generate strong field enhancement (top and bottom). Then, the intensity of the resulting field can be written as

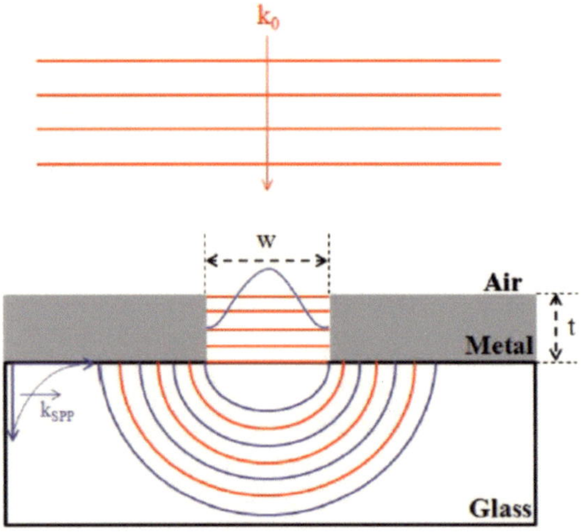

Figure 8. Illustration of the adopted model. A single frequency incoming plane wave with wave vector k_0 in air is linearly polarized perpendicular to a slit of subwavelength width w, milled in a metallic film with thickness t deposited on a BK7 glass substrate. Here, k_{SPP} is the wavevector of the SPP mode. (Ferri et al., 2011)

$$E \approx \frac{E_0}{\pi}\frac{w}{t}\cos\left(k_{SPP}t + \frac{\pi}{2}\right) \tag{4}$$

where $k_{SPP} = 2\pi/\lambda_{SPP}$, $\lambda_{SPP} = 2\pi/\mathrm{Re}\left[\beta\right]$, $\beta = k_0 n_{SPP}$, $n_{SPP} = \left(\varepsilon_{metal}\varepsilon_{glass}/\varepsilon_{metal} + \varepsilon_{glass}\right)^{1/2}$, and $k_0 = 2\pi/\lambda_0$ (Gay et al., 2006b; Pacifici et al., 2008; Maier, 2007). Here, E_0 represents the electrical field of the incoming plane wave, where λ_0 is its wavelength. Also, k_{SPP} and λ_{SPP} are the wavevector and wavelength of the SPP, β is the propagation constant of the superposed travelling wave, and n_{SPP} is the effective index of the SPP, which is for the interface between the metal and dielectric. In addition, ε_{metal} and ε_{glass} are the dielectric permittivities of metal and glass, respectively, and are functions of the excitation wavelength. In this sense, ε_{Ag} = - 7.89 + 0.74i and ε_{glass} = 2.31 are the tabulated dielectric constants of Ag and BK7 glass in the wavelength of 488.0 nm. In the same way, ε_{Au} = - 9.49 + 1.23i and ε_{glass} = 2.29 are the corresponding dielectric constants of Au and BK7 in 632.8 nm (Palik, 1985). Here it is important to point out that the wavelength values of 488.0 and 632.8 nm are known to be close to the plasmon excitation wavelength of Ag and Au, respectively. Increasing the metallic film thickness leads to decoupling of the top and bottom antenna.

For illustration purposes, Fig. 9 shows corresponding 2D numerical simulations carried out with Comsol Multiphysics® for TM-polarized waves for a 150 nm slit fabricated in an Ag film when illuminated by the line at 488.0 nm of an Ar ion laser. Fig. 9(a) shows the amplitude of the magnetic H field (along the z direction). Figs. 9(b) and 9(c) shows the amplitude of the electric E field (in the y direction), with its vector representation in the x-y plane. Fig. 9(a) shows how the incident plane wave is modified by the existing subwavelength slit. It is possible to see that the considered wave is almost completely reflected from the unstructured part of the film. Around the slit entrance, the amplitude of the standing wave is markedly attenuated, where some lightwave transmission to the exit facet is apparent. On the dielectric/metal interface, a train of surface waves (SPPs) is evident together with waves propagating into space. In the rims, different charge configurations can be obtained, which can be symmetric or antisymmetrically coupled (Prodan et al., 2003). This coupling leads to determined charge configurations in each rim of the slit (top and bottom). From Figs. 9(b) and 9(c), these surface modes are clearly seen. It is possible to notice from the figures these resonances [antisymmetrically and symmetrically coupled in Figs. 9(b) and 9(c), respectively] on the facets of the slit. These modes are associated with localized SPs, which are nonpropagating excitations due to direct light illumination of the conduction electrons of the metallic nanostructure coupled to the electromagnetic field (Maier, 2007). A similar behaviour was observed in the simulations for Au films when

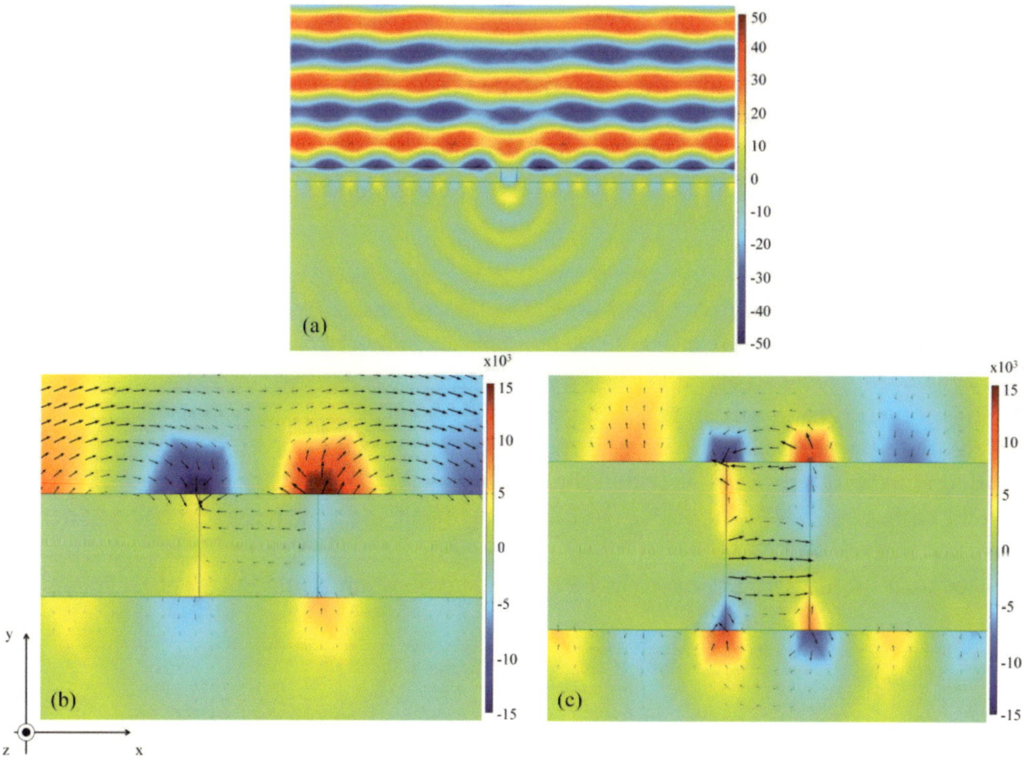

Figure 9. 2D numerical simulations of a 150 nm slit fabricated in an Ag film when illuminated by the line at 488.0 nm of an Ar ion laser. (a) Amplitude of the magnetic H field (along the z direction). (b) and (c) Amplitude of the electric E field (in the y direction), and its vector representation in the x-y plane. The value of the Ag film thickness in (a) and (b) is 120 nm, and in (c) is 270 nm. Length spans: (a) x = 4 μm and y = 2 μm, (b) and (c) x = 600 nm and y = 400 nm

excited by the line at 632.8 nm of an HeNe laser. The main apparent difference is the lower transmitted intensity due to a higher absorption loss attributable to the particular characteristics of Au (Palik, 1985), i.e., $Re[\varepsilon_{Au}] = - 9.49$ in contrast to $Re[\varepsilon_{Ag}] = - 7.89$. This is verified in Fig. 10, where it can be seen that the normalized transmission intensity for the Ag films is improved more than for the Au samples.

In general, the present numerical simulations qualitatively show appreciable light transmission through the slits. Actually, it was experimentally observed that the transmission sensitively depends on the metallic film thickness and slit width. As a first approximation, the theoretical slit transmission intensity can be given simply by the square modulus of Eq. (4). In this way, Fig. 10 plots as predicted [from Eq. (4)] and measured (using the setup shown in Fig. 7) transmission intensities as a function of the film thickness for the various slit widths milled in the Ag and Au samples. Also, the insets of Fig. 10 show the measured transmission versus slit width for certain film thicknesses. The relative slit transmission intensities are obtained by subtracting the background originating from the metal film and normalizing to the intensity from the wider slit structures. It is valuable to notice from Fig. 10 the very good correspondence between the theoretical estimate and the experiment. Therefore, taking into account the errors associated with the experimental determination of film thickness, slit width, and optical transmission intensities, it is possible to affirm that (1) the slits' transmission varies with metallic film thickness and presents a damped oscillatory behaviour as the film thickness increases, and (2) the transmission increases linearly with increasing slit width for a fixed metallic film thickness. Although the general behaviour is similar, distinct optical properties [see Fig. 4(a)] lead to perceptible differences in the transmitted intensity and the position of maxima and minima between the Ag and Au films.

To help in elucidating the first observation, it is valuable to note that Fabry–Perot (FP) resonances are expected to contribute to the enhanced transmission of subwavelength slit arrays (Porto et al., 1999; Takahura, 2001; Pang et al., 2007; Garcia-Vidal et al., 2002). In this way, FP modes related to the finite depth of the slits in the present films should give rise to transmission maxima at certain wavelengths (Garcia-Vidal et al., 2002). Actually, an accurate analysis recently published shows that the two maxima observed in Figs. 10(a) and 10(b) correspond to FP-like resonances within the slit volume for the first half-wave and full wave of the light within the slit (Weiner, 2011). In this sense, the FP multiple reflection effect within the slits leads to significant modulation of the transmission as a function of metal film thickness. These transmission maxima occur if the FP resonance condition is fulfilled (Li et al., 2009):

$$2k_0 Re\left[\frac{\lambda_0}{\lambda_{SPP}}\right]t + \arg(\phi_1\phi_2) = 2m_y\pi \tag{5}$$

where m_y (the FP mode) is an integer and ϕ_1 and ϕ_2 denote the phase of the reflection coefficients of the slit at the incident and output interfaces, respectively. Thus, the effect of slit depth on the transmission enhancement can be easily understood. When the incident wavelength and slit depth are satisfied by Eq. (5), a transmission maximum will occur. Furthermore, from Eq. (5), it is expected that the transmission under a certain incident wavelength has a period of $\lambda_{SPP}/2$ as a function of slit depth. The results of Fig. 10 can now be simply explained for the present Ag and Au films. Therefore, when the film thickness is near half- or full-integer wavelengths of the guided mode within the slit "cavity", optimal transmission is achieved, which implies a field enhancement inside the slit.

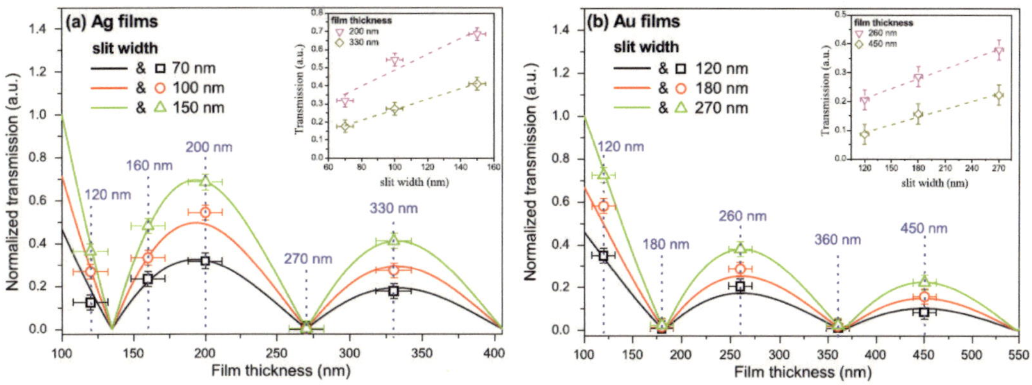

Figure 10. Theoretically estimated (lines) and experimental (symbols) normalized slit transmission intensities versus film thickness for the slits of the (a) 120, 160, 200, 270, and 330 nm thick Ag films, and the (b) 120, 180, 260, 360, and 450 nm thick Au samples. The dotted straight lines point out the thickness of the considered samples. The insets show the measured transmission versus slit width for some film thicknesses. Here, the dashed straight lines are linear fittings of the experimental points (Ferri et al., 2011)

We can apply an FP analysis to obtain the finesse F from Fig. 10, given by

$$F = \frac{\pi}{2\sin^{-1}\left(1/\sqrt{f}\right)} \tag{6}$$

where $f = 4R(1 - R)^{-2}$ is the finesse factor. Here, R is the reflectivity, given by $R = 1 - T$, where T is the transmission (Born et al., 1993). We determine from Fig. 10(a) for the Ag samples with 100 nm of slit width and thicknesses of 120, 160, 200, 270, and 330 nm, the reflectivities R = 0.73, 0.67, 0.46, 0.99, and 0.73, and the corresponding finesses F = 9.98, 7.74, 3.84, 312.58, and 9.89, respectively. From Fig. 10(b) for the Au samples with 120nm of slit width and thicknesses of 120, 180, 260, 360, and 450 nm, we determine the reflectivities R = 0.66, 0.99, 0.79, 0.98, and 0.91, and the resultant finesses F = 7.45, 312.58, 13.27, 155.49, and 33.29, respectively. Here it is important to point out that for an FP cavity, the definition of quality factor (Q factor) is equivalent to the finesse (Shyu et al., 2011). Therefore, we can clearly see that both the R and Q factor values are significantly affected by the film thickness for a fixed slit width. Also, it is interesting to notice that near-zero transmission is a sign of high reflectivity values and high Q factors. For the maximum transmission points, a backward reasoning applies.

Finally, the fact that the transmission increases linearly with increasing slit width is in accordance with literature (Kihm et al., 2008), where it was observed that the far-field transmitted intensity from a single slit shows a monotonic increase with the width, as expected from macroscopic intuition. In other words, the physical cavity length L_z and the optical cavity length $L_c(\lambda)$ are related, such that $L_c(\lambda) = L_z + 2\delta(\lambda)|r(\lambda)|^2$, where $r(\lambda) = |r(\lambda)|\exp(i\phi)$ is the Fresnel coefficient, describes the shift of resonance wavelength from a perfect metal reflector due to field penetration $\delta(\lambda)$ into the metal mirror. But L_z is constant for all studied samples (20 μm), resulting in ϕ_z constant for a determined depth of metallic film. Nevertheless, the monotonic increase with the width w (= L_x) can also be explained considering FP resonances, for which we will use a simple analytical model to investigate the experimental results based on geometric arguments. Considering the standing wave mode in the cavity, when the penetration depth is ignored, the resonant condition of the slits can be written as

$$\frac{1}{\lambda_{SPP}^2} = \left(\frac{m_z + \phi_z}{2L_z}\right)^2 + \left(\frac{m_x + \phi_x}{2L_x}\right)^2 \tag{7}$$

We have applied Eq. (7) using the data of 200 nm of depth with widths L_x = 70, 100, and 150 nm for an Ag film, and 260 nm of depth with widths L_x = 120, 180, and 270 nm for an Au film. L_z = 20 μm for both samples. Also, we considered in our calculations that ϕ_z and ϕ_x are practically constant, since we did not observe any peak shift in the transmission spectra of the samples in analysis. In the interface, values of $\lambda_{SPP\text{-}Ag}$ = 269.9 nm and $\lambda_{SPP\text{-}Au}$ = 361.6 nm were used. It was obtained for the Ag film m_z = 78 and m_x = 1, 1.4, and 2, and for the Au film m_z = 145 and m_x = 1, 1.5, and 2, i.e., an increase in L_x allows an increase in the transmission intensity spectra for a fixed depth [see insets of Figs. 10(a) and 10(b)].

6.2. Multilayered metallic thin films

6.2.1. Motivation

Although the detailed picture of the transmission enhancement is still being investigated, the excitation of SPPs on the two surfaces of the metal film has been proposed to be involved in the process (Moreno et al., 2004; Lezec et al., 2004). In fact, it was already shown that when two perforated metal films are spaced by a dielectric layer (cascaded metallic structure), the transmission is further increased compared to a single perforated metal film (Ye et al., 2005). Additionally, bimetallic structures, such as films and nanoparticles, have attracted considerable attention for plasmon resonance excitation (Zynio et al., 2002; Gupta et al., 2005; Tan et al., 2007; Chen et al., 2010). These works are focused on improving sensitivity and evanescent field enhancement by optimization of the thickness of Ag/Au layers. Recently, optimization in terms of spectral characteristics was also theoretically demonstrated (Dyankov et al., 2011).

Here, we propose novel structures providing a unique opportunity to generate plasmonic modes. The structures are based on subwavelength slits fabricated in multilayered metallic thin films. The main feature of the novel structure is that the metal film consists of alternating layers of Ag and Au. In this Section, we demonstrate that the slits transmission can be augmented by increasing their widths, with the advantage to offer minor losses in comparison with a single perforated metal film.

6.2.2. Experimental considerations

The experimental procedure adopted here is similar to that shown in Section 6.1.2. Multilayered Ag/Au/Ag/Au and Au/Ag/Au/Ag films with total thickness t = 200 nm (d = 50 nm for each layer) were thermally evaporated onto BK7 glass substrates. Slits with widths in the range of approximately 60–600 nm were milled in the films using FIB lithography with the same already mentioned conditions. The slit length was fixed at 5 μm. For example, the right panel of Fig. 11 shows a scanning electron micrograph of a slit with width w about 180 nm fabricated in the Ag/Au/Ag/Au film.

The transmission measurement setup is identical to that shown in the left panel of Fig. 7. Here, the 488.0 nm wavelength light beam from the Ar ion laser was used for the Ag/Au/Ag/Au film, and the 632.8 nm wavelength light beam from the HeNe laser was used for the Au/Ag/Au/Ag sample. The left panel of Fig. 11 shows the basic schematic of the experimental setup.

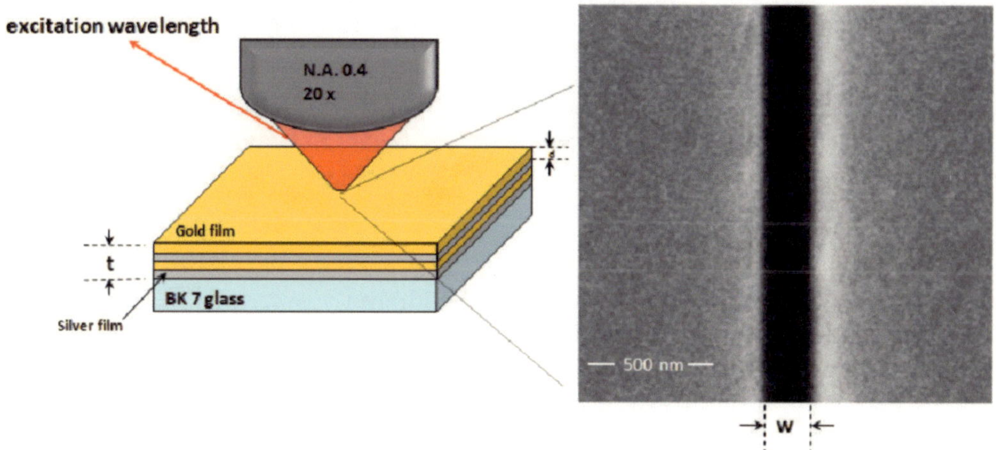

Figure 11. Left panel shows a simplified schematic of the optical transmission experiment. Similarly to the left panel of Fig. 7, a 488.0 nm (for the Ag/Au/Ag/Au film) and a 632.8 nm (for the Au/Ag/Au/Ag sample) Ar ion and HeNe laser light sources, respectively, are normally focused onto the sample surface by a 20× microscope objective lens. Right panel shows a scanning electron micrograph (taken with 60000× magnification) of a typical structure. The considered slit has approximately 180 nm of width and was focused-ion beam milled through the Ag/Au/Ag/Au sample. In the experiments, the total thickness t of the Ag/Au/Ag/Au and Au/Ag/Au/Ag films was fixed at 200 nm. The width w is varied from 60 nm to 600 nm (Ferri et al., 2012)

6.2.3. Results and discussion

Fig. 12(a) shows the physical picture adopted in this work to investigate the light transmission through the subwavelength slits fabricated in the multilayered metallic films. The essential elements of the model are represented (Gay et al., 2006b). Basically, an incident monochromatic light beam in air is linearly polarized perpendicular to the slit of subwavelength width w, milled in a multilayered metallic film with thickness t, deposited on a dielectric substrate (BK7 glass). Each metallic layer has a thickness d (= 50 nm).

The physical processes involved in the far-field intensity enhancement for the present subwavelength slits are similar to those discussed in Section 6.1.3, in the sense that the first metallic layer is responsible to start the generation of plasmonic excitations. Form our simulation results we can see that at the interface between the adjacent metallic layers the condition for generation of SPPs is not fulfilled, Fig. 12(b). Nevertheless, there is an induced charge current due to the plasmonic surface excitations from the first to the last metallic layer, see Fig. 12(c), resulting in an asymmetric distribution [similar to Fig. 9(b)]. Additionally, the transmission over each layer is given by the Beer-Lambert law, i.e., $I_i = I_{i-1}\exp(-\alpha d)$ (Born et al., 1993), where α is the absorption coefficient of the corresponding metal layer, d is the thickness of the layers, and i is the layer number. Then, the resulting transmitted intensity can be written as (Gay et al., 2006b; Maier, 2007; Pacifici et al., 2008; Ferri et al., 2011),

$$I_{out} = I_i + \left| \frac{E_0}{\pi} \frac{w}{t} \cos\left(k_{SPP} t + \frac{\pi}{2} \right) \right|^2, \ i = 1,\ldots,4 \quad (8)$$

Here, the physical quantities of the second term are identical with those in Eq. (4). The only difference is that the considered dielectric medium is the air, which tabulated dielectric constant is ε_{air} = 1.00 (Palik, 1985). In this sense, the first layer is assumed to govern the extraordinary transmission (I_{SPP}) of the subwavelength slits fabricated in the present multilayered metallic films.

Numerical simulations carried out with Comsol Multiphysics® were also performed for the multilayered metallic samples. A similar behaviour was observed in comparison to that shown in Section 6.1.3. For illustration purposes, Figs. 12(b) and 12(c) show simulations for a 50 nm slit fabricated in the Au/Ag/Au/Ag film when illuminated by the line at 632.8 nm typical of an HeNe laser. Fig. 12(b) shows the amplitude of the magnetic H field (along the z direction). Fig. 12(c) shows the amplitude of the electric E field (in the y direction), with its vector representation in the x-y plane.

The theoretical slit transmission intensity can be given by Eq. (8). In this way, Fig. 13 plots as predicted and measured transmission intensities as a function of the slit width for the various subwavelength structures milled in the Ag/Au/Ag/Au and Au/Ag/Au/Ag samples. For comparison purposes, the insets of Fig. 13 also show the simulated slit optical transmission obtained from Comsol Multiphysics® versus slit width for the considered multilayered metallic films and single perforated Ag and Au films with 200 nm of thickness. The relative slit transmission intensities are obtained by subtracting the background originating from the metal film and normalizing to the intensity from the wider slit structures. It is valuable to notice from Fig. 13 the very good correspondence between the theoretical estimative and experiment. Therefore, it is possible to affirm that: (1) the transmission increases linearly with increasing slit width, and (2) for a fixed width, the transmission of the multilayered structures is augmented in comparison with a single perforated metal film of the same thickness. It is evident that this last observation is more apparent for the Au/Ag/Au/Ag film.

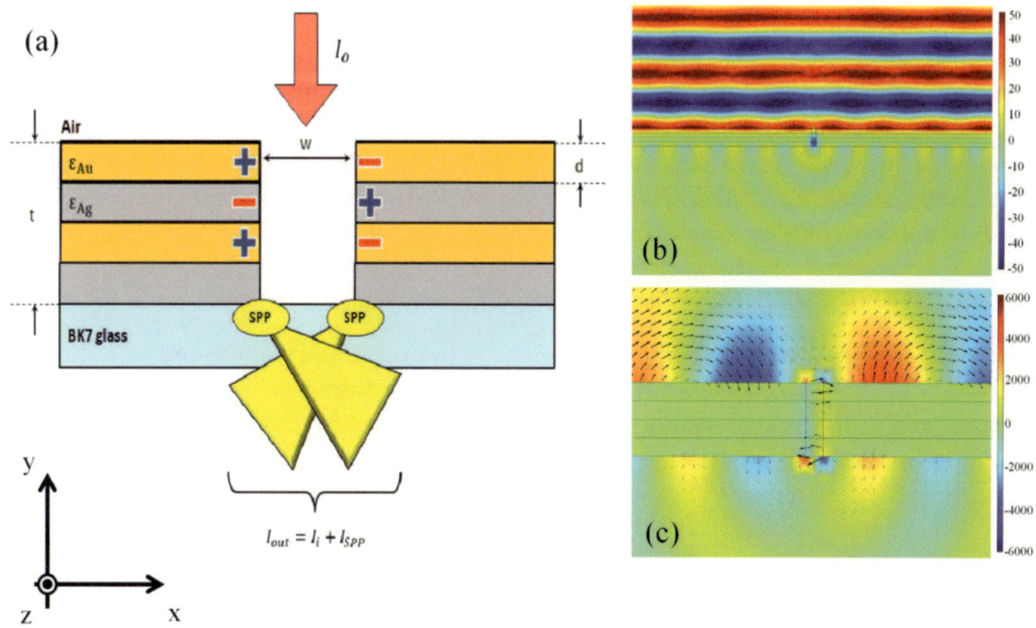

Figure 12. (a) Illustration of the adopted model. A single frequency incoming plane wave with intensity I_0 in air is linearly polarized perpendicular to a slit of subwavelength width w, milled in a multilayered metallic film with thickness t deposited on a BK7 glass substrate. Each metallic layer has a thickness d. The overall transmitted intensity I_{out} is a sum of the transmission of each layer I_i with the contribution due to the SPPs created in the air/metallic film interface I_{SPP}. (b) and (c) 2D simulations of a slit fabricated in the Au/Ag/Au/Ag film. (b) Amplitude of the magnetic H field (along the z direction). (c) Amplitude of the electric E field (in the y direction), and its vector representation in the x-y plane. Length spans: (b) x = 4 μm and y = 2 μm, and (c) x = 600 nm and y = 400 nm (adapted from Ferri et al., 2012)

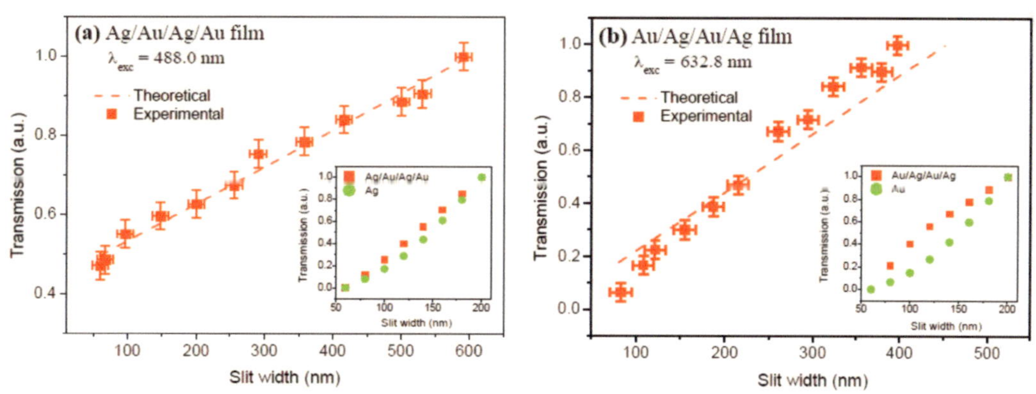

Figure 13. Theoretically estimated (dashed straight lines) and experimentally obtained (squares) normalized slit transmission intensities versus slit width for the various subwavelength structures milled in the (a) Ag/Au/Ag/Au and (b) Au/Ag/Au/Ag samples. The insets show details of the simulated slit optical transmission versus slit width for the considered metallic multilayered films and single perforated Ag and Au films with 200 nm of thickness (Ferri et al., 2012)

The fact that the far-field transmitted intensity from the present slits shows a monotonic increase with their widths is in the same trend of that previously observed in single perforated metal films (Ferri et al., 2011; Kihm et al., 2008), as expected from macroscopic intuition. In that case, the dependence could be explained considering Fabry-Perot resonances of the standing wave mode in the slit "cavity", in conjunction with the generation of SPPs. However, in the present case, such a resonant condition (Ferri et al., 2011; Kihm et al., 2008; Li et al., 2009) cannot be applied, since we have distinct reflection coefficients due to the existence of different materials in the slits. Nevertheless, the monotonic increase with the width is expected simply by considering the dependence of the transmitted intensity with the w parameter in Eq. (8).

Finally, the observation that the transmission of the metallic multilayered structures is augmented in comparison with a metal film of the same thickness when perforated with subwavelength slits, can be elucidated considering that, for the present multilayered films, the optical transmission profile is assumed to be mainly governed by the first metallic layer,

given that it is responsible to start the generation of plasmonic excitations. Subsequently, we just have electronic conduction to the underlying metallic layers, as already pointed out. Furthermore, each metallic layer (with 50 nm of thickness) additionally contribute to the overall transmission according to the Beer-Lambert law, in contrast to a single perforated metal film with the same total thickness of 200 nm of the proposed multilayered films.

Independently of the preceding discussion about multilayered metallic thin films, it is valuable to mention that gain-assisted propagation of SPPs at the interface between a metal and a dielectric with optical gain have been the focus of much research activity (Avrutsky, 2004; Nezhad et al., 2004). In this context, Er^{3+}-doped tellurite glasses as the dielectric medium is very attractive (Wang et al., 1994). Very recently, our research group have gave significant contributions regarding the excitation and/or improvement of the luminescence of Er^{3+} ions embedded in these glassy matrices through plasmonic nanostructures (Rivera et al., 2012a; Rivera et al., 2012b).

7. Conclusion

The physics of the transmission of light through subwavelength apertures in metallic films has been a topic of intense research in recent times. In this chapter, we have presented a review of this field, showing some essential subjects involved in this phenomenon. Although the current understanding of this phenomenon is not complete or even not substantially correct, the materials presented in many literatures are useful and clue us on how to go ahead. In particular, we presented in this chapter some contributions of our research group regarding the optical transmission through subwavelength single slits in metallic thin films. The simulations qualitatively reveal that the transmission profile is controlled by interference between the incident standing wave and plasmonic surface excitations. It was possible to observe that the slits' transmission is significantly affected by the metallic film thickness, presenting a damped oscillatory behavior as the film thickness is augmented. In addition, for a fixed metallic film thickness, the transmission increases linearly with increasing slit width. For a fixed wavelength and slit width, FP modes within the slits lead to significant modulation of the transmission as a function of metal film thickness. As well, it was shown that the transmission of multilayered structures is augmented in comparison with a single perforated metal film with a similar thickness. In this sense, we have demonstrated that metallic multilayered structures have the advantage to offer minor losses in comparison with a single perforated metal film.

Author details

V. A. G. Rivera, F. A. Ferri, O. B. Silva, F. W. A. Sobreira and E. Marega Jr.
Instituto de Física de São Carlos, Universidade de São Paulo, São Carlos, Brazil

Acknowledgement

The authors are indebted to Prof. J. Weiner, Prof. A. R. Zanatta and Dr. M. A. Pereira-da-Silva (all at the Instituto de Física de São Carlos, USP, Brazil) for the helpful discussions, optical transmission experiments and atomic force microscopy measurements, respectively. We also would like to thanks Prof. B.-H. V. Borges (Departamento de Engenharia Elétrica, EESC, USP, Brazil) for the support with Comsol Multiphysics®. This work was financially supported by the Brazilian agencies FAPESP and CNPq under CEPOF/INOF.

8. References

Avrutsky, I. (2004). Surface Plasmons at Nanoscale Relief Gratings between a Metal and a Dielectric medium with Optical Gain. *Physical Review B*, Vol.70, No.15, pp. 155416-1 – 155416-6

Bethe, H.A. (1944). Theory of diffraction by small holes. *Physical Review*, Vol.66, No.7–8, (October), pp. 163–182.

Born, M. & Wolf, E. (1993). Principles of Optics: Electromagnetic Theory of Propagation Interference and Diffraction of Light (Pergamon), ISBN 978-008-0139-87-6, Cambridge, United Kingdom.

Bouhelier, A., Huser, T., Tamaru, T. et al. (2001). Plasmon optics of structured silver films. *Physical Review B*, Vol. 63, No. 15, (March) pp. 155404-1 – 155404-9.

Bouwkamp, C.J. (1950). On Bethe's Theory of Diffraction by Small Holes. *Philips Research Reports*, Vol.5, No.5, pp. 321–332.

Bouwkamp, C.J. (1950). On the Diffraction of Electromagnetic Waves by Small Circular Disks and Holes. *Philips Research Reports*, Vol.5, No.6, pp. 401–422.

Bouwkamp, C.J. (1954). Diffraction theory. *Reports on Progress in Physics*, Vol.17, No.1 pp. 35–100.

Chen, X. & Jiang, K. (2010). Effect of aging on optical properties of bimetallic sensor chips. *Optics Express*, Vol.18, No.2, pp.1105-1112.

Degiron, A., et al.(2004). Optical Transmission Properties of a Single Subwavelength Aperture in a Real Metal. *Optics Communications*, Vol.239, No.1-3, pp.61–66.

Ditlbacher, H., et al. (2002). Fluorescence Imaging of Surface Plasmon Fields. *Applied Physics Letters*, Vol.80, No.3, pp. 404–406.

Ditlbacher, H., et al.(2002). Two-dimensional Optics with Surface Plasmon Polaritons. *Applied Physics Letters*, Vol.81, No.10, pp. 1762–1764.

Dyankov, G., et al.(2011). Plasmon Modes Management. *Plasmonics*, Vol.6, No.4, pp. 643–650

Ebbesen, T.W., et al. (1998). Extraordinary optical transmission through sub-wavelength hole arrays. *Nature*, Vol.391, pp. 667–669.

Ferri, F.A., et al. (2011). Influence of Film Thickness on the Optical Transmission Through Subwavelength Single Slits in Metallic Thin Films. *Applied Optics*, Vol.50, No.31, pp. G11-G16.

Ferri, F.A. et al. (2012). Surface Plasmon Propagation in Novel Multilayered Metallic Thin Films. *Proceedings of SPIE*, Vol.8269, pp. 826923-1-826923-6, San Francisco, USA, January 23-26, 2012.

Fontana, E. (2006). Thickness Optimization of Metal Films for the Development of Surface-Plasmon-based Sensors for Nonabsorbing Media. *Applied Optics*, Vol.45, No.29, pp.7632–7642.

Forbes, R. & Djuric, Z (1996). Progress in understanding liquid-metal ion source operation *9th International Vacuum Microelectronics Conference*, pp.468-472, Saint Petersburg, Russia, July 7-12, 1996.

Garcia-Vidal, F.J. & Martin-Moreno, L. (2002). Transmission and Focusing of Light in One-Dimensional Periodically Nanostructured Metals. *Physical Review B*, Vol.66, No.15, pp. 155412-1 – 155412-10.

Garcia-Vidal, F.J. et al.(2010). Light Passing Through Subwavelength Apertures, *Reviews of Modern Physics*, Vol.82, No.1, pp. 729-787.

Gay, G. et al.(2006). Surface Wave Generation and Propagation on Metallic Subwavelength Structures Measured by Far-field Interferometry. *Physical Review Letters*, Vol.96, No.21, pp. 213901-1 – 213901-4.

Gay, G. et al.(2006). The Optical Response of Nanostructured Surfaces and the Composite Diffracted Evanescent Wave Model. *Nature Physics*, Vol.2, No.4, pp. 262–267.

Ghaemi, H.F., et al.(1998). Surface Plasmons Enhance Optical Transmission through Subwavelength Holes. *Physical Review B*, Vol.58, No.11, pp. 6779–6782.

Gupta, B.D. & Sharma, A.K.(2005). Sensitivity Evaluation of a Multi-layered Surface Plasmon Resonance-based Fiber Optic Sensor: a Theoretical Study. *Sensors and Actuators B:Chemical* Vol.107, No.1, pp. 40-46.

Hecht, B., et al.(1996). Local Excitation, Scattering, and Interference of Surface Plasmons. *Physical Review Letters*, Vol. 77, No. 9, pp. 1889–1892.

Huang, C.P., et al.(2007). Plasmonics:Manipulating Light at the Subwavelength Scale. *Active and Passive Electronic Components*, Vol.2007, No.30946, pp.1-13.

Hutter, E. & Fendler, J.H.(2004). Exploitation of Localized Surface Plasmon Resonance. *Advanced Materials*, Vol. 16, No. 19, pp. 1685–1706.

Jackson, J.D. (1999). Classical Electrodynamics. (John Wiley & Sons, Inc., New York, NY, 3rd edition), ISBN 978-047-1309-32-1.

Janssen, O.T.A., et al.(2006). On the Phase of Plasmons Excited by Slits in a Metal Film. *Optics Express*, Vol.14, No.24, pp.11823–11832.

Kalkum, F., et al.(2007). Surface-wave Interferometry on Single Subwavelength Slit-groove Structures Fabricated on Gold Films. *Optics Express*, Vol.15, No.5, pp. 2613–2621.

Kihm, H.W., et al.(2008). Control of Surface Plasmon Generation Efficiency by Slit-width Tuning. *Applied Physics Letters*, Vol.92, No.5, pp.051115-1 – 051115-3.

Kim, J.H. & Moyer, P.J.(2006). Thickness Effects on the Optical Transmission Characteristics of Small Hole Arrays on Thin Gold Films. *Optics Express*, Vol.14, No.15, pp.6595–6603.

Lalanne, P., et al.(2005). Theory of Surface Plasmon Generation at Nanoslit Apertures. *Physical Review Letters*, Vol.95, No.26, pp. 263902-1 – 263902-4.

Lal, S., et al.(2007). Nano-optics from Sensing to Waveguiding. *Nature Photonics*, Vol.1, No.11, pp. 641-648.

Laluet, J.-Y., et al.(2008). Generation of Surface Plasmons at Single Subwavelength Slits: from Slit to Ridge Plasmon. *New Journal of Physics*, Vol.10, No.10, pp. 105014-1 – 105401-9.

Langford R, et al.(2007). Focused Ion Beam Micro- and Nanoengineering. *MRS Bulletin*, Vol.32, No.5, pp. 417–423.

Lezec, H.J. & Thio, T.(2004). Diffracted Evanescent Wave Model for Enhanced and Suppressed Optical Transmission through Subwavelength Hole Arrays. *Optics Express*, Vol.12, No.16, pp. 3629–3651.

Li, Z.-B, et al.(2009). Fabry–Perot Resonance in Slit and Grooves to Enhance the Transmission through a Single Subwavelength Slit. *Journal of Optics A: Pure and Applied Optics*, Vol.11, No.10, pp. 1-4.

Lindquist, N.C. et al.(2012). Engineering Metallic Nanostructures for Plasmonics and Nanophotonics. *Reports on Progress in Physics*, Vol.75, No.3, pp. 1-61.

Maier, S. A. (2007). Plasmonics: Fundamentals and Applications. (Springer). ISBN 978-0387-33150-8.

Melngailis, J. (1987). Focused Ion Beam Technology and Applications. *Journal of Vacuum Science and Technology B: Microelectronics and Nanometers Structures*, Vol.5, No.2, pp.469-495.

MoberlyChan, W., et al.(2007). Fundamentals of Focused Ion Beam Nanostructural Processing: Below, at, and Above the Surface. *MRS Bulletin*, Vol.32, No.5, pp. 424–432.

Moreno, L. M. & Garcia-Vidal, F. J. (2004). Optical Transmission through Circular Hole arrays in Optically Thick Metal Films. *Optics Express*, Vol.12(16), No.16, pp. 3619-3628.

Nezhad, M., et al.(2004). Gain Assisted Propagation of Surface Plasmon Polaritons on Planar Metallic Waveguides. *Optics Express*, Vol.12, No.17, pp. 4072-4079.

Orloff, J. (1993). High-Resolution Focused Ion Beams. *Review of Scientific Instruments*, Vol.64, No.5, pp. 1-26.

Pacifici, D., et al.(2008). Quantitative Determination of Optical Transmission through Subwavelength Slit Arrays in Ag films: Role of Surface Wave Interference and Local Coupling between adjacent Slits. *Physical Review B*, Vol.77, No.11, pp. 1-5.

Palik, E.D.(1985). Handbook of Optical Constants of Solids. (Academic Press). ISBN 0125444206

Pang, Y., et al.(2007). Optical Transmission through Subwavelength Slit Apertures in Metallic Films. *Optics Communications*, Vol.280, No.1 pp. 10–15.

Porto, J.A., et al.(1999). Transmission Resonances on Metallic Gratings with very Narrow Slits. *Physical Review Letters*, Vol.83, No.14, pp. 2845–2848.

Prodan, E, et al.(2003). A Hybridization Model for the Plasmon Response of Complex Nanostructures. *Science*, Vol.302, No.5644, pp. 419–422.

Reyntjens, S. & Puers, R.(2001). A Review of Focused Ion Beam Applications in Microsystem Technology. *Journal of Micromechanics and Microengineering*, Vol.11, No.4, pp.287-300.

Rivera, V.A.G., et al.(2012). Focusing Surface Plasmons on Er 3+ Ions with Convex/Concave Plasmonic lenses. *Proceedings of SPIE*, Vol.8269, pp.82692I-1 – 8269I-6, San Francisco, USA, January 23-26, 2012.

Rivera, V.A.G., et al.(2012). Luminescence enhancement of Er 3+ Ions from Electric Multipole Nanostructure Arrays. *Proceedings of SPIE*, Vol.8269, pp.82692H-1 – 8269H-7, San Francisco, USA, January 23-26, 2012.

Sarid, D.(1981). Long-range Surface-plasma waves on very thin metal films. *Physical Review Letters*, Vol. 47, No. 26, pp. 1927–1930.

Shou, X., et al.(2005). Role of Metal Film Thickness on the Enhanced Transmission Properties of a Periodic Array of Subwavelength Apertures. *Optics Express*, Vol.13, No.24, pp.9834–9840.

Shyu, L.-H., et al.(2011). Influence of Intensity loss in the Cavity of a Folded Fabry–Perot Interferometer on Interferometric Signals. *Review of Scientific Instruments*, Vol.82, No.6, pp. 063103-1 – 063103-3.

Takakura, Y.(2001). Optical Resonance in a Narrow Slit in a Thick Metallic Screen. *Physical Review Letters*, Vol.86, No.24, pp. 5601–5603.

Tan, Y.Y., et al.(2007). Two-layered Metallic Film Induced Surface Plasmons for Enhanced Optical Propulsion of Microparticles. *Applied Physics Letters*. Vol.91, No.14, pp.141108-1-141108-3.

Tao T, et al.(1990). Focused ion beam Induced Deposition of Platinum. *Journal of Vacuum Science and Technology B: Microelectronics and nanometer structures*. Vol.8, No.6, pp.1826-1829.

Treacy, M.M.J.,(1999). Dynamical Diffraction in Metallic Optical Gratings. *Applied Physics Letters*, Vol.75, No.5, pp. 606–608.

Treacy, M.M.J.,(2002) Dynamical Diffraction Explanation of the Anomalous Transmission of Light through metallic gratings. *Physical Review B*, Vol.66, No.19. pp.195105-1–191505-11.

Volkert, C.A. & Minor, A.M.(2007). Focused Ion Beam Microscopy and Micromachining. *MRS Bulleting*, Vol.32, No.5, pp. 389-395.

Wang, J.S., et al.(1994). Tellurite glass: a new candidate for fiber devices. *Optical Materials*, Vol.3, No.3, pp.187-203

Weiner, J.(2009). The Physics of Light Transmission through Subwavelength Apertures and Aperture Arrays. *Reports on Progress in Physics*, Vol.72, No.6, pp. 1-19.

Weiner, J.(2011). The Electromagnetics of Light Transmission through Subwavelength Slits in Metallic Films. *Optics Express*, Vol.19, No.17, pp. 16139–16153.

Xie, Y., et al.(2005). Transmission of Light through a Periodic Array of Slits in a Thick Metallic Film. *Optics Express*, Vol.13, No.12, pp.4485–4491.

Ye, Y.H. & Zhang, J.Y.(2005). Enhanced Light Transmission through Cascaded Metal Films Perforated with Periodic Hole Arrays. *Optics Letters*, Vol.30, No.12, pp.1521-1523.

Yin, L., et al.(2004). Surface Plasmons at single nanoholes in Au films. *Applied Physics Letters*, Vol.85, No.3, pp. 467–469.

Zayats, A.V., et al.(2005). Nano-optics of Surface Plasmon Polaritons. *Physics Reports*, Vol.408, No.3-4, pp. 131–314.

Zynio, S.A., et al.(2002). Bimetallic layers increase sensitivity of affinity sensors based on surface plasmon resonance. *Sensors*, Vol.2, No.2, pp. 62-70.

Chapter 8

Plasmonic Lenses

Yongqi Fu, Jun Wang and Daohua Zhang

Additional information is available at the end of the chapter

http://dx.doi.org/10.5772/50029

1. Introduction

The resolution of almost all conventional optical system is indispensably governed by the diffraction limit. This resolution limit can be overcome by use of focusing the evanescent waves in the near field region. The concept of "superlens" was proposed firstly by Pendry in 2000 [1]. When ε= -1and μ= -1, the negative refractive index material plate can be a perfect lens [2-4]. Because of the dispersion and absorption in the materials, the conditions of ε= -1and μ= -1 is hard to satisfy for the natural materials. Although the perfect lens may not exist, the superlens which can provide higher resolution beyond the diffraction limit have been proved. And focusing by means of surface plasmon polarisons (SPPs) by plasmonic lens is attracting much interest recently due to its unique feature of extraordinary enhanced transmission [5-8]. It means that we can focus the evanescent components of an illuminated object in the near-field region with subdiffraction-limit resolution [9]. This allows them to break the conventional barrier of diffraction limit, and leads to the formation of concentrated sub-wavelength light spots on the order of nanometers. Plasmonic lens is always consisted by metal and dielectric and can excite SPPs and always can be used for focusing, imaging, and beam shaping and so on.

In this paper, a literature review is given for the purpose of displaying a physical picture of plasmonic lenses for the relevant reader. Firstly, the basic theory about the plasmonic lens is presented. Then several examples of plasmonic lens are given. Here we mainly focus on the typical concepts of the plasmonic lens reported so far.

2. Plasmonic lens on the basis of negative refractive index materials

2.1. Superlens

Although perfect lens proposed by Pendry may not exist, superlens is realized and proved by Zhang's group in 2003 [10-13] and other research groups [14-21]. Here we mainly introduce the typical works which were done by Zhang's group. They showed that optical evanescent waves could indeed be enhanced as they passed through a sliver superlens. Figure 1 below shows configuration of the superlens they designed.

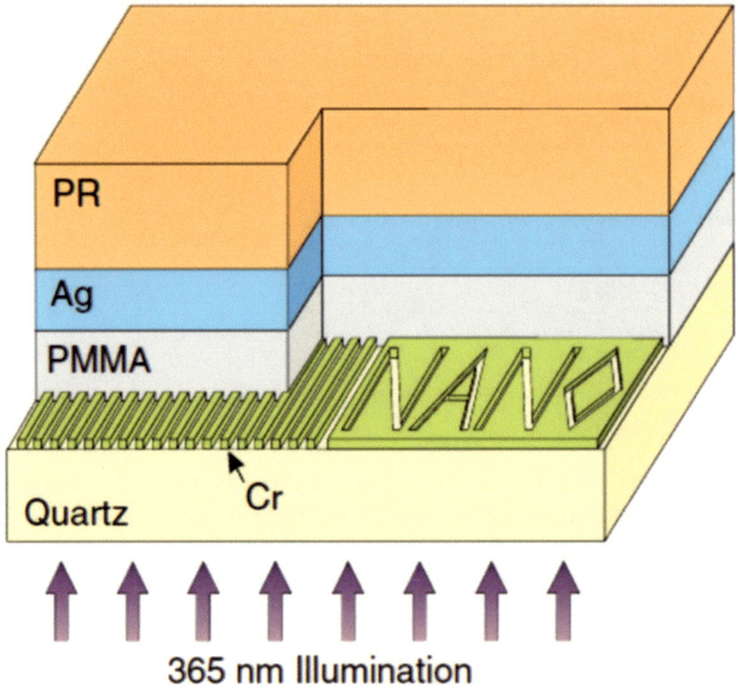

Figure 1. Optical superlensing designed by Zhang's group. Reprinted with permission from "N. Fang, H. Lee, C. Sun and X. Zhang, Science 308, 534-537 (2005)." of copyright ©2005 American Institute of Physics.

As can be seen, a set of objects were inscribed onto a chrome screen. The objects were designed to be placed about 40 nm away from the silver film which is 35 nm in thickness. And the chrome objects were patterned on quartz by using focused ion beam (FIB) technique, a 40 nm thick layer of polymethyl methacrylate (PMMA) was used to planarized them. The objects are imaged onto the photoresist on the other side of the silver film under ultraviolet (UV) illumination (at a wavelength of 365 nm). The negative photoresist which is 120 nm thick is used to record the near-field image. The substrate is food-exposed under an I-line (365 nm) mercury lamp. The exposure flux is 8 mW/cm^2, and an optimal exposure time of 60 s is applied to reduce the surface root mean square modulation below 1 nm for both the silver and PMMA surfaced; otherwise, the dispersion characteristics of the superlens would be modified and would in turn smear the details of the recorded image. The optical image is converted into topographic modulations by developing the negative photoresist and is mapped using atomic force microscopy (AFM).

Because the electric and magnetic responses of materials were decoupled in the near field, only the permittivity needs to be considered for transverse magnetic (TM) waves. This makes noble metals such as silver natural candidates for optical superlensing. Silver is chosen here. As surface charges accumulate at the interface between the silver and the imaging medium, the normal component of an electric of silver is selected and the permittivity of the silver and that of the adjacent medium are equal and of opposite sign. Such a delicate resance is essential to ensure the evanescent enhancement across the slab. For enhanced transmission of evanescent waves, it is found that an asymptotic impedance match ($k_{zi} / \varepsilon_i + k_{zj} / \varepsilon_j = 0$) has to be met at the surface of the silver, known as the surface plasmon excitation condition (k_{zi}, cross-plane, wave vector in silver; ε_i, permittivity of silver; k_{zj}, cross-plane wave vector in dielectric; and ε_j, permittivity of dielectric). It is widely known in metal optics that when the two media take the opposite sign in permittivity and $|\varepsilon_i| >> \varepsilon_j$, only surface plasmons at the narrow range of in-plane wave vector (k_x) that are close to ko can be resonantly coupled. However, less well known is that when $|\varepsilon_i| \sim \varepsilon_j$ and we are of opposite sign, the excitable surface plasmon band of k_x is significantly broadened, resulting in the superlensing effect.

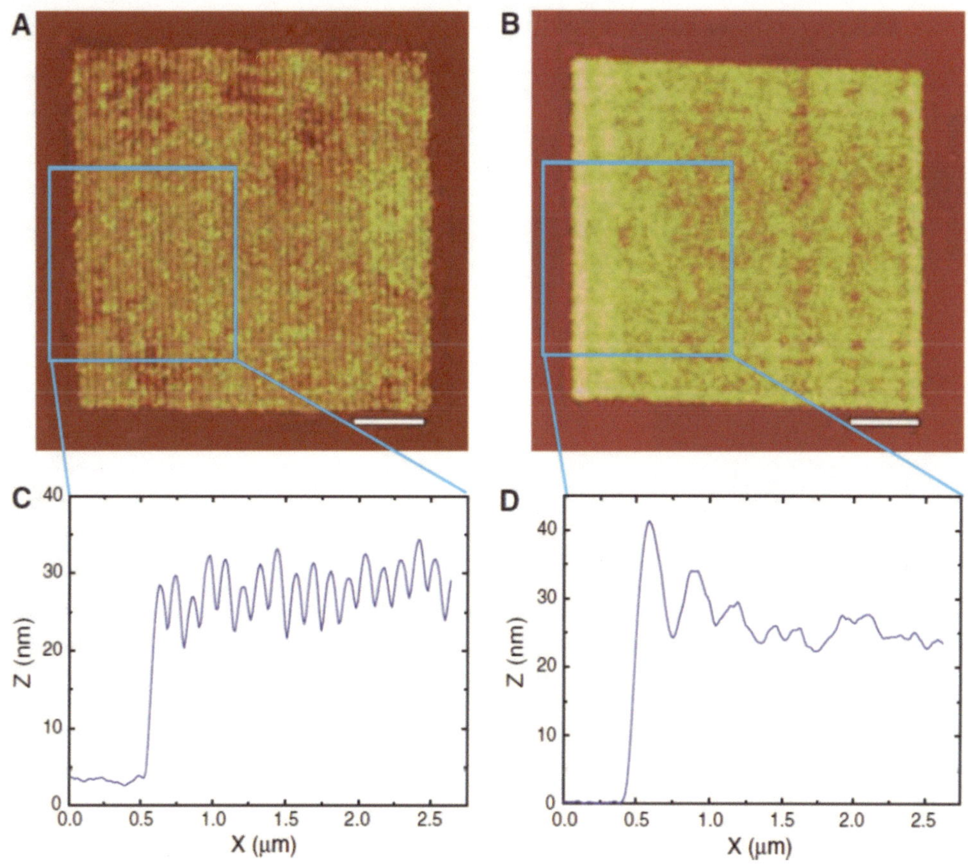

Figure 2. (A) AFM of the developed image (scale bar, 1 µm; color scale from dark red to bright yellow, 0 to 150 nm). (B) Control experiments were carried out, in which the silver superlens layer was replaced by a 35 nm thick PMMA layer, for a total PMMA thickness of 75nm. (C) The blue solid curve shows the clearly demonstrating the 63 ± 4 nm half-pitch resolved with a 35 nm silver superlens. X direction is relative displacement along the cross-section direction. (D) The blue solid curve shows the average cross section of Fig. 2B (control sample). Reprinted with permission from "N. Fang, H. Lee, C. Sun and X. Zhang, Science 308, 534-537 (2005)." of copyright ©2005 American Institute of Physics.

The intensity of evanescent waves decays with a characteristic length Z, and

$$Z^{-1} = 4\pi\sqrt{a^{-2} - \varepsilon\lambda^{-2}}\,, \tag{1}$$

where a is the period of a line array, and ε is the permittivity of the surrounding media. In Zhang's experiment, for the 60 nm half-pitch and $\varepsilon = 2.4$, the decay length is estimated to be 11 nm. Thus it is obviously difficult to resolve a 60 nm half-pitch object from a distance of 75 nm away if there isn't a superlens to enhance and transmit the evanescent waves. So we could find the photoresist images with typical average height modulations of 5 nm to 10 nm from Fig. 2 C. And this is assisted by careful control of the surface morphology of the PMMA and silver surface. In addition, Zhang also proved that the silver superlens can also image arbitrary nanostructures with sub-diffraction-limited resolution. The recorded image "NANO" in Fig. 3 B shows that the fine features from the mask showed in Fig. 3 A in all directions with good fidelity can be faithfully produced.

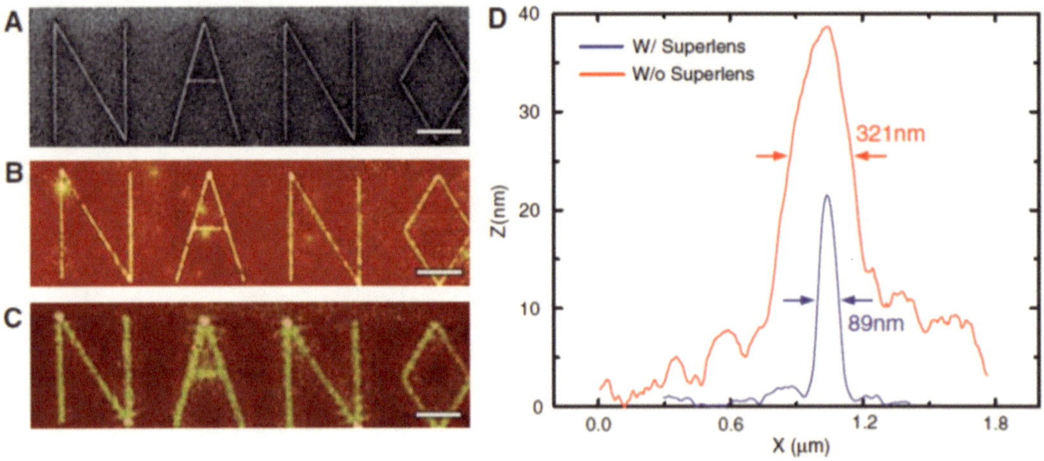

Figure 3. An arbitrary object "NANO" was imaged by silver superlens. (A) FIB image of the object. The linewidth of the "NANO" object was 40 nm. Scale bar in (A) to (C), 2 μm. (B) AFM of the developed image on photoresist with a silver superlens. (C) AFM of the developed image on photoresist when the 35 nm thick layer of silver was replaced by PMMA spacer as a control experiment. (D) The average cross section of letter "A" shows an exposed line width of 89 nm (blue line), whereas in the control experiment, we measured a diffraction-limited full width at half-maximum ling width of 321 ± 10 nm (red line). Reprinted with permission from "N. Fang, H. Lee, C. Sun and X. Zhang, Science 308, 534-537 (2005)." of copyright ©2005 American Institute of Physics.

2.2. Hyperlens

The images imaged by the superlens we talked about above are the same size as the objects. And there is no working distance. The hyperlens here was also proposed by Zhang's group [22]. It can magnify a sub-diffraction-limited image and projects it into the far field. Figure 4 A is the schematic of the hyperlens. It consists of a curved periodic stack of Ag (35 nm) and Al_2O_3 (35 nm) deposited on a half-cylindrical cavity fabricated on a quartz substrate. Sub-diffracion-limited objects were inscribed into a 50-nm-thick chrome layer located at the inner surface (air side). The anisotropic metamaterial was designed so that the radial and tangential permittivities have different signs.

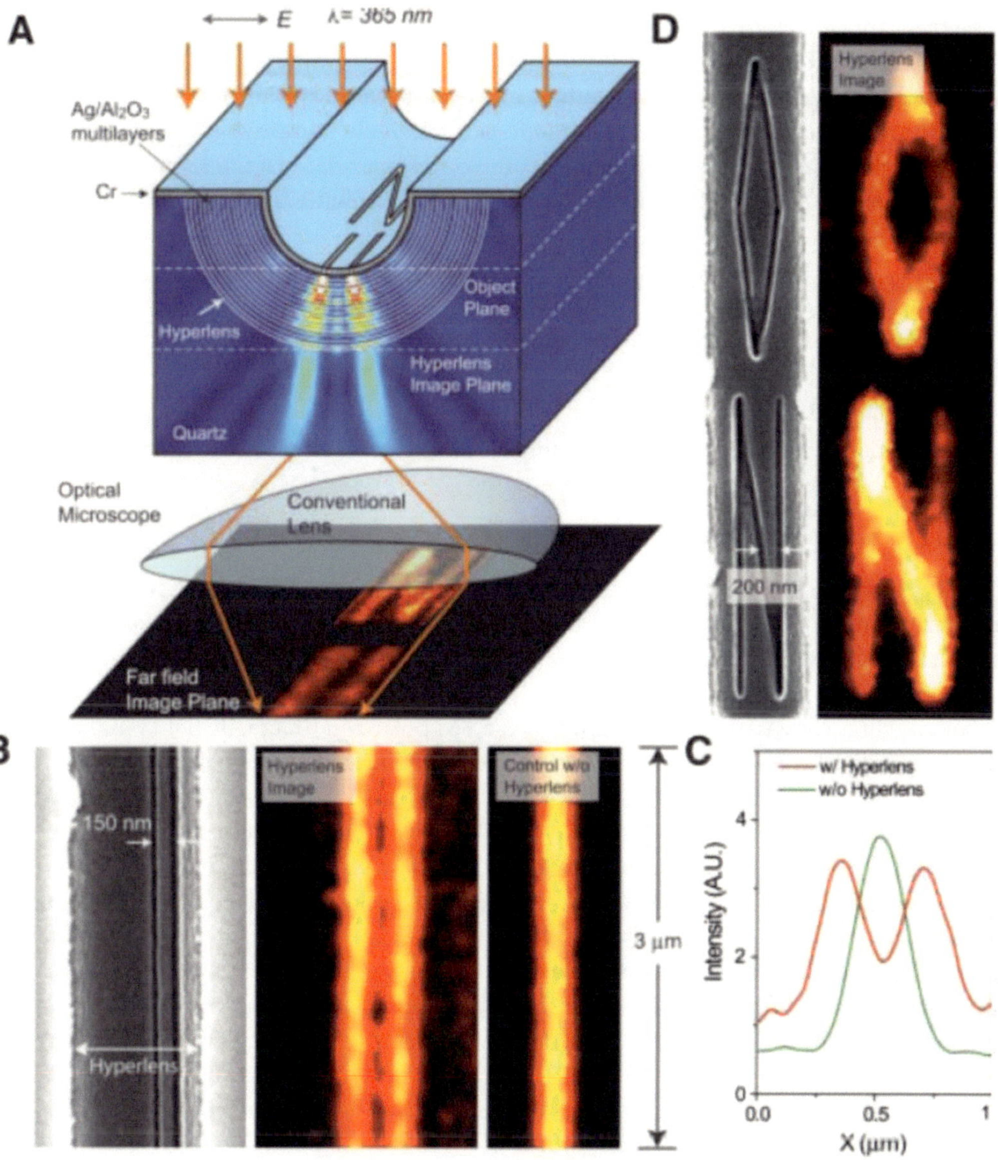

Figure 4. Optical hyperlens. (A) Schematic of heperlens and nymerical simulation of imaging of sub-diffraction-limited objects. (B) Hyperlens imaging of line pair object with line width of 35 nm and spacing of 150 nm. From left to right, scanning electron microscope image of the line pair object fabricated near the inner side of the hyperlens, magnified hyperlens image showing that the 150-nm-spaced line pair object can be clearly resolved, and the resulting diffraction-limited image from a control experiment without the hyperlens. (C) The averaged cross section of hyperlens image of the line pair object with 150-nm spacing (red), whereas a diffraction-limited image obtained in the control experiment (green). A.U.: arbitrary units. (D) An arbitrary object "ON" imaged with subdiffraction resolution. Line width of the object is about 40 nm. The hyperlens is made of 16 layers of Ag/Al_2O_3. Reprinted with permission from "Zhaowei Liu, Hyesog Lee, Yi Xiong, Cheng Sun, Xiang Zhang, Science 315. 1686 (2007)" of copyright ©2007 of AAAS.

The object imaged with hyperlens was a pair of 35-nm-wide lines spaced 150 nm apart. Upon illumination, the scattered evanescent field from the object enters the anisotropic medium and propagates along the radial direction. Because of the conservation of angular momentum, the tangential wave vectors are progressively compressed as the waves travel outward, resulting in a magnified image at the outer boundary of the hyperlens. Hence the magnified image (350-nm spacing) can be clearly resolved with an optical microscopy.

3. Plasmonic lens on the basis of subwavelength metallic structures

3.1. Subwavelength metallic structure for superfocusing

3.1.1. One-dimensional structures for focusing

In this section, we presented two types of tuning methods for the purpose of phase modulation: depth tuning [23] and width tuning [24-26] approaches.

3.1.1.1. Depth-tuned. strctures

Three types of plasmonic slits (convex, concave, and flat/constant groove depth) with different stepped grooves have been designed and fabricated to achieve efficient plasmonic focusing and focal depth modulation of the transmitted beam. Figure 5 shows the fabricated depth-tuned plasmonic lens using focused ion beam milling [27]. The general design of the plasmonic slit is shown in Fig. 6 (a) [28]. When a TM polarized (magnetic field parallel to the y-direction) monochromatic plane wave impinges on the slit, it excites collective oscillations of the electrons at the surface, which is known as SPPs. The SPPs propagate along the surface of the metal film and are diffracted to the far-field by the periodic grooves, which are designed with a width smaller than half of the incident wavelength to allow a high diffraction efficiency [29]. Constructing interference of such diffracted beams leads to the focusing effect at a certain point on the beam axis [30]. Since the diffracted beams are modulated by the nanometric grooves, through adjusting the parameters of the grooves (such as our width, depth, period and number), the diffracted beams can be fully manipulated resulting in a tailored ultra-compact lens with subwavelength resolution and nanometer accuracy [31]. Most interestingly, it has been numerically found [32] that the relative phase at the exit end of the slit increases steadily with the increasing groove depth, making it possible to achieve continuous phase retardation by simply designing surrounding grooves with stepped depths as shown in Figs. 6 (b) and 6 (c). This has led to a great simplification of the plasmonic lens design without increasing the groove number or generating a bump on the metal film [33].

Figure 7 (a) presents a detailed comparison of the measured intensity distribution with simulation using FDTD at the slit cross section (along x-direction, as indicated by the white dot line in Fig. 2 (a) [23]). A good agreement has been found between the experiment and the theoretical prediction except that the measured full-width at a half-maximum (FWHM) of the central lobe (approximately 281 nm) is slightly larger than the calculated value of about 230 nm. This is because the measured intensity distribution approximately equals to the convolution of the finite probe size (30-80nm) and the actual intensity distribution of the transmitted light.

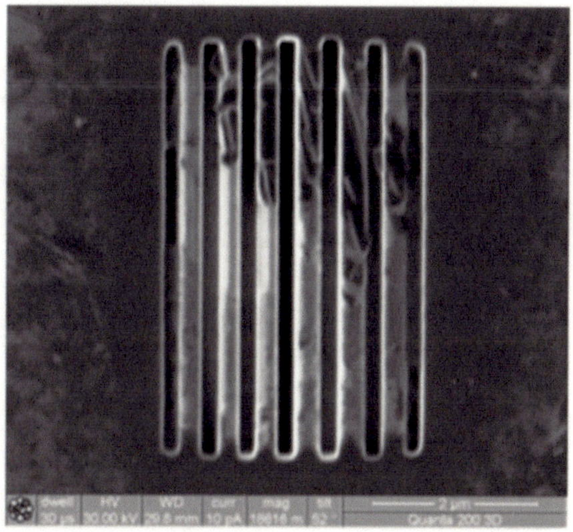

Figure 5. FIB image of the fabricated depth-tuned nanostructure (type of concave) on the Ag thin film with thickness of 200 nm.

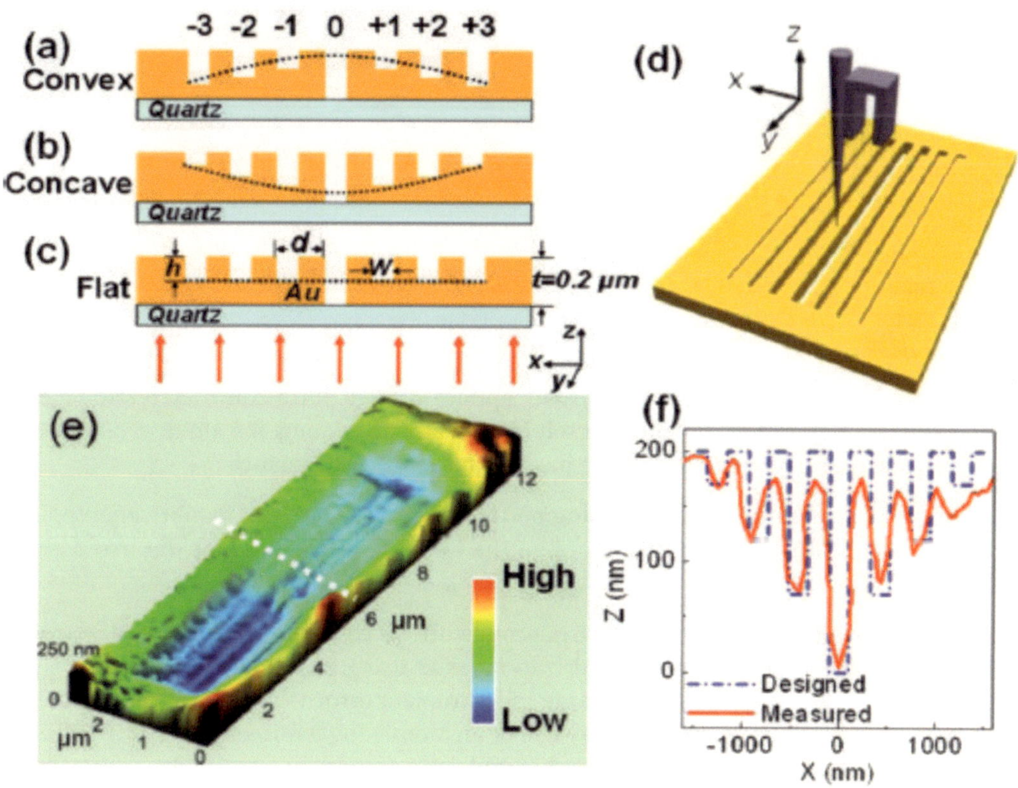

Figure 6. Schematic drawing of the nanoplasmonic slits with (a) convex, (b) concave and (c) flat shaped profiles. (d) Schematic drawing of near-field measurement setup. (e) Measured topographic image of the slit with concave corrugations. The marked area 'A' shows a larger overall depth than that of the area 'B'. (f) Cross section of the concave groove-slit at the position indicated by the dashed line in (e).

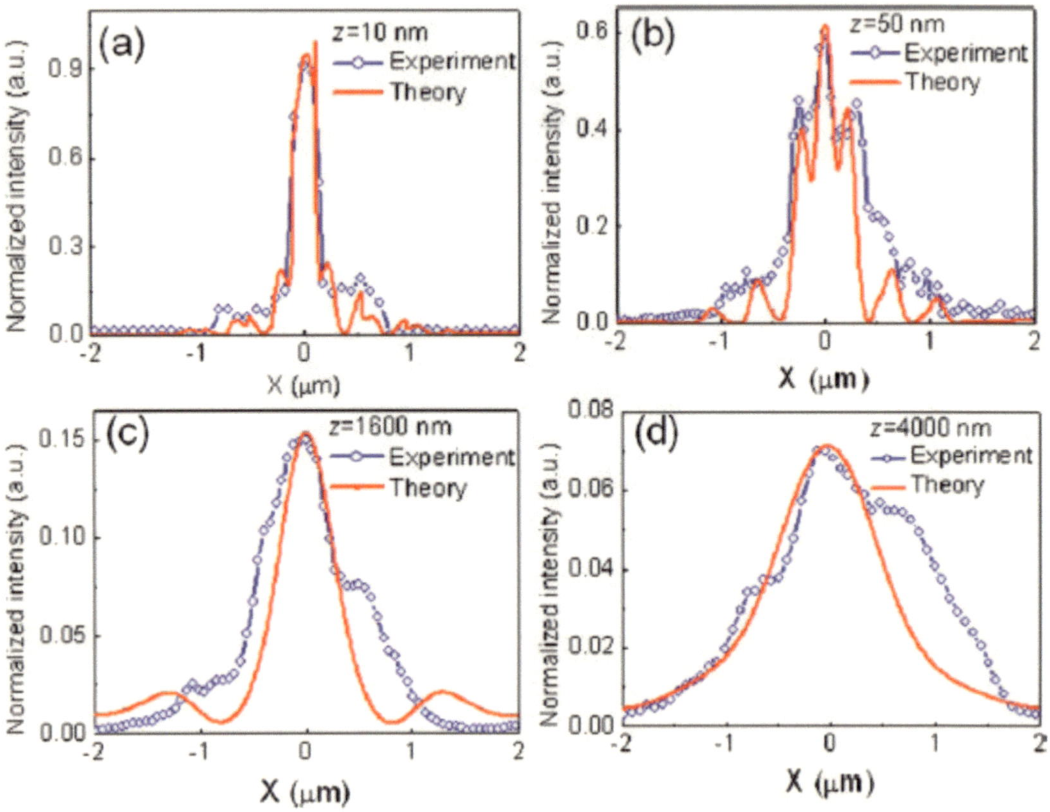

Figure 7. Comparison of measured and theoretical cross sections at x=0 in Fig. 2 of the reference paper for (a) z=10 nm, (b) z=50 nm (c) z=1600 nm and (d) z=4000 nm.

Near-field measurement reveals unambiguously the light interaction with the slits and confirms the functionalities of the nanoplasmonic lens. The simple plasmonic lens demonstrated in this paper can find broad applications in ultra-compact photonic chips particularly for biosensing and high-resolution imaging. Among the three types metallic structures, the type of convave structure has best focusing performance.

In adition, V-shaped influence on focusing performance was analyzed in fabrication point of view [34]. The incident angle dependance on the focusing properties was discussed also in Ref. [35].

Regarding fabrication of the metallic structures with depth-tuned grooves, it is worthy to point out that the geometrical characrerization issue using atomic force microscope after focused ion beam direct milling [31]. Large measurement error is found during geometrical characterization of the nanostructures by use of an atomic force microscope (AFM) working in tapping mode. Apex wearing and 34° full cone angle of the probe generate the measurement errors during characterizing the nanostructures with the feature size of 200 nm and below. To solve this problem, a FIB trimmed AFM probe is employed in the geometrical characterization, as shown in Fig. 8. The results show that the error is improved greatly using the trimmed probe.

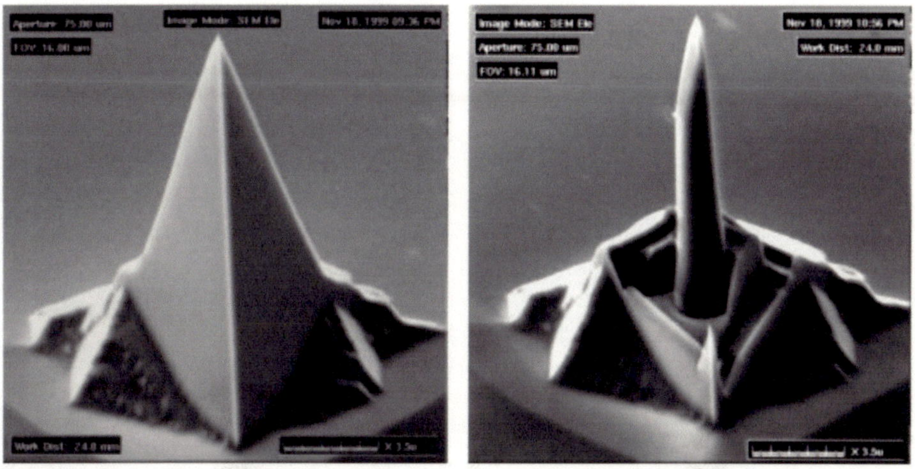

(a) (b)

Figure 8. AFM probe for tapping mode. (a) the commercial probe with half cone angle of 17° and material of Si_3N_4. (b) the FIB trimmed probe with high aspect ratio.

Influence of polarization states on focusing properties of the depth-tuned metallic structures was reported [31]. The structure was designed with geometrical parameteres shown as Fig. 9. Figure 10 shows the total electric-field intensity $|E|^2=|Ex|^2+|Ey|^2+|Ez|^2$ at x-y plane along the longitudinal direction at z = 1.35 μm at λ=420 nm for the (a) elliptical polarization (EP), (b) circular polarization (CP), and (c) radial polarization (RP) cases. The intensity of transverse electric field, $|Ex|^2+|Ey|^2$, is significantly tuned. In the figure, the intensity along the horizontal (x) is equal to that along the vertical (y), while the intensity along the

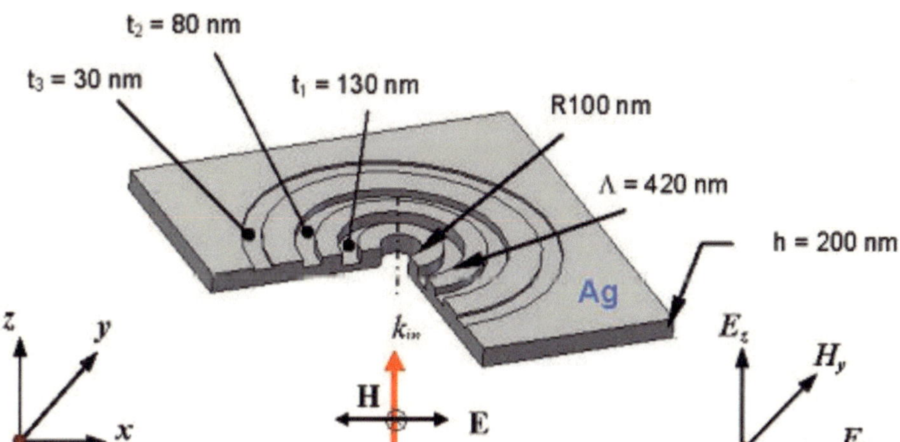

Figure 9. An annular plasmonic lens having a depth-tuned structure (groove depths, t_1= 130 nm, t_2= 80 nm and t_3=30 nm) milled in the output side of a Ag thin film (thickness, h = 200nm). Other structure parameters are: central hole diameter=200 nm, groove width=200 nm, and groove period=420 nm. The structure is incident with TM-polarized light having electric vector along the x direction. Reprinted with permission from "Jun Wang, Wei Zhou and Anand K. Asundi, Opt. Express 17, 8137-8143 (2009)." of copyright ©2009 Optical Society of American.

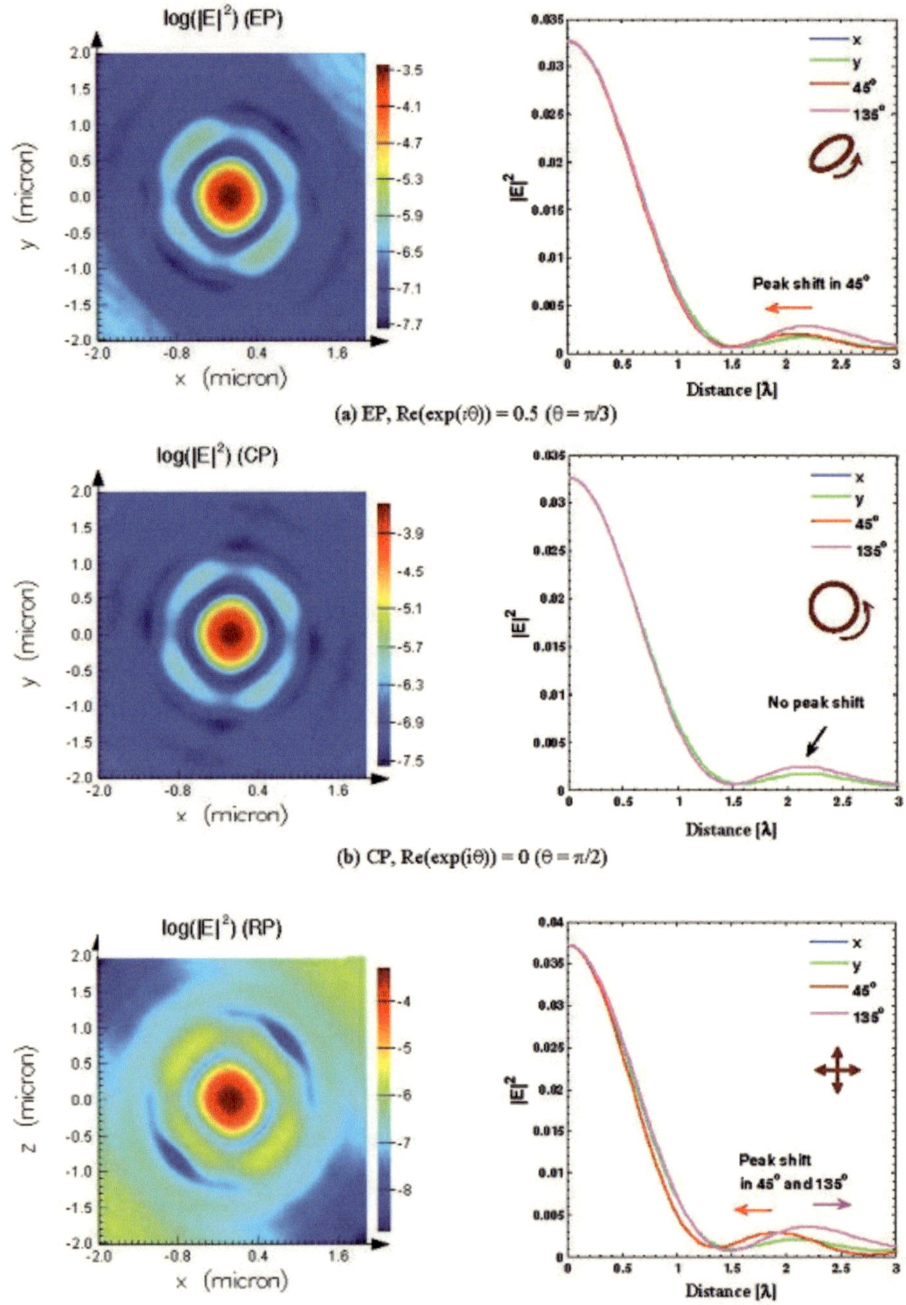

Figure 10. Total electric field (left) transmitted through the structure under illumination using different polarization states, including (a) CP, (b) EP, and (c) RP, showing the phase modulation effect on the beam profile (right) along the transverse direction in x and y and diagonal directions along 45° and 135° with respect to the x. Refer to Fig. 1 for the directions. Reprinted with permission from "Jun Wang, Wei Zhou and Anand K. Asundi, Opt. Express 17, 8137-8143 (2009)." of copyright ©2009 Optical Society of American.

diagonal directions (45° and 135° with respect to the x is tuned), the peak shift is observed at the side lobe, which is $<0.1\lambda$ for the EP case. In Fig. 10 (a), in the direction of 45°. And the beam in 45° is narrower than that in the x- or y-directions. In addition, the phase function Re(exp(iθ)) indicates tuning capability. Re (exp(iθ)) = 0.5, where $\theta = \pi/3$, for the EP case, and the phase function becomes 0 for the CP and 1 for the RP case. For example, in Fig. 10 (b), the uniformly distributed total-electric-field intensity is observed in the x-y plane, while, in Fig. 10 (c), the peak shifts 0.2λ in the 45° larger than that for the EP case, and much narrow beam is observed in the same direction. The same plasmonic modes are observed for CP, EP, or RP polarization cases as for TM case. Using a polarized plane wave the transverse electric field is tuned; the tuning 45° and 135° with respect to effect on focus spot is observed along the diagonal directions in 45° the x-direction. Of the cases, RP approach forms the smallest focus spot along the 45° using Re[exp(iθ)]=1, showing maximum tuning capability, while CP approach the phase function Re[exp(iθ)]=0 forms a symmetrically electric field distribution in the focal plane. Phase function indicates the tuning capability.

3.1.1.2. Width-tuned structures

A novel method is proposed to manipulate beam by modulating light phase through a metallic film with arrayed nano-slits, which have constant depth but variant widths. The slits transport electro-magnetic energy in the form of surface plasmon polaritons (SPPs) in nanometric waveguides and provide desired phase retardations of beam manipulating with variant phase propagation constant. Numerical simulation of an illustrative lens design example is performed through finite-difference time-domain (FDTD) method and shows agreement with theory analysis result. In addition, extraordinary optical transmission of SPPs through sub-wavelength metallic slits is observed in the simulation and helps to improve elements's energy using factor.

To illustrate the above idea of modulating phase, a metallic nano-slits lens is designed [24]. The parameters of the lens are as follows: $D = 4$ μm, $f = 0.6$ μm, $\lambda = 0.65$ μm, $d = 0.5$ μm, where D is the diameter of the lens aperture, f the focus length, the wavelength and d the thickness of the film. The two sides of the lens is air. The schematic of lens is given in Fig. 11, where a metallic film is perforated with a great number of nano-slits with specifically designed widths and transmitted light from slits is modulated and converges in free space.

After numerous iterations of calculation using the FDTD algorithm, the resulting Poyinting vector is obtained and showed in Fig. 12 (a). A clear-cut focus appears about 0.6 micron away from the exit surface, which agrees with our design. The cross section of focus spot in x direction is given in Fig. 12 (b), indicating a full-width at half-maximum (FWHM) of 270 nm. The extraordinary light transmission effects of SPPs through sub-wavelength slits is also observed in the simulation with a transmission enhance factor of about 1.8 times. These advantages promise this method to find various potential applications in nano-scale beam shaping, integrate optics, date storage, and near-field imaging ect.

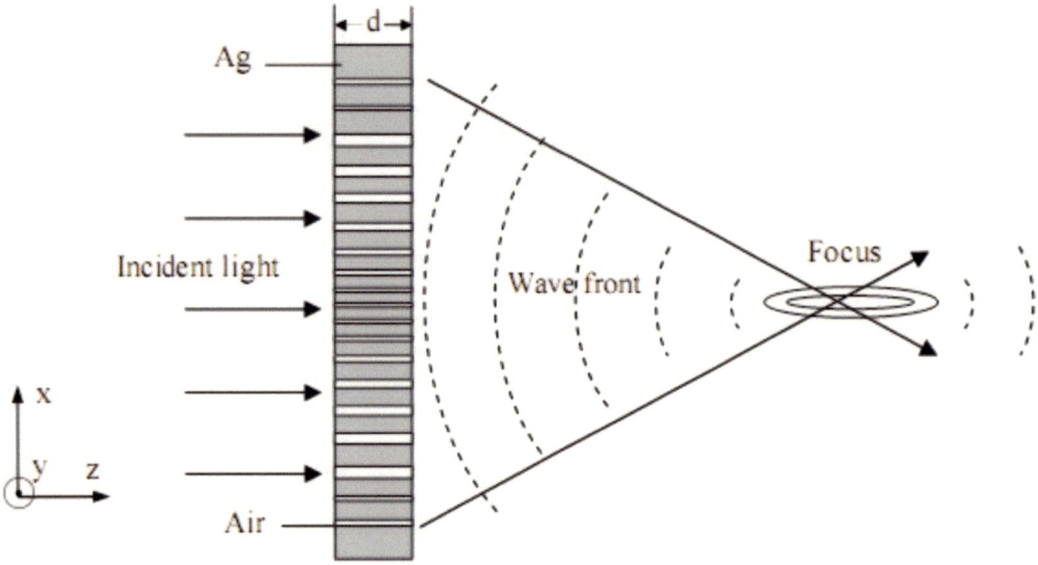

Figure 11. A schematic of a nano-slit array with different width formed on thin metallic film. Metal thickness in this configuration is *d*, and each slit width is determined for required phase distribution on the exit side, respectively. A TM-polarized plane wave (consists of Ex, Hy and Ez field component, and Hy component parallel to the y-axis) is incident to the slit array from the left side. Reprinted with permission from "H. F. Shi, C. T. Wang, C. L. Du, X. G. Luo, X. C. Dong, and H. T. Gao, Opt. Express 13, 6815-6820 (2005)." of copyright ©2009 Optical Society of American.

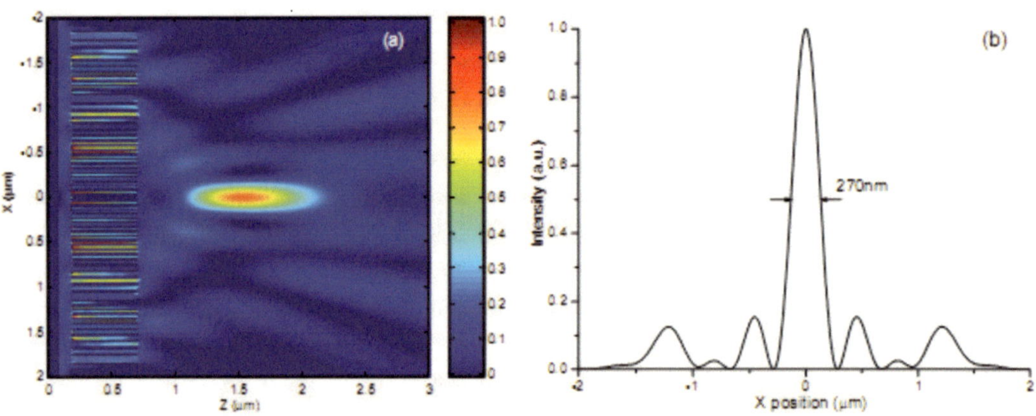

Figure 12. (a) FDTD calculated result of normalized Poynting Vector Sz for designed metallic nano-slits lens. Film thickness is 500nm, and the total slits number is 65. The structure's exit side is posited at z=0.7 μm. (b) Cross m.section of the focus at z=1.5 μm. Reprinted with permission from "H. F. Shi, C. T. Wang, C. L. Du, X. G. Luo, X. C. Dong, and H. T. Gao, Opt. Express 13, 6815-6820 (2005)." of copyright ©2009 Optical Society of American.

As an experimentlal verification example, Lieven et. al. [37] experimentally demonstrated planar lenses based on nanoscale slit arrays in a metallic film. The lens structures consist of optically thick gold films with micron-size arrays of closely spaced, nanoscale slits of varying widths milled using a focused ion beam. They found an excellent agreement between electromagnetic simulations of the design and confocal measurements on manufactured structures. They provide guidelines for lens design and show how actual lens behavior deviates from simple theory.

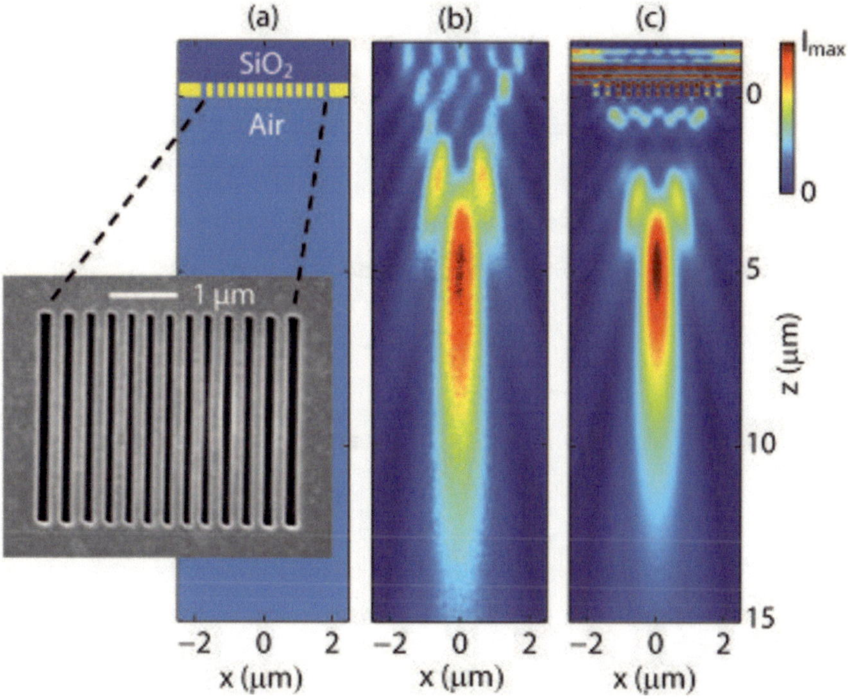

Figure 13. Planar lens based on nanoscale slit array in metallic film. (a) Geometry of the lens consisting of a 400nm optically thick gold film (yellow) with air slits of different widths (80 to150 nm) (light blue) milled therein on a fused silica substrate (dark blue). The inset shows a scanning electron micrograph of the structure as viewed from the air-side. (b) Focusing pattern measured by confocal scanning optical microscopy (CSOM). (c) Finite-difference and frequency-domain (FDFD) simulated focusing pattern of the field intensity through the center of the slits. In order to show the features of the focus spotclearly, the field intensity inside the slits is saturated. Reprinted with permission from "LievenVerslegers et. al., Nano Lett. 9, 235-238 (2009)" of copyright ©2009 Chemical Society of American.

The basic geometry consists of an array of nanoscale slits in an otherwise opaque metallic film (see Fig. 13 (a)). Figure13 shows the main results of the work, which combines fabrication, characterization, andsimulation. Pane l (a) shows the fabricated structure, while panels (b) and (c) represent the measured and simulated field intensity in across section through the center of the slits (along the x-direction). Both the measurement and the simulation clearly demonstrate focusing of the wave. The agreement between experiment and simulation is excellent. Moreover, the simulation image is generated using the designed parameters as the slit width rather than the actual slit width measured in the SEM, as is commonly done when comparing nanophotonics simulation and experiments. The agreemen there thus indicates the robustness in design and the fault tolerance of this approach for focusing. The effect of lens size can be exploited to control the focusing behavior as is shown clearly in Fig. 14. Both lenses introduce the same curvature to the incident plane wave as the lens from Fig. 13, since we consist of slits with the same width as the original design (2.5 μm long). By omitting one outer slit on each side for the lens in Fig. 14 (a), one gets the lens, as shown in Fig. 14 (b).

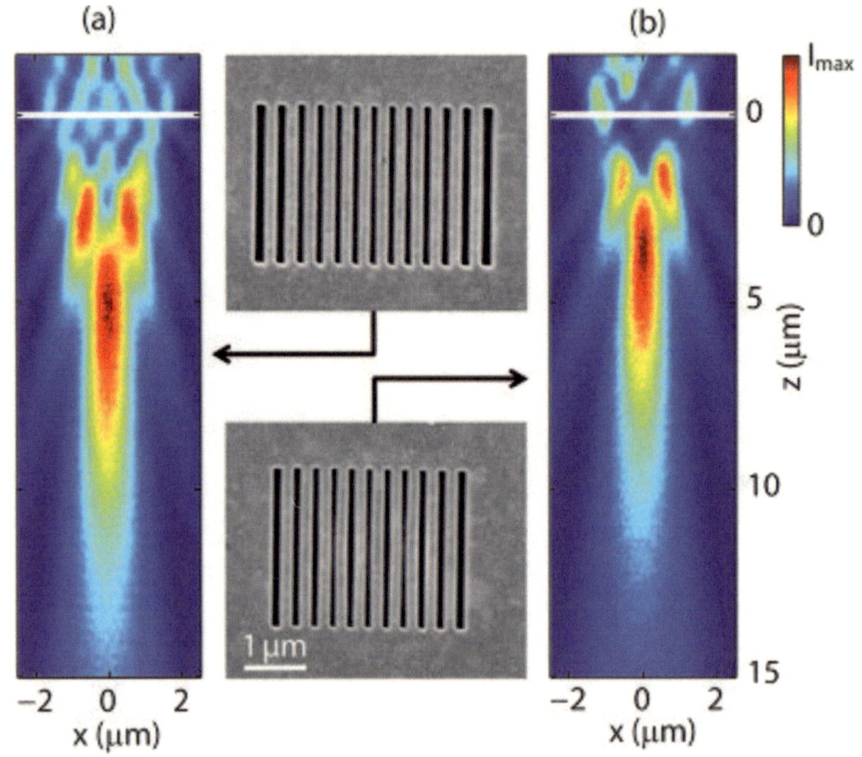

Figure 14. Control of the cylindrical lens behavior by design of nanoscale slit array parameters. Effect of lens size on focusing for (a) a lens with 13 slits (80-150nm by 2.5 μm) and (b) a lens with 11 slits (80-120nm by 2.5 μm). The white line gives an estimate of the lens position. Both scanning electron micrographs are on the same saclew. Reprinted with permission from "LievenVerslegers et. al., Nano Lett. 9, 235-238 (2009)" of copyright ©2009 Chemical Society of American.

This first experimental demonstration is a crucial step in the realization of this potentially important technology form any applications in optoelectronics. Moreover, the design principles presented here for the specil case of a lens can be applied to construct a wide range of optical components that rely on tailoring of the optical phase front.

3.1.2. Two-dimensional structures for focusing

3.1.2.1. Circular grating-based metallic structures for focusing

Recently, Jennifer et. al. reported the generation and focusing of surface plasmon polariton (SPP) waves from normally incident light on a planar circular grating (constant slits width and period) milled into a silver film [38]. The focusing mechanism is explained by using a simple coherent interference model of SPP generation on the circular grating by the incident field. Experimental results concur well with theoretical predictions and highlight the requirement for the phase matching of SPP sources in the grating to achieve the maximum enhancement of the SPP wave at the focal point. NSOM measurements show that the plasmonic lens achieves more than a 10-fold intensity enhancement over the intensity of a single ring of the in-plane field components at the focus when the grating design is tuned to the SPP wavelength.

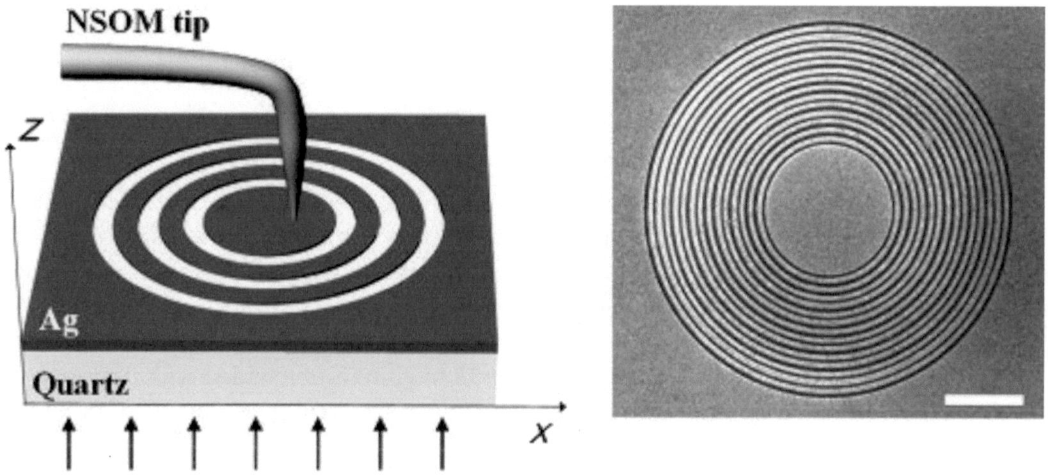

Figure 15. (a) Experimental scheme for near-field measurements. Circular gratings are cut into a silver film deposited on a quartz substrate. Laser light is normally incident from the quartz side, and the electromagnetic near-field is monitored with a metal coated NSOM tip. (b) SEM image of a sample with 15 rings. The scale bar is 5 microns. Reprinted with permission from "Jennifer M. Steele , Zhaowei Liu , Yuan Wang, and Xiang Zhang, Opt. Express 14, 5664-5670 (2006)." of copyright ©2006 Optical Society of American.

To investigate this focusing experimentally, rings with different periods were cut into 150 nm thick silver films. Silver was evaporated onto a quartz plate at a high rate to ensure a surface with minimal roughness. Rings were milled into the metal using an FEI Strata 201 XP focused ion beam (FIB), with the inner most ring having a diameter of 8 microns. Additional rings were added with a period either close to or far from resonance with the excited SPP waves. The surface plasmons were excited with linearly polarized laser light incident from the quartz side. The electromagnetic near-field of these structures was recorded using near-field scanning optical microscopy (NSOM) in collection mode using a metal coated NSOM tip. A metal coated tip was chosen over an uncoated tip to increase the resolution of the scan. Previous experimental results on samples with similar geometry compare favorably with computer simulations, indicating the interaction of the SPP near field with the metal tip is negligible. The measurement scheme can be seen in Fig. 15 (a) with an SEM image of a typical sample shown in Fig. 15 (b). The phase change of SPP waves across a barrier is an interesting issue that has received very little attention. If the slit width is much smaller than the SPP wavelength, the slit will have very little effect on the SPP and the phase change should be very small. However, if the slit width is on the order of the SPP wavelength, as the SPP waves cross a slit opposite charges will be induced on opposite sides of the slit, providing quasi-electrostatic coupling across the barrier. The authors calimed that it is possible that the phase change will be sensitive to the slit width. The number and period of rings, film material, and slit geometries provide experimental handles to tune the plasmonic lens to accommodate specific applications, making this technique a flexible plasmonic tool for sensing applications.

The following two sections below introduce the metallic subwavelength structures with chirped (variant periods) slits and nanopinholes acted as the plasmonic lenses for the purpose of superfocusing.

3.1.2.2. Illumination under linear polarization state

A novel structure called plasmonic micro-zone plate-like (PMZP) or plasmonic lens with chirped slits is put forth to realize superfocusing. It was proposed by Fu's group [39,40]. Unlike conventional Fresnel zone plate (CFZP), a plasmonic structure was used and combined with a CFZP. Configuration of the PMZP is an asymmetric structure with variant periods in which a thin film of Ag is sandwiched between air and glass. The PMZP is a device that a quartz substrate coated with Ag thin film which is embedded with a zone plate structure with the zone number N < 10. Figure 16 is an example of schematic diagram of the structure.

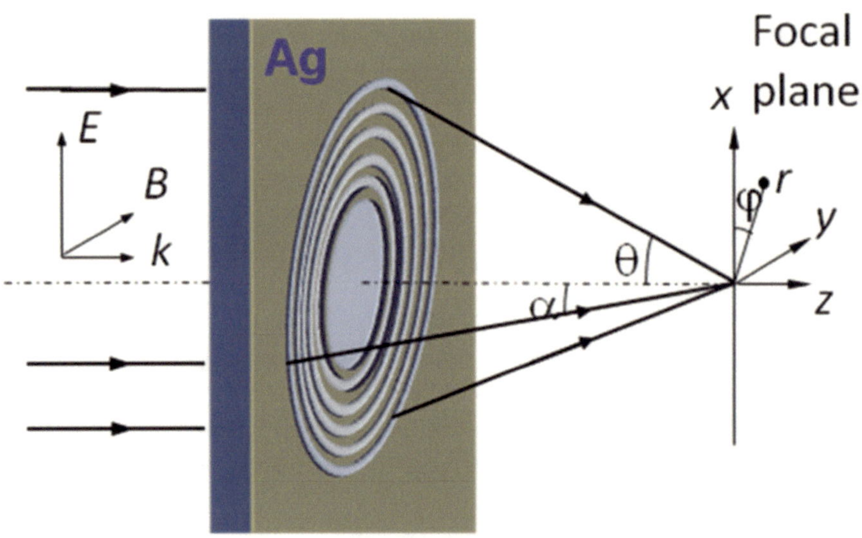

Figure 16. Schematic of the plasmonic micro-zone plate for super-focusing.

Following the electromagnetic focusing theory of Richard and Wolf (Richards & Wolf, 1959)[41], the electric field vector in the focal region is given by,

$$\mathbf{E}(r, z, \varphi) = -i\,[I_0(r, z) + I_2(r, z)\cos 2\varphi]\,\mathbf{i} - i\,I_2(r, z)\sin 2\varphi\,\mathbf{j} - 2I_1(r, z)\cos\varphi\,\mathbf{k}. \quad (2)$$

Let **i**, **j**, **k** be the unit vectors in the direction of the co-ordinate axes. To be integral over the individual zones, I_0, I_1, I_2 are expressed as,

$$I_0(r,z)=\sum_n T_n \int_{\alpha_{n-1}}^{\alpha_n} \sqrt{\cos\theta}(1+\cos\theta)J_0(kr\sin\theta)\exp(ikz\cos\theta)\sin\theta d\theta,$$

$$I_1(r,z)=\sum_n T_n \int_{\alpha_{n-1}}^{\alpha_n} \sqrt{\cos\theta}\sin\theta J_1(kr\sin\theta)\exp(ikz\cos\theta)\sin\theta d\theta,$$

$$I_2(r,z)=\sum_n T_n \int_{\alpha_{n-1}}^{\alpha_n} \sqrt{\cos\theta}(1-\cos\theta)J_2(kr\sin\theta)\exp(ikz\cos\theta)\sin\theta d\theta.$$

The wave vector $k = 2\pi/\lambda$.

According to the equation, the intensity of lateral electric field component, $|E_x|^2$, follows the zero-order Bessel function J_0 of the first kind, while the intensity of longitudinal electric field component, $|E_z|^2$, follows the first-order Bessel function J_1 of the first kind. In the total electric field intensity distribution, all the field components add up. With high numerical aperture, this leads to not only asymmetry of the focus spot but also an enlarged focus spot. As shown in Figs. 17 (a)~(b), the total electric field and individual electric field components, $|E|^2$, $|E_x|^2$, &$|E_z|^2$, emerged from λ_{sp}-launched FZP lens are in comparison with that the total electric field and individual electric field components from a λ_{in}-launched PMZP lens. It is found that the intensity ratio, $|E_x|^2/|E_z|^2$, can increase up to 10 times for the λsp-launched PMZP lens. A focus spot having the polarization direction along the x direction is obtained.

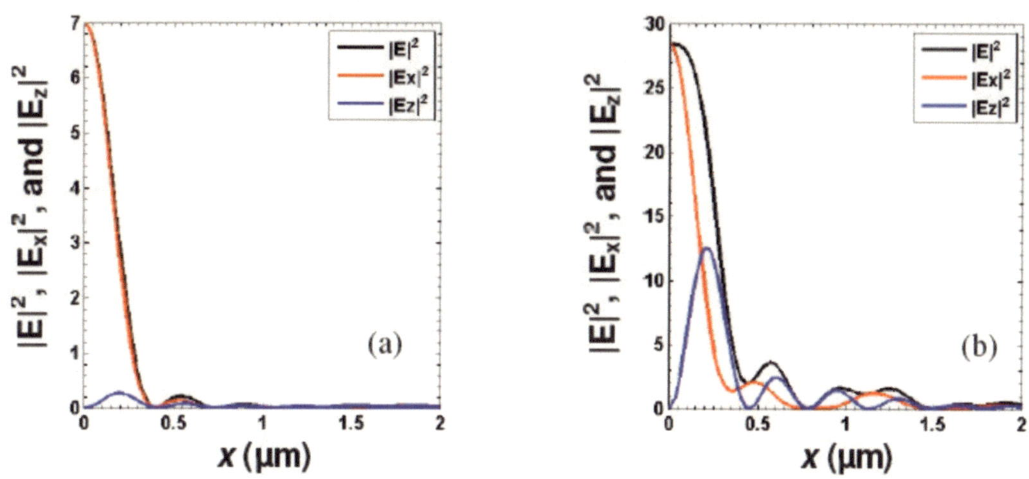

Figure 17. The total electric field and electric field components, $|E|^2$, $|E_x|^2$, and $|E_z|^2$, emerged from (a) a λ_{sp}-launched FZP lens and (b) a λ_{in}-launched FZP lens. It is observed that the proposed superlens enables to restrict the depolarization effect and produces a linearly- polarized focus spot having the polarization direction in the x direction.

The chirped slits can form a focal region in free space after the exit plane. The final intensity at the focal point is synthesized by iteration of each zone focusing and interference each other, and can be expressed as

$$I = \alpha \sum_{i=1}^{N} C I_0 \frac{4r_i}{\lambda_{SP}} e^{-(r_i / l_{SP})} \tag{3}$$

where I_0 is the incident intensity, r_i is the inner radius of each zone, i is the number of the zones, l_{SP} is the propagation length for the SPP wave, α is interference factor, and C is the coupling efficiency of the slits. C is a complicated function of the slit geometry and will likely have a different functional form when the slit width is much larger or much smaller than the incident wavelength

The PMZPs is an asymmetric structure. For an evanescent wave with given k_x, we have $k_{zj} = +[\varepsilon_j(\omega / c)^2 - k_x^2]^{1/2}$ for j=1 (air) and j=3 (glass) and $k_{zj} = +i[k_x^2 - \varepsilon_j(\omega / c)^2]^{1/2}$ for j=2 (Ag film). Superfocusing requires regenerating the evanescent waves. Thus the PMZP needs to be operated with the condition $| k_{z1} / \varepsilon_1 + k_{z2} / \varepsilon_2 | | k_{z2} / \varepsilon_2 + k_{z3} / \varepsilon_3 | \to 0$. Physically, this would require exciting a surface plasmon at either the air or the glass side. For $E_\perp$ wave, a negative permittivity is sufficient for focusing evanescent waves if the metal film thickness and object are much smaller than the incident wavelength. Because electric permittivity $\varepsilon < 0$ occurs naturally in silver and other noble metals at visible wavelengths, a thin metallic film can act as an optical super lens. In the electrostatic limit, the p-polarized light, dependence on permeability μ is eliminated and only permittivity ε is relevant. In addition, diffraction and interference contribute to the transition from the evanescent waves to the propagation waves in the quasi-far-field region. Above all, the PMZPs form super focusing by interference of the localized SPP wave which is excited from the zones. This makes it possibly work at near and quasi-far-field with lateral resolution beyond diffraction limit. Also the PMZP has several zones only, its dimension is decreased greatly compared the CFZP.

As an example, an appropriate numerical computational analysis of a PMZP structure's electromagnetic field is carried out using finite-difference and time-damain (FDTD). It is illuminated by a plane wave with a 633 nm incident where Ag film has permittivity $\varepsilon_m = \varepsilon_m' + i\varepsilon_m'' = -17.6235 + 0.4204i$. An Ag film with thickness $h_{Ag} = 300nm$ centered at z=150 nm has an embedded micro-zone-plate structure. Zone number *N*=8, and outer diameter *OD*=11.93 μm. The widths of each zone from first ring to last ring, calculated by using the conventional zone plate equations, are 245, 155, 116, 93, 78, 67, 59, and 52 nm. In the FDTD simulations, the perfectly

matched layer boundary condition was applied at the grid boundaries. Figure 17 is the simulation result. From the result, the simulated focal length of the PMZP, f_{PMZP} and depth of focus (*DOF*) are larger than those of the designed values using the classical equations $r_n = (n\lambda f_{FZP} + n^2\lambda^2 / 4)^{1/2}$ and $DOF = \pm 2\Delta r^2 / \lambda$, where *n*=1, 2, 3,..., f_{PMZP} is the designed principal focal length of Fresnel zone plates and given in terms of radius *R* of the inner ring

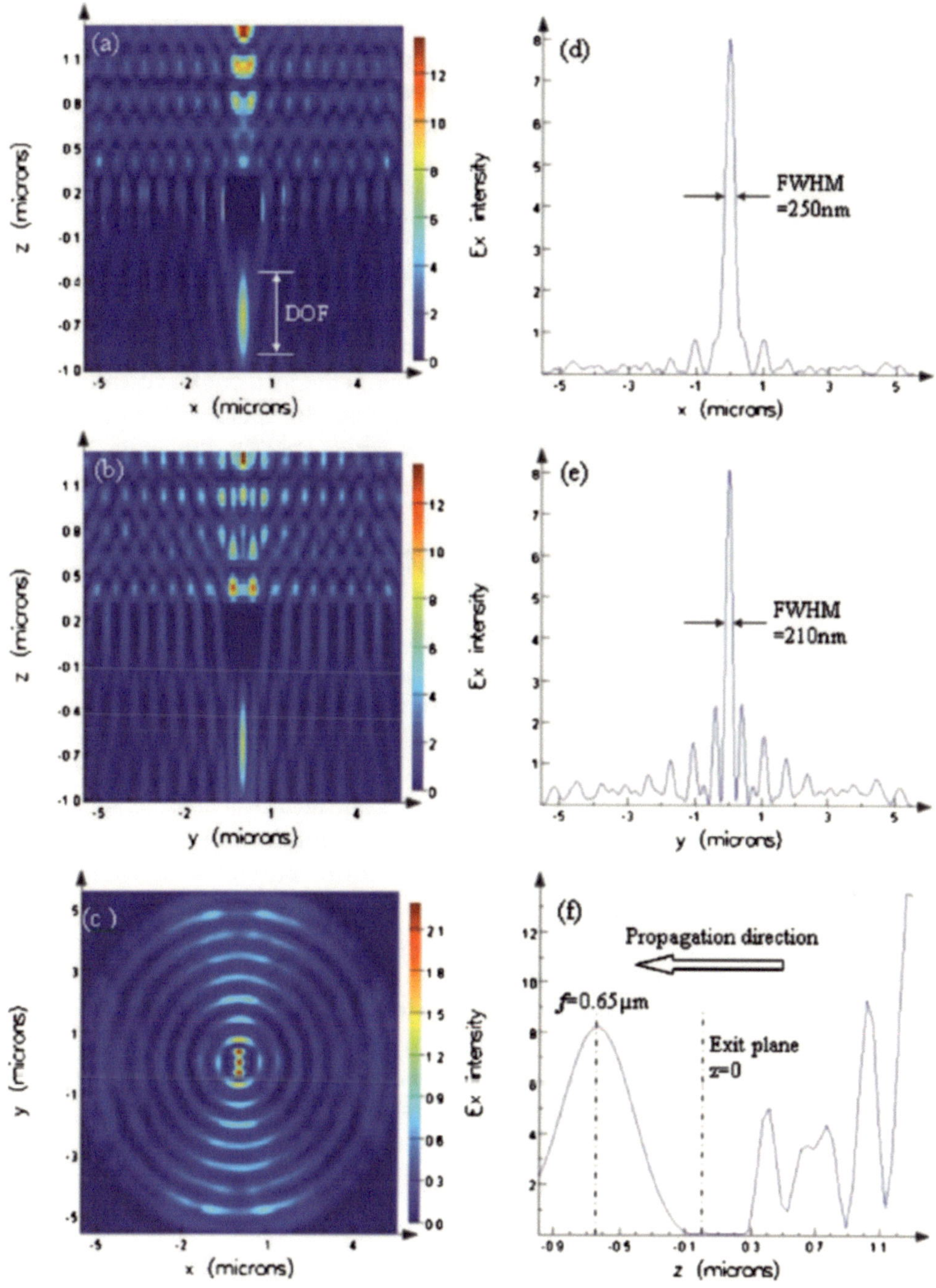

Figure 18. The simulation result of the example of the PMZP. The propagation direction is z. Electric field intensity $|E_x|^2$ at (a) y-z plane, (b) x-z plane, and (c) x-y plane. Electric field transmission in the line z=-0.65 μm (calculated focal plane) at (d) x-z plane, y=0; (e) y-z plane, x=0; and (f) y-z plane, x=0. The designed focal length and outmost zone width using scalar theory is f=1μm and 53 nm, respectively. The calculated DOF is ~700 nm (scalar theory designed value is 8.85 nm) the site z=0 is the exit plane of the Ag film.

and incident wavelength by $f_{FZP} = R^2 / \lambda$, Δr is the outmost zone width, and λ is the incident wavelength. It may be attributed to the SPPs wave coupling through the cavity mode and is involved for the contribution of the beam focusing. The focusing is formed by interference between the SPPs wave and the diffraction waves from the zones.

Figures 18 (a)~(f) are electric intensity distribution $|E_x|^2$ for $E_\perp$ wave in z-y, x-z, and x-y plane, respectively. The numerical computational analysis of the electromagnetic field is carried out using finite-difference and time-domain (FDTD) algorithm. It can be seen that focused spot size [full width at half maximum (FWHM)] at y-z plane is smaller than the one at x-y plane. For conventional Fresenl zone plate, when zone numbers are few, the first sidelobe is large. In contrast, our PMZP has much lower first sidelobe. Suppressed sidelobe at y-z plane is higher than the one at XOZ plane due to the incident wave with $E_\perp$ wave. Transmission with the $E_\perp$ wave illumination is extraordinarily enhanced due to the excited SPP wave coupling which is then converted to propagation wave by diffraction. Figure 2 (f) shows that there is only one peak transmission after exit plane of the Ag film. Furthermore, Fabry-Pérot-like phenomenon is found through the central aperture during the SPP wave coupling and propagation in cavity mode, as shown in Figs.17 (a), (b) and (f). It plays a positive role for the enhancement transmission.

Further characterization of the plasmonic lens was done,as shown in Figs. 19 (a)-(d) [42]. The Au thin film of 200 nm in thickness was coated on quart substrate using e-beam evaporation technique. The lens was fabricated using focused ion beam (FEI Quanta 200 3D dual beam system) direct milling technique, as shown in Fig. 19 (b). Geometrical characterization was performed using an atomic force microscope (Nanoscope 2000 from DI company). Figure 19 (c) shows topography of the FIB fabricated plasmonic lens. The optical measurement was performed with a near-field optical microscope (MultiView 2000^{TS} from Nanonics Inc. in Israel) where a tapered single mode fiber probe, with an aperture diameter of 100 nm, was used working in collection mode. The fiber tip was raster scanned at a discrete constant height of 500 nm, 1.0 μm, 1.5 μm, 2.0 μm, 2.5 μm, 3.0 μm, 3.2 μm, 3.5 μm, 3.7 μm, 4 μm, 4.5 μm, and 5 μm, respectively, above the sample surface, and allowing us to map the optical intensity distribution over a grid of 256×256 points spanning an area of 20×20 μm^2. Working wavelength of the light source is 532 nm (Nd: YAG laser with power of 20 mW). Additionally, a typical lock-in amplifier and optical chopper were utilized to maximize the signal-to-noise ratio. Figure 19 (d) shows the measured three-dimensional (3D) electric field intensity distribution of the lens at propagation distance of 2.5 μm.

Figure 20 is a re-plotted 3D image of the NSOM measured intensity profiles along *x*-axis probed at the different propagation distance *z* ranging from 5 nm to 5 μm. It intuitively shows the intensity distribution along propagation distance. It can be seen that the peak intensity is significantly enhanced from 0.01 μm to 1 μm, and then degraded gradually in near-field region because of SPP-enhanced wave propagation on Au surface vanished in free space when z >1 μm. Only the interference-formed beam focusing region exits in near-field region. It is also in agreement with our calculated results. For more information, please see Ref. [43].

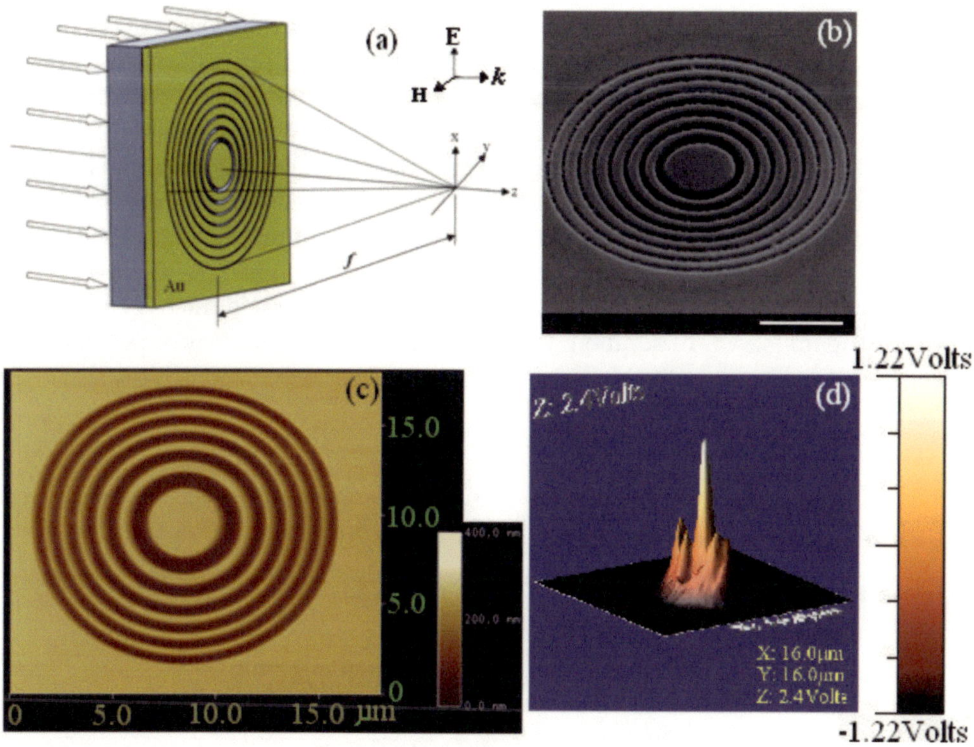

Figure 19. (a) Schematic diagram of the sandwiched plasmonic lens with chirped circular slits corrugated on Au film. Width of the outmost circular slit is 95 nm. Lens dimension (outer diameter) is 12 µm. (b) Scanning electron microscope image of the lens fabricated using focused ion beam milling technique. The scale bar is 4 µm. (c) AFM measurement result: topography of the fabricated lens. (d) NSOM characterization result of the lens: 2D E-field intensity distribution at propagation distance of 2.5 µm

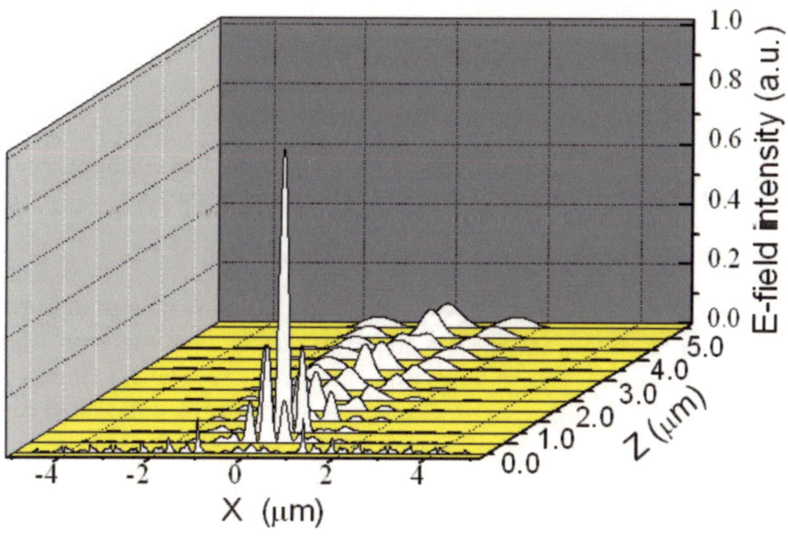

Figure 20. Measured 3D E-field intensity distribution of the plasmonic lens vs. lateral x and propagation distance z using NSOM. The figure was replotted using the NSOM probed data.

A hybrid Au-Ag subwavelength metallic zone plate-like structure was put forth for the purpose of preventing oxidation and sulfuration of Ag film, as well as realizing superfocusing, as shown in Fig. 21 [43]. The Au film acts as both a protector and modulator in the structure. Focusing performance is analyzed by means of three-dimensional (3D) finite-difference and time-domain (FDTD) algorithm-based computational numerical calculation. It can be tuned by varying thicknesses of both Au and Ag thin films. The calculated results show that thickness difference between the Au and Ag thin films plays an important role for transmission spectra. The ratio of Au to Ag film thicknesses, h_{Au}/h_{Ag}, is proportional to the relevant peak transmission intensity. In case of $h_{Au} \approx h_{Ag}$=50 nm, both transmission intensity and focusing performance are improved. In addition, the ratio h_{Au}/h_{Ag} strongly influences position of peak wavelengths λ_{Au} and λ_{Ag} generated from beaming through the metallic structures.

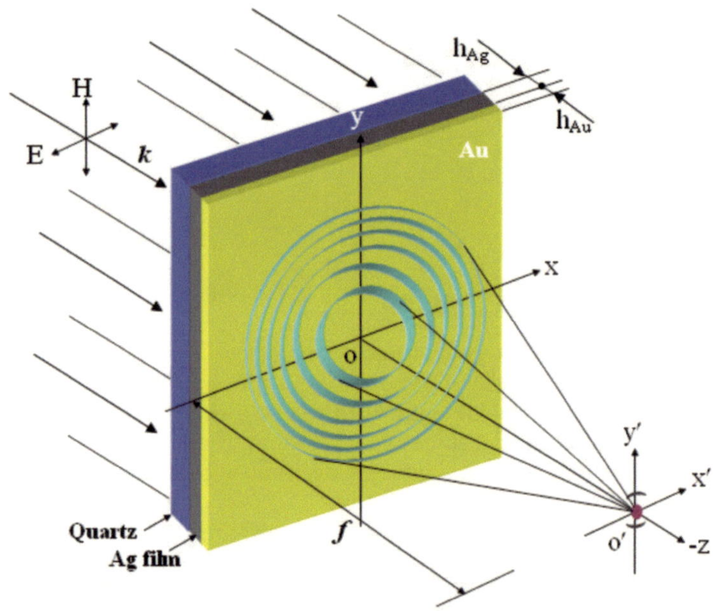

Figure 21. Schematic of the plasmonic micro-zone plate super-focusing with focal length *f*. It is illuminated by a plane wave with 633 nm incident wavelength. In our FDTD simulations the perfectly matched layer (PML) boundary condition was applied at the grid boundaries.

The following features were found from the calculation results:

1. Thickness difference between the Au and Ag thin films plays important role for transmission spectra. It determines peak transmission intensity. Ratio of film thicknesses $\alpha = h_{Au}/h_{Ag}$ is proportional to the relevant peak intensity. It may attributed to the resonant wavelength of the hybrid structure which is proportional to α.
2. In the case of $h_{Au} \approx h_{Ag}$ = 50 nm, $\alpha \approx 1$, both transmission intensity and focusing performance are improved in comparison to the other cases (fixing h_{Au} = 50 nm and varying h_{Ag} from 10 nm to 200 nm).
3. For h_{Ag} = 200 nm, both transmission intensity and focusing performance are improved gradually with increasing h_{Au}. However, unlike the fixed 200 nm thickness of the Ag

film, fixing h_{Au} = 50 nm and 200 nm respectively and varying h_{Ag}, the corresponding optical performances are not improved gradually with increasing h_{Ag}.

4. Ratio $\alpha = h_{Au}/h_{Ag}$ strongly influences position of peak wavelengths λ_{Au} and λ_{Ag} generated from the metallic subwavelength structures.

The calculation results show that thickness of both the Au and Ag thin films has significant tailoring function due to the great contribution to superfocusing and transmission. Improved focusing performance and enhanced transmission can be obtained if h_{Au} and h_{Ag} match each other. This hybrid subwavelength structure has potential applications in data storage, nanophotolithography, nanometrology, and bio-imaging etc.

However, the rings-based structures have higher sidelobes. To supress the sidelobes, a circular holes-based plasmonic lens was reported, as shown in Fig. 22 [44]. In the plasmonic lens with fixed pinhole diameters, propagation waves still exist for much reduced periodicity of pinholes due to the SPPs wave coupling, which interferes with the diffraction wavelets from the pinholes to form a focusing region in free space. Increasing incident wavelength is equivalent to reducing the pinhole diameters, and rapid decay of the EM field intensity will occur accordingly. The superlens proposed by the authors has the advantages of possessing micron scale focal length and large depth of focus along the propagation direction. It should be especially noted that the structure of the superlens can be easily fabricated using the current nanofabrication techniques, e.g. focused ion beam milling and e-beam lithography.

To further improve focusing quality of the circular holes-based plasmonic lens, an elliptical nanoholes-based plasmonic lens was put forth, as shown in Fig. 9 [45]. The plasmonic lens is

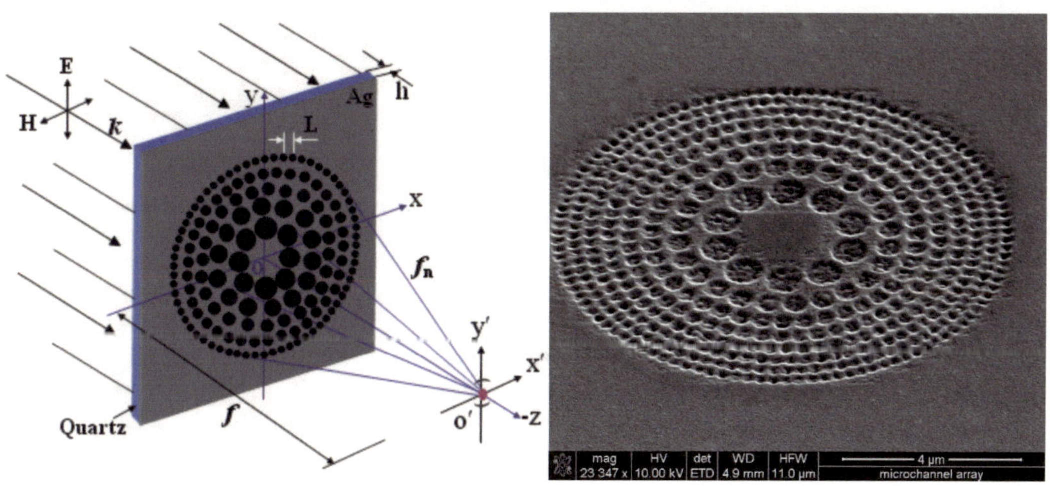

(a) (b)

Figure 22. (a) Schematic of the pinhole array with focal length *f*. Lateral central distance *L* determines wave coupling between the neighboring holes. The pinholes are uniformly distributed in the zones. It is illuminated by a plane wave with 633 nm incident wavelength and p-polarization (transverse-magnetic field with components of E_x, H_y, and E_z). The perfectly matched layer (PML) boundary condition was applied at the grid boundaries in the three-dimensional FDTD simulation. (b) Nanofabrication of the lens using focused ion beam.

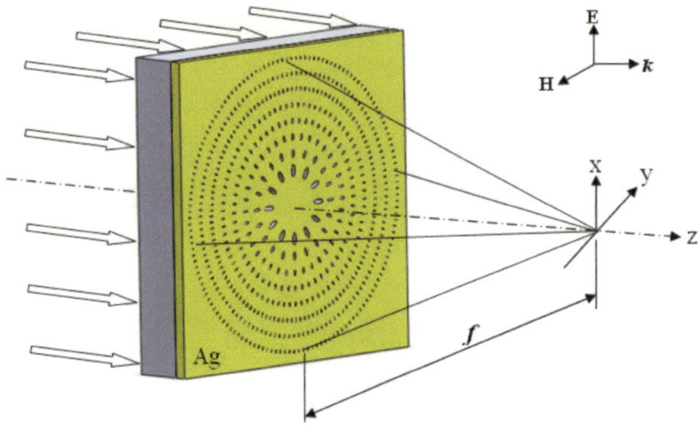

(a)

(b)

Figure 23. (a) Schematization of the pinhole array with focal length *f*. Lateral central distance *L* determines of wave coupling between the neighbored holes. The pinholes are uniformly distributed along the zones. It is illuminated by various waves with 633 nm incident wavelength. And we have different polarization states such as TM, EP and RP. (b) layout of elliptical pinholes with total 8 rings $\delta = 0.6$ being used in our computational numerical calculation.

composed of elliptical pinholes with different sizes distributed in different rings with variant periods. Long-axis of the ellipse is defined as $a_n=3w_n$, whereas w_n is width of the corresponding ring width, and *n* is the number of rings. A thin film of Ag coated on the glass substrate is perforated by the pinholes. The numbers of w_n and radius for different rings are listed in Table1.

Ring No.	1	3	3	4	5	6	7	8
Ring radius(μm)	1.41	2.11	2.78	3.43	4.08	4.73	5.37	6.01
w_n (nm)	245	155	116	93	78	67	59	53

Table 1. The numbers of w_n for different rings with orders from inner to outer (designed *f*=1μm)

As an example, the authors studied the case of 200 nm thickness Ag film coated on quartz substrate and designed a nanostructure with 8 rings on the metal film (see Fig. 23). The pinholes are completely penetrated through the Ag film. The number of pinholes from inner to outer rings is 8, 20, 36, 55, 70, 96, 107, and 140, respectively. Outer diameter of the ring is 12.05 μm. Radius of the rings can be calculated by the formula $r_n^2 = 2nf\lambda + n^2\lambda^2$, where f is the focal length for working wavelength of λ=633 nm which we used in our works. For simplicity, we define a ratio of short-axis to long-axis δ= *b*/*a* (where *a* is the length of long-axis of the elliptical pinholes, and *b* the short-axis). The used metal here is Ag with dielectric constant of $\varepsilon_m = -17.24 + i0.498$ at λ=633 nm, and ε_d=1.243 for glass. The incident angle θ is 0° (normal incidence). In our analysis, we simulated the cases of the ratios: δ = 0.1, 0.2, and 0.4, respectively. The ultra-enhanced lasing effect disappears when the ratio δ→1 (circular pinholes). Orientation of the pinholes is along radial direction. The pinholes symmetrically distribute in different rings with variant periods. It can generate ultra-enhanced lasing effect and realize a long focal length in free space accordingly with extraordinarily elongated depth of focus (*DOF*) of as long as 13 μm under illumination of plane wave in linear y-polarization.

Figure 24 shows our computational results: E-field intensity distribution $|Ey|^2$ at x-z and y-z planes for (a) and (b) δ=0.1; (c) and (d) δ=0.2; and (e) and (f) δ=0.4, respectively. It can be seen that E-field intensity distribution is symmetric due to linear y-polarization which is formed by uniformly rotating linear polarization radiating along radial directions. By means of interfering constructively, and the Ez component interferes destructively and vanishes at the focus. Thus we have $|E|^2=|Ex|^2+|Ey|^2+|Ez|^2=|Ey|^2$, whereas Ex=0, and Ez=0. The lasing effect-induced ultra-long *DOF* is 7 μm, 12 μm, and 13 μm for δ=0.1, 0.2, and 0.4, respectively. It is three orders of magnitude in comparison to that of the conventional microlenses. The ultra-enhanced lasing effect may attribute to the surface plasmon (SP) wave coupling in the micro- and nano-cavity which form Fabry-Pérot resonance while the beam passing through the constructive pinholes. Calculated full-width and half-maximum (FWHM) at propagation distance z= -7 μm in free space is 330 nm, 510 nm, and 526 nm for δ=0.1, 0.2, and 0.4, respectively.

Lasing effect of the plasmonic lens with extraordinarily elongated *DOF* has the following unique features in practical applications:

1. In bioimaging systems such as confocal optical microscope, three-dimensional (3D) image of cells or molecular is possible to be obtained and conventional multilayer focal plane scanning is unnecessary.
2. In online optical metrology systems, feed-back control system can be omitted because height of surface topography of the measured samples as large as ten micron is still within the extraordinarily elongated *DOF* of the plasmonic lenses.
3. In plasmonic structures-based photolithography systems, the reported experimental results were obtained at near-field with tens nanometers gap between the structure and substrate surface. Apparently, it is difficult to control the gap in practical operation. However, using this plasmonic lens, the working distance between the structure and substrate surface can be as long as 12 μm and even longer. Practical control and operation process will be much easier and simplified than the approaches before in case of using this lens.

Like the discussion above regarding illumination with different polarization states, influence of polarization states on focusing properties of the the plasmonic lenses with both chirped circular slits and elliptical nanopinholes were reported [46, 47].

The lens was fabricated using focused ion beam directly milling technique, and characterization of the elliptical nanopinholes-based plasmonic lens was carried out using NSOM, as shown in Figs. 25 (a) and (b). Currently, this work is still in progress in the research group.

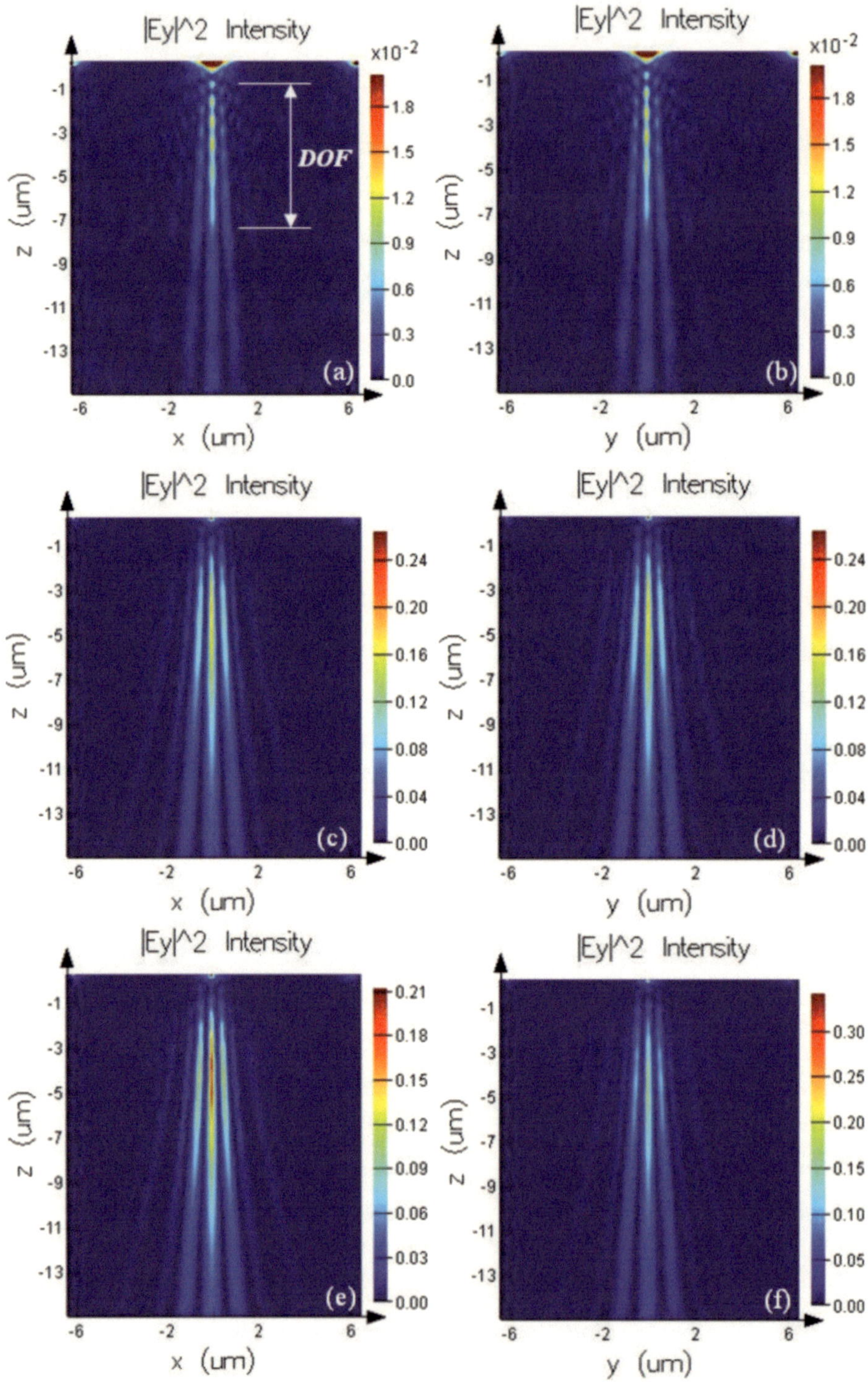

Figure 24. E-field intensity distribution $|Ey|^2$ at x-z and y-z planes for (a) and (b) δ=0.1; (c) and (d) δ=0.2; and (e) and (f) δ=0.4, respectively.

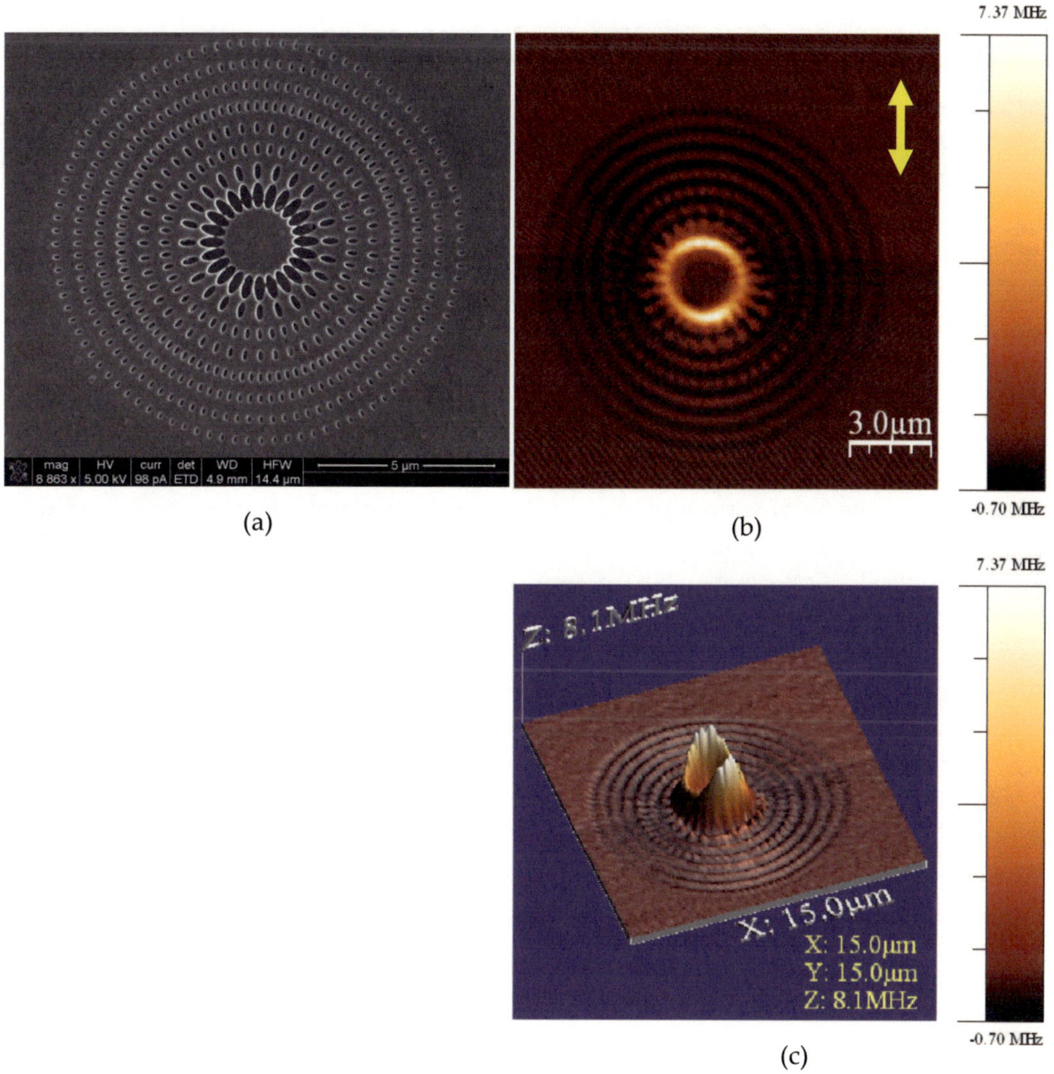

Figure 25. (a) SEM micrograph of the elliptical nanopinholes-based plasmonic lens. (b) 2D image of NSOM probing along propagation distance at *z*=20nm in free space. The arrow indicates direction of linear polarization. (c) 3D image of NSOM probing.

Here we addressed an issue here from fabrication point of view. Real part of Au permittivity ε_{Au} will be increased due to Ga^+ implantation of the FIB directly etching process. Theoretically, propagation constant k_{SP} will be increased due to the large real part of dielectric constant as $k_{SP} = \frac{\omega}{c}\sqrt{\frac{\varepsilon_m \varepsilon_d}{\varepsilon_m + \varepsilon_d}}$, whereas ω is the incident frequency, *c* is the speed of light in vacuum, ε_m and ε_d is dielectric constant of metal and dielectric, respectively. The increased k_{SP} will cause strong transmission enhancement and generate extended skin depth in free space which is helpful for formation of the focusing region and makes a positive contribution on the plasmonic focusing accordingly.

An elliptical nano-pinholes-based plasmonic lens was studied experimentally by means of FIB nanofabrication, AFM imaging, and NSOM characterization for the purpose of proof of plasmonic finely focusing. Both modes of sample scan and tip scan were employed for the lens probing. For the NSOM-based optical characterization of the plasmonic lenses, both of them have their own characteristics. The former can generate a bright-field like image with strong and uniform illumination; and the latter can produce a dark-field like image with high contrast which is helpful for checking focusing performance of the lenses. Our experimental results demonstrated that the lens is capable of realizing a subwavelength focusing with elongated depth of focus.

3.1.2.3. Illumination under radial polarization state

Most recently, the circular rings-based plasmonic lens was experimentally demonstrated [48, 49]. The focusing of surface plasmon polaritons by a plasmonic lens illuminated with radially polarized light was investigated. The field distribution is characterized by near-field scanning optical microscope. A sharp focal spot corresponding to a zero-order Bessel function is observed. For comparison, the plasmonic lens is also measured with linearly polarized light illumination, resulting in two separated lobes. Finally, the authors verify that the focal spot maintain sits width along the optical axis of the plasmonic lens. The results demonstrate the advantage of using radially polarized light for nanofocusing applications involving surface plasmon polaritons. Figures 26 (a) and (b) are reported NSOM characterization results. For comparison purposes, this theoretical cross section is also shownin Figure 26 (b). The profile of the Bessel function can be clearly observed. Neglecting the contribution of the E component, the theoretical spot size (based on full width half-maximum criterion) is 380nm. The measured spot size is slightly larger, 410 ± 39nm (error was estimated by taking several cross sections through the center of the PL along different directions).

3.2. Subwavelength metallic structure for imaging

A planar lens based on nanoscale slit arrays in a metallic film is present here for subwavelength imaging in the far field. Figure 27 is the schematic of the optical imaging with metallic slab lens [51]. To illustrate the design, both object distance a and image distance b are set to be 1 μm. The aperture and the thickness of silver slab lens are 3μm and 300 nm respectively. The permittivity ε_m=-29.26+i1.348 is used for silver at 810 nm.

When the light with magnetic field polarized in the y direction impinged on the surface of silver slab, SPPs can be excited at the slit entrance. The SPPs propagate inside the slits in the specific waveguide modes until reaching the export where we radiate into free space and form the optical image at the desired position. For the imaging of object localized on the axis x=0, the phase retardation of light transmitted through the lens is giver by

$$\Delta\varphi(x) = 2n\pi + \Delta\varphi(0) + \frac{2\pi}{\lambda}(a + b - \sqrt{a^2 + x^2} - \sqrt{b^2 + x^2}) \quad (4)$$

where n is an integer number. Therefore, the key point of designing the metallic lens is to determine the width and position of slits for appropriate phase retardation.

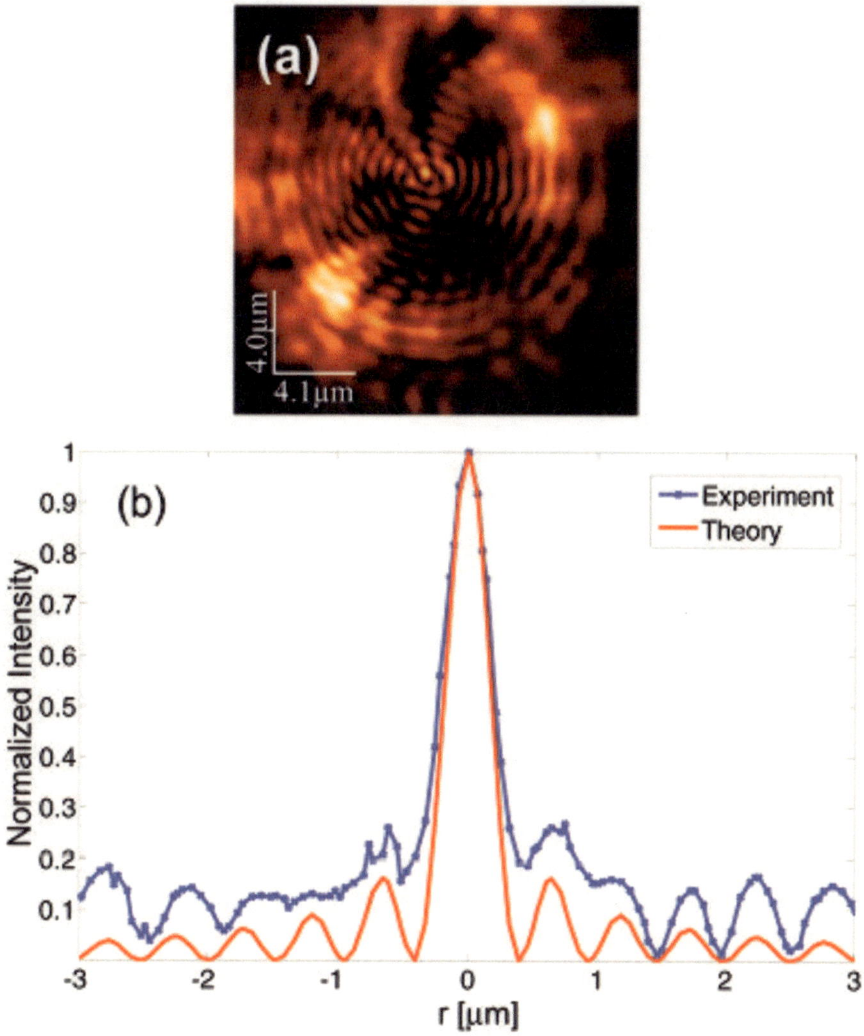

Figure 26. NSOM measurement showing SPP focusing in the plasmonic lens illuminated by radially polarized light. The NSOM probe was at a constant height of 2 μm above the PL. (a) 2D NSOM scan. Bright regions correspond to high intensity. (b) Normalized experimental (blue, cross markers) and theoretical (red, solid line) cross sections through the center of the PL. The sharp focus can be clearly observed. Reprinted with permission from "Gilad M.Lerman, Avner Yanai,and Uriel Levy, Nano Lett. 9, 2139-2143 (2009)." of copyright ©2009 Chemical Society of American.

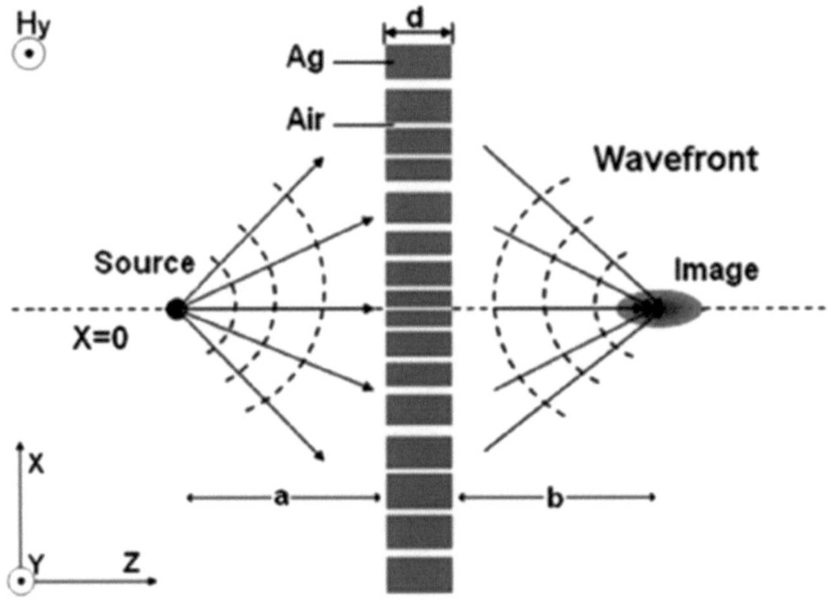

Figure 27. A schematic of optical imaging by a metallic slab lens with nanoslits. Reprinted with permission from "Ting Xu, et. al. Appl. Phys. Lett. 91,201501 (2007)" of copyright ©2007 American Institute of Physics.

Because of the complexity of the accurate description of the processes of SPPs' excitation, propagation, and coupling in metallic slits, the convenient and effective way is the approximation of these processed and making sure of good accuracy simultaneously. The coupling of SPPs during the propagation in slits is neglected, provided that the metallic wall between any two adjacent slits is larger than the skin depth in metal, about 24 nm for silver at a wavelength of 810 nm giver by

$$Z = \frac{1}{k_0}\left[\frac{\mathrm{Re}(\varepsilon_m) + \varepsilon_d}{\mathrm{Re}(\varepsilon_m)^2}\right]^{1/2} \tag{5}$$

The coupling effect occurred at the exit surface from neighboring slits is also omitted, compared with the intensity of directly radiated light from slits. The design above also displays that only a slight deviation is produced by coupling effect and it affects the image profile insignificantly. Assuming that the slit width is much smaller than a wavelength, it is justified to only consider the fundamental mode in the slit. Its complex propagation constant β in the slit is determined by the following equation:

$$\tanh(\sqrt{\beta^2 - k_0^2\varepsilon_d}\,w/2) = \frac{-\varepsilon_d\sqrt{\beta^2 - k_0^2\varepsilon_m}}{\varepsilon_m\sqrt{\beta^2 - k_0^2\varepsilon_d}} \tag{6}$$

where k_0 is the wave vector of light in free space, ε_m and ε_d are the permittivity of the metal and dielectric material inside the slits, and w is the slit width. The real and imaginary parts of β determine the phase velocity and the propagation loss of SPPs inside the metallic slit, respectively.

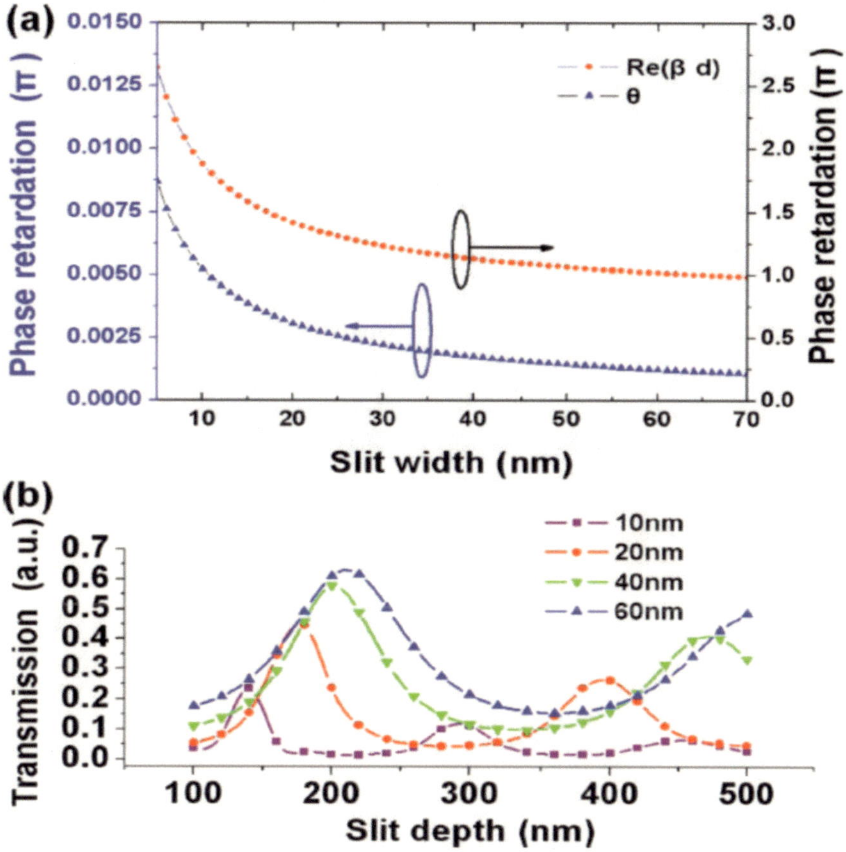

Figure 28. (a) Dependence of phase retardation on the slit width. Red and blue tags represent the contributions for phase retardation from the real part of propagation constant and multiple reflections. The wavelength of incident light is 810 nm. (b) Transmittance of optical field (810 nm wavelength) through a nanoslit vs the slit depth ranging from 100 to 500 nm. Slit widths are 10, 20, 40, and 60 nm. Reprinted with permission from "Ting Xu, et. al. Appl. Phys. Lett. 91,201501 (2007)" of copyright ©2007 American Institute of Physics.

To illustrate the validity of metallic slab lens, finite-difference time-domain simulations are performed. Figure 28 (a) illuminates the dependence of phase retardation on slit width. Figure 28 (b) plots the finite-difference time-domain (FDTD) simulated transmittance of light through a slit with variant thickness as ranging from 100 to 500 nm by normal incidence at the wavelength of 810 nm. Calculated steady optical field (the magnetic field $|Hy|^2$) of the simulation result is shown in Fig. 29. Obvious image spot can be seen at the position around Z=2.43 μm with the full width at half maximum of 396 nm approximately half of the incident wavelength. The slight focal shift of about 60 nm. Above all, the metallic slab lens displays a considerably good performance for imaging objects in the far field region.

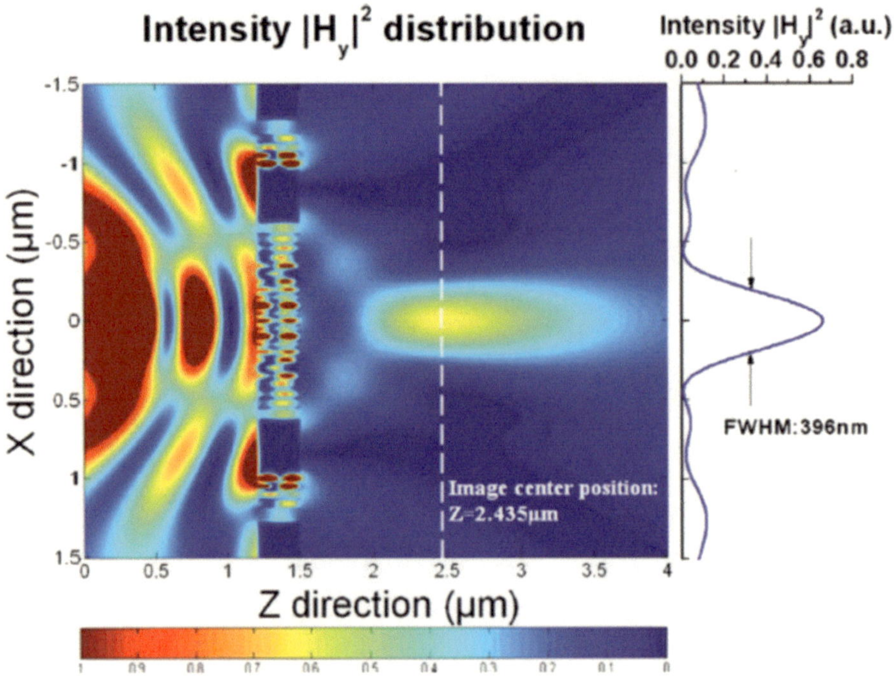

Figure 29. Calculated steady magnetic field intensity |Hy|2 of the simulation results using FDTD method. The spirce is localized at X=0 and Z=0.2 μm. The metallic slab lens ranged from Z=1.2 μm to Z=1.5 μm. The radiated light from the source is TM polarized with a wavelength of 810nm. The curve at the right side represents the cross section of image plane at Z=2.435 μm. Reprinted with permission from "Ting Xu, et. al. Appl. Phys. Lett. 91,201501 (2007)" of copyright ©2007 American Institute of Physics.

In addition, nanorod array was reported being used for imaging [51]. Mark et. al reported parallel conducting wires as a lens for subwavelength microwave imaging [52].

4. Plasmonic Lens on the basis of waveguide modes

Another structure that can realize nanofocusing was theoretically reported [53]. SPPs propagating toward the tip of a tapered plasmonic waveguide are slowed down and asymptotically stopped when we tend to the tip, never actually reaching it (the travel time to the tip is logarithmically divergent). This phenomenon causes accumulation of energy and giant local fields at the tip. Focusing of fundamental cylindrical SPP wave is formed at apex of the taper tip, as shown in Fig. 30. Figure 31 displays the amplitudes of the local optical fields in the cross section of the system for the normal and longitudinal (with respect to the axis) components of the optical electric field. As SPP's move toward the tip, the SPP fields start to localize at the metal surface, and simultaneously, our wavelength is progressively reducing and amplitude growing. The field magnitudes grow significantly at small |Z|. The transverse x component grows by an order of magnitude as the SPP's approach the tip of the guide, while the longitudinal z component, which is very small far from the tip, grows relatively much stronger. The 3D energy concentration occurs at the tip of a smoothly tapered metal nanoplasmonic waveguide. This causes the local field increase by 3 orders of magnitude in intensity and four orders in energy density.

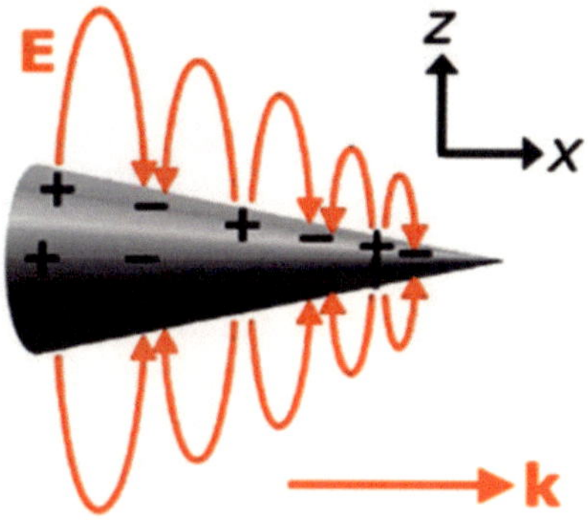

Figure 30. Geometry of the nanoplasmonic waveguide. The radius of the waveguide gradually decreases from 50 nm to 2 nm. Reprinted with permission from "Mark I. Stockman, Phys. Rev. Lett. 93, 137404 (2004)" with copyright © 2004 of American Society of Physics.

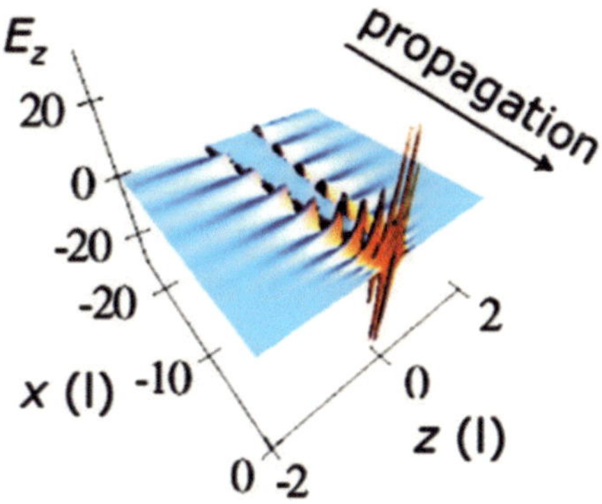

Figure 31. Snapshot of instanteneous fields (at some arbitrary moment t=0): longitudinal component Ez of the local optical electric field are shown in the longitudinal cross section (xz) plane of the system. The fields are in the units of the far-zone (excitation) field. Reprinted with permission from "Mark I. Stockman, Phys. Rev. Lett. 93, 137404 (2004)" with copyright © 2004 of American Society of Physics.

Most recently, Ewold *et. al.* reported our tapered waveguid structure for nanofocusing, as shown in Fig. 32 [54]. It was used for focusing of surface plasmon polaritons (SPPs) excited with 1.5 μm light in a tapered Au waveguide on a planar dielectric substrate by experiments and simulations. We find that nanofocusing can be obtained when the asymmetric bound mode at the substrate side of the metal film is excited. The propagation and concentration of this mode to the tip is demonstrated. No sign of a cutoff waveguide width is observed as the SPPs propagate along the tapered waveguide. Simulations show that such concentrating behavior is not possible for excitation of the mode at the low-index side of the film. The mode that enables the focusing exhibits a strong resemblance to the asymmetric mode responsible for focusing in conical waveguides. This work demonstrates a practical implementation of plasmonic nanofocusing on a planar substrate.

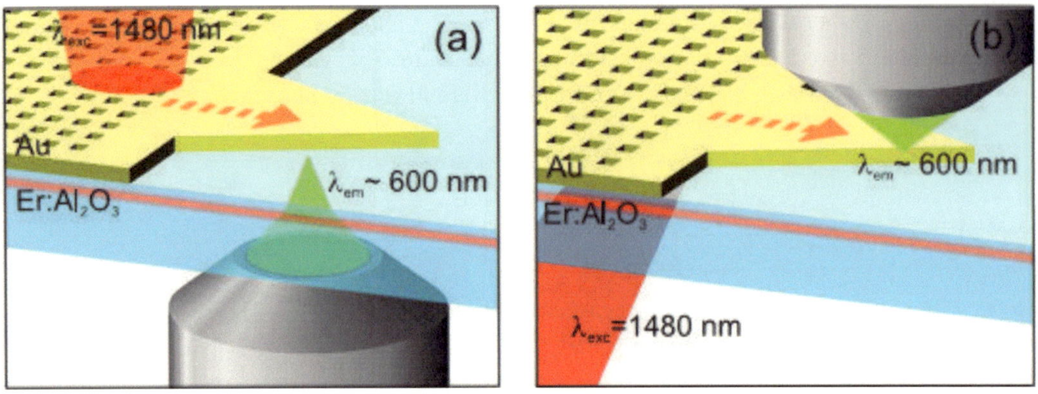

Figure 32. Schematic of the experimental geometry in the case of upconversion luminescence detection through the substrate (a), or from the air side of the sample (b). In both cases the SPPs are excited with infrared light at the Au/Al_2O_3 interface in the direction of the arrow. The red line schematically indicates the Er depth profile. Reprinted with permission from "Ewold Verhagen, Albert Polman, and L. (Kobus) Kuipers, Opt. Express 16, 45-57 (2008)." of copyright ©2008 Optical Society of American.

A fiber-pigtailed 1.48 μm diode pump laser is used as excitation source. Figure 32 shows a schematic of the hole array/taper geometry. The pitch of the hole array is chosen such that p-polarized light with a wavelength of 1.48 μm is diffracted to generate SPPs propagatin at the substrate side of the film. To maximize the excitation of the desired SPP mode, the excitation beam is focused to a 10 μm wide spot near the edge of the array. The triangularl shaped tapered waveguide starts at a distance of 6 μm from the edge of the excitation arra and has a base width of 12 μm and a length of 60 μm (taper angle 11°).

5. Plasmonic lens on the basis of curved chains of nanoparticles

Focusing of surface plasmon polaritons (SPPs) beams with parabolic chains of gold nanoparticles fabricated on thin gold films was reported [55]. SPP focusing with different parabolic chains is investigated in the wavelength range of 700-860nm, both experimentally and thoretically. Mapping of SPP fields is accomplished by making use of leakage radiation microscopy, demonstrating robust and efficient SPP focusing into submicron spots. Numerical simulations based on the Green's tenor formalism show very good agreement with the experimental results, suggesting the usage of elliptical corrections for parabolic structures to improve our focusing of slightly divergent SPP beams.

Shortly after the above work, excitation,focusing and directing of surface plasmon polaritons (SPPs) with curved chains of nanoparticles located on a metal surface is investigated both experimentally and theoretically by Evlyukhin et. al., as shown in Fig. 33 [56]. We demonstrate that, by using a relatively narrow laser beam (at normal incidence) interacting only with a portion of a curved chain of nanoparticles, one can excite an SPPs beam whose divergence and propagation direction are dictated by the incident light spot size and its position along the chain. It is also found that the SPPs focusing regime is strongly influenced by the chain inter-particle distance. Extensive numerical simulations of

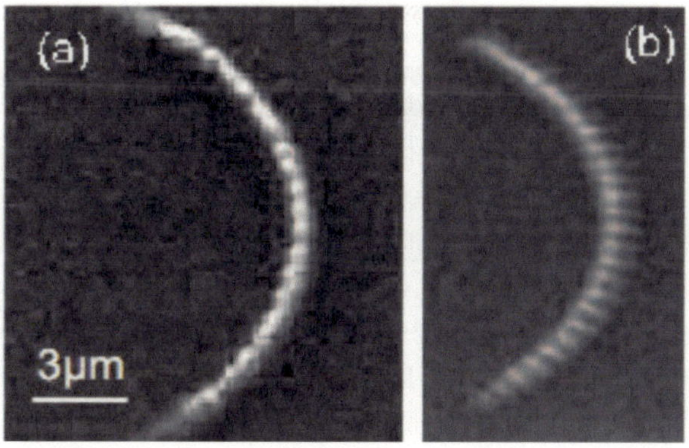

Figure 33. Scanning electron microscope (SEM) (a) the top image of the structure, (b) the image of the structure obtained with view angle 45°. The radius of curved chains of nanoparticles is equal to 10μm. The particle in-plane size (diameter) and inter-particle distance are estimated to be about 350nm and 850nm, respectively, the particle height is 300nm. Reprinted with permission from "A. B. Evlyukhin, et. al, Opt. Exp. 15, 16667-16680 (2007)." of copyright ©2007 Optical Society of American.

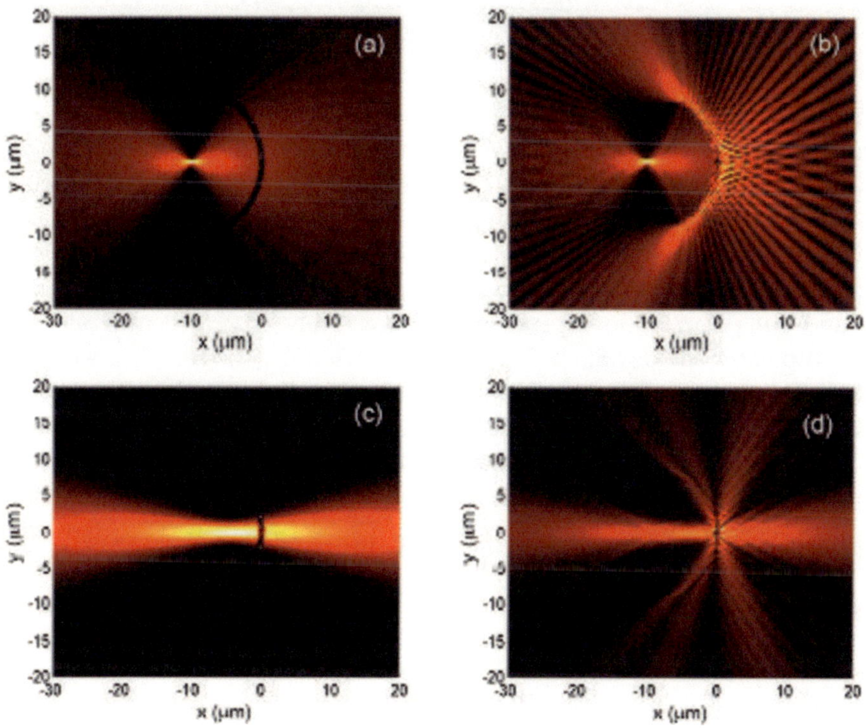

Figure 34. Magnitude of scattered electric field calculated above the gold surface with a curved chain (with R=10 μm and β=60°) of spheroid gold nanoparticles illuminated by a light beam at the wavelength of 800nm being incident perpendicular to the gold surface and polarized along x-direction. The waist W of the incident beam and the inter-particle (center-to-center) spacing D in the chain are (a) W=10μm, D=400nm; (b)W=10μm, D=800nm; (c)W =1.5 μm, D=400nm; (d)W =1.5 μm, D=800nm. Reprinted with permission from "A. B. Evlyukhin, et. al, Opt. Exp. 15, 16667-16680 (2007)." of copyright ©2007 Optical Society of American.

the configuration investigated experimentally are carried out for a wide set of system parameters by making use of the Green's tensor formalism and dipole approximation. Comparison of numerical results with experimental data shows good agreement with respect to the observed features in SPP focusing and directing, providing the guidelines for aproper choice of the system parameters.

When the inter-particle distance is smaller than the light wavelength, the pattern of the field magnitude distribution is relatively smooth [Fig. 34 (a)]. In this case, the illuminated part of the chain exhibits scattering properties that are similar to those of a continuous ridge. Note that straight ridges are frequently used for excitation of a divergent SPP beam on a metal surface in SPP experiments. When increasing the inter-particle distance, the individual particles of the chain become relatively independent sources of the scattered waves whose phases differ considerably, resulting in a complex interference pattern [Fig. 34 (b)] – a system of divergent SPPs rays. A similar trend is also seen for a relatively narrow incident light beam [see Figs. 34 (c) and (d)]. If the light spot size being determined by *W* is sufficiently small in comparison with the chain curvature radius *R* so that the diffraction angle of a SPPs beam is approximately equal to the focusing angle *W*/*R*, the focusing effect decreases and the maximum of SPPs intensity moves toward the nanoparticle chain. Strong SPPs focusing effects have been obtained for relatively lager *W*/*R*.

Apart from the chain particle-based plasmonic structures for nanofocusing, the reverse pattern: nanoholes, constructed in curved chains can also realize focusing of SPP wave, as shown in Fig. 35 [57]. The focused SPPs can be directly coupled into a waveguide located at the focal plane. The constructive interference of SPPs launched by nanometric holes allows us to focus SPP into a spot of high near-field intensity having subwavelength width. Near-field scanning optical microscopy is used to map the local SPP intensity. The resulting SPPs patterns and our polarization dependence are accurately described in model calculations based on a dipolar model for the SPP emission at each hole. Furthermore, we show that the high SPPs intensity in the focal spot can be launched and propagated on a Ag strip guide with a 250 × 50 nm^2 cross section, thus overcoming the diffraction limit of conventional optics. The combination of focusing arrays and nano-waveguides may serve as a basic element in planar nano-photonic circuits.

Not only focusing and imaging, the similar plasmonic structures can be used as beam sppliters [58] and beam shaping [59]. Nanofabrication of the plasmonic lenses were reported in Ref. [60-62].

6. Plasmonic Talbot effect of nanolenses

Previous introduction in Section 3 shows a common phenomenon which there several focal points exist along central axis in free space after exit plane of the nanolenses.

To explore its physical mechanism, we compared the phenomenon to another well-known story: Talbot effect, *i.e.* self-imaging. The self-imaging means that when a one-dimensional (1D) periodic structure is illuminated by the monochromatic plane wave, the image of that structure can be observed at the periodical distance from the back side of the structure. It is

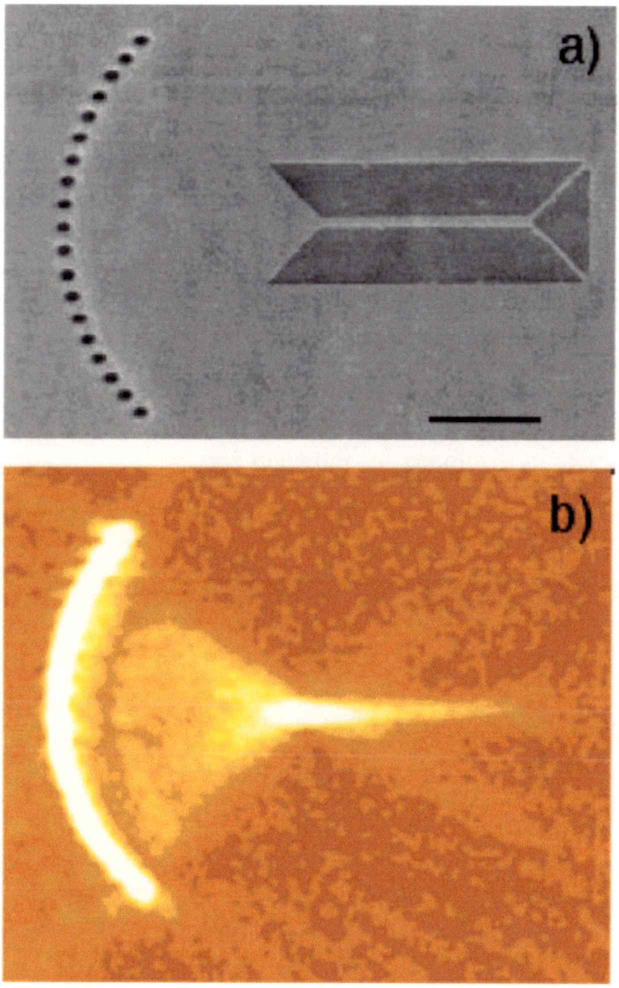

Figure 35. (a) SEM image of the focusing array coupled to a 250-nm-wide Ag strip guide; light gray, Ag; dark gray, Cr; scale bar, 2 μm.(b) NSOM image of the SPPs intensity showing focusing and guiding. Reprinted with permission from "Leilei Yin, et. al., Nano Lett. 2005, *5*, 1399–1402." of copyright ©2005 Chemical Society of American.

a type of imaging by diffraction rather than an ordinary imaging of a lens. Actually, some researchers are still interested in the self-imaging effect since H. F. Talbot discovered it in 1836. And intensive theoretical and experimental studies regarding Talbot effect have been done since then. For example, the plasmon analogue of the self-imaging Talbot effect of a row of holes drilled in a metal film was described and theoretically analyzed, and suggested the potential applications in sensing, imaging, and optical interconnects on the basis of plasmon focal spots aimed at plasmon waveguides [63]. Subsequently, the Talbot effect regarding SPPs imaging on the basis of a rather different system by a quite different approach was studied theoretically and experimentally [62-64]. Furthermore, the Talbot effect for volume electromagnetic waves has been used in a variety of applications. And it is expected that the analogue for SPPs will be found applications in numerous nanoscale plasmonic devices.

Here, as an example, the Talbot effect of an Ag nanolens with five periodic concentric through the rings illuminated by a radially polarized light was computationally studied. Rigorous finite-difference and time-domain (FDTD) algorithm was employed in the computational numerical calculation. The results indicate that several focal points can be obtained at different locations due to the SPPs-related Talbot effect at $\lambda_{inc.}$=248 nm. The positions are quite different from that of values calculated by the Talbot distance equation reported in Ref. [65]. A minimum diameter of 100 nm at site of full width and half maximum (FWHM) was derived at the propagation distance of Z=396 nm. To further study the phenomenon in physics, it was compared with the traditional Talbot distance calculated using the scalar diffraction theory in the sections below.

For simplification of the analyses, the structure is simplified that is equivalent to a one-dimensional grating structure with the same geometrical parameters as the nanolens mentioned before due to its symmetry, as shown in Fig. 36.

The scalar field immediately behind an infinite grating at the original position (Z = 0) when it is illuminated by an unit intensity plane wave can be described by a Fourier series representing a weighted set of plane wave components as:

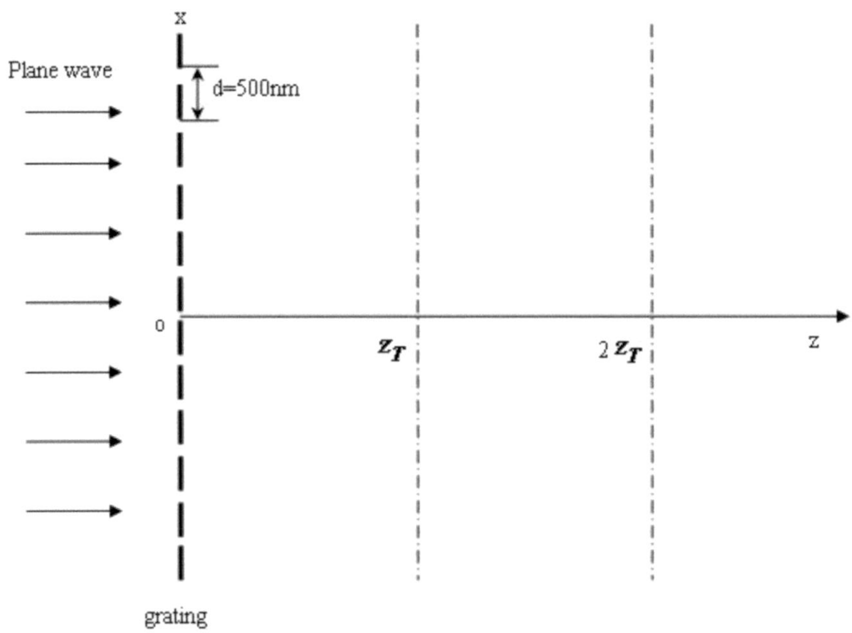

Figure 36. Self-imaging of the grating structure with the same geometrical parameters as the nanolens mentioned before.

$$U(x) = \sum_{n=-\infty}^{\infty} c_n \exp\left(j2\pi \frac{n}{d} x \right) \tag{7}$$

where d is the grating period (is 500 nm here) and c_n is the nth Fourier coefficient. The coefficients represent the complex intensity and the phase.

According to the frequency domain analysis method, the diffraction field distribution U_z at a free-space propagation distance Z is given by

$$U_z(x)=\sum_{n=-\infty}^{\infty} c_n \exp\left(j2\pi\frac{n}{d}x\right)\exp\left[-j\pi\lambda z\left(\frac{n}{d}\right)^2\right]\exp(jkz) \tag{8}$$

Note that when

$$z=\frac{2md^2}{\lambda} \qquad (\mathrm{m}=1,2,3,\ldots) \tag{9}$$

then

$$\exp\left[-j\pi\lambda z\left(\frac{n}{d}\right)^2\right]=1 \tag{10}$$

U_z can be written as

$$U_z(x)=\sum_{n=-\infty}^{\infty} c_n \exp\left(j2\pi\frac{n}{d}x\right)\exp(jkz)=U(x)\exp(jkz) \tag{11}$$

Intensity distribution equals to the original intensity

$$I=\left|U_z(x)\right|^2=\left|U(x)\right|^2 \tag{12}$$

The Talbot distance of a grating with a period of d is

$$z_T=\frac{2d^2}{\lambda} \tag{13}$$

Substituting d and λ with their corresponding values of 500 nm and 248 nm, respectively, we can get the Talbot distance of z_T=2.016 μm. It indicates that self-imaging of the periodic grating structures can be observed at positions of Z=2.016 μm, 4.032 μm, and 6.048 μm, etc., which is integral number times of z_T.

It was found that the result analyzed above by the scalar diffraction theory is quite different from the results calculated by the FDTD numerical analysis method. The former is far field diffraction, but the latter is SPPs coupling and interfering at near field. Thus the focal points do not repeat at the same positions as that of the scalar one along the propagation direction. The phenomenon described before can be attributed to that of the surface plasmons (SP). SPPs is excited at all azimuthal directions that interferes each other constructively and creates a strongly enhanced localized fields at the focal points. The size of the focal spots is less than half a wavelength and the focal positions are not determined by the Talbot distance.

In comparison to the results of above mentioned Talbot distances from scalar theory, only the contribution from the field of SPPs is taken into account in the calculations of the interference pattern. In the geometry considered in this paper, the conversion of the incident SPPs into the volume electromagnetic waves is weak. The Talbot distance of 2.016 μm (first order) calculated by formula $z_T = k\Lambda^2/\pi$ (it is the same as Eq. (13) actually. They were written in two different forms.) from Maradudin *et. al.* [63] and van Oosten [64] is different also from our calculated value of 0.3962 μm (first order) here. The reason we get a different answer compared to the other two methods is because the other two methods use the paraxial approximation in far-field regime, which state that the pitch of the structure is much greater than the incident wavelength. However, the near-field regime in which we operate is when the wavelength is about half the period. For our structures, the paraxial approximation does not hold anymore. Therefore, we can draw a conclusion that the theoretical equations deduced for the structures of metallic dot arrays [63, 65] are not suitable to our nanolens structures. The approach of computational numerical calculation must be employed for the study of the Talbot distance. For more information regarding theoretical calculation, readers can see Ref.[65].

To compare the experimental results with the theoretically calculated results, we plotted E-field intensity profiles versus propagation distance z together with that of the numerically calculated in the same figure, as shown in Fig. 37. It can be seen that variation tendency of the E-field intensity of the measured results is in agreement with the theoretically calculated results in general. As can be seen from Fig. 36, the positions of the measured first three focal points are slightly in advance of the calculated values due to the interaction of the metal surfaces between the NSOM probe and nanolenses at near-field. But for the last two ones,

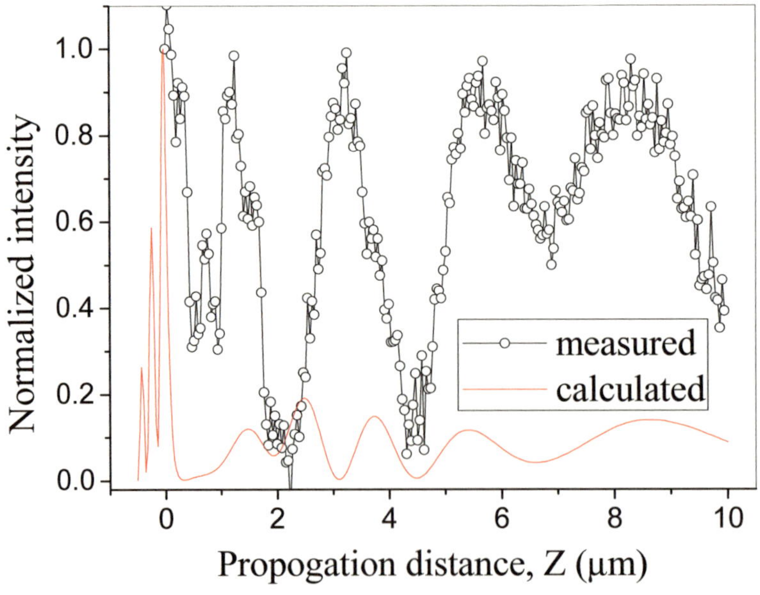

Figure 37. Comparison of E-field intensity distribution along propagation axis between experimental and theoretical results.

the measured and calculated results coincide with each other very well. It can be explained that the interaction between the NSOM probe and the nanolenses surface is weak at the far-field. The influence on the probing is significantly degraded and can be ignored accordingly.

In comparison to a standard diffraction experiment, if the wavelength is large the slit is subwavelength, surely there will be no diffracted waves (also known as cylindrical waves) and therefore, the distinction between short wavelength and long wavelength is obvious. For the standard diffraction, transmission intensity decays exponentially with slit size (Bethe's theory was put forth in 1944). However, for our plasmonic lenses, the transmission intensity can be enhanced due to inherent advantage of plasmonic resonance which is well known already. Here, we used metallic subwavelength structure to generate the self-imaging. It is more apparent in near field region. The contribution of plasmonic resonance here is that enhancing the intensity of the plasmonic Talbot-based focusing (see Fig. 36). It is especially important for nano-photolithography for which large working distance and high exposure intensity are crucial issues. Our previous FDTD calculation results [66] demonstrated that only one focal point can be observed for the case of λ>300 nm, which is a threshold value for the SPPs-based Talbot effect. For the short wavelength, it is smaller than the slit width. Thus it is consistent to the Talbot effect condition. For more information regarding experimental characterization, readers can see Ref.[67].

7. Summary

In summary, the main reason of the diffraction limit of a conventional optical lens is that a conventional lens is only capable of transmitting the propagating. The evanescent carry subwavelength information about the object decay, but it decay exponentially and cannot be collected in far field region. In contrast, plasmonic lens is an alternative to these problems. There are two key concepts about plasmonic lens. One is the concept of negative refractive index. The other is the transmission enhancement of evanescent waves. The concept of negative refractive index is the basic issue of plasmonic lens. And this is very common for noble metal in specific frequency.

The scalar Talbot effect cannot interpret the Talbot effect phenomenon for the metallic nanolenses. It may attribute to the paraxial approximation applied in the Talbot effect theory in far-field region. However, the approximation does not hold in our nanolenses structures during the light propagation. In addition, the Talbot effect appears at the short wavelength regime only, especially in the ultra-violet wavelength region.

Also plasmonic lenses provide excellent optical property for us. We can shape the beam beyond the diffraction limit, such as superfocusing, imaging and so on. Also we can exist in various forms. And in this paper, we show them in three form generally. Some new ideas will be produced while we design the plasmonic lenses to resolve practical problems. Currently, applications of the plasmonic lenses reported include photodetector [67] and ring resonator [68]. With rapid development of the plasmonic lenses, it is reasonable to believe that extensive applications will be found in the near future.

Author details

Yongqi Fu
School of Physical Electronics, University of Electronic Science and Technology of China, China

Jun Wang and Daohua Zhang
School of Electronic and Electrical Engineering, Nanyang Technological University, Republic of Singapore

Acknowledgement

Some reported research works in this chapter were supported by National Natural Science Foundation of China (No. 11079014 and 61077010).

8. References

[1] J. B. Pendry, "Negative Refraction Makes a Perfect Lens", Phys. Rev. Lett. 5, 3966-3969 (2000).
[2] Anthony Grbic and George V. Eleftheriades, "Overcoming the diffraction Limit with a planar Left-Handed Transmission-Line Lens", Phys. Rev. Lett. 92, No. 11, 117403 (2004).
[3] D.R. Smith, J.B. Pendry, M.C.K. Wiltshire, "Metamaterials and negative refractive index", Science 305, 788-792 (2004).
[4] S. Anantha Ramakrishna, "Physics of negative refractive index materials", Rep. Prog. Phys. 68 (2), 449-521 (2005).
[5] T. W. Ebbesen, H. J. Lezec, H. F. Ghaemi, T. Thio and P. A Wolff, "Extraordinary optical transmission through subwavelength hole array", Nature 391, 667-669 (1998).
[6] H.J. Lezec, A. Degiron, E. Devaux, R.A. Linke, F. Martín-Moreno, L.J. García-Vidal, and T.W. Ebbesen, "Beaming light from a subwavelength aperture", Science 297, 820 (2002).
[7] Liang-Bin Yu, Ding-Zheng Lin et al., "Physical origin of directional beaming emitted from a subwavelength slit", Phys. Rev. B 71, 041405 (2005).
[8] W. L. Barnes, A. Dereux, and T. W. Ebbesen, "Surface plasmon subwavelength optics", Nature 424, 824-830 (2003).
[9] N. Fang, X. Zhang, "Imaging properties of a metamaterial superlens", Appl. Phys. Lett. 82(2), 161-163 (2003).
[10] N. Fang, H. Lee, C. Sun and X. Zhang, "Sub-Diffraction-Limited Optical Imaging with a Silver Superlens", Science 308, 534-537 (2005).
[11] Z. Liu, S. Durant, H. Lee, Y. Pikus, N. Fang, Y. Xiong, C. Sun, and X. Zhang, "Far-Field Optical Superlens", Nano Lett. Vol. 7, No. 2, 403-408 (2007).
[12] N. Fang, Z. W. Liu, T.J. Yen, X. Zhang, "Regenerating evanescent waves from a silver superlens", Opt. Express 11(7), 682-687 (2003).
[13] Z. W. Liu, J. M. Steele, W. Srituravanich, Y. Pikus, C. Sun, and X. Zhang, "Focusing surface plasmons with a plasmonic lens", Nano Lett. 5, 1726-1729 (2005).
[14] D.O.S. Melville, R.J. Blaike, C.R. Wolf, "Submicron imaging with a planar silver lens", Appl. Phys. Lett. 84(22), 4403-4405 (2007).

[15] Salandrino and N. Engheta, "Far-field Subdiffraction Optical Microscopy Using Metamaterial Crystals: Theory and Simulations", Phys. Rev. B 74, 075103 (2006).

[16] W. Nomura, M. Ohtsu, and T. Yatsui, "Nanodot coupler with a surface plasmon polariton condenser for optical far/near-field conversion", Appl. Phys. Lett. 86, 181108 (2005).

[17] Liang Feng, A. Kevin Tetz, Boris Slutsky, Vitaliy Lomakin, and Yeshaiahu Fainman, "Fourier plasmonics: diffractive focusing of in-plane surface plasmon polariton waves", Appl. Phys. Lett. 91, 081101 (2007).

[18] Z. Liu, H. Lee, Y. Xiong, C. Sun and X. Zhang, "Far-Field Optical Hyperlens Magnifying Sub-Diffraction-Limited Object", Science 315, 1686 (2007).

[19] A. V. Zayats, I. I. Smolyaninov, and A. A. Maradudin, "Nano-optics of surface plasmon polaritons", Phys. Rep. 408, 131-314 (2005).

[20] L. Zhou and C. T. Chan, "Relaxation mechanisms in three-dimensional metamaterial lens focusing", Opt. Lett. 30, 1812 (2005).

[21] H. Ditlbacher, J. R. Krenn, G. Schider, A. Leitner, F. R. Aussenegg, "Two-dimensional optics with surface plasmon polaritions", Appl. Phys. Lett. 81, 762-764 (2002).

[22] Yongqi Fu, W. Zhou, L.E.N. Lim, C. Du, H. Shi, C.T Wang and X. Luo, "Transmission and reflection navigated optical probe with depth-tuned serface corrugations", Appl. Phys. B 86, 155-158 (2007).

[23] H. F. Shi, C. T. Wang, C. L. Du, X. G. Luo, X. C. Dong, and H. T. Gao, "Beam manipulating by metallic nano-slits with variant widths", Opt. Express 13, 6815-6820 (2005).

[24] Avner Yanai and Uriel Levy, "The role of short and long range surface plasmons for plasmonic focusing applications", Opt. Express 17, No. 16 (2009).

[25] M. H. Wong, C. D. Sarris, and G. V. Eleftheriades, "Metallic Transmission Screen For Sub-Wavelength Focusing", Electronics Letters, 43, 1402-1404 (2007).

[26] Yongqi Fu, Wei Zhou, Lennie E.N. Lim, Chunlei Du, Haofei Shi, Changtao Wang, "Geometrical characterization issues of plasmonic nanostructures with depth-tuned grooves for beam shaping". Opt. Eng. 45, 108001 (2006).

[27] Baohua Jia, Haofei Shi, Jiafang Li, Yongqi Fu, Chunlei Du, and Min Gu, "Near-field visualization of focal depth modulation by step corrugated plasmonic slits", Appl. Phys. Lett. 94, 151912 (2009).

[28] Y. Xie, A.R. Zakharian, J.V. Moloney, and M. Mansuripur, Transmission of light through slit apertures in metallic films. Optics Express. 12. 6106-6121, (2004).

[29] L.B. Yu, D.Z. Lin, Y.C. Chen, Y.C. Chang, K.T. Huang, J.W. Liaw, J.T. Yeh, J.M. Liu, C.S. Yeh, and C.K. Lee, Physical origin of directional beaming emitted from a subwavelength slit. Physical Review B - Condensed Matter and Materials Physics. 71. 1-4, (2005).

[30] B. Ung and Y. Sheng, "Optical surface waves over metallo-dielectric nanostructures: Sommerfeld integrals revisited". Optics Express. 16. 9073-9086, (2008).

[31] H. Shi, C. Du, and X. Luo, "Focal length modulation based on a metallic slit surrounded with grooves in curved depths". Applied Physics Letters. 91, (2007).

[32] Z. Sun and H.K. Kim, "Refractive transmission of light and beam shaping with metallic nano-optic lenses". Applied Physics Letters. 85. 642-644, (2004).
[33] Yongqi Fu, W. Zhou, L.E.N Lim, C. Du, H. Shi, C. Wang and X. Luo, "Influence of V-shaped plasmonic nanostructures on beam propagation", Appl. Phys. B 86, 461-466(2007).
[34] Zhaowei Liu, Jennifer M. Steele, Hyesog Lee, and Xiang Zhang, "Tuning the focus of a plasmonic lens by the incident angle", Appl. Phy. Lett. 88, 171108 (2006).
[35] Jun Wang, Wei Zhou and Anand K. Asundi, "Effect of polarization on symmetry of focal spot of a plasmonic lens". Opt. Express 17, 8137-8143 (2009).
[36] LievenVerslegers, PeterB.Catrysse, ZongfuYu, JustinS.White, Edward S.Barnard, Mark L.Brongersma, and Shanhui Fan, "Planar Lenses Based on Nanoscale Slit Arrays in a Metallic Film". Nano Lett. 9, 235-238 (2009).
[37] Jennifer M. Steele , Zhaowei Liu , Yuan Wang, and Xiang Zhang, Resonant and non-resonant generation and focusing of surface plasmons using circular gratings. Opt. Express 14, 5664-5670 (2006).
[38] Yongqi Fu, W. Zhou, L.E.N.Lim, C.L. Du, X.G.Luo, Plasmonic microzone plate: Superfocusing at visible regime, Appl. Phys. Lett. 91, 061124 (2007).
[39] Yongqi Fu, Yu Liu, Xiuli Zhou, Zong Weixu, Fengzhou Fang, Experimental demonstration of focusing and lasing of plasmonic lens with chirped circular slits, Opt. Express 18 (4), 3438–3443 (2010).
[40] Wang, J., Zhou, W., Li, E. P. & Zhang, D. H., Subwavelength focusing using plasmonic wavelength-launched zone plate lenses. Plasmonics, 6, 269-272 (2011).
[41] Richards, B. & Wolf, E. Electromagnetic Diffraction in Optical Systems. II. Structure of the Image Field in an Aplanatic System. Proc. Roy. Soc. Lond A, Vol.253, No.1274, 358-379 (1959).
[42] Yongqi Fu, Wei Zhou, Hybrid Au-Ag subwavelength metallic structures with variant periods for superfocusing, J. Nanophotonics 3, 033504 (22 June 2009).
[43] Yongqi Fu, Wei Zhou, Lim Enk Ng Lennie, Nano-pinhole-based optical superlens, Research Letter in Physics, Vol.2008, 148505 (2008).
[44] Yongqi Fu, Xiuli Zhou, Yu Liu, Ultra-enhanced lasing effect of plasmonic lens structured with elliptical nano-pinholes distributed in variant period. Plasmonics, 5 (2), 111-116 (2010).
[45] Zhenkui Shi, Yongqi Fu, Xiuli Zhou, Shaoli Zhu, Polarization effect on focusing of a plasmonic lens structured with radialized and chirped elliptical nanopinholes. Plasmonnics 5(2), 175-182 (2010).
[46] Yu Liu, Yongqi Fu, Xiuli Zhou, Polarization dependent plasmonic lenses with variant periods for superfocusing. Plasmonics, 5(2), 117-123 (2010).
[47] Gilad M.Lerman, Avner Yanai,and Uriel Levy, "Demonstration of nanofocusing by use of plasmonic lens illuminated with radial polarized light". Nano Lett. 9, 2139-2143 (2009).
[48] Qiwen Zhan, "Cylindrical vector beams: from mathematical concepts to applications", Advances in Optics and Photonics 1, 1-57 (2009).

[49] T. Xu, C. L. Du, C. T. Wang, and X. G. Luo, "Subwavelength imaging by metallic slab lens with nanoslits", Appl. Phys. Lett. 91, 201501 (2007).
[50] A. Ono, J. Kato, S. Kawata, "Subwavelength optical imaging through a metallic nanorod array", Phys. Rev. Lett. 95, 267407 (2005).
[51] P. A. Belov, Y. Hao, S. Sudhakaran, "Subwavelength microwave imaging using an array of parallel conducting wires as a lens", Phys. Rev. B 73, 033108 (2006).
[52] Mark I. Stockman, "Nanofocusing of Optical Energy in Tapered Plasmonic Waveguides", Phys. Rev. Lett. 93, 137404 (2004).
[53] Ewold Verhagen, Albert Polman, and L. (Kobus) Kuipers, "Nanofocusing in laterally tapered plasmonic waveguides", Opt. Express 16, 45-57 (2008).
[54] A. B. Evlyukhin, S. I. Bozhevolny, A. L. Stepanov, R. Kiyan, C. Reinhardt, S. Passinger, and B. N. Chichkov, "Focusing and directing of surface plasmon polaritons by curved chains of nanoparticles", Opt. Exp. 15, 16667-16680 (2007).
[55] I. P. Radko, S. I. Bozhevolnyi, A. B. Evlyukhin, and A. Boltasseva, "Surface plasmon polariton beam focusing with parabolic nanoparticle chains", Opt. Express 15, 6576-6582 (2007).
[56] Leilei Yin, Vitali K. Vlasko-Vlasov, John Pearson, Jon M. Hiller, Jiong Hua, Ulrich Welp, Dennis E. Brown, and Clyde W. Kimball, "Subwavelength Focusing and Guiding of Surface Plasmons", Nano Lett. 2005, *5*, 1399–1402.
[57] Z. J. Sun and H. K. Kim, "Refractive transmission of light and beam shaping with metallic nano-optic lenses", Appl. Phys. Lett. 85, 642 (2004).
[58] Z. J. Sun, "Beam splitting with a modified metallic nano-optics lens", Appl. Phys. Lett. 89, 26119 (2006).
[59] H. Ko, H. C. Kim, and M. Cheng, "Light transmission through a metallic/dielectric nano-optic lens", J. Vac. Sci. Technol. B 26, 62188-2191 (2008).
[60] S. Vedantam, H. Lee, J. Tang, J. Conway, M. Staffaroni, J. Lu and E. Yablonovitch, "Nanoscale Fabrication of a Plasmonic Dimple Lens for Nano-focusing of Light", Proceeding of SPIE 6641, 6641J (2007).
[61] James A. Shackleford, Richard Grote, Marc Currie, Jonathan E. Spanier, and Bahram Nabet, "Integrated plasmonic lens photodetector", Appl. Phys. Lett. 94, 083501 (2009).
[62] W. Srituravanich, L. Pan, Y. Wang, C. Sun, C. Bogy, and X. Zhang, "Flying plasmonic lens in the near field for high-speed nanolithography", Nature Nanotech. 3, 733-737 (2008).
[63] Dennis M R, Zheludev N I and García de Abajo F J , "The plasmon Talbot effect," Opt. Express, 15, 9692-700 (2007).
[64] D. van Oosten, M. Spasenovi, and L. Kuipers, "Nanohole Chains for Directional and Localized Surface Plasmon Excitation." Nano Lett. 10, 286-290 (2010).
[65] A A Maradudin and T A Leskova, "The Talbot effect for a surface plasmon polariton," New J. Phys. 11, 033004 (2009).
[66] Lingli Li, Yongqi Fu, Hongsheng Wu, Ligong Zheng, Hongxin Zhang, Zhenwu Lu, Qiang Sun, Weixing Yu, The Talbot effect of plasmonic nanolens, Opt. Express, 19(20), 19365-19373 (2011).

[67] Yiwei Zhang, Yongqi Fu, Yu Liu, Xiuli Zhou, Experimental study of metallic elliptical nanopinhole structure-based plasmonic lenses. Plasmonics, 6(2), 219-226 (2011).
[68] V. S. Volkov, S. I. Bozhevolnyi, E. Devaux, J.-Y. Laluet, and T. W. Ebbesen, "Wavelength selective nanophotonic components utilizing channel plasmon polaritons", Nano Lett. 7, 880 (2007).

Chapter 9

Fundamental Role of Periodicity and Geometric Shape to Resonant Terahertz Transmission

Joong Wook Lee and DaiSik Kim

Additional information is available at the end of the chapter

http://dx.doi.org/10.5772/48243

1. Introduction

Optical properties of the plasmonic structures based on the surface of variously structured metal plates became a subject of intense research ever since Ebbesen et al. reported the enhanced optical transmission through subwavelength hole arrays [1]. Their experimental results provided evidence that the enhanced optical properties originate from the coupling between the incident light and surface plasmon modes. Studies of such plasmonic structures and its mechanisms have a long history in the development of electromagnetism, though it has been an intense subject of research in recent years.

In the 1900s, Wood and Rayleigh reported the anomalous optical properties of reflection gratings [2,3], which are intimately connected with the excitation of surface bound waves on the metal-dielectric interfaces [4]. Hessel and Oliner presented two types of the anomalies occurring in the reflection grating surface: a Rayleigh wavelength type attributed to the usual propagating surface plasmon modes and a resonance type which is related to the guided waves supported by the grating itself [5]. On the other hand, since the 1970s, the optical properties of transmission gratings or grids have been occasionally studied in the microwave, terahertz (THz), and infrared regions [6-10]. In particular, Lochbihler and Depine in 1993 presented a theoretical approach for calculating the fields diffracted by the gratings made of highly conducting wires [11]. Here they calculated Maxwell's equations based on modal expansions inside the metal and Rayleigh expansions outside the metal, applying the theoretical treatments by means of surface impedance boundary condition method [11,12].

This research field reinvigorated after the Ebbesen's pioneering work which was focused on the understanding of mechanisms of the enhanced transmission through periodically perforated holes or slits on metallic plates. The one-dimensional plasmonic structures with periodic arrays of slits have been studied by some groups with theoretical and experimental approaches in the optical region [13-18]. These results strongly support the existence of two mechanisms of generating enhanced optical transmission: the geometric shape controlled mainly by the slit thickness and the periodicity determining the excitation of coupled surface plasmon polaritons (SPPs) on both surfaces of the perforated metallic structures. In particular the former called as a Fabry-Perot-like behavior was studied theoretically in a single narrow slit and clearly verified by strong resonant transmission of microwave radiation [19,20].

On the other hand in two-dimensional plasmonic structures with periodic arrays of holes, the geometric shape effect has been underestimated in spite of many researches about the strong transmission enhancement in the optical region [21-26]. Recently, the geometric shape resonance however was emerged as one of the dominant mechanisms of the strong transmission enhancement [27-29]. In particular, Koerkamp et al. already noted that the strong optical transmission is strongly influenced by a hole shape and a shape change from circular to rectangular [30]. The importance of the geometric shape was also confirmed by studies of strong enhancement of the light transmission through random arrays of holes [30-32] and single apertures [33-36], in particular a single rectangular hole theoretically studied by Garcia-Vidal et al. [37].

Unlike most typical systems in optical and infrared regions, THz time-domain spectroscopy system based on coherent THz radiation is ideally suited for phase-sensitive, broadband transmission measurements [38-41]. Furthermore, long wavelengths in THz region, 3000 μm to 30 μm (0.1 THz to 10 THz) in general, permit relatively simple fabrication of free-standing samples which can exclude complex surface modes by the substrate. For these reasons, the coherent THz waves have been generally used to study SPP behaviors [42-55]. The SPPs in the strictest sense however do not exist on perfectly conducting metal-dielectric interfaces since surface electromagnetic fields on the dielectric side cannot be strongly confined to the interface. However, recently the concept of the SPP has been extended even to include perfectly good metals with corrugations, which is referred to simply as designer (or spoof) SPP [56-58]. Pendry et al. reported the electromagnetic surface excitations localized near the surface which are governed by an effective permittivity of the same plasma form [56].

In the THz region, the most important thing is that the perfect transmission, of up to near-unity, at specific frequencies can be achieved since most metals become perfect conductor and therefore have extremely small ohmic loss fraction. Until now, a few groups have theoretically predicted the perfect transmission [59-64] and experimentally observed the near-unity transmission [65-67] in the plasmonic structures with periodic arrays of slits and holes. However, despite great progresses in realizing the perfect transmission, we do not have the clear understanding of what are the relative contributions of the geometric shape and the periodicity effects toward the perfect transmission in specific conditions.

Here, we first report on the importance of the combined effects of the geometric shape and periodicity in enhancing THz transmission through one-dimensional periodic arrays of slits. Theoretical predictions based on perfect conductor model show that the measured transmission peaks lie within the broad geometric shape band. We also report that the perfect transmission can be realized by the geometric shape resonance appearing in spectral region below first Rayleigh minimum. Calculated near-field distributions confirm that the perfect transmission is caused by the concentration of the electric and magnetic fields within the silts. Furthermore, we find that, despite the existence of only the geometric shape effect in random arrays of slits of relatively thick thicknesses, the enhanced transmission comes to the perfect transmission in the long wavelength region. Angle dependent transmission spectra and relative phase measurements clarify the relative roles of the geometric shape and periodicity toward the perfect transmission.

2. Experimental setup

In our experiments, we used a standard THz time-domain spectroscopy system with a spectral range from 0.1 to 2.5 THz, based on a femtosecond Ti:sapphire laser which generates optical pulses with average power of 600 mW and temporal width of about 120 fs set at a center wavelength of 760 nm. Experimental setup is shown in Fig. 1. A photoconducting antenna method is used to generate THz pulses. A crystal for the emitter is undoped, <110> orientated semi-insulating GaAs (INGCRYS Laser System Ltd.) with dark resistivity in range 5.3×10^7 to 5.6×10^7 Ohm cm. Two metallic electrodes are painted by silver paint on one side of the crystal, having a separation of about 0.4 mm which is among a large aperture emitter. Unlike a typical biased system of the photoconductive antenna, an ac bias voltage of 300 V and 50 kHz square wave was applied to the electrodes of the crystal [68]. To make this system, we used a high voltage pulse generator (DEI Model PVX-4150) which generates the squared high-voltage sources by combining DC high voltage source from a power supply (F. u. G. Model MCL 140-650) with square function generated using a pulse/function generator.

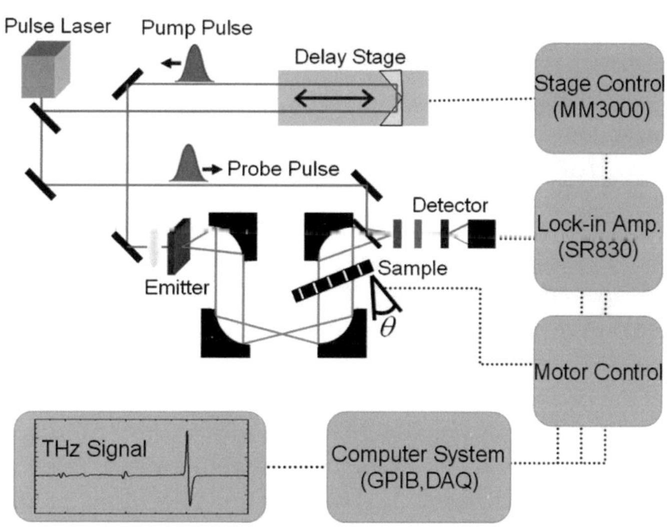

Figure 1. Schematic diagram of the THz time-domain spectroscopy setup. The angle dependent transmission spectra are measured by tilting the sample stage.

Optical pulses generated from the femtosecond Ti:sapphire laser are divided into two parts, pump and probe beams, by a beamsplitter. The pump beam is focused on the surface of the emitter by using a 25 mm focal lens and immediately generates the THz wave which is essentially single cycle electromagnetic pulses. The generated THz wave is collimated by using an off-axis parabolic mirror at a focal distance. The degree of collimation is of the order of $\Delta\theta \sim \lambda/D \sim 1/100$ radian because our beam size D is about 5 cm and the THz wave is a point-like source. The collimated THz wave with the beam diameter of about 5 cm impinges on a reference box or samples. The angle-dependent transmission spectra are measured by tilting the sample stage in range -10 to 50 degrees. The THz pulses transmitted through the samples meet the optical probe beam again on an electro-optic crystal (ZnTe). The polarization change of the optical probe beam due to the Pockels effect is detected by a home-built differential photo-detector and provides us information of both amplitude and phase of the THz wave [69,70]. The signals are automatically acquired by a synchronized system (as shown in Fig. 1) of three components, a motion controller which tilts the sample stage, a data acquisition system, and a delay stage which controls time delay between pump and probe optical pulses. The whole setup is also enclosed in a box which is purged with dry nitrogen gas to reduce the absorption effects of water vapor.

We performed our experiments as follows. First, the collimated THz wave impinges on a reference box made of a 2 cm by 2 cm square hole punctured on a piece of aluminum. We measure the transmission spectrum through the reference box. We then mount our samples with the plasmonic structures of periodic or random arrays of slits perforated on aluminum plates, right on top of the reference box, and measure the transmission spectrum. Our system provides the time-resolved trace of the transmitted THz electric field with subpicosecond temporal resolution as an original signal. The fast Fourier transforms of the measured time-domain THz pulses give us information on both spectral amplitude and phase. The normalization is carried out over the entire frequency simply by dividing the spectral amplitude of the transmitted THz pulse with one of the reference beam transmitting through the reference box.

3. Sample fabrication

The samples were fabricated by a micro-drilling and a femtosecond laser machining methods. The latter is based on laser ablation which is performed by amplified femtosecond pulses. The pulses have energy of up to 1 mJ centered at 800 nm and a repetition rate of 1 kHz. The positions of the focused laser pulses are accurately controlled by using a galvanometer scanner (Scanlab AG, Germany) and a mechanical shutter system. The line edge roughness of the perforated slits is controlled to make successful slit structures, having the LER 3σ value of less than 3 μm which is two orders of magnitude less than the THz wavelengths of interest. The achieved high quality originates from the superiority of femtosecond laser machining system. The samples manufactured by this method have a fixed period d of 500 μm and different thicknesses h of 17, 50, 75 and 153 μm respectively. The silt widths a of these samples are 78, 80, 85, and 83 μm respectively, which are properly designed to be subwavelength [as shown in Fig. 2(a)].

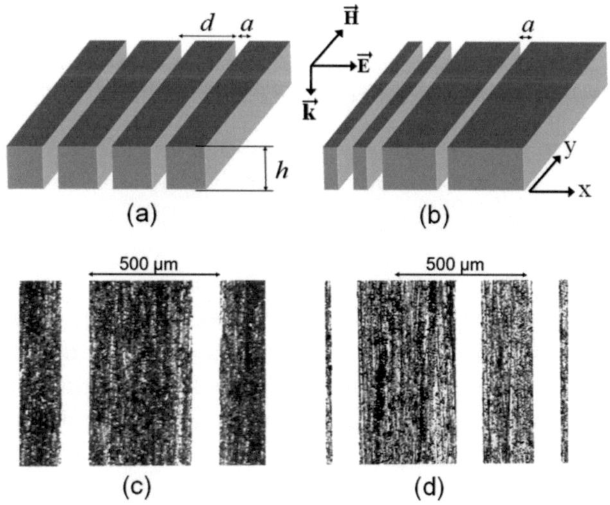

Figure 2. (a), (b) Schematics of the periodic and random arrays of slits of width a, period d, and thickness h, respectively. (c), (d) Microscopic images of the periodic and random arrays of 400 μm thick slits, respectively (Fig. (1) in reference [71]).

On the other hand, the samples with thicknesses of 200 and 400 μm were fabricated by the micro-drilling method by using a screw with a diameter of 100 μm which is a minimum size in diameter of available and commercial screw. These thick plates were chosen for explicitly observing Fabry-Perot-like modes inside the slits. In particular the samples with thickness of 400 μm were considered suitable for randomization of the slit structures, based on the different frozen-phonon fashion of neighboring silts [as shown in Fig. 2(b)]. The position of the *i*th slit by the frozen-phonon fashion is given by

$$x_i = x_0 + id + \Delta x_i \tag{1}$$

where x_0 is the position of the basis of coordinate and Δx_i is the random amount of the *i*th slit which is getting by generating random values [72]. We displaced the position of each slit with random amounts of ± *d/4* and ± *d/2* which, in this chapter, are called as *Random Type I* and *Random Type II.*

Comparison between transmission properties through random and periodic arrays of slits reveal that the transmission spectra strongly affected by the geometric shape show similar spectral waveforms in both random and periodic arrays, since the geometric shape resonance depends on intrinsic properties of the structure factors not the structural arrangement of the slits. The geometric shape in the one-dimensional plasmonic structures is determined by the thickness *h* of the slits, providing the Fabry-Perot-like modes $f_c = nc/2h$ where *c* is the speed of the light in vacuum and *n* is the integer mode index. In contrast, the periodicity effect might be found in transmission spectra through the periodic arrays not random, having different constructive interference patterns and different far-field transmissions. Therefore the relative contribution of the geometric shape and periodicity is directly compared in this chapter.

4. Theoretical framework

Herein, we are first interested in theoretical approach of THz electromagnetic wave transmission through the periodic arrays of slits, which may allow us to improve our intuitive understanding of the underlying physics and to find the mechanisms enhancing the transmission. To calculate the transmittance, the metallic structures are considered as a perfect conductor so that finite conductivity effects are neglected. This is a good approximation in the THz region, since the real and imaginary parts of dielectric constant of most metals are of the order of -30000 and 100000 respectively. We also consider a simple model system with rectangular shaped slits as shown in Fig. 2(a). In this system, the slits are along the y axis and a direction of propagation of the incident THz wave is the z axis which is normal to the sample surface. The polarization of the incident plane wave is placed along the x axis (TM polarization). Here, there are three different regions: the incident and reflection region (region I), the metal and perforated slit region (region II), and the transmission region (region III).

Let $f(x,z)$ be the spatial function of the component along the z axis of the magnetic field. The fields can be expressed in terms of Rayleigh expansions in region I and III and modal expansion in region II [11,46]. The boundary matching is only carried out at the metal-dielectric interfaces of I-II and II-III since which are semi-infinite. Here we use the single mode approximation inside the slits since the slit width is much smaller than the wavelength. The magnetic fields are described as

$$f_1(x,z) = \sum_{-\infty}^{\infty}(R_n e^{i\chi_n(z+h/2)} + \delta_{n0}e^{-ik(z+h/2)})e^{i\alpha_x x}, \text{ region I} \tag{2}$$

$$f_3(x,z) = \sum_{-\infty}^{\infty} T_n e^{-i\chi_n(z-h/2)}e^{i\alpha_x x}, \text{ region III} \tag{3}$$

and

$$f_2(z) = A\frac{\sin(kz)}{\sin(kh/2)} + B\frac{\cos(kz)}{\cos(kh/2)}, \text{ region II} \tag{4}$$

where $\alpha_n = 2\pi n/d$, $\chi_n = \sqrt{k^2 - \alpha_n^2}$, $k = \omega/c = 2\pi/\lambda$, and R_n and T_n are the complex amplitudes of the reflected and transmitted diffracted waves respectively. The continuity of the tangential components of the fields implies the continuity of $f(x,z)$ and its normal derivative at $z = \pm h/2$ along $0 \le x \le a$. These conditions are

$$A + B = \sum_{n=-\infty}^{\infty}(R_n + \delta_{n0})e^{i\alpha_n x} \tag{5}$$

for the continuity of the field at $z = -h/2$,

$$Ak\cot(\frac{kh}{2}) - Bk\tan(\frac{kh}{2}) = \sum_{n=-\infty}^{\infty}(i\chi_n R_n - ik\delta_{n0})e^{i\alpha_n x} \tag{6}$$

for the continuity of the normal derivative of the field at $z = -h/2$,

$$-A + B = \sum_{n=-\infty}^{\infty} T_n e^{i\alpha_n x} \tag{7}$$

for the continuity of the field at $z = h/2$, and

$$Ak\cot(\frac{kh}{2}) + Bk\tan(\frac{kh}{2}) = \sum_{n=-\infty}^{\infty} (-i\chi_n T_n) e^{i\alpha_n x} \tag{8}$$

for the continuity of the normal derivative of the field at $z = h/2$.

With the orthogonality, we apply some of integral equation to obtain simple solution with proper projection and integration. Multiplying a term $e^{-i\alpha_m x}$ and integrating its results of Eqs. (5)-(8) give

$$A + B = \sum_{n=-\infty}^{\infty} (R_n + \delta_{n0}) P_n \tag{9}$$

$$\left[Ak\cot(\frac{kh}{2}) - Bk\tan(\frac{kh}{2}) \right] Q_n = i\chi_n R_n - ik\delta_{n0} \tag{10}$$

$$-A + B = \sum_{n=-\infty}^{\infty} T_n P_n \tag{11}$$

and

$$\left[Ak\cot(\frac{kh}{2}) + Bk\tan(\frac{kh}{2}) \right] Q_n = -i\chi_n T_n \tag{12}$$

where the quantities P_n and Q_n appearing in Eqs. (9)-(12) are defined as

$$P_n = \frac{1}{a}\int_0^a e^{i\alpha_n x} dx \tag{13}$$

and

$$Q_n = \frac{1}{d}\int_0^a e^{-i\alpha_n x} dx \tag{14}$$

respectively. By using Eqs. (9)-(12), we obtain

$$A = \frac{P_0}{1 + ikW\cot(kh/2)} \tag{15}$$

$$B = \frac{P_0}{1 - ikW\tan(kh/2)} \tag{16}$$

$$R_n = 1 - \frac{2ikQ_nP_0}{\chi_n}\left[\frac{\cot(kh) - ikW}{1 + k^2W^2 + 2ikW\cot(kh)}\right] \tag{17}$$

and

$$T_n = \frac{2ikQ_nP_0}{\chi_n\sin(kh)}\left[\frac{1}{1 + k^2W^2 + 2ikW\cot(kh)}\right] \tag{18}$$

where W is defined as

$$W = \sum_{n=-\infty}^{\infty}\frac{P_nQ_n}{\chi_n} \tag{19}$$

For normal incidence, zero-order transmittance for the periodic arrays of slits is expressed as follows

$$T_0 = \frac{2ia}{d\sin(kh)}\left[\frac{1}{1 + k^2W^2 + 2ikW\cot(kh)}\right] \tag{20}$$

5. Experimental results

5.1. Enhanced transmission in thin metal plates

Theoretical calculations are first compared with experimental results of samples with the thicknesses of 17, 75 and 153 μm, which are carefully chosen so that the peak positions due to the geometric shape resonance are placed on the higher spectral region than the first Rayleigh minimum. Measured THz time traces and corresponding spectral amplitudes obtained by the fast Fourier transform method are shown in Fig. 3(a) and 3(b) respectively. The top curves in Fig. 3(a) and 3(b) show the reference signals transmitted through the reference box. The single cycle THz pulses with a temporal width of about 2 ps in time domain are transformed to the spectral pulses with a well-formed shape centered at 0.5 THz and a broad spectral width. In our experimental setup, we effectively removed multiple reflections of the THz pulses occurring from the emitter, the electro-optic crystal and a beamsplitter, which can help assure good quality of the measured time-resolved THz signals.

The thinnest sample with the thickness of 17 μm does not have the peak position of the enhanced transmission locating inside the presented spectral region. However, the two curves for the thicknesses of 75 and 153 μm in Fig. 3(a) have significant long-time oscillations with the periods of 0.87 and 1.80 picoseconds respectively, which also appears in the Fourier transformed spectra as peaks at 1.14 and 0.56 THz respectively (Fig. 3(b), gray arrows in bottom two curves). Here the peak positions of the enhanced transmission appear at the frequency regions slightly below the Rayleigh minima $f_R = c/d$, which indicate that the transmission properties are related to the periodicity effect. On the other hand, the peak positions shift toward longer wavelength region as the sample thickness increases, keeping the typical characteristics of Fabry-Perot-like resonance. These observations indicate that the enhanced transmissions are attributed to the combined effects of the geometric shape resonance and the periodicity.

Theoretical calculations shown in Fig. 3(c) are in well agreement with the experimental results except for extremely sharp transmission peaks immediately below the first Rayleigh wavelength. In particular in the cases of the two thinnest samples, the peak linewidths have almost infinitesimal values less than several gigahertz, which cannot be experimentally observed because of the finite temporal range of time domain signals and finite sample size. Theoretically predicted peak suddenly appears at 0.56 THz for h=153 μm as shown in bottom curve of Fig. 3(b). This appearance results from the approach of the geometric shape resonance toward the first Rayleigh minimum at which the first diffracted orders become evanescent.

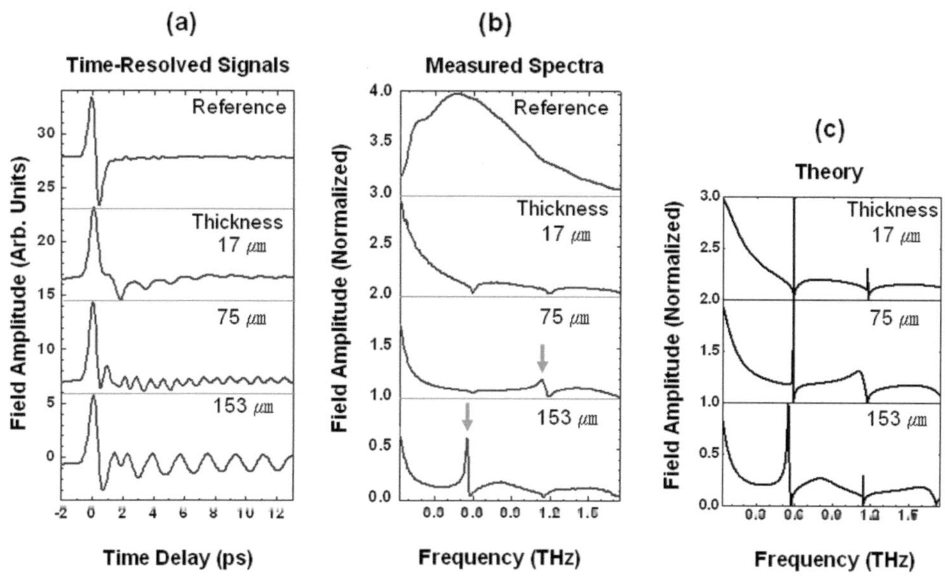

Figure 3. (a) Time traces of the incident terahertz wave (the upper figure) and the transmitted waves for three samples with a fixed period of 500 μm and different thicknesses of 17, 75, and 153 μm. The slit widths of the three samples are 78, 80, and 83 μm respectively. (b) Fourier transforms for four time traces in (a). The gray arrows represent the resonant peak positions. (c) Theoretical calculations for three samples in (b).

The independent contribution of Fabry-Perot-like resonance can draw out from single-slit approximation which can be achieved by carrying the distance between slits to be infinite. Figure 4(a) and 4(c) show the resonant peaks of transmission spectra through the samples with the thicknesses of 75 and 153 μm respectively, theoretically predicted by the single-slit approximation. The red-shifted Fabry-Perot conditions predicted by Takakura in a single narrow slit of a thick metallic plate assure the precision of the theoretical predictions. These peak positions induced only by the Fabry-Perot-like resonance are compared with the measured angle-dependent transmission amplitudes [Fig. 4(b) and 4(d)], since the geometric shape resonance is angle-independent. For the case of the 75 μm thickness, the experimental results show a resonance band over all incident angles around 1.3 THz, which is reasonably predicted by the first Fabry-Perot-like resonance band centered at 1.2 THz [Fig. 4(a), shadowed area]. The theoretical prediction and experimental results for the sample of the 153 μm thickness also show the corresponding two spectral bands centered at 0.65 and 1.45 THz. From these theoretical prediction and experimental results, we note that the peak positions of resonant transmission can be experimentally determined by broad Fabry-Perot-like resonance band existing nearby.

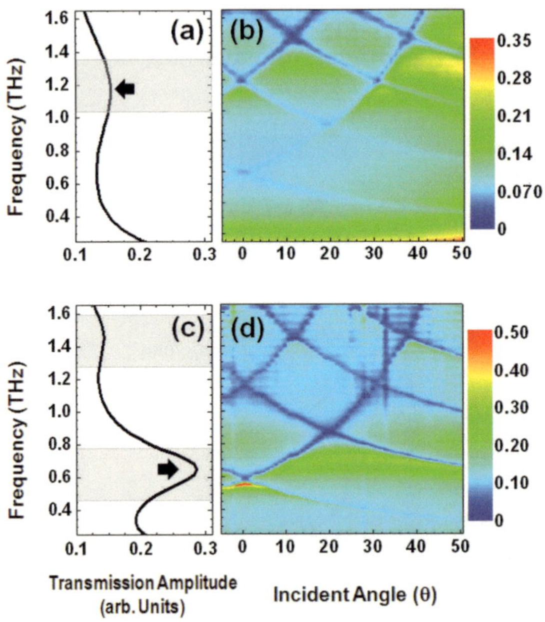

Figure 4. (a), (c) Transmission spectra, at normal incidence, predicted by single-slit approximation for the samples with the thicknesses of 75 and 153 μm respectively. Arrows represent the red-shifted first Fabry-Perot positions calculated by Takakura (Eq. (9), ref. [19]). (b), (d) Measured angle-dependent transmission spectra for the samples with the thicknesses of 75 and 153 μm respectively.

At this point, we can expect that perfect transmission phenomenon will be appeared under the condition of overlapping between the Fabry-Perot-like resonance band and totally evanescent spectral regions below the first Rayleigh wavelength. The first diffracted waves touch down on the metal surface at the wavelength of the first Rayleigh minimum, at which wavelength eventually all higher diffracted orders become evanescent which is the reason why the total incident energy can be converged to the zero-order transmission amplitude. We therefore design a thicker sample to realize the perfect transmission.

5.2. Perfect transmission in thick metallic plates

The sample with the thickness of 200 μm eventually has broad Fabry-Perot-like resonance band centered at 0.45 THz which is located in the spectral region below the first Rayleigh minimum appearing at 0.6 THz as shown in Fig. 5(a). Angle dependent transmission amplitudes [Fig. 5(b)] show the perfect transmission (in this sample, over 99% at normal incidence) centered at 0.5 THz. Despite the two transmission minima lines caused by Rayleigh frequencies for the ±1 diffraction orders, the perfect transmission appears at a broad incident angle range up to 10° because of the strong contribution of the Fabry-Perot-like resonance. Theoretical calculation with Eq. (20) is in good agreement with experimental result at normal incidence.

The perfect transmission phenomenon necessarily appears when the peak position of the enhanced transmission is located anywhere in the spectral region below the first Rayleigh frequencies, since all the diffracted orders of the transmittance T_n except for zeroth order become zero. A very useful theoretical condition for the perfect transmission can be therefore derived by setting the zeroth order reflection coefficient at zero as following

$$|R_0| = \left| 1 - \frac{2iQ_0P_0}{1}\left[\frac{\cot(kh) - ikW}{1 + k^2W^2 + 2ikW\cot(kh)}\right]\right| = 0 \tag{21}$$

Below the first Rayleigh frequencies, W can be considered as

$$W = \frac{P_0Q_0}{k} + \frac{i\gamma}{k} \tag{22}$$

since all the reflected diffracted orders except for zeroth order become evanescent. Where γ is an imaginary part of kW. We can then write the condition of the perfect transmission as

$$1 - (P_0Q_0)^2 - \gamma^2 - 2\gamma\cot(kh) = 0 \tag{23}$$

For the theoretical calculation of the first Fabry-Perot-like resonance centered at 0.48 THz [Fig. 5(c)], a magnetic field profile inside the slits becomes symmetric [Fig. 5(d)] which is proved by the condition for the perfect transmission given by Eq. (23). Indeed, the magnetic field inside the slits given by Eq. (4) converges into the symmetric waveform at the first Fabry-Perot-like resonance as

$$f_2(z) \cong B\frac{\cos(kz)}{\cos(kh/2)} \tag{24}$$

since a denominator term of B is minimized under the condition of Eq. (23). It can be also intuitively understood that, under the condition of the perfect transmission at the first Fabry-Perot-like resonance, the electric field strengths at both sides of entrance and exit of the slits are the same despite opposite orientations of the electric fields at each position. Because the light coupling between resonant modes of the structured metal surfaces in the transmission and reflection regions is related to the matching condition of the phase of the electric and the magnetic fields, the electric field strength at the entrance of the slits cannot be reproduced by the asymmetric profiles shown in Fig. 5(d).

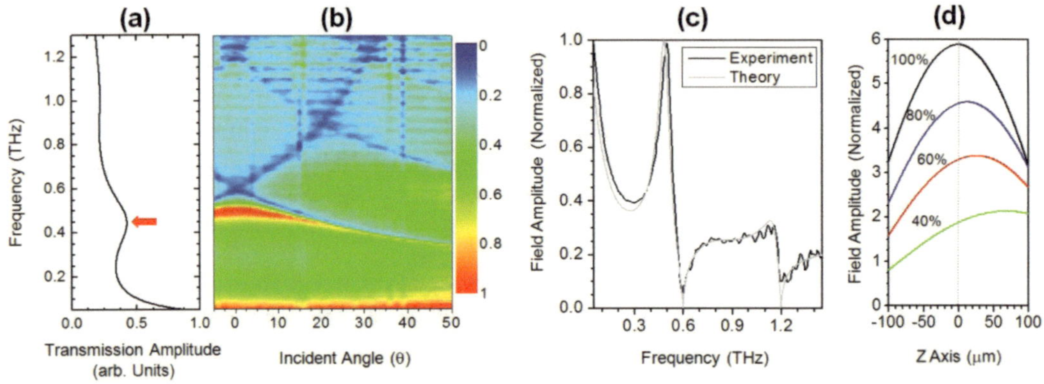

Figure 5. (a) Transmission spectra, at normal incidence, predicted by single-slit approximation for the sample with the thickness of 200 μm. An arrow represents the peak position of first Fabry-Perot-like resonance mode. (b) Measured angle-dependent transmission spectra for the sample with the thickness of 200 μm. The period is 500 μm and the slit width is 100 μm. (c) Experimental and theoretical results, at normal incidence, of the sample with the thickness of 200 μm. (d) The magnetic field profiles along the z direction inside the slits calculated by the perfect conductor model. Under the condition of the perfect transmission appearing at the first Fabry-Perot-like mode, the field profile is symmetric (black line). At neighboring frequencies with the transmission amplitudes of 80 (blue line), 60 (red line) and 40% (green line) respectively, the field profiles become asymmetric.

Near-field distributions of THz waves above the metal surfaces can also give a key for understanding the perfect transmission phenomenon. The enhancement factor of the near electric fields onto the slits can be intuitively predicted by using the language of the surface impedance as

$$Z = \frac{a}{d}\frac{E_{slit}}{H_0} + \frac{(d-a)}{d}\frac{E_{metal}}{H_0} \tag{25}$$

where E_{slit}, E_{metal} and H_0 are the electric fields along the x direction at the slits and the metal and the incident magnetic field respectively. Since the effective surface impedance becomes one under the condition of the perfect transmission and there is no tangential component of the electric field on the metal surface, we can write Eq. (25) as

$$Z_{eff} = \frac{a}{d}\frac{E_{slit}}{H_0} = 1 \tag{26}$$

The degree of the electric field enhancement is not only proportional to the incident magnetic field, but also inversely proportional to the areal sample coverage $\beta = a/d$ as

$$E_{slit} = \beta^{-1} H_0 \tag{27}$$

For the case of the areal sample coverage of $\beta = 0.2$ ($a = 100$ and $d = 500$ μm), the theoretical results show the enhancement factor of 5 on the entrance of the silts as shown in Fig. 6(a). At a specific frequency at which the perfect transmission is realized, the incident waves are focused on the entrance of the slits due to the strong evanescent waves built up by a touchdown of all orders of diffracted modes on metal surface, followed by coupled with the Fabry-Perot-like resonant modes inside the slits, and then the focused waves reached at the exit of the slits are completely converted to the transmission.

These results suggest a possibility of realizing even stronger enhancement of the electric field strength onto the slits. We therefore fabricated a sample with the areal sample coverage of $\beta = 0.1$ by increasing the period. (Reducing the slit width is actually more effective, but the technical limit of the micro-drilling method to fabricate thick metallic plates is 100 μm.) Calculated near-field distribution of the sample is given by Fig. 6(b), showing the enhancement factor of 10 corresponding to the inverse of the areal sample coverage. The near-field enhancement necessarily induces the transmission enhancement into the far-field. Measured angle dependent transmission spectra shown in Fig. 7 show not only angle-independent, broad spectral bands near 0.3 and 0.6 THz caused by the geometric shape resonance, but also a strong transmission peak with the value of about 0.85 at 0.26 THz at normal incidence. Though the peak value does not reach the THz transparency because of a narrow spectral bandwidth, the enhancement factor compared with the areal sample coverage of 0.1 is over eight. Controlling the areal sample coverage by using the period or width is therefore extremely important in realizing a high power near-field THz source.

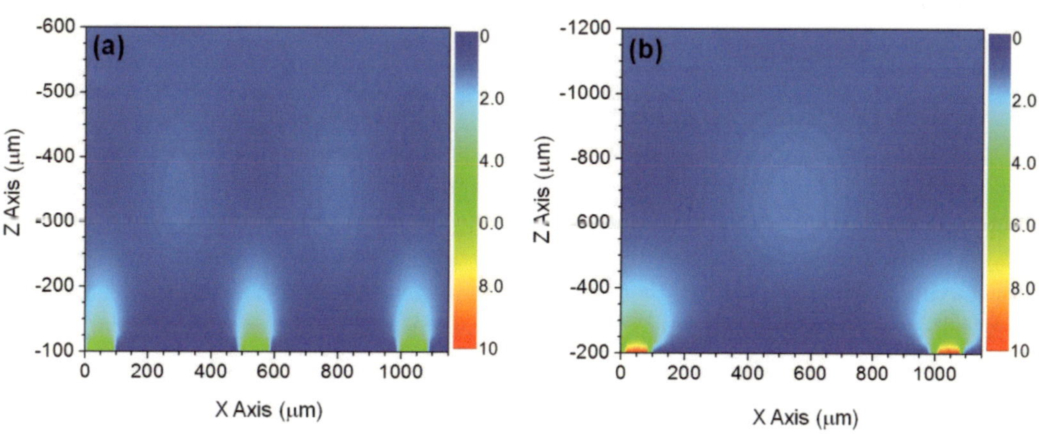

Figure 6. Electric field amplitudes for the samples with different periods of 500 (a) and 1000 μm (b) respectively. The thickness h is 400 μm and the width a is 100 μm in both cases. Enhancement factors at the entrance of the slits are 5 and 10 respectively.

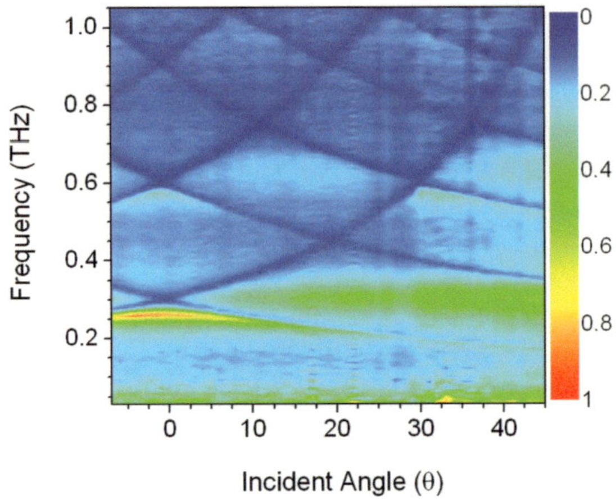

Figure 7. Measured angle-dependent transmission spectra for the sample with the period of 1000 μm. The thickness is 400 μm and the slit width is 100 μm.

5.3. THz transparency by geometric shape

At this point, it is important to understand that what the relative contributions of the periodicity and the geometric shape resonance toward the perfect transmission are in a given situation. From the experimental viewpoint, the best method to compare the difference between the relative contributions of two mechanisms is to fabricate the samples with random arrays and to investigate its transmission properties. The calculated transmission spectra as shown in Fig. 8 play an important role in determining the appropriate sample parameters. For the convenience of comparison, the transmission spectra are plotted versus the first Fabry-Perot frequency, $f_c = c/2h$ instead of the slit thickness.

In the cases of the relatively thinner metal plates, there are two peak maxima of the transmission spectra: one corresponds to the first Fabry-Perot-like mode, $f_c = c/2h$, and the other corresponds to the so-called zeroth Fabry-Perot-like mode appearing at the long wavelength region near 0 THz, which is originated from the geometric shape effect of the semi-infinite slits in a thin metal film. In the long wavelength region near 0 THz, an aluminum plate with the thickness of several hundreds of microns can be actually regarded as an optically thin metal film. With increasing the thickness of the aluminum plates, the peak maxima originating from higher Fabry-Perot-like modes such as $f_c = 2c/2h$, $3c/2h$, and etc are added consecutively, maintaining the perfect transmission.

All curves of the Fabry-Perot-like modes show that the curves approach to straight lines of the classical Fabry-Perot positions (Fig. 8, dotted lines) as getting away from the wavelength of the first Rayleigh minimum of 0.6 THz. The calculated results may therefore present that, when the peak maxima approach to the long wavelength limit, the contribution by the coupled evanescent surface modes presented in Eq. (19) becomes weak. In our experiments, we therefore fabricated the samples of periodic and random arrays of slits with the thickness of 400 μm since in which there are three different peak maxima originating from the zeroth

(Fig. 8, C), first (Fig. 8, B) and second (Fig. 8, A) Fabry-Perot-like modes and each peak maximum may be stimulated by different contributions of the periodicity and the geometric shape. In fact, as the peak position is getting away from the wavelength of the first Rayleigh minimum toward longer wavelength region, the geometric shape effect becomes dominant.

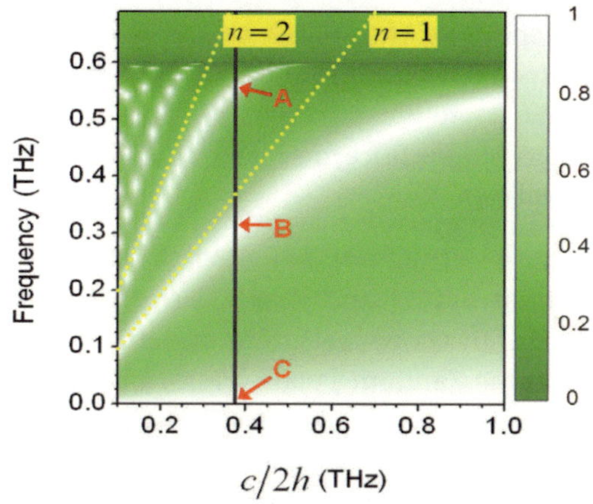

Figure 8. Calculated transmission spectra plotted versus the frequency of the first Fabry-Perot-like resonance. The dotted lines represent the first and second classical Fabry-Perot resonances. The sample with the thickness of 400 μm has three Fabry-Perot resonance modes (A, B, and C) appearing at the frequency region below the first Rayleigh minimum.

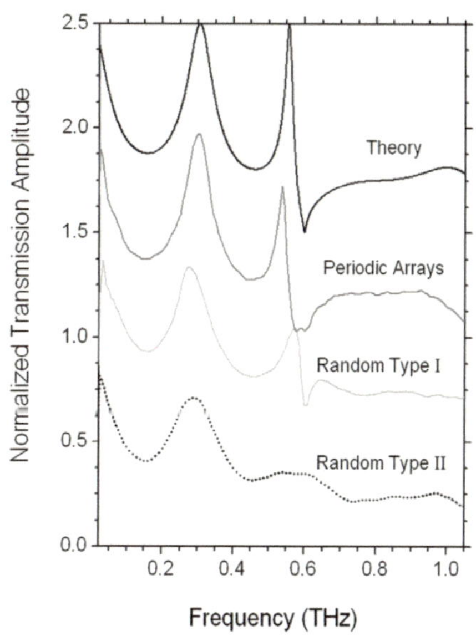

Figure 9. Normalized transmission spectra at normal incidence for three types of the samples: periodic arrays (dark gray line), random type I (bright gray line), and random type II (dotted line). The black line is the calculated transmission spectra for the case of the periodic arrays. These samples have same areal sample coverage of 0.2 and same thickness of 400 μm. The randomization of the slit structures is based on the different frozen-phonon fashion.

Figure 9 shows the normalized transmission amplitudes of periodic and random arrays of silts at normal incidence. Here the samples have same slit width *a* of about 100 μm and same areal sample coverage of β=0.2. As predicted in theoretical calculations, the calculated transmission spectrum (Fig. 9, a solid line) shows three peak maxima at 0.56, 0.30 and near zero THz, corresponding to the second, first and zeroth Fabry-Perot-like modes respectively. The measured transmission spectrum (Fig. 9, a dark gray line) of the periodic arrays of slits agrees well with the calculated one, except slight disagreement at the peak maximum of the second Fabry-Perot-like mode. Here the experimental linewidth of the second Fabry-Perot-like mode does not completely reproduce the theoretical one since which is too sharp to experimentally observe in real sample.

We then consider two random structures with the different frozen-phonon fashion of neighboring slits. The Random Type I and II samples were fabricated by displacing the position of each slit of periodic arrays with random amounts of ± *d*/4 and ± *d*/2 respectively. The transmission spectrum of the Random Type I (Fig. 9, a bright gray line) shows a slightly lower peak value at 0.30 THz and a considerably decreased peak value at 0.56 THz, but overall the spectral shape of the transmission spectra are well matched, particularly reproducing the peak value due to the zeroth Fabry-Perot-like mode. The measured transmission spectrum of the Random Type II (Fig. 9, a dotted line) clearly exhibits the spectral properties of the perfectly random arrays of slits, which does not show any more the transmission minimum at 0.6 THz originating from the first Rayleigh minimum due to the periodicity effect only. On the other hand, although the peak value appearing near 0.3 THz is slightly decreased with further increasing the fraction of random amounts, the spectral shape and peak position can be not only considered unchangeable, the resonant peak also exhibits an enhancement factor larger than 3.5. Furthermore, the resonant peaks of the zeroth Fabry-Perot-like mode are more dramatic where all the transmission spectra in theory and experiments are perfectly matched, showing the transmittance more than 80%.

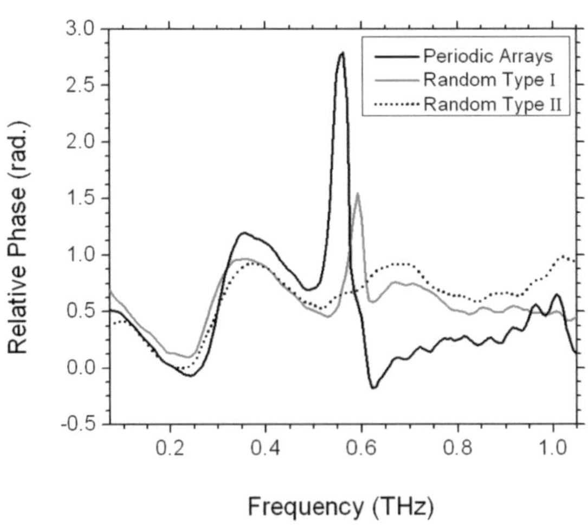

Figure 10. Relative phase changes in radians at normal incidence for three types of the samples: periodic arrays (black line), random type I (gray line), and random type II (dotted line).

The relative phase of transmitted THz waveforms not only improves the understanding for the relative contributions of the geometric shape and the periodicity but also makes more clearly what the dominant mechanism in enhancing the THz transmission is. When the geometric shape effect becomes dominant, there will be a little difference between the random and periodic arrays of slits, as shown in Fig. 10. In addition, the geometric shape does not depend on the incident angle of the THz waves. Shown in Fig. 11 are the angle dependent transmission amplitudes for the samples. For the sample of periodic arrays of slits, there are two dominant transmission minima lines crossing at 0.6 THz at zero angle which is the Rayleigh wavelengths corresponding to the ±1 diffracted orders as

$$f_{\pm 1}(\theta) = \frac{c}{d[1 \mp \sin(\theta)]} \tag{28}$$

The Rayleigh wavelengths determined by the grating period distinctly appear as transmission minima lines over all the incident angle of the THz waves, generating an angle-dependent and strongly enhanced transmission peak near 5.4 THz. With increasing the randomness, the transmission minima lines of the Rayleigh wavelengths become indistinct as shown in Fig. 11(b) and eventually disappear as shown in Fig. 11(c). Similarly, the angle dependent transmission peak appearing at the sample of periodic arrays of slits also disappears since its strong enhancement is caused by the strong field accumulation of the evanescent surface waves by the periodic structures. On the contrary, the enhanced peaks appearing near 0.3 THz are essentially angle independent, making the strong enhancement.

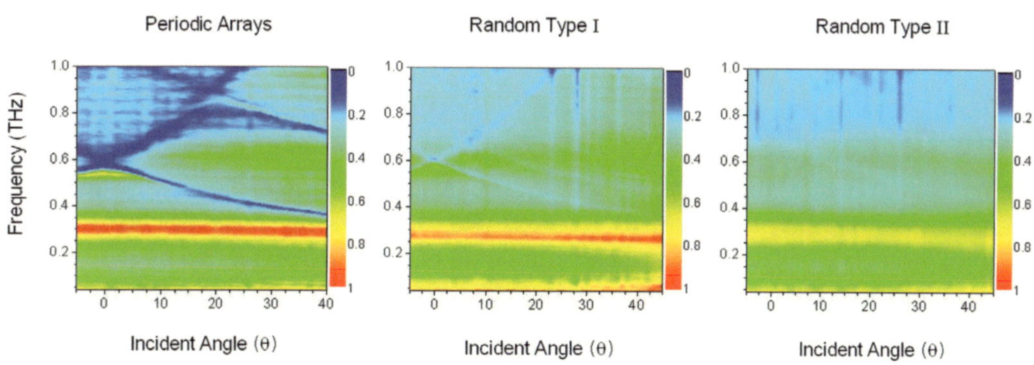

Figure 11. Normalized angle dependent transmission amplitudes for three types of the samples: periodic arrays (a), random type I (b), and random type II (c).

6. Conclusion

In conclusion, we have demonstrated two mechanisms for perfect transmission appearing at specific frequencies through the periodic and random arrays of slits. The theoretical results show that the perfect transmission can be realized in spectral region below the first Rayleigh minimum determined by the periodicity. Under the condition for the perfect transmission, the symmetric electric- and magnetic-field profiles inside the slits are excited and the incident energy is funneled onto the slits with the enhancement factor proportional to the inverse of the areal sample coverage. The measured transmission spectra through the slits of thick metal plates show the enhanced transmission of perfect transmission, which is attributed by the increase of the spectral linewidth resulting from the approach of the geometric shape resonance to the spectral region below the first Rayleigh minimum. The resonant bands of the geometric shape presented by the single silt approximation and the measured angle dependent transmission spectra not only are angle independent but also determine the spectral peak positions of the perfect transmission.

The random arrays of slits designed to remove the effect of the periodicity clearly show the dominant contribution of the geometric shape resonance toward the enhanced transmission. In the long-wavelength region far from the first Rayleigh minimum, the geometric shape resonance becomes dominant and the transmission enhancement factors are almost equal to the cases of the periodic arrays of slits. The strong transmission and local field enhancements open a possibility of potential applications in perfect transmission of solid materials such as semiconductors and biological materials requiring the high power THz sources and in designing photonic devices such as transmission filters, low-pass frequency filters, and transmissive waveguides requiring high transmission at desired frequencies.

Author details

Joong Wook Lee*
Advanced Photonics Research Institute, GIST, Gwangju, Republic of Korea

DaiSik Kim
Department of Physics and Astronomy, Seoul National University, Seoul, Republic of Korea

Acknowledgement

The authors gratefully thank Q. H. Park and M. A. Seo for useful discussions and S. C. Jeoung for sample preparation. This research was supported by Basic Science Research Program through the National Research Foundation of Korea (NRF) funded by the Ministry of Education, Science and Technology (2010-0021181) and by the APRI Research Program of GIST.

7. References

[1] Ebbesen T. W., Lezec H. L., Ghaemi H. F., Thio T., and Wolff P. A. (1998) Extraordinary optical transmission through sub-wavelength hole arrays. Nature 391: 391, 667-669.
[2] Wood R. W. (1902) On a remarkable case of uneven distribution of light in a diffraction grating spectrum. Phil. Mag. 4: 396-408.
[3] Rayleigh L. (1907) On the passage of electric waves through tubes, or the vibrations of dielectric cylinders. Phil. Mag. 14: 60-65.

* Corresponding Author

[4] Fano U. (1941) The theory of anomalous diffraction gratings and of quasi-stationary waves on metallic surfaces (Sommerfeld's waves). J. Opt. Soc. Am. 31: 213-222.

[5] Hessel A. and Oliner A. A. (1965) A New Theory of Wood's Anomalies on Optical Gratings. Appl. Opt. 4: 1275-1297.

[6] Ulrich R. (1968) Interference filters for the far infrared. Appl. Opt. 7: 1987-1996.

[7] Chen C. C. (1970) Transmission Through a Conducting Screen Perforated Periodically with Apertures. IEEE trans. Microwave Theory Tech. 18: 627-632.

[8] Bliek P. J., Botten L. C., Deleuil R., Paedran R. C. M., and Maystre D. (1980) Inductive grids in the region of diffraction anomalies: theory, experiment, and applications. IEEE trans. Microwave Theory Tech. 28: 1119-1125.

[9] Keilmann F. (1981) Infrared high-pass filter with high contrast. Int. J. Infrared Millimeter Waves 2: 259-272.

[10] Ulrich R. (1974) Modes of propagation on an open periodic waveguide for the far-infrared, in optical and acoustical micro-electronics, ed. by J. Fox, Micr. Res. Inst. Symp. Series 23: Polytec. Press, New York 359-376.

[11] Lochbihler H. and Depine R. (1993) Highly conducting wire gratings in the resonance region. Appl. Opt. 32: 3459-3465.

[12] Lochbihler H. (1994). Surface polaritons on gold-wire gratings. Phys. Rev. B 50: 4795-4801.

[13] Porto J. A., Garcia-Vidal F. J., and Pendry J. B. (1999) Transmission resonances on metallic gratings with very narrow slits. Phys. Rev. Lett. 83: 2845-2848.

[14] Cao Q. and Lalanne P. (2002) Negative role of surface plasmons in the transmission of metallic gratings with very narrow slits. Phys. Rev. Lett. 88: 057403.

[15] Collin S., Pardo F., Teissier R., and Pelouard J. L. (2002) Horizontal and vertical surface resonances in transmission metallic gratings. J. Opt. A: Pure Appl. Opt. 4: 154-160.

[16] Astilean S., Lalanne P., and Palamaru M. (2000) Light transmission through metallic channels much smaller than the wavelength. Opt. Commun.175: 265-273.

[17] Treacy M. M. J. (1999) Dynamical diffraction in metallic optical gratings. Appl. Phys. Lett. 75: 606-608.

[18] Lee K. G. and Park Q. H. (2005) Coupling of Surface Plasmon Polaritons and Light in Metallic Nanoslits. Phys. Rev. Lett. 95: 103902.

[19] Takakura Y. (2001) Optical resonance in a narrow slit in a thick metallic screen. Phys. Rev. Lett. 86: 5601-5603.

[20] Yang F. and Sambles J. R. (2002) Resonant Transmission of Microwaves through a Narrow Metallic Slit. Phys. Rev. Lett. 89: 063901.

[21] Martin-Moreno L., Garcia-Vidal F. J., Lezec H. J., Pellerin K. M., Thio T., Pendry J. B., and Ebbesen T. W. (2001) Theory of extraordinary optical transmission through subwavelength hole arrays. Phys. Rev. Lett. 86: 1114-1117.

[22] Salomon L., Grillot F., Zayats A. V., and Fornel F. (2001) Near-field distribution of optical transmission of periodic subwavelength holes in a metal film. Phys. Rev. Lett. 86: 1110-1113.

[23] Martin-Moreno L. and Garcia-Vidal F. J. (2004) Optical transmission through circular hole arrays in optically thick metal films. Opt. Express 12: 3619-3628.

[24] Hohng S. C., Yoon Y. C., Kim D. S., Malyarchuk V., Muller R., Lienau C., Park J. W., Yoo K. H., Kim J., Ryu H. Y., and Park Q. H. (2002) Light emission from the shadows: Surface plasmon nano-optics at near and far fields. Appl. Phys. Lett. 81: 3239-3242.

[25] Kim D. S., Hohng S. C., Malyarchuk V., Yoon Y. C., Ahn Y. H., Yee K. J., Park J. W., Kim J., Park Q. H., and Lienau C. (2003) Microscopic Origin of Surface-Plasmon Radiation in Plasmonic Band-Gap Nanostructures. Phys. Rev. Lett. 91: 143901.

[26] Ropers C., Park D. J., Stibenz G., Steinmeyer G., Kim J., Kim D. S., and Lienau C. (2005) Femtosecond light transmission and subradiant damping in plasmonic crystals. Phys. Rev. Lett. 94: 113901.

[27] Gordon R. and Brolo A. G. (2005) Increased cut-off wavelength for a subwavelength hole in a real metal. Opt. Express 13: 1933-1938.

[28] Klein Koerkamp K. Enoch J., S., Segerink F. B., Van Hulst N. F., and Kuipers L. (2004) Strong influence of hole shape on extraordinary transmission through periodic arrays of subwavelength holes. Phys. Rev. Lett. 92: 183901.

[29] Degiron A. and Ebbesen T. W. (2005) The role of localized surface plasmon modes in the enhanced transmission of periodic subwavelength apertures. J. Opt. A: Pure Appl. Opt. 7: S90-S96.

[30] van der Molen K. L., Klein Koerkamp K. J., Enoch S., Segerink F. B., Van Hulst N. F., and Kuipers L. (2005) Role of shape and localized resonances in extraordinary transmission through periodic arrays of subwavelength holes: Experiment and theory. Phys. Rev. B 72: 045421.

[31] Ruan Z. and Qiu M. (2006) Ehanced transmission through periodic arrays of subwavelength holes: the role of localized waveguide resonances. Phys. Rev. Lett. 96: 233901.

[32] Lee J. W., Seo M. A., Kang D. H., Khim K. S., Jeoung S. C., and Kim D. S. (2007) Terahertz electromagnetic wave transmission through single rectangular holes and slits in thin metallic sheets. Phys. Rev. Lett. 99: 137401.

[33] Bethe H. A. (1944) Theory of diffraction by small holes. Phys. Rev. 66: 163-182.

[34] Boukamp C. J. (1954) Diffraction Theory. Rep. Prog. Phys. 17 : 35-100.

[35] García de Abajo F. J. (2002) Light transmission through a single cylindrical hole in a metallic film. Opt. Express 10: 1475-1484.

[36] Degiron A., Lezec H. J., Yamamoto N., and Ebbesen T. W. (2004) Optical transmission properties of a single subwavelength aperture in a real metal. Opt. Commun. 239: 61-66.

[37] Garcia-Vidal F. J., Moreno E., Porto J. A., and Martin-Moreno L. (2005) Transmission of light through a single rectangular hole. Phys. Rev. Lett. 95: 103901.

[38] Van Exter M. and Grischkowsky D. (1990) Optical and electronic properties of doped silicon from 0.1 to 2 THz. Appl. Phys. Lett. 56: 1694-1696.

[39] Jiang Z., Li M., and Zhang X. C. (2000) Dielectric constant measurement of thin films by differential time-domain spectroscopy. Appl. Phys. Lett. 76: 3221-3223.

[40] Kang C., Maeng I. H., Oh S. J., Son J. H., Jeon T. I., An K. H., Lim S. C., and Lee Y. H. (2005) Frequency-dependent optical constants and conductivities of hydrogenfunctionalized single-walled carbon nanotubes. Appl. Phys. Lett. 87: 041908.

[41] Han H., Park H., Cho M., and Kim J. (2002) THz pulse propagation in plastic photonic crystal fiber. Appl. Phys. Lett. 80: 2634-2636.

[42] Hara J. F., Averitt R. D., and Taylor A. J. (2004) Terahertz surface plasmon polariton coupling on metallic gratings. Opt. Express 12: 6397-6402.

[43] Torosyan G., Rau C., Pradarutti B., and Beigang R. (2004) Generation and propagation of surface plasmons in periodic metallic structures. Appl. Phys. Lett. 85: 3372-3374.

[44] Gomez Rivas J., Schotsch C., Haring Bolivar P., and Kurz H. (2003) Enhanced transmission of THz through subwavelength holes. Phys. Rev. B 68: 201306.

[45] Gomez Rivas J., Kuttge M., Haring Bolivar P., and Kurz H. (2004) Propagation of surface plasmon polaritons on semiconductor gratings. Phys. Rev. Lett. 93: 256804.

[46] Lee J. W., Seo M. A., Kim D. S., Jeoung S. C., Lienau Ch., Kang J. H. and Park Q. H. (2006) Fabry-Perot effects in THz time-domain spectroscopy of plasmonic band-gap structures. Appl. Phys. Lett. 88: 071114.

[47] Azad A. K., Zhao Y., and Zhang W. (2005) Transmission properties of terahertz pulses through an ultrathin subwavelength silicon hole array. Appl. Phys. Lett. 86: 141102.

[48] Qu D., Grischkowsky D., and Zhang W. (2004) Terahertz transmission properties of thin, subwavelength metallic hole arrays. Opt. Lett. 29: 896-898.

[49] Cao H. and Nahata A. (2004) Influence of aperture shape on the transmission properties of a periodic array of subwavelength apertures. Opt. Express 12: 3664-3672.

[50] Shin Y. M., So J. K., Jang K. H., Won J. H., Srivastava A., and Park G. S. (2007) Superradiant terahertz smith-purcell radiation from surface-plasmon excited by counter-streaming electron beams. Appl. Phys. Lett. 90: 031502.

[51] Jeon T. I., Zhang J., and Grischkowsky D. (2005) THz Sommerfeld wave on a single metal wire. Appl. Phys. Lett. 86: 161904.

[52] Jeon T. I. and Grischkowsky D. (2006) THz Zenneck surface wave propagation on a metal sheet. Appl. Phys. Lett. 88: 061113.

[53] Seo M. A., Park H. R., Koo S. M., Park D. J., Kang J. H., Suwal O. K., Choi S. S., Planken P. C. M., Park G. S., Park N. K., Park Q. H. and Kim D. S. (2009) Terahertz field enhancement by a metallic nano slit operating beyond the skin-depth limit. Nature Pnoton. 3: 152-156.

[54] Matsui T., Agrawal A., Nahata A. and Vardeny Z. V. (2007) Transmission resonances through aperiodic arrays of subwavelength apertures. Nature 446: 517-521.

[55] Williams C. R., Andrews S. R., Maier S. A. Fernández-Domínguez A. I., Martin-Moreno L., and García-Vidal F. J. (2007) Highly confined guiding of terahertz surface plasmon polaritons on structured metal surfaces. Nature Photon 2: 175-179.

[56] Pendry J. B., Martin-Moreno L., and Garcia-Vidal F. J. (2004) Mimicking Surface Plasmons with Structured Surfaces. Science 305: 847-848.

[57] Garcia-Vidal F. J., Martin-Moreno L., and Pendry J. B. (2005) Surfaces with holes in them: new plasmonic metamaterials. J. Opt. A: Pure Appl. Opt. 7: S97-S101.

[58] Hibbins A. P., Evans B. R., and Sambles J. R. (2005) Experimental verification of designer surface plasmons. Science 308: 670-672.

[59] Garica de Abajo F. J., Gomez-Santos G., Blanco L. A., Borisov A. G., and Shabanov S. V. (2005) Tunneling mechanism of light transmission through metallic films. Phys. Rev. Lett. 95: 067403.

[60] Garcia de Abajo F. J., Gomez-Medina R., and Saenz J. J. (2005) Full transmission through perfect-conductor subwavelength hole arrays. Phys. Rev. E 72: 016608.

[61] Treacy M. M. J. (2002) Dynamical diffraction explanation of the anomalous transmission of light through mettallic gratings. Phys. Rev. B 66: 195105.

[62] Garcia-Vidal F. J. and Martin-Moreno L. (2002) Transmission and focusing of light in one-dimensional periodically nanostructured metals. Phys. Rev. B 66: 155412.

[63] Kang J. H., Park Q. H., Lee J. W., Seo M. A., and Kim D. S. (2006) Perfect transmission of THz waves in structured metals. J. Korean Phys. Soc. 49: 881-884.

[64] Lee J. W., Seo M. A., Park D. J., Jeoung S. C., Park Q. H., Lienau Ch., and Kim D. S. (2006) Terahertz transparency at Fabry-Perot resonances of periodic slit arrays in a metal plate: experiment and theory. Opt. Express 14: 12637-12643.

[65] Tanaka M., Miyamaru F., Hangyo M., Tanaka T., Akazawa M., and Sano E. (2005) Effect of a thin dielectric layer on terahertz transmission characteristics for metal hole arrays. Opt. Lett. 30: 1210-1212.

[66] Lee J. W., Seo M. A., Park D. J., Kim D. S., Jeoung S. C., Lienau Ch., Park Q-Han, and Planken P. C. M. (2006) Shape resonance omni-directional terahertz filters with near-unity transmittance. Opt. Express 14: 1253-1259.

[67] Lee J. W., Seo M. A., Sohn J. Y., Ahn Y. H., Kim D. S., Jeoung S. C., Lienau Ch., and Park Q-Han (2005) Invisible plasmonic metamaterials through impedance matching to vacuum. Opt. Express 13: 10681-10687.

[68] Zhao G., Schouten R. N., Van der Valk N., Wenckebach W. Th., and Planken P. C. M. (2002) Design and performance of a THz emission and detection setup based on a semi-insulating GaAs emitter. Rev. Sci. Instrum. 73: 1715-1719.

[69] Saleh B. E. A. and Teich M. C. (1991) Fundamentals of Photonics. John Wiley & Sons, Inc.

[70] Wu Q. and Zhang X. C. (1995) Free-space electro-optic sampling of terahertz beams. Appl. Phys. Lett. 67: 3523-3525.

[71] Lee J. W. and Kim D. S. (2009) Relative contribution of geometric shape and periodicity to resonant terahertz transmission. J. Appl. Phys. 107: 113109.

[72] Nau D., Schönhardt A., Bauer Ch., Christ A. I., Zentgraf T., Kuhl J., Klein M. W., and Giessen H. (2007) Correlation effects in disordered metallic photonic crystal slabs. Phys. Rev. Lett. 98: 133902.

Section 3

Emerging Concepts with Plasmonics

Chapter 10

Application of Surface Plasmon Resonance Based on a Metal Nanoparticle

Amir Reza Sadrolhosseini, A. S. M. Noor and Mohd. Maarof Moksin

Additional information is available at the end of the chapter

http://dx.doi.org/10.5772/51219

1. Introduction

Several researchers have focused on biosensor for monitoring of biological interaction. Moreover, the detection of bimolecular is an extremely significant problem. Hence, the development of optical biosensors based on optical properties of noble metal nanoparticles using Surface Plasmon Resonance was considered.

Surface plasmon resonance (SPR) is a powerful technique to retrieve information on optical properties of biomaterial and nanomaterials. Biosensor based on SPR is a versatile technique for biological analysis applications. Essentially, SPR depends on the optical properties of metal layer [1] and enviromental changes so it is related to charge density oscillation at the interface between them [2]. Hence, biomolecular possess an extreme sensitivity to plasmon resonance and they remove the requirement for extrinsic biomolecular labeling [3]. One advantage of SPR is, the light beam never passes through the dielectric medium of interest and hence the effect of absorption of the light in the analyte can be ignored. Hence, the main potential of surface plasmon resonance is characterization of medium after the metal layer.

Sometimes polymer layer was coated on gold later for detection and improvement the sensor performence. So, the properties of nanostructures gave rised to interest in biosensing. Nano particles and Nano tube have improved the sensitivity, selectivity, and multiplexing capacity in biosensors. Different types of nanostructure-based biosensors including carbon nanotube, SnO_2 nanoparticles, nanowires, and nanogap biosensors have emerged..

Biomolecular interaction are determined and predicted via angular modulation sensor; hence, the out put of SPR biosensor sensor is angle shift (θ_{spr}), associated with the point of minimum reflected light intensity and is very sensitive to the changes in the dielectric constant of the medium. So, the sensitivity of modern SPR sensing systems based on the Kretschmann configuration is such that they are capable of detecting refractive indexes as high as approximately 5 × 10^{-7} refractive index units, which corresponds to a 1 pg/mm^2 surface coverage of biomolecules [4, 5].

Unfortunately, when sensing biomolecules of low molecular weight or in low concentration, the variation in SPR angle is very small and this causes difficulties in detecting the acquired signal [6]. Hence, several researchers have focused on enhancement of sensitivity of biomolecular sensor. Hence, the properties of metal nanoparticles (gold and silver) have contemplate as a result of the physical characteristics of surface-enhanced Raman scattering (SERS). Zhu at al., (1998) utilized colloidal gold nanoparticle films to enhance the the Raman scattering of the underlying molecules [7]. On the other hand the colloidal gold nanoparticles were used as a immobilized element on the thin gold film[8, 9]. Moreover, to improve the roughness of the film, the inclusion of colloidal Au nanoparticles was added to the gold film. SPR biosensors using colloidal Au-modified gold films has the large angle shift and increase in minimum reflectivity suggesting that SPR spectrum depend on size and volume fraction of the Au nanoparticles [10]. To improve the detection of SPR biosensors, He et al., (2000)[11] and; Lyon et al., (1998)[10] reported using biomolecule-tagged colloidal Au nanoparticles. In this case, the SPW damped strongly because the intraction of the biomolecules -colloidal Au were occouured with the immobilized bio-monolayer sensing. However, using this method to generate a biomolecular interaction signal involves complex procedures for the preparation of nano-sized colloidal Au particles and chemical immobilizations to tag the biomolecules. This process, eliminates the major advantage of SPR technology, namely that biomolecular labeling is not required [6]. On the methods for improving the the sensitivity of SPR biosensors was reported by Gestwicki et. al. [12]. They enhanced the sensitivity and response of biosensor based on absorption spectroscopy techniques [13, 14]. They shown the resonance angle can be improved by caused by the conformational changes of proteins and enzymes when they bind with analytes of low molecular weights and the response of sensor can be increased by adopting Au nanoshells or nanoclusters.

The Au nanocluster-embedded dielectric film is fabricated through a simple co-sputtering method utilizing a multi-target sputtering system. In the proposed method, the resolution of the SPR biosensor is enhanced by precisely regulating the size and volume fraction of the Au nanoclusters in order to control the surface plasmon effect [6].

Carbon nanotubes (CNTs)[15] have been used as both electrode and transducer components in biosensors becuse of their interesting electrochemical and electrical properties including thier flexibility for varying the electrical property (from metallic to semiconductor) and their, small diameters, high mechanical strength, and easy fabrication/integration. Wang [16] and Balasubramanian and Burghard [17], presented the design methodologies, performance characteristics, and potential applications of CNT biosensors, as well as the factors limiting their practical use. It is obvious that many challenges still do remain, especially regarding the sensitivity of the nanotube response to environmental effects, and also the chosen type of raw CNT material, and its functionalization.

The other nanostructured sensors is based on nano ceramic such as ZnO and SnO_2. For example, the SnO_2 sensor has demonstrated a potential for gas sensing applications. Liu et al. [18] developed a highly sensitive and nanostructured SnO_2 thin-film sensor for the detection of ethanol vapor. Its lower LOD (below 1 ppm) demonstrates an improvement over that of other works reported in the literature. Recently, Choi et al. [19] implemented a functionalized ZnO nanowire field effect transistor (FET) biosensor for the low level detection of biomolecular interactions. This biosensor was fabricated without any doping which showed an enhanced sensitivity compared to other nanowire biosensors.

Fundamentally, the base of the biosensor is the coupling of a ligand-receptor binding reaction [1] to a signal transducer. Moreover, angular, phase and polarization modulations are the various methods applicable for distinguishing the interaction of bimolecular with the receptor. Essentially, SPR is a quantum electromagnetic phenomenon that appears at the interface of the dielectric and the metal. Under certain conditions, the energy of the light beam is absorbed by collective excitation of the free electrons called surface Plasmon (SP), which lies between the interface of the dielectric and the metal. On the other hand, when the momentum of the photon matches that of the Plasmon, the resonance appears as an interface of two two media with dielectric constants of opposite signs, and the SP wave propagates along the interface. In accordance with the SP wave properties, the SPR is classified as propagating the SPR, the long-range SPR and the localized SPR.

In the case of propagating SPR (PSPR) [20], the amplitude of the SP wave is decaying exponentially normal to the interface, which the Plasmon propagates, for a distance of about one to ten microns. The interactions between the surface and molecules in the analyte lead to the SPR signal shift which can be observed in shift of the angle of resonance and the wavelength. A second type of Plasmon's wave is Long-range SPR (LRSPR) that occurs in thin metal films with low attenuation; hence, the traveling distance of the Plasmon wave is about a millimeter in the visible range.

The localized SPR (LSPR) [20, 21] is exited on a nano-scale. On the other hand, the LSPR appears in the metal structure as a lateral size that is less than half the wavelength of the excitation photon. Hence, the light beam will interacts with the nanoparticles, so as to leads to a locally Plasmon oscillation occurring around the particle with a specific frequency similar to LSPR. However, the LSPR is more sensitive to changes in the local dielectric analyte [20]. Typically, researchers will measure the changes in the local dielectric environment or analyte via making shifts in the LSPR wavelength measurements. Moreover a variant of angle-resolved sensing that is possible for LSPR is also possible, and LSPR can provide real-time kinetic data for the binding processes. In addition, the PSPR signal is more sensitive to changes in the bulk refractive index than are the LSPR signals, so when the LSPR and PSPR signals are examined the interaction with the molecular adsorption layer and, the response of the two methods becomes comparable in the short-range[20]. Consequently, both sensitivity and flexibility are the capability of the LSPR. Moreover, the decay length is shorter than PSPR; hence, the LSPR are offered for the sensor.

2. Optical properties of metal

Assume that the material such as metal consist of charges which can be set in motion by an oscillating electric field of light that is polarized in one direction. In accordence with Newton's second law of motion, displacement of the electron and damping factor which is a function of mass of electron and relaxation time, the polarization density (P_c) of the conduction electrons is given as

$$P_c = -nex_0(t) = -\frac{ne^2\tau}{m_e\omega}E\frac{\omega\tau - i}{\omega^2\tau^2+1} \tag{1}$$

where $x_0(t)$, m_e and τ are the displacement , mass of electron respectively and a decay or relaxation time [2]. Hence, the total polarization is

$$P = P_b + P_c \tag{2}$$

Wher P_b is the background polarization, and the susceptibility ($\chi = \frac{P}{E}$) therefore the dielectric function (ε) is

$$\varepsilon = 1 + 4\pi\chi \Rightarrow \varepsilon = 1 + 4\pi\frac{P_b + P_c}{E} \tag{3}$$

By using the equations (1-3) the dielectric function can be written as [2, 21],

$$\varepsilon = 1 + \frac{4\pi P_b}{E} - \frac{4\pi ne^2\tau}{m_e\omega}\cdot\frac{\omega\tau - i}{\omega^2\tau^2+1} \tag{4}$$

or

$$\varepsilon = \varepsilon' - i\varepsilon'' \tag{5}$$

In the case of high frequency of $\omega \to \infty$

$$\varepsilon_\infty = 1 + \frac{4\pi P_b}{E} \tag{6}$$

Then by using equations (4) and (6),

$$\varepsilon = \varepsilon_\infty - \frac{4\pi ne^2\tau^2}{m_e(\omega^2\tau^2+1)} + \frac{4\pi ne^2\tau}{\omega\, m_e(\omega^2\tau^2+1)}i \tag{7}$$

Two frequency regimes of interest in equation (7) are to be considered for studying the material response [2].

i. Low frequency regime

For the case of the frequency ω is less than $1/\tau$, $\varepsilon'' = 4\tau\pi ne^2/(\omega m_e)$ the real terms of equation (7) can be neglected [2]. In this case the media is absorbent and the absorption coefficient (α) is

$$\alpha = \frac{\omega}{c}\sqrt{2\varepsilon''} \ , \ \alpha = \frac{\omega}{c}\sqrt{\frac{8\pi ne^2\tau}{m_e\omega}} \tag{8}$$

ii. High frequency regime

Since the dielectric function is real in the high frequency regime, the imaginary part of Eq. (7) can be ignored. In this case the frequency ω is more then $1/\tau$, the dielectric function is therefore,

$$\varepsilon = \varepsilon_\infty - \frac{4\pi ne^2\tau^2}{m_e(\omega^2\tau^2 + 1)} \tag{9}$$

or,

$$\varepsilon = \varepsilon_\infty(1 - \frac{\omega_p^2}{\omega^2}) \tag{10}$$

Where, the plasma frequency (ω_p) is defined as

$$\omega_p = \frac{4\pi e^2 n}{\varepsilon_\infty m_e} \tag{11}$$

As can be seen from equation (10) the dielectric function vanishes at the plasma frequency. However, for $\omega > \omega_p$ the medium behaves like non-absorbing and the light beam traverses through the medium without attenuation. On the other hand when $\omega < \omega_p$, the medium is of high reflectivity behaviour. Therefore the plasma frequency is the frequency of collective oscillations of the electron gas (plasma). The electron density oscillates at the plasma frequency and plasma oscillations are longitudinal type excitation [2]. The quantum of the plasma oscillation is called *plasmon* which can be excited by X ray and light or by an inelastic electron scattering [22].

To obtain the conditions of excitation to result in resonance, let's consider the electrical field component in the *x* direction,

$$\vec{E} = \vec{E}e^{i(k_x x - \omega t)}e^{-\alpha\ z} \tag{12}$$

where α is the absorption coefficient and k_x is the wave number. The absorption coefficients and dielectric constants of two adjacent media is related by

$$\frac{\alpha_1}{\alpha_2} = -\frac{\varepsilon_1}{\varepsilon_2} \tag{13}$$

By using the transverse wave equation ($\nabla^2\vec{E} = \frac{\varepsilon}{c^2}\frac{d^2\vec{E}}{d^2t}$) on the components of the electrical field, a dispersion relation for surface plasmon can be obtained as

$$k^2_{palasmon} = \frac{\omega}{c}\left[\frac{\varepsilon_1 \cdot \varepsilon_2}{\varepsilon_1 + \varepsilon_2}\right] \tag{14}$$

The plot of this equation is shown in Figure 1a. In the figure the wave number of the surface plasmon wave is always larger than the photon wave number and hence the photon would not be able to excite surface plasmon. The dispersion curve of the photon and surface plasmon wave must cross each other as shown in Figure 1.b [2, 20] for surface plasmon resonance to occur when the momentum of incoming light is equal to momentum of the plasmons. This can be done by using prism coupling method as in the present work.

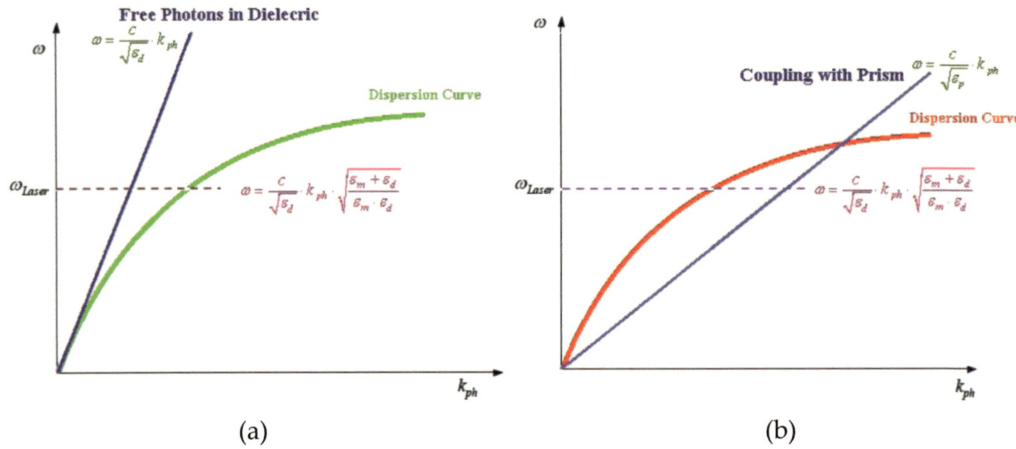

Figure 1. a) The wave number of photon is always less than the wave number of Plasmon at all frequencies and b) The wave number of photon may equal the wave number of number of plasmon.

3. Optical properties of the metal nano particles

The nanoparticles of interest are typically large enough to accurately apply classical electromagnetic theory to describe their interaction with light [23, 24]. However, they are also small enough to allow observation strong variations in the optical properties depending on the particle size, shape, and local environment. Because of the complexity of the systems being studied, efficient computational methods are the capable of treating large size materials are essential. In the past few years, several numerical methods have been developed to determine the optical properties of small particles, including the discrete dipole approximation (DDA) [23], T-matrix, and spectral representation methods (SR) [25]. We employed the DDA, which is a computational procedure suitable for studying the scattering and absorption of electromagnetic radiation by particles with sizes in the order or less of the wavelength than the incident light. DDA has been applied to a broad range of problems, including interstellar and interplanetary dust grains, ice crystals in the atmosphere, human blood cells, surface features of semiconductors, metal nanoparticles and their aggregates, and more. The DDA was first introduced by Purcell and Pennypacker [26] and has been subjected to several improvements, in particular those made by Draine et. al [27]. Below, we briefly describe the main characteristics of DDA and its numerical implementation, namely, the DDSCAT cod[1]. For a full description of DDA and DDSCAT [23], the reader can consult refs [24-28].

The main idea behind DDA is to approximate a scatter, in our case the nanoparticle, by large enough array of polarizable point dipoles. Once the location and polarizability of each dipole are specified, the calculation of the scattering and absorption efficiencies by the dipole array can be determined, depending on the accuracy of the mathematical algorithms and the capabilities of the computational hardware. Although the calculation of the radiated fields in DDA is, in theory and principle, also possible, that calculation is actually beyond the computational capabilities of any of the current systems.

Let us assume an array of N polarizable point dipoles located at $\{r_i\}$, $i = 1, 2, ..., N$, each one characterized by a polarizability α_i . The system is excited by a monochromatic incident plane wave $E_{inc}(r,t) = E_0 e^{ik.r-i\omega t}$ where r , t , ω, $k = \omega / c = 2\pi / \lambda$, c and λ are the position vector, time, the angular frequency, the wave vector, the speed of light and the wavelength of the incident light. Each dipole of the system is subjected to an electric field that can be split in two contributions: (i) the incident radiation field, plus (ii) the field radiated by all of the other induced dipoles. The sum of both fields is the so-called local field at each dipole and is given by

$$E_{i,loc} = E_{i,inc} + E_{i,dip} = E_0 e^{ik.r_i} - \sum_{i \neq j} A_{ij}.P_j \tag{15}$$

where P_i is the dipole moment of the ith element, and A_{ij} with $i \neq j$ is an interaction matrix with 3×3 matrixes as elements, such that [23]

$$\vec{A}_{ij}.\vec{P}_i = \frac{e^{ikr_{ij}}}{r_{ij}^3} \left\{ k^2 \vec{r}_{ij} \times (\vec{r}_{ij} \times \vec{P}_j) + \frac{(1 - ikr_{ij})}{r_{ij}^2} \left[r_{ij}^2 \vec{P}_j - 3\vec{r}_{ij}(\vec{r}_{ij}.\vec{P}_j) \right] \right\} \tag{16}$$

Here $r_{ij} = \left| \vec{r}_i - \vec{r}_j \right|$, and $\vec{r}_{ij} = \vec{r}_i - \vec{r}_j$, and we are using *cgs* units. Once we solve the 3N-coupled complex linear equations given by the relation

$$\vec{P}_i = \alpha_i.\vec{E}_{i,loc} \tag{17}$$

and determined each dipole moment $\vec{P}_i$, we can then find the extinction and absorption cross sections for a target, C_{ext} and C_{abs} in terms of the dipole moments as [23]

$$C_{ext} = \frac{4\pi k}{\left| \vec{E}_0 \right|^2} \sum_{i=1}^{N} (\vec{E}_{i,inc}^{*}.\vec{P}_i) \tag{18}$$

[1] Discrete Dipole Scattering (DDSCAT) is a Fortran code for simulation and calculation of scattering and absorption of light by particles.

$$C_{abs} = \frac{4\pi k}{\left|\vec{E}_0\right|^2} \sum_{i=1}^{N} \left\{ \text{Im}\left[\vec{P}_i \cdot (\alpha_i^{-1})^{\dagger} P_i^{\dagger} \right] - \frac{2}{3} k^3 \left| P_i \right|^2 \right\} \tag{19}$$

where † means complex conjugate. The scattering cross section can be obtained using the following relation:

$$C_{ext} = C_{sca} + C_{abs} \tag{20}$$

Certain arbitrariness in the construction of the array of dipole points represents a solid target of a given geometry. For example, the geometry of the grid where the dipoles have to be located is not uniquely determined, and a cubic grid is thus usually chosen. Also, it is not obvious how many dipoles are actually required to adequately approximate the target or which choice is the best for the dipole polarization. If one chooses the separation between dipoles d such that $d \ll \lambda$, then one can assign the polarize for each particle i in a vacuum, using the lattice dispersion relation (LDR) polarization, α_i^{LDR} at a third order in k, given by [23, 27]

$$\alpha_i^{LDR} = \frac{\alpha_i^{CM}}{1 + \alpha_i^{CM} \left[b_1 + b_2 \varepsilon_i + b_3 S \varepsilon_i \right] (k^2/d)} \tag{21}$$

where ε_i is the macroscopic dielectric function of the particle, α_i^{CM} is the polarizability given by the well-known Clausius- Mossotti relation

$$\alpha_i^{CM} = \left(\frac{d}{3} \right)^3 \frac{\varepsilon_i - 1}{\varepsilon_i + 2} \tag{22}$$

and S, b_1, b_2 and b_3 are coefficients of the expansion. Now the question is, how many dipoles do we need to mimic the continuum macroscopic particle with an array of discrete dipoles? The answer is not straightforward, because we have to consider the convergence of the physical quantities as a function of the dipole number. It has been found that $N \geq 10^4$ for an arbitrary geometry is a good starting number, as shown in the Appendix. However, we do have a matrix of $(3N)^2$ complex elements that require a large amount of computational effort. In this work, we have employed the code adapted by Draine and Flatau to solve the complex linear equations found in DDA. To solve the complex linear equations directly would require tremendous computer capabilities; however, one can use iterative techniques to compute the vector $\vec{P} \equiv \{\vec{P}_i\}$. In this case, each iteration involves the evaluation of matrix-vector products such as $\vec{A}.\vec{P}^{(n)}$, where n is the number of the iteration. The algorithm, named DDSCAT, locates the dipoles in a periodic cubic lattice, and then uses fast Fourier transform techniques to evaluate matrix-vector products such as $\vec{A}.\vec{P}$, which thus allows the whole computation of the final vector $\vec{P}$ for a large number of dipoles [28]. For a detailed description of DDA and DDSCAT code, the reader can look at refs [24-28].

In 2003, Sosa et. al. [23] defined the extinction, absorption and scattering efficiencies or coefficients, Q_{ext}, Q_{sca} and Q_{abs} as

$$Q_{ext} = \frac{C_{ext}}{A},\ Q_{abs} = \frac{C_{abs}}{A},\ Q_{sca} = \frac{C_{sca}}{A} \tag{23}$$

Where $A = \pi a_{eff}^2$ and a_{eff} is/are defined through the concept of an effective volume equal to $4\pi a_{eff}^3 / 3$.

In Figure 2 [23], we show Q_{ext}, Q_{sca} and Q_{abs} in dotted, dashed, and solid lines, respectively, as a function of the wavelength of the incident light, λ for nanometric-sized particles. The calculations were done for nanoparticles with $a_{eff} = 50nm$ and dielectric functions as measured on bulk silver and gold by Johnson and Christy [29]. The nanoparticles were represented or mimic by around 65000 point dipoles in order to have a good convergence on their optical properties, as discussed below. The number of these dipoles is quite large in comparison with the numbers used in previous studies on isolated and supported small metallic nanoparticles [30, 31] where, incidentally, only an extinction efficiency was reported. Our use of a large number of dipoles is in agreement with the results found in a previous study where we learned that even for small metallic nanoparticles with radii of about a few nanometers, one needed more than 12000 dipoles to achieve convergence on their optical properties [32]. A larger number of dipoles is necessary to achieve that convergence. In particular, we found that extinction efficiency converges very rapidly as a function of the number of dipoles. However, this is not the case for the absorption or scattering efficiencies where a very large number of dipoles were necessary to achieve that convergence. It is also well known that for small nanoparticles *(<20 nm)* the light absorption process dominates the extinction spectrum, whereas for very large nanoparticles *(>100 nm)*, the light-scattering process dominates. As we show in this paper, to elucidate the optical properties of medium-size nanoparticles, it is indispensable to undertake an in-depth study of the scattering and absorption efficiencies, and not only of the extinction one, because both phenomena, namely, scattering and absorption, are the same order of magnitude. Figure 2 shows the optical efficiencies for a sphere with a radius of *50 nm*. In the spectra, we can observe that at about *320 nm* all the efficiencies have a local minimum that corresponds to the wavelength at which the dielectric functions of silver. Therefore, this feature of the spectra is inherent to the material properties and, as we observe below, that feature independent of the particle geometry. Below *320 nm*, the absorption of light is mainly due to the intra-band electronic transitions of silver; therefore, this feature of the spectra should also be quite independent of the shape and size of the particles, as it is actually corroborated in all of the graphs shown below that corresponding to a silver particle. At about 350 nm, the spectrum of Q_{abs} shows a peak is related to the excitation of the surface Plasmon of the sphere; therefore, this feature is inherent to the geometry of the particle, although the position depends on its material properties. At larger wavelengths the spectrum of Q_{abs} shows specific features from *380 nm* to about *500 nm* that corresponds to plasmon excitations due to higher multi-polar charge distributions [23]. If we look only at the Q_{ext}, it is not possible to observe the same features because scattering effects hide them. The Q_{sca}

spectrum shows a broad structure from *320 nm* to *750 nm*, with a maximum being about *400 nm*, three times more intense than the maximum of Q_{abs}. The characteristics of the Q_{sca} spectrum are mainly due to the size of the particle. In a previous work [33], we found that this maximum is less pronounced as the radius of the sphere increases. Also, the Q_{sca} spectra of nanospheres decay slowly as the radius increases, which means that as the sphere becomes larger it scatters light at longer wavelengths, just as expected. The long tail in the extinction spectrum corresponds to Rayleigh scattering which also observed in the extinction spectra shown below.

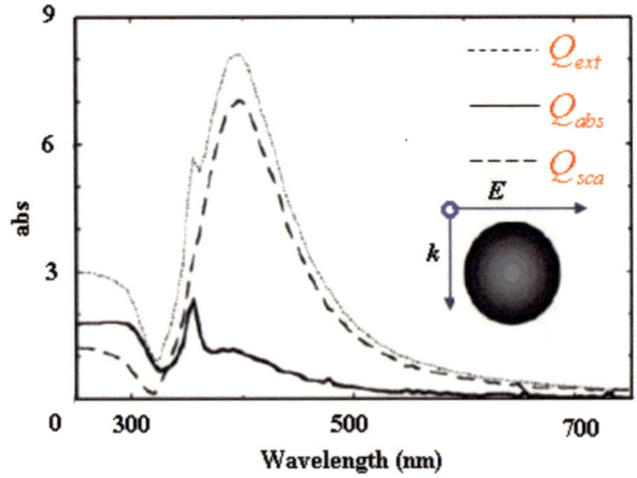

Figure 2. Optical coefficients for a silver nanosphere[23]

4. Theory

4.1. Surface plasmon resonance with prism coupling

We study SPR in the classical Kretschmann's configuration shown in Figure 3, where one side of the glass prism is coated thin gold or silver film in contact with the probed sample. Surface plasmons, which are collective oscillations of free electrons, are propagating along the interface between the sample and the metal film, and are thus sensitive to the optical properties of both the sample and the metal film. These plasmons can only be excited with TM-polarized light, which has an electric field component normal to the surface that generates the required surface charge distribution for the excitation [34].

The incident light is totally internally reflected at the interface and its intensity decreases with incident angle which is also called as coupling angle. By using the momentum and energy conservation laws [1, 35] the condition of resonance is obtained as follows:

$$k_{glass} \sin\theta_i = \frac{\omega}{c} n_{glass} \sin\theta_i = k_{plasmon} \tag{24}$$

where θ_i is the coupling angle.

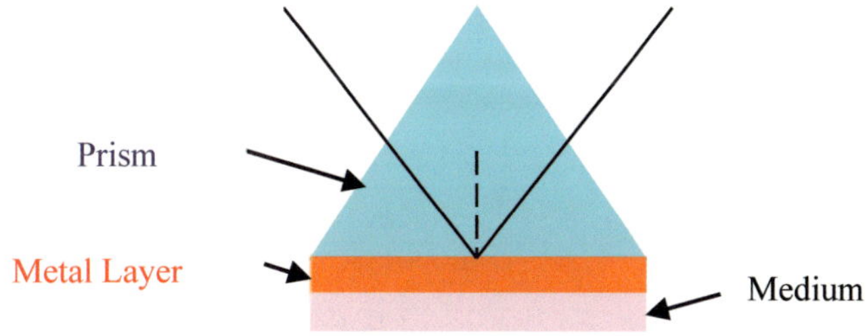

Figure 3. Kretschmann configuration

To excite the SPW the electrical and magnatisam fields are parallel and perpendicular in the incident plane which is defined by light propagation vector and the normal to the interface. Thus, in accordance with Figure 4 and the boundary conditions, the relationship between electrical and magnetism fields can be written as follows [2, 22]:

$$B_a = n_0\sqrt{\varepsilon_0\mu_0}(E_0 + E_{r1}) = n_1\sqrt{\varepsilon_0\mu_0}(E_{t1} + E_{i1}) \tag{25}$$

$$B_b = n_1\sqrt{\varepsilon_0\mu_0}(E_{i2} + E_{r2}) = n_2\sqrt{\varepsilon_0\mu_0}E_{t1} \tag{26}$$

$$E_a = (E_0 - E_{r1})\cos\theta_0 = (E_{t1} - E_{i1})\cos\theta_{t1} \tag{27}$$

$$E_b = (E_{i2} - E_{r2})\cos\theta_{t1} = E_{t2}os\theta_{t2} \tag{28}$$

The relationship between E_a, B_a and E_b, B_b are obtained by using phase change due to light passing through the different layers and $E_{i1} = E_{r2}e^{-i\delta}$, $E_{i2} = E_{t1}e^{-i\delta}$, $B_{i1} = B_{r2}e^{-i\delta}$, $B_{i2} = B_{t1}e^{-i\delta}$, $B = \frac{n}{c}E$ and $c = \frac{1}{\sqrt{\varepsilon_0\mu_0}}$ [2, 22], as follows [36]:

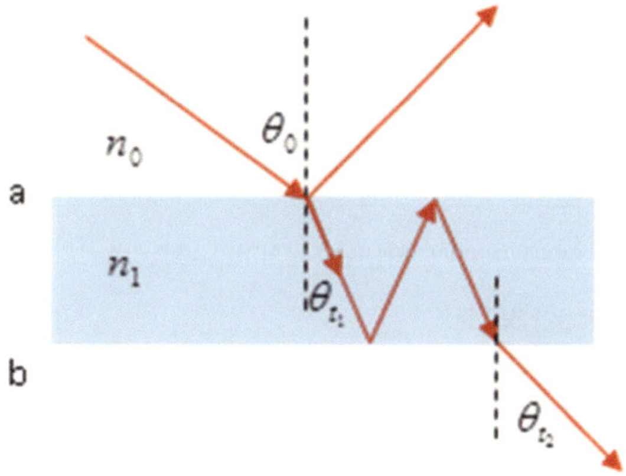

Figure 4. Reflection and refraction at the interface

$$E_a = (e^{-i\delta}\cos\delta)E_b - (i\frac{e^{-i\delta}}{\tau_1}\sin\delta)B_b \tag{29}$$

$$B_a = (-\tau_1 i e^{-i\delta} \sin\delta) E_b + (e^{-i\delta} \cos\delta) B_b \tag{30}$$

as τ_n is

$$\tau_N = \frac{n_N}{\cos\theta_{t_N}} \sqrt{\varepsilon_0 \mu_0} \tag{31}$$

Consequently, the layers cause the phase change of light and the matrix layers obtain from the coefficient of equations (30) and (31) as follows:

$$\begin{bmatrix} E_a \\ B_a \end{bmatrix} = M \begin{bmatrix} E_b \\ B_b \end{bmatrix}$$

where M is a matrix consist of

$$M = \begin{bmatrix} \cos\delta & -i\dfrac{\sin\delta}{\tau_1} \\ -i\tau_1 \sin\delta & \cos\delta \end{bmatrix}, \tag{32}$$

where δ is the phase shift due to the beam passing through different layer given as

$$\delta = \frac{2\pi}{\lambda} d n_1 \cos\theta_{t_1} \tag{33}$$

where d is the thickness of layer. Considering more than one layer is involved for description and simulation of SPR, the matrix for N-layer can be achieved from the relation between tangential fields at the boundary of the first layer up to the N_{th} layer as follows [2, 36, 37]

$$\begin{bmatrix} E_a \\ B_a \end{bmatrix} = M_1 M_2 M_3 .....M_N \begin{bmatrix} E_b \\ B_b \end{bmatrix} \tag{34}$$

Substituting,

$$M = M_1 M_2 M_3 ....M_N \tag{35}$$

and the characteristic matrix [38] is

$$M = \begin{bmatrix} m_{11} & m_{12} \\ m_{21} & m_{22} \end{bmatrix} \tag{36}$$

If one layer is deposited on the prism, the reflection and transmission coefficients are defined respectively as

$$r = \frac{E_{r_1}}{E_0} \tag{37}$$

and

$$\Gamma = \frac{E_{t_2}}{E_0} \tag{38}$$

and from equations (25, 26, 27, 28 and 36), it can be obtained that

$$\begin{bmatrix} (E_0 - E_{r_1})\cos\theta_0 \\ n_0\sqrt{\varepsilon_0\mu_0}(E_0 + E_{r_1}) \end{bmatrix} = \begin{bmatrix} m_{11} & m_{12} \\ m_{21} & m_{22} \end{bmatrix} \begin{bmatrix} E_{t_2}\cos\theta_{t_1} \\ n_2\sqrt{\varepsilon_0\mu_0}E_{t_2} \end{bmatrix} \tag{39}$$

By substituting equations (37 and 38) in equation (39) and after simplification, the reflection coefficient is as follows:

$$r = \frac{m_{21} + m_{22}\tau_2 - m_{11}\tau_0 - m_{12}\tau_2\tau_0}{m_{21} + m_{22}\tau_2 + m_{11}\tau_0 + m_{12}\tau_2\tau_0} \tag{40}$$

whereby the reflectivity is

$$R = rr^* \tag{41}$$

If there are three media of prism, n_0, metal layer, n_1 and air ($n_2 = 1$); and the thickness of metal layer, t, is less than wavelength ($\lambda = 632.8nm$) then the characteristic matrix is

$$M = \begin{pmatrix} 1 & -i(2\pi/\lambda)d\cos^2\theta_{t1} \\ -i(2\pi/\lambda)dn_1^2 & 1 \end{pmatrix} \tag{42}$$

According to equations (37, 40) and (41), the reflection coefficient is

$$r = \frac{\overbrace{\left(\tau_0 n_2\sqrt{\varepsilon_0\mu_0} - \cos\theta_{t_2}\right)}^{A} - i\alpha\overbrace{\left(n_1^2\cos\theta_{t_2}\tau_0 - \cos^2\theta_{t_1}\sqrt{\varepsilon_0\mu_0}\right)}^{B}}{\underbrace{\left(\tau_0 n_2\sqrt{\varepsilon_0\mu_0} + \cos\theta_{t_2}\right)}_{C} - i\alpha\underbrace{\left(n_1^2\cos\theta_{t_2}\tau_0 + \cos^2\theta_{t_1}\sqrt{\varepsilon_0\mu_0}\right)}_{D}} \tag{43}$$

Substituting α for $2\pi t/\lambda$, and after multiplying the numerator and denominator of equation (42) with the complex conjugate of the denominator, the reflection coefficient is derived as follows:

$$r = \frac{AC + \alpha^2 BD - i\alpha(BC - DA)}{C^2 + \alpha^2 D^2} \tag{44}$$

and the phase change [39] is

$$\tan\phi = \frac{\alpha(BC - DA)}{AC + \alpha^2 BD} \tag{45}$$

4.2. Theory of localized SPR

In a general and simple case, we can assume that the nanoparticle is as metallic sphere with the radius R. It is embedded in a liquid medium. So, the SP wave just can propagate less than the 100 nm or the size of particle; hence, becomes localized.

The electrical field outside the sphere is [20]:

$$E_{out}(x,y,z) = E_0\hat{z} - \left|\frac{\varepsilon_m - \varepsilon_d}{\varepsilon_m + 2\varepsilon_d}\right| \times a^3 E_0 \left[\frac{\hat{z}}{r^3} - \frac{3z}{r^5}(x\hat{x} + y\hat{y} + z\hat{z})\right] \quad (46)$$

where E_0 is the applied field magnitude polarized in the z direction. The free electron clouds are dislocate by the electromagnetic wave field and produces uncompensated charges near the particle surface which produces corresponding opposing forces (as in Figure 5 [20]).

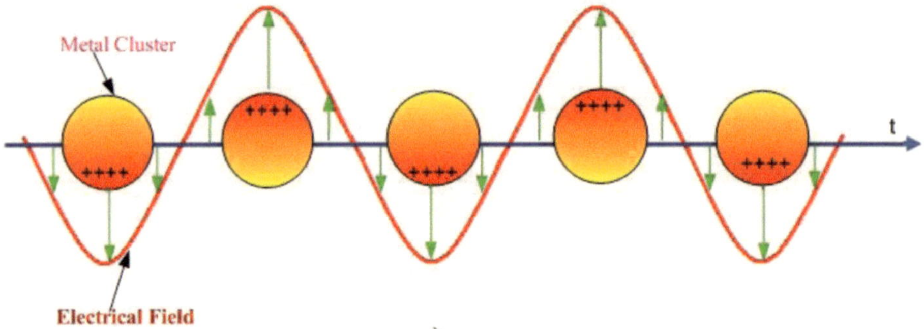

Figure 5. Collective Oscillations of free electrons due to applied electric field

The optical resonance related to these oscillations is called the localized SPR (LSPR) [20]. The origin of the term "surface" comes from the knowledge that the oscillations are caused by the polarization of the particle surface, and because the generated electric field is larger near the particle surface and decays farther away far from the surface, which makes that field similar to the evanescent field at a flat interface in the propagating SPR case. In the general case, the frequency of the collective oscillations does not coincide with the applied wave frequency and is determined by many factors, electron concentration, electron effective mass, the particle shape and size, interaction between the particles, and the influence of the environment. However, for an elementary description of the nanoparticles plasmon resonance it is sufficient to use the usual dipole approximation, and the Drude model. The electrostatic polarizability of particle α_0 [20] assigns the scattering and absorption of light by a small particle. The electrostatic polarizability of the particle can be calculated using the metal optical permittivity ε_m, the medium dielectric constant ε_d, and the particle geometrical dimensions [20, 40]:

$$\alpha_0 = a^3\left(\frac{\varepsilon_m - \varepsilon_d}{\varepsilon_m + 2\varepsilon_d}\right) \quad (47)$$

Moreover, static polarizability is [20]

$$\alpha = \frac{a\alpha_0}{a - (k^2 + 0.67ik^3 a)\alpha_0} \tag{48}$$

Where

$$k = \frac{2\pi\sqrt{\varepsilon_d}}{\lambda}$$

is the medium wave number. The fundamental assumption in the theoretical model is that of a spherical particle shape, which allows us to utilize extended the Mie theory. Furthermore, we can assume that the particles are distributed in a triangular two-dimensional array instead of in a disordered layer. This assumption allow us to perform efficient estimates of the inter particle coupling effects. The calculated quantity is the extinction cross-section, which is formally obtained from the integral [41]:

$$C_{ext} = \frac{1}{2I_0}\mathrm{Re}\int_A \left[E_i \times H_s^{\dagger} + E_s \times H_i^{\dagger} \right] \cdot \hat{n} dA \tag{49}$$

Here, ($E_i(r)$, $H_i(r)$) and ($E_s(r)$, $H_s(r)$ are the incident and total scattered fields, respectively, and I_0 is the incident irradiance. The integral is evaluated over an arbitrary surface A that encloses the entire nanoparticle system. In the case of an isolated spherical nanoparticle illuminated by a plane wave, the scattering cross-sections C_{ext} are calculated based on the Mie scattering theory [42]. From Eqs.(47) and (49) and badi the scattering cross-sections is[20]

$$C_{ext} \cong 4\pi k \,\mathrm{Im}(\alpha) \tag{50}$$

In the static case the extinction cross-section will be [43] explained by Eq (50). One can see from the latter expression that the extinction spectrum has a strong resonance when:

$$\varepsilon_m(\omega = \omega_{res}) = -2\varepsilon_d \tag{51}$$

The permittivity of bulky metal is explained by the Drude model; hence, we utilize it to obtain the resonance frequency:

$$\varepsilon_m(\omega) = \varepsilon_{ib} - \frac{\omega_p^2}{\omega(\omega + i\gamma_b)} \tag{52}$$

where ε_{ib} is the contribution of the inter band electronic transitions and γ_b is the volume decay constant. The metal permittivity is a main parameter for determining the LSPR frequency. So, we consider the exact Drude model. From Eq. (51) and Eq. (52) making the assumption that the significant contribution to the extinction spectrum is caused by absorption, the effective absorption cross-section can be expressed by the resonance frequency as follows

$$C_{abs} = \frac{12\pi k}{a^3}\frac{\varepsilon_d \operatorname{Im}(\varepsilon_m)}{|\varepsilon_m - \varepsilon_d|^2} \times |\alpha| = \pi a^2 \frac{12 \cdot k \cdot a \cdot \varepsilon_d}{(2\varepsilon_d + \varepsilon_{ib}^2)} \times \frac{\omega_p^2 (\gamma_b / \omega)(\omega + \gamma_b)^2}{(\omega^2 + \gamma_b^2 - \omega_{res}^4)^2 + \omega_{res}^4 \times (\gamma_b^2 / \omega^2)} \quad (53)$$

The resonance frequency and resonance wavelength are given by [20]

$$\omega_{res} = \frac{\omega_p}{\sqrt{\varepsilon_{ib} + 2\varepsilon_d}} \quad \lambda_{res} = \lambda\sqrt{\varepsilon_{ib} + 2\varepsilon_d} \quad (54)$$

where

$$\omega_{res},\ \lambda_{res} \text{ and } \lambda_p = \frac{2\pi c}{\omega_p}$$

are the resonance frequency, resonance wavelength and plasma wavelength respectively, demonstrated as [44, 45] that is near the resonance, and sometimes for comparison between the metallic bulk permittivity and the metallic nano particle permittivity, the absorption spectrum can be approximately reduced to a Lorenzian profile.

The wavelength modulation is the common method used for LSPR sensor. So, it is the shift in the maximum (or minimum) of the spectrum curve that is monitored as a function of changes in the local dielectric environment that is caused by analyte absorption. The mechanism of this method is described in Eq. (48), and it has been demonstrated for variation of refractive index or the length of a molecular which was adsorbed to receptor. In this equation, the sensitivity to the bulk refractive index ($S_{\lambda \to bulk}$) plays an important role in determining the total response of the LSPR. This sensitivity depends on various factors such as the size and shape of the nanoparticles. In the following section, we discuss the effect of the shape and design of the particle on the sensing process. Apparently, the resonance position is not dependent on the particle size as can be seen by Eq. (54) for a sphere and Eq. (50) for an arbitrary smooth particle and only variation in the intensity of the SPR band with particle size is observed. However, experimentally the SPR bandwidth as well as the band position is observed depending on the nanoparticle size. In the modification to the Mie theory, Drude and Sommerfeld considered that the relative permittivity of the nanoparticle depends on the size $\varepsilon(\omega, R)$ in lieu of $\varepsilon(\omega)$. The real and the imaginary parts of the modified relative permittivity are given by [46, 47]:

$$\varepsilon'(\omega) = \varepsilon_{ib} - \frac{\omega_p^2}{\omega^2 - \omega_d^2} \text{ and } \varepsilon''(\omega) = \frac{\omega_p^2 \times \omega_d}{\omega(\omega^2 + \omega_d^2)} \quad (55)$$

Where ε_{ib} is the high-frequency limit of $\varepsilon(\omega)$, arising from the response of the core electrons (electrons in completely filled bands); ω_d is the relaxation or damping frequency that represents collisions of electrons with the lattice (phonons) and defects; and ω_d is the bulk plasmon frequency. Dependence of the relative permittivity on the particle radius R

was introduced by assuming that the particle size is smaller than the mean free path of the conduction electrons. The damping frequency ω_d is related to the mean free path of the conduction electrons in the bulk metal R_{bulk} and the Fermi velocity of electrons υ_f by:

$$\omega_d = \frac{\upsilon_f}{R_{bulk}} \tag{56}$$

When R becomes smaller than the mean free path, surface scattering is dominant. Such surface scattering results in the peak broadening and induces a $1/R$ dependence of the SPR bandwidth. In this case, the mean free path R_{eff} becomes size-dependent according to:

$$\frac{1}{R_{eff}} = \frac{1}{R} + \frac{1}{R_{bulk}} \tag{57}$$

One can see from the above equations that the metal particle's shape dictates the spectral signature of its plasmon resonance; the ability to change this parameter and study the effect on LSPR response is a very important experimental challenge. The development of increasingly sophisticated lithographic and chemical methods now allows for the routine production of a wide variety of complex NPs and their assemblies [50].

4.3. Metal cluster

The intraction of light with the metal cluster is the atractive aspect in nanotechnology. The electrons between metal atoms in small clusters are localized, and the plasmon resonance is a size dependent phenomenon.

The absorption and scattering cross section C_{abs} and C_{sca} are the common parameter to express the optical properties. Hence, absorption and scattering particles are not purely, so extinction cross section should be considered, which is the sum of both C_{abs} and C_{sca} or and $C_{ext} = C_{abs} + C_{sca}$. In the quasi-static the metal cluster is a spherical form with the radius is smaller than the wavelength ($R << \lambda$). On the other hand, extinction cross section may be written in other form as follows:

$$C_{ext}(\omega) = 9\frac{\omega}{c}\varepsilon_m^{3/2} V_0 \frac{\varepsilon_2(\omega)}{\left[\varepsilon_1(\omega) + 2\varepsilon_m\right]^2 + \varepsilon_2^2(\omega)} \tag{58}$$

Where ω / c, V_0, ε_m and $\varepsilon(\omega)$ are the wave vector, cluster volume, dielectric function of the embedding medium and dielectric function of the particle metal. The extinction cross section is due to dipolar absorption.

A multidipolar contribution, quadrupole extinction and quadrupole scattering are conquered in the scattering cross section in this region. Consequantly, the position and shape of the plasmon resonance will depend on the size, the dielectric functions of the metal clusters, the dielectric function of the surrounding medium and the shape of the particles, its aggregations and, interactions with the substrate etc..

4.4. The nanoparticle in liquid matrix

Today, any investigation of nano-fluid is considerable. Hence, in this instance, the complex refractive index of liquids containing isotropic nanoparticles and optical the response of medium are investigated. The diameter of the particles is much smaller than the wavelength of the incident light. The response of a medium to an external electromagnetic field cannot be completely described using macroscopic fields as local field effects play an important role. This pont was first demonstrated by Lorentz who showed an impressive accuracy when predicting the dielectric function of the medium [49]. The external field induces a collection of dipole moments [50] and the local field factor relates the macroscopic fields to the local ones. These local field factors are important in the derivation of various effective medium theories (EMTs) considered next here. Regardless of the composition of the nanocomposite, the effective dielectric function can always be estimated by Wiener [51]

$$\varepsilon_{eff}(\omega) \le \sum_{i=1}^{N} f_i \varepsilon_i(\omega) \ , \ \frac{1}{\varepsilon_{eff}(\omega)} \le \sum_{i=1}^{N} \frac{f_i}{\varepsilon_i(\omega)} \tag{59}$$

where the summation is more than the number of constituents N, and their corresponding volume fractions f_i and complex dielectric functions $\varepsilon_i(\omega)$. These limits correspond to capacitors that are connected in parallel or as a series *i.e.* the Wiener bounds give upper and lower limits for the effective complex dielectric function for a certain volume fraction irrespective of the geometry of the nanostructure. Even tighter limits, such as Hashin–Shtrikman and Bergman–Milton, can be utilized in the estimation of the effective dielectric function for two-phase nanocomposites with known volume fractions of the constituents [51–54]. For a more detailed description of the different models see, *e.g.*, a review article by Hale [55].

The Maxwell Garnet (MG) EMT can be used to describe the effective dielectric function of a solution containing a small number of spherical nanoparticles. This dilute limit must be assumed as the MG model itself does not include interaction between nanoparticles or agglomeration. In such a case the effective dielectric function of a medium is given by

$$\frac{\varepsilon_{eff}(\omega) - \varepsilon_h(\omega)}{\varepsilon_{eff}(\omega) + 2\varepsilon_h(\omega)} = f_i \frac{\varepsilon_i(\omega) - \varepsilon_h(\omega)}{\varepsilon_i(\omega) + 2\varepsilon_h(\omega)} \tag{60}$$

where $\varepsilon_i(\omega)$ and $\varepsilon_h(\omega)$ are the frequency-dependent complex dielectric functions of the inclusions and the host, respectively. Naturally the corresponding volume fractions f_i and satisfy $f_i + f_h = 1$. The predicted optical properties of an MG medium are in fine agreement with the measured ones/properties for relatively low volume fractions of the inclusions.

Mathematically this low volume fraction limit of the MG formalism arises from the asymmetric equation for the effective dielectric function since the effective dielectric function would become dependent on the choice of the host material for a two-phase MG system with comparable volume fractions. The Bruggeman effective medium theory (BMT) [56] is based on the assumption that the inclusions are embedded in the effective medium itself, which is invariant if the constituents are replaced by each other. The Bruggeman formalism has been extended to inclusions with non-spherical shape by introducing a geometric (or depolarization) factor g [57], which depends on the shape of the inclusions. This generalized BMT is

$$f_h \frac{\varepsilon_h(\omega) - \varepsilon_{eff}(\omega)}{\varepsilon_h(\omega) + g[\varepsilon_h(\omega) - \varepsilon_{eff}(\omega)]} + f_i \frac{\varepsilon_i(\omega) - \varepsilon_{eff}(\omega)}{\varepsilon_i(\omega) + g[\varepsilon_i(\omega) - \varepsilon_{eff}(\omega)]} = 0 \tag{61}$$

For example, the classical Bruggeman formula is obtained with a value of g = 1/3 which corresponds to spherical inclusions whereas for two-dimensional circular inclusions the value of g = 1/2. Thus by extracting the effective dielectric function of a two-phase nanocomposite with known volume fractions and dielectric functions of the constituents, we can solve 10 for g and then obtain the geometrical shape of the investigated particles.

The quadratic Eq. (61) has two solutions for the complex effective dielectric function but only the positive branch of the square root is physically reasonable and is given by

$$\varepsilon_{eff}(\omega) = \frac{-c(\omega) + [c^2(\omega) + 4g(1-g)\varepsilon_h(\omega)\varepsilon_i(\omega)]^{1/2}}{2(1-g)} \tag{62}$$

Where

$$c(\omega) = 3(f_h - 1)\varepsilon_h(\omega) + (3f_i - 1)\varepsilon_i(\omega) \tag{63}$$

The frequency-dependent effective refractive index $n_{eff}(\omega)$ and the effective extinction coefficient $k_{eff}(\omega)$ keff(ω) of a nanoliquid are connected to effective dielectric function $\varepsilon_{eff}(\omega)$ as follows:

$$n_{eff}(\omega) = \frac{1}{\sqrt{2}}[\mathrm{Re}\{\varepsilon_{eff}\} + (\mathrm{Re}\{\varepsilon_{eff}\}^2 + \mathrm{Im}\{\varepsilon_{eff}\}^2)^{1/2}]^{1/2} \quad k_{eff}(\omega) = \frac{\mathrm{Im}\{\varepsilon_{eff}\}}{2n_{eff}} \tag{64}$$

We show in Figure 4.4 the effective dielectric functions for MG and BR type curves with geometric factors g = 1/3 and g = 1/2, which are inside the Wiener bounds as expected.

Further, at low volume fractions these three curves are almost overlap, whereas there is a rather larger difference for larger volume fractions. Hence it is important to know how to model the medium with nanoparticles to gain an accurate retrieval of the complex refractive index of an individual nanoparticle. Before closing this section we want to mention that the MG model can be generalized to hold for ellipsoids [58] by using a geometric factor such as that for BR [59]. The calculation of this geometric factor g for ellipsoids can be found from [60].

In the Rayleigh limit $r_0 << \lambda$, the optical intensity at SPR resonance decays approximately as $I \propto \left(\frac{r_0}{r}\right)^6$ where r is the distance from the center of the metal particle. We can use this dependence in order to define an effective refractive index of the medium surrounding the metal core according to:

$$n_{eff} = \frac{n_2 \int_{r_0}^{r_0+d} 4\pi r^2 \left(\frac{r_0}{r}\right)^6 \partial r + n_1 \int_{r_0+d}^{\infty} 4\pi r^2 \left(\frac{r_0}{r}\right)^6 \partial r}{\int_{r_0}^{\infty} 4\pi r^2 \left(\frac{r_0}{r}\right)^6 \partial r} = n_2 - \frac{n_2 - n_1}{\left(1 + \left(\frac{d}{r_0}\right)^3\right)} \tag{65}$$

5. SPR sensor parameters

The main performance parameters of SPR sensors for nano application are sensitivity, linearity, resolution, accuracy, reproducibility and limitation of detection.

Sensitivity

Sensor sensitivity is the ratio of the change in the sensor output to the change in the refractive index [1].

$$S = \frac{\partial Y}{\partial n} \tag{66}$$

S, Y and n are the sensitivity, sensor output and refractive index of the probe medium respectively.

Sensor linearity

Sensor linearity is related to linear relationship of output sensor to quantity which should be measured.

Resolution and Reproducibility

The resolution of a sensor is the smallest change in the measured quantity which produces a detectable change in the sensor output. The resolution of SPR sensor usually refers to a bulk refractive index resolution. In SPR biosensor, an equivalent of this term is the limit of detection. Sensor accuracy describes the closeness of agreement between the measured quantity and the true value of the quantity such as refractive index or concentration of the analyte. It is usually expressed in absolute terms or as a percentage of the error/output ratio.

Reproducibility is the ability of a sensor to provide the same output when measuring the same value of the quantity measurement such as concentration or refractive index under the same operating condition over a period of time. It is expressed as the percentage of full range. The limitation of detection is an important parameter for evaluation of sensor. It is related to minimum detection of concentration of analyte. The minimum variation of quantity which measured with the sensor is related to minimum concentration value which is expressed the limitation detection.

6. Simulation of SPR Curve for various refractive index

Figure 6 shows the simulation of SPR curve for variation of the refractive index of analyte. The refractive index of prism and gold layer are 1.83 and .235+3.31i respectively. The wavelength of incident light was 632.8 nm (He-Ne Laser). When the incident angle was changed from 30º to 90º the refractive index of analyte shifted from 1.1 up to 1.6. Figure 7 shows the resonance angle shifted from 39.76º to 80.24º. Moreover, in accordance with Figures 8 and 9, the changes in refractive index are 0.01 and 0.001 respectively. As a result, the resonance angle shifts for small variation of refractive index of analyte are 10.944 and 0.08 respectively [61].

Figure 10 shows the variation of reflectivity with refractive index of analyte. Consequently, the reflectivity increased with increasing refractive index of analyte.

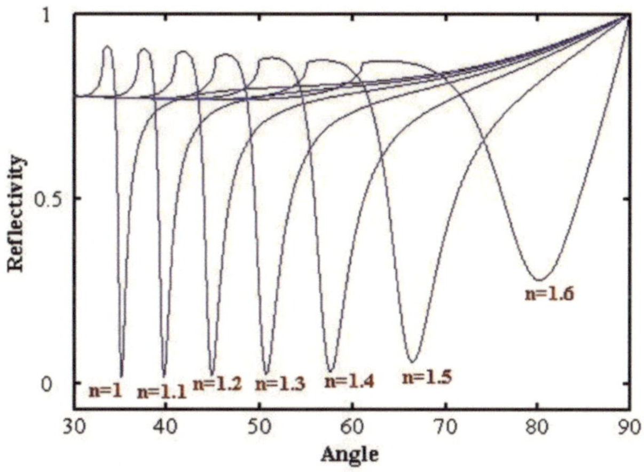

Figure 6. Simulation of SPR curve for various refractive index of analyte from 1.1 to 1.6[61]

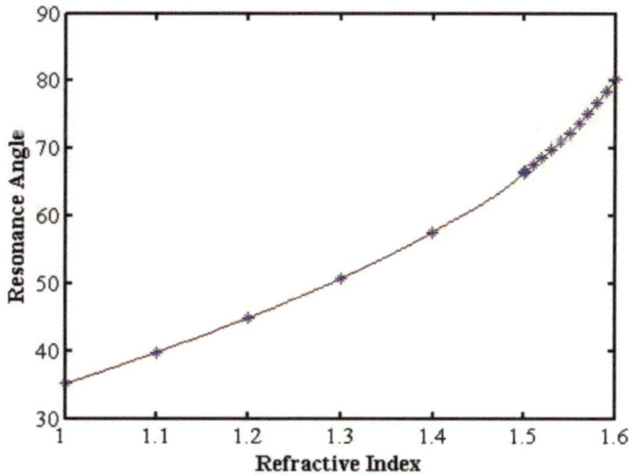

Figure 7. The variation of resonance angle with refractive index of analyte[61]

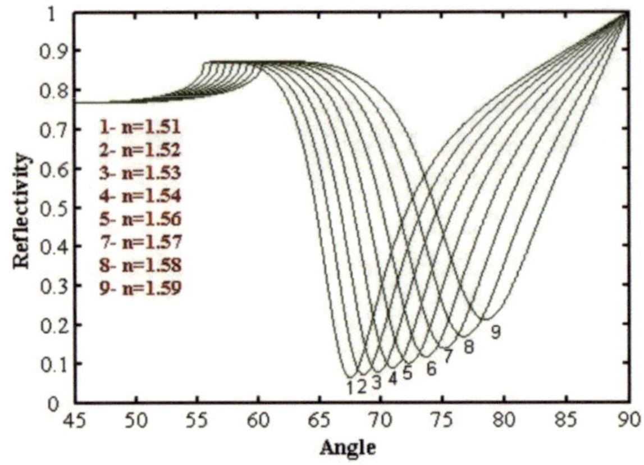

Figure 8. Simulation of SPR curve for various refractive index of analyte from 1.51 to 1.59[60]

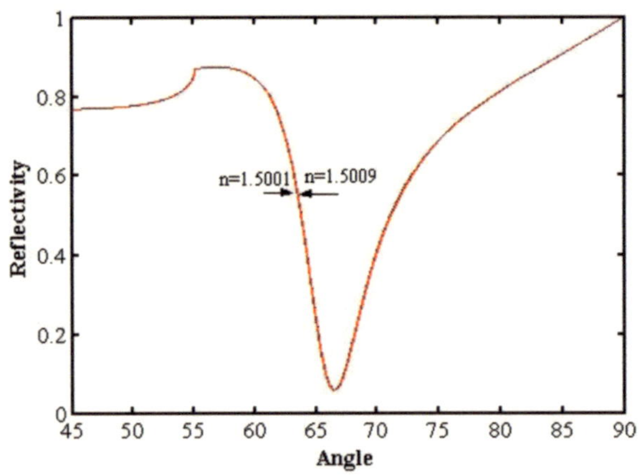

Figure 9. Simulation of SPR curve for variation of refractive index of analyte for 1.5001 and 1.5009[61]

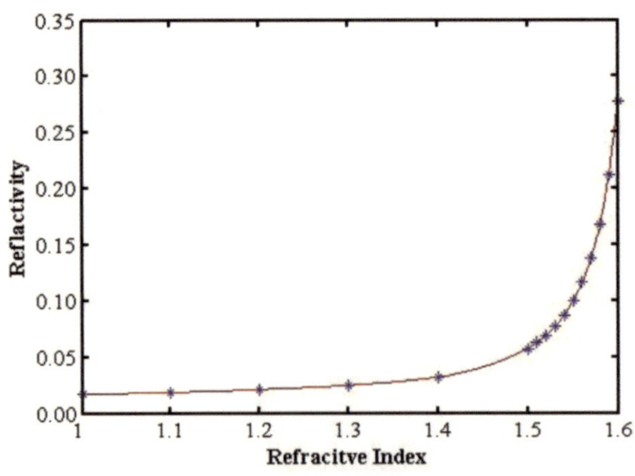

Figure 10. Reflectivity with refractive index of analyte[61]

If the nano-fluid is considered, the refractive index of analyte has an imaginary part which is changed from 0.001 to 0.2, and the resonance angle and reflectivity are shifted from 52.34º to 54.628º and .0252 to 0.6587 respectively, that are depicted in figure 11.

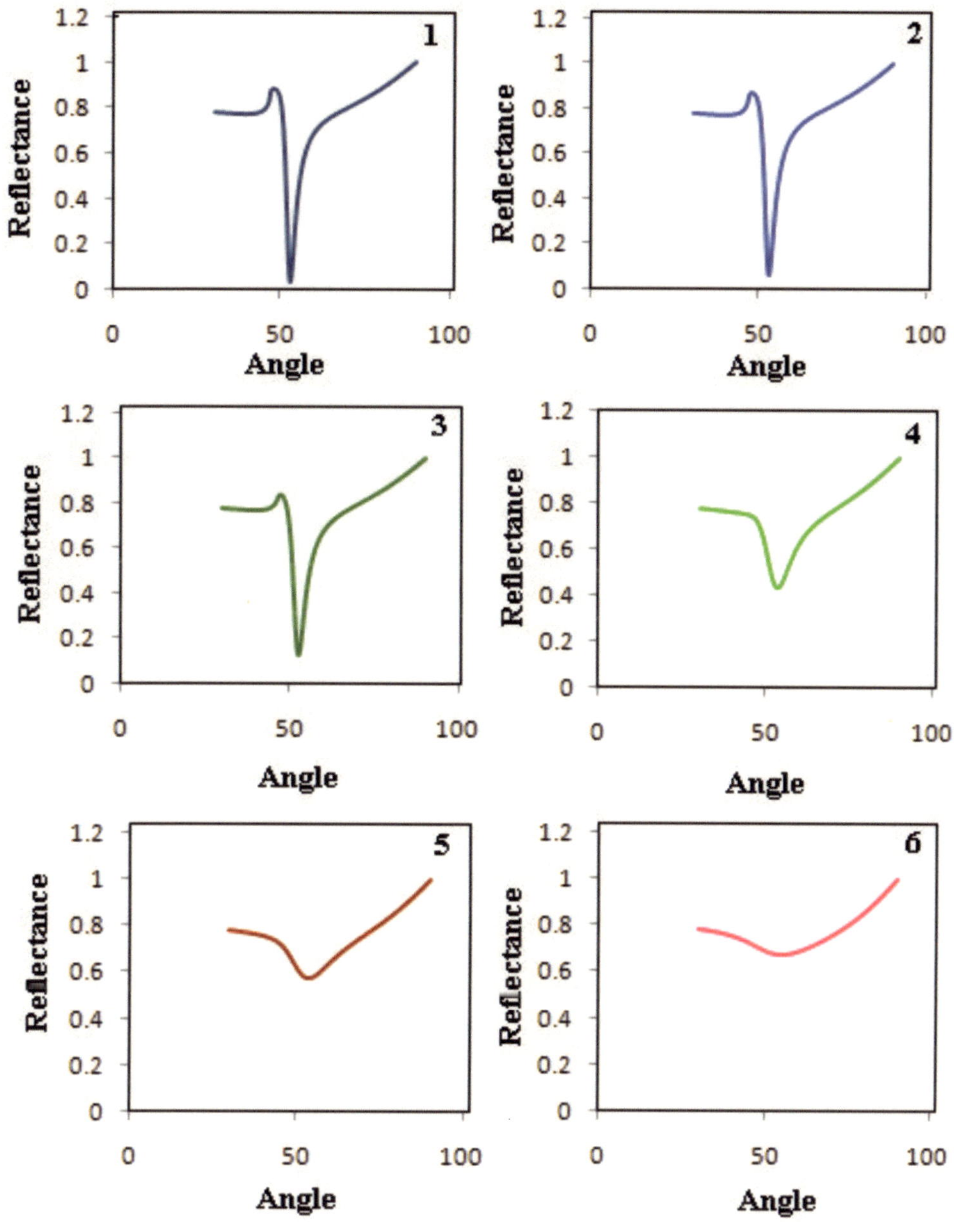

Figure 11. SPR signals related to simulation of nano fluid

The pertinent parameters were sorted in Table 1. Figure 12a and 12b show the variation of resonance angle and reflectivity versus the imaginary part.

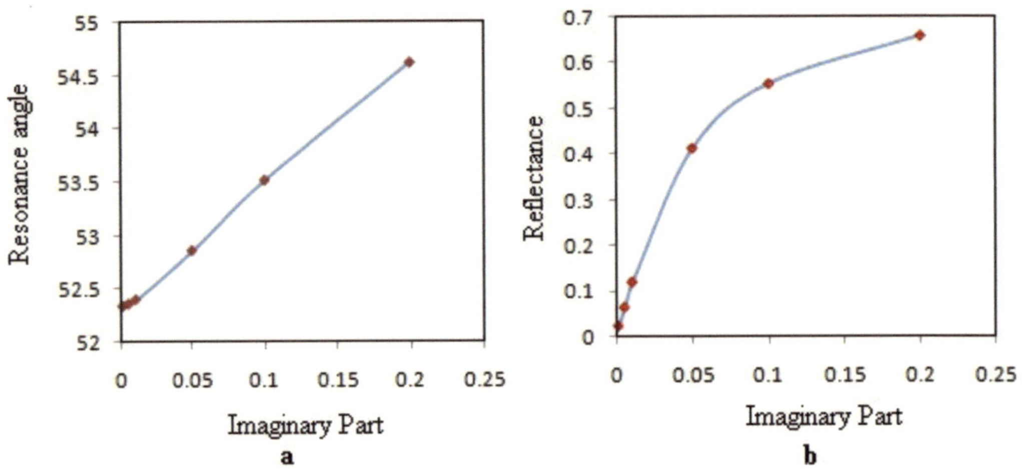

Figure 12. Variation of resonance angle and reflectance versus the imaginary part of liquid

Refractive index	Resonance angle	Reflectance
1.331+0.001i	52.34	0.0252
1.331+0.005i	52.364	0.0663
1.331+0.01i	52.405	0.1221
1.331+0.05i	52.864	0.4140
1.331+0.1i	53.523	0.5546
1.331+0.2i	54.628	0.6587

Table 1. Pertinent parameters for simulation of nanofluid

7. Application of SPR nano-metal

One of the most attractive aspects of collective excitations of Plasmon is their use to concentrate light in sub wavelength structures and enhance transmission through periodic arrays of sub wavelength holes in optically thick metal films. SPs are tightly bound to metal–dielectric interfaces penetrating around 10 nm into the metal (the so-called skin depth) and typically more than 100 nm into the dielectric (depending on the wavelength). SPs at the optical wavelengths concentrate light in a region that is considerably smaller than their wavelength, a feature that suggests the possibility of using surface – plasmon polaritons for the fabrication of nanoscale photonic circuits operating at optical frequencies. In addition the possibility of manufacturing such nano beds in planar form makes them remarkably important for biosensing towards biochip application.

Aslan et. al., reported on Nanogold-plasmon-resonance-based glucose sensing in 2004 [62]. Their sensor based on the aggregation and disassociation of 20-nm gold particles. The plasmon absorption was shifted due to interaction of nanoparticle and glucose and the response range is $\mu M \rightarrow mM$ glucose levels in many biological sample.

Application of Au nanocluster-embedded film was reported by Hu, et. al. in 2004 [6]. The used the mentioned dielectric film (SiO_2) in biosensor and prepared a novel ultrahigh-resolution surface plasmon resonance biosensor. The limitation of sensor was reported 0.1 pg/mm^2 and this sensor can detect the interaction of small molecules in low concentrations. The sensitivity of sensor depend on the size and volume fraction of the embedded Au nanoclusters [6].

Another application of gold nanoparticle in SPR sensor was reported by Chau et al in 2006 [63]. A novel localized surface plasmon resonance based on fiber-optic was used for sensing Ni^{2+} ion and label-free detection of streptavidin and staphylococcal enterotoxin B at the picomolar level using self-assembled gold nanoparticles [63].

In order to develop the fully integrated portable surface plasmon resonance (SPR) system for detection of explosives, the amplification strategy of SPR signal was investigated. Indirect competitive inhibition method allowed the middle-sized SPR sensor to detect trinitrotoluene (TNT) at ppt level [64]. However, this enhanced SPR signal was not high enough to detect TNT at ppt level by a miniaturized SPR sensor. Therefore, localized surface plasmon resonance (LSPR) effect using Au nanoparticle as further signal amplification approach was used. The amplification method of indirect competitive inhibition and LSPR were combined together for fabrication of the immunosurface using Au nanoparticle. TNT detectable range of this immune surface was from 10 ppt (10 pg/ml) to 100 ppb (100 ng/ml), which was almost comparable to that without Au nanoparticle. The observed resonance angle change due to binding monoclonal TNT antibody (M-TNT Ab) with the immune surface modified with Au nanoparticlewas amplified to four times higher than that in absence of Au nanoparticle [64].

LSPR sensors and silver nanoparticle were applied with various thiolate self-assembled monolayers (SAM) to provide chemical selectivity for detection of volatile organic compounds (VOCs) [65]. LSPR spectrums of silver nanoparticles were shifted due to binding and for detection should be measured as the response signal. The limitation of sensor detection was as low as 18–30ppm for heptanone, depending on the surface modification of Ag nanoparticles [65].

Moreover for colorimetric hydrogen peroxide, the LSPR of poly(vinylalcohol) capped silver nanoparticles was used [66]. The silver nanoparticles are directly synthesized in the PVA matrix treatment in aqueous medium. No other reagent is used. . The silver nanoparticles have the catalytic ability for the decomposition of hydrogen peroxide; then the decomposition of hydrogen peroxide induces the degradation of silver nanoparticles. Hence, a remarkable change in the localized surface plasmon resonance absorbance strength could be observed. As a result, the yellow color of the silver nanoparticle–polymer solution was gradually changed to transparent colour. Furthermore, when this transparent solution was subjected to thermal treatment, it became again yellow and the UV–vis spectroscopy confirmed that nanoparticles were again formed, suggesting the renewability of this sensor. The determination of reactive oxygen species such as hydrogen peroxide has possibilities for applying to medical and environmental applications [66].

Salmonella was detected using LSPR and Au nanoparticles [67]. In this case, the small contact area between the nanoparticle and the bacteria was the cause of the short range interaction of the local electric field and bacteria. Hence, the SPR signals were shifted to higher value that can be explained by the Mie theory and effective medium theory. So, this is alternative methods for detection of Salmonella [67].

Author details

Amir Reza Sadrolhosseini
Center of Excellence for Wireless and Photonics Networks (WiPNet), Faculty of Engineering Universiti Putra Malaysia, UPM Serdang, Malaysia

A. S. M. Noor
Center of Excellence for Wireless and Photonics Networks (WiPNet), Faculty of Engineering Universiti Putra Malaysia, UPM Serdang, Malaysia
Department of Computer and Communication Systems Engineering, Faculty of Engineering, Universiti Putra Malaysia, UPM Serdang, Malaysia

Mohd. Maarof Moksin
Department of Physics, Faculty of Science, Universiti Putra Malaysia, Serdang, Malaysia

8. References

[1] Homola J. Surface plasmon resonance based sensors. Verlag Berlin Heidelberg: Springer; 2006.

[2] Peyghambarian N, Koch S W, Mysyrowicz A. Introduction to Semiconductor Optics. New Jersey: Prentice Hall; 1993.

[3] Liedberg, B., Nylander, C., Lundstrom, I. 1983. Surface plasma resonance for gas detection and biosensing. Sensors and Actuators B 1983; 4 299.

[4] Stenberg E, Persson B, Roos H, Urbaniczky, C. Quantitative determination of surface concentration of proteins with surface plasmon resonance using radiolabeled protein. J. Colloid. Interf. Sci. 1991; 143 513–526.

[5] Homola J, Yee S, Gauglitz G. Surface plasmon resonance sensors: review. Sens. Actuat. B 1999; 54 3–15.

[6] Hu W, Chenb S-J, Huangc K-T, Hsud J H, Chend W Y, Changa G L, Lai K-A. A novel ultrahigh-resolution surface plasmon resonance biosensor with an Au nanocluster-mbedded dielectric film. Biosensors and Bioelectronics 2004;19 1465–1471.

[7] Zhu T, Zhang X, Wang J, Fu X, Liu Z . Assembling colloidal Au nanoparticles with functionalized self-assembled monolayers. Thin Solid Films 1998; 327–329 595–598.

[8] Gradar K C, Brown K R, Keating C D, Stranick S J, Tang S-L, Natan M J. Nanoscale characterization of gold colloid monolayers: a comparison of four techniques. Anal. Chem. 1997;69 471–477.

[9] Brown K R, Lyon L A, Fox A P, Reiss B D, Natan M J. Hydroxylamine seeding of colloidal Au nanoparticles. Part 3. Controlled formation of conductive Au films. Chem. Mater. 2000; 12 314–323.

[10] Lyon L A, Musick M D, Natan M J. Colloidal Au-enhanced surface plasmon resonance immunosensing. Anal. Chem. 1998; 70 5177–5183.

[11] He L, Musick M D, Nicewarner S R, Salinas F G, Benkovic S J, Natan M J, Keating C D. Colloidal Au-enhanced surface Plasmon resonance for ultrasensitive detection of DNA hybridization. J. Am. Chem. Soc. 2000; 122 9071–9077.

[12] Gestwicki J E, Hsieh H V, Pitner J B. Using receptor conformational change to detect low molecular weight analytes by surface plasmon resonance. Anal. Chem. 2001; 73 5732–5737.

[13] Sun Y, Xia Y. Increased sensitivity of surface plasmon resonance of gold nanoshells compared to that of gold solid colloids in response to environmental changes. Anal. Chem. 2002; 74 5297–5305.

[14] Mayer C, Stich N, Schalkhammer T, Bauer G. Slide-format proteomic biochips based on surface-enhanced nanocluster-resonance. Fresenius J. Anal. Chem. 2001; 371 238–245.

[15] Kouzani A Z, Dai X J, Michalski W P. Investigation of the effect of design parameters on sensitivity of surface plasmon resonance biosensors. Biomedical Signal Processing and Control - Biomed Signal Process Control 2011 6(2) 147-156, 2011.

[16] Wang J. Carbon-nanotube based electrochemical biosensors: a review. Electroanalysis 2005; 17 7–14.

[17] Balasubramanian K, Burghard M. Biosensors based on carbon nanotubes, Analytical and Bioanalytical Chemistry 2006; 385 452–468.

[18] Liu Y, Koep E, Liu M. A highly sensitive and fast-responding SnO2 sensor fabricated by combustion chemical vapor deposition. Chemistry of Materials 2005; 17 3997–4000.

[19] Choi A, Kim K, Jung H-I, Lee S Y. ZnO nanowire biosensors for detection of biomolecular interactions in enhancement mode. Sensors and Actuators B: Chemical 2010; 148 577–582.

[20] Zalevsky Z, Abdulhalim I. Integrated nanophotonics devices Elsevier Science 2010.

[21] Willets K A, Van Duyne R P. Localized Surface Plasmon Resonance Spectroscopy and Sensing. Annu. Rev. Phys. Chem. 2007; 58:267–97.

[22] Kittle C. Solid State Physics New York: Wiley; 1996.

[23] Sosa I O, Noguez C, Barrera R G. Optical Properties of Metal Nanoparticles with Arbitrary Shapes. J. Phys. Chem. B 2003; 107 6269-6275.

[24] Kelly K L, Lazarides A A, Schatz G C. Computational electromagnetics of metal nanoparticles and their aggregates. Computing in Science and Engineering, 2001; 3(4) 67.

[25] Mishchenko M I, Hovenier J W, Travis L D. Light Scattering by Nonspherical Particles. San Diego: Academic Press; 2000.

[26] Toma H E, Zamarion V M, Toma S H, Araki K. The coordination chemistry at gold nanoparticles J. Braz. Chem. Soc. 2010; 21(7).

[27] Draine B T. Astrophys. J. 1998; 333 848.
[28] Draine B T, Flatau P J. Program DDSCAT; University of California at San Diego: San Diego, CA.
[29] Johnson P B, Christy, R. W. Phys. ReV. B 1972; 6 4370.
[30] Yang W-H, Schatz G C, van Duyne R P. Discrete dipole approximation for small particles with arbitrary shapes. J. Chem. Phys. 1995; 103 869.
[31] Jensen T R, Duval M L, Kelly K L, Lazarides A A, Schatz G C, Van Duyne, R P. Nanosphere lithography: Effect of the external dielectric medium on the surface plasmon resonance spectrum of a periodic array of silver nanoparticles, Journal of Physical Chemistry B, 1999; 103(45) 9846 – 9853.
[32] Noguez C, Sosa I, Barrera R G. Optical characterization of isolated nanoparticles with arbitrary shapes MRS Proc. 2002; 704 275.
[33] Noguez C, Sosa I, Barrera R G. MRS Proc. 2002, 704, 275; Singh, R. K., Partch, R., Muhammed, M., Senna, M., Hofmann, H., Eds.
[34] Fontana E. Thickness Optimization of Metal Films for the Development of Surface Plasmon Based Sensors for non-absorbing media. Journal of Applied optics 2006.
[35] Schasfoort, R. and Tudos, A. 2008, Handbook of Surface Plasmon Resonance. RSC Publishing, Cambridge, U.K.
[36] Sharma K. 2006. Optics. Academic Press, California, 301-316.
[37] Beketov G V, Shirshov Y M, Shynkarenko O V, Chegel V I. Surface plasmon resonance spectroscopy: prospects of superstrate refractive index variation for separate extraction of molecular layer parameters. Sensors and Actuators B. 1998; 48 432.
[38] Sadrolhosseini A R, Moksin M M, Yunus W M M, Talib Z A, Abdi M M. Surface Plasmon Resonance Detection of Copper Corrosion of Biodiesel Using Polypyrrole-Chitosan Layer Sensor. Optical review 2011 18(4) 331-337.
[39] Yuan X-C, Ong B, Tan Y, Zhang D, Irawan R, Tjin S. Sensitivity-stability-optimized surface plasmon resonance sensing with double metal layers. Journal of optics 2006; 959 963-8.
[40] Stewart ME, Anderton CR, Thompson LB, Maria J, Gray SK, Rogers JA . Nanostructured plasmonic sensors. Chem Rev 2008;108: 494–521.
[41] Xu H, Kall M. Modeling the optical response of nanoparticle- based surface Plasmon resonance sensors. Sensors and Actuators B 2002; 87 244-249.
[42] Mie G. Beitrage zer optik truber meiden speziell kolloidaler metallosungen. Ann. Phys. (Leipzig) 1908; 25 377–445.
[43] Gupta S, Huda S, Kilpatrick PK, Velev OD. Characterization and optimization of gold nanoparticle-based silver-enhanced immu- noassays. Anal Chem 2007; 79 3810–20.
[44] Baptista P, Pereira E, Eaton P, Doria G, Miranda A, Gomes I. Gold nanoparticles for the development of clinical diagnosis methods. Anal Bioanal Chem 2008; 391 943–50.
[45] Luo PG, Stutzenberger FJ. Nanotechnology in the detection and control of microorganisms. Adv Appl Microbiol 2008;63 145–81.

[46] Bruzzone S, Arrighini GP, Guidotti C. Some spectroscopic proper- ties of gold nanorods according to a schematic quantum model founded on the dielectric behavior of the electron-gas confined in a box. Chem Phys 2003; 291 125–40.
[47] Prodan E, Radloff C, Halas NJ, Nordlander P. A hybridization model for the plasmon response of complex nanostructures. Science 2003; 302 419–22.
[48] Wang H, Brandl DW, Nordlander P, Halas NJ. Plasmonic nanos- tructures: artificial molecules. Acc Chem Res 2007; 40 53–62.
[49] Lorentz H A. -ber die beziehungzwischen der fortpflanzungsgeschwindigkeit des lichtes der körperdichte. Ann. Phys. 1880; 9 641–665.
[50] Sutherland R L. Handbook of Nonlinear Optics New York: Marcel Dekker; 1996.
[51] Hashin Z, Shtrikman S. A variational approach to the theory of the effective magnetic permeability of multiphase materials. J. Appl. Phys. 1963; 33 3125–3131.
[52] Milton G W. Bounds on the complex dielectric constant of a composite material. Appl. Phys. Lett. 1980; 37 300–302.
[53] Bergman D J. Exactly solvable microscopic geometries and rigorous bounds for the complex dielectric constant of a two-component composite material. Phys. Rev. Lett. 1980; 44 1285–1287.
[54] Aspnes D E. Bounds to average internal fields in two-component composites. Phys. Rev. Lett. 1982; 48 1629–1632.
[55] Hale D K. The physical properties of composite materials. J. Mater. Sci. 1976; 11 2105–2141.
[56] Sipe J E, Boyd R W. Nonlinear susceptibility of composite optical materials in the Maxwell Garnett model. Phys. Rev. A 1992; 46 1614–1629.
[57] Zeng X C, Bergman D J, Hui P M, Stroud D. Effective-medium theory for weakly nonlinear composites. Phys. Rev. B 1988; 38 10970–10973.
[58] Cohen R W, Cody G D, Coutts M D, Abeles B. Optical properties of granular silver and gold films. Phys. Rev. B 1973; 8 3689–3701.
[59] Lagarkov A N, Sarychev A K. Electromagnetic properties of composites containing elongated conducting inclusions. Phys. Rev. B 1996; 53 6318–6336.
[60] Grosse P, Offermann V. Analysis of reflectance data using the Kramers–Kronig relations. Appl. Phys. A 1991; 52 138–144.
[61] Sadrolhosseini A R. Surface Plasmon Resonance Characterization Of Biodiesel. PhD thesis. Universiti Putra Malaysia; 2011.
[62] Kadir Aslan K, Lakowicz J R, Geddes C D. Nanogold-Plasmon-Resonance-based glucose sensing. Analytical Biochemistry 2004; 330 145-155.
[63] Chau L K, Lin Y F, Cheng S F, Lin T J. Fiber-optic chemical and biochemical probes based on localized surface plasmon resonance. Sensors and Actuators B 2006; 113 100–105.
[64] Kawaguchi T, Shankaran D R, Kim S J, Matsumoto K, Toko K, Miura N. Surface plasmon resonance immune sensor using Au nanoparticle for detection of TNT. Sensors and Actuators B 2008; 133 467–472.

[65] Chena Y Q, Lub C J. Surface modification on silver nanoparticles for enhancing vapor selectivity of localized surface plasmon resonance sensors. Sensors and Actuators B 2009; 135 492–498.

[66] Filippo E, Serra A, Manno D. Poly(vinyl alcohol) capped silver nanoparticles as localized surface Plasmon resonance-based hydrogen peroxide sensor. Sensors and Actuators B 2009 138 625–630.

[67] Fua J, Park B, Zhaoa Y. Limitation of a localized surface plasmon resonance sensor for Salmonella detection. Sensors and Actuators B 2009; 141 276–283.

Chapter 11

Localized Surface Plasmon Resonances: Noble Metal Nanoparticle Interaction with Rare-Earth Ions

V.A.G. Rivera, F.A. Ferri and E. Marega Jr.

Additional information is available at the end of the chapter

http://dx.doi.org/10.5772/50753

1. Introduction

Particles of sizes between 1 and 100 nm show fascinating properties with unusual characteristics that lead to the formation of unique properties in nanosystems, which are not observed in ordinary materials. These are considered hereby as nanoparticles (NPs). Additionally, metallic NPs with sizes smaller than the wavelength of light show strong dipolar excitations in the form of localized surface plasmon resonances (LSPR). LSPRs are non-propagating excitations of the conduction electrons of metallic NPs coupled to the electromagnetic field [1]. This effect has been the subject of extensive research, both fundamental and with a view to applications [2,3]. The resonance frequency of the oscillation, i.e., the surface plasmon (SP) energy, it is essentially determined by the dielectric properties of the metal and the surrounding medium, and by the particle size and shape. The collective charge oscillation causes a large resonant enhancement of the local field inside aand near the NP. This field enhancement is used in surface-enhanced Raman scattering (SERS) [4] and is currently discussed for potential applications in nonlinear optical devices [5], in optical tweezers [6], and generally for the manipulation of the local photonic density of states.

Rare-earth (RE) elements are a group of chemical elements known as Lanthanides that occur together in the periodic table. These elements are used in common consumer goods such as: computer memory, DVD's, rechargeable batteries, cell phones, car catalytic converters, magnets, fluorescent lighting and much more. Furthermore, play an essential role in modern national defense, e.g.: lanthanum in night-vision goggles, neodymium in laser range-finders, guidance systems, communications; europium in fluorescents and phosphors in lamps and monitors; erbium amplifiers in fiber-optics data transmission; samarium in permanent magnets stable at high temperatures and, others technological applications. We center our attention in the RE ions and their optical properties that are widely employed in photonic. Devices of general interest span RE ions concentrations of tens to several thousand parts per million (ppm), resulting in devices of one to tens of meters long such as channel waveguides and optical fiber, respectively. In optical devices, the RE should ideally be confined as a delta function in the center of the core for maximum gain per unit pump power. Thereby, there is a necessary tradeoff between the confinement and the RE ions concentration, such that the more confined structures require a higher rare-earth ions concentration. An important feature that distinguishes them from other optically active ions: they emit and absorb over narrow wavelength ranges, the wavelengths of the emission and absorption transitions depend the host material, the intensities of these transitions are weak, the lifetimes of metastable states are long, and the quantum efficiencies tend to be high, except in aqueous solutions. These properties lead to excellent performance of RE ions in many optical applications. Devices that provide gain (e.g. lasers and amplifiers), must have low scattering losses, and one is restricted to using single-crystal or glass hosts. Whereas in many applications crystalline materials are preferred for reasons that include higher peak cross sections or better thermal conductivities, the versatility of glasses and the broader emission and absorption spectra they provide have led to the use of RE doped glasses in many applications, eventually running into the clustering limit for the particular host glass composition [7].

Most of existing and potential future applications of NPs suffer from damping caused by metal absorption. Sudarkin and Demkovich [8] suggested increasing the propagation length of the surface plasmon polariton[1] (SPP) by creating the population inversion in the dielectric medium adjacent to the metallic surface (film). Recently, gain-assisted propagation of the SPPs at the interface between a metal and a dielectric with optical gain has been analyzed theoretically [9,10]. Thus, NPs embedded in a gain media represent a field enhancement sustained that resonant excitation can lead to a reduction in the threshold for achieving inversion in the optically active surrounding medium, and the presence of gain can counteract the inherent absorption losses in the NP [1]. While this enhancement of plasmon resonances in gain medium not is experimentally confirmed, amplification of fluorescence due to field enhancement in gain medium with metal NPs has been observed [11,12,13,14].

In the present chapter, we explore the physics of localized surface plasmons by first considering the interaction of metal NPs with an electromagnetic wave in order to arrive at the resonance condition (LSPR). After, we show studies of plasmon resonances in particles with different shapes and sizes, and the effects of interactions between particles in the ensemble. Then we will focus on the RE ions, discussing the optical properties of the trivalent ions through the principles of quantum mechanics (in terms of oscillator strengths) with special interest in the interactions between the 4f electrons themselves. Since all the other electronic shells are spherically symmetric. Finally we present diverse experimental results of the interaction of RE ions interaction with NPs, resulting in an enhancement of the

[1] A SPP is an electromagnetic wave propagating along the interface between two media possessing permittivities with opposite signs, such as the metal–dielectric interface.

intensity emission of the RE ions due to long-range electromagnetic interaction between LSPR and the RE ions.

In this context, it is worth mentioning that the main purpose of this chapter is to show some of concrete concept of *noble metallic nanoparticle interaction with rare-earth ions* exhibiting a field enhancement in transparent medium with noble metal NPs embedded via a *localized surface plasmon resonance.*

2. Localized surface plasmon resonance and metallic nanoparticles

First, we know that a surface plasmon resonance (SPR) can be described as the resonance collective oscillation of valence electrons in a solid stimulated by beam of light incident (electromagnetic field – see Figure 1). The resonance condition is established when the frequency of light matches the natural frequency of valence electrons oscillating against from this restoring force. SPR in nanometer-sized structures is called LSPR. For gold and silver NPs, the resonance falls into the visible region of the electromagnetic spectrum. A striking consequence of this is the bright colors exhibited by particles both in transmitted and reflected light, due to resonantly enhanced absorption and scattering. This effect has found applications for many hundreds of years, for example in the staining of glass for windows or ornamental cups[2].

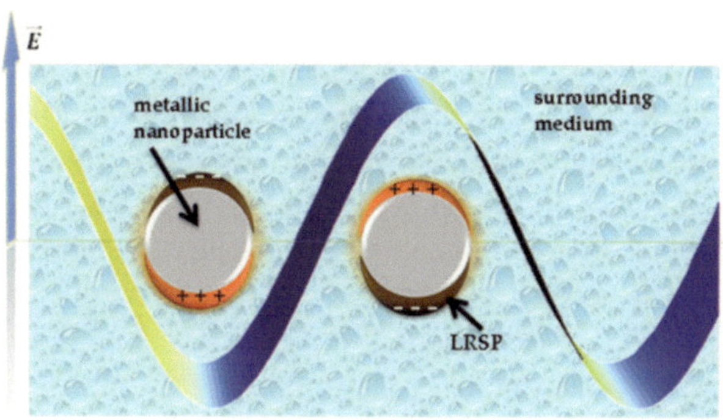

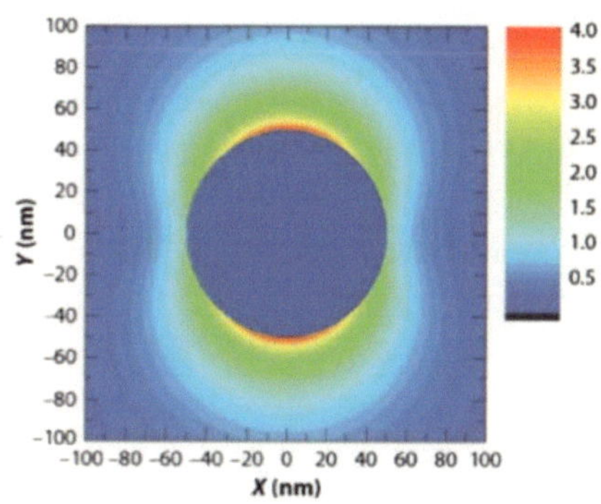

Figure 1. Left: Schematic of the electron charge displacement (valence electrons) in a metallic NP (LSPR) interacting with an incident plane wave, with electric field polarized *E* into a host matrix. Right: The corresponding electric field strength pattern for a 100 nm silver sphere, irradiated at a wavelength of 514 nm, this is reproduced from X. Lu et al. [15].

2.1. Noble metal nanoparticles in an electric field

For spherical NPs smaller than compared to local variations of the involved electromagnetic fields ($d<<\lambda$) in the surrounding medium, the quasi-static approximation [16], we lead to a

[2] Localized surface plasmons have been observed since the Romans who used gold and silver nanoparticles to create colored glass objects such as the Lycurgus Cup (4th Century A.D.). A gold sol in the British museum, created by Michael Faraday in 1857, is still exhibiting its red color due to the plasmon resonance at ,530 nm [L. M. Liz-Marzan, Mater. Today **7**, 26 (2004).].

good agreement between theory and experiment. So, one can calculate the spatial field distribution (assuming as a problem of a particle in an electrostatic field), from the Laplace equation for the potential $\nabla^2\Phi = 0$, therefore the electric field $E = -\nabla\Phi$. Besides, the harmonic time dependence can then be added to the solution once the field distributions are well known. The solution this problem was obtained by Jackson [17]:

$$\Phi_{in} = -\frac{3\varepsilon_{NP}}{\varepsilon_{NP} + 2\varepsilon_m} E_0 r \cos\theta \tag{1}$$

$$\Phi_{out} = -E_0 r \cos\theta + \frac{\varepsilon_{NP} - \varepsilon_m}{\varepsilon_{NP} + 2\varepsilon_m} E_0 \left(\frac{d}{2}\right)^3 \frac{\cos\theta}{r^2} \tag{2}$$

E_0 is the amplitude of the electric field, ε_{NP} and ε_m are the dielectric permittivity of the NP and of the surrounding medium respectively, both are function the excitation frequency ω, r is the module of the position vector. Here Φ_{out} describes the superposition of the applied field and that of a dipole located at the NP center, i.e., an applied field induces dipole moment inside the sphere proportional to $|E_0|$. Additionally, we can arrive at

$$\alpha = 4\pi \left(\frac{d}{2}\right)^3 \frac{\varepsilon_{NP} - \varepsilon_m}{\varepsilon_{NP} + 2\varepsilon_m} \tag{3}$$

Here α is the (complex) polarizability of the NP in the electrostatic approximation. Under the condition that $|\varepsilon_{NP} + 2\varepsilon_m|$ is a minimum, the polarizability shows a resonant enhancement. Thus, we can also write the Frohlich condition $\mathrm{Re}\left[\varepsilon(\omega)_{NP}\right] = -2\varepsilon_m$ that is the associated mode the dipole surface plasmon of the NP (in an oscillating field). Nevertheless, the distribution of the electric field E_{in} inside and E_{out} outside the sphere can be written as:

$$E_{in} = \frac{3\varepsilon_m}{\varepsilon_{NP} + 2\varepsilon_m} E_0 \tag{4}$$

$$E_{out} = E_0 + \frac{3\mathrm{n}(\mathrm{n}\cdot\mathrm{p}) - \mathrm{p}}{4\pi\varepsilon_0\varepsilon_m} \left(\frac{1}{r}\right)^3 \tag{5}$$

Here, $\mathrm{p} = \varepsilon_0\varepsilon_{\mathrm{m}}\alpha E_0$ is the dipole moment. Therefore, a resonance in α implies a resonant improvement of both the internal and dipolar fields, resulting in prominent applications of NPs in optical devices. Now consider a plane wave incident with $E(r,t) = E_0 e^{-i\omega t}$, this induce on NP an oscillating dipole moment $\mathrm{p}(t) = \varepsilon_0\varepsilon_{\mathrm{m}}\alpha E_0 e^{-i\omega t}$, i.e., we have a scattering of the plane wave by the NP. In this sense, the electromagnetic fields associated with an electric dipole in the near, intermediate and radiation zones are [17]:

$$H = \frac{ck^2}{4\pi}(\mathrm{n}\times\mathrm{p})\frac{e^{ikr}}{r}\left(1 - \frac{1}{ikr}\right) \tag{6}$$

$$E = \frac{1}{4\pi\varepsilon_0\varepsilon_m}\left\{k^2(\mathrm{n}\times\mathrm{p})\times\mathrm{n}\frac{e^{ikr}}{r} + (3\mathrm{n}(\mathrm{n}\cdot\mathrm{p})-\mathrm{p})\left(\frac{1}{r^3}-\frac{ik}{r^2}\right)e^{ikr}\right\} \tag{7}$$

With $k = 2\pi/\lambda$ and n is the unit vector in the direction of the point P of interest. For *kr*<<1 (near zone) we have the electrostatic result of (5). The magnetic field present has the form $H = \frac{i\omega}{4\pi}\frac{(\mathrm{n}\times\mathrm{p})}{r^2}$. Hence, in the near field the fields are predominantly electric and for static field $(kr \to 0)$, the magnetic field vanishes. For *kr*>>1, the dipole fields have spherical-wave form: $H = \frac{ck^2}{4\pi}\frac{(\mathrm{n}\times\mathrm{p})e^{ikr}}{r}$ and $E = \sqrt{\frac{\mu_0}{\varepsilon_0\varepsilon_m}}H\times\mathrm{n}$.

From the viewpoint of optics, it is much interesting to note that another consequence of the enhanced polarization α in which a NP scatters and absorbs light [18]. For a sphere of volume *V* and dielectric function $\varepsilon_{NP} = \varepsilon_1 + i\varepsilon_2$ in the quasi-static limit, the extinction cross section $C_{ext} = C_{abs} + C_{sca}$ is:

$$C_{ext} = 9\frac{\omega}{c}\varepsilon_m^{3/2}\frac{\varepsilon_2}{(\varepsilon_1 + 2\varepsilon_m)^2 + \varepsilon_2^2} \tag{8}$$

2.2. Mie theory

For particles with larger dimensions, where the quasi-static approximation is not justified due to significant phase-changes of the driving field over the particle volume, a rigorous electrodynamics approach is required. This way, Gustav Mie solved Maxwell's equations for the case of an incoming plane interacting with a spherical particle [19]. In essence, the electromagnetic fields are expanded in multipole contributions and the expansion coefficients are found by applying the correct boundary conditions for electromagnetic fields at the interface between the metallic NP and its surrounding.

The extinction cross section of a spherical NP is given by the following expression:

$$\sigma_{ext} = \frac{\lambda^2}{2\pi}\sum_{n=0}^{\infty}(2n+1)\mathrm{Re}\{a_n + b_n\} \tag{9}$$

Here the parameters a_n and b_n are defined as:

$$a_n = \frac{\Psi_n(\beta)\Psi'_n(m\beta) - m\Psi_n(m\beta)\Psi'_n(\beta)}{\xi_n(\beta)\Psi'_n(m\beta) - m\Psi_n(m\beta)\xi'_n(\beta)} \tag{10}$$

$$b_n = \frac{m\Psi_n(\beta)\Psi'_n(m\beta) - \Psi_n(m\beta)\Psi'_n(\beta)}{m\xi_n(\beta)\Psi'_n(m\beta) - \Psi_n(m\beta)\xi'_n(\beta)} \tag{11}$$

The size parameter β is defined as $\beta = \dfrac{\pi d m_0}{\lambda_0}$, where λ_0 is the incident wavelength with respect to vacuum, and m_0 represents the refractive index of the surrounding medium. The Ricatti-Bessel functions Ψ and ξ are defined in terms of the half-integer-order Bessel function of the first kind ($J_{n+1/2}(z)$), $\Psi_n(x) = \left(\dfrac{rx}{2}\right)^{0.5} J_{n+1/2}(x)$ and $\xi_n(x) = \left(\dfrac{rx}{2}\right)^{0.5} H_{n+1/2}(x)$, $H_{n+1/2}(x)$ is the half-integer-order Hankel function of the second kind.

We will focus our attention in silver and gold, since the localized plasmon resonance condition mentioned above is satisfied at visible light frequencies. Additional advantages of these metal NPs include simple preparation methods for a wide range of sizes and shapes and easy surface conjugation to a variety of ligands.

We now evaluate the extinction cross section using the Mie theory from equation (9), Figure 2, we take dielectric constants for silver and gold from Palik [20], and the medium dielectric constant is assumed to be 1.0 (i.e., a particle in a vacuum) and 2.0 for NPs with sizes different.

We can see from Figure 2 the dependence of resonance frequency with size of NPs (silver and gold) and the refraction index (surrounding medium). K. Lance Kelly et al. (2003), show that for spherical NPs clear differences between the quasistatic and the Mie theory results. However, the important features are retained, e.g. the frequency resonance. Although Mie theory is not a very expensive calculation, the quasistatic expressions are convenient to use when only qualitative information is needed [21].

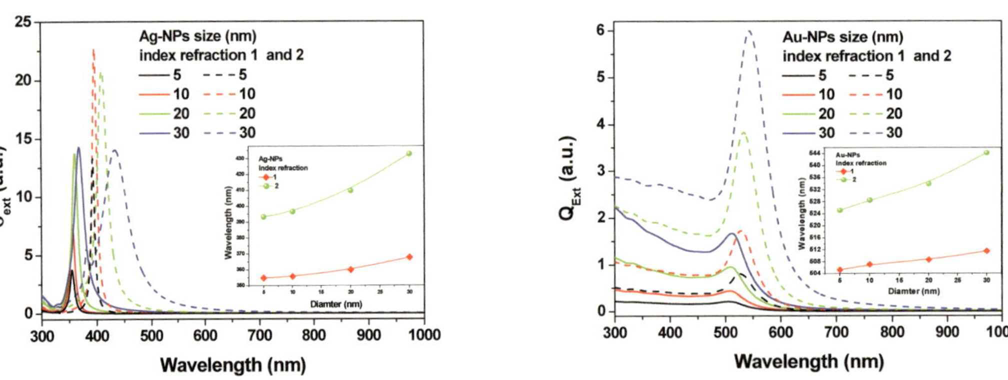

(a) (b)

Figure 2. Extinction cross section from Mie theory for (a) silver and (b) gold NPs. Both in function of size with refraction index 1 (line curves) and 2 (dot lines). Inset figure shows a red-shift with the increment of size NPs and refraction index.

Now, Figure 3 shows clearly the dependence of resonance frequency embedded in a glass (with refraction index of: 1.99 for 400 nm, 1.88 for 630 nm, 1.75 for 900 nm and 1.70 for 1000 nm) and in a bulk with refraction index 2, in both cases with 20 nm size and a size distribution of 20 % STD, this simulation was extracted from program MiePlot v4.2.11 of Philip Laven. Also, it is possible to observe an increment of bandwidth and the intensity of peak for NPs embedded in the glass when compared with the material bulk. Therefore, LSPR results in enhanced local electromagnetic fields near the surface of the NP (Novotny & Hecht, 2006) [22], see Figure 1.

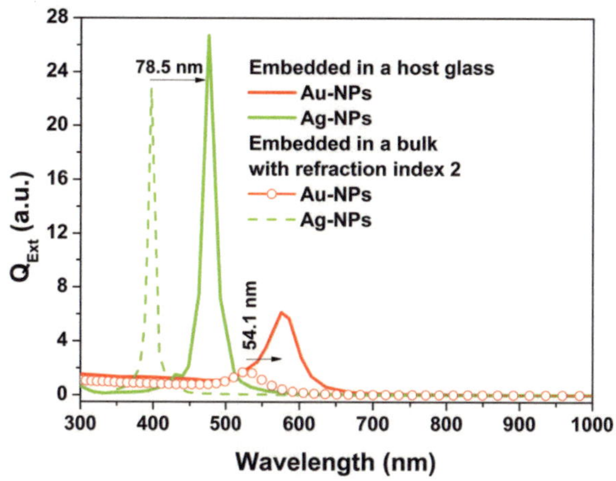

Figure 3. Extinction cross section from Mie theory via the program MiePlot v4.2.11 of Philip Laven, for silver and gold NPs with a size distribution of 20 % STD of 20 nm size embedded in a glass with refraction index in function of the wavelength and a bulk with refraction index 2 constant. In both cases, we observed a red-shift (see arrow). Inset figure shows the size distribution of NPs for 50 NPs in both cases.

Hideki Nabika and Shigehito Deki [23], show an experimental evidence of this dependence above mentioned for silver NPs of different sizes were synthesized by reducing $AgNO_3$ with *N,N*-dimethylformamide (DMF), Poly(vinylpyrrolidone) (PVP) aqueous solution (10 mL, 10 wt %) was added to DMF (80 mL), followed by an addition of $AgNO_3$ aqueous solution (10 mL). They obtained three sets of spherical silver NPs with a particle diameter varying from 9.7 to 27.1 nm and rod-shaped silver NPs with an aspect ratio of 1.79, its results are show Figure 4.

Thereby, the LSPR properties are sensitive to its environment in the order of its shape and size [24], the dielectric function of the glass host containing the NPs, changes the LSPR bandwidth and resonance frequency, see Figure 2-4.

We can be written as the sum of the bound and quasi-free (conduction) electron contributions [25]: $\varepsilon(\omega)=\varepsilon^{ib}(\omega)-\omega_p^{\,2}/\omega[\omega+i\gamma]$, γ is the damping the resonance, ω the excitation frequency, ω_p the plasma frequency. The bound electron contribution ε^{ib} remains unchanged [26]. A similar expression can be used for the contribution of conduction electrons [27]: $\gamma=1/\tau^{NP}=1/\tau_0+2g_sV_F/d$. Where the first term, $1/\tau_0$ is associated to bulklike electron scattering process in the particle and the second term is a consequence of quasi-electron-free interaction with the surface and, for a sphere, V_F is the Fermi velocity, and g_s is the surface factor [25]. Nevertheless, the LSPR dependence on the matrix refractive index (n_λ) also can be calculated by [12]: $\omega_p = \sqrt{4\pi n' e^2 / \varepsilon_0 \varepsilon_d(\omega) m^*}$, where n' is the electrons density, e the electron charge, $\left(\varepsilon_d(\omega) = n_\lambda^{\,2}\right)$ the dielectric permittivity, *and* m^* the electron mass.

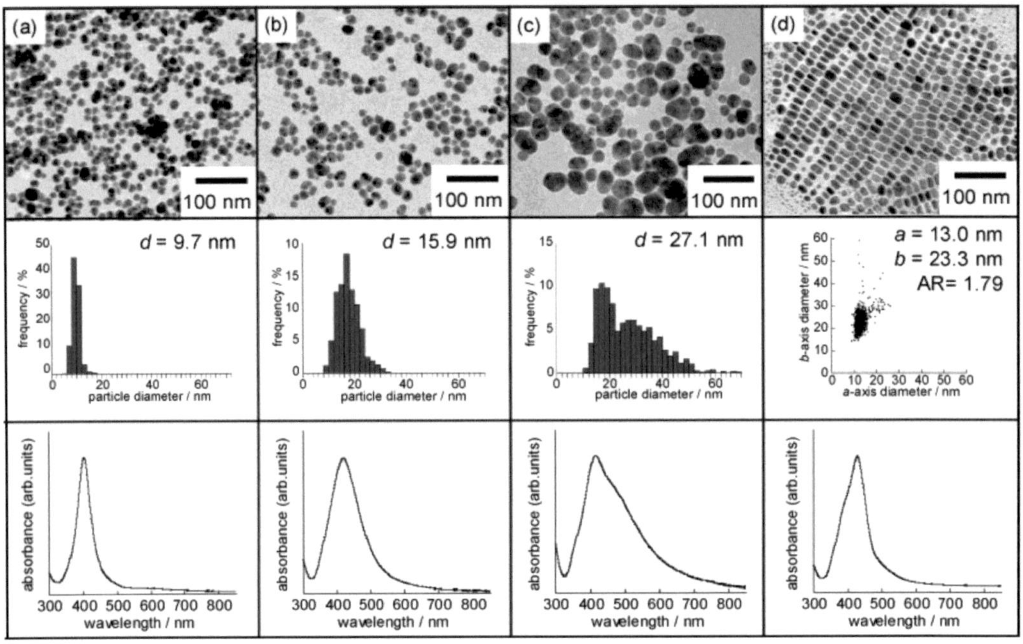

Figure 4. TEM micrographs, size distributions and optical absorption spectra of spherical silver NPs. Hideki Nabika and Shigehito Deki [23]. (a) 90 °C, $AgNO_3$ = 0.125 M, (b) 90 °C, $AgNO_3$ = 0.250 M, and (c) 110 °C, $AgNO_3$ = 0.250 M and (d) of the rod-shaped silver NPs.

On the other hand, in the Gans theory, the LSPR is only a function of the aspect ratio and refractive index. Thus, in certain conditions, a linear relationship between them can be resulted [28]. Nevertheless, numerical results suggest that, even when the aspect ratio is fixed and the retardation effect is weak, the position of longitudinal resonance can still depend strongly with the aspect ratio [29, 30] . Using the model of Cheng - ping Huang et al [31], we can write:

$$\lambda_p = \pi n \sqrt{10\kappa\left(2\delta^2 + r^2 \ln\left[\kappa\right]\right)} \tag{12}$$

Where $\kappa = l/2r$ is the aspect ratio of the NPs (with the inner radius r and in a cylindrical region with the length l), and δ is the skin depth. Cheng - ping Huang considered the problem as using an LC circuit model without solving the Laplace or Maxwell equations, and show that its results overcomes the deficiency of Gans theory and provides a new insight into the phenomenon. This way, we can observe resonance wavelength change with the aspect ratio of NP, Figure 5. This means a breakdown of the linear behavior presenting oscillations electrons originating from the amorphous geometry of the NPs associated with the inertia of electrons, see inset Figure 2 as well.

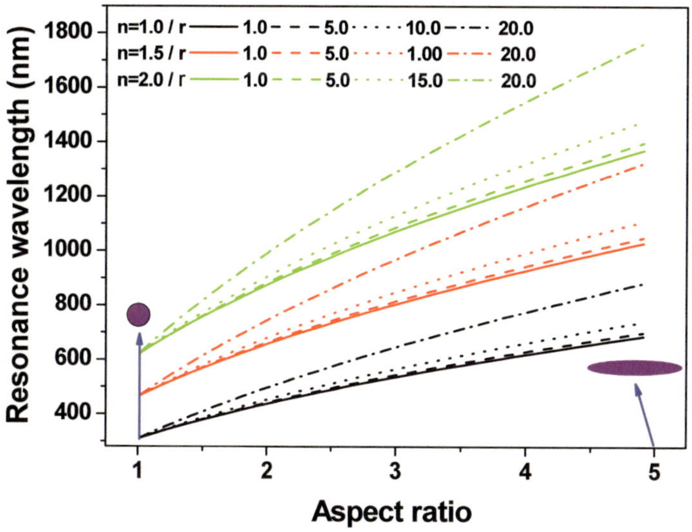

Figure 5. Resonance wavelength with aspect ratio from Cheng - ping Huang prediction for radius different and embedded in a bulk with refraction index 1.0 , 1.5 and 2.0.

Figure 6 illustrates this further, via experiments of X. Lu et al. [15] and Mock et al. [32]. Panel (a) the nanobar Ag NPs and the corresponding dark-field light scattering spectra, in broad agreement with the simulated results shown in [15]. Panel (b) shows the dipolar plasmon lineshapes of colloidal silver NPs of different shapes [32].

2.3. Coupling between metallic nanoparticles

We can say that the localized plasmon resonance frequency of a single metallic NP can be shifted through of alterations in shape, size and surrounding medium from the Frohlich condition, section 2.1. Nevertheless, in a NP ensembles we can obtain additional shifts due to electromagnetic interactions between the localized modes, see Figure 3 and 4 (c). These interactions are basically of a dipolar nature (when $d \ll \lambda$). So, the NP ensemble can be treated as an ensemble of interacting dipoles (in a first approximation). Those NPs can be embedded into a host matrix ordered or random, in one-, two- or three-dimensional arrays with interparticle spacing D. Electromagnetic coupling of those arrays shows interesting localization effects for closely spaced particles such as, enhancement process due to field localization in NP junctions. Assuming a dipolar approximation the NPs can be treated as point dipoles. In this sense, two regimes have to be distinguished (i) For $D \ll \lambda$, near-field interactions with a distance dependence of D^{-3} dominate, equation (5), and the NP array is described as an array of point dipoles coupling via their near-field, see section 2.1 and Figure 7. These arrays can serve as hot-spots for field enhancement, e.g. in a context of surface-enhanced Raman scattering (SERS). (ii) For larger NP separations, far-field dipolar coupling with a distance dependence of D^{-1} dominates, see section 2.1.

Therefore, the maximum field enhancement is determined by the shortest distance between two equipotential particles.

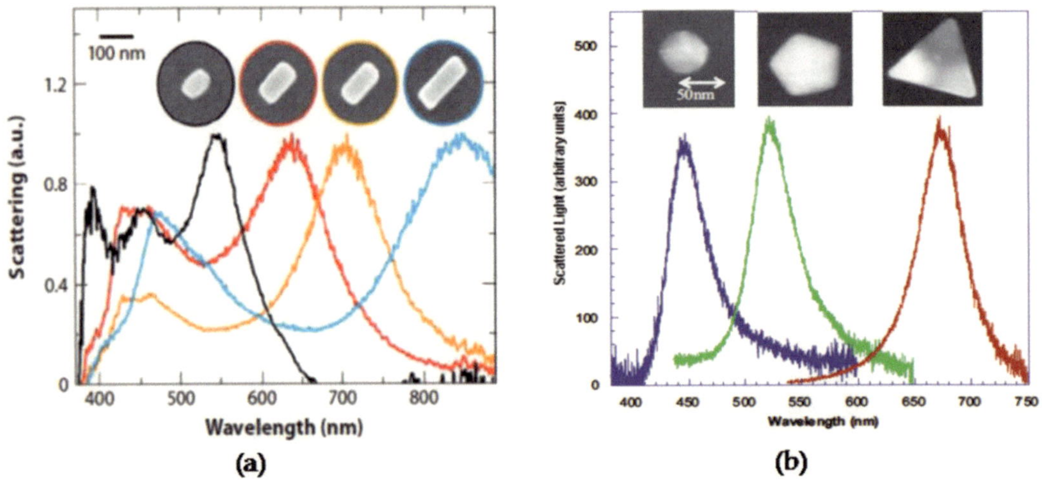

Figure 6. Panel (a) scattering spectra. In the insets are SEM images of the individual Ag nanobars. X. Lu et al. [15]. For simple, highly symmetric shapes (sphere, cube and triangular plate), the spectra are dominated by a single peak, but with the peak position sensitive to the shape, and ranging from 400 to up 800 nm. Panel (b) Scattering spectra of single silver nanoparticles of different shapes obtained in dark-field configuration. Mock et al. [32].

Figure 7 shows the near field coupling between NPs. Here, the restoring force acting on the oscillating electrons of each NP in the chain is either increased or decreased by the charge distribution of neighboring particles. Depending on the polarization direction of the exciting light, this leads to a blue-shift of the plasmon resonance for the excitation of transverse modes, Figure 7 (a), and a red-shift for longitudinal modes, Figure 7 (b).

Figure 7. Schematic of near-field coupling between metallic NPs for two different polarizations.

Today, there are several theoretical methodologies available to describe the interaction among the metallic NPs with electromagnetic radiation [33,34,35,36,37,38,39,40].

3. Rare-earths

Rare earths ions in solids are either divalent or trivalent. Their electronic configuration is $4f^{N}5s^{2}5p^{6}$ or $4f^{N-1}5s^{2}5p^{6}$, respectively. By far the most common valence state of the RE ions in solids is the trivalent one. Those ions have a long history in optical and magnetic applications. We have special interest in the devices luminescent using crystal, powders, and glasses. On the other hand, divalent RE ions have also been used in laser devices, but only in relatively exotic ones for cryogenic operation [41,42]. The most frequently used laser-active RE ions and host media together with typical emission wavelength ranges are shown in the table 1:

Ion	Common host media	Important emission wavelengths (μm)
Neodymium (Nd^{3+})	YAG, YVO_4, YLF, silica	1.03–1.1, 0.9–0.95, 1.32–1.35
Ytterbium (Yb^{3+})	YAG, tungstates, silica	1.0–1.1
Erbium (Er^{3+})	YAG, silica, tellurite, chacogenetos glasses	1.5–1.6, 2.7, 0.55
Thulium (Tm^{3+})	YAG, silica, fluoride glasses	1.7–2.1, 1.45–1.53, 0.48, 0.8
Holmium (Ho^{3+})	YAG, YLF, silica	2.1, 2.8–2.9
Praseodymium (Pr^{3+})	silica, fluoride glasses	1.3, 0.635, 0.6, 0.52, 0.49
Cerium (Ce^{3+})	YLF, LiCAF, LiLuF, LiSAF, and similar fluorides	0.28–0.33

Table 1. Common laser-active rare earth ions and host media and important emission wavelengths.

The 4f electrons are not the outermost ones. They are shielded from external fields by two electronic shells with larger radial extension ($5s^2 5p^6$), which explains the atomic nature of their spectra absorption/emission, Figure 8. Thus the 4f electrons are only weakly perturbed by the charges of the surrounding atoms. Important characteristics that distinguish them from other optically active ions are: (i) the wavelengths of the emission and absorption transitions are relatively insensitive to host material, (ii) the intensities of these transitions are weak, (iii) the lifetimes of metastable states are long and, (iv) the quantum efficiencies tend to be high. This is why the RE ions are such a useful probe in a solid; the crystal environment constitutes only a small perturbation on the atomic energy levels, and many of the solid state, and hence spectroscopic, properties can be understood from a consideration of the free ions. In turn, the wavefunctions of the free ions constitute a good zero order approximation for a description of solid-state properties, Figure 8.

The solutions to this problem can then befactored into a product of a radial and angular function. Whereas the radial function depends on the details of the potential the spherical symmetry ensures that the angular component is identicalwith that of a hydrogen atom and can be expressed as spherical harmonics. Except for Ce^{3+} and $Yb^{3+,}$ which have only one electron, the solutions of the central-field problem are products of one-electron states that are antisymmetric under the interchange of a pair of electrons, as required by the Pauli exclusion principle.

In describing the state of a multielectron atom, the orbital angular momenta and the spin angular momenta are added separately. The sum of the orbital angular momenta is designated by the letter L, and the total spin is characterized by S. The total angular momentum J of the atom may then be obtained by vector addition of L and S. The collection of energy states with common values of J, L, and S is called a term. Here, the Russell-

Saunders notation for the energy levels (terms) is used, $^{2S+1}L_J$, this takes into account the spin-spin coupling, orbit-orbit coupling and spin-orbit coupling. The energy levels of a free RE ion are usually interpreted by considering only interactions between the 4f electrons themselves. Since all the other electronic shells are spherically symmetric, Figure 8, their effect on all the terms of a configuration is the same in first order, and therefore do not contribute significantly to the relative positions of the 4f energy levels, we can write:

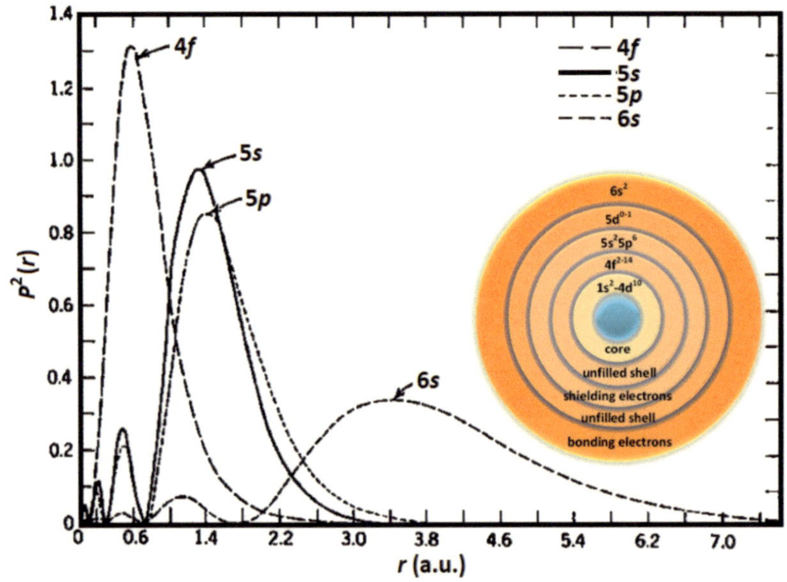

Figure 8. Square of the radial wavefunctions for the 4f, 5s, 5p and 6s energy levels from Hartree-Fock. It was a calculation for Gd⁺ by Freeman e Watson (1962) [46]. Picture inset schematically shows that the 4f orbital is within the 6s, 5p and 5s levels.

$$H = -\frac{\hbar^2}{2m}\sum_{i=1}^{N}\Delta_i - \sum_{i=1}^{N}\frac{Z^* e^2}{r_i} + \underbrace{\sum_{i<j}^{N}\frac{e^2}{r_{ij}}}_{H_C} + \underbrace{\sum_{i=1}^{N}\zeta(r_i) s_i \cdot l_i}_{H_{SO}} + V_{EF} \tag{13}$$

Where N=1…14 is the number of the 4f electrons, Z^*e the screened charge of the nucleus because we have neglected the closed electronic shell, V_{EF} treats the interaction of the ion with the electromagnetic field, and $\zeta(r_i)$ the spin-orbit coupling function [43],

$$\zeta(r_i) = \frac{\hbar^2}{2m^2c^2 r_i}\frac{dU(r_i)}{dr_i} \tag{14}$$

Where $U(r_i)$ is the potential in which the electron i is moving. The first two terms of the Hamiltonian (equation (13)) are spherically symmetric and therefore do not remove any of the degeneracies within the configuration of the 4f electrons, therefore we can neglect. The next two terms, which represent the mutual Coulomb interaction of the 4f electrons (H_C) and their spin-orbit interaction (H_{SO}) are responsible for the energy level structure of the 4f electrons. Details of the matrix element calculations of *Hc* and *Hso* can be found in ref. [44,45].

Figure 8 shows the radial distribution functions 4f, 5s, 5p, and 6s electrons for Gd^+ as obtained from Harteww-Fock calculation by Freeman and Watson [46]. We can see that the 4f electrons are inner electrons with relatively small $<r^n>$ values. As we shall see, the crystal field interaction will be small in REs. It is also evident that the 4f wavefunctions do not extend very far beyond the 5s and 5p shells. Thus, all quantities of a solid depend on an overlap of the 4f wavefunctions with those of a neighboring ion have to be small in RE compounds.

3.1. Trivalent ions in a static crystal field

The 4f shell of the RE ions are an unfilled shell and therefore have a spherical charge distribution. If the ion is introduced into a crystal, the ion experiences an inhomogeneous electrostatic field, the so-called crystal field, which is produced by the charge distribution into the crystal. This crystal field distorts the closed shells of the RE ion. Producing an effect on the energy level 4f, i.e., removes to a certain degree the degeneracy of the free ion 4f levels, thus producing a major modification of the energy levels (but this depends on the crystal symmetry). In the luminescence spectrum are observed additional transitions, which originate from the excited crystal field levels of the ground term is evident, and the spectra look more complicated. On the other hand, these spectrums can be used to determine the crystal field energies of the ground term.

We shall now try to understand the behavior of these additional transitions. To do so we draw on our treatment of the Coulomb interaction of the 4f electrons. The crystal field interaction comes by the interaction of the 4f electrons with all the charges of the crystal with all charges of the crystal, except for trivial factors, it is therefore given by one over the radius vector between the 4f electrons and the crystal charges. This function is given, apart from radial factors, of two spherical harmonics Y_{kq}, one containing the coordinates of the 4f electrons, the other containing coordinates of the crystal charge. The latter has to be integrated over the whole crystal, whereupon it gives the strength of the crystal potential at the site of the RE ion. Finally, this shows the existence of the splitting telling us that the development of the crystal potential into spherical harmonics contains a second order term.

3.1.1. Crystal field splitting

Crystal field splitting has two aspects: (i) symmetry, namely, the number of the levels into which an ion is free the *J* term are split in a crystal field of a given symmetry; (ii) the actual size of the crystal field splitting.

This way, the point charge model can be used for illustrative purposes but not for a quantitative description of the crystal field interaction. A realistic description of this interaction must take into account that the crystal field is built up of the spatially extended charge clouds of the individual ions. Thus, these charge clouds can penetrate each other and thereby interact. Hence, we can considered the following elements:

1. The ions are considered static in the crystal, i.e., we can neglected the lattice vibrations and their effect on the energy levels.
2. We regard the 4f electrons of one RE ion, as representative of those of all the RE ions in the crystal and, thus the interaction of 4f electrons of adjacent ions is neglected.
3. The crystal consists of the extended charge distributions. This produces an overlap of the charge distributions of the neighboring ions and the 4f electrons. Aditionally, a charge transfer between 4f electrons and the electrons of the ligands can take place. Both contribute to the crystal field interaction.
4. The 4f electrons of one ion are considered to be independent of each other, that is correlation effects play no significant role.

Therewith, we can calculate the crystal potential $\Phi(r_i,\varphi_i,\theta_i)$ at the site of the 4f electrons and the potential energy of the 4f electrons in this potential. If the crystal has charge density $\rho(R)$ and the 4f electrons have radius r_i, we have:

$$V=-\sum_i \int \frac{e_i\rho(R)}{|R-r_i|}d\tau=-\sum_{k,i} e_i \int \rho(R)P_k(\cos[R,r_i])\frac{r_<^k}{r_>^{k+1}}d\tau \tag{15}$$

Where $r_<$ and $r_>$ are, respectively, the smaller and larger value of r and r_i. Here $P_k(\cos[x])$ are the Legendre polynomials. Also, the equation (15) can be re-write as:

$$V=-\sum_{k,q,i} B_{k,q}C_{k,q}(\theta_i,\varphi_i) \tag{16}$$

Where the crystal field parameter $B_{k,q}$ ($k \leq 6$ for f electrons) and the summation i is carried out over all the 4f electrons of the ions. $B_{k,q}$ have the form:

$$B_{k,q}=-e\int (-1)^q \rho(R)C_{k-q}(\theta,\varphi)\frac{r_<^k}{r_>^{k+1}}d\tau \tag{17}$$

Here $C_{k,q}$ is called a tensor operator and is defined as: $C_{k,q}=\left(\frac{4\pi}{2k+1}\right)^{0.5} Y_{k,q}$

In early treatments of the crystal fiel interaction it was assumed that in RE compounds the point charge contribution would be the dominant part of the crystal field interaction. Then, in equation (16) the integral over the lattice can be replaced by a sum over all lattice points and $r_<^k$ can be replaced by r_i^k. The latter replacement can be performed as long as the charge distribution of the crystal does not enter that of the 4f electrons (as long as $r_i < R$), which implies that the potential acting on the 4f electrons obeys the Laplace equation $(\Delta\Phi(r,\varphi,\theta)=0)$ at the position of the 4f electrons.

Figure 9 further illustrates the effect of spin–orbit and crystal field interactions on the energy levels of the Er^{3+} ion. This figure were extracted from A.J. Kenyon [47]. A further splitting of the energy levels comes about when the ion is placed in a silica host. The $4f^N$ configuration is composed for a number of states where the quantum numbers (L, S, J, and another arbitrary one) define the terms of the configuration, all of which are degenerate in the central-field approximation, as illustrated in Figure 9. Next in the hierarchy is spin–orbit, the strongest of the magnetic interactions. Spin–orbit lifts the degeneracy in total angular momentum and splits the *LS* terms into *J* levels, Figure 9.

We can see that the environment provided by silica hosts destroys the spherically symmetric environment that Er^{3+} ions enjoy in the vapor phase. Thus the degeneracy of the 4f atomic states will be lifted to some degree. This splitting is also referred to as stark splitting, and the resulting states are called Stark components (of the parent manifold). The even-*k* terms in the expansion split the free ion *J* multiplets into Stark components generally separated by 10–100 cm^{-1} (see Figure 9). The ion–lattice interaction can mix multiplets with different *J* values (*J* mixing), although it usually remains a good quantum number. The odd-*k* terms admix higher lying states of opposite parity [e.g., $4f^N15d^1$] into the $4f^N$ configuration. This admixture does not affect the positions of the energy levels, but it has a very important effect on the strengths of the optical transitions between levels.

3.2. Optical properties

In recent years most of the interest in luminescent RE ions has concentrated on the trivalent erbium (Er^{3+}), and in particular its emission band around 1550 nm. This emission is within the telecommunication windows, and fortuitously coincides with the 1550 nm intra-4f $^4I_{13/2}\rightarrow{}^4I_{15/2}$ transition of the Er^{3+} ion.

For this reason exists today a great interest in utilizing erbium-doped materials for gain elements and sources in telecommunications systems [48,49,50]. The development of the erbium-doped fiber amplifier (EDFA) in the late 1980s [51,52] exploited the $^4I_{13/2}\rightarrow{}^4I_{15/2}$ transition and allowed the transmission and amplification of signals in the 1530–1560 nm region without the necessity for expensive optical to electrical conversion [53]. Hence, it is originated a driving force for research in RE doped fibers and integrated optics waveguides has been their use for amplifying weak signals in optical communications systems at 1300 and 1600 nm. This may be achieved by simply splicing a section of RE-doped fiber into the transmission one and injecting pump light through a fiber coupler. The signal generated within the RE emission band stimulates emission of radiation at the same frequency, amplifying the optical communication signal with high gain, high efficiency and low noise, which is highly advantageous for optical communications [54]. There are five main RE candidates for use as dopants in fiber or waveguide amplifiers for optical communications systems: Er^{3+}, Tm^{3+}, Nd^{3+}, Pr^{3+} and Dy^{3+}. The Er^{3+} and Tm^{3+} ions are the choice for the 1400–1600 nm window centered at 1550 nm, based on the $^4I_{13/2}\rightarrow{}^4I_{15/2}$ transition of Er^{3+} ion and the $^3H_4\rightarrow{}^3F_4$ transition of Tm^{3+} ion. The $^4F_{3/2}\rightarrow{}^4I_{13/2}$ emission of Nd^{3+} ion, the $^1G_4\rightarrow{}^3H_5$ transition of Pr^{3+} ion and the $^6F_{11/2}(^6H_{9/2})\rightarrow{}^6H_{15/2}$ transition of Dy^{3+} ion are all potentially useful for the 1300 nm telecommunication window.

Further, the 3+ ions all exhibit intense narrow-band intra-4f luminescence in a wide variety of hosts, and the shielding provided by the $5s^2$ and $5p^6$ electrons (see Figure 8), which means that rare-earth radiative transitions in solid hosts resemble those of the free ions and electron–phonon coupling is weak. Although some of the divalent species also exhibit luminescence (e.g., samarium and europium), it is the trivalent ions that are of most interest.

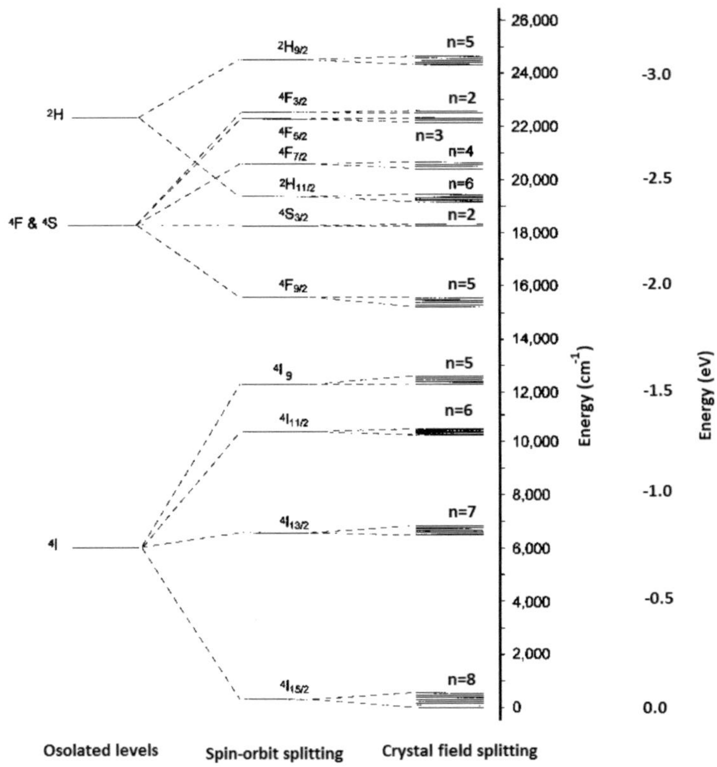

Figure 9. The effect of spin-orbit and crystal field splitting on the energy levels of the Er^{3+} ion in silica matrix. The energy diagram shows the hierarchy of splitting resulting from electron-electron and electron-host interactions. Figure extracted from A.J. Kenyon [47].

As mentioned previously, the intra-4f transitions are parity forbidden and are made partially allowed by crystal field interactions mixing opposite parity wavefunctions, resulting in luminescence lifetimes long (often in the millisecond range), and linewidths narrow. The selection of an appropriate ion with intense and narrow-band emission can be obtained across much of the visible region and into the near-infrared. Figure 10 shows energy level diagrams for the isolated 3+ ions of each of the 13 lanthanides with partially filled 4f orbitals.

Almost all the RE ions their emission is due to optical transitions within the f-manifold (e.g. Tb^{3+} ($4f^8$), Gd^{3+} ($4f^7$) and Eu^{3+} ($4f^7$). The 4f-electrons are well shielded from the chemical environment and therefore have almost retained their atomic character. Nevertheless, for a number of RE ions, also broad emission bands are known. Prominent examples are Eu^{2+} and Ce^{3+}. Here, the emission is from to 5d-4f optical transitions. As electrons participate in the chemical bonding, the d-f emission spectra consist of broad bands. These transition types are allowed and are consequently very fast (a few μs or less).

Some line emission is not a specific property of RE ions and in addition to that, in the case of RE ions, broad emission spectra can be obtained as well, depending on the optical transitions involved. In this respect, RE ions are not unique. The unique properties of the RE ions originate from the fact that the spectral position of the emission lines is almost independent of the host lattice, in contrast to line emission generate by the emission of metal ions.

3.2.1. *Radiative transitions: Intensities of optical transitions*

The terms shown in equation (13) are time-dependent; thereby, they do not lead to stationary states of the system. This way, they are treated using time-dependent perturbation theory resulting in transitions between the states established by the static interactions. In luminescent devices the most important term is V_{EF} which gives rise to the emission and absorption of photons from decay radiative of the RE ions. This involves the interaction between the electron charge and the electric field and, the interaction between the electron spin and the magnetic field.

The experimental data on spectra of RE ions show that the radiation is mostly electric dipole (ED) nature, though some cases of the magnetic dipole radiation are also observed. Since the optical transitions take place between levels of a particular $4f^N$ configuration, the electric dipole radiation is forbidden in first order, because the electric dipole operator has uneven parity and the transition matrix element must have even parity (Laporte selection rules). Van Vleck [55] pointed out the dipole electric radiation can only occur because the $4f^N$ states have admixtures of $4f^{N-1}nl$ configuration (nl will be mostly 5d), thus the $4f^{N-1}nl$ has to be chosen such that it has opposite parity from $4f^N$. The admixture is produced via interactios that have odd parity and this depend of the host matrix of the RE ions. However, we can have four dominant sources of optical radiation in RE ions spectra:

i. Forced electric dipole induced by odd terms of the crystal field.
ii. Forced electric dipole radiation induiced by lattices radiation.
iii. Allowed magnetic dipole transition.
iv. Allowed electric quadrupole radiation.

In free atoms, magnetic dipole is about six orders of magnitude weaker than the electric dipole radiation. This latter occurs only as a consequence of a perturbation, both kinds of radiation shows up in the RE spectra with about intensity. Then, quadrupole radiation are less probable still in comparison with the magnetic dipole.

There are different types of transitions between levels (called electric dipole, electric quadrupole, and magnetic dipole) and therefore the transitions are divided into allowed ones (with a high probability) and so-called forbidden transitions (for transitions where the electric dipole transition is quantum mechanically forbidden).

For the first order allowed magnetic dipole radiation the free atom selection rules are still quite valid: $\Delta S = \Delta L = 0$ and $\Delta J = 0, \ \pm 1 \ \left(0 \neq 0\right)$. In the electric dipole case we have: $\Delta l = \pm 1$, $\Delta S = 0$, $\left|\Delta L\right|$, $\left|\Delta J\right| \leq 2l$. However, since ED transitions is induced by the crystal field, the free

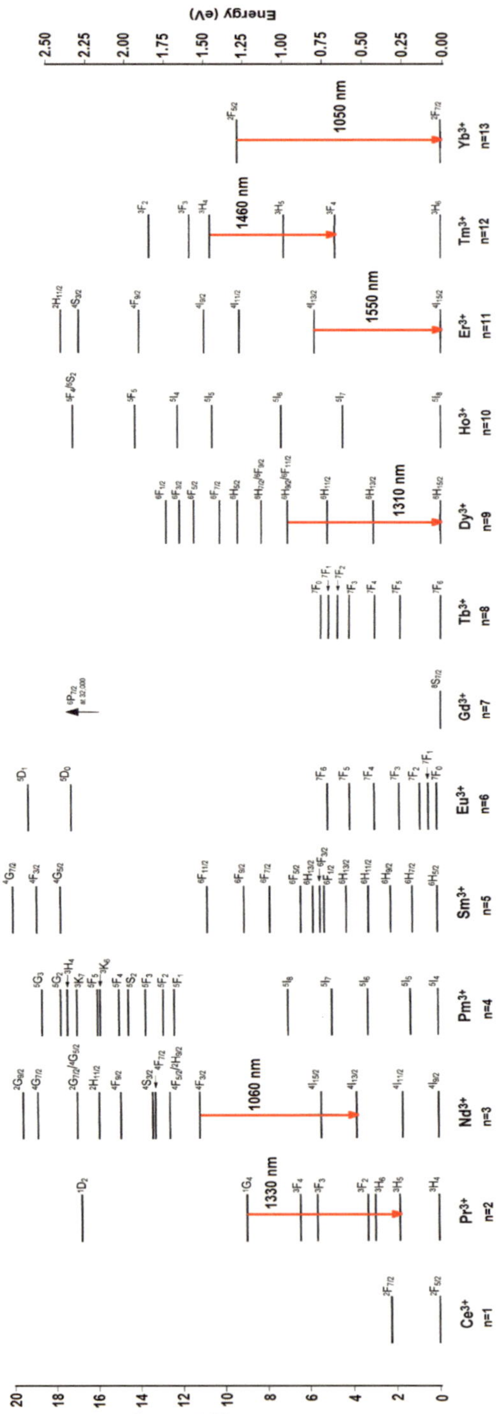

Figure 10. Energy levels of the triply charged lanthanide ions. Besides, the most technologically important radiative transitions are labeled. Figure adapted from A.J. Kenyon [47].

atom selection rules break down almost completely, i.e., the selection rules with regard to the *L*, *S* and *J* quantum numbers are now essential governed by the crystal field interaction yielding $|\Delta J| \leq 6$ (approximately). For the electric quadrupole case the rules selection is: $\Delta S = 0, \ |\Delta L|, |\Delta J| \leq 2$.

We now outline the Judd-Ofelt theory for the determination of the intensities in RE crystal spectra [56,57]. It is in essence a quantification of the ideas formulated by van Vleck [55] under optical radiation sources. Intensities are often expressed in terms of oscillator strengths, where the total oscillator strength for a transition from an energy level *a* to an energy level *b* is given:

$$f(a,b) = \frac{8\pi^2 m\upsilon}{3h(2J+1)}\left[\frac{\left(n^2+2\right)^2}{9n}S_{ed} + nS_{md}\right] \tag{18}$$

Here *h* is the Planck constant, υ the frequency of the transition $a \rightarrow b$, *n* the refraction index of host. S_{ed} and S_{md} are the electric dipole and magnetic dipole intensities respectively and are defined as:

$$\begin{aligned} S_{ed}(\alpha J,\alpha' J') &= \sum_{\lambda=2,4,6} \Omega_\lambda \left\langle f^N\alpha[SL]J \left\| U^{(\lambda)} \right\| f^N\alpha'[S'L']J' \right\rangle^2 \\ S_{md}(\alpha J,\alpha' J') &= \beta^2 \sum_{\lambda=2,4,6} \Omega_\lambda \left\langle f^N\alpha[SL]J \left\| L+2S \right\| f^N\alpha'[S'L']J' \right\rangle^2 \end{aligned} \tag{19}$$

$U^{(\lambda)}$ is a tensor operator of rank λ, and the sum runs over the three terms values 2, 4 and 6 of λ. With:

$$\Omega_\lambda = (2\lambda+1)\sum_{k,q}\frac{\left|B_{k,q}^{\ *}\right|^2 \left|Y_{k,\lambda}\right|^2}{2k+1} \tag{20}$$

The Ω_λ parameters have so far been assumed to arise solely from crystal field; however, they also contain contributions from admixtures by the lattice vibrations. The asterisk means that for $B_{k,q}^{\ *}$ the radial integral is to be taken between states 4f and nl instead of 4f and 4f.

Also, it is interesting to determine the coefficient for spontaneous light from state $a(\alpha J)$ to state $b(\alpha' J')$. Those are defined as:

$$A(\alpha J,\alpha' J') = A_{ed} + A_{md} = \frac{64\pi^4 v^3}{3hc^2(2J+1)}\left[\frac{n\left(n^2+2\right)^2}{9}S_{ed} + n^3 S_{md}\right] \tag{21}$$

Here *c* velocity of light in vacuum. The Judd-Ofelt formalism has been applied to the analysis of a number of systems. In most of these analyses the crystal field splitting of the terms is neglected; therefore, the total absorption intensities between the ground term and the excited terms are analyzed with only three empirical parameters $\Omega_\lambda(\lambda = 2,4 \text{ and } 6)$.

If b is an excited state that decays only by the emission of photons, its observed that relaxation rate is the sum of the probabilities for transitions to all possible final states. The total rate is the reciprocal of the excited-state lifetime τ_a

$$\tau_a = \frac{1}{A(\alpha J, \alpha' J')} \tag{22}$$

The branching ratio $\beta_{a,b}$, for the transition $b \to a$ is the fraction of all spontaneous decay processes that occur through that channel and is defined as follows:

$$\beta_{a,b} = \frac{A(a,b)}{\sum_c A(a,c)} = A(a,b)\tau_a \tag{23}$$

The branching ratio, which has an important influence on the performance of a device based on a particular transition, appears often in the discussion of specific ions. It has a significant effect on the threshold of a laser and the efficiency of an amplifier.

4. Metallic nanoparticle embedded in a gain media: LSPR interaction radiative transitions

The RE-doped laser crystals and glasses are among the most popular solid-state gain media. In order of maintaining the efficiency of these materials in which the RE dopant is uniformily dispersed, as in multi-component glasses. In this sense, glasses have good optical, mechanical, and thermal properties to withstand the severe operating conditions of optical amplifier. Desirable properties include hardness, chemical inertness, absence of internal strain and refractive index variations, resistance to radiation-induced color centers, and ease of fabrication. Furthermore, the matrix host composition affects the solubility of the RE dopant, this affect the lifetime, absorption, emission, and excited state absorption cross sections of the dopant transitions [47]. This competing absorption phenomenon can seriously diminish the efficiency of an optical device. For example, a decrease in the excited-state absorption for Er^{3+}-doped fibers going from a germane-silicate host to an alumina-silicate host has been verified [58], demonstrating the importance of host selection for a given RE ion or laser transition. Host glasses compatible with this relatively high concentration of RE without clustering require the open, chain-like structure of phosphate glasses or the addition of modifier ions (Ca, Na, K, Li, or other) to open the silicate structure and increase solubility [48]. The limitation owing to clustering in a predominantly silica host without modifier ions has been well documented [19]. The maximum erbium concentration in silica for optimum amplifier performance has been suggested to be lesser to 100 ppm [59]. However a 14.4-dB gain, 900-ppm erbium-doped silica fiber amplifier has been reported, indicating that higher concentrations can produce useful devices [48]. Nevertheless, the interactions between host and RE ions, it is necessary to consider background losses from impurity absorption and scattering mechanisms that decrease the efficiency of the optical device. For example, depending on the phonon energies of the host matrix, some of the level lifetimes can be strongly quenched by multi-phonon transitions. Such effects are minimized in low-phonon-energy host media such as fluoride fibers. The effect of internal loss is most dramatic in distributed amplifiers, where pump light must travel long distances in the process of distributing gain.

Additionally, into the glass matrix can happen diverse kinds of interactions, in particular dipole–dipole interactions between the RE ions similar or of different species, allow energy transfer between those. This is exploited e.g. in Er^{3+}–Yb^{3+}-codoped fibers, where the pump radiation is dominantly absorbed by Yb^{3+} ions and mostly transferred to Er^{3+} ions [48].

In this way, RE-doped glasses shown are excellent materials that provide gain media. This makes them ideal candidates for embedding metallic NPs, resulting in new materials for the fabrication of devices optical with large enhancements even when the gain is saturated due light localization effects. It has been shown theoretically and experimentally that homogeneous aggregates of structures supporting LSPR can lead to extremely large enhancement of local field amplitudes exceeding those of single structures [60,61]. Besides, inorganic glasses are the host matrix for metallic NPs formation. The wide temperature range of glass viscosity growth provides the possibility to control over the NPs size within the wide range by means of modifying the temperature and duration of thermal processing. In fact, only such kind of matrix makes it possible to control and investigate all the stages of NPs formation, including the starting stage [14,62].

The LSPR in metallic NPs is predicted to exhibit a singularity when the surrounding dielectric medium has a critical value of optical gain [63]. This singularity is obtained from the equation (4), for $\mathrm{Re}\left[\varepsilon_{NP}(\omega)\right] \approx -2\varepsilon_m$, the NP absorption presents a maximum. This is so-called the Frohlich condition is associated with dipole mode [1].

Therefore, the strong local electric field induced by the NP (LSPR) can enhance the total electric field, and can also improve the quantum yield of the luminenescence of the RE ions from the NP (this strong enhancement of electromagnetic fields is essential in nonlinear optics effects) [64,65]. Thus, the enhancement obtained with these NPs is due to the formation of EDs which generates a polarization given by $\vec{P}_{NP} = qeN'\vec{x}$, where $\vec{x}$ is the distance between each NP , $q = \pi d^3/6$ is the NP specific volume, N' is the conduction electrons density, and e is the electronic charge [12]. Therefore, we obtained a modification in the local electric field by these dipoles (local field correction). Using the results obtained by O.L. Malta et al [66], the effective electric field can be written as:

$$\vec{E}_{eff} = (\varepsilon_0 + 2)\left[1 + q\omega_P^2 / \left\{3\varepsilon_0\left[(1-q)(\omega_P^2 / 3\varepsilon_0) - \omega^2 + i\gamma\omega\right]\right\}\right]\vec{E}_0 / 3 \tag{24}$$

Where ε_0 is the dielectric constant in the presence of an external electromagnetic field of amplitude E_0. In the presence of an electromagnetic wave we have LSPR which forms EDs separated by different distance $\vec{r}$, some of which will contribute to the luminescence enhancement.

Assuming that the RE ions may occupy different sites in the host, a direct coupling between the excited states of the RE ions and the NPs modifies the Stark levels energies [12,13,67]. As the NPs just give a contribution to the local field when the light is present, the oscillator strength of a spectral line, corresponding to transition from the ion ground level i to the component f of the excited level can be re-write as:

$$f(a,b)=\frac{8\pi^2 m\upsilon}{3h(2J+1)}\left[\frac{\left(n^2+2\right)^2}{9n}S_{ed}+\sum_{\lambda}\left|\langle i|D_{\lambda\ NP}^{(1)}|f\rangle\right|^2+nS_{md}\right] \tag{25}$$

The second term into parenthesis is added to the equation from the Judd-Ofelt theory [56,57], and represents the ED transition due to LSPR of NP. To obtain non vanishing matrix elements of the components $D_{\lambda\ NP}^{(1)}$ it is necessary to admix into $\langle i|$ and $|f\rangle$ other states of opposite parity.

Thus, the initial and final states are: $\langle i|=\langle\phi_i|+\sum_{\beta}\langle\phi_i|V|\phi_\beta\rangle/(E_i-E_\beta)\langle\phi_\beta|$, and $|f\rangle=|\phi_f\rangle+\sum_{\beta}\langle\phi_\beta|V|\phi_f\rangle/(E_f-E_\beta)|\phi_\beta\rangle$, then:

$$\langle i|\vec{P}+\vec{P}_{NP}|f\rangle=\sum_{\beta}\left\{\frac{\langle\phi_i|V|\phi_\beta\rangle\langle\phi_\beta|\vec{P}|\phi_f\rangle}{E_i-E_\beta}+\frac{\langle\phi_i|\vec{P}|\phi_\beta\rangle\langle\phi_\beta|V|\phi_f\rangle}{E_f-E_\beta}+\frac{\langle\phi_i|V|\phi_\beta\rangle\langle\phi_\beta|\vec{P}_{NP}|\phi_f\rangle}{E_i-E_\beta}\ldots\right.$$
$$\left.\ldots+\frac{\langle\phi_i|\vec{P}_{NP}|\phi_\beta\rangle\langle\phi_\beta|V|\phi_f\rangle}{E_f-E_\beta}\right\} \tag{26}$$

Where V is the crystalline field, equation (16), responsible for the Stark levels and treated as a perturbation, ϕ_i and ϕ_f have the same parity, ϕ_β has opposite parity in relation of ϕ_i and ϕ_f, $\vec{P}$ is the ED from the light interaction. For example, Er^{3+} ions presented an electronic configuration [Xe]4f^{11}, with J =15/2 been the spin-orbit ground state because the f shell is more than half-filled. The order of these Stark levels depends on the strength of the crystal field and the influence of $\vec{P}_{NP}$.

This can be observed experimentally in the luminescence spectra of different RE ions shown in Figure 11 interaction with silver or gold NPs in different host matrix, as a blue- or red-shift (see equation 26) and a pronounced increase in luminescence (see equation 25).

In this scenario, the influence of $\vec{P}_{NP}$ appeared in the luminescence spectra as a blue or a red -shift of the peaks, e.g. a transfer energy from the NPs dipoles to RE ions, thus modifying the Stark energy level's bandwidth, see Figure 11. This energy transfer is obtained from the second term added in equation (25) and can be understood through the following definition: $D_{q'\ NP}^{(1)}=\sum_{m}r_m C_{\lambda\ NP}^{(1)}(\theta_m,\phi_m)$, where $C_{\lambda\ NP}^{(1)}(\theta_m,\phi_m)$ is the Racah tensor and where r_m is the distance between the NPs and the RE ions ($r_m=|\vec{x}|$).

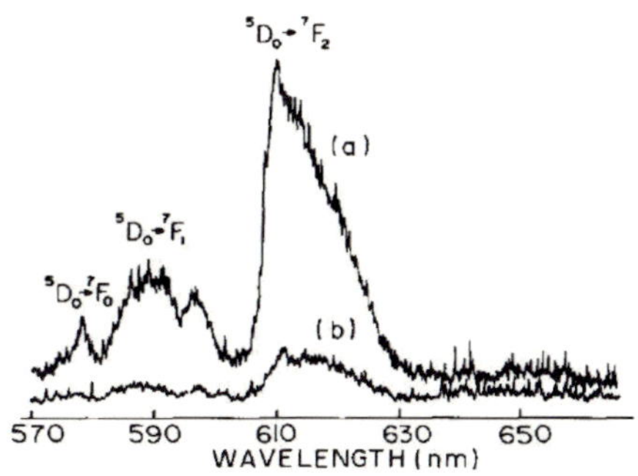

5D0→7F2
(a)
5D0→7F1
5D0→7F0
(b)
570 590 610 630 650
WAVELENGTH (nm)

(a)

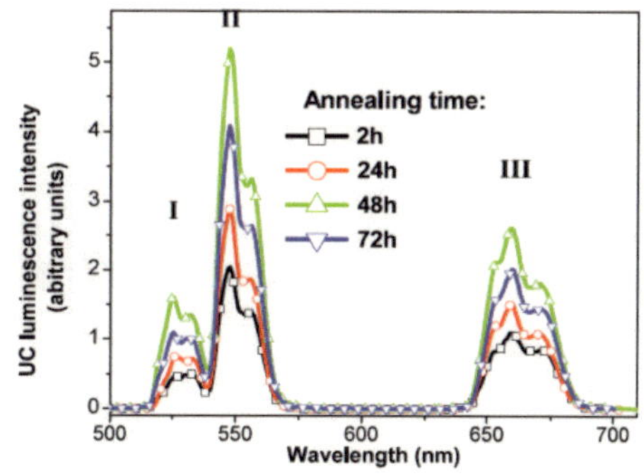

II
I
III
Annealing time:
2h
24h
48h
72h
UC luminescence intensity (abitrary units)
Wavelength (nm)

(b)

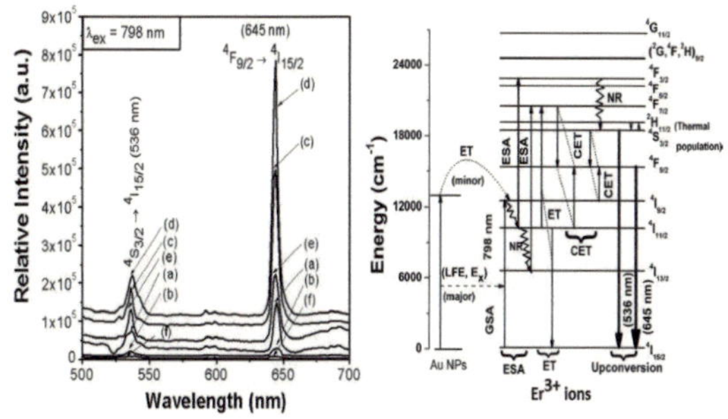

λex = 798 nm
(645 nm)
Relative Intensity (a.u.)
Wavelength (nm)
Energy (cm-1)
Au NPs
ESA
ET
Upconversion
Er3+ ions

(c)

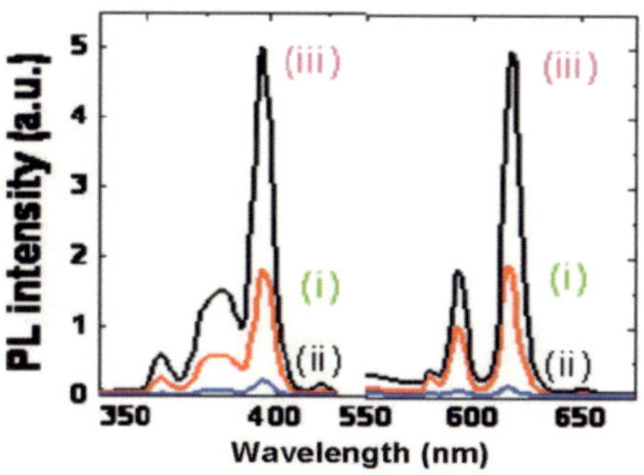

PL intensity (a.u.)
(iii)
(i)
(ii)
Wavelength (nm)

(d)

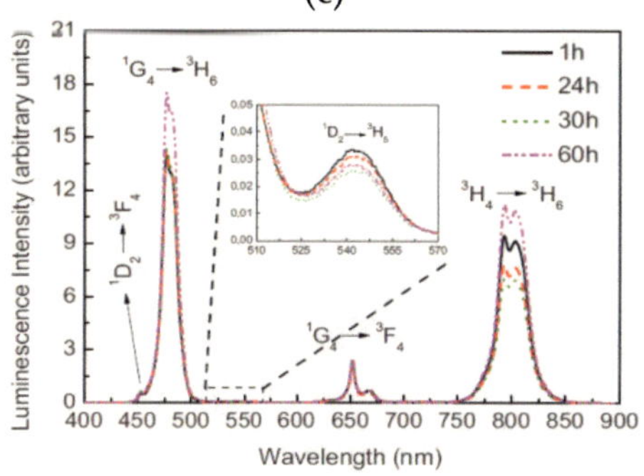

1h
24h
30h
60h
Luminescence Intensity (arbitrary units)
Wavelength (nm)

(e)

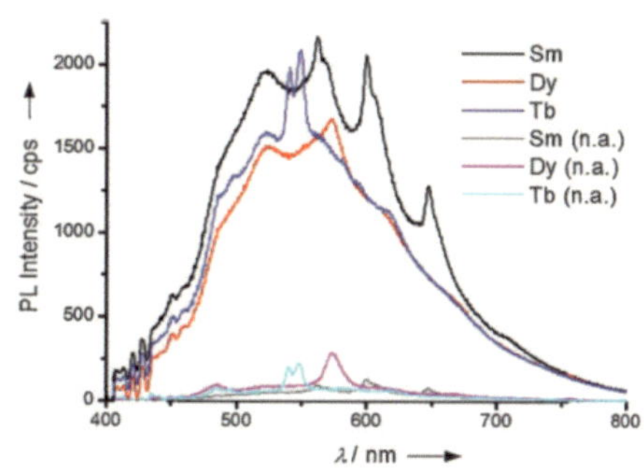

Sm
Dy
Tb
Sm (n.a.)
Dy (n.a.)
Tb (n.a.)
PL Intensity / cps
λ / nm

(f)

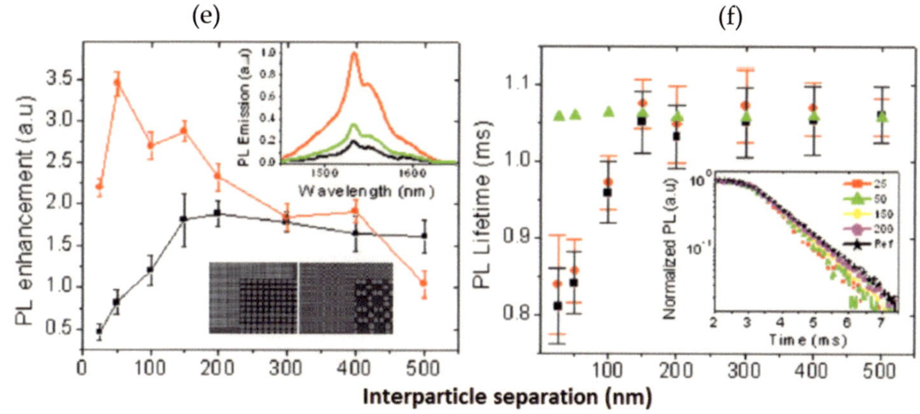

PL enhancement (a.u)
PL Emission (a.u)
Wavelength (nm)
PL Lifetime (ms)
Normalized PL (a.u)
Time (ms)
Interparticle separation (nm)

(g)

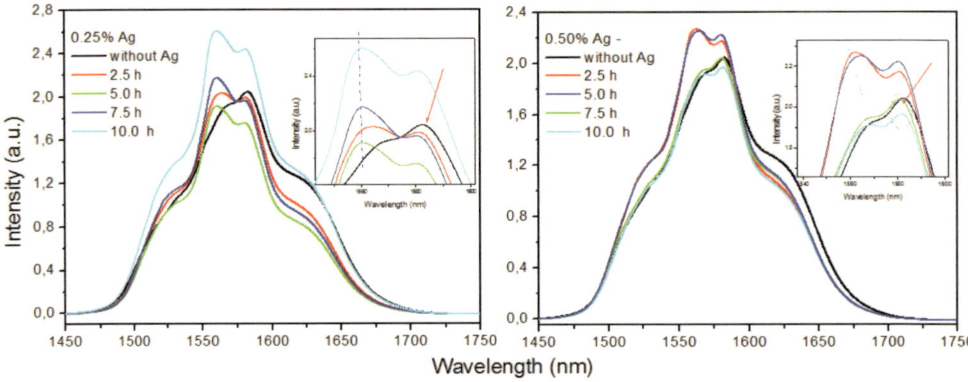

(h)

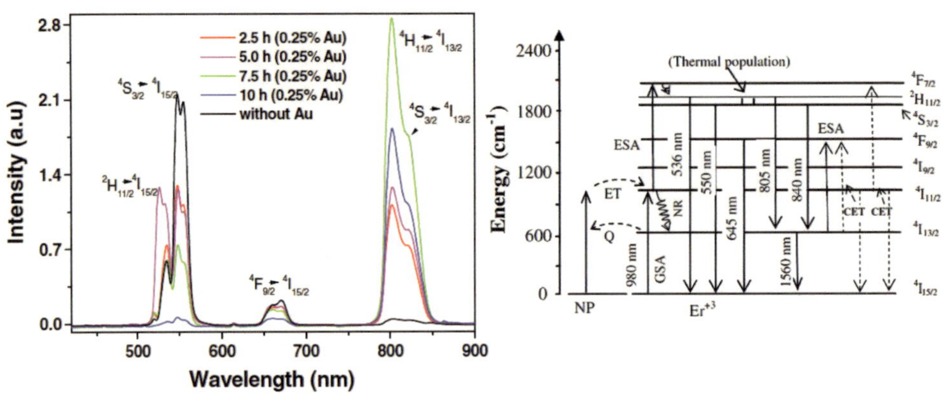

(i)

Figure 11. (a). Fluorescence spectra of the Eu^{3+} ions in the presence (a) and in the absence (b) of Ag NPs.The silver concentration is 7.5 in weight percent. The matrix glass composition can be found in [66].
(b). Frequency UC spectra for excitation at 980 nm, (for composition see Ref. [68]. After the cooling the samples were annealed for different durations (24, 48, and 72 h) in order to nucleate silver NPs [68].
(c). Upconversion spectra extracted from [69] under excitation wavelength at λ_{ex}=798 nm (for composition and for amplification ratio see Ref. [69]. The bases of the emission curves (c), (d), and (e) have been uplifted for better visibility.
(d). PL spectra extracted from [70] (i) Eu(III)EDTA, $3H_2O$ complex, (ii) Eu complex with Au nanoparticles and (iii) Eu complex with Au–ZnO nanoparticles.
(e). Upconversion spectra of Tm^{3+}/Yb^{3+} codoped PbO–GeO_2 samples containing silver NPs obtained by pumping the samples with a diode laser operating at 980 nm. For more details see [71].
(f). PL spectra ($\lambda_{exc}\approx$337 nm) of sy-activated with 0.037 mol% Ag containing glasses co-doped with 0.19 mol% Sm (black curve), Dy (red), and Tb (blue), respectively, and of non-activated and non-annealed (n.a.) 0.037 mol% Ag containing samples co-doped with 0.19 mol% Sm (grey), Dy (magenta), and Tb (light blue) [72].
(g). Picture on the left, Integrated PL enhancement in periodic (circle) and Fibonacci (square) nanoparticle arrays of various interparticle separations. The top inset shows the representative PL spectrum of periodic (bottom), Fibonacci (top) nanoparticle arrays with Δ_{min}= 50 nm and unpatterned area (middle),and the inset bottom shown the SEM micrograph of (left) periodic, (right) Fibonacci array Au nanocylinders. Picture on the right, PL lifetime of periodic (circle), Fibonacci (square), unpatterned

(triangle). Inset Er decay of unpatterned (star) and Fibonacci arrays with different Δ_{min} as specified in the legend. For more details see Ref. [73].
(h). Picture on the left; PL of the TE025-Y samples pumped with diode laser at 980 nm, showing PL enhancement. The inserted figure shows a zoom of the peaks. The vertical dashed line is a reference for showing the blue-shift of the peaks, and the arrow indicates the enhancement due to transfer energy from EDs to Er^{3+}. Picture on the right; same for the samples TE050-Y. The enhancement of luminescence was found to be reproducible for all ours samples. These pictures was extracted and modified from Ref. [12]. These spectrums are of Er^{3+}-doped tellurite glass, more details see [12].
(i). Luminescence spectra of Er^{3+}:Au-doped tellurite glass for several annealing times, pumped at 980 nm. Picture on the left; Upconversion, for the range $400<\lambda< 900$ nm. Picture on the right; Energy-levels diagram of Er^{3+} for the luminescence spectra. ET stands for energy transfer, CET for cooperative energy transfer, Q for quenching, GSA for ground state absorption, NR for non-radiative decay, and ESA for excited state absorption. For more details see Ref. [13].

Nevertheless, an interesting question arises: how these NPs are excited?

As explained above, those NPs can be excited by a predefined incident radiation through a direct coupling between the excited states of the both RE ions and NPs, resulting in: (i) a local field increase (Frohlich condition), at ω_p, (ii) a nonradiatively decay (heat generation by Joule effect) or (iii) a radiative energy release which depends on the albedo of the NPs. Thereby, the exact response of LSPR will depend on the details of the physical system (e.g. arrangement, shape, host matrix), and usually not strictly symmetric about the resonant frequency. A schematic representation of the interaction process within the RE:NP system is depicted in Figure 12 (a).

Also we can elucidate the enhancement (quenching) from the process of energy transfer as following: Er^{3+} emission promotes energy transfers into a plasmonic mode, which can decay nonradiatively by heat generation (Joule effect) or radiatively by releasing energy that depends on the albedo of the NPs. We consider two types of emission from a system of identical dipoles [74]:

$$I_1 \propto I_p \eta_p \eta_0 \tag{27}$$

$$I_2 \propto I_1 \eta_0 \eta_{LSPR} Q_{scatt} \tag{28}$$

Here, I_1 is the intensity emission for a single emitter, I_2 is the intensity emission of the NP, I_p the pump intensity, η_p is the pump efficiency, η_0 is the internal quantum efficiency of the energy transfer, η_{LSPR} is the efficiency of the energy transfer (nonradiative) to LSPR modes and the scattering efficiency of the plasmon at the emission wavelength Q_{scatt} is:

$$Q_{scatt} = Q_{scatt}^{(Ray)} = \frac{8}{3} \frac{\omega_p{}^4 y^4}{\left(\omega^2 - \omega_p{}^2\right)^2 + \frac{4}{9} y^6 \omega_p{}^4} \tag{29}$$

Where $Q_{scatt}^{(Ray)}$ is the Rayleigh scattering [67,68] and, $y = \omega D/c$, with c the speed of light. Therefore the total intensity I_T, can be written as [62]:

$$I_T = I_1(1-\eta_{LSPR}) + \eta_0 I_2 \tag{30}$$

We consider I_1 as the contribution from the emitter dipole (RE ions) uncoupled to the metal and I_2 the plasmon enhanced emission from coupled dipoles, see Figure 12 (b).

Finally, we can obtain the following conditions: ${\eta_0}^2 Q_{scatt}^{(Ray)} \geq 1$, where $Q_{scatt}^{(Ray)}(\omega, D)$, i.e. the luminescence enhancement depends on the incident radiation frequency and the NP size. But, as were mentioned, the incident radiation for the NP activation comes from the emitter dipole $(I_1\eta_{LSPR})$ and not from the pump radiation (I_p), i.e., from transition radiative from RE. The coupling efficiency η_{LSPR} is defined by: $\eta_{LSPR} = \Gamma_{LSPR} / (\Gamma_{rad} + \Gamma_{nonrad} + \Gamma_{LSPR})$, where: Γ_{LSPR} is the energy transfer rate to the LSPR mode [73,74].

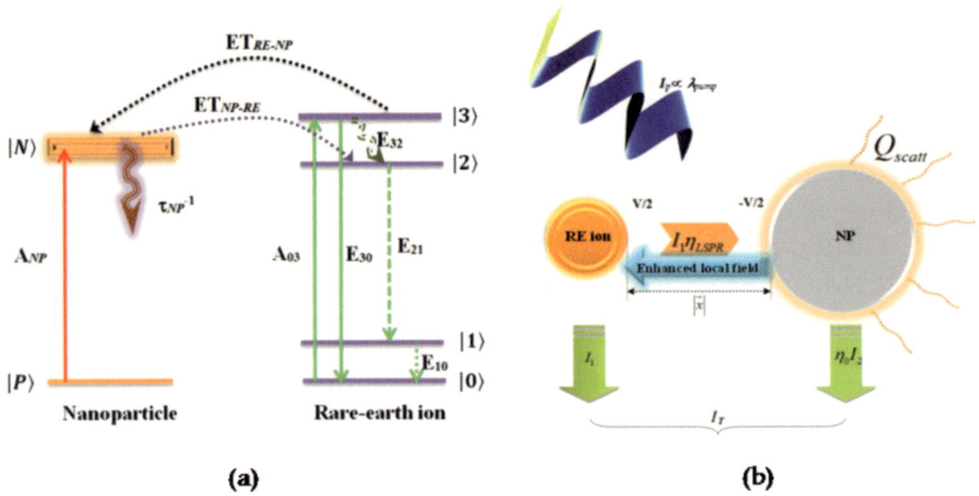

Figure 12. (a) Energy level scheme for a resonant and RE ion absorption. The dot line indicates the energy transfer between RE→NP or NP→ RE, and the vertical dot line shows the transition radiative under consideration. The curved arrows indicate non-radiative transitions. (b) Schematic representation of the system RE:NP. A monochromatic plane wave with pump intensity (I_p) which is proportional to the pump wavelength λ_{pump}, induces the following processes: (i) absorption of RE ion, I_p, (ii) activation of NPs, due the coupling RE:NP (transitions levels), $I_1\eta_{LSPR}$ (iii) NP transmitter, via electric dipole. Such coupling depends on the coupling efficiency η_{LSPR}. Where I_1 the intensity emission of emitter, I_2 the intensity emission of the NP, η_{LSPR} the efficiency of the energy transfer (nonradiative) to LSPR modes. Equipotential surfaces (electric dipole coupling) with electric potentials -V/2 and V/2 for the NPs and the Er^{3+} ions respectively.

The strong local electric field induced by NPs (ED), increases the quantum yield η of the RE luminescence, defined by the ratio of the local field $\vec{E}_{loc}$ and the incident field $\vec{E}_i$, $\eta = (r_m + d)/d = |\vec{E}_{loc}| / |\vec{E}_i|$ [14], here $\vec{E}_{eff} = \vec{E}_{loc} + \vec{E}_i$. Thus, the maximum field enhancement is determined by the shortest distance between two equipotential particles. It corroborates well with the demonstration in the references [12,13,67,75].

Another possible mechanism for the energy transfer is due to that part of silver or gold (small silver or gold aggregates) probably remained under the form of ions, atoms, charged or neutral dimmers and multimers. Consequently, the contribution of the latter species to the RE luminescence enhancement and band shape features cannot be excluded [14], Figure 11. Moreover, the insertion of silver or gold in the samples leads to strong modification of the glassy network and consequently on the RE local environment, this is verified in reference [62].

5. Conclusion

This chapter present results where it was demonstrated the simultaneous exploitation of the enhanced local field due to NPs and energy transfer processes in order to enhance the luminescence spectra of a glassy composite material, and others. Besides, from these nanoparticles it is possible to modulate the down/up-conversion emission of the REs with applicability in areas such as optical telecommunication, including biomedical imaging and energy conversion. On the other hand, this significant enhanced fluorescence has high potential for application in photonics, optical displays, lasers and optical memory devices, amongst others. However, the success of new applications of nanoparticles depends on improvement in the understanding of the properties of LSPR and the environment around them. Further efforts and systematic studies must be realized in order to offer new developments to extend the analytical applications field of metallic nanoparticle interaction with rare-earth ions.

We hope that this chapter stimulated our readers for the development of theoretical and experimental work on plasmonics and rare-earths.

Author details

V.A.G. Rivera, F.A. Ferri and E. Marega Jr.
Instituto de Física de São Carlos, INOF/CEPOF, USP, São Carlos – São Paulo, Brazil

Acknowledgement

This work was financially supported by the Brazilian agencies FAPESP, CNPq and CEPOF/INOF. V.A.G. Rivera thanks to FAPESP for financial support (project 2009/08978-4 and 2011/21293-0) that allowed my pos-doctoral and my gratefully to Dr. Luiz Antonio Nunes of the Instituto de Fisica de São Carlos – University São Paulo - Brazil and the Dr. Yannick Ledemi and the Dr. Younnes Messaddeq of the Centre d'Optique, Photonique et laser – University Laval- Canada for the discussions on this issue.

6. References

[1] Stefan Alexander Maier. Plasmonics Funadamentals and Applications. Springer Science+Business Media LLC; 2007.
[2] Mark I. Stockman. Opt. Exp. 2011;19(22): 22029-22106.
[3] Y. Wang, E. W. Plummer and K. Kempa. Advances in Physics 2011;60(5): 799-898.

[4] K. Kneipp, H. Kneipp, I. Itzkan, R. R. Dasari, and M. S. Feld. J. Phys.: Condens. Matter 2002;14:R597–R624.
[5] Y. Takeda and N. Kishimoto. Nuc. Inst. Met. Phys. Res. B 2003;206: 620–623.
[6] Peter J. Reece. Nat. Phot. 2008;2:333-334.
[7] Desurvire E., J. L. Zyskind and C. R. Giles. IEEE/OSA J. Ligh. Technol. 1990;LT8: 1730–1741.
[8] A. N. Sudarkin and P. A. Demkovich. Sov. Phys. Tech. Phys. 1989;34: 764-766.
[9] I. Avrutsky. Phys. Rev. B 2004;70: 155416-155421.
[10] M. P. Nezhad, K. Tetz, and Y. Fainman. Opt. Express 2004;12: 4072-4079.
[11] Chau, K. J., Dice, G. D., and Elezzabi, A. Y. Phys. Rev. Lett. 2005;94: 173904-173907.
[12] V.A.G. Rivera, S. P. A. Osorio, Y. Ledemi, D. Manzani, Y. Messaddeq, L. A. O. Nunes, and E. Marega. Opt. Exp. 2010;18: 25321-25328.
[13] S.P.A. Osorio, V.A.G. Rivera, L.A.O. Nunes, E. Marega, D. Manzani, Y. Messaddeq. Plasmonics 2012;7: 53-58.
[14] V.A.G. Rivera, Y. Ledemi, S.P.A. Osorio, D. Manzani, Y. Messaddeq, L.A.O. Nunes and E. Marega. J. Non-Crys. Sol. 2012;358: 399-405.
[15] X. Lu, M. Rycenga, S.E. Skrabalak, B.Wiley, and Y. Xia. Annu. Rev. Phys. Chem. 2009;60: 167-192.
[16] U. Kreibig, M. Vollmer. Optical Properties of Metal Clusters. Springer-Verlag, Berlin; 1995.
[17] Jackson John D. Classical Electrodynamics. John Wiley & Sons, Inc., New York, NY, 3rd edition; 1999.
[18] Bohren, Craig F. and Huffman Donald R. Absorption and scattering of light by small particles. John Wiley & Sons, Inc., New York, NY, 1 edition; 1983.
[19] Mie Gustav. Ann. Phys. 1908;25: 377-345.
[20] E. D. Palik. Handbook of Optical Constants of Solids Academic. Elsevier, Orlando, FL. ISBN: 978-0-12-544415-6; 1985.
[21] K. Lance Kelly, Eduardo Coronado, Lin Lin Zhao, and George C. Schatz. J. Phys. Chem. B 2003;107: 668-677.
[22] Novotny L. and Hecht B. Principles of nano-optics, Cambridge University Press, United Kingdom, ISBN-13 968-0-521-83224-3; 2006.
[23] Hideki Nabika and Shigehito Deki. J. Phys. Chem. B 2003;107(35): 9161-9164.
[24] D. D. Evanoff, R. L. White and G. Chumanov. J. Phys. Chem. B 2004;108(37): 1522-1524.
[25] H. Baida, P. Billaud, S. Marhaba, D. Christofilos, E. Cottancin, A. Crut, J. Lermé, P. Maioli, M. Pellarin, M. Broyer, N. Del Fatti, and F. Vallé. Nano Lett. 2009;9(10): 3463-3469.
[26] C. Voisin, N. Del Fatti, D. Christofilos and F.J. Valleé. Phys. Chem. B 2001;105: 2264-2280.
[27] F. Hache, D. Ricard and C.J. Flytzanis. J. Opt. Soc. Am. B 1986;3(12): 1647-1655.
[28] S. Link, M. B. Mohamed and M. A. El-Sayed. J. Phys. Chem. B 1999;103: 3073-3077.
[29] H. Kuwata, H. Tamaru, K. Esumi and K. Miyano. Appl. Phys. Lett. 2003;83: 4625-4627.
[30] S.W. Prescott and P. Mulvaney. J. Appl. Phys. 2006;99: 123504-123510.

[31] Cheng-ping Huang, Xiao-gang Yin, Huang Huang and Yong-yuan Zhu. Opt. Exp. 2009;17(8): 6407-6413.
[32] Mock J. J., Barbic M., Smith D. R., Schultz D. A., and Schultz S. J. Chem. Phys. 2002;116(15): 6755–6759.
[33] Purcell E. M. and Pennypacker C. R. Astrophys. 1973;186: 705-714.
[34] Ruppin R. Phys. Rev. B 1982;26: 3440-3444.
[35] Pinchuk A., Kalsin A., Kowalczuk B., Schatz G. and Grzybowski B. J. Phys. Chem. C 2007;111: 11816-11822.
[36] Gerardy J. M. and Ausloos M. Phys. Rev. B 1982;25: 4204-4229.
[37] Claro F. Phys. Rev. B 1984;30: 4989-4999.
[38] Rojas R. and Claro F. Phys. Rev. B 1986;34: 3730-3736.
[39] Olivares I., Rojas R., Claro F. Phys. Rev. B 1987;35: 2453-2455.
[40] M. Chergui, A. Melikian , H. Minassian. J. Phys. Chem. C 2009;113: 6463–6471.
[41] P. P. Sorokin and M. J. Stevenson. Phys. Rev. Lett. 1960;5: 557-559.
[42] D.C. Brown. IEEE J. Sel. Top. Quantum Electron. 2005;11: 587-599.
[43] S. Hufner. Optical Spectra of Transparent Rare Earth Compounds. Academic press New York – San Francisco – London; 1978.
[44] B.R. Judd. Operator Techniques in Atomic Spectroscopy. McGraw-Hill, New York; 1963.
[45] B.G. Wybourne. Spectroscopic properties of Rare Earths, Wiley, New York; 1965.
[46] A.J. Freeman and R.E. Watson. Phys. Rev. 1962;127: 2058-2075.
[47] A.J. Kenyon. Prog. Quan. Elec. 2002;26: 225–284
[48] Michel J.F. Digonnet and Marcel Dekker. Rare-earth-doped fiber lasers and amplifiers. 2d ed., edited by, Inc. New York – Basel; 2001.
[49] V.A.G. Rivera, E.F. Chillce, E.G. Rodrigues, C.L. Cesar, L.C. Barbosa. J. Non-Crys. Sol. 2006;353: 125-130.
[50] V.A.G. Rivera, E.F. Chillce, E.G. Rodrigues, C.L. Cesar, L.C. Barbosa. Proc. SPIE 2006;6116: 190-193.
[51] P.J. Mears, L. Reekie, I.M. Jauncey and D.N. Payne. Elec. Lett. 1987;23: 1026-1028.
[52] E. Desurvire, R.J. Simpson and P.C. Becker. Opt. Lett. 1987;12: 888-890.
[53] E. Desurvire. Phys. Today 1994;47: 20-27.
[54] M. Yamane and Y. Asahara. Glasses for Photonics. Cambridge - University Press, Cambridge, United Kingdom; 2002.
[55] J.H. van Vleck. J. Phys. Chem. 1937;41: 67-80.
[56] B.R. Judd. Phys. Rev. 1962;127: 750-761.
[57] G. S. Ofelt. J. Chem. Phys. 1962;37: 511-520.
[58] Arai K., H. Namikawa, K. Kumata, T. Honda, Y. Ishii, T. Handa. J. Appl. Phys. 1986;59: 3430–3436.
[59] Shimizu M., M. Yamada, M. Horigucho and E. Sugita. IEEE Pho. Tech. Lett. 1990;2: 43–45.
[60] V.A. Markel, V. M. Shalaev, E. B. Stechel, W. Kim, and R. L. Armstrong, Phys. Rev. B 1996;53: 2425.
[61] V. M. Shalaev, E. Y. Poliakov and V. A. Markel. Phys. Rev. B 1996;53: 2425-2449.

[62] V.A.G. Rivera, S.P.A. Osorio, D. Manzani, Y. Messaddeq, L.A.O. Nunes, E. Marega Jr. Opt. Mat. 2011;33: 888-892.

[63] N. M. Lawandy. Appl. Phys. Lett. 2004;85: 5040-5042.

[64] S. Kim, J. Jin, Y. Kim, I. Park and Y. Kim. Nature 2008;453: 757-760.

[65] S. Kuhn, U. Hakanson, L. Rogobete and V. Sandoghdar. Phys. Rev. Lett. 2006;97: 017402.

[66] O.L. Malta, P.O. Santa-Cruz, G.F. de Sá and F. Auzel. J. of Lumin. 1985;33: 261-272.

[67] T. Som and B. Karmakar. J. Appl. Phys. 2009;105: 013102.

[68] L.R.P. Kassab, F.A.Bonfim, J.R. Martinelli, N.U. Wetter, J.J.Neto and Cid B. Araujo. Appl. Phys. B 2009;94: 239-242.

[69] T. Som and B. karmakar. J. Opt. Soc. Am. B 2009;26(12): B21-B27.

[70] Krishna Kanta Haldar and Amitava Patra. App. Phys. Lett. 2009;95: 063103.

[71] Thiago A. A. Assumpção, Davinson M. da Silva, Luciana R. P. Kassab and Cid B. de Araújo. J. App. Phys. 2009;106: 063522.

[72] Maik Eichelbaum and Klaus Rademann. Adv. Funct. Mater. 2009;19: 2045–2052.

[73] A. Gopinath, S. V. Boriskina, S. Yerci, R. Li and L. Dal Negro. Appl. Phys. Lett. 2010;96: 071113.

[74] J. R. Lakowicz. Anal. Biochem. 2005;337: 171-194.

[75] F. Le, D. W. Brandl, Y. A. Urzhumov, H. Wang, J. Kundu, N. J. Halas, J. Aizpurua, and P. Nordlander. ACS Nano 2008;2: 707-718.

Chapter 12

Resonant Excitation of Plasmons in Bi-Gratings

Taikei Suyama, Akira Matsushima, Yoichi Okuno and Toyonori Matsuda

Additional information is available at the end of the chapter

http://dx.doi.org/10.5772/50754

1. Introduction

Metal gratings have an interesting property known as resonance absorption in the optics region [Raeter 1982], which causes partial or total absorption of incident light energy. This absorption is associated with the resonant excitation of plasmons on a grating surface; incident light couples with surface plasmons via an evanescent spectral order generated by the grating [Nevièr 1982]. Resonance absorption in metal film gratings has been the subject of many theoretical [Nevièr 1982] and experimental investigations focused on various applications including chemical sensing [DeGrandpre 1990, Zoran 2009], surface enhanced phenomena such as Raman scattering [Nemetz 1994], and photonic bandgaps [Barnes 1995, Tan 1998].

A thin-film metal grating, which is a corrugated thin metal film, also results in absorption similar to that observed for thick gratings [Inagaki, Motosuga 1985, Chen 2008, Bryan-Brown 1991, Davis 2009]. Absorption in thin-film metal gratings, however, is much more complicated than in thick gratings because of the existence of coupled plasmon modes in addition to those observed in thick gratings. If the metal film is sufficiently thick, single-interface surface plasmons (SISPs) alone are excited [Raeter 1977, Okuno 2006, Suyama 2009]. However, if the film is sufficiently thin, simultaneous excitation of surface plasmons occurs on both surfaces; these plasmons interfere with each other and produce two coupled plasmon modes, short-range surface plasmons (SRSPs) and long-range surface plasmons (LRSPs) [Chen 1988, Hibbins 2006].

Most previous studies on resonance absorption have mainly dealt with metal gratings whose surfaces are periodic in one direction. Metal bi-gratings, which are periodic in two directions, also yield plasmon resonance absorption, similar to singly periodic gratings [Glass 1982, Glass 1983, Inagaki, Goudonnet 1985, Harris 1996]. In this work, we therefore investigated coupled plasmon modes excited in multilayered bi-gratings [Matsuda 1993, Matsuda 1996, Suyama 2010]. We anticipated interesting behavior in the resonance phenomenon due to the presence of the double periodicity. Further, in view of the fact that layered gratings are interesting structures for optical device applications, we investigated a multilayered bi-gratings, which is a stack of thin-film bi-gratings made of a dielectric or metal. This paper is structured as follows. After formulating the problem in Section **2**, we briefly describe a method for obtaining a solution in Section **3**. Focusing our attention on the resonant excitation of plasmon modes, we then show the computational results in Section **4**, before presenting the conclusions of this study.

2. Formulation of the problem

Here, we formulate the problem of diffraction from multilayered bi-gratings when an electromagnetic plane wave is incident on it. The time-dependent factor, exp($-i\omega t$), is suppressed throughout this paper as customary.

2.1. Geometry of the gratings

Figure 1 shows the schematic representation of multilayered sinusoidal gratings with double periodicity. The grating, with L-1 laminated grating layers, has a period d in both the X- and Y-directions. The semi-infinite regions corresponding to the medium above the

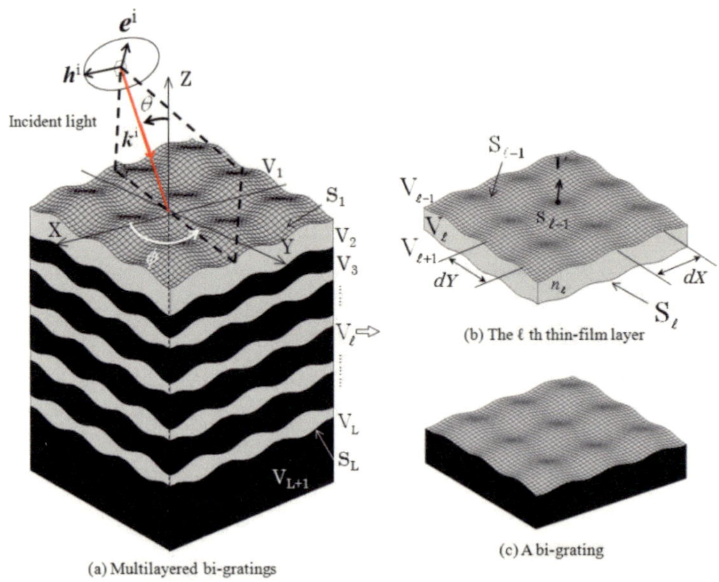

Figure 1. Schematic representation of bi-gratings. (a) Multilayered bi-gratings; (b) The l th thin-film bi-grating; (c) A bi-grating

grating and the substrate are denoted by V_1 and V_{L+1}, respectively. The individual layers in the grating, beginning from the upper layer (light-incidence side), are denoted by $V_\ell(\ell=2,3,\ldots,L)$. All of the regions $V_\ell(\ell=1,2,\ldots,L+1)$ are filled with isotropic and homogeneous media with refractive indices n_ℓ, and the permeability of each region is equal to that of vacuum, μ_0. The interface between V_ℓ and $\mathrm{V}_{\ell+1}$ is denoted by S_ℓ ($\ell = 1,2,\ldots,L$). The profile of S_ℓ is sinusoidal and is given by

$$z = \eta_\ell(x,y) = \frac{h_l}{4}\left\{\sin\left(\frac{2\pi x}{d} + p_\ell\right) + \sin\left(\frac{2\pi y}{d} + p_\ell\right)\right\} - \sum_{i=1}^{\ell-1} e_i \tag{1}$$

Here, p_ℓ represents the phase of S_ℓ in each direction, and e_ℓ denotes the average distance between S_ℓ and S_{l+1}. The value h_ℓ is regarded as the groove depth of the boundary S_ℓ.

It should be noted that we can easily generalize the shape of S_ℓ by distinguishing between *d* and p_ℓ in the *X*-and *Y*-directions, by writing them as d_x, d_y, $p_{x\ell}$ and $p_{y\ell}$, respectively. A singly periodic grating is a special case where $d_y \to \infty$ in the doubly periodic case. In the present paper, we concentrate our attention on describing the analysis only for the doubly periodic case, since it reduces to the singly periodic case through a simple procedure.

2.2. Incident wave

The electric and magnetic fields of an incident wave are given by

$$\begin{pmatrix} E^{\mathrm{i}} \\ H^{\mathrm{i}} \end{pmatrix}(P) = \begin{pmatrix} e^{\mathrm{i}} \\ h^{\mathrm{i}} \end{pmatrix} \exp\left(ik^{\mathrm{i}} \cdot P \right) \tag{2}$$

with

$$\boldsymbol{h}^{\mathrm{i}} = \left(1/\omega\mu_0\right) \boldsymbol{k}^{\mathrm{i}} \times \boldsymbol{e}^{\mathrm{i}} \tag{3}$$

Here, *P* is the position vector for an observation point P(*X*, *Y*, *Z*), and $\boldsymbol{k}^{\mathrm{i}}$ is the wave vector of the incident wave defined by

$$\boldsymbol{k}^{\mathrm{i}} = \left[\alpha, \beta, -\gamma\right]^{\mathrm{T}} \tag{4}$$

where $\alpha = n_0 k \sin\theta\cos\varphi$, $\beta = n_0 k \sin\theta \sin\varphi$, and $\gamma = n_0 k \cos\theta$. The symbol *k* (= $2\pi/\lambda$) is the wave number in vacuum and λ is the wavelength of the incident wave. We define θ and φ as polar and azimuth angles, respectively, as shown in Fig. 1(a), and the superscript "T" denotes a transposition.

The amplitude of the incident electric field can be decomposed into TE and TM modes [Chen 1973] (with respect to the *Z*-axis) and is written as

$$\boldsymbol{e}^{\mathrm{i}} = \cos\delta\, \boldsymbol{e}^{\mathrm{TE}} + \sin\delta\, \boldsymbol{e}^{\mathrm{TM}} \tag{5}$$

$$\boldsymbol{e}^{\mathrm{TE}} = \left[\sin\varphi, -\cos\varphi, 0\right]^{\mathrm{T}} \tag{6}$$

$$\boldsymbol{e}^{\mathrm{TM}} = \left[\cos\theta\cos\varphi, \cos\theta\sin\varphi, \sin\theta\right]^{\mathrm{T}} \tag{7}$$

Here, the superscript TE (TM) indicates the absence of a Z component of the electric (magnetic) field.

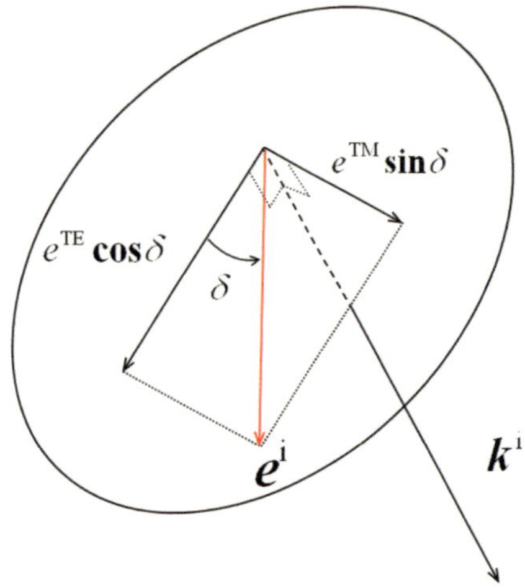

Figure 2. Definition of a polarization angle.

The symbol δ is the polarization angle between $\boldsymbol{e}^{\mathrm{TE}}$ and $\boldsymbol{e}^{\mathrm{i}}$ as shown in Fig. 2; for $\delta = 0^{\circ}$ ($\delta = 90^{\circ}$), this represents TE-mode (TM-mode) incidence.

2.3. Diffracted wave

We denote the diffracted fields as $E_{\ell}(P)$ and $H_{\ell}(P)$ in the region $V_{\ell}(\ell = 1,2,\ldots L, L+1)$. These satisfy the following conditions:

(C1) Helmholtz equation:

$$\left(\nabla^2 + n_\ell^2 k_\ell^2\right)\begin{pmatrix}\boldsymbol{E}_\ell \\ \boldsymbol{H}_\ell\end{pmatrix}(\boldsymbol{P}) = 0 \qquad (\ell = 1,2,\ldots L, L+1) \tag{8}$$

(C2) Radiation conditions:

E_1 and H_1 propagate or attenuate in the positive Z-direction.

E_{L+1} and E_{L+1} propagate or attenuate in the negative Z-direction.

(C3) Periodicity conditions:

$$f(X + d, Y, Z) = \exp(i\alpha d) f(X, Y, Z) \tag{9}$$

$$f(X, Y + d, Z) = \exp(i\beta d) f(X, Y, Z) \tag{10}$$

Here, *f* denotes any component of $E_{\ell}(P)$ or $H_{\ell}(P)$.

(C4) Boundary condition ($0 < X < d\,; 0 < Y < d\,; \ell = 1,2,\ldots,L$):

$$\boldsymbol{\nu}\times\left[\begin{pmatrix}\boldsymbol{E}_{\ell-1}\\ \boldsymbol{H}_{\ell-1}\end{pmatrix}+\delta_{\ell 1}\begin{pmatrix}\boldsymbol{E}^{\mathrm{i}}\\ \boldsymbol{H}^{\mathrm{i}}\end{pmatrix}-\begin{pmatrix}\boldsymbol{E}_{\ell}\\ \boldsymbol{H}_{\ell}\end{pmatrix}\right]=0 \quad (\text{on } \boldsymbol{S}_{\ell}) \tag{11}$$

where $\delta_{\ell 1}$ is the Kronecker delta, and ν_{ℓ} denotes the unit vector normal to the surface S_{ℓ}, which is given by

$$\nu_{\ell}=\frac{\left[-\dfrac{\partial\eta_{\ell}(X,Y)}{\partial X},-\dfrac{\partial\eta_{\ell}(X,Y)}{\partial Y},1\right]}{\sqrt{\left(\dfrac{\partial\eta_{\ell}(X,Y)}{\partial X}\right)^{2}+\left(\dfrac{\partial\eta_{\ell}(X,Y)}{\partial Y}\right)^{2}+1}} \tag{12}$$

3. Mode-matching method

We next explain the mode-matching method [Yasuura 1965, Yasuura 1971, Okuno 1990] for determining the diffracted field produced by the multilayered bi-grating. We introduce vector modal functions in the region $V_{\ell}\,(\ell=1,\ldots,L+1)$ to express the diffracted field in each individual region. To construct the wave functions $E^{\mathrm{d}}_{\ell N}(\boldsymbol{P})$ and $H^{\mathrm{d}}_{\ell N}(\boldsymbol{P})$, we define the electric modal function $\phi^{\mathrm{TE,TM}\pm}_{\ell mn}(\boldsymbol{P})$ and the magnetic modal function $\psi^{\mathrm{TE,TM}\pm}_{\ell mn}(\boldsymbol{P})$ as

$$\varphi^{\mathrm{TE,TM}\pm}_{\ell mn}(\boldsymbol{P})=e^{\mathrm{TE,TM}\pm}_{\ell mn}\phi^{\pm}_{\ell mn}(\boldsymbol{P}) \tag{13}$$

$$e^{\mathrm{TE}\pm}_{\ell mn}=\frac{k^{\pm}_{\ell mn}\times i_{Z}}{\left|k^{\pm}_{\ell mn}\times i_{Z}\right|},\ e^{\mathrm{TM}\pm}_{\ell mn}=\frac{e^{\mathrm{TE}\pm}_{\ell mn}\times k^{\pm}_{\ell mn}}{\left|e^{\mathrm{TE}\pm}_{\ell mn}\times k^{\pm}_{\ell mn}\right|} \tag{14}$$

$$\psi^{\mathrm{TE,TM}\pm}_{\ell mn}(\boldsymbol{P})=\frac{1}{\omega\mu_{0}}k^{\pm}_{\ell mn}\times\varphi^{\mathrm{TE,TM}\pm}_{\ell mn}(\boldsymbol{P}) \tag{15}$$

where $\phi^{\pm}_{\ell mn}(\boldsymbol{P})$ is the solution of the Helmholtz equation satisfying the periodic condition in the region $V_{\ell}\,(\ell=1,.2,\ldots,L+1)$. It is written as

$$\varphi^{\pm}_{\ell mn}(\boldsymbol{P})=\exp\left(ik^{\pm}_{\ell mn}\cdot\boldsymbol{P}\right)\quad(m,n=0,\pm1,\pm2,\ldots) \tag{16}$$

where the positive and negative signs match on either side of the equation, and $k^{\pm}_{\ell mn}$ is the wave vector of the (m,n)th order diffracted wave given by

$$k^{\pm}_{\ell mn}=\left[\alpha_{m},\beta_{n},\pm\gamma_{\ell mn}\right]^{\mathrm{T}} \tag{17}$$

$$\alpha_{m}=\alpha+\frac{2m\pi}{d},\ \beta_{n}=\beta+\frac{2n\pi}{d} \tag{18}$$

$$\gamma_{\ell mn} = \left(n_\ell^2 k^2 - \alpha_m^2 - \beta_n^2\right)^{1/2}, \ \mathrm{Re}\left(\gamma_{\ell mn}\right) \geq 0 \text{ and } \mathrm{Im}\left(\gamma_{\ell mn}\right) \geq 0 \tag{19}$$

Note that the superscripts + and - represent upwardly and downwardly propagating waves in the positive and negative Z-direction, respectively.

In terms of the linear combinations of the vector modal functions, we form approximate solutions for the diffracted electric and magnetic fields in V_ℓ :

$$\begin{aligned}\begin{pmatrix} \boldsymbol{E}_{\ell N}^{\mathrm{d}} \\ \boldsymbol{H}_{\ell N}^{\mathrm{d}} \end{pmatrix}(\boldsymbol{P}) = & \sum_{m,n=-N}^{N} A_{\ell mn}^{\mathrm{TE+}}(N) \begin{pmatrix} \varphi_{\ell mn}^{\mathrm{TE+}} \\ \psi_{\ell mn}^{\mathrm{TE+}} \end{pmatrix}(\boldsymbol{P}) \\ & + \sum_{m,n=-N}^{N} A_{\ell mn}^{\mathrm{TM+}}(N) \begin{pmatrix} \varphi_{\ell mn}^{\mathrm{TM+}} \\ \psi_{\ell mn}^{\mathrm{TM+}} \end{pmatrix}(\boldsymbol{P}) \\ & + \sum_{m,n=-N}^{N} A_{\ell mn}^{\mathrm{TE-}}(N) \begin{pmatrix} \varphi_{\ell mn}^{\mathrm{TE-}} \\ \psi_{\ell mn}^{\mathrm{TE-}} \end{pmatrix}(\boldsymbol{P}) \\ & + \sum_{m,n=-N}^{N} A_{\ell mn}^{\mathrm{TM-}}(N) \begin{pmatrix} \varphi_{\ell mn}^{\mathrm{TM-}} \\ \psi_{\ell mn}^{\mathrm{TM-}} \end{pmatrix}(\boldsymbol{P}) \\ & \left(\ell = 1, 2, \ldots, L+1\right)\end{aligned} \tag{20}$$

with

$$\psi_{\ell mn}^{\mathrm{TE,TM}\pm}(\boldsymbol{P}) = \frac{1}{\omega\mu_0} k_{\ell mn}^{\pm} \times \phi_{\ell mn}^{\mathrm{TE,TM}\pm}. \tag{21}$$

Here, $A_{1mn}^{\mathrm{TE-}}(N) = A_{1mn}^{\mathrm{TM-}}(N) = 0$ and $A_{L+1mn}^{\mathrm{TE+}}(N) = A_{L+1mn}^{\mathrm{TM+}}(N) = 0$ because of the radiation conditions (C3) stated in **Section 2.3**.

The approximate solutions $\boldsymbol{E}_{\ell N}^{\mathrm{d}}(\boldsymbol{P})$ and $\boldsymbol{H}_{\ell N}^{\mathrm{d}}(\boldsymbol{P})$ already satisfy the Helmholtz equation (C1), the periodicity conditions (C2), and the radiation conditions (C3). The unknown coefficients $A_{\ell mn}^{\mathrm{TE}\pm}(N)$ and $A_{\ell mn}^{\mathrm{TM}\pm}(N)$ are therefore determined such that the solutions approximately satisfy the boundary conditions (C4). In the mode-matching method [Yasuura 1996, Yasuura 1971, Okuno 1990], the least-squares method is employed to fit the solution to the boundary conditions [Hugonin 1981]. That is, we find coefficients that minimize the weighted mean-square error by

$$\begin{aligned} I_N = & \int_{S_1'} \left| \nu \times \left[\mathrm{E}_{1N}^{\mathrm{d}} + \mathrm{E}^{\mathrm{i}} - \mathrm{E}_{2N}^{\mathrm{d}} \right](s_1) \right|^2 ds \\ & + \left|\Gamma_1\right|^2 \int_{S_1'} \left| \nu \times \left[\mathrm{H}_{1N}^{\mathrm{d}} + \mathrm{H}^{\mathrm{i}} - \mathrm{H}_{2N}^{\mathrm{d}} \right](s_1) \right|^2 ds \\ & + \sum_{\ell=2}^{L} \left\{ \int_{S_\ell'} \left| \nu \times \left[\mathrm{E}_{\ell N}^{\mathrm{d}} - \mathrm{E}_{\ell+1N}^{\mathrm{d}} \right](s_\ell) \right|^2 ds + \right. \\ & \left. \left|\Gamma_\ell\right|^2 \int_{S_\ell'} \left| \nu \times \left[\mathrm{H}_{\ell N}^{\mathrm{d}} - \mathrm{H}_{\ell+1N}^{\mathrm{d}} \right](s_\ell)^2 \right| ds \right\} \end{aligned} \tag{22}$$

Here, $S_\ell^{'}$ is a one-period cell of the interface S_ℓ, and Γ_ℓ is the intrinsic impedance of the medium in V_ℓ.

To solve the least-squares problem on a computer, we first discretize the weighted mean-square error I_N by applying a two-dimensional trapezoidal rule where the number of divisions in the x-and y-direction is chosen as $2(2N+1)$ [Yasuura 1965, Yasuura 1971, Okuno 1990]. We then solve the discretized least-squares problem by the QR decomposition method. Computational implementation of the least-squares problem is detailed in the literature [Lawson 1974, Matsuda 1966, Suyama 2008].

4. Numerical results

Here we show some numerical results obtained by the method described in the preceding section. After making necessary preparation, we show the results for three the bi-grating, thin-film bi-grating and multilayered thin-film bi-gratings cases.

4.1. Preparation

It is known that the solutions obtained by mode-matching method [Yasuura 1965, Yasuura 1971, Okuno 1990] have proof of convergence. We, therefore, can employ the coefficient $A_{\ell mn}^{\mathrm{TM,TE}\pm}(N)$ with sufficiently large N for which the coefficients are stable in evaluating diffracted fields.

The power reflection and transmission coefficient of the (m, n) order propagating mode in V_1 and $\mathrm{V}_{\mathrm{L}+1}$ are given by

$$\rho_{\mathrm{mn}} = \rho_{mn}^{\mathrm{TE}} + \rho_{mn}^{\mathrm{TM}} \; [\mathrm{Re}(\gamma_{1mn}) \geq 0], \; \rho_{mn}^{\mathrm{TE}} = \frac{\gamma_{1mn}}{\gamma}\left|A_{1mn}^{\mathrm{TE}+}\right|^2, \; \rho_{mn}^{\mathrm{TM}} = \frac{\gamma_{1mn}}{\gamma}\left|A_{1mn}^{\mathrm{TM}+}\right|^2 \tag{23}$$

$$\tau_{\mathrm{mn}} = \tau_{mn}^{\mathrm{TE}} + \tau_{mn}^{\mathrm{TM}} \; [\mathrm{Re}(\gamma_{L+1mn}) \geq 0], \; \tau_{mn}^{\mathrm{TE}} = \frac{\gamma_{L+1mn}}{\gamma}\left|A_{L+1mn}^{\mathrm{TE}-}\right|^2, \; \tau_{mn}^{\mathrm{TM}} = \frac{\gamma_{L+1mn}}{\gamma}\left|A_{L+1mn}^{\mathrm{TM}-}\right|^2 \tag{24}$$

The coefficient defined above is the power carried away by propagating diffraction orders normalized by the incident power. We calculate the total diffraction efficiency $\rho^{\mathrm{total}} = \sum' \rho_{mn}$ where $\sum'$ denotes a summation over the propagating orders.

Although it is known that the solutions obtained by mode-matching modal expansion method [Yasuura 1965, Yasuura 1971, Okuno 1990] have proof of convergence for problems of diffraction by gratings, we compare our results with other existing theoretical [Glass 1983] and experimental results [Inagaki, Goudonnet 1985] on plasmon resonance absorption in bi-gratings to show the validity of the present method. Figure 3 shows the reflectivity curves calculated by the present method and those from Rayleigh's method [Glass 1983] for three sinusoidal silver bi-gratings with different corrugation amplitudes. As confirmed in Fig. 3, the reflectivity curves from the present method are coincident with those of the

Rayleigh's method [Glass 1983]. Next, we make comparison with the experimental results [Inagaki, Goudonnet 1985] in which the resonance angle θ_d, i.e., the polar angle at which the dip of reflectivity occurs, is observed near 55° for a sinusoidal silver bi-grating with $h = 0.048$ μm, $d = 2.186$ μm, $\lambda = 0.633$ μm, and $\phi = 45°$. The resonance angle calculated from the present metod for these parameters is $\theta_d = 54.1°$, which is close to the experimental data. These examples show that the present method gives reliable results for the analysis of plasmon-resonance absorption in metal bi-gratings.

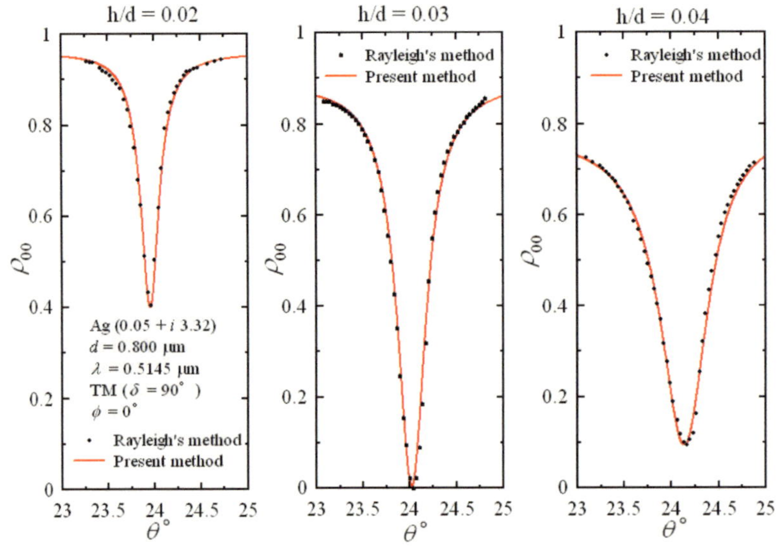

Figure 3. Comparison of resonance absorption curves calculated by the present method with other existing theoretical results. Solid curves show our results, and dotted curves are taken from Figure 2. of Ref. [Glass 1983].

In the numerical examples presented here, we deal with a shallow sinusoidal silver bi-grating with height $h = 0.030$ μm and period $d = 0.556$ μm. The wavelength of an incident light is chosen as $\lambda = 0.650$ μm. We take $n_2 = 0.07+i4.20$ as the refractive index of silver at this wavelength [Hass 1963].

It should be noted, however, that the index of a metal film depends not only on the wavelength but also on the thickness of the film, in particular when the film is extremely thin it may take unusual values if circumstances require. When dealing with a thin metal structure, hence, we should be careful in using the index value given in the literature. As for the value taken in our computation, we assume that $n_2 = 0.07+i4.20$ is available even for the case of $e/d = 0.02$. This is because a similar assumption was supported by experimental data in a problem of diffraction by an aluminum grating with a thin gold over-coating.

4.2. A bi-grating case

Using the numerical algorithm stated in the previous section, we first investigate the absorption in a metal bi-gratings by $L = 1$ as shown in Fig. 1(c). The semi-infinite regions corresponding to the medium above the grating and the substrate are denoted by V_1 and V_2, respectively. V_1 is vacuum (V) with a relative refractive index n_1=1 and V_2 consists of a lossy metal characterized by a complex refractive index n_2.

4.2.1. *Diffraction efficiency*

Figure 4 shows the total diffraction efficiency of a sinusoidal silver bi-grating as functions of a polar angle θ when the azimuthal angle ϕ = 30° is fixed. In the efficiency curves we observe four dips which occur at the same angles of incidence for both TE and TM polarized incident light. In this subsection, we demonstrate that the dips are associated with absorption that is caused by the coupling of surface plasmons with an evanescent mode diffracted by a sinusoidal bi-grating. For convenience, the four dips in Fig. 4 are labeled as A, B, C, and D.

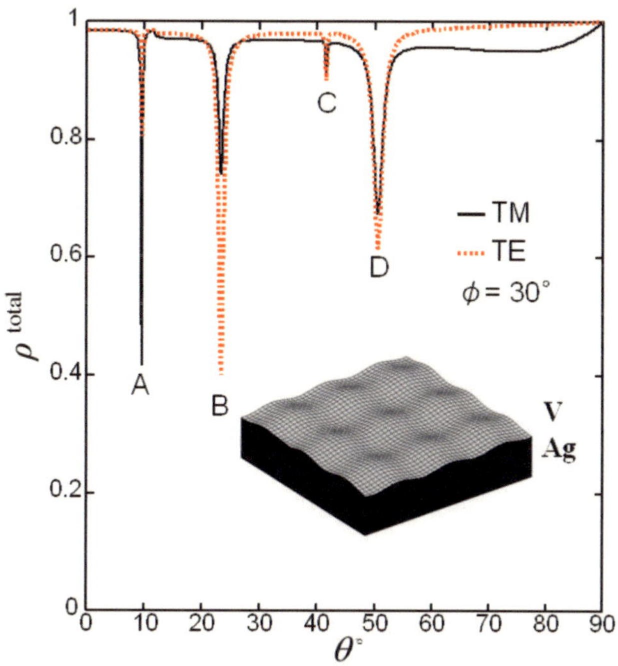

Figure 4. Total diffraction efficiencies ρ^{total} as functions of θ (L=1).

4.2.2. *Expansion coefficients*

In Fig. 5 we plot the expansion coefficients of the (0, -1)st-order and (-1, 0)th-order TM vector modal function, which are two evanescent modes, as a functions of θ under the same parameters as in Fig. 4. Solid curves in Fig. 5 represent the real part of the expansion coefficient, and dashed curves for the imaginary part. We observe a resonance curve of the expansion coefficient $A^{TM}_{1(-1,0)}$ at the angles of incidence θ = 9.5°, i.e., dip A, and $A^{TM}_{1(0,-1)}$ at the angles of incidence θ = 23.3°, 41.5° and 49.5°, i.e., dips B, C, and D in Fig. 4 for both TE and TM incidence.

This implies that the TM component of the (0, -1)st-order and (-1, 0)th-order evanescent mode couples with surface plasmons at dips B and D. We can similarly confirm that dips A and C are associated with the coupling of the TM component of the (-1, 0)th- and (-1, -1)st-order evanescent mode, respectively, with surface plasmons, although we do not include any numerical example here.

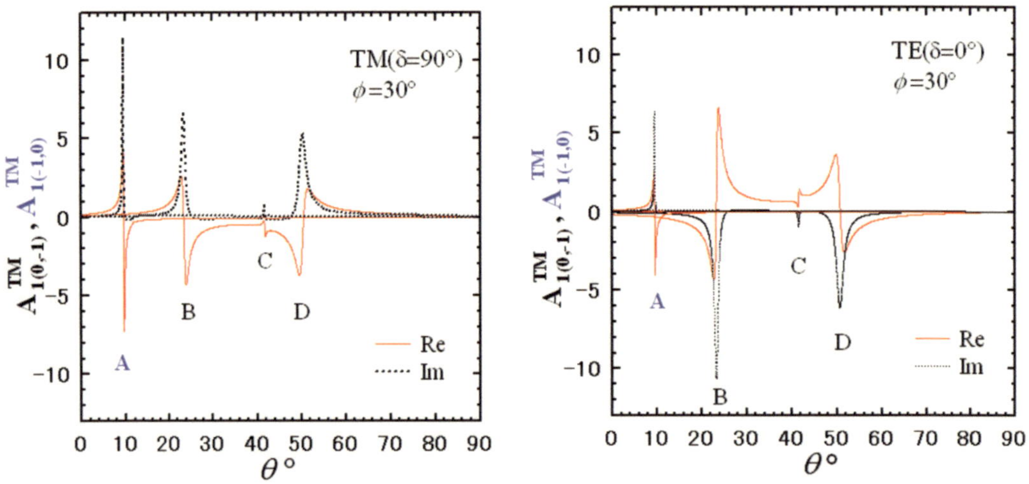

Figure 5. $A^{TM}_{1(0,-1)}$ and $A^{TM}_{1(-1,0)}$ as functions of θ for both TM- (a) and TE- (b) polarized incident light.

4.2.3. Field distributions and energy flows

In order to investigate the resonant excitation of surface plasmons, we study field distributions and energy flows in the vicinity of the grating surface when the absorption occurs. Here we consider the case of the dip B at which the TM component of the (0, -1)st-order evanescent mode couples with surface plasmons. We calculate the electric field of the TM component of the (0, -1)st-order evanescent mode $E^{TM}_{\ell 0-1} = A^{TM}_{\ell 0-1}(N)\ \varphi^{TM}_{\ell 0-1}(\ell = 1,2)$ and the total electric field $E^{\,t}$. The magnitude of these fields along the Z-axis where $Y = (\beta_{-1}/\alpha_0) * X$ is plotted in Fig. 6(a). We observe in this figure that very strong electric fields are induced at the grating metal surface and the fields exponentially decay away from the surface.

Next we show the energy flows **S** that are the real part of Poynting's vectors for the total field. The X and Y components of the energy flows **S** are plotted as the vector (S_X, S_Y) in Fig. 6(b). The energy flows are calculated over the region close to the grating surface:

$$\{P(x,y,z): 0 \le x \le d, 0 \le y \le d, \\ z = (h/4)[\sin(2\pi x/d) + \sin(2\pi y/d)] + 0.01d\}. \tag{25}$$

The energy flow at a point P is given by Re[$S(P)$], where $S(P)=(1/2)\, E^t(P)\times \overline{H^t}(P)$ stands for Poynting's vector, E^t and H^t denote total fields, and the over-bar means complex conjugate.

We calculate the energy flow at each point located densely near the grating surface and show the results in Fig. 6(b).

Figure 6(b) shows that the energy of electromagnetic fields in the vicinity of the grating surface flows uniformly in the direction that the (0, -1)st-order evanescent mode travels in the *XY* plane. We thus confirm that surface plasmons are excited on the grating surface through the coupling of the TM component of an evanescent mode.

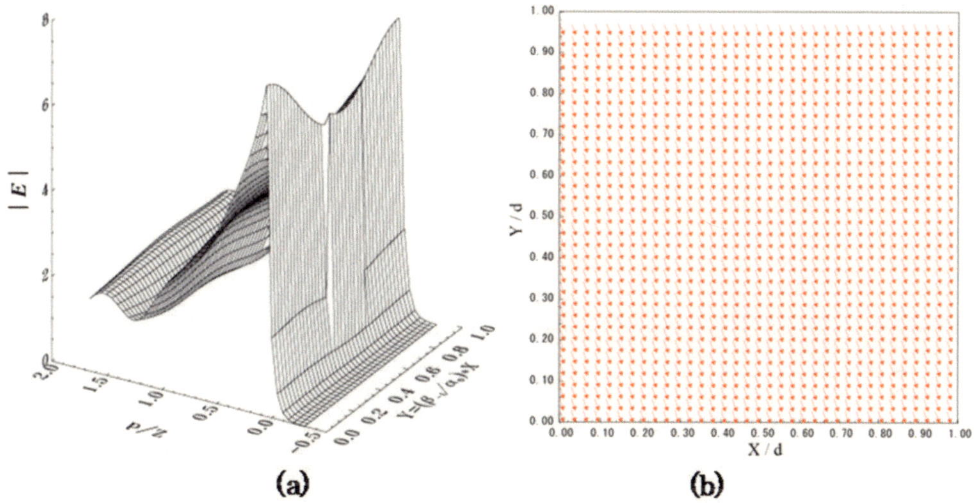

Figure 6. Field distribution (a) and Energy flow (b) for the total field when plasmon resonance absorption occurs at $\phi = 30°$, $\theta = 23.3°$.

4.2.4. *Polarization conversion through plasmon resonance absorption*

Diffracted fields from a sinusoidal metal bi-grating have both TE and TM component for an arbitrary polarized incident light. We therefore observe polarization conversion that a TM (or TE) component of the incident light is converted into a TE (or TM) component of the reflected light. It has been pointed out [Chen 1973, Inagaki, Goudonnet 1985] that the polarization conversion [Elston 1991, Matsuda 1999, Suyama 2007] is strongly enhanced when the plasmon-resonance absorption occurs in a sinusoidal metal bi-grating. Our study confirms the enhancement of polarization conversion through plasmon-resonance absorption. In Fig. 7, the TE and TM component of the diffraction efficiency ρ_{00} ($\rho_{00} = \rho_{00}^{\mathrm{TE}} + \rho_{00}^{\mathrm{TM}}$) of Fig. 4 are shown for the case of the TM incidence. The TM component ρ_{00}^{TM} is decreased at the position of the plasmon-resonance absorption, but the TE component ρ_{00}^{TE} is contrary increased there. That is, the resonant excitation of surface plasmons causes the enhancement of polarization conversion. On the other hand, in the case of the TE incidence the conversion from a TE to a TM component occurs through the plasmon-resonance absorption. It should be noted that the conversion efficiency depends on the azimuthal angle φ and the depth of the grating surface h.

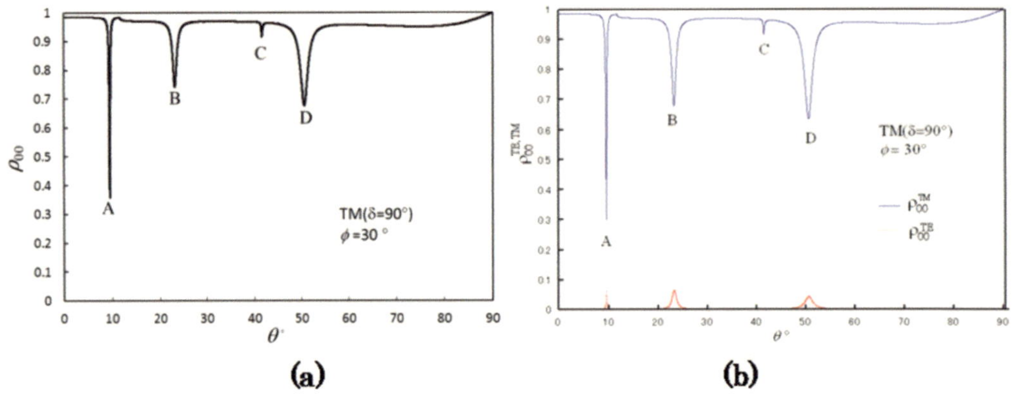

Figure 7. Diffraction efficiencies of ρ_{00} (a) ρ_{00}^{TE} and ρ_{00}^{TM} (b) as functions of θ; parameters are the same as in Figure 4.

4.2.5. Prediction of resonance angles

We seek to determine a complex incidence angle θ_c for which total or partial absorption occurs, i.e., $\rho_{\mathrm{total}}\left(=\rho_{00}\right)$ takes a minimum. This angle relates to the propagation constant of the surface plasmon on the corrugated surface. Here, we denote by $\hat{\alpha}_{\mathrm{SP}}$ and $\hat{\beta}_{\mathrm{SP}}$ the X and Y components of the surface plasmon wave vector normalized by the wave number k_1:

$$\begin{aligned} \hat{\alpha}_{SP} &= \sin\theta_{\mathrm{c}}\cos\phi + m\lambda / d, \\ \hat{\beta}_{SP} &= \sin\theta_{\mathrm{c}}\sin\phi + n\lambda / d. \end{aligned} \tag{26}$$

In reality, we cannot realize a complex incidence angle θ_{c}. We can, however, estimate the real angle of incidence at which the absorption occurs by taking the real part of Eq. (26).

If the wavevector of the surface plasmon ($\hat{\alpha}_{\mathrm{SP}}, \hat{\beta}_{\mathrm{SP}}$) is obtained, we can estimate the resonance angle θ_{SP} for each azimuthal angle φ from the phase-matching condition for coupling of a surface plasmon wave with the (m, n)th-order evanescent mode:

$$\begin{aligned} \mathrm{Re}\{\hat{\alpha}_{\mathrm{SP}}\} &= \sin\theta_{\mathrm{SP}}\cos\varphi + m\lambda / d, \\ \mathrm{Re}\{\hat{\beta}_{\mathrm{SP}}\} &= \sin\theta_{\mathrm{SP}}\sin\varphi + n\lambda / d. \end{aligned} \tag{27}$$

We solve the homogenous problem [Nevièr 1982] for a sinusoidal metal bi-grating by present method and then obtain the surface-plasmon wave vector. Table 1 shows the propagation constants of the surface plasmon and the resonance angles θ_{SP}. The data demonstrate that the estimated resonance angle θ_{SP} agrees with θ_{d} which is the absorption peak in Fig. 4. Figure 8 shows the estimated resonance angle θ_{SP} as a function of the azimuthal angle φ for the sinusoidal bi-grating considered in Fig. 4. Note that points A, B, C, and D in Fig. 8 are results obtained from the absorption peak of total-efficiency curves in Fig. 4. From this figure, we can find the resonance angle for each azimuthal angle.

Dip	Mode	Re($\hat{\alpha}_{SP}$)	Im($\hat{\alpha}_{SP}$)	Re($\hat{\beta}_{SP}$)	Im($\hat{\beta}_{SP}$)	θ°_{SP}	θ°_{d}
A	(-1, 0)	-1.026277	-0.000804	0.087099	-0.001279	9.49	9.5
B	(0, -1)	0.343795	-0.004910	-0.971021	-0.002498	23.39	23.3
C	(0, -1)	-0.595897	-0.001282	-0.837168	-0.000549	41.44	41.5
D	(0, -1)	0.668265	-0.004365	-0.778336	-0.005224	50.50	50.5
E	(0, -1)	0.149639	-0.000679	-1.019425	-0.000679	12.22	12.2
	(-1, 0)	-1.019426	-0.000679	0.149639	-0.000679	12.22	

[a] $h = 0.03\mu\text{m},\ d = 0.556\mu\text{m},\ \lambda = 0.650\mu\text{m}$

Table 1. Propagation Constants and Estimated Resonance Angles[a]

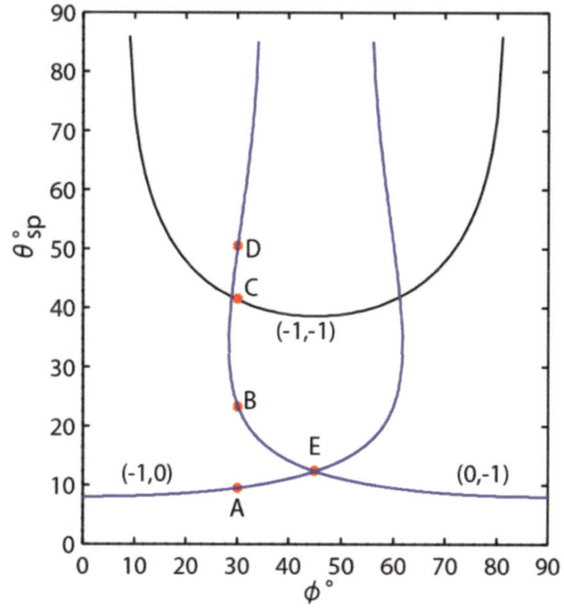

Figure 8. Resonance angles θ_{SP} as functions of azimuthal angle ϕ .

4.2.6. Simultaneous resonance absorption

From Fig. 8, it is predicted the angle of incidence at which the simultaneous resonance absorption occurs from the position of the intersection of the (-1,0) and (0,-1) curve. At the intersection **E**, the (0,-1)st- and (-1,0)th-order evanescent modes couple simultaneously with two surface-plasmon waves at the same angle of incidence. Thus, two surface-plasmon waves are excited simultaneously in directions symmetric with respect to the plane of incidence and interact with each other. The interference of the surface-plasmon waves causes the standing wave in the vicinity of the grating surface. This is confirmed from Figs. 9, the strong fields along the Z-axis where Y=X, and where the *X* and *Y* components of Poynting's vectors **S** on a surface 0.01*d* above the one-unit cell of the grating surface are plotted as the vector (S_X, S_Y). We further observe in Fig. 10 that the simultaneous excitation of the surface plasmons waves causes the strong absorption for both TE- and TM-polarized incident light.

It has been reported [Barnes 1995, Ritchie 1968] that surface-plasmon band gaps exist at the angles of incidence at which simultaneous excitation of plasmon waves occurs, and that the appearance of the band gaps depends strongly on the surface profile. Hence, there is a possibility that a band gap will be observed at the point E in Figs. 9 and 10 provided that the grating profile is appropriately chosen, because two plasmon modes are excited at that point.

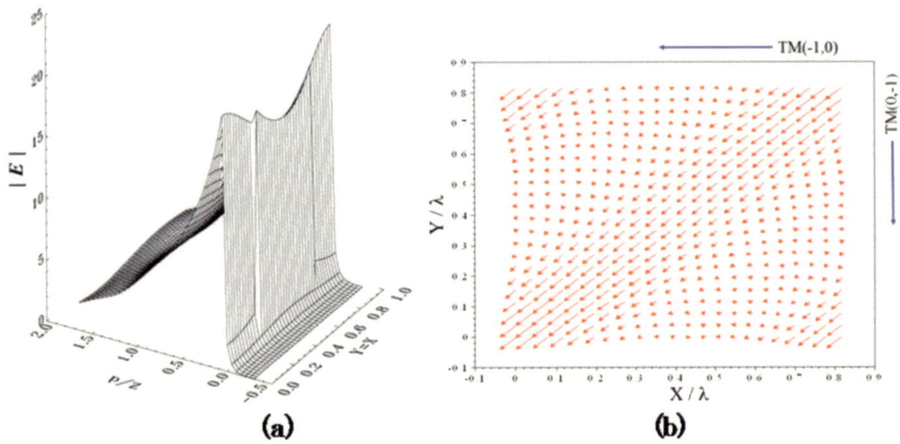

Figure 9. Field distributions and Energy flows for the total field when plasmon resonance absorption occurs.

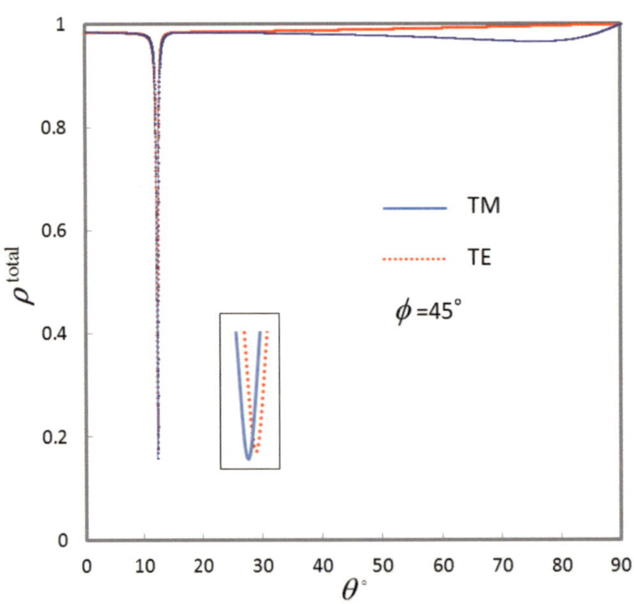

Figure 10. Total diffraction efficiencies ρ^{total} as functions of θ.

4.3. A Thin-film Bi-grating case

As a numerical example, we consider a sinusoidal silver (Ag) film bi-grating having a common period as shown in Fig. 1(b). The values of the parameters are the same as those in Fig. 4 except for the thickness of the silver film. Using the present algorithm, we calculated the diffraction efficiencies and field distributions to clarify the properties of the coupled surface plasmon modes.

4.3.1. Diffraction efficiency

First, we consider a sinusoidal silver-film bi-grating. The bi-grating is denoted by $L = 2$ (V/Ag/V). Figure 11 shows the (0,0)th order power reflection ρ_{00} in V_1 (Vacuum) (a) and the transmission coefficient τ_{00} in V_3 (Vacuum) (b) as functions of the incident angle θ for two values of e/d when the azimuth angle $\phi = 0^\circ$ is fixed; e is the thickness of the silver film. We observe partial absorption of the incident light as dips in the efficiency curves in Fig. 11(a), in addition to the constant absorption corresponding to the reflectivity of silver. We assume that the dips are caused by resonant excitation of surface plasmons. If this is the case, each of the dips can be related to one of the three types of plasmon modes: a SISP that is observed as a single dip at $\theta = 8.0^\circ$ on the $e/d = 0.4$ curve, and a SRSP and LRSP corresponding to the dips in the $e/d = 0.08$ curve at $\theta = 6.54^\circ$ and 8.8°, respectively.

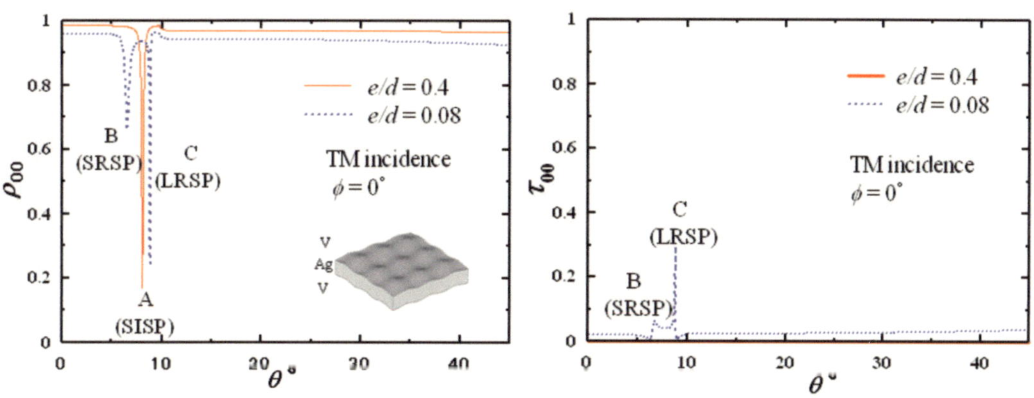

Figure 11. The (0, 0)-th order diffraction efficiencies ρ_{00} (a) and τ_{00} (b) as functions of θ for two values of e/d (*L*=2).

When the grating is thick ($e/d = 0.4$), the power can be seen in V_1 alone and no transmitted power exists in V_3 in Fig. 11. Increasing θ from 0°, we first observe the dip at $\theta = 8.0^\circ$ corresponding to absorption in Fig. 11(a). If the grating is relatively thin ($e/d = 0.08$), the power exists in both V_1 and V_3. Although the power in V_3 is generally small, it becomes large at the incidence angles for which absorption was observed in Fig. 11. This suggests that coupled oscillations occur on the upper and lower surfaces of the grating.

4.3.2. Expansion coefficients

We examined the same phenomena observing the modal expansion coefficients in $V_l(l=1,3)$. Figures 12 and 13 illustrate the (-1,0)th-order coefficients $A_{1(-1,0)}^{TM+}$ for $e/d=0.4$ and 0.08, respectively. We observe, in Fig. 12(a) ($e/d=0.4$), the resonance characteristics (enhancement and rapid change in phase) of the coefficient $A_{1(-1,0)}^{TM+}$ near $\theta=8.0^{\circ}$. The coefficient $A_{3(-1,0)}^{TM-}=0$ remains unchanged, as seen in Fig. 12(b). This means that the incident wave illuminating the grating at this angle causes coupling between the (-1,0)th-order evanescent mode and some oscillation excited on the upper surface of the grating. The oscillation exists locally in the vicinity of the illuminated surface and hence does not have any influence on the field in V_3.

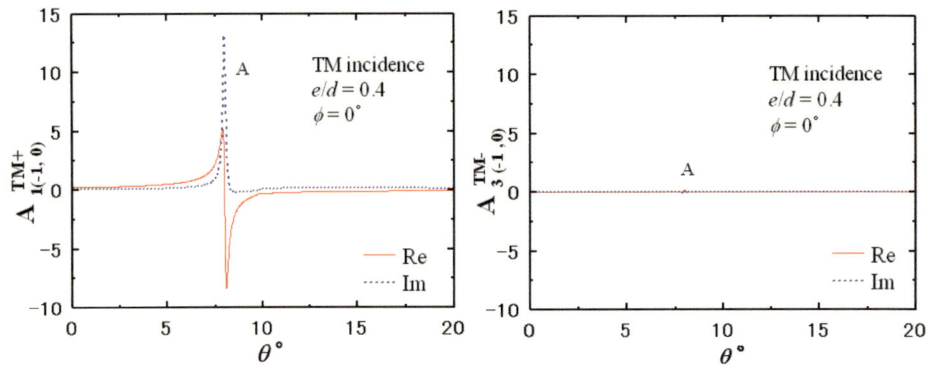

Figure 12. The (-1, 0)th order modal coefficients $A_{1(-1,0)}^{TM+}$ (a) and $A_{3(-1,0)}^{TM-}$ (b) as functions of θ at $e/d=0.4$.

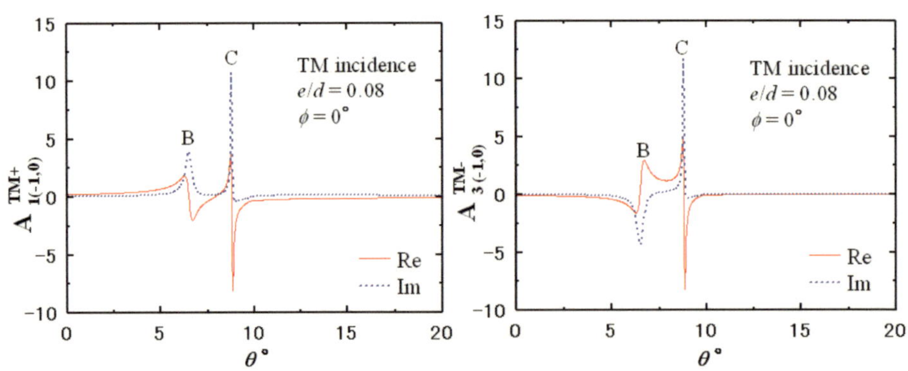

Figure 13. The (-1, 0)th order modal coefficients $A_{1(-1,0)}^{TM+}$ (a) and $A_{3(-1,0)}^{TM-}$ (b) as functions of θ at $e/d=0.08$.

In Fig. 13 ($e/d=0.08$), we find the resonance characteristics in both $A_{1(-1,0)}^{TM+}$ and $A_{3(-1,0)}^{TM-}$. In addition, they appear around two incidence angles: $\theta=6.54^{\circ}$ and $\theta=8.8^{\circ}$. This means that the oscillation in the vicinity of the upper surface causes another oscillation on the lower surface at this thickness. The oscillations interfere with each other and result in two coupled oscillating modes: the SRSP and LRSP. This means that the TM component of the (-1,0)th order evanescent mode couples with the surface plasmons simultaneously excited on the upper and lower surface of the film grating. The two surface plasmons interfere with each other and result in symmetric and antisymmetric coupled modes, SRSP and LRSP, as we will see next.

4.3.3. Field distributions and energy flows

We consider the same phenomena observing the field distributions and energy flows near the grating surfaces. In the former we find that the total field is enhanced. In the latter we observe the symmetric (even) and anti-symmetric (odd) nature of the oscillations, which correspond to the LRSP and SRSP [Raeter 1977].

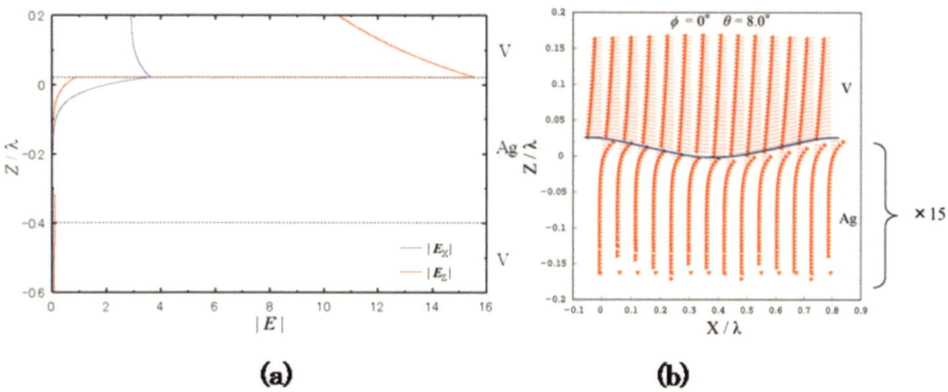

Figure 14. Field distributions (a) and energy flows (b) at $\theta = 8.0^{\circ}$, which corresponds to the single dip on the $e/d = 0.4$ curve in Fig. 11.

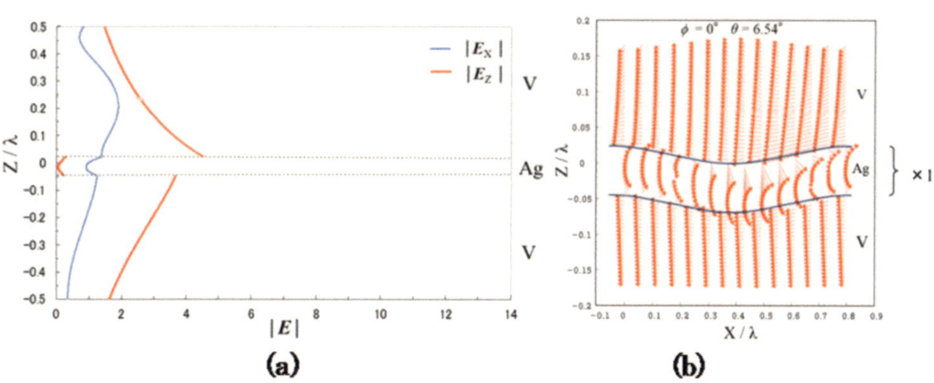

Figure 15. Field distributions (a) and energy flows (b) at $\theta = 6.54^{\circ}$, which corresponds to the left dip on the $e/d = 0.08$ curve in Fig. 11.

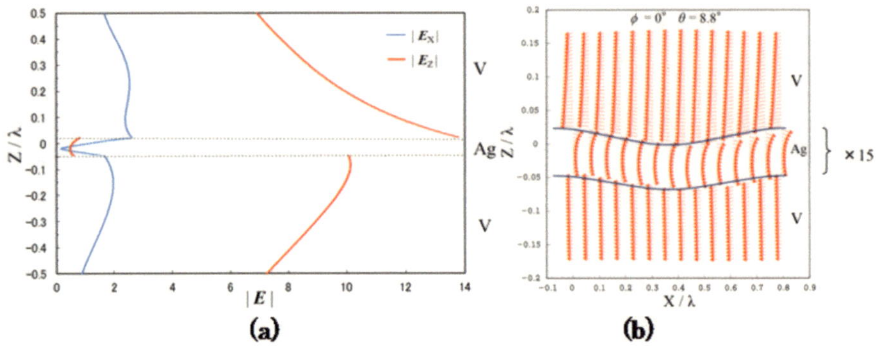

Figure 16. Field distributions (a) and energy flows (b) at $\theta = 8.8^{\circ}$, which corresponds to the right dip on the $e/d = 0.08$ curve in Fig. 11.

Figures 14, 15, and 16 show the field distributions of the *X*- and *Z*-components of the total electric fields (a) and energy flows (b) in the vicinity of the silver-film grating at the incidence angles at which absorption was observed in Fig. 11. The abscissa and ordinate show the magnitude and distance in the *Z* direction normalized by the wavelength λ. The parallel broken lines represent the grating surfaces.

Figure 14 shows $\boldsymbol{E_X}$ and $\boldsymbol{E_Z}$ (a), *Sx* and *Sz* (b) for the case of $e/d = 0.4$ at $\theta = 8.0^{\circ}$, which corresponds to the single dip in Fig. 11(a). Figures 15 and 16 show the same thing for the $e/d = 0.08$ case. Figure 15 illustrates the results at $\theta = 6.54^{\circ}$, where the left dip is observed in Fig. 11(a). On the other hand, Fig. 16 depicts the results at $\theta = 8.8^{\circ}$, corresponding to the right dip in Fig. 11(a). In Figs. 14 to 16, we observe strong enhancement of $\boldsymbol{E_X}$ and $\boldsymbol{E_Z}$ (note that the magnitude of the incident radiation is 1), which is observed at the incidence angles where absorption occurs.

We observed that the total field above the grating surface decays exponentially in the *Z* direction and the magnitude of the total field is almost E_Z in Fig. 14(a). The state of affairs is nearly the same in the metal region except for the rapid decay. Because the grating is thick, the oscillation near the upper surface does not reach the lower surface and, hence, the field below the grating is zero. Figure 14(b) illustrates the energy flow **S**, which is magnified by 15 in the metal region. We see that the energy flow is almost in the *X* direction and that it goes in opposite directions in vacuum and in metal. This is commonly observed when a SISP is excited.

In Figs. 15(a) and 16(a), we again see the enhancement of E_X and E_Z on the upper and lower surface of the silver-film grating, respectively. The rate of enhancement in Fig. 15(a) is not as large as that in Fig. 16(a). E_Z is strongly enhanced at both the upper and lower surfaces of the grating and exponentially decays away from each surface. We thus observe the simultaneous excitation of surface plasmons at the surfaces. We can understand the difference of the field distributions assuming that the former and the latter refer to Figs. 15 and 16 are the results of the SRSP and the LRSP mode excitation. Figures 15(b) and 16(b) complement the understanding showing the even and odd nature of relevant oscillations.

4.4. Multilayered thin-film bi-gratings case

Next, we consider multilayered thin-film bi-gratings as shown in Fig. 1(a) indicated by $L = 4$ ($V/Ag/SiO_2/Ag/V$) that consists of a stack of silver and SiO_2 films pairs. As listed in the figure, the values of the parameters are the same as those in Fig. 11 except for $L = 4$, $n_{SiO_2} = 1.5$, $e_{SiO_2} / \mathrm{d} = 0.3$ or 0.08.

4.4.1. *Diffraction efficiency*

Figure 17 shows the (0,0)th order power reflection ρ_{00} in V_1 (Vacuum) (a) and the transmission coefficient τ_{00} in V_5 (Vacuum) (b) as functions of the incident angle θ for two different values of e_{SiO_2} / d. The curve for $e_{SiO_2} / d = 0.08$ is almost the same as for a sinusoidal silver film bi-grating (Fig. 11): the coupled surface plasmon modes, SRSP and LRSP, are excited. On the other hand, in the curve for $e_{SiO_2} / d = 0.3$, we find a new type of absorption of incident light besides the plasmon resonance absorption associated with SRSP or LRSP.

This is related to the resonant excitation of a guided wave supported by the SiO_2 film. This absorption is characterized by its occurrence over a wider range of θ. For example, in the case of $e_{SiO_2} / d = 0.3$, the extinction power is more than 50% for all angles of incidence θ ranging from 0° to 12°.

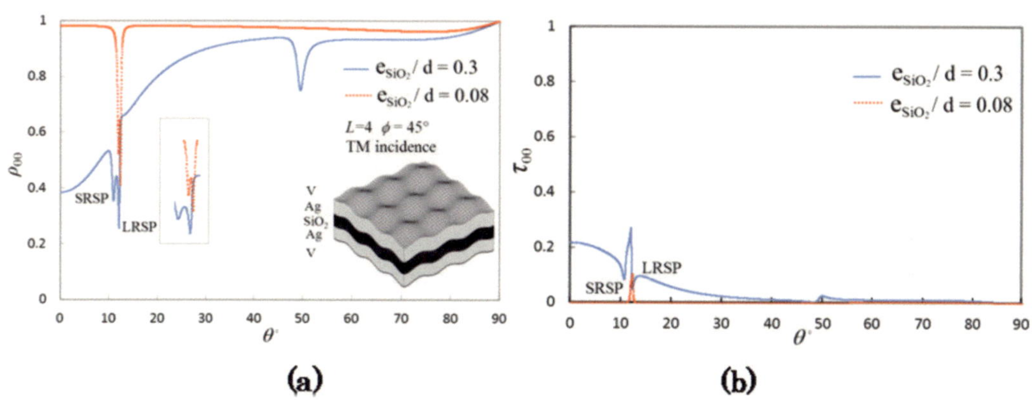

Figure 17. The (0, 0)-th order diffraction efficiencies ρ_{00} (a) and τ_{00} (b) as functions of θ for two values of e_{SiO_2}/d (L=4).

4.4.2. *Field distributions*

In order to examine the properties of the wide absorption found in Fig. 17, we investigated the field distributions of the total electric field E^{total} and the TM component of the (0, 0)th-order diffracted electric field $E^{TM}_{l(0,0)}$ in the vicinity of the SiO_2 film. The magnitude of E^{total} and $E^{TM}_{l(0,0)}$ (l =1,2,...5) along the Z-axis are plotted in Fig. 18 where $\theta = 0^\circ$ and $e_{SiO_2} / d = 0.3$.

We observe in the figure that the field distributions of E^{total} inside the SiO_2 film indicates a standing wave pattern corresponding to the normal mode of a one dimensional cavity resonator, and that the distribution is almost close to that of $E^{TM}_{3(0,0)}$. Hence, we conclude that the wide absorption observed in the multilayered grating V/Ag/SiO_2/Ag/V is associated with resonance of the (0, 0)th-order diffracted wave $E^{TM}_{3(0,0)}$ in the SiO_2 film sandwiched by a sinusoidal silver film grating.

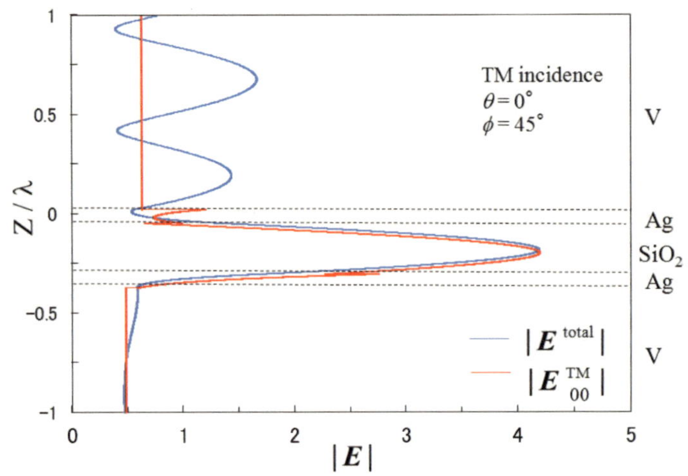

Figure 18. Standing wave pattern of the electric filed in the SiO_2 film.

5. Conclusions

We have investigated the resonance absorption associated with the resonant excitation of surface plasmons in bi-gratings. Calculating diffraction efficiency, expansion coefficients, field profiles, and energy flows, we examined the characteristics of the resonant excitation of surface plasmons in detail. Interesting phenomena were revealed, including the conversion of a TM (or TE) component of the incident light into a TE (or TM) component at several different incidence angles, strong field enhancement on the grating surface where surface plasmons are excited, and simultaneous resonance absorption that does not occur in the case of a singly periodic grating in general. The results presented here facilitate a clear understanding of the coupled plasmon modes, SISP, SRSP and LRSP, excited in a thin film doubly periodic metal grating.

Author details

Taikei Suyama, Akira Matsushima and Yoichi Okuno
Graduate School of Science and Technology, Kumamoto University, Kurokami, Kumamoto, Japan

Toyonori Matsuda
Kumamoto National College of Technology, Suya, Nishigoshi, Japan

Acknowledgement

This work was supported in part by Grant-in-Aid for Scientific Research from Japan Society for the Promotion of Science (Grant number 23560404). The authors thank Shi Bai and Qi Zhao for help in numerical computation and in preparation of the manuscript.

6. References

Raeter, H., (1982) Surface Plasmon and Roughness, *Surface Polaritons*, V. M. Argranovich and D. L. Mills, (Ed.), Chap. 9, 331-403, North-Holland, New York

Nevièr, M., (1980) The Homogenous Problem, *in Electromagnetic Theory of Gratings* R. Petit (Ed.), Chap. 5, 123-157, Springer-Verlag, Berlin

DeGrandpre, M.D., and L.W. Burgess, (1990) Thin film planar waveguide sensor for liquid phase absorbance measurement, *Anal. Chem.*, 62, 2012-2017

Zoran, J. and M. Jovan, (2009) Nanomembrane-Enabled MEMS Sensors: Case of Plasmonic Devices for Chemical and Biological Sensing, *Micro Electronic and Mechanical Systems*, Kenichi Takahata (Ed.), ISBN: 978-953-307-027-8, InTech

Nemetz, A., U. Fernandez and W. Knoll, (1994) Surface plasmon field-enhanced Raman spectroscopy with double gratings, *J. Appl. Phys*, 75, 1582-1585

Barnes,W. L., T. W. Preist, S. C. Kitson, J. R. Sambles, N. P. K. Cotter, and D. J. Nash, (1995) Photonic gaps in the dispersion of surface plasmons on gratings, *Phys. Rev. B*, 51, 11164 -11167

Tan, W.-C., T. W. Preist, J. R. Sambles, M. B. Sobnack, and N. P. Wanstall, (1998) Calculation of photonic band structures of periodic multilayer grating systems by use of a curvilinear coordinate transformation, *J. Opt. Soc. Am. A*, 15, 9, 2365–2372

Inagaki, T., M. Motosuga, E.T. Arakawa and J.P. Goudonnet, (1985) Coupled surface plasmons in periodically corrugated thin silver films, *Phys. Rev. B*, 32, 6238-6245

Chen, Z., I. R. Hooper, and J. R. Sambles, (2008) Strongly coupled surface plasmons on thin shallow metallic gratings, *Phys. Rev. B*, 77, 161405

Bryan-Brown, G.P., S.J. Elston, J.R. Sambles, (1991) Coupled surface plasmons on a silver coated grating, *Opt. Commu.*, 82, No. 1-2, 1-5

Davis, T. J., (2009) Surface plasmon modes in multi-layer thin-films, *Opt. Commu.*, 282, 135-140

Raeter, H., (1977) Surface plasma ocillations and their applications, *Physics of Thin Films*, G. Hass and M.H. Francombe (Ed.), Chap. 9, 145-261, Academic Press, New York

Okuno, Y. and T. Suyama, (2006) Numerical analysis of surface plasmons excited on a thin metal grating, *Journal of Zhejiang University -Science A*, 7, No. 1, 55-70

Suyama, T., Y. Okuno and T. Matsuda, (2009) Plasmon resonance-absorption in a metal grating and its application for refractive-index measurement, *Journal of Electromagnetic Waves and Applications*, 20 , No. 2, 159-168

Chen, Z. and H. J. Simon, (1988) Attenuated total reflection from a layered silver grating with coupled surafce waves, *J. Opt. Soc. Am. B*, 5, No. 7, 1396-1400

Hibbins, A. P., W. A. Murray, J. Tyler, S. Wedge, W. L. Barnes and J. R. Sambles, (2006) Resonant absorption of electromagnetic fields by surface plasmons buried in a multilayered plasmonic nanostructure, *Phys. Rev. B*, 74, 073408

Glass, N. E., A. A. Maradudin and V. Celli, (1982) Surface plasmons on a large-amplitude doubly periodic corrugated surface, *Phys. Rev. B*, 26, 5357-5365

Glass, N. E., A. A. Maradudin, and V. Celli, (1983) Theory of surface-polariton resonances and field enhancements in light scattering from doubly periodic gratings, *J. Opt Soc. Am.*, 73, 1240-1248

Inagaki, T., J. P. Goudonnet, J. W. Little and E. T. Arakawa, (1985) Photoacoustic study of plasmon-resonance absorption in a doubly periodic grating, *J. Opt Soc. Am. B*, 2, 432-439

Harris, J. B., T. W. Preist, J. R. Sambles, R. N. Thorpe and R. A. Watts, (1996) Optical response of doubly periodic gratings, *J. Opt Soc. Am. A*, 13, 2041-2049

Chen, C. C., (1973) Transmission of microwave through perforated flat plates of finite thickness, *IEEE Trans. Microwave Theory Tech.*, MTT-21, 1-6

Yasuura, K. and T. Itakura, (1965) Approximation method for wave functions (I),(II), and (III), *Kyushu Univ. Tech. Rep*, 38, 72-77, 1965; 38, 378-385, 1966; 39, 51-56, 1966

Yasuura, K.(1971). A view of numerical methods in diffraction problems, *Progress in Radio Science 1966-1969*, W. V. Tilson and M. Sauzade (Ed.), 257-270, URSI, Brussels

Okuno, Y., (1990) Mode-matching Method, *Analysis Methods for Electromagnetic Wave Problems*. E. Yamashita (Ed.), 107-138, Artech House, Boston

Hugonin, J. P., R. Petit, and M. Cadilhac, (1981) Plane-wave expansions used to describe the field diffracted by a grating, *J. Opt. Soc. Am.*, 71, No. 5, 593–598

Lawson, C. L. and R. J. Hanson, (1974) Solving Least Squares Problem, Prentice-Hall, Englewood Cliffs, N. J.

Matsuda, T. and Y. Okuno, (1993) A numerical analysis of planewave diffraction from a multilayer-overcoated grating, *IEICE*, J76-C-I, No. 6, 206–214

Matsuda, T. and Y. Okuno, (1996) Numerical evaluation of plane-wave diffraction by a doubly periodic grating, *Radio Sci.*, 31, 1791-1798

Suyama, T., Y. Zhang, Y. Okuno, Z. Q. Luo and T. Matsuda, (2010) Surface Plasmon Resonance Absorption in a Multilayered Bigrating, *PIERS Online*, 6, No. 1, 76-80

Suyama, T., Y. Okuno, A. Matsushima and M. Ohtsu, (2008) A numerical analysis of stop band characteristics by multilayered dielectric gratings with sinusoidal profile, *Progress In Electromagnetics Research B*, 2, 83-102

Hass, G. and L. Hadley, (1963) Optical properties of metals, *American Institute of Physics Handbook* , D. E. Gray, (2nd Ed.), 6-107, McGraw-Hill, New York.

Elston, S. J., G. P. Bryan-Brown and J. R. Sambles, (1991) Polarisation conversion from diffraction gratings, *Phys. Rev. B*, 44, 6393-6399

Matsuda, T., D. Zhou, and Y. Okuno, (1999) Numerical analysis of TE-TM mode conversion in a metal grating placed in conical mounting, *IEICE C-I*, J82-C-I, No. 2, 42-49 (in Japanese).

Suyama, T., Y. Okuno and T. Matsuda, (2007) Enhancement of TM-TE mode conversion caused by excitation of surface plasmons on a metal grating and its application for refractive index measurement, *Progress In Electromagnetics Research*, 72, 91-103

Ritchie, R. H., E. T. Arakawa, J. J. Cowan and R. N. Hamm, (1968) Surface-plasmon resonance effect in grating diffraction, *Phys. Rev. Lett.*, 21, 1530-1533

A Treatise on Magnetic Surface Polaritons: Theoretical Background, Numerical Verification and Experimental Realization

Yu-Hang Yang and Ta-Jen Yen

Additional information is available at the end of the chapter

http://dx.doi.org/50752

1. Introduction

The polariton is a kind of coupling between electromagnetic waves (photons) and elementary excitations such as phonons, plasmons and magnons, It includes two modes of surface and bulk polariton that can be excited by means of semiconductors, metals, ferromagnets, antiferromagnets, and so on. The coupling causes an intensity enhancement of the electromagnetic field, which is very useful for nano-technology such as biosensing, waveguide applications, nano-antenna device. The majority of the literature is concentrating on the magnetic surface polariton supported by magnetic materials. The outline of paper is as follows. In sections 2 we introduce the fundamental of the polaritons first, and then the general dispersion equation of the magnetic surface polaritons (MSPs) mode by considering the full form of Maxwell's equations in section 3. Next, in sections 4 and 5, we present numerical and experimental results of realizing the MSPs, by naturally existing materials – in section 4, by ferromagnetic and antiferromagnetic materials and in section 5, by the effective media of ferromagnetic and anti-ferromagnetic superlattices, respectively. By employing metamaterials, artificially constructed materials whose properties mainly stem from structures rather than their constitutive elements, one can also achieve MSPs mode with a greater engineering freedom; thus, in section 6, we start from interpreting what metamaterials and their operation rationales are, and then how metamaterials support the MSPs mode. Finally, in section 7, there comes a conclusion. Note that in this study we moderately modify the quoted definitions in the diagrams to correspond to our definitions.

2. Fundamental of polaritons

The well-known surface plasmon polaritons (SPPs) denote that a collective oscillation of electrons couples with the transverse magnetic wave, and they propagate along the interface between two media (ex: a metal and a dielectric) with an exponential decay into two media. Due to the polarization dependence of SPPs, it is a reasonable perception to realize a magnetic analog by a transverse electric wave. In addition, we have known the fact that the wave equation can be expressed by either electric or magnetic component of the electromagnetic wave, so that it implies that all electromagnetic phenomena chould be symmetric. In recent years, more and more researchers have paid intensive attention to the electric SPPs and their applications, in particular for bio-sensing [15] and nanophotonic applications [27].

In contrast, the magnetic surface polaritons (MSPs) mode did not attract much attention yet, and that is because magnetic responses are typically weak and their resonant frequencies are usually below the far infrared region. In our opinion, it is useful and interesting to understand the mechanism of the MSPs even it still does not yet become a protagonist. Throughout this study, we will show there is a potential material to promote the inherent magnetism in natural materials.

At first, let us briefly define the well-known magnetic polariton. In general, the magnetically ordered media support three kinds of elementary modes, spin waves (or call magnons), magnetostatic modes, and magnetic polaritons. Three modes are classified by the dominant restoring force as shown in Fig. 1 [21]. When the wavevector (k) in vacuum is greater than $10^8\ m^{-1}$, the exchange interaction is important and one calls this mode as magnon. The equation of motion is usually needed to solve this mode. For magnetostatic mode (magnetostatic limit, $k \ll \frac{\omega}{c}$), $10^7 \leqslant k \leqslant 10^8\ m^{-1}$, the exchange and dipolar interactions may be both important; $3 \times 10^3 \leqslant k \leqslant 10^7\ m^{-1}$, the dipolar terms are mainly dominant. For this region, we need a Hamiltonian treatment for such a magnetostatic mode. For magnetic polaritons, $k \leqslant 3 \times 10^3\ m^{-1}$, it is recognized as electromagnetic region where the full form of Maxwell's equations including retardation corrections have to be considered. As an example, the dispersion relation of a ferromagnetic insulator [21] is plotted in Fig. 1, in where ω_0 and ω_m will be defined later. In this literature, we just concentrate on MSPs, and study three situations: a pure ferromagnets (or antiferromagnets), ferromagnetic (or antiferromagnetic) superlattices, and magnetic metamaterials.

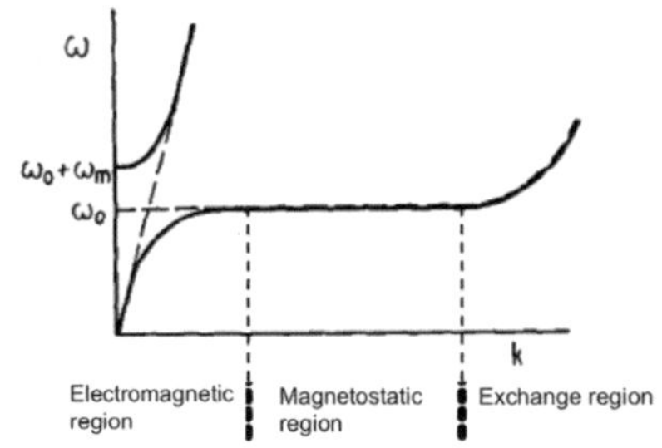

Figure 1. The dispersion diagram. In general, $k > 10^8\ m^{-1}$ is an exchange region; $3 \times 10^3 \leqslant k \leqslant 10^8\ m^{-1}$ is a magnetostatic region; $k \leqslant 3 \times 10^3\ m^{-1}$is an electromagnetic region.

3. General dispersion equation for magnetic surface polaritons

It is worthy to derive a general dispersion relation in advance before we begin studying the real cases. It should help one to understand all definitions that are used in our whole literature. First, the geometry considered in this study consists of two semi-infinite linear media, including the magnetic and nonmagnetic media shown in Fig. 2. Second, we assume

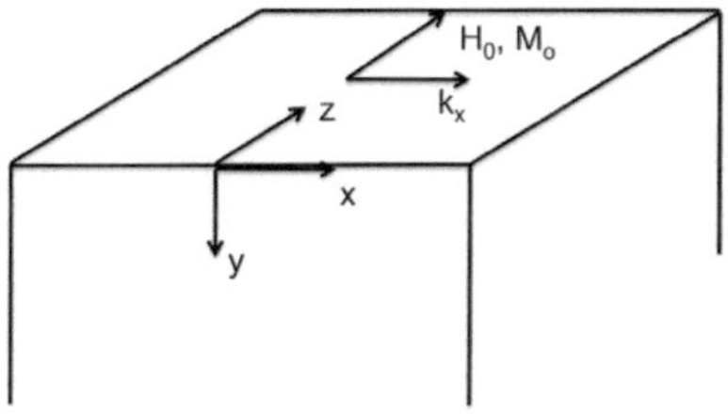

Figure 2. The geometry considered in this study. The half space above the $y > 0$ plane is a magnetic medium; below the $y < 0$ plane is a nonmagnetic medium. The external static magnetic field (H_0) is parallel to the easy axis, $\hat{z}$.

that the magnetic permeability tensor of the magnetic medium can be written by

$$\overleftrightarrow{\mu}_{eff}(\omega) = \begin{pmatrix} \mu_{xx} & i\mu_{xy} & 0 \\ -i\mu_{xy} & \mu_{yy} & 0 \\ 0 & 0 & \mu_{zz} \end{pmatrix}, \tag{1}$$

and the electric permittivity can be given by

$$\overleftrightarrow{\varepsilon}_{eff}(\omega) = \begin{pmatrix} \varepsilon_{xx} & 0 & 0 \\ 0 & \varepsilon_{yy} & 0 \\ 0 & 0 & \varepsilon_{zz} \end{pmatrix}. \tag{2}$$

Note that we are interesting in the dispersion relation itself so that the damping and spatial dispersion can be ignored here. It will be worthy to consider the damping only when the reflectivity calculation is executed; spatial dispersion is meaningful only for magnons. The constitutive relations for the linear medium read

$$\vec{D}(\omega) = \overleftrightarrow{\varepsilon}_{eff}(\omega)\vec{E}(\omega), \quad \vec{B}(\omega) = \overleftrightarrow{\mu}_{eff}(\omega)\vec{H}(\omega). \tag{3}$$

Next, we follow our previous work [37] and ready to derive rigorously a general dispersion equation. We are looking for MSPs mode. Therefore, without loss of generality, the transverse electric fields of the MSPs mode above and below the plane $y = 0$ can be respectively written by

$$\begin{aligned} \vec{E}^{+}(\omega) &= \hat{z}\,\mathrm{E}_z^{+} e^{ik_x x} e^{-\alpha^{+} y}, \\ \vec{E}^{-}(\omega) &= \hat{z}\,\mathrm{E}_z^{-} e^{ik_x x} e^{\alpha^{-} y}. \end{aligned} \tag{4}$$

where k_x is the direction of the MSPs mode in vacuum. α^+ and α^- stand for attenuation coefficients of the MSPs mode for $y > 0$ and $y < 0$, respectively. Faraday's law can solve the corresponding magnetic induction as follows,

$$\begin{aligned} \vec{B}^{+}(\omega) &= \frac{1}{i\omega}\left[(-\alpha^{+})\hat{x} - (ik_x)\hat{y}\right]\vec{E}^{+}(\omega), \\ \vec{B}^{-}(\omega) &= \frac{1}{i\omega}\left[(\alpha^{-})\hat{x} - (ik_x)\hat{y}\right]\vec{E}^{-}(\omega). \end{aligned} \tag{5}$$

Next, the magnetic fields can be solved by Eq. 3:

$$\begin{aligned} \vec{H}^{+}(\omega) &= \frac{-1}{\mu_{xy}^2 - \mu_{xx}\mu_{yy}}\left[(-\mu_{yy}\alpha^{+} + \mu_{xy}k_x)\hat{x} + (-i\mu_{xy}\alpha^{+} - i\mu_{xx}k_x)\hat{y}\right], \\ \vec{H}^{-}(\omega) &= (-\alpha^{+})\hat{x} + (-ik_x)\hat{y}. \end{aligned} \tag{6}$$

Applying Ampere's law with Eqs. 1-3, we can solve the following wave equations:

$$\begin{aligned}
&\left[\mu_{xx}\frac{\partial^2}{\partial x^2}+\mu_{yy}\frac{\partial^2}{\partial y^2}+\left(\mu_{xx}\mu_{yy}-\mu_{xy}^2\right)\frac{\omega^2}{c^2}\varepsilon_{zz}\right]\vec{E}^+(\omega)=0,\\
&\left(\frac{\partial^2}{\partial x^2}+\frac{\partial^2}{\partial y^2}+\frac{\omega^2}{c^2}\varepsilon_{zz}\right)\vec{E}^-(\omega)=0.
\end{aligned} \tag{7}$$

Using Eqs. 4 and 7, we can solve the attenuation coefficients as follows,

$$\begin{aligned}
\alpha^+ &= \sqrt{\frac{\mu_{xx}}{\mu_{yy}}k_x^2-\left(\frac{\mu_{xx}\mu_{yy}-\mu_{xy}^2}{\mu_{yy}}\right)\frac{\omega^2}{c^2}\varepsilon_{zz}},\\
\alpha^- &= \sqrt{k_x^2-\frac{\omega^2}{c^2}}.
\end{aligned} \tag{8}$$

Note that it is also straightforward to solve the bulk polaritons, propagating perpendicular to the $\hat{z}$ axis. All we do is just to replace $\alpha^{\pm}$ by ik_y. Then, one is given by

$$\begin{aligned}
\frac{c^2\left(\mu_{xx}k_x^2+\mu_{yy}k_y^2\right)}{\omega^2} &= \left(\mu_{xx}\mu_{yy}-\mu_{xy}^2\right)\varepsilon_{zz},\\
k_x^2+k_y^2 &= \frac{\omega^2}{c^2}.
\end{aligned} \tag{9}$$

The continuity of the tangential components of the electric field (E_x) and the magnetic filed (H_x) at the interface ($y=0$) yields

$$E_x^+=E_x^-,\quad \frac{\mu_{yy}B_x^+ + i\mu_{xy}B_y^+}{\mu_{xx}\mu_{yy}-\mu_{xy}^2}=\alpha^-. \tag{10}$$

Using the Eqs. 5, 6 and 10, the general dispersion equation of the MSP's mode reads

$$\frac{-\alpha^+\mu_{yy}+\mu_{xy}k_x}{\mu_{xx}\mu_{yy}-\mu_{xy}^2}=\alpha^-. \tag{11}$$

Note that Eq. 11 itself implies a non-reciprocal dispersion relation due to $\pm|k_x|$. Now, we have enough equations to begin our discussion for the MSPs mode.

4. Realization of MSPs mode by magnetic materials

Over the past 50 years, Damon and Eshbach [8] first reported the surface magnetostatic modes without retardation effect for the case of a ferromagnetic slab; effective media for superlattices have also done [17]. Later Hartstein et al. reported more detailed discussion of the MSPs for semi-infinite medium of a pure magnetic materials [1]. Before discussing the semi-infinite gyromagnetic uniaxial medium (gyromagnetic ratio,$\gamma \neq 0$), we have to assume several limitations for this study: (a) we neglect the local effects of the surface on the spin wave (i.e. pinning effects) and (b) the exchange interaction is ignored as well.

4.1. Semi-infinite ferromagnets

In this part, we consider the case, a pure ferromagnet and vacuum. For simplicity, we treat the electric permittivity of a ferrromagnet $[\overleftrightarrow{\varepsilon}(\omega)]$ as scalar value, 1. Also, we neglect damping of $\overleftrightarrow{\mu}(\omega)$ so that μ_{xx}, μ_{zz}, and μ_{xy} in Eq. 1 are real. The external static magnetic filed (H_0) is along the easy axis $+\hat{z}$ shown in Fig. 2. The half space $y < 0$ is vacuum, and $y > 0$ is a pure ferromagnet. In the Voigt geometry, the magnetic permeability satisfies the equation, $\mu_v(\omega) = \mu_{xx} - \mu_{xy}^2 / \mu_{xx}$ plotted in Fig. 3 [1]. We also define $\omega_v = \gamma\sqrt{B_o H_0}$ at resonance and $\omega_1 = \gamma B_0$ at $\mu_v(\omega) = 0$. For a uniaxial ferromagnet $(\mu_{xx} = \mu_{yy})$, the Eqs. 8 and 11 are rewritten by

$$\alpha^+ = \sqrt{k_x^2 - \mu_v \frac{\omega^2}{c^2}}, \quad \alpha^- = \sqrt{k_x^2 - \frac{\omega^2}{c^2}}. \tag{12}$$

and

$$\frac{-\alpha^+ \mu_{xx} + \mu_{xy} k_x}{\mu_{xx}^2 - \mu_{xy}^2} = \alpha^-. \tag{13}$$

For the MSPs mode, the α^+ and α^- in Eq. 12 have to be positive that yield to an exponential

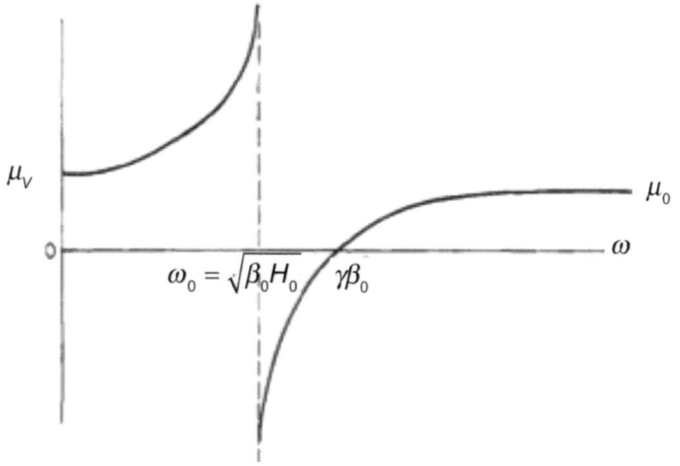

Figure 3. Frequency dependence of magnetic permeability in the Voigt geometry. [1]

decay into both magnetic and nonmagnetic materials (vacuum). The dispersion diagram [1] is plotted in Fig. 4 in where the shaded region makes the existence of the MSPs mode possible. For a uniaxial ferromagnet, the components of the magnetic permeability tensor in Eq. 1 can be written by

$$\begin{aligned} \mu_{xx} &= \mu_{yy} = \mu_0 \left(1 + \frac{\omega_m \omega_0}{\omega_0^2 - \omega^2}\right), \\ \mu_{zz} &= \mu_0, \\ \mu_{xy} &= \mu_0 \frac{\omega_m \omega}{\omega_0^2 - \omega^2}. \end{aligned} \tag{14}$$

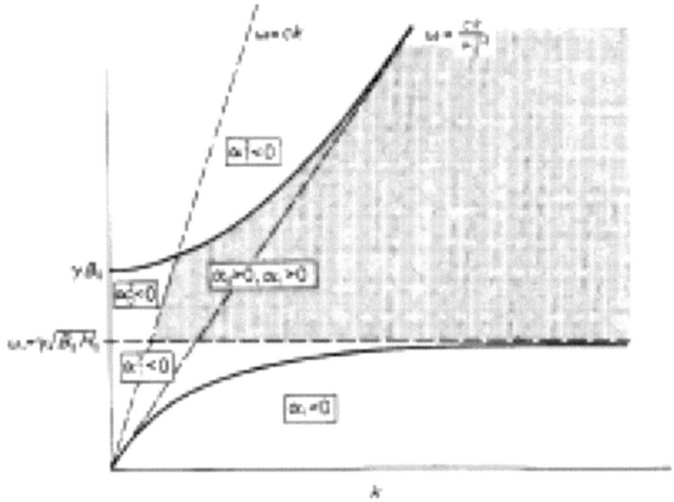

Figure 4. The dispersion relation of a pure ferromagnet. [1]

Here, two angular frequencies (ω_m and ω_0) are written in terms of the magnetization (M_0) and external field (H_0):

$$\omega_m = 4\pi\gamma M_0, \ \ \omega_0 = \gamma H_0, \tag{15}$$

where ω_0 is angular frequency of the ferromagnetic resonance, and γ is the gyromagnetic ratio. Note that μ_0 in Eq. 14 is caused by other magnetic dipole excitation (such as optical magnons) at higher frequency (ex: $\mu_0 = 1.25$ in YIG). Before discussing a detailed calculation of dispersion equation, we would like to return to Eq. 6 that can simply predict a non-reciprocal dispersion relation. One write down the polarization of the magnetic field of the MSPs at a fixed point by investigating the components in Eq. 6:

$$\frac{H_y^+}{H_x^+} = i\frac{\mu_{xy}\alpha^+ + \mu_{xx}k_x}{\mu_{yy}\alpha^+ - \mu_{xy}k_x}. \tag{16}$$

When we take the magnetostatic limits ($k_x \to \pm\infty$), $\alpha^\pm$ will be very close to $|k_x|$ [see Eq. 12]. Then, the Eq. 16 reduces to $H_y^+ = iH_x^+$ for the $+k_x$ direction and $H_y^+ = -iH_x^+$ for the $-k_x$. Therefore, the polarization of magnetic field of the MSPs mode has a different rotation for different $\pm k_x$ directions, leading to a non-reciprocal result. Next, we are ready to investigate it in detail by solving the Eq. 13 at the magnetostatic limit $k_x \to \pm\infty$. One is given by

$$\mu_{xx} \pm \mu_{xy} \equiv \mu_{\rho\pm}(\omega) = -1. \tag{17}$$

Using Eq. 14, we write

$$\mu_{\rho\pm} = \mu_0\left(1 + \frac{\omega_m}{\omega_0 \mp \omega}\right) = -1. \tag{18}$$

Eq. 18 describes a condition to which the MSPs mode have a magnetostatic analog. Plotting a dispersion relation of the surface polariton [Eq. 13] is a good method to observe the condition shown in Fig. 5. Note that an applied static field is already considered in Fig. 5. It is very clear to observe the non-reciprocal nature of the MSPs mode (solid curves). Next, we consider

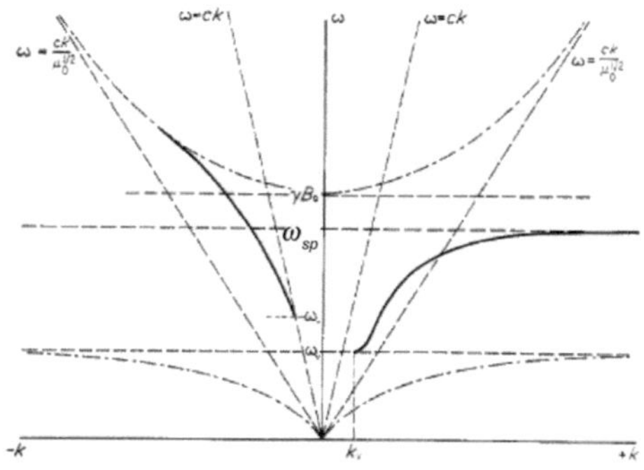

Figure 5. Non-reciprocal relation. The solid curves are the surface polaritons, and dot-dash curves are bulk polaritons. [1].

four situations: (a) For the magnetostatic limit $k_x \to +\infty$, we take $\mu_{\rho+}$ in Eq. 18, and then the frequency of the asymptotic surface polaritons reads

$$\omega_{sp} = \gamma \left(\frac{H_0 + \mu_0 B_0}{1 + \mu_0} \right), \tag{19}$$

where $B_0 \equiv H_0 + 4\pi\gamma M_0$. This is just the frequency of the unretardated surface magnon mode (also termed as DE mode [8]). (b) For the magnetostatic limit $k_x \to -\infty$, we find that $\mu_{\rho-}$ is always positive, yielding to no magnetostatic angalog. In general, the MSPs mode is called as real mode if magnetostatic limit is valid; otherwise, it is a virtual mode. Therefore, a non-reciprocal nature of the MSPs mode in a pure ferromagnet is acknowledged according to a magnetostatic mode. The investigation is in a good agreement with abovementioned polarization of magnetic field. (c) Next, let us consider a special case, $\mu_0 = 1$, and for the moment we concentrate on the $+k_x$ solution. The frequency of the MSPs mode starts at

$$\omega_v = \gamma \sqrt{B_0 H_0}, \tag{20}$$

and from Eq. 13 one obtains a corresponding wavevector

$$k_v = \frac{\omega_v}{c} \sqrt{\frac{B_0}{B_0 - H_0}} > \frac{\omega_v}{c}, \tag{21}$$

where the retardation corrections are already involved. The ω_v is also the frequency for the bulk magnon excitation under a condition, $\mu_v(\omega) \to \pm\infty$ [see Fig. 3]. If there is no external field, the ω_v reduces to zero so that the dispersion relation of the MSPs mode begins from the zero wavevector. The situation just likes electric SPPs supported by a dielectric and a metal [24]. (d) Now, let us consider the $-k_x$ solution including the retardation corrections. The retarded MSPs mode begins at the light line where $\mu_{xx}(\omega_-) = 1$ [see Fig. 5]. We solve Eq. 14 to yield the beginning frequency:

$$\omega_- = \gamma \sqrt{\frac{H_0^2 + \mu_0 H_0 B_0}{\mu_0 - 1}}. \tag{22}$$

Such a virtual mode terminates on the upper branch of the bulk polaritons. In addition, we see that from Eq. 22 the MSPs mode never starts at $\mu_0 \to 1$ (i.e. $\omega_- \to \infty$). Compared

Fig. 3 with Fig. 5, we note that the MSPs mode in the $-k_x$ direction still can exist even if $\mu_v > 1$. It is a very different from isotropic medium ($\mu_{xx} = \mu_{yy} = \mu_{zz}$) because its MSPs mode is only found within a gap where $\mu_v < 1$. From Eq. 21, the light line cannot excite any MSPs mode due to lack of the momentum conservation. In order to excite the MSPs mode in a pure ferromagnet, the optical coupler is need to transfer the momentum of the light to MSPs mode. Here, we quote the reference [[14]] to show the numerical calculations of the reflectivity spectra of the MSPs mode. Note that the damping factor is no longer neglected when one calculates a reflectivity spectrum. In Fig. 6, the calculated reflectivity spectra result from Otto ATR [1] configuration (optical coupler) for a pure ferromagnet (YIG). The figures 6(a) and 6(b) correspond to $+k_x$ and $-k_x$ directions, respectively. "h" is the air gap between a prism and a YIG. In Fig. 6(a), it shows that a larger damping ($1/\tau$) of the MSPs has larger loss, leading to much difficulty for determining the resonance frequency. In Fig. 6(b), the resonance frequency of the MSPs mode is blue-shift among increasing incidence angles. The asymmetric reflectivity deep results from an appreciable photon content in both figures 6. In general, the resonance frequency of the MSPs mode for a pure ferromagnet lies in a microwave region so that one needs a big sample for measuring the MSPs mode. There is, as we know, not a direct experimental verification.

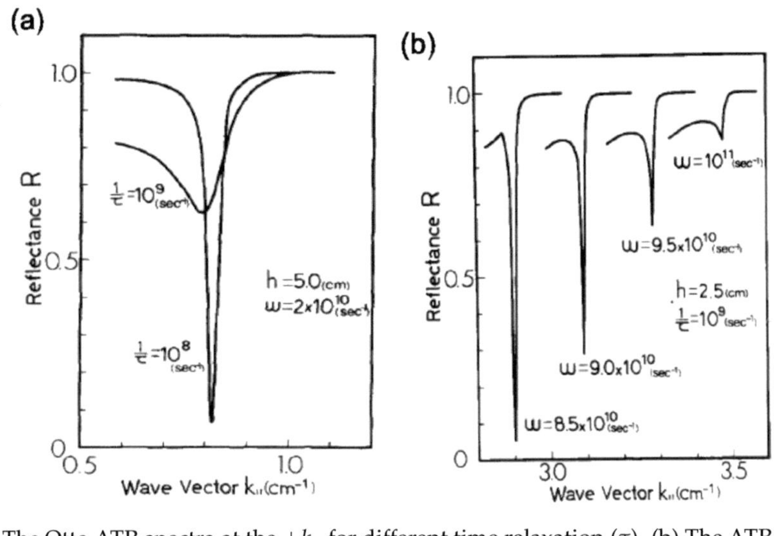

Figure 6. (a) The Otto ATR spectra at the $+k_x$ for different time relaxation (τ). (b) The ATR at the $-k_x$ for different incidence angle within the prism. [14]

4.2. Semi-infinite antiferromagnets

In this case, we focus on a gyromagnetic uniaxial antiferromagnetic medium in the Voigt configuration. Again, we neglect the exchange interactions and damping of $\overleftrightarrow{\varepsilon}(\omega)$ and $\overleftrightarrow{\mu}(\omega)$, and the external static magnetic field is applied in the $\hat{z}$ direction. The magnetization (M_0) splits the ferromagnetic resonant frequency (ω_0) into two separable frequencies given as $\omega_{\pm} = \omega \pm \omega_0$. The effective magnetic permeability can be derived by considering Bloch's

[1] attenuated total reflectance

equations [20], which reads

$$\begin{aligned}
\mu_{xx} &= \mu_{yy} = \mu_0 \left(1 + \frac{\omega_m \omega_A}{\omega_{an}^2 - \omega_+^2} + \frac{\omega_m \omega_A}{\omega_{an}^2 - \omega_-^2}\right), \\
\mu_{zz} &= \mu_0, \\
\mu_{xy} &= \mu_0 \left(\frac{\omega_m \omega_A}{\omega_{an}^2 - \omega_+^2} - \frac{\omega_m \omega_A}{\omega_{an}^2 - \omega_-^2}\right),
\end{aligned} \tag{23}$$

where ω_{an} is the antiferromagnetic resonance frequency and is determined by the anisotropy field (H_A) and exchange field (H_E),

$$\omega_{an} = \sqrt{\omega_A \left(2\omega_E + \omega_A\right)} = \gamma \sqrt{H_A \left(2H_E + H_A\right)}, \tag{24}$$

where $\omega_A = \gamma H_A$ and $\omega_E = \gamma H_E$. Next, one just follows the Eq. 8, yielding to Eq. 25,

$$\begin{aligned}
\alpha^+ &= \sqrt{k_x^2 - \left(\frac{\mu_{xx}^2 - \mu_{xy}^2}{\mu_{xx}}\right) \frac{\omega^2}{c^2} \varepsilon_{zz}}, \\
\alpha^- &= \sqrt{k_x^2 - \frac{\omega^2}{c^2}}.
\end{aligned} \tag{25}$$

Also, the dispersion relation from Eq. 11 can read

$$\frac{-\alpha^+ \mu_{xx} + \mu_{xy} k_x}{\mu_{xx}^2 - \mu_{xy}^2} = \alpha^-. \tag{26}$$

In the absence of the applied field ($H_0 = 0$), one has $\mu_{xy} = 0$ [see Eqs. 15 and 23]. Then, Eqs. 25 and 26 reduce more compact forms:

$$\begin{aligned}
\alpha^+ &= \sqrt{k_x^2 - \varepsilon_{zz} \mu_{xx} \frac{\omega^2}{c^2}}, \\
\alpha^- &= \sqrt{k_x^2 - \frac{\omega^2}{c^2}}.
\end{aligned} \tag{27}$$

$$\frac{-\alpha^+}{\mu_{xx}} = \alpha^-. \tag{28}$$

Using Eqs. 27 and 28 together, a simpler dispersion relation of surface polariton can be given by

$$k_x^2 \left(1 - \mu_{xx}^2\right) = \frac{\omega^2}{c^2} \mu_{xx} \left(\varepsilon_{zz} - \mu_{xx}\right). \tag{29}$$

The simpler dispersion relation [5] is plotted in Fig. 7. In Fig. 7, we can observe the reciprocal property of the MSPs mode. Also, it is straightforward to calculate the frequency of the

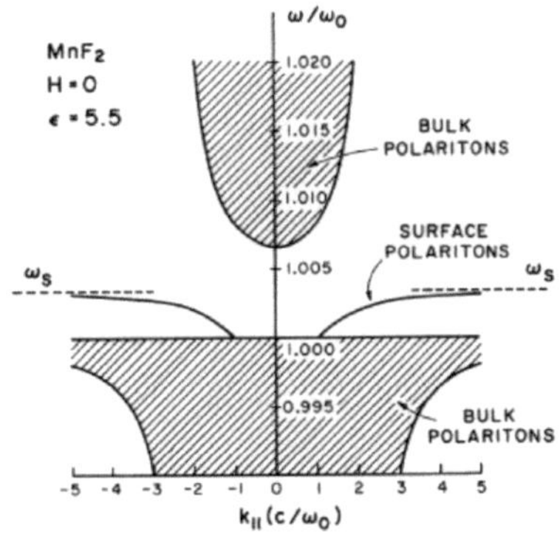

Figure 7. The shaped areas are bulk polaritons, and two solid curves are surface polaritons. Without an external static magnetic field, the dispersion relation is reciprocal. ($H_0 = 0$, $\varepsilon_{zz} = 5.5$) [5]

MSPs mode at the magnetostatic limit. One just solves $\mu_{xx} = -1$ [see Eq. 28] and yields the asymptotic frequency of MSPs shown in Fig. 7 as follows,

$$\omega_s = \sqrt{\omega_{an}^2 + \omega_m \omega_A}. \tag{30}$$

On other hand, with an applied static field, Eq. 26 has to be solved numerically shown in Fig. 8 [5]. Interestingly, one can clearly see the non-reciprocal dispersion relation of the MSPs mode that is just like a pure ferromagnet case with an applied static field. Note that the third MSPs mode starts at near $\omega/\omega_0 = 1.01$ and has no magnetostatic analog (i.e. virtual mode). However, two MSPs modes at lower frequency region have magnetostatic limits (i.e. ω_{sm+} and ω_{sm-}).

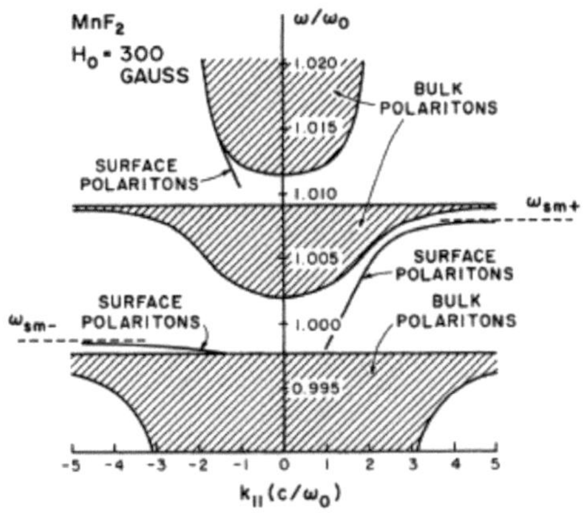

Figure 8. Dispersion relation for MSPs mode on MnF_2 at an applied field of 300 G. There are two real modes and virtual mode that are all non-reciprocal modes. [5]

Now, we would like to talk about the experimental verification for the surface polaritons. Let's simply describe the numerical results. There are typically two kinds of couplers: ATR and diffraction gratings. Here, we quote reference [32] in where they considered the diffraction gratings. In Fig. 9, the sample is MnF_2, a uniaxial antiferromagnet with the parameters $H_E = 7.87\ KG$, $H_A = 550\ HG$, $M_0 = 0.6\ KG$, and $\varepsilon_{zz} = 5.5$. Here, the Ω is our previous definition of antiferromagnetic resonance frequency, ω_{an}. There two solid lines are light lines ($k_x = \omega_0/c$), and two dotted vertical lines correspond to the wavevectors induced by the gratings ($q = k_x \pm s$). It is clear to observe that light lines never interact any of the branch of MSPs mode (dash curves). However, wavevector induced by gratings ($q = k_x + s$) indeed interacts a MSPs mode at the direction ($+k_x$), leading to the excitation of the MSPs mode. Note that when Eq. 1 includes damping terms, the leaky MSPs modes appear as shown by solid curves in Fig. 9. The difficulty to observe the surface polaritons for a pure antiferromagnet should be easier than a pure ferromagnet. The reason is that the resonance frequency of the MSPs mode lies at far infrared regions and the sample need not be so large. M.R.F. Jensen et al reported the first direct experimental evidence [18] for a pure antiferromagnet (FeF_2, $\omega_{an} \sim 1.57\ THz$).

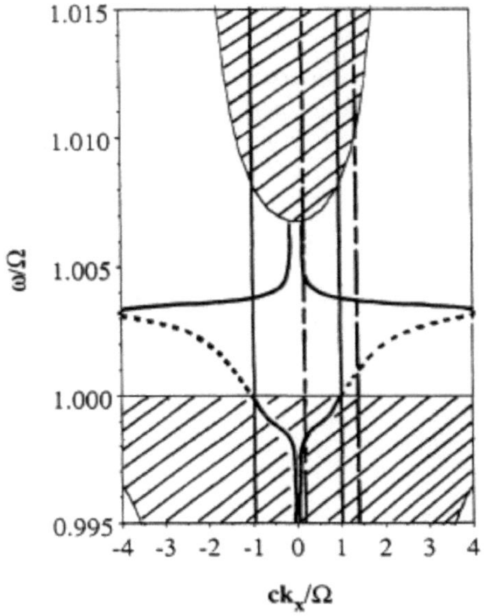

Figure 9. The shaped areas are bulk modes and the dashed lines are surface modes at no applied static field. The dotted vertical lines are the grating induced lines ($q = k_x \pm s$), and the light line is solid line ($q = k_x \pm s$). "s" is a reciprocal spatial period of the gratings. [[32]]

5. Realization of MSPs by effective media of magnetic superlattices

There are many researches on superlattice's surface polariton such as lateral superlattices [23, 36]. However, the lateral superlattice cases will be excluded in this study. In this section, the period of the superlattices will only to be along $\hat{y}$ axis. The superlattices considered here is composed of two alternating uniaxial magnetic materials and uniaxial magnetic (or nonmagnetic) materials ($\mu_{xx}^{(j)} = \mu_{yy}^{(j)}$, j = medium 1 or 2). Proposed that the condition, $kd << 1$ (period, $d = a + b$, see Fig. 10) is valid, and then the effective medium theory can adequately approximate the bulk and surface polaritons.

The magnetic permeability of each layer in a magnetic superlattices still can be respectively described by Eq. 1, and then we quote references [2, 25] that rigorously derived the effective magnetic permeability and electric permittivity of the magnetic superlattices. Therefore, one

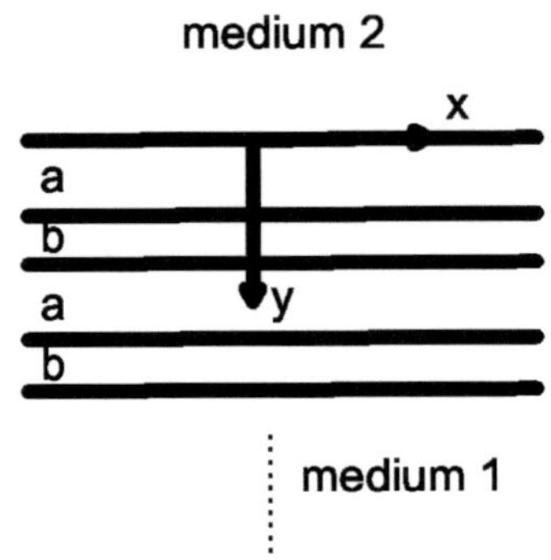

Figure 10. The region $y > 0$ is the magnetic superlattices composed of two alternating materials. The symbol "a" stands for the thickness of a ferromagnetic (or antiferromagnetic) material; "b" stands for the thickness of a nonmagnetic materials. $d = a + b$ is a period of two alternating magnetic materials.

is given by

$$\begin{aligned} \bar{\mu}_{xx} &= \frac{(a+b)^2 \mu_{xx}^{(1)} \mu_{xx}^{(2)} + ab \left[\left(\mu_{xx}^{(1)} - \mu_{xx}^{(2)} \right)^2 - \left(\mu_{xy}^{(1)} - \mu_{xy}^{(2)} \right)^2 \right]}{(a+b) \left(a\mu_{xx}^{(2)} + b\mu_{xx}^{(1)} \right)}, \\ \bar{\mu}_{yy} &= \frac{(a+b)\, \mu_{xx}^{(1)} \mu_{xx}^{(2)}}{a\mu_{xx}^{(2)} + b\mu_{xx}^{(1)}}, \\ \bar{\mu}_{xy} &= \frac{a\mu_{xy}^{(1)} \mu_{xx}^{(2)} + b\mu_{xy}^{(2)} \mu_{xx}^{(1)}}{a\mu_{xx}^{(2)} + b\mu_{xx}^{(1)}}, \\ \bar{\varepsilon}_{zz} &= \frac{a\varepsilon_{zz}^{(1)} + b\varepsilon_{zz}^{(2)}}{a+b}. \end{aligned} \tag{31}$$

Most works on magnetic superlattices have almost considered the special case, that is, medium 2 is nonmagnetic ($\mu_{xx}^{(2)} = 1$ and $\mu_{xy}^{(2)} = 0$). Here, we also concentrate on such special case. Accordingly, we substitute Eq. 14 into $\mu_{xx}^{(1)}$ and $\mu_{xy}^{(1)}$ in Eq. 31 for ferromagnetic/nonmagnetic superlattices; for antiferromagnetic/nonmagnetic superlattices, Eq. 23 may be chosen. For both cases, an external static magnetic filed (H_0) is applied along the easy axis $+\hat{z}$.

5.1. Ferromagnetic superlattices

In the ferromagnetic superlattices case, we see $\bar{\mu}_{xx} \neq \bar{\mu}_{yy}$ and $\bar{\mu}_{xy} = a\mu_{xy}^{(1)} / \left(a + b\mu_{xx}^{(1)} \right)$ so that the attenuation coefficients and dispersion equation of the MSPs mode can be given by [see Eqs. 8 and 11]

$$\begin{aligned} \alpha^+ &= \sqrt{\frac{\bar{\mu}_{xx}}{\bar{\mu}_{yy}} k_x^2 - \left(\frac{\bar{\mu}_{xx}\bar{\mu}_{yy} - \bar{\mu}_{xy}^2}{\bar{\mu}_{yy}} \right) \frac{\omega^2}{c^2} \bar{\varepsilon}_{zz}}, \\ \alpha^- &= \sqrt{k_x^2 - \frac{\omega^2}{c^2}}, \end{aligned} \tag{32}$$

and

$$\frac{-\alpha^{+}\bar{\mu}_{yy} + \bar{\mu}_{xy}k_x}{\bar{\mu}_{xx}\bar{\mu}_{yy} - \bar{\mu}_{xy}^2} = \alpha^{-}. \tag{33}$$

Again, Eq. 33 directly depends on the direction of k_x, leading to non-reciprocal dispersion relation. We would like to quote the results of the reference [4], and then the dispersion relation (Eq. 33) is rewritten as follows,

$$\left(\alpha^{-}\right)^2 \bar{\mu}_{xx}\bar{\mu}_v - 2\left(\alpha^{-}\right)\bar{\mu}_{xx}\bar{\mu}_v - k_x^2 + \bar{\mu}_{xx}\frac{\omega^2}{c^2} = 0. \tag{34}$$

where we substitute α^{+} in Eq. 32. The numerical calculation of Eq. 34 is shown in Fig. 11 at the condition, $a = b$. Note that we re-define the expression in the Eq. 15 for the reason to be consistent with reference [4]:

$$\omega_0 = g\mu_0\gamma_0 H_0, \;\; \omega_m = g\mu_0\gamma_0 M_0, \tag{35}$$

where g is a Lande factor, μ_0 is a magnetic permeability in the vacuum, and $\gamma_0 = e/2m$ with $e > 0$ and m being the electron charge and electron mass. The numerical results show no magnetostatic analog (real mode) for this configuration (i.e. $a = b$) at $k_x \to \pm\infty$, and only a virtual mode exists for $-k_x$. For $a > b$, the magnetostatic analog can occurs at $-k_x$ (the evidence is not shown here). In addition, if the retardation corrections are included, then the virtual MSPs mode still exists for $a < b$. Note that Fig. 11 shows that the surface polaritons exist only in a restricted range.

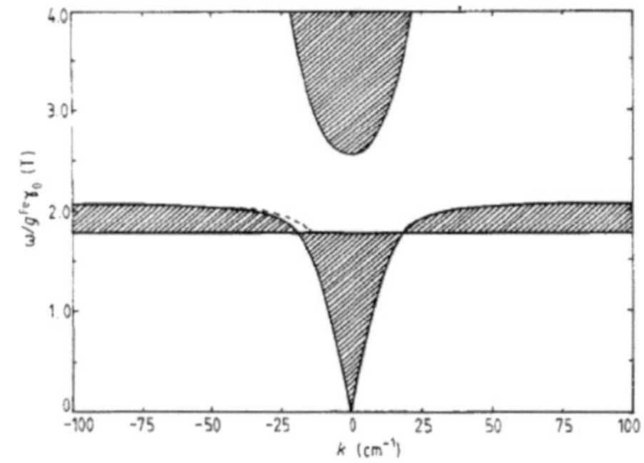

Figure 11. Ferromagnetic (Fe)/nonmagnetic superlattices. $a = b = 5 \times 10^{-4}$ cm. $\mu_0 H_0 = 1$ T. $\mu_0 M_0 = 2.15$ T. $g^{Fe} = 2.15$. The shaped region is the bulk mode; the dotted line is the surface mode. [4]

5.2. Antiferromagnetic superlattices

Let us consider the superlattices that are composed of medium 1 (uniaxial antiferromagnetic) alternating with medium 2 (non-magnetic). In the absence of an applied static magnetic field ($H_0 = 0$), the component of the magnetic permeability of two media can read from Eq. 23,

$$\begin{aligned}
\mu_{xx}^{(1)} &= \mu_{yy}^{(1)} = 1 + \frac{2\omega_m\omega_A}{\omega_{an}^2 - \omega^2} = \frac{\omega_1^2 - \omega^2}{\omega_{an}^2 - \omega^2} \equiv \mu_1, \\
\mu_{xx}^{(2)} &= \mu_{yy}^{(2)} = 1 \equiv \mu_2, \\
\mu_{zz}^{(1)} &= \mu_{zz}^{(2)} = \mu_0 = 1, \\
\mu_{xy}^{(1)} &= \mu_{xy}^{(2)} = 0.
\end{aligned} \tag{36}$$

where $\omega_1 = \sqrt{\omega_{an}^2 + 2\omega_m\omega_A}$ at $\mu_1 = 0$. Here, we do not attempt to derive the non-reciprocal dispersion equation at an applied static field, but just show the numerical results later. Using Eqs. 31 and 36, the component of the effective magnetic permeability can be given by

$$\bar{\mu}_{xx} = \frac{\bar{\omega}_1^2 - \omega^2}{\omega_{an}^2 - \omega^2}, \qquad \bar{\mu}_{yy} = \frac{\omega_1^2 - \omega^2}{\bar{\omega}_{an}^2 - \omega^2}, \tag{37}$$

where $\bar{\omega}_1^2 = \left(b\omega_{an}^2 + a\omega_1^2\right)/(a+b)$ and $\bar{\omega}_{an}^2 = \left(a\omega_{an}^2 + b\omega_1^2\right)/(a+b)$. Reference [3] showed the results of Eq. 37 in Fig. 12 for $a > b$.

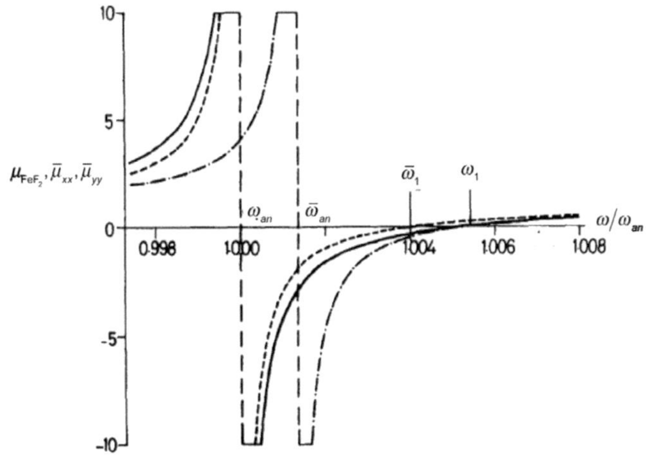

Figure 12. A FeF_2/ZnF_2 superlattices with $\frac{a}{a+b} = 0.75$. μ_{FeF_2} (solid line), $\bar{\mu}_{xx}$ (dash line), and $\bar{\mu}_{yy}$ (dash-dot line). ω_{an}, $\bar{\omega}_{an}$, $\bar{\omega}_1$ and $\bar{\omega}_1$ are marked. [3]

Now the attenuation coefficients can be given by Eq. 8

$$\alpha^+ = \sqrt{\frac{\bar{\mu}_{xx}}{\bar{\mu}_{yy}}k_x^2 - \bar{\varepsilon}_{zz} \cdot \bar{\mu}_{xx}\frac{\omega^2}{c^2}}, \qquad \alpha^- = \sqrt{k_x^2 - \frac{\omega^2}{c^2}}. \tag{38}$$

And, the general dispersion equation of the MSPs mode reads from Eq. 11

$$\frac{-\alpha^+}{\bar{\mu}_{xx}} = \alpha^-, \tag{39}$$

Note that Eq. 39 implies a reciprocal dispersion relation of the surface polaritons. Substitution of Eq. 38 into Eq. 39 can solve the explicit form

$$k_x = \frac{\omega}{c}\sqrt{\frac{\bar{\mu}_{yy}\left(\bar{\mu}_{xx} - \bar{\varepsilon}_{zz}\right)}{\bar{\mu}_{xx}\bar{\mu}_{yy} - 1}}. \tag{40}$$

Eqs. 39 and 40 describes the condition for the existence of the MSPs mode. First, we investigate the magnetostatic limit ($k_x \to \pm\infty$), yielding to

$$\bar{\mu}_{xx} = -1, \; \bar{\mu}_{xx}\bar{\mu}_{yy} = 1. \tag{41}$$

We define $\bar{\mu}_{yy} < 0$ as real mode and $\bar{\mu}_{yy} > 0$ as virtual mode. In a result, only the real mode has a magnetostatic analog, and the virtual mode does not. Here, we simply persent two kinds of situations: (1) If $a > b$, the real mode lies in a frequency region, $\bar{\omega}_{an} < \omega < \omega_{sp}$ (i.e. $\bar{\mu}_{yy} < 0$) and virtual mode in another frequency region, $\omega_{an} < \omega < \bar{\omega}_{an}$ (i.e. $\bar{\mu}_{yy} > 0$) [see Fig. 12]. At $k_x \to \pm\infty$, Eq. 41 yields the asymptotic frequency of the MSPs mode

$$\omega_{sp} = \sqrt{\frac{\omega_{an}^2 + \omega_1^2}{2}}, \tag{42}$$

where $\omega_1 = \sqrt{\omega_{an}^2 + 2\omega_m\omega_A}$ at $\mu_1 = 0$ [see Eq. 36]. Note that the resonance frequency of the MSPs mode is independent of a and b and, consequently, is just the same as a pure antiferromagnet (see Eq. 30). (2) For $a < b$, there is only the virtual mode to exist at the frequencies, $\omega_{an} < \omega < \bar{\omega}_1$.

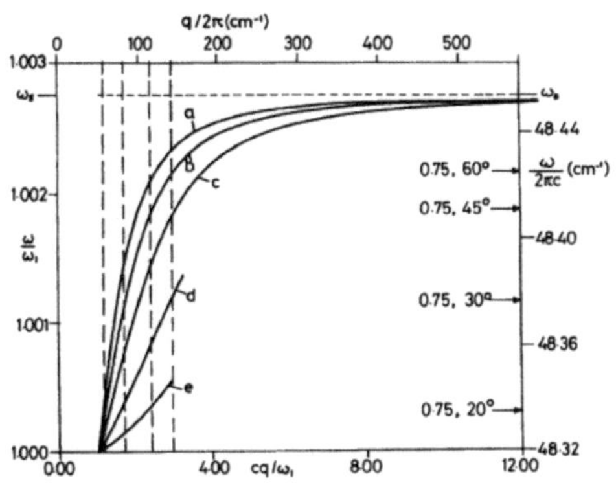

Figure 13. The dispersion relation versus f_1 at no applied static magnetic field ($H_0 = 0$). The curves from a to e represent $f_1 = 1$, 0.75, 0.5, 0.25, and 0.1, respectively. [3]

For the FeF_2/ZnF_2 system, reference [3] has summarized the results of the dispersion curves of the MSPs modes versus $f_1 \equiv a/(a+b)$ various at $+k_x$ shown in Fig. 13. Figure 13 marks four black arrows for $f_1 = 0.75$ that respectively corresponds to four incidence angle: 20°, 30°, 45°, and and 60°. These angles are in a position to transfer momentum of light into the MSPs mode, leading to a reflectivity deep in the Otto ATR spectra (like Fig. 6). Note that one can clearly observe that only real modes (magnetostatic analog) can exist at the conditions, $a > b$.

Under an applied static magnetic field, on other hand, the numerical calculations should be required for this discussion shown in Fig. 14 [17]. Figure 14(a) shows a pure antiferromagnet (i.e. $f_1 = 1$); Fig. 14(b) shows $f_1 = 0.75$; Fig. 14(c) is $f_1 = 0.25$. For $f_1 = 1$ and $f_1 = 0.75$, one can observe that there are two magnetostatic modes and one is virtual mode at $\pm k_x$. As for $f_1 = 0.25$, there are all virtual modes at $\pm k_x$.

Finally, If $a = b$ ($f_1 = 0.5$), the dispersion relation of the antiferromagnetic superlattices (MnF_2) [4] is shown in Fig. 15. It shows the zero and non-zero applied static magnetic field, respectively. Fig. 15(a) shows the reciprocal dispersion relation; Fig. 15(b) shows

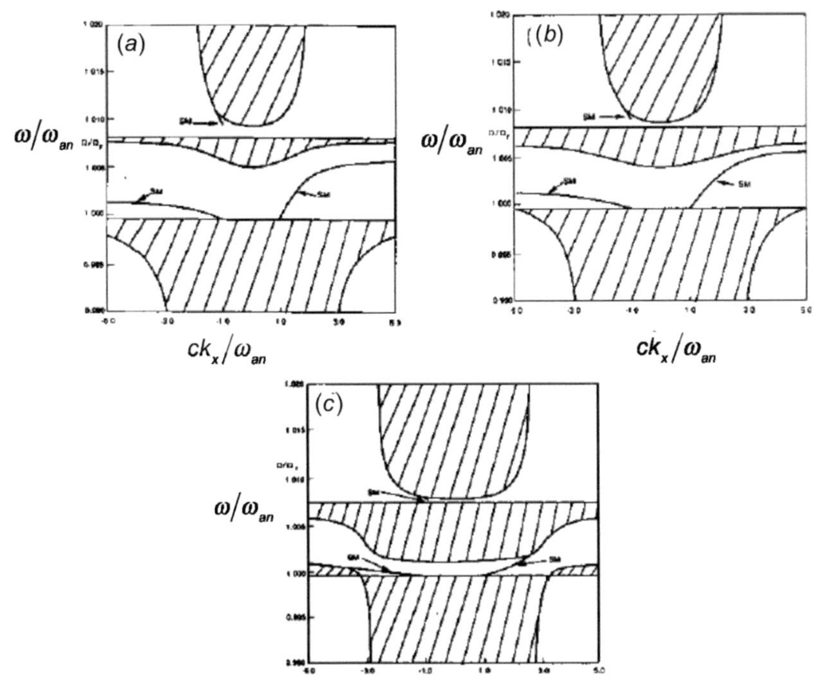

Figure 14. The dispersion relation at an applied static magnetic field. The solid curves represent the surface modes, and shaded regions are bulk modes. (a) a pure antiferromagnet (MnF_2) (b) antiferromagnetic superlattices, MnF_2/ZnF_2 ($f_1 = 0.75$). (c) antiferromagnetic superlattices, MnF_2/ZnF_2 ($f_1 = 0.25$). $\omega_{an} = 260\ GHz$. $H_0 = 200\ G$, $H_A = 7.85\ kG$, $H_E = 550\ kG$, $M_0 = 0.6\ kG$, $\varepsilon_{zz}^{(1)} = 5.5$, and $\varepsilon_{zz}^{(2)} = 8$. [3]

non-reciprocal dispersion relation. The MSPs modes only exist in a restricted region that bulk modes are not excited. Here, for consistence with reference [4], we recall the Eq. 35 and change the expression in Eq. 24 as following:

$$\omega_{an} = g\mu_0\gamma_0\sqrt{H_A\left(H_A + 2H_E\right)}\,. \tag{43}$$

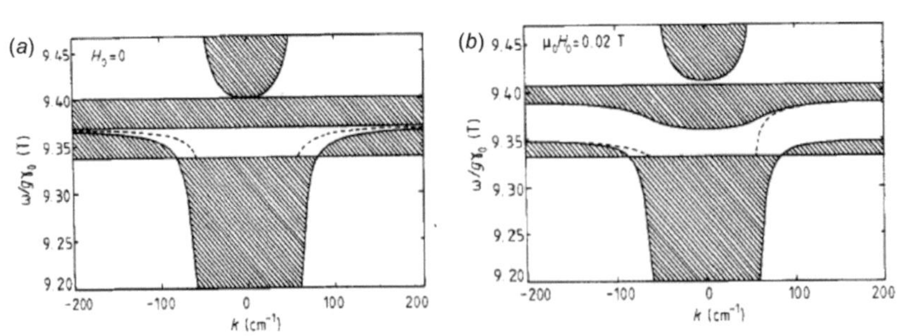

Figure 15. The antiferromagnetic (MnF_2)/nonmagnetic superlatticess. $a = b = 5 \times 10^{-4}\ cm$, $\mu_0 H_E = 55\ T$, $\mu_0 H_A = 0.787\ T$, $\mu_0 M_0 = 0.754\ T$, and $g = 2$. (a) $\mu_0 H_0 = 0$. (b) $\mu_0 H_0 = 0.02\ T$. The shaped region is the bulk polaritons; the dotted curve is the surface polaritons mode. [4]

6. Realization of MSPs by metamaterials

6.1. Introduction of metamaterials

In 1968, a Russian physicist V. G. Veselago theoretically calculated that a medium might bend light to the "wrong" pathway whenever the medium could "give" a negative refractive index [35]. In his theory, the negative refractive index medium (NRIM) is composed of both negative electric permittivity and negative magnetic permeability at the same time but unfortunately, Veselago never experimentaly demonstrated such an NRIM mainly because he could not find out a material with negative magnetic permeability. Three decades later, British physicist J.B. Pendry suggested that the two-dimensional metallic wires [12] and metallic split-ring resonators (SRRs) [11] yield to negative permittivity and negative permeability, respectively. These innovative electromagnetic responses were artificially created by a periodic array of sub-wavelength unit cells that afterwards, were termed as metamaterials and led to a variety of rare and even unprecedented electromagnetic properties such magnetic magnetism at terahertz region [33], inverse optical rules [e.g., inverse Snell's law [22], inverse Doppler shift [30] and inverse Cerenkov radiation [28]] , superlensing effect [31], invisibility cloaking [7], and others.

In 2001, Shelby et al. integrated the metallic wires and metallic SRRs into a prism structure [see Fig. 16], and for the first time demonstrated the negative refractive index [22]. This result demonstrated that metamaterials were recognized to be an effective medium due to the sub-wavelength characteristics. Therefore, the magnetic metamaterials (like SRRs) can be expected to process the MSPs mode as same as that for the aforementioned magnetic superlattices. In fact, some groups have already theoretically predicated the surface mode of the ferromagnetic metals [1] and the metamaterials [26].

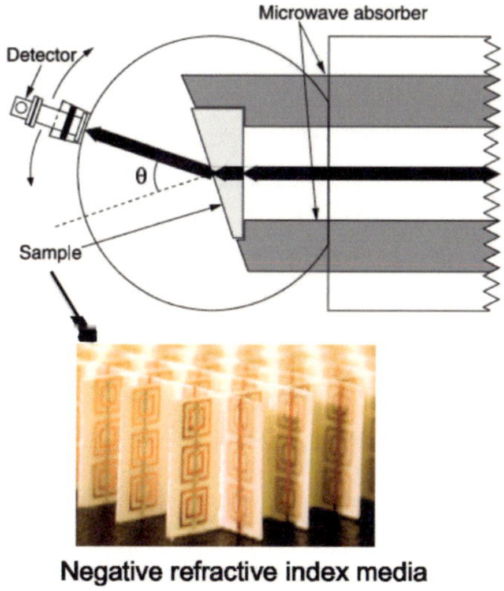

Figure 16. Experimental verification of the negative refractive index medium. The sample consisted of a periodic array of the metallic wires and the SRRs is so called a NRIM whose geometrical shape is a prism now. One can observe that the NRIM bends light into a "wrong" pathway, and Snell's law verifies the negative refractive index. [22]

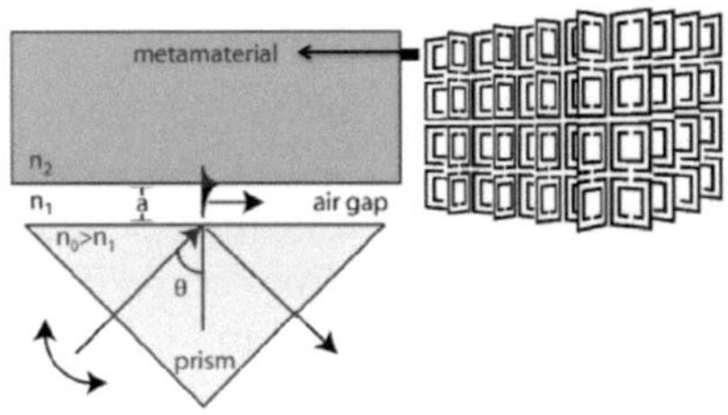

Figure 17. The Otto ATR configuration. The magnetic metamaterials is composed of a two-dimensional grid of the SRRs. "a" is gap between prism and metamaterials. [16]

6.2. Uniaxial magnetic metamaterials

Now, we quote an interesting work done by J. N. Gollub et at. who constructed a two-dimensional array of the SRRs into a physical three-dimensional bulk [16] shown in Fig. 17. Such the physical three-dimensional bulk metamaterials is considered as a semi-infinite medium due to the subwavelength SRR structure. According to their geometry, we regard the effective electric permittivity and magnetic permeability of the bulk as anisotropic and uniaxial, respectively. Therefore, it is straightforward to write

$$\overleftrightarrow{\varepsilon}_{eff}(\omega) = \begin{pmatrix} \bar{\varepsilon}_{xx} & 0 & 0 \\ 0 & \bar{\varepsilon}_{yy} & 0 \\ 0 & 0 & \bar{\varepsilon}_{zz} \end{pmatrix}, \quad \overleftrightarrow{\mu}_{eff}(\omega) = \begin{pmatrix} \bar{\mu}_{xx} & 0 & 0 \\ 0 & \bar{\mu}_{xx} & 0 \\ 0 & 0 & \bar{\mu}_{zz} \end{pmatrix}. \tag{44}$$

Also, the attenuation coefficients and dispersion equation of the surface mode of this bulk medium can be written by setting $\bar{\mu}_{xy} = 0$ and $\bar{\mu}_{xx} = \bar{\mu}_{yy}$ into Eqs. 8 and 13:

$$\alpha^+ = \sqrt{k_x^2 - \bar{\varepsilon}_{zz}\bar{\mu}_{xx}\frac{\omega^2}{c^2}}, \quad \alpha^- = \sqrt{k_x^2 - \frac{\omega^2}{c^2}}, \tag{45}$$

and

$$\frac{-\alpha^+}{\bar{\mu}_{xx}} = \alpha^-. \tag{46}$$

They are as same as those for a pure antiferromagnet without an applied static field [Eqs. 27 and 28]. The magnetic permeability of the bulk medium is calculated by means of the standard retrieval process [6, 34] shown in Fig. 18 [16]. There exists a $\bar{\mu}_{xx} < 0$ between 12 and 14 GHz in where the MSPs mode is expected. From Eq. 46, one knows that the momentum never be conserved without an optical coupler (diffraction gratings or prism). The experimental result of the Otto ATR spectra [16] is shown in Fig. 19 , where the dash curve is numerical calculation and the solid curve is experimental measurement. The experimental result is in a good agreement with numerical calculation, and demonstrated successfully that the MSPs mode is excited occurring at reflectivity deep (12.2 GHz).

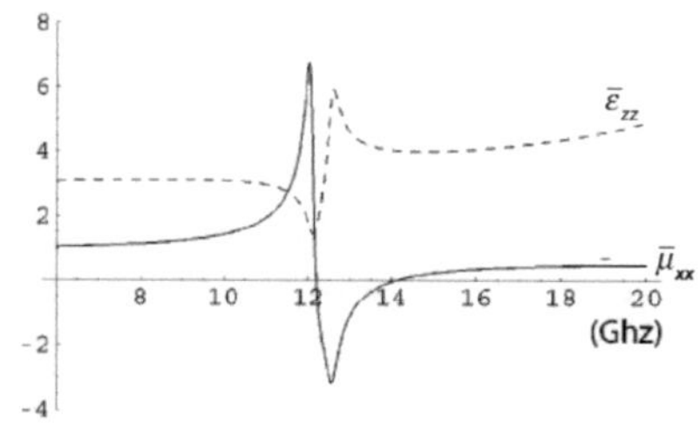

Figure 18. The effective permittivity and permeability of the bulk metamaterials. $\bar{\mu}_{xx} < 0$ for 12 and 14 GHz. [16]

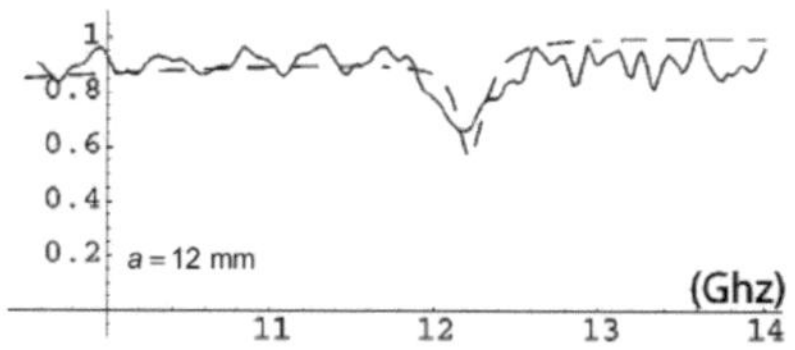

Figure 19. The experimental results of the ATR spectra, and $a = 12\ mm$. The dash curve is numerical calculation, and the solid curve is experimental measurement. [16]

6.3. Biaxial magnetic metamaterials

Here, we shortly re-describe what we just talked about the conditions for an excitation of the MSPs at magnetostatic limit. First, for a pure uniaxial ferromagnet, the MSPs can be excited at the condition, $\mu_{xx} \pm \mu_{xy} = -1$ [Eq. 17] with an applied static magnetic field. Note that the surface mode might exist whether or not the μ_v in Fig. 3 is positive or negative. Second, for a pure antiferromagnet without an applied static field, the condition is $\mu_{xx} = -1$ at magnetostatic limit, which is also for uniaxial magnetic metamaterials. Third, for a ferromagnetic superlattices, the condition is not considered in this chapter, but we present a special numerical result shown in Fig. 11. Fourth, for an antiferromagnetic superlattices, the necessary condition is $\bar{\mu}_{xx} < 0$. It collocates with $\bar{\mu}_{yy} < 0$ being a real mode and $\bar{\mu}_{xy} > 0$ being a virtual mode, respectively.

Now, we would like to introduce our previous work [37] that considered a situation as same as the antiferromagnetic superlattices. In a word, the magnetic metamaterials have an effect biaxial tensors are given by

$$\overleftrightarrow{\varepsilon}_{eff}(\omega) = \begin{pmatrix} \bar{\varepsilon}_{xx} & 0 & 0 \\ 0 & \bar{\varepsilon}_{yy} & 0 \\ 0 & 0 & \bar{\varepsilon}_{zz} \end{pmatrix}, \quad \overleftrightarrow{\mu}_{eff}(\omega) = \begin{pmatrix} \bar{\mu}_{xx} & 0 & 0 \\ 0 & \bar{\mu}_{yy} & 0 \\ 0 & 0 & \bar{\mu}_{zz} \end{pmatrix}. \tag{47}$$

The proposal structure of our work [shown in Fig. 20] is a periodic array of the sandwich unit cells. Fig. 20(b) shows there are three layers in a sandwich unit cell that includes two SRRs and one spacer (Roger RT5800). The dimension of the unit cell is $6 \times 6\ mm^2$ and thickness is $2\ mm$, which is smaller than our operating wavelength (23 mm or 13.22 GHz). Therefore, it is safe to consider a sandwich unit cell as an effective medium. Its effective electric permittivity and magnetic permeability are retrieved shown in Fig. 21. At 13.22 *GHz*, we have

$$\bar{\mu}_{xx} = -4.20,\ \bar{\mu}_{yy} = -0.252,\ \bar{\varepsilon}_{zz} = 3.46. \tag{48}$$

Therefore, our design only possesses the real mode. Using COMSOL Multiphysics finite element based electromagnetic solver [2], we simulate the field distribution of the MSPs mode shown in Fig. 22 where one can clearly see the exponential decay of the MSPs propagating at the interface. By substituting Eq. 48 into Eq. 40, the wavelength of the MSPs mode can be evaluated as 3.947 *mm* that is in a good agreement with the simulation result (not shown). Note that such a biaxial magnetic metamaterials depends on designed geometry to determine the resonance frequency of the MSPs mode, instead of inherent magnetism.

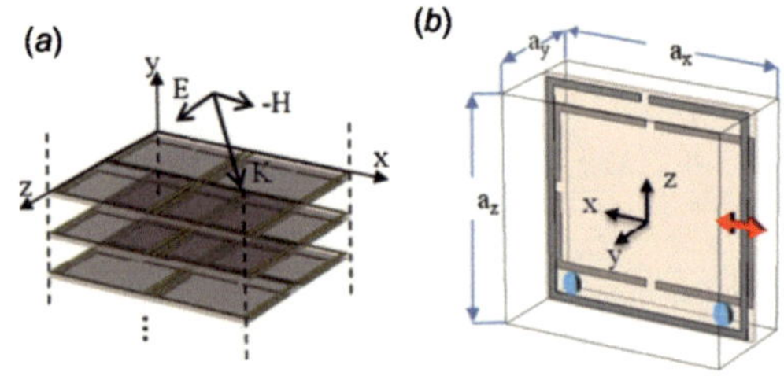

Figure 20. The sandwich unit cell of the biaxial magnetic metamaterials. $a_x = a_y = 6\ mm, a_z = 2\ mm$. [37]

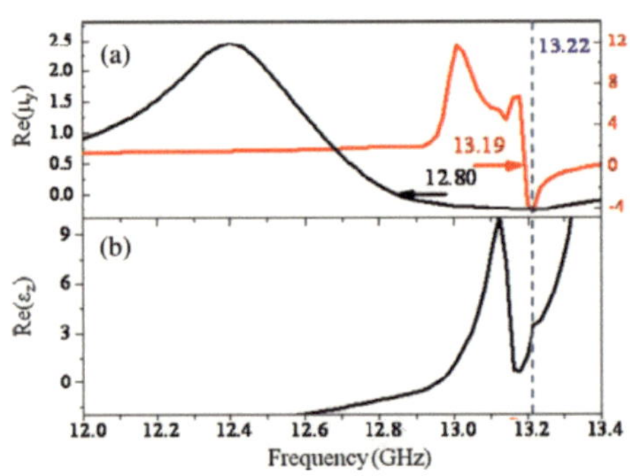

Figure 21. The retrieval result of the biaxial magnetic metamaterials. (a) Red curve is $\bar{\mu}_{xx}$; black curve is. (b) Black curve is $\bar{\varepsilon}_{zz}$. At 13.22 GHz, we have $\bar{\mu}_{xx} = -4.2, and \bar{\mu}_{yy} = -0.252, \bar{\varepsilon}_{zz} = 3.46$. [37]

6.4. One-dimensional magnetic photonic crystal

In last two sections, the samples are not composed of magnetic materials, and so there need not be an applied static magnetic field. Now, we basically introduce that a magnetic metamaterials is composed of YIG rods [29] shown in Fig. 23(a). The structure is like a one-dimensional photonic crystal (PC), and arranged periodically in a square lattice embedded within the air. The lattice constant is $a = 8\ mm$ and the radius of the YIG rod is $r = 2\ mm$ where axis of the rods is at $\hat{z}$ axis. Again, an applied magnetic static field is along the easy axis $\hat{z}$. The operating frequency is 5 *GHz* (i.e. wavelength $= 60\ mm$) that is enough

[2] http://www.comsol.com/

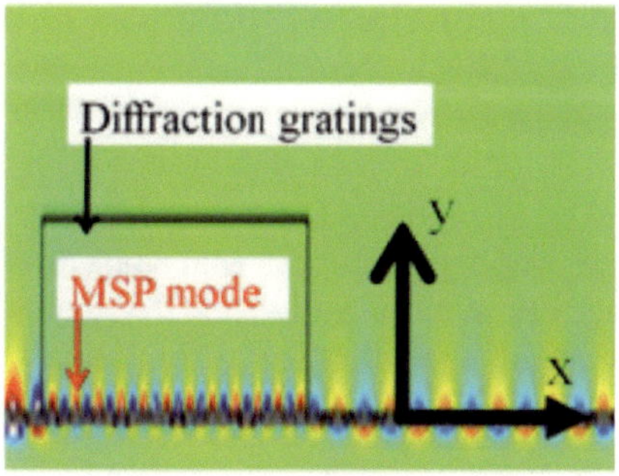

Figure 22. The field distribution of the MSPs mode for a biaxial magnetic metamaterials at 13.22 GHz. [37]

lager to be regarded this structure as an effective medium. For the reflectivity simulation, the magnetic permeability of the YIG needs a damping factor that can be modified from Eq. 14:

$$\omega_0 \to \omega_0 + i\beta\omega \tag{49}$$

The retrieved effective permittivity and permeability, and band diagram of the bulk polaritons are plotted in Figs. 23(c) and 23(b), respectively. Note that the condition of the magnetostatic mode is as same as that for a pure ferromagnet [i.e. see Eq. 17]. In a word, the condition is $\mu_{eff} = -1$ [see in Fig. 23(c)]. In this work, Liu et al. planed to apply a static magnetic field upon the PC, which yields to the modeling of the reflectivity as shown in Fig. 23(a). One can observe the reflectivity has "own" direction at the resonance frequency of the

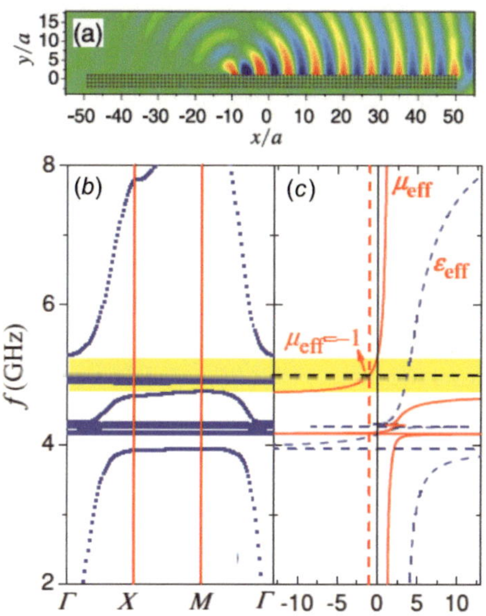

Figure 23. (a) The line source is transverse magnetic wave with the electric field polarized along the rod, and is placed near the surface of the magnetic metamaterials ($\frac{y}{a} = 1$; $\frac{x}{a} = -10$) (b) The photonic band diagram of the magnetic metamaterials with $H_0 = 900\ Oe$. The yellow region is photonic band gap. (c) The retrivevd effective constitutive parameters. The blue dashed curve is ε_{eff} and red solid curve is μ_{eff}. Note that $\mu_{eff} = 1$ at $5\ GHz$. [29]

MSPs. The main mechanism is non-reciprocal dispersion relation of the MSPs because the left-propagating direction of the MSPs is forbidden.

7. Conclusions

In this study, we review a few important papers that help one to understand the conditions about the existence of the magnetic surface polaritons (MSPs) modes. Note that ferromagnetic resonance frequency is typically in the microwave frequency region, and antiferromagnetic resonance frequency in the far infrared. However, Metamaterials resonance frequency can be artificially determined.

The MSPs mode can further be considered as two kinds of modes, real mode (magnetostatic analog) and virtual mode. First, for real mode, we basically summarize those aforementioned conditions. (a) For a pure uniaxial ferromagnet with $\overleftrightarrow{\varepsilon} = 1$, at an external static magnetic field (H_0), the condition is $\mu_0 \left[1 + \omega_m/(\omega_0 - \omega)\right] = -1$ where ω_{msp} is locating within the region, $\mu_V < 1$. In addition, the dispersion relation of the MSPs mode is non-reciprocal. On the other hand, without H_0, the condition is $\mu_0 \left[1 - \omega_m/\omega\right] = -1$ and magnetostatic analog starts at $k_x = 0$ with non-reciprocal dispersion relation. (b) For a pure uniaxial anti-ferromagnet, without H_0, the condition is $\mu_{xx} = -1$ where ω_{msp} is living within the region, $\mu_V < 1$. The dispersion relation of the MSPs is reciprocal. On other hand, with H_0, numerical calculation shows there is a possibility to have a magnetostatic analog with non-reciprocal dispersion relation. (c) For a ferromagnetic superlattices, with H_0, there is a magnetostatic analogs for $a/b > 0.5$. The dispersion relation of the MSPs is non-reciprocal. (d) For an anti-ferromagnetic superlattices, without H_0, the condition is $\bar{\mu}_{xx} = -1$ and $\bar{\mu}_{xx}\bar{\mu}_{yy} = 1$. The dispersion relation of the MSPs mode is reciprocal for $a > b$. On other hand, with H_0, numerical calculation shows that the magnetostatic analog can exist for $a > b$.

Second, for the virtual modes, the numerical calculation is necessary, and it is not easy to give a exact condition. However, for magnetic superlattices, if the ratio of "a" to "b" is less than 0.5, then the virtual modes should exist. Finally, let us comment the conditions for magnetic metamaterials. The magnetic metamaterials are ability to design their constitutive parameters by means of a specific the geometry of the sub-wavelength unit cell. Theoretically, one can well control the real modes or virtual modes, yielding to the versatile devices such as optical bistability [9, 10, 13], second harmonic generation [19]. Although over the past researches have less attention to the MSPs modes due to the experimental difficulty compared to electric SPPs, we expect the MSPs mode will get more attention by means of the magnetic metamaterials in the future.

Acknowledgments

The authors would like to gratefully acknowledge the financial support from the National Science Council (98-2112-M-007-002MY3, 99-2923-M-007-003-MY2, 99-ET-E-007-002-ET, 100-ET-E-007-002-ET, 100-2120-M-002-008, 100-2120-M010-001).

Author details

Yu-Hang Yang and Ta-Jen Yen
Department of Materials Science and Engineering, National Tsing Hua University, Hsinchu, Taiwan

8. References

[1] A. Hartstei, E. Burstein, A. M. R. B. & Wallis, R. F. [1973]. Surface polaritons on semi-infinite gyromagnetic media, *Journal of Physics C-Solid State Physics* 6(7): 1266–1276.
[2] Almeida, N. S. & Mills, D. L. [1988]. Effective-medium theory of long-wavelength spin-waves in magnetic superlattices, *Physical Review B* 38(10): 6698–6710.
[3] Almeida, N. S. & Tilley, D. R. [1990]. Surface-polaritons on antiferromagnetic superlattices, *Solid State Communications* 73(1): 23–27.
[4] Barnas, J. [1990]. Spin-waves in superlattices .3. magnetic polaritons in the voigt configuration, *Journal of Physics-Condensed Matter* 2(34): 7173–7180.
[5] Camley, R. E. & Mills, D. L. [1982]. Surface polaritons on uniaxial antiferromagnets, *Physical Review B* 26(3).
[6] D. R. Smith, S. Schultz, P. M. & Soukoulis, C. M. [2002]. Determination of effective permittivity and permeability of metamaterials from reflection and transmission coefficients., *Phys. Rev. B* 65: 195104.
[7] D. Schurig, J. J. Mock, B. J. J. S. A. C. J. B. P. A. F. S. & Smith, D. R. [2006]. Metamaterial electromagnetic cloak at microwave frequencies, *Science* 314(5801): 977–980.
[8] Damon, R. W. & Eshbach, J. R. [1961]. Magnetostatic modes of a ferromagnet slab, *Journal of Physics and Chemistry of Solids* 19(3-4): 308–320.
[9] G. D. Xu, T. Pan, T. C. Z. & Sun, J. [2008]. Optical bistability with surface polaritons in layered structures containing left-handed metallic magnetic composites, *Applied Physics B-Lasers and Optics* 93(2-3): 551–557.
[10] I. L. Lyubchanskii, N. N. Dadoenkova, A. E. Z. Y. P. L. & Rasing, T. [2008]. Optical bistability in one-dimensional magnetic photonic crystal with two defect layers, *Journal of Applied Physics* 103(7).
[11] J. B. Pendry, A. J. Holden, D. J. R. & Stewart, W. J. [1999]. Magnetism from conductors and enhanced nonlinear phenomena, *Ieee Transactions on Microwave Theory and Techniques* 47(11): 2075–2084.
[12] J. B. Pendry, A. J. Holden, W. J. S. & Youngs, I. [1996]. Extremely low frequency plasmons in metallic mesostructures, *Physical Review Letters* 76(25): 4773–4776.
[13] J. Bai, S. F. Fu, S. Z. & Wang, X. Z. [2011]. Reflective optical bi-stability of antiferromagnetic films, *European Physical Journal B* 83(3): 343–348.
[14] J. Matsuura, M. F. & Tada, O. [1983]. Atr mode of surface magnon polaritons on yig, *Solid State Communications* 45(2).
[15] J. N. Anker, W. P. Hall, O. L. N. C. S. J. Z. & Duyne, R. P. V. [2008]. Biosensing with plasmonic nanosensors, *Nature Materials* 7(6): 442–453.
[16] J. N. Gollub, D. R. Smith, D. C. V. T. P. & Mock, J. J. [2005]. Experimental characterization of magnetic surface plasmons on metamaterials with negative permeability, *Physical Review B* 71(19).
[17] M. C. Oliveros, N. S. Almeida, D. R. T. J. T. & Camley, R. E. [1992]. Magnetostatic modes and polaritons in antiferromagnetic nonmagnetic superlattices, *Journal of Physics-Condensed Matter* 4(44): 8497–8510.
[18] M. R. F. Jensen, T. J. Parker, K. A. & Tilley, D. R. [1995]. Experimental-observation of magnetic surface-polaritons in fef2 by attenuated total-reflection, *Physical Review Letters* 75(20): 3756–3759.
[19] M. W. Klein, C. Enkrich, M. W. & Linden, S. [2006]. Second-harmonic generation from magnetic metamaterials, *Science* 313(5786): 502–504.

[20] Mills, D. L. & Burstein, E. [1974]. Polaritons - electromagnetic modes of media, *Reports on Progress in Physics* 37(7): 817–926.

[21] Pincus, P. [1962]. Propagation effects on ferromagnetic resonance in dielectric slabs, *Journal of Applied Physics* 33(2): 553.

[22] R. A. Shelby, D. R. S. & Schultz, S. [2001]. Experimental verification of a negative index of refraction, *Science* 292: 77.

[23] R. E. Camley, M. G. C. & Tilley, D. R. [1992]. Surface-polaritons in antiferromagnetic superlattices with ordering perpendicular to the surface, *Solid State Communications* 81(7): 571–574.

[24] Raether, H. [1988]. *Surface plasmons on smooth and rough surfaces and on gratings*, Springer-Verlag.

[25] Raj, N. & Tilley, D. R. [1987]. Polariton and effective-medium theory of magnetic superlattices, *Physical Review B* 36(13): 7003–7007.

[26] Ruppin, R. [2000]. Surface polaritons of a left-handed medium, *Physics Letters A* 277(1): 61–64.

[27] S. Lal, S. L. & Halas, N. J. [2007]. Nano-optics from sensing to waveguiding, *Nature Photonics* 1(11): 641–648.

[28] S. Xi, H. S. Chen, T. J. L. X. R. J. T. H. B. I. W. J. A. K. & Chen, M. [n.d.]. Experimental verification of reversed cherenkov radiation in left-handed metamaterial, *Physical Review Letters* 103(19).

[29] S. Y. Liu, W. L. Lu, Z. F. L. & Chui, S. T. [2011]. Molding reflection from metamaterials based on magnetic surface plasmons, *Physical Review B* 84(4).

[30] Seddon, N. & Bearpark, T. [2003]. Observation of the inverse doppler effect, *Science* 302(5650): 1537–1540.

[31] Smolyaninov, II, Y. J. H. & Davis, C. C. [2007]. Magnifying superlens in the visible frequency range, *Science* 315(5819): 1699–1701.

[32] Stamps, R. L. & Camley, R. E. [1989]. Greens-functions for antiferromagnetic polaritons .2. scattering from rough surfaces, *Physical Review B* 40(1): 609–621.

[33] T. J. Yen, W. J. Padilla, N. F. D. C. V. D. R. S. J. B. P. D. N. B. & Zhang, X. [2004]. Terahertz magnetic response from artificial materials, *Science* 303(5663): 1494–1496.

[34] T. M. G. Xudong Chen, B.-I. W., Pacheco, J. J. & Kong, J. A. [2004]. Robust method to retrieve the constitutive effective parameters of metamaterials, *PHYSICAL REVIEW E* 70: 016608.

[35] Veselago, V. G. [1968]. Electrodynamics of substances with simultaneously negative values of sigma and mu, *Soviet Physics Uspekhi-Ussr* 10(4): 509.

[36] Wang, X. Z. & Tilley, D. R. [1995]. Retarded modes of a lateral antiferromagnetic nonmagnetic superlattice, *Physical Review B* 52(18): 13353–13357.

[37] Y.H. Yang, I. W., U. H. L. & Yen, T. [2012]. Magnetic surface polariton in a planar biaxial metamaterial with dual negative magnetic permeabilities, *Plasmonics* 7(1): 87–92.

Chapter 14

Photoacoustic Based Surface Plasmon Resonance Spectroscopy: An Investigation

K. Sathiyamoorthy, C. Vijayan and V.M. Murukeshan

Additional information is available at the end of the chapter

http://dx.doi.org/10.5772/52545

1. Introduction

Surface Plasmon resonance effect (SPR) has very important applications in many fields. [1-3] The most significant of them perhaps is in the field of real-time bio-sensing. It is important for bio-sensors to be sensitive enough to detect even the smallest amounts of bio-molecules. There are several instances when conventional spectroscopy becomes inadequate, even for clearer and transparent materials. Such a situation arises when one attempts to measure a very weak absorption, which in turn involves the measurement of significantly small changes in the intensity of the strong and unattenuated transmitted signal.

Surface plasmons (SP) are the quanta of elementary excitations involving the oscillations of the electron cloud against the background ionic arrangement in a metal surface. SP can be excited at the metal-dielectric interface using light by matching the momenta between the incident photon and the surface plasmon. Momentum matching can be achieved by various techniques.[4-7] The simplest way to achieve momentum matching between excitation photon and plasmon is utilization of high index prism over the metal medium. SP excitation using prism coupling technique is based on either Kretschmann or Otto configuration.[8, 9] The present study is based on Kretschmann configuration in which the metal film has direct contact with the prism base. The magnitude of the wavevector of SP depends on the dielectric constants of both the metal and the surrounding dielectric medium. It is extremely sensitive to properties of the dielectric medium which is in contact with the metal. As SP circumvents diffraction, it can be confined in nanosize domain and hence have interaction with material structures of dimension from sub-micrometer to a few nanometers. The most common technique widely examined for investigating the effects of the excitation of SP is the measurement of reflection coefficient of incident optical beam. But such an approach is mostly not suitable for the sample with weak absorption characteristics. Higher detection sensitivities than those possible with such an approach are always desirable for improving sensing performance. The measurement of phase variation during SP excitation is expected to provide higher sensitivity to SPR sensors. However, the complexity of the technique associated with phase detection impedes the targeted development. Moreover, most of the optical detection systems (such as linear photo arrays, CCD etc.) currently employed for measuring the reflection remain off the interrogation region due to their sophistication thereby increasing the size and the cost of the system.

The significance of the photo acoustics techniques is in this context that it may pave the way for the design of bio-sensors with greater sensitivity as they are capable of detecting temperature rises of 10^{-6} to $10^{-5}$of a degree in the sample, corresponding to a thermal power generation of about 10^{-6}*calories/cm*^{3}. Furthermore, since the sample itself constitutes the electromagnetic radiation detector enabling studies over a wide range of optical and electromagnetic wavelengths without the need to change detector systems. Besides, the compactness of the PA detection system enables easy integration with the existing Kretschmann configuration based SPR system for applications such as on-chip detection.

Photoacoustic (PA) spectroscopy is a very powerful analytical tool for examining the optical absorption properties of solids as it measures directly the energy absorbed by the material on exposure to light. Conventional optical absorption/transmission spectroscopy requires a homogenous and partially transparent sample and further, it cannot be used with highly scattering samples, where the scattering affects significantly the accuracy of the measurement of optical absorption coefficient. The only other method of obtaining some spectroscopic information from opaque samples is diffuse optical reflection spectroscopy where special treatment is required for sample surface preparation. Such problems are absent in the case of photoacoustic spectroscopy and spectral information can be obtained from a variety of samples including opaque materials, powders etc.

The photoacoustic effect was first reported by Alexander Graham Bell in 1880 though the wide use of this technique for spectroscopic purposes started only almost after a century. From then, there has been a steady progress in the development of PA spectrometers in view of their potential application particularly in biological and environmental studies.[10, 11] A major improvement in the sensitivity of photoacoustic method has been brought about by the availability of laser sources, highly sensitive microphones and other efficient acoustic detection systems.[10-12] Markus Nagele et al developed a novel type of highly sensitive multipass resonant PA cell for trace-gas detection making use of new compact low-power laser sources such as quantum-cascade lasers.[12] Frank Muller et al presented a transportable, highly sensitive photoacoustic spectrometer based on a continuous-wave dual-cavity optical parametric oscillator.[13] Innovations in the design of inexpensive and high performance photoacoustic spectrometers continue to attract recent attention, rendering this form of spectroscopy increasingly versatile.[10, 11, 14-16]

The principle and theory of standard PA technique are well established by Rosencwaig who described the fundamental principles and investigated several possible applications of the PA technique.[17] Patel and Tam reviewed the physics of the technique and extended the technique to many other kinds of applications.[18-21] The most frequently used methods are microphonic detection of the periodic pressure variations developed due to nonradiative deexcitation of the absorbed light by the sample enclosed in an air tight cell. In the case of solid samples, the sample is placed inside a specially designed and acoustically insulated gas filled chamber. The sample is then illuminated with monochromatic light, whose intensity is modulated by means of a mechanical chopper. Light absorption by a material usually leads to two types of deexcitation processes namely radiative and nonradiative. The nonradiative deexcitation processes result in heating the sample. Since the incident radiation is intensity modulated by the chopper, the sample gets heated up repetitively. The resulting periodic heat flow from the solid absorber to the surrounding gas medium creates pressure fluctuations in the cell, which are detected by the microphone. Therefore the depth of specimen responsible for the PA signal is restricted within a thermal diffusion length defined by.[22]

$$\mu = \sqrt{\frac{k}{\pi \rho f c}} \tag{1}$$

Where k is the thermal conductivity, ρ the density, c the specific heat and f is the chopping frequency. The effect of scattered light on measurement is completely avoided by placing the microphone at remote place. This high sensitive microphone is connected to the main chamber containing experimental sample through a small tunnel. Thus PAS is immune towards the stray light from the highly scattering sample. The photo-exciting system consists of a high power optical source, a monochromator and light focusing set up. The detection system consists of a high sensitive microphone, preamplifier and a dual phase lock-in amplifier. The uses of lock-in amplifier effectively increases S/N ratio and enables the simultaneous measurement of phase and amplitude of the PA signal with respect to light modulation frequency. Other than spectral measurements, the photoacoustic technique can also be used in several other applications such as measurement of thermal diffusivity, detection of phase transition, depth profiling, subsurface imaging etc [7, 8, 9, 10].[23-25] The main aspect of cell design is the positioning of the microphone with respect to incident light to avoid scattered light and to improve performance [11, 12, 13].[26-28] In the present study the PA cell designed to study SP excitation and provision is made to study samples of different properties in different gas environments.

Both PA and reflectance techniques are established in view of simultaneous measurement of SP excitation. The conventional way of detection of reflectance is by either a linear array of photodiodes or a goniometer. A single photodetector can also be used to measure reflectance by using a second rotation stage moving in conjunction with rotation of the prism. The system employing these types of configurations is found to be expensive and hence two-prism concepts has been proposed in the recent past.[29, 30] The aim of the earlier reported two-prism configuration is to avoid above mentioned expensive detection techniques.[29] The second prism is integrated to counter compensate the displacement that caused due to the beam walk over the metal coupled face of the first prism during SPR angular measurement.[29] The main aim of usage of second prism in the earlier study is to make the reflected beam stationary and incident normally with respect to a single photodiode over wide angular rotation.[29] But in the case of SPR measurement, it is imminent that the surface plasmon excitation spot i.e., the interrogation spot should remain stationary during SPR angular measurement as the beam walk over the metal face of the prism affects the SPR measurement sensitivity significantly. A two-prism technique with a modified optical configuration is designed to improve SPR measurement. The proposed optical configuration involves two prisms which are positioned in such a way that the SP excitation spot and reflected light remain stationary during angular displacement and rotation. Finally these two concepts such as two prisms configuration and PA measurement are integrated to find application as SPR sensor based on photoacoustic detection

This chapter in this context describes objectives, scope and methodology for this photoacoustic based surface plasmon resonance spectroscopy. The main objective is to carry out photoacoustic investigation of surface plasmons, so as to explore the possibility of employing photoacoustic techniques in real time sensing based on plasmon resonance spectroscopy. Such sensors exploit the high sensitivity of the surface plasmon frequency to the refractive index of layers of adsorbed molecules on the surface. Generally the resonance conditions are influenced by the material adsorbed onto the thin metal film, which will be explored in detail. An in house developed photoacoustic technique is employed as the detection system for spectroscopic investigations. Two types of set-ups are proposed to investigate SP assisted sensing applications. The first set up is based on a closed photoacoustic cell in which dye coated glass and Au substrates used as sensor element. The second sensor is based on photoacoustic based measurement of resonance angle of prism-based SP excitation. The setup is also equipped with photodetecting system to measure the reflectance of the incident light. The applications of two proposed methodologies for possible sensor application are explained by taking porphyrin as sensor element as it finds wide applications in the biomedical field. There has been a lot of recent interest in supramolecular aggregation effects of porphyrins leading to the formation of self organized nanorods, nanowheels and nanotubes in view of their scope in sensor applications as well as the interesting modifications in their physical properties. The use of metalloporphyrin arrays as sensors is a new concept in artificial olfaction. Its application as 'Electronic nose' applications based on visualizing the color changes associated with the interaction of vapors have been reported. Besides, it properties on Au substrate is also investigated to examine the effect of SP on its optical properties and on the improvement of the sensitivity of the measurement.

2. Review of theory

2.1. Theoretical investigation on prism based SP excitation

The principle of prism based SPR sensor is the measurement of attenuated total reflection (ATR) which is an experimental technique often used with infrared (IR) spectroscopy.

Attenuated total reflection spectroscopy utilizes total internal reflection (TIR) which occurs when a beam of radiation from a denser medium enters a rarer medium above the critical angle θc. The ATR method in IR spectroscopy involves using a crystal as the high refractive index material to excite the test sample. Infra-red light is passed through the ATR crystal in such a way that it should reflect at least once off the internal surface in contact with the sample. This reflection forms an evanescent wave which extends into the sample, typically by a few micrometres. The attenuated energy from each evanescent wave is passed back to the infra-red beam, which then exits the opposite end of the crystal. A detector can be used to measure the exiting beam to detect any changes to the beam.

In the same context SPR sensor also uses a high index prism to excite surface plasmon in the metal film. Here surface plasmon not an evanescent wave as in FT-IR spectroscopy interacts with the test sample but the measurement method is based on same ATR principle. In the SPR sensor, ATR is due to excitation of SP at the SPR angle. It is observed above the critical angle when the momentum of the excitation light matches with that of the surface plasmon i.e., when $k_{sp} = k_o n_p \sin\theta_{sp}$, where k_{sp} and $k_o$are the wavevectors of the incident photon and excited plasmon respectively, n_p is the refractive index of the prism and θ is the angle of incident beam with respect to normal to the prism surface]. As the excitation of surface plasmon depends on the nature of the surrounding dielectric medium, any change in property of the surrounding dielectric will be reflected as shift in SPR angle.

Fresnel's equations are used to explain the excitation of SP using high index prism. The Fresnel's equations describe how light behaves when it moves through media of differing refractive indices. It describes the reflection and transmission of electromagnetic waves at a dielectric interface. As the SPR experiment setup utilizes an ATR method and the ATR method is applied by using a prism with Kretschmann configuration, where the metal layer is in direct contact with the ATR crystal, Fresnel equations for a three layer model (prism-metal-air) is applicable for the present configuration.

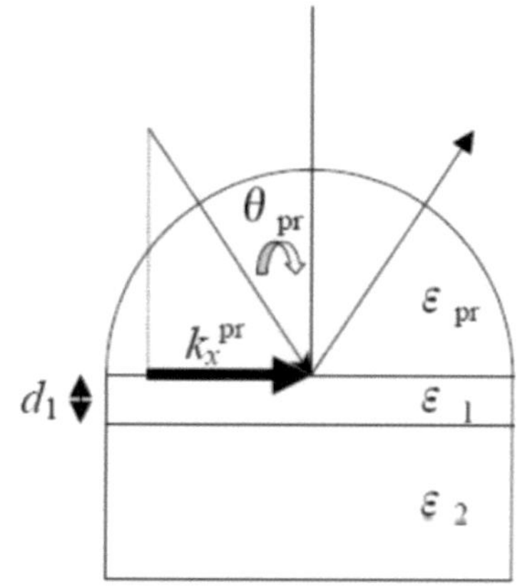

Figure 1. Schematic of three-layer prism setup[31]

Fresnel's equation of three layer model, i.e. prism-metal-air, is defined as

$$R = \left| r_{pr12}^{p} \right|^2 \tag{2}$$

where R is the reflectance of the incident light, and r_{pr12}^{p} is the reflected portion of incident light, the superscript P denotes a P-polarized model, and subscript pr12 denotes a prism-medium 1-medium 2 model.

The reflected portion of incident light is given by[31]

$$r^{p}_{pr12} = \frac{r^{p}_{pr1} + r^{p}_{12} e^{2ikz1d1}}{1 + r^{p}_{pr1} r^{p}_{12} e^{2ikz1d1}} \tag{3}$$

where r^{p}_{pr1} denotes the reflected potion of light from prism to medium 1, r^{p}_{12} denotes the reflected portion of light from medium 1 to medium 2, $kz1d1$ denotes the phase factor.

The reflected portion of light from prism to medium 1 is given by[31]

$$r^{p}_{pr1} = \frac{\cos\theta_{pr} / n_{pr} - (\varepsilon_1 - n_{pr}{}^2 \sin^2\theta_{pr})^{1/2} / \varepsilon_1}{\cos\theta_{pr} / n_{pr} + (\varepsilon_1 - n_{pr}{}^2 \sin^2\theta_{pr})^{1/2} / \varepsilon_1} \tag{4}$$

where θ_{pr} denotes the incident angle of the light entering the prism, n_{pr} denotes the refractive index of the prism, and ϵ_1 denotes the dielectric constant of medium 1.

The reflected portion of light from medium 1 to medium 2 is given by[31]

$$r^{p}_{12} = \frac{(\varepsilon_1 - n_{pr}{}^2 \sin^2\theta_{pr})^{1/2} / \varepsilon_1 - (\varepsilon_2 - n_{pr}{}^2 \sin^2\theta_{pr})^{1/2} / \varepsilon_2}{(\varepsilon_1 - n_{pr}{}^2 \sin^2\theta_{pr})^{1/2} / \varepsilon_1 + (\varepsilon_2 - n_{pr}{}^2 \sin^2\theta_{pr})^{1/2} / \varepsilon_2} \tag{5}$$

where ϵ_1 denotes the dielectric constant of medium 1, n_{pr} denotes the refractive index of the prism, θ_{pr} denotes the incident angle of the light entering the prism and $\tilde{\epsilon}_2$ denotes the dielectric constant of medium 2.

Phase factor is given by(Yamamoto, 2008)

$$kz1d1 = \frac{\omega}{c} (\varepsilon_1 - n_{pr}{}^2 \sin^2\theta_{pr})^{1/2} \tag{6}$$

where $\omega = \frac{2\pi c}{\lambda}$ is the frequency, c denotes the speed of light, λ denotes the wavelength of incident light, ϵ_1 denotes the dielectric constant of medium 1, n_{pr} denotes the refractive index of the prism and θ_{pr} denotes the incident angle of the light entering the prism.

These equations relate reflectance (intensity of exiting beam) to the refractive indices of the prism, metal and air, thickness of metal layer, wavelength of incident light and the dielectric constants of the metal and air.

2.2. Theory of photoacoustic spectroscopy

The PA effect is based on the generation of acoustic signal when a sample enclosed in an airtight chamber is heated periodically on irradiation with an intensity modulated beam. The corresponding pressure fluctuation produced by the sample in the ambient gas is easily detected by a sensitive microphone. Assuming a optically opaque sample and negligible heat flux into the air in contact with the irradiated surface of the sample, the pressure fluctuation inside the chamber is given by[22, 32]

$$Q = \frac{\nu P_0 I_0 (\alpha_s \alpha_g)^{1/2}}{2\pi l_g T_0 k_s f \sinh(l_s \sigma_s)} e^{j(\omega t - \frac{\pi}{2})} \tag{7}$$

where γ is the ratio of specific heat capacity of air, P_0 is the ambient pressure, T_0 is the ambient temperature, I_0 is the radiation intensity, f is the modulation frequency, $\alpha_i, k_i l_i$ are thermal diffusivity, thermal conductivity and length of the medium $i = g$ refers to gas and $i = s$ refers to solid sample. The complex quantity$\sigma_i = (1 + j)a_i$, where $a_i = \left(\frac{\pi f}{\alpha}\right)^{\frac{1}{2}}$ is the thermal diffusion coefficient of the medium i. The expression in equation (7) can be solved based on whether the sample is thermally thick or thermally thin. A sample is considered to be thermally thick when the heat generated due to the absorption of light in the sample reaches the surface and enters the surrounding air column of the photo acoustic cell. For thermally thick sample, the product $l_s a_s$ will become greater than $1(i.e., (l_s a_s \gg 1))$. In that case equation (7) of the PA signal reduces to,

$$Q = \frac{\nu P_0 I_0 (\alpha_s \alpha_g)^{1/2}}{\pi l_g T_0 k_s} \frac{\exp[-l_s(\frac{\pi f}{\alpha_g})^{\frac{1}{2}}]}{f} e^{j(\omega t - \frac{\pi}{2} - l_s \alpha_s)} \tag{8}$$

From the above expression it is clear that the photo acoustic signal varies exponentially with the modulation frequency according to $[\left(\frac{1}{f}\right)\exp\left(-b\sqrt{f}\right))]$, where $b = l_s\left(\frac{\pi}{\alpha_s}\right)^{\frac{1}{2}}$ and the phase decreases linearly with$b\sqrt{f}$. Thus, thermal diffusivity α_s can be obtained either from the amplitude or phase data with respect to the modulation frequency for the optically opaque and thermally thick sample.

For the thermally thin sample, the product $l_s a_s$ will become much lesser than 1(i.e., $(l_s a_s \ll 1)$. The equation 7 in this case simplifies to

$$Q = \frac{\nu P_0 I_0 (\alpha_g)^{1/2} \alpha_s}{(2\pi)^{3/2} l_g l_s T_0 k_s} \frac{e^{j(\omega t - 3\pi/4)}}{f^{3/2}} \tag{9}$$

The above expression implies that the PA signal amplitude from a thermally thin sample under the heat transmission configuration varies as $f^{-1.5}$ and the phase is insensitive to the variation in the modulation frequency. Hence in the case of thermally thin and optically opaque samples, the PA signal is independent of the optical absorption coefficient of the sample, but depends on the wavelength of the incident radiation.

3. Experimental set-up and characterization

3.1. Protocol for sample preparation

Fe(III) tetraphenylporphyrin is prepared according to standard procedures reported in literature.[33] About 50 *mg* of 5,10,15,20 tetraphenyl porphine Fe(III) chloride procured from

Aldrich, is dissolved in chloroform and shaken with 1*M* hydrochloric acid for four hours. The chloroform layer was separated and dried with unhydrous sodium sulphate. The iron porphyrin was purified by recrystallization from chloroform-ethanol mixture and the optical absorption spectrum was checked with standard spectrum. The HCl treatment is to remove any μ-oxo compounds which are present in iron porphyrin samples. The samples are prepared in different physical forms, namely, liquid, powder and film. The porphyrin solutions are prepared at two different concentrations 10^{-4} *M* and 10^{-6} *M* respectively in chloroform. The glass substrate of dimension 1 *cm* x 1*cm* is ultrasonically cleaned using acetone and then double distilled water a couple of times and vacuum dried. The solution with higher concentration is used to coat a film on the glass plate and Au coated BK7 prism. A few drops of the porphyrin solution are deposited over the glassplate using a 2 ml syringe to form a porphyrin film. Similarly phorphyrin film is formed on Au film coated on the prism base. The deposited film is allowed to dry under vacuum. As the solvent evaporates, the porphyrin film shrinks resulting in the formation islands of porphyrin microclusters on the glass substrate at scattered positions. This approach has been used recently to prepare porphyrin samples where aggregation effects have been observed.[34, 35] Size characterization of the aggregates is done with Atomic Force Microscopy. Contact AFM measurements were done using a Digital Instruments Multimode head attached to a Nanoscope-IV controller. Standard Si_3N_4 cantilevers were used for the normal topography and friction mode imaging. Photoacoustic (PA) studies on powder samples are done by smearing the porphyrin on the grease coated glass substrate.

3.2. Two-prism configuration to excite and measure SP excitation

A schematic diagram representing the experimental set-up is shown in figure 2. Laser beam is focused on to the prism base using the lens and reflected light is measured using an ultrasensitive photodiode. The polarizer is set to polarize the incident beam as P polarized in the direction parallel to the plane of incidence, in this case in the horizontal plane. SP is excited using Kretschmann configuration in which the metal film is in direct contact with the prism base. Lock-in-amplifier and chopper are integrated with the proposed set-up to improve signal to noise ratio (SNR) of the measurement. A chopper is used to modulate the intensity of the incident light and the lock-in amplifier is employed to measure the data with reference to the chopping frequency. Au film is used as the metal film, which is coated on the base facet of BK7 prism.

The prism is fixed on the rotation table and rotated in steps of 0.1° to change the momentum of incident light with respect to dielectric-metal interface. SP will be excited once the momentum between the incident light and surface plasmon is matched. Maximum energy transfer between photon to surface plasmon takes place at SPR angle and is very sensitive to dielectric property of the material adjacent to Au film and is given by $K_{sp} = \frac{\omega}{c}\sqrt{\frac{\varepsilon_m \varepsilon_d}{\varepsilon_m + \varepsilon_d}}$,

where ω is the frequency, c is the velocity of the light, and ε_m and ε_d are dielectric constants of metal and dielectric medium respectively.

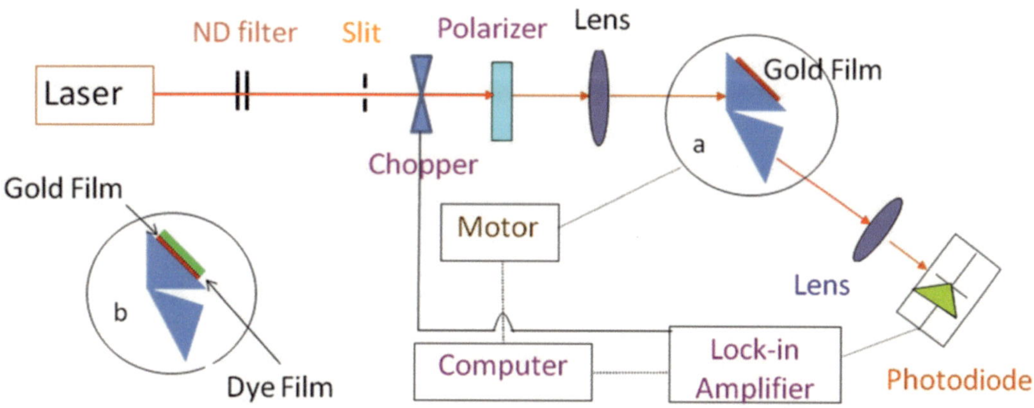

Figure 2. Schematic diagram of experimental set-up for SPR sensor

The point at which SP is excited is called the interrogation point/spot. It is represented by a symbol S in figure 3. This spot should not walk over the prism base during SPR measurement and is avoided by positioning the prism with respect to rotation axis (P) of the prism table.[36]

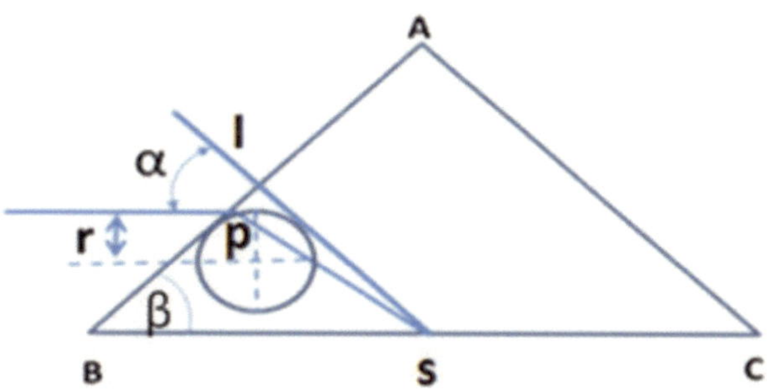

Figure 3. The configuration for incident ray with respect to axis of rotation of the prism table (P) to keep the coupling spot remains practically stationary during SPR measurement

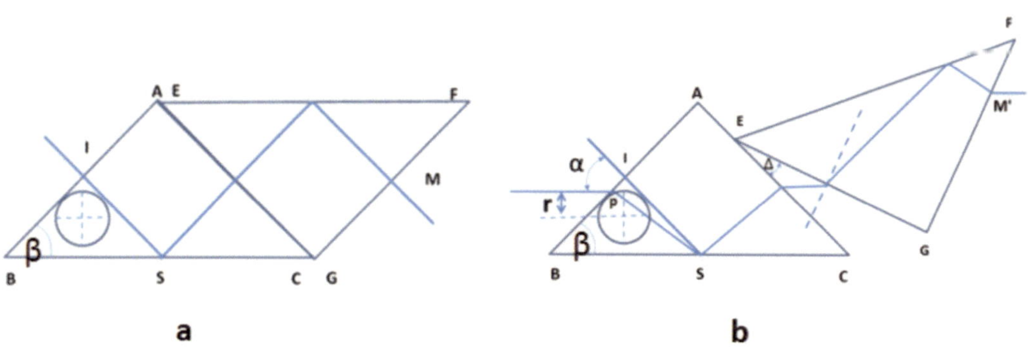

Figure 4. The two-prism configuration with (a) angle = 0° and (b) angle = Δ° between two prisms and ray traces within the prisms

Figure 3 represents a single prism marked by ABC. The solid line (IS) represents the normal incidence of the excitation beam. The centre of the dotted cross represents the axis of the rotation. A circle is drawn with axis of rotation as center and $r = l_p / n_p$ radius, where l_p is the length of the light path inside the prism from the entrance face to the coupling spot and n_p is the refractive index of the prism. It touches the prism surface at I and the beam which strikes at the point of contact (I) always ensures stationary SP excitation spot (S) during SPR angular rotation.

Though the coupling spot remains stationary at the prism base, the reflected light from it always rotates with respect to rotation axis. In such configuration, a linear photodiode array or goniometer is used to collect reflected light from the prism base and is quite expensive. Hence the significance of the concept of employing the two prism configuration.[29, 30] In the two prism technique, the first prism (ABC) is used to excite surface plasmon and the second prism (GEF) is employed to counter the orientation of reflected light from the first prism to opposite direction. This configuration always ensures that the reflected light from both prisms always remains fixed in orientation with respect to the incident beam direction and hence a single photodiode is employed to record the reflected light. Figure 4 represents the two-prism configuration with ABC and GEF as the two prisms. In figure 4 (a) each prism has been rotated by $90°$ ($\Delta = 0$) and the combination act as a parallel plate. The beam which is incident normally (IS) on two prism combination will also emerge out normally at **M**. Though the interrogation spot remains stationery during rotation, the light getting reflected from the second prism (GEF) will makes a walk-over on the prism base. Hence there will be a small shift (δ) in the exit beam (**M'**) with respect to fixed point M and the amount of shift depends on angular separation Δ between the two prisms. Figure 4 (b) represents the shift of the exit beam M ' due to angular separation Δ and is given by

$$\delta = \frac{l}{2} \cos\left(\sin^{-1}\left(\frac{1}{n_p} \sin\left(\frac{1}{n_p} (\sin\alpha \cos\beta + \left(n_p{}^2 - \sin^2\alpha\right)^{1/2} \sin\beta) + \Delta \right)\right)\right) \quad (10)$$

where α is the angle of rotation of the prism with normal incident angle, β is the angle of the prism and Δ is angle between the prisms, n_p is the refractive index and l is the length of the prism facet,.

Figure 5 represents the shift (δ) as a function of incident angle α. The plot A represents the displacement curve for the prism configuration $\Delta = 0$. It shows minimum at 90° rotation and exhibits displacement of about 0.035 mm with respect to **M** during the angular rotation between 40° to 120° which is found to be lower than the earlier reported two prism configuration.[29] The shifts in displacement for different angular separations (Δ) are also investigated. The shift in minimum towards higher degrees due to increase in angular separation (Δ) is observed. Displacement studies for different angular separations (Δ) are represented by the graphs A, B, C and D. The shift is about 2° per 3° angular separation is observed.

In the present SPR sensor study, the angle (Δ) between the two prisms is selected as 3° to get the reflected beam well deviated from the direction of the incident beam for interference free detection. Besides this, it is ensured that the reflected light remains well within the active area of the photodiode during the SPR angular detection. Both amplitude and phase of the reflected light are measured simultaneously (figure 6). The figure 6(a) represents the measurement of reflectance as the function of the angular rotation. The dip in the graph is due to excitation of SP and has an SPR angle at *43.6* °. The narrowness of the resonance dip indicates the good confinement of SP at the metal-dielectric interface. The experiment is repeated for improved signal to noise ratio and the obtained signal is theoretically analyzed.

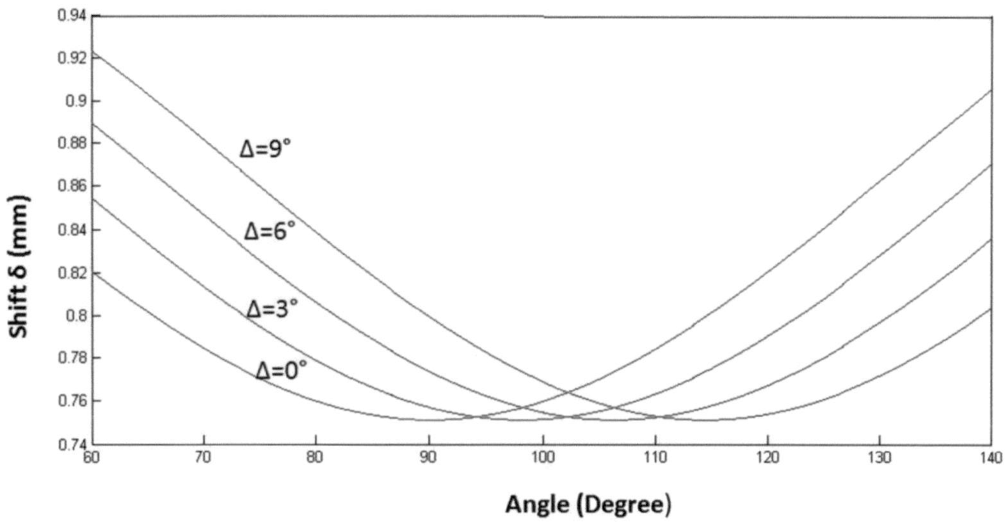

Figure 5. Theoretical simulation of the expression (10), Shift (δ) in terms of Angle (α) for various angular separations (Δ)

The quantitative description of the reflectivity R for the P wave is given by Fresnel's equations for systems comprising three or more layers as,

$$R = \left| r^{p}{}_{012} \right|^2 = \left| \frac{r^{p}{}_{01} + r^{p}{}_{12} \exp(2ik_{z1}d)}{1 + r^{p}{}_{01} r^{p}{}_{12} \exp(2ik_{z1}d)} \right|^2 \tag{11}$$

where d is the thickness of the metal film, and the reflection coefficient of the P wave is given by

$$r^{p}{}_{ik} = \left(\frac{k_{zi}}{\varepsilon_i} - \frac{k_{zk}}{\varepsilon_k} \right) \Big/ \left(\frac{k_{zi}}{\varepsilon_i} + \frac{k_{zk}}{\varepsilon_k} \right) \tag{12}$$

$r^{p}{}_{ik}$ is the Fresnel coefficient of p- polarized light between the ith layer and the kth layer in the glass prism(layer 1)/metal (layer 2)/sample (layer 3) configuration (refer figure 1b).

Moreover $r^{p}{}_{012} = \mathrm{Re}\left(r^{p}{}_{012}\right) + \mathrm{Im}\left(r^{p}{}_{012}\right)$. Hence, the phase of the reflected light can be expressed as

$$\phi = \tan^{-1}\left(r^{p}{}_{012}\right) = \tan^{-1}\left(\text{Im}\left(r^{p}{}_{012}\right) / \text{Re}\left(r^{p}{}_{012}\right)\right) \tag{13}$$

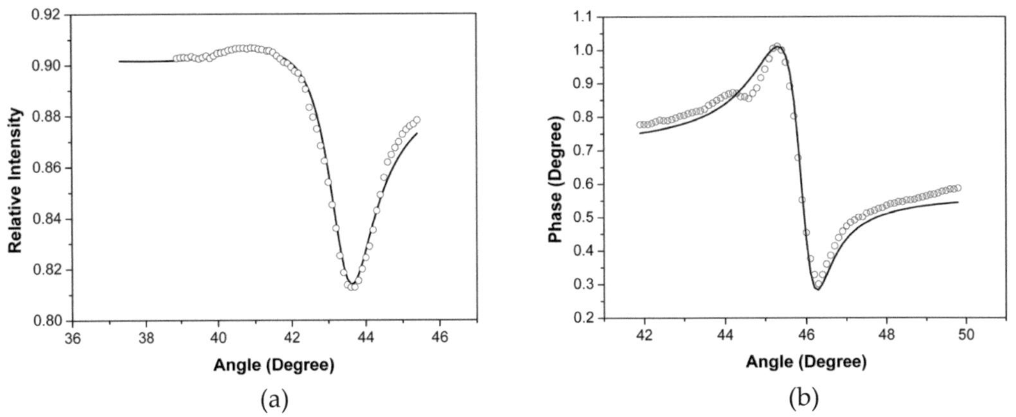

Figure 6. (a) the reflectivity and (b) the phase shift of the reflected light of a SPR sensor. Circles represent the experimental value and solid lines represent the theoretical simulation

Using the expression (11), theoretical fit on the experimental data (circles) is done to get the refractive index and thickness of the Au film. The solid line in figure 6(a) represents the theoretical fit. The refractive index and thickness of Au film are found to be0.1726 + i3.3418 and *85*nm respectively at wavelength 532 nm.

Theoretical simulation is also performed on the obtained phase data (figure 6(b)) by using expression (13) to validate the results obtained through amplitude data (figure 6(a)). The solid line in figure 6(b) represents the theoretical fit done on the measured phase data pointed represented by circles. It may be noted from figure 6(b) that the theoretical fit matches exactly with the experimentally obtained phase data.

3.3. Variable volume closed photoacoustic cell

Figure 7 shows a sketch of the designed variable volume closed PA cell and corresponding experimental set-up. The cell is made from an aluminium rod of diameter 80 *mm*. The main chamber is cylindrical in shape with a volume that can be varied between 160to 640 mm^3. The microphone is connected to the main chamber through a narrow tunnel of diameter 1.2*mm*. A half inch 50*mV*/*Pa* sensitive microphone from G.R.A.S is used as PA detecting element. The head of the microphone is acoustically isolated from the surroundings by a proper protective cover made of aluminium, screwed to the side of the PA cell. This cover also prevents dust particles from reaching the microphone diaphragm. PA cell is designed in such a way that its cell volume could be changed to study samples of different properties in different gas environments. The sample holder is made of a brass rod of 1*cm* thick and 1 *cm* diameter with a depression on the flat surface. Sample holders with depression of various depths are also made to accommodate samples of different absorption capacities. Particularly samples with low absorption coefficient are taken in relatively large amounts in the case of PA studies. Since the PA signal varies inversely as the gas volume and the thermal conductivity of the gas ($\sqrt{k_g}\frac{P_0}{T_0}$ where k_g is the thermal conductivity of the sample, P_0 the pressure and T_0 temperature of the coupling gas medium), the distance between the sample and the cell window must be greater than the thermal diffusion of the gas since the boundary layer of the gas acts as an acoustic piston generating the pressure wave. The distance of the sample surface from the optical widow is varied by screwing the threaded cylindrical rod, on which sample holder rests, connected at the base of the PA cell. (see figure.1). This alters the cell volume, which is an important design parameter. The source used to excite optically the sample is xenon lamp of 500 *W* from M. Watnabe & Co Ltd. An image of the source is condensed within the size of the entrance slit of the monochromator using a proper combination of optical components to maximize the amount of light entering the monochromator to increase intensity of the monochromatised light. Since the intensity of the PA signal is proportional to the intensity of the incident light, the monochromatised light from the exit face of the monochromator is tight focused on the sample surface. The intensity of the light falling on the sample is modulated by a mechanical chopper of frequency 10 Hz (SR540 chopper) and the reference output of the chopper controller is connected to the lock-in amplifier (SR830 dual-phase digital lock-in amplifier). The lock-in amplifier analyses the PA signal with respect to chopping frequency. The designed PA spectroscopic setup is characterized by measuring frequency response of the spectrometer using carbon black and the result agrees well with the Rosencwaig's theory [figure8]. Then the photoacoustic spectrometer is standardized by recording the spectrum of Holmium oxide.[37]

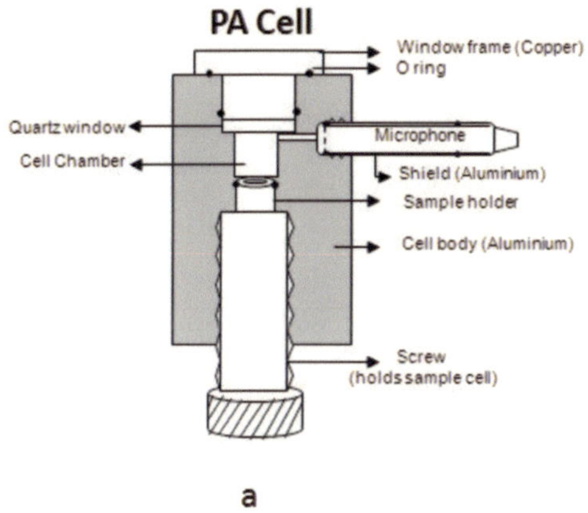

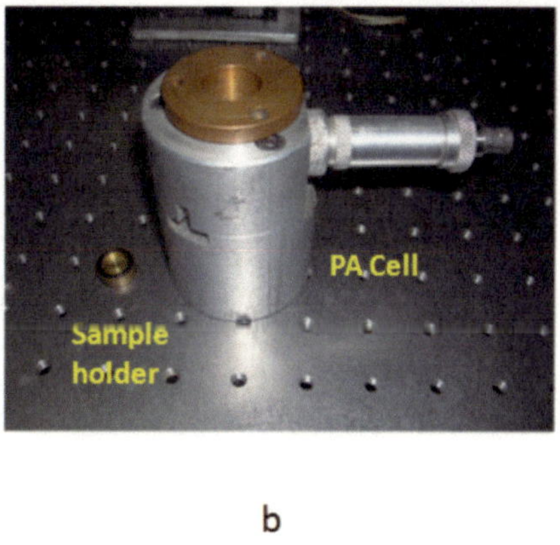

Figure 7. (a) Schematic diagram of Photoacoustic cell and (b) Experimental set-up.

The entire PA system is computer interfaced. Wavelength tuning of the monochromator is achieved by a stepper motor rotated by a driving circuit which contains switching circuit powered by 2N3055 power transistor and speed of the scanning is controlled through LPT1 port of the computer. The lock-in amplifier is controlled through the RS232 port.

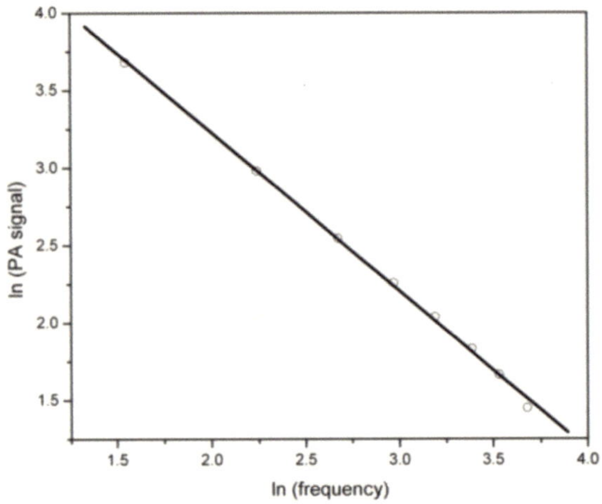

Figure 8. Frequency response of the PA cell.

3.4. Photoacoustic based surface plasmon resonance spectroscopy

The photoacoustic set-up to measure resonance angle of SP excitation in prism coupling technique is shown in the following figure.[38] The light from the source [DPSSL, SUWTECH LDC-1500] was coupled to an optical fibre and guided to an assembly consisting of a polarizer and an optical chopper. Another optical fibre is used to collect the polarised beam to guide through the focusing lens onto the prism base. The prism was mounted on the PA chamber containing microphone. The PA chamber was fabricated in such a way that a small air cell separates the gold-layer and the transducer chamber so as to enable the creation of sound waves.

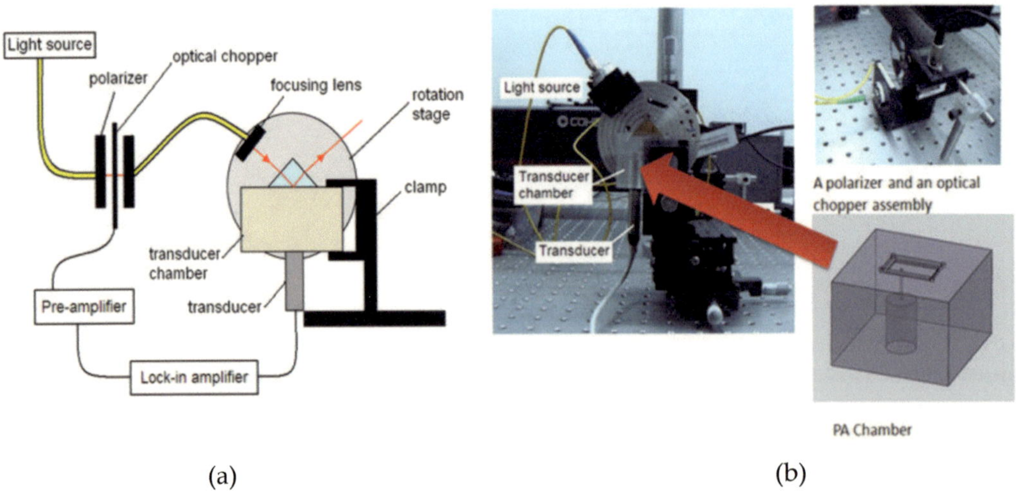

Figure 9. (a) Schematic of SPR experiment setup measuring photoacoustic signal (side view) and (b) the corresponding experimental set-up

The technical drawings for the transducer chamber can be found in figure 9 (b). It encloses chamber of depth 1 mm. PA chamber is connected to the microphone by a tunnel of diameter 1.2 mm and a length of about 14 *mm*. The contact areas between the prism and transducer chamber as well as between the transducer chamber and the transducer were sealed using vacuum grease to prevent any leakage of the acoustic piston effect out of the air pocket. The transducer chamber is positioned in such a way that the beam focuses on a stationary spot on the gold nano-layer on the prism throughout the scanning range of incident angles. The beam spot on the gold nano-layer is ensured to be stationary for the same reasons mentioned earlier in SPR experiment conducted for reflectivity measurement. For this investigation, a pre-amplifier with filter is used in addition to the lock-in amplifier to further improve the signal to noise ratio. The transducer is connected to the source input, while the optical chopper is connected to the reference input channels of the lock-in amplifier. The rotation stage is again jogged at intervals of 0.1°, and the photoacoustic signal at each jog step is recorded. Due to the sensitivity of the transducer, Fast Fourier Transform (FFT) filter smoothing was done on the photoacoustic signals to eliminate high frequency noise.

Carbon black is a strong absorber of the light where $(1/\alpha) < 1$ and $\mu > 1$ (where αisthe optical absorption coefficient) and hence a strong PA signal is expected. The strength of the signal is inversely proportional to the chopping frequency according to Rosencwaig's theory.[22, 39] Thus a log-log plot of the PA signal versus chopping frequency is expected to be a straight line of slope-1.

The PA cell is tested by recording the spectrum of carbon black as the standard sample. A Helium-Neon laser is used as the source of excitation. The PA signal is recorded as the function of modulation frequency obtained using a mechanical chopper. Figure 10 represents the measured frequency response of the PA cell. The data fit well to a straight line of slope -1.

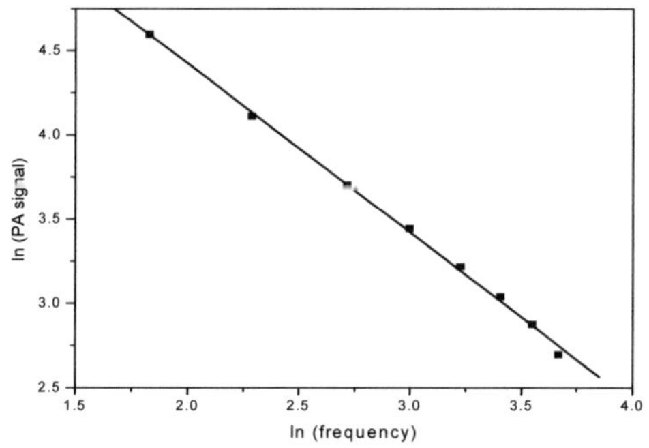

(a)

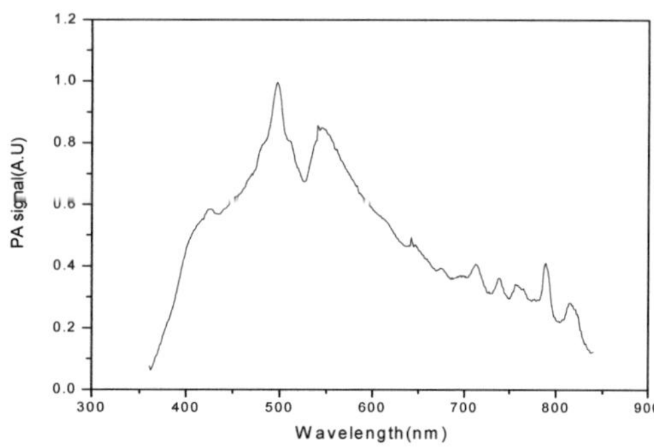

(b)

Figure 10. (a) Frequency response of PA cell and (b) PA Spectrum of carbon black.

Fig 10 (b) shows the PA spectrum of carbon black recorded with the Xenon lamp as the source. In the case of thermally thin and optically opaque samples, the PA signal is independent of the optical absorption coefficient of the sample, which in turn depends on the wavelength of the incident radiation. Thus the carbon black sample, which satisfies the above condition, should exhibit the spectrum of the source used. The observed spectrum shows the intensity variation of the xenon lamp as a function of wavelength.

4. Results and discussion

The application of the proposed configuration is illustrated by taking porphyrin as sample (dielectric medium). There has been a lot of recent interest in supramolecular aggregation effects of porphyrins in view of their scope in sensor applications. The use of metalloporphyrin arrays as sensors is a new concept in artificial olfaction.

The metalloporphyrin selected for the present study is FeTPPCl. Figure 11 shows the AFM pictures of the film containing FeTPP aggregates on the glass plate. Microclusters of the FeTPP in the size range of 0.8 to 1.6 *microns* are clearly seen in the picture. The AFM pictures indicate an inhomogeneous distribution of porphyrin molecules obtained on the glass surface.

FeTPP aggregates on the glass plate are studied using optical absorption (OA) and photoacoustic spectroscopy. Figure 12 shows the PA spectrum of FeTPP in three physical forms, namely, solution at low concentration (curve A), powder (curve B) and film on glass plate (curve C). The PA spectrum of the solution at low concentration is obtained by depositing a couple of drops of porphyrin solution on the tissue paper. In the case of powder samples, only a few micrograms are taken to avoid signal saturation. The PA spectrum of aggregates on the glass plate is compared with those of porphyrin solution at low concentration deposited on tissue paper and porphyrin powder in order to distinguish between the effects due to aggregation and concentration.

The Soret band is seen in figure 12 at around 450 nm and the Q bands in the region of 500-600 *nm*. These features are similar to those reported for the solution spectrum.[40] The peak corresponding to the Soret band is found to be red shifted to 450 nm in the case of films from the corresponding value of 444 nm for the spectra of the powder as well as the dilute solution. This shift and the accompanying broadening occur only in the film and not in the cases of the dilute solution. These effects are not observed even in the case of the powder sample where the concentration is high. Thus the observed spectral features appear to result from aggregation effects rather than concentration effects. The aggregates could be J-aggregates which are characterized by shifting of bands to larger wavelengths with respect to long wavelength absorption band of the monomers at about 444nm. Figure 12(i) shows the OA spectrum of the films on the glass substrate. No specific feature is seen in the Q band region in the optical spectrum. However, the PA spectrum, shown in figure 12ii(C) clearly exhibits a pronounced two-peak structure between 500-600 nm. Such features are usually attributed to hyper spectra due to metal to ligand charge transfer.[33, 41] These arise from the considerable mixing of the metal d_π orbitals with the LUMO of the Porphyrin, which renders Fe porphyrins of considerable interest for photochemical studies. The broadening, almost nonzero extinction coefficient throughout the visible region and spectral shifts of the bands are indicative of formation of aggregates during the preparation of films on glass substrate. [35]

The Q bands in OA spectrum are not well pronounced as compared to PA spectrum of the film, apparently due to the inhomogeneous distribution of film on the glass substrate and low absorption of the sample. The broad spectral features of FeTPP in liquid form at low concentration, film form and powder form appear to be similar. However, the process of preparing a solution-dried film on glass plates leads to some amount of aggregation as judged by the shift and the broadening in the bands. The hyper spectral features in Q band region of the FeTPP are clearly resolved in PA spectra, particularly in the case of aggregates. This could be due to the fact that the efficiency of nonradiative thermal processes is much larger in aggregates (than in monomeric species) to which PAS is very sensitive. The increased metal-ligand energy transfer in the case of aggregate films could be the reason for the observation of well pronounced hyper spectra.

PA studies at low concentration solution and powder are done to bring out the effect of concentration on shifting and broadening of the peaks in PA spectra. The PA spectrum of porphyrin solution at low concentration is studied by depositing a few drops of porphyrin on the tissue paper, which is a convenient way of recording the spectra of liquid samples using the microphone-based PA spectrometer. The experiment is also repeated by depositing low concentrated solution on a glass plate to avoid aggregation. However, such samples did not produce any appreciable PA signal due to the low concentration. Aggregation processes are also absent in powder samples and occur when the porphyrin films are deposited on the glass substrate. The PA spectra reveal the aggregation effects in a more detailed manner compared to OA spectra. PA spectra of powder samples can be obtained using only at a few micrograms of the sample smeared over the grease coated glass plate as signal saturation effects are observed at higher density.[42]

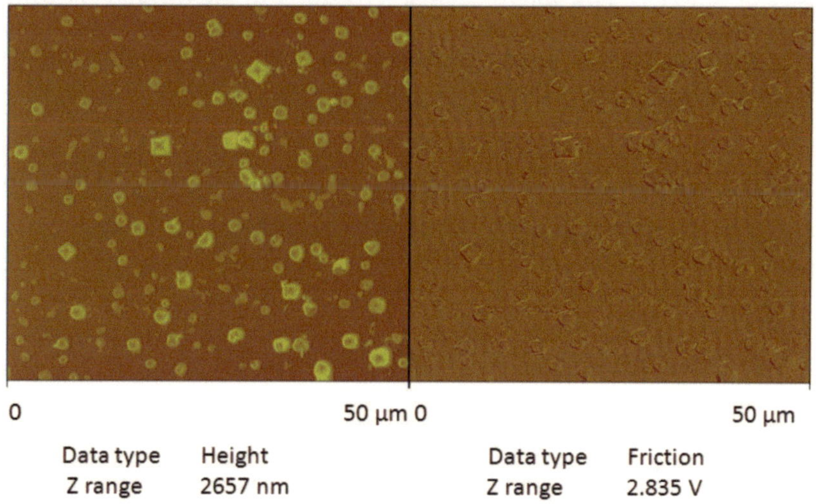

Figure 11. AFM picture of FeTPP aggregates on glass substrate

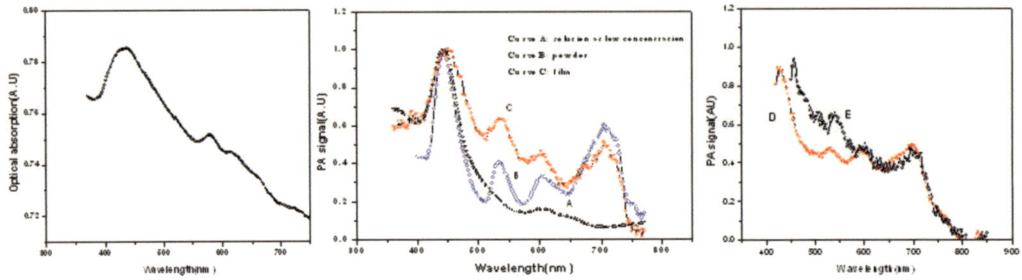

i *ii* *iii*

Figure 12. (i) Optical absorption spectrum of FeTPP on glass substrate, (ii) Photoacoustic spectrum of FeTPP.[(a) curve A -solution at low concentration, curve B -powder and curve C- film on glass substrate.] and (iii) Photoacoustic spectrum of FeTPP.[(d) Curve D -dye film on glass plate and curve E –dye film on Au film.]

AFM studies on porphyrin deposited on Au film show no distinct features as observed in the case of porphyrin film formed on glass substrate. PA studies on porphyrin film on Au substrate {Figure 12(iii) show improved PA signal as compared to porphyrin film on glass substrate [Figure 12(iii) D]. The proposed concept based on PA cell with porphyrin-glass and porphyrin-Au substrates is expected to find applications as gas sensor.

Investigation of SPR excitation is performed using the two-prism technique. Figure 13shows surface plasmon excitation in Au film of thickness 60 nm using prism in Kretschmann configuration. The graph is a plot of reflectivity against angular rotation of the prism. It exhibits a dip at 43.95 degrees due to excitation of surface plasmon. The angle at which SP is excited is called surface plasmon resonance angle. From the experimental data, it can be observed that a peak in reflectivity occurs before a dip. At the peak, the ratio of incident light intensity to reflected light intensity is at a maximum. This would mean that there is minimum loss of incident light and the peak is a result of TIR. The peak is at the angle where the highest amount of energy of the incident light is passed, through the evanescent wave along the metal and crystal interface, back to the reflecting light.

Therefore, the subsequent minimum reflectivity would give the SPR angle, since the angle is after the critical angle for TIR, and is where the greatest amount of energy is lost due to maximum excitation of surface plasmons. After the SPR angle, reflectivity increases as incident angle increases. This indicates that the amount of energy lost through surface plasmon excitation is decreasing as the consequence of deviation from momentum match between incident photon and excited SP, with the light intensity returning to initial levels. The profile of reflectivity against incident angle forms a basis for comparison with photoacoustic signals.

The experimental set-up used to study SPR is described in the experimental section titled photoacoustic based surface plasmon resonance spectroscopy. Figure 13 shows the graph of photoacoustic signal obtained against incident angle for gold nano-layer of 60 nm thickness. The graph contains two photoacoustic profiles for two modulation frequencies such as 7 Hz and 11 Hz trespectively. From the data, it can be seen that at both frequencies, a peak occurs at the SPR angle. This result is consistent with earlier reported data by different groups.[43-46] For 7 Hz modulation frequency, the photoacoustic signal is larger at every incident angle as compared to 11 Hz. These indicate that lower frequencies resulted in larger photoacoustic signals.

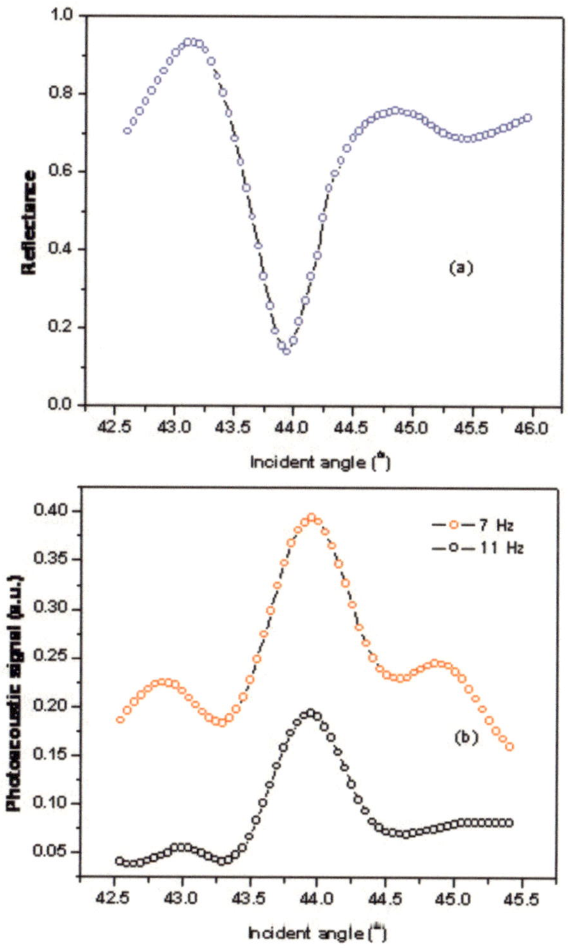

Figure 13. Comparison of (a) reflectivity profile and (b) photoacoustic profile

Figure 13 shows the comparison of the reflectivity profile with photoacoustic signal profile against incident angle. As can be seen from the graph, peaks in reflectivity correspond with dips in photoacoustic signal, and vice versa. It is thus established that reflectivity and photoacoustic signal have an inverse relationship. According to Inagaki et al[43], the absorptivity for *p*-polarized photons is affected at each angle of incidence into two parts, one of which contributes to the photoacoustic signal by decaying nonradiatively into heat, and the other part which escapes from the sample by decaying into photons or into emission of photoelectrons without contributing to the photoacoustic signal.

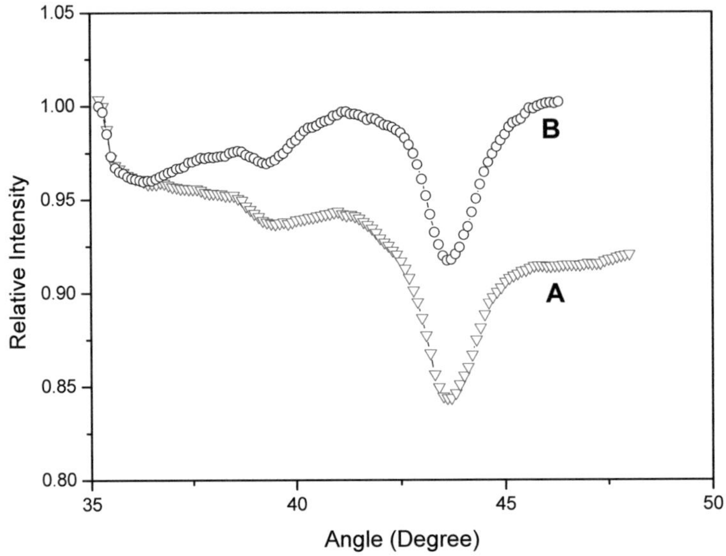

Figure 14. Reflectivity response of a Kretschmann surface plasmon arrangement as a function of angle for 532 nm wavelength. (A) Bare Au film and (B) Au-Dielectric (porphyrin medium).

At the critical angle where reflectivity is the highest, the absorption of *p*-polarized photons is the lowest since there is little difference between the incident intensity of light and reflected intensity of light. As such, the least amount of *p*-polarized photons decays into heat. Fluctuation of thermal expansion is then at its least, resulting in low amplitudes of the photoacoustic piston effect, i.e. minimal photoacoustic signal. Conversely, At the SPR angle, reflectivity is at its lowest due to maximum absorption of *p*-polarized photons. The resulting amount of p-polarized photons decaying into to heat is at a maximum. This fluctuation of thermal expansion would then be at its maximum, resulting in high amplitudes of the photoacoustic piston effect, i.e. maximum photoacoustic signal.

SPR measurement is repeated on Au-porphyrin dielectric medium to study the response of SPR angle due to adsorption of dye molecules for possible sensor applications. The plot B in Figure 14 represents the corresponding SPR measurement. The result shows a lowering in the reflection minimum of SP excitation as compared bare Au film (plot A in figure 14). The excited surface plasmon in the metallic film suffers significantly from strong damping caused by internal absorption and radiation losses. The damping of surface plasmon is characterized by an exponential decay of its intensity away from the metal surface. The radiative loss is due to plasmon re-emission of light. The metal dissipation loss in this case is given by[47]

$$\gamma_m = \frac{2\pi}{\lambda_e} \frac{\varepsilon_m^{''}}{(\varepsilon_m^{'})^2} \left(\frac{\varepsilon_m^{'} \varepsilon_d^{'}}{\varepsilon_m^{'} + \varepsilon_d^{'}} \right)^{3/2} \tag{14}$$

where and denote the real and imaginary part of the dielectric constant of the metal, where is the real part of surrounding dielectric medium. This loss can be overcome by gain medium such as dye at an optimum concentration. The gain γ_g needed to compensate for the SP loss is given by[47]

$$\gamma_g = \frac{2\pi}{\lambda_e}\frac{\varepsilon_d''}{(\varepsilon_d')^2}\left(\frac{\varepsilon_m'\varepsilon_d'}{\varepsilon_m'+\varepsilon_d'}\right)^{3/2} \tag{15}$$

here ε_d'' is the imaginary part of the dielectric constant of dye medium which is the function of dye concentration (*N*) and emission cross section (σ_e). Hence the lowering in the reflection minimum of SP excitation could be due to dye assisted enhancement of SP.

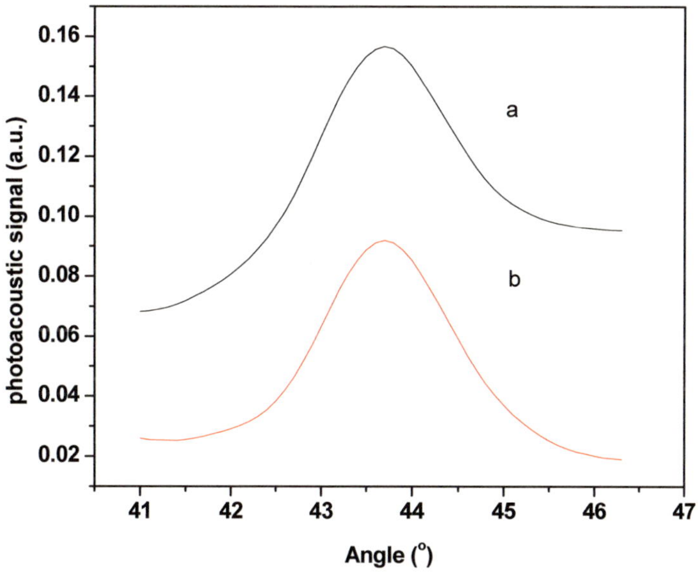

Figure 15. Photoacoustic investigation of (A) Bare Au film and (B) Au-Dielectric (porphyrin medium).

The photoacoustic investigation of SP assisted porphyrin excitation shows a decrease in PA signal (figure 15), which signal could be due to enhanced SP excitation in the presence porphyrin gain medium. The excited fluorophore in the gain medium emits plasmons instead of photons due to strong interaction between the excited dye molecules and electron vibration energy of SP. Any change in the property of the dye due to physio-adsorption and chemisorption will influence the excited SP, which will be reflected in reflective and photoacoustic measurements.

5. Conclusion

Photoacoustic investigation of surface plasmons was thus realized, establishing the relationship between SPR and photoacoustics. An in house developed photoacoustic technique was designed and demonstrated to be feasible as a real time sensor based on plasmon resonance spectroscopy through the repeatability of photoacoustic results using different modulation frequencies. Theoretical simulations done verify the experimental results for SPR, thereby providing a basis of comparison with the results of the photoacoustic experiment.

There is ample scope for future investigation on realizing photoacoustic bio-sensor of various types. The main aim is the identification of a target molecule by having a suitable biological recognition element attached to the surface of the sensor (be it metal, polymer or glass).

Author details

K. Sathiyamoorthy and V.M. Murukeshan
Nanyang Technological University, School of Mechanical and Aerospace Engineering, Singapore

C. Vijayan
Department of Physics Indian Institute of Technology Madras Chennai, India

6. References

[1] G. Klenkar and B. Liedberg, Analytical and Bioanalytical Chemistry 391, 1679 (2008).
[2] J. Borejdo, Z. Gryczynski, N. Calander, P. Muthu, and I. Gryczynski, Biophysical journal 91, 2626 (2006).
[3] X. Chen, M. C. Davies, C. J. Roberts, K. M. Shakesheff, S. J. B. Tendler, and P. M. Williams, Analytical Chemistry 68, 1451 (1996).
[4] S. Y. Wu, H. P. Ho, W. C. Law, C. Lin, and S. K. Kong, Opt. Lett. 29, 2378 (2004).
[5] Y. H. Joo, S. H. Song, and R. Magnusson, Opt. Express 17, 10606 (2009).
[6] J. D. Wright, C. von Bultzingslowen, T. J. N. Carter, F. Colin, P. D. Shepherd, J. V. Oliver, S. J. Holder, and R. J. M. Nolte, Journal of Materials Chemistry 10, 175 (2000).
[7] L. Pang, G. M. Hwang, B. Slutsky, and Y. Fainman, APPLIED PHYSICS LETTERS 91, 123112 (2007).
[8] Z. Genchev, N. Nedelchev, E. Mateev, and H. Stoyanov, Plasmonics 3, 21 (2008).
[9] H.-S. Leong, J. Guo, R. G. Lindquist, and Q. H. Liu, Journal of Applied Physics 106, 124314 (2009).
[10] S. T. Persijn, E. Santosa, and F. J. M. Harren, Applied Physics B: Lasers and Optics 75, 335 (2002).
[11] J. Li, X. Gao, W. Li, Z. Cao, L. Deng, W. Zhao, M. Huang, and W. Zhang, Spectrochimica Acta Part A: Molecular and Biomolecular Spectroscopy 64, 338 (2006).
[12] M. Nägele, D. Hofstetter, J. Faist, and M. W. Sigrist, Analytical Sciences 17, 497 (2001).
[13] F. Müller, A. Popp, F. Kühnemann, and S. Schiller, Opt. Express 11, 2820 (2003).
[14] L.-y. Hao, J.-x. Han, Q. Shi, J.-h. Zhang, J.-j. Zheng, and Q.-s. Zhu, Review of Scientific Instruments 71, 1975 (2000).

[15] J. A. Balderas-Lopez, Review of Scientific Instruments 77, 064902 (2006).
[16] J. A. Balderas-Lopez, Review of Scientific Instruments 77, 086104 (2006).
[17] A. Rosencwaig, in *Advances in Electronics and Electron Physics*, edited by L. Marton (Academic Press, 1978), Vol. Volume 46, p. 207.
[18] C. K. N. Patel and A. C. Tam, Reviews of Modern Physics 53, 517 (1981).
[19] A. C. Tam and C. K. N. Patel, Appl. Opt. 18, 3348 (1979).
[20] C. K. N. Patel and A. C. Tam, Nature 280, 302 (1979).
[21] C. K. N. Patel and A. C. Tam, Applied Physics Letters 34, 467 (1979).
[22] A. Rosencwaig and A. Gersho, Journal of Applied Physics 47, 64 (1976).
[23] K. A. Azez, Journal of Alloys and Compounds 424, 4 (2006).
[24] Q. Shen, M. Inoguchi, and T. Toyoda, Thin Solid Films 499, 161 (2006).
[25] T. S. Silva, A. S. Alves, I. Pepe, H. Tsuzuki, O. Nakamura, M. M. F. d. A. Neto, A. F. da Silva, N. Veissid, and C. Y. An, Journal of Applied Physics 83, 6193 (1998).
[26] N. C. Fernelius, Appl. Opt. 18, 1784 (1979).
[27] L. C. Aamodt, J. C. Murphy, and J. G. Parker, Journal of Applied Physics 48, 927 (1977).
[28] J. W. G. Ferrell and Y. Haven, Journal of Applied Physics 48, 3984 (1977).
[29] B. C. Mohanty and S. Kasiviswanathan, Review of Scientific Instruments 76, 033103 (2005).
[30] K. Sathiyamoorthy, P. A. Kurian, C. Vijayan, and M. P. Kothiyal, edited by R. Guo, S. S. Yin and F. T. S. Yu (SPIE, San Diego, CA, USA, 2007), p. 669818.
[31] M. Yamamoto, Department of Energy and Hydrocarbon Chemistry, Kyoto University, Kyoto-Daigaku-Katsura, Nishikyo-ku, 615-8510, JAPAN 48, 1 (2008).
[32] N. A. George, C. P. G. Vallabhan, V. P. N. Nampoori, and P. Radhakrishnan, Optical Engineering 41, 251 (2002).
[33] F.-P. Montforts, Angewandte Chemie International Edition 43, 5431 (2004).
[34] M. L. Merlau, W. J. Grande, S. T. Nguyen, and J. T. Hupp, Journal of Molecular Catalysis A: Chemical 156, 79 (2000).
[35] N. A. Rakow and K. S. Suslick, Nature 406, 710 (2000).
[36] R. Ulrich and R. Torge, Appl. Opt. 12, 2901 (1973).
[37] K. Sathiyamoorthy, C. Vijayan, and M. P. Kothiyal, Review of Scientific Instruments 78, 043102 (2007).
[38] K. Sathiyamoorthy, J. Joseph, C. J. Hon, and M. V. Matham, edited by M. F. Costa (SPIE, Braga, Portugal, 2011), p. 80010K.
[39] A. Rosencwaig, Optics Communications 7, 305 (1973).
[40] V. Lepentsiotis and R. van Eldik, Journal of the Chemical Society, Dalton Transactions, 999 (1998).
[41] K. S. Suslick, F. V. Acholla, and B. R. Cook, Journal of the American Chemical Society 109, 2818 (1987).
[42] J. W. Lin and L. P. Dudek, Analytical Chemistry 51, 1627 (1979).
[43] T. Inagaki, K. Kagami, and E. T. Arakawa, Appl. Opt. 21, 949 (1982).

[44] T. A. El-Brolossy, T. Abdallah, M. B. Mohamed, S. Abdallah, K. Easawi, S. Negm, and H. Talaat, The European Physical Journal - Special Topics 153, 361 (2008).
[45] S. Negm and H. Talaat, Solid State Communications 84, 133 (1992).
[46] Q. Shen and T. Toyoda, Jpn. J. Appl. Phys. 39, 511 (2000).
[47] J. Seidel, S. Grafström, and L. Eng, Physical Review Letters 94, 177401 (2005).

Chapter 15

Electrically-Driven Active Plasmonic Devices

Young Chul Jun

Additional information is available at the end of the chapter

http://dx.doi.org/10.5772/ 50756

1. Introduction

Enhanced light-matter interactions in light-confining structures (such as optical cavities) have been extensively investigated for both fundamental studies and practical applications. Plasmonic nanostructures, which can confine and manipulate light down to the nanometer scale, are becoming increasingly important (Atwater 2007, Brongersma 2009). Plasmonic resonators and antennas can convert free-space light into intense, localized fields or enable coupling into deep-subwavelength-guided modes. Wherever subwavelength control over light is required, plasmonic structures are likely to play a vital role. Strong field enhancement in those structures can also alter light-matter interactions at a very fundamental level. Many areas of optical physics and devices can benefit from such extreme light concentration and manipulation (Schuller 2010). Moreover, these metal nanostructures can be simultaneously used as electrical contacts for current injection or application of electric fields (MacDonald 2010, Cai 2012). Thus, plasmonic structures are naturally suited for electrically driven devices. These active plasmonic devices can enable a broad range of new opto-electronic applications.

In this chapter, we review device concepts for electrically-driven active plasmonic devices, including the author's own work. We divide the following sections based on the physical mechanism leading to electrical control. In each section, the device operation and physics are explained. Finally, the conclusions and future prospects are presented.

2. Electric field induced liquid crystal reorientation

One possible way to obtain electrical control of plasmonic devices is to manipulate the refractive index of the dielectric medium adjacent to the metal surface. Liquid crystals (LCs) are an excellent example of such a tunable medium. The large broadband optical anisotropy of LCs makes them ideal for electrically tunable plasmonic structures (Kossyrev 2005, Chu 2006). For example, a E7 LC layer was employed in contact with metallic hole arrays, and electrical tuning of surface plasmon dispersion and optical transmission properties has been recently demonstrated (Dickson 2008). Plastic spacers (~ 60 um thick) were used to create the LC cell between the perforated gold film and the indium tin oxide (ITO) glass used as a top electrode. The static electric field controlled the LC's orientation in the cell. When the electric field was applied, the LC preferentially aligned with the field which was perpendicular to the interface. The maximum refractive index variation corresponding to a 90° reorientation of the LC was $n_e - n_o = 0.21$. (Ordinary and extraordinary refractive indexes, perpendicular and along the LC orientation, are $n_o = 1.525$ and $n_e = 1.734$.) This, in turn, induced a change in the SPP effective index at the Au/LC interface (given by Eq. 1) (Raether 1988) and optical response of the plasmonic crystals.

$$n_{SPP} = \sqrt{\varepsilon_{Au}\varepsilon_{LC} / (\varepsilon_{Au} + \varepsilon_{LC})} \tag{1}$$

Changes in both transmission amplitude and resonance frequency were observed. The changes in the transmission near the band-edges of the plasmonic crystals were most prominent, as expected.

3. Electrically induced thermal heating or phase transition

Another approach for electrical control is based on electrically induced thermal heating or phase transition. Figure 1a shows a 'thermo-plasmonic' device based on electrically-driven thermal heating and refractive index changes in a thermo-optic polymer (Nikolajsen 2004). Long-range (LR) surface plasmon polariton (SPP) waveguides for a wavelength of 1.55 μm were utilized to implement Mach-Zender interferometric modulators (MZIMs) and directional-coupler switches (DCSs) (Fig. 1a). First, 15 nm thin and 8 μm wide gold stripes were sandwiched between 15 μm thick layers of a thermo-optic polymer (BCB, benzocyclobutene) and heated by electrical signal currents. The LRSPP mode was excited by end-fire coupling with a single mode fiber. Gold stripes worked simultaneously as plasmonic waveguides and electrical heating elements. This configuration maximizes the influence of applied electrical signals. Low driving powers (< 10 mW for modulators and < 100 mW for switches), high extinction ratios (> 30 dB), and moderate response times (~ 1 ms) were demonstrated.

The dissipated power in the device for the temperature increase ΔT can be estimated as $P \sim 2\kappa \Delta T L w/d$, where $\kappa \sim 0.2$ W/mK is the polymer thermal conductivity, $w = 8$ μm is the stripe width, and $d = 15$ μm is the cladding thickness. The temperature increase needed for complete extinction at the modulator output is given as $\Delta T = (\partial n/\partial T)^{-1}(\lambda/2L)$, where $\partial n/\partial T \sim -2.5 \times 10^{-5}$ °C^{-1} is the polymer thermo-optic coefficient. Thus, the estimate for the modulator driving power is obtained as $P \sim 2\kappa \Delta T L w/d = (\partial n/\partial T)^{-1}(\kappa w \lambda/d) \approx 7$ mW, which is close to the measured value.

More recently, a semiconductor based electrical tuning of extraordinary optical transmission (EOT) through a metal hole array was reported (Shaner 2007). The gold aperture array, designed to operate in the mid-infrared spectral range (~ 1200 cm^{-1}), was fabricated on a doped GaAs epilayer (Fig. 1b). When the current flowed under the metal grating layer, the epilayer was resistively heated, and frequency tuning of the EOT spectrum (over 25 cm^{-1}) was achieved (Fig. 1c).

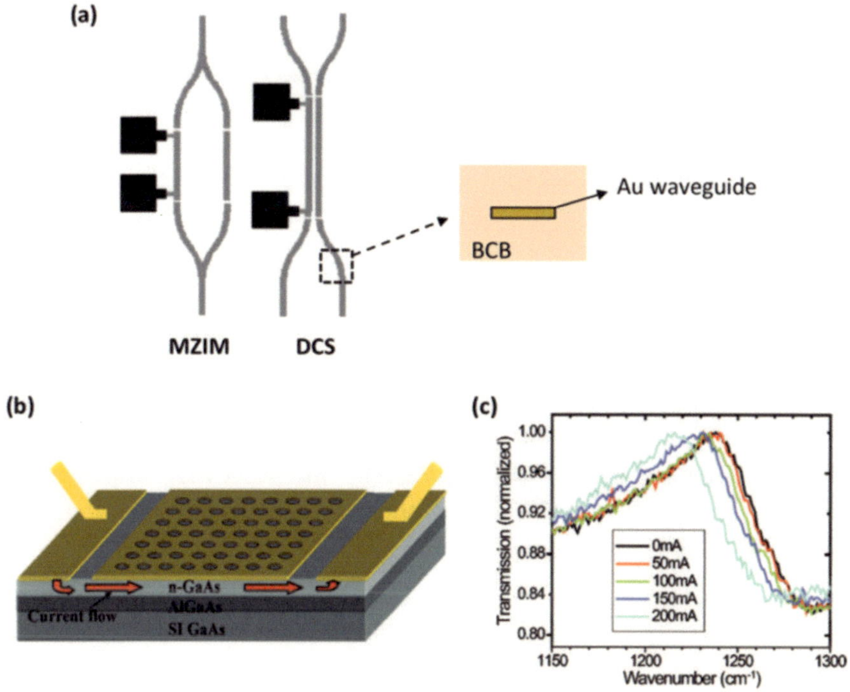

Figure 1. Active plasmonic devices based on electrically induced thermal heating (Nikolajsen 2004, Shaner 2007). (a) Schematic of LRSPP-based plasmonic modulator and switch using an electrically-driven refractive index change in a thermo-optic polymer (BCB). (b) Schematic of a tunable EOT device. The current path through n-doped GaAs epilayer is shown. (c) Transmission spectra through the EOT device. A redshift is observed as the current level increases.

Light transmission through such a structure relies on surface plasmon (SP) excitations at the air/metal and metal/semiconductor interfaces. The SP resonance condition for the metal/semiconductor interface is approximately given by

$$\sqrt{i^2 + j^2}\,\lambda = a_0 \,\mathrm{Re}\left(\sqrt{\frac{\varepsilon_s \varepsilon_m}{\varepsilon_s + \varepsilon_m}}\right) \tag{2}$$

Here, λ is the free space wavelength, a_0 is the lattice constant, i and j are integers (related to reciprocal lattice vectors), and ε_s and ε_m are the permittivities of the semiconductor and metal. Thermal heating of the semiconductor layer affects both a_0 and ε_s, and thus induces the spectral tuning of the transmission peak. Thermal expansion of the heated sample can change a_0, which would lead to a red-shift in the SP resonance (Eq. 2). However, this is a relatively small effect. The semiconductor dielectric constant ε_s can be affected by temperature in several ways. The temperature dependence of the dielectric function can be modelled using the Drude approximation:

$$\varepsilon(\omega,T,N) = \varepsilon_{\infty}(T)\left(1 - \frac{\omega_p(T,N)^2}{\omega^2 + i\omega\Gamma(T,N)}\right) \quad \text{where} \quad \omega_p(T,N)^2 = \frac{N(T)q^2}{\varepsilon_0\varepsilon_{\infty}(T)m^*(T,N)} \tag{3}$$

Here, ε_∞ is the background high frequency dielectric constant, ω_p is the plasma frequency, N is the electron density, m^* is the electron effective mass, and Γ is the damping term (= $1/\tau$, τ is the electron scattering time). The sample temperature T can be determined by monitoring the GaAs bandgap change (e.g. measuring the band edge PL as a function of current). High sample temperature can increase thermal generation of free carriers and affect the dielectric constant. This also affects the electron scattering time and effective mass.

Electrically induced domain switching or phase transition can modify the substrate refractive index too. Recently, domain switching in the ferroelectric barium-titanate ($BaTiO_2$) thin film was employed for active plasmonic devices (Dicken 2008). $BaTiO_2$ is a perovskite ferroelectric material that exhibits a large electrooptic coefficient (on the order of r ~ 100 pm/V) and large birefringence $\Delta n = 0.05$ (ordinary and extraordinary indexes are $n_o = 2.412$ and $n_e = 2.36$, respectively). The application of a bias across an Ag/$BaTiO_2$ SPP waveguide induced domain switching in the $BaTiO_2$ film. This modified the refractive index in the ferroelectric layer and modulated the optical output of a plasmonic interferometric device (working at $\lambda_0 = 688$ nm) by up to 15%.

Related works on tunable terahertz metamaterials (MMs) use phase transition in vanadium dioxide (VO_2) (Driscoll 2009). VO_2 is a correlated electron material that exhibits an insulator-to-metal (IMT) phase transition which can be controlled thermally, electrically, or optically. Local heating by electrical current through the VO_2 film induced the IMT and a change in the refractive index of the film. With increased temperature, the refractive index of VO_2 increased and the MM resonance frequency red-shifted. This electrically controlled frequency tuning is highly hysteretic and persistent; thus, this MM/VO_2 structure can also be used as a memory device. Similar electro-optic switching of plasmonic metamaterials was demonstrated in the near-infrared region, using the transition between crystalline and amorphous phases in a chalcogenide glass layer (Samson 2010).

4. Electrical carrier injection or depletion in semiconductor structures

We have so far considered electrical control based on liquid crystal reorientation, thermal heating, phase transition, etc. These are attractive enough for some applications, but they are rather slow processes. Therefore, for other applications, faster electrical control based on conventional semiconductor device technology may be preferred. Electrical tuning via carrier injection or depletion in semiconductor structures has been widely studied for dielectric photonics structures, e.g. for high speed, compact silicon photonic modulators (Soref 1987, Reed 2010). Carrier injection or depletion modifies the refractive index of a medium and can induce a change in the behavior of dielectric resonators.

A similar concept has been applied to active tuning of plasmonic MMs at terahertz frequencies (Chen 2006, Chan 2009). The free carrier absorption in a doped-GaAs substrate was dynamically controlled with an electric bias, by changing the carrier concentration in the substrate; this causes a strong amplitude modulation in the MM transmission. However, at higher frequencies (such as mid-infrared or mid-IR), the free carrier absorption is much smaller, so the transmission amplitude cannot be modulated in this way.

Nevertheless, using highly-doped semiconductor layers, we can still induce spectral tuning of mid-IR resonances by changing the refractive index of the substrate. Recently, electrically tunable mid-IR MMs based on depletion-type semiconductor devices was demonstrated (Jun 2012-1). Gold split ring resonator (SRR) arrays work simultaneously as an optical MM layer and electrical metal gate. With a reverse bias applied to the metal gate, the refractive index of the substrate directly underneath the plasmonic resonators varies through changes in the depletion width in a highly doped semiconductor. This results in frequency tuning of MM resonances. The mid-IR spectral range is technologically important for a number of applications, including chemical sensing and thermal imaging.

The images and schematics of the device are shown in Figs. 2a and 2b. The gold SRRs are connected to a metal gate via electrical bus lines and the whole MM layer works as an electrical gate. The MM layer is placed on top of an n-doped GaAs epilayer to form a metal-semiconductor junction. We need a highly doped n+ layer in order to induce a large dielectric constant change. An insulating barrier (e.g. undoped $Al_{0.3}Ga_{0.7}As$ layer) is also included to reduce leakage current. A modified SRR geometry was chosen because of its strong field enhancement in the two gaps and its compatibility with electrical connectivity, but other MM geometries can be used too.

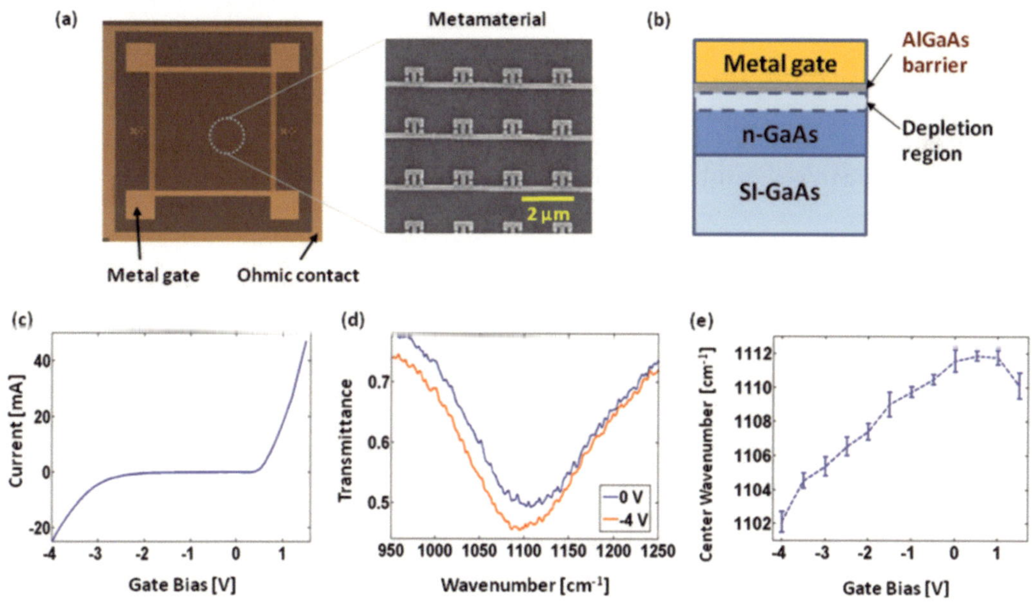

Figure 2. Fig. 2. Device structure and Measurement data (Jun 2012-1). (a) Image showing the device structure. The metamaterial layer (1 mm x 1 mm in size) is connected to the metal gate. These are surrounded by the outer Ohmic contact pad which contacts the n-doped layer. (b) Schematic of the substrate. It is composed of the barrier (30 nm $Al_{0.3}Ga_{0.7}As$) and doped semiconductor (n-GaAs, N_D = 5 x

10^{18} cm^{-3}) regions. The depletion region width in n-GaAs is varied by applying an external voltage. (c) IV curve exhibits a diode contact behavior. (d) Fourier transform infrared spectroscopy (FTIR) transmission measurement at room temperature for gate biases $V_G = 0$ V and -4 V. (e) The center frequency of the metamaterial resonance as a function of gate bias.

The dielectric constant ε of a semiconductor substrate can be modeled using the Drude approximation (Eq. 3). In the high doping regime (> 10^{18} cm^{-3}), the dielectric constant decreases rapidly with doping density (Fig. 3a). So, starting from an n+ GaAs doped layer, we can remove carriers with a reverse bias by increasing the depletion region to obtain a large dielectric constant change (e.g. $\Delta\varepsilon \sim 5.5$ for $\Delta N_D = 5 \times 10^{18}$ cm^{-3}, $\lambda_0 = 10$ μm).

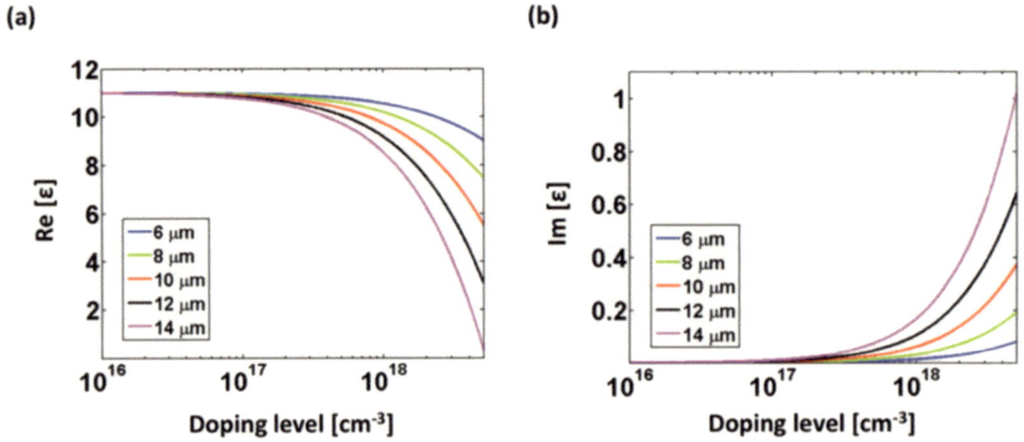

Figure 3. Drude model calculations of dielectric constants in n-doped GaAs for several mid-IR wavelengths (Jun 2012-1). (a) is the real part and (b) is the imaginary part of the dielectric constant.

The depletion width changes in an n-doped GaAs layer can be estimated using a textbook formula for a MIS (metal-insulator-semiconductor) capacitor:

$$W = \left[\frac{2\varepsilon_{GaAs}\varepsilon_0}{qN_D}(-\phi_s) \right]^{1/2} \tag{4}$$

where the surface potential ϕ_s is related to the gate voltage V_G as follows:

$$V_G = \phi_s - \frac{\varepsilon_{GaAs}}{\varepsilon_{AlGaAs}} W_{barrier} \left[\frac{2qN_D}{\varepsilon_{GaAs}\varepsilon_0} |\phi_s| \right]^{1/2} + \phi_{MS} \tag{5}$$

Here, ϕ_{MS} is the flat-band voltage. Note that the dielectric constants here are static values, which are different from the high frequency ones in Eq. 1. The surface potential ϕ_s is introduced to consider a voltage drop across a barrier layer and obtain the actual voltage bias applied in the doped semiconductor region. The depletion width gradually increases with a reverse bias, and the width change depends on the doping density, barrier material/thickness, etc.

We can model the MM resonance shifts using electromagnetic simulations. Here, finite difference time domain (FDTD) simulations are used to simulate the transmission spectra. We utilize dielectric constants and depletion widths obtained by previous equations. The wavelength-dependent dielectric constants from the Drude model are used for a doped semiconductor layer. We employ a different depletion width for each bias and repeat the numerical simulation. A broadband light pulse is incident from the top (normal to the MM plane) and polarized orthogonal to the SRR gap to excite the lowest MM resonance. The transmission is measured on the substrate side.

Figure 4 shows the calculated transmission spectrum. The transmission dip shows a spectral shift upon the application of a bias (Fig. 4a) and gradually red-shifts with increasing reverse bias (Fig. 4b) in agreement with our measurements. This red-shift can be easily understood from a LC resonator model. From $\omega_0 = 1/\sqrt{LC}$, we have $\omega_0 \propto 1/\sqrt{\varepsilon} = 1/n$, where n is the refractive index of a substrate. When the depletion width increases, the substrate refractive index increases (Fig. 3a) and the resonance red-shifts.

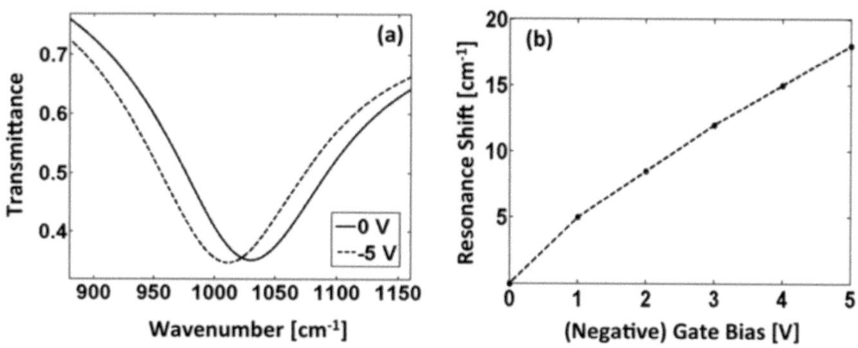

Figure 4. FDTD simulations of MM transmission spectra (Jun 2012-1). (a) Transmission spectra for the gate bias $V_G = 0$ V and -5 V. (b) Resonance red-shift as a function of reverse bias to the gate.

Electrical tunability can be further increased using asymmetric metamaterial designs (Jun 2012-2), which can be more sensitive to a substrate permittivity change. Recently, a classical analog of electromagnetically induced transparency (EIT) was demonstrated in several plasmonic and metamaterial structures. One such structure consists of two metal arcs with different lengths in order to have resonances at slightly different frequencies (Fig. 5, inset). The two metal arms are out-of-phase within a narrow frequency range, and the transmission spectrum has a transparency window due to the destructive interference from two resonator arms. Because this transparency window is caused by interference from two coupled resonators, it can be more sensitive to a substrate refractive index change. Figure 5a shows the simulated transmission spectrum for several different gate biases. We find that the interference part has larger tuning than other frequencies, as expected.

Depletion-type devices are also good for local index control. We can electrically address each resonator arm separately (Figs. 5a-5c). Depending on the bias scheme, we obtain quite different tuning behaviors. Here, we show a two-coupled resonator case. However, more sophisticated structures can be designed to enable a tailored spectrum change with a voltage bias. Electrically tunable metamaterials can be used for novel active infrared devices, such as optical filters, switches, modulators, and phase shifters.

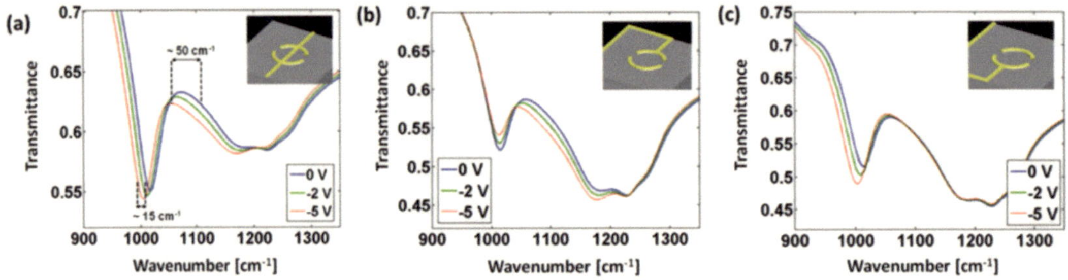

Figure 5. Asymetric metamaterial design based on coupled resonators (Jun 2012-2). Simulated transmission spectrum for several different gate biases, where (a) both resonator arms are connected to an electrical bus line – the straight gold line in the inset – and electrically biased, (b) only the smaller arc is connected to the electrical bus line, (c) only the larger arc is connected. The inset shows the unit cell of the FDTD simulation.

Other semiconductor structures or materials can be also adopted. For example, a metamaterial layer can be patterned on quantum well or quantum dot substrates to induce optical coupling between them (Gabbay 2011). The frequency and strength of this optical coupling can be tailored by designing proper metamaterial/substrate structures. Furthermore, this coupling can be electrically tuned by the stark shift of the quantum well or quantum dot transitions (Gabbay 2012), for instance. Graphene can be also employed for tunable plasmonic metamaterials. Due to its high polarizability, a single atomic layer of graphene can induce a noticeable change in the optical response of metamaterials (Papasimakis 2010). Electrostatic gating can be used to actively tune the optical response of graphene (Wang 2008), thus inducing a change in metamaterial responses. ITO can be also used for electrically carrier injection to induce a change in plasmonic mode indexes at visible and near-infrared wavelengths (Feigenbaum 2010).

A conventional metal-oxide-semiconductor (MOS) capacitor can be utilized as a plasmonic waveguide. Recently, a MOS field-effect plasmonic modulator was reported, termed a 'plasMostor' (Dionne 2009). That employs a metal-MOS-metal (Ag-SiO2-Si-Ag) waveguide structure, operating at 1.55 μm. It supports both photonic and plasmonic modes and has a transmission coefficient determined by interference between them; e.g., application of a gate bias (> 0.7 V) drives the MOS into accumulation and changes the Si index, which cut off the photonic mode and induces the transmission change. This plasmonic modulator can achieve modulation ratios approaching 10 dB in device volumes of half a cubic wavelength with femtojoule switching energies and the potential for gigahertz modulation frequencies.

5. Electrically pumped plasmon-emitting diode

The excitation of SPs with free-space light requires a special momentum matching technique (such as prism coupling, gratings, or scatterers). However, if the metal surface is very close to a light emitter (e.g. fluorescent molecules or quantum dots), the SPs can be excited directly via near-field coupling. Metal layers that support and guide the SP mode can also serve as contact electrodes. This brings an interesting opportunity for electrically pumped SP sources. The development of such electrical SP sources is important for miniaturized photonic circuitry and integrated sensing platforms.

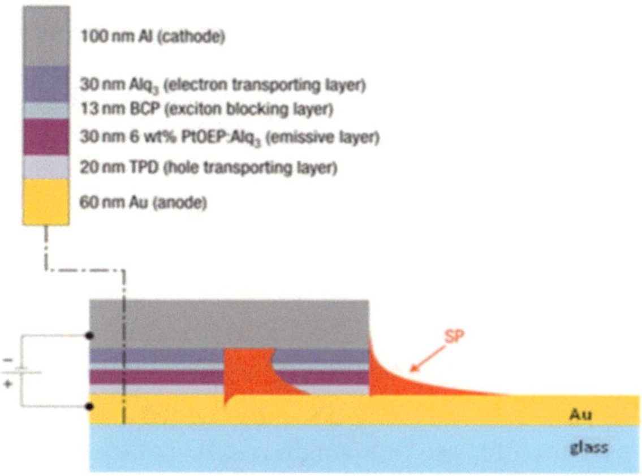

Figure 6. Schematic of an organic plasmon-emitting diode (Koller 2008). SPs are excited electrically through exciton recombination inside the MDM waveguide and later outcoupled to single surface SP wave as shown above.

Figure 6 shows an electrically pumped SP source using organic light-emitting diode (OLED) structures (Koller 2008). The organic active layers are placed in between two gold layers (i.e., a metal-dielectric-metal structure, or MDM). Metal layers work simultaneously as the SP waveguide and metal electrode. Electrically injected electrons and holes recombine in the active organic layer and generate the SP wave. The excited SP mode in the MDM waveguide can extend further into a single metal surface as shown in Fig. 6. The strong electromagnetic field intensity of the SP mode can enable a compact sensing platform without external illumination optics.

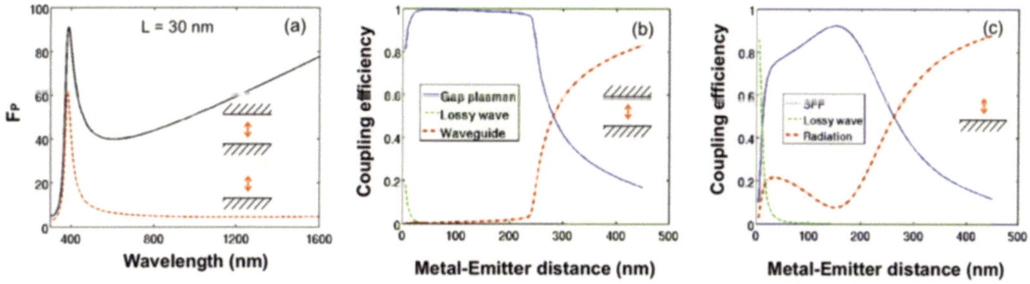

Figure 7. Light emission of a dipolar emitter inside an Ag/SiO_2/Ag MDM slab waveguide or near an Ag/SiO_2 interface (Jun 2008). We assume that the emitter has unit quantum efficiency. (a) Spontaneous emission enhancement as a function of free-space wavelength. In the MDM case, the emitter is in the center of the gap. The gap size L = 30 nm. (b) Fraction of dissipated energy into different decay pathways as a function of the metal-emitter distance. The emission wavelength λ_0 = 800 nm. (c) same as (b) but for the single metal/dielectric interface case.

A similar electrical SP source based on silicon CMOS processes has been reported (Walters 2010): gold cladding layers surround a semi-insulating layer of alumina that contains silicon quantum dots and form a MDM plasmonic waveguide. When a sufficient voltage is applied across the insulator layer, tunnelling electrons can excite the embedded quantum dots through impact ionization processes, producing the SP wave. The device fabrication and materials are compatible with well-established silicon microelectronics processes.

Because of their small optical mode volume, MDM structures can achieve large spontaneous emission enhancement and high coupling ratio into well-defined SPP waveguide modes (Jun 2008). Figure 7 explains this for a single dipolar emitter which is oriented normal to the interfaces. The spontaneous emission enhancement factor, F_P, is obtained by calculating the work done on the dipole by its own reflected field from the metal surfaces. Figure 7a shows F_P as a function of the free-space wavelength. In addition to a resonant enhancement peak around the SP resonance wavelength of silver, there is a strong non-resonant enhancement due to the tight confinement of modes between the two metallic films. In the non-resonant region, propagation lengths of SP modes become longer, which is desirable for many applications. For comparison, the case of a single metal surface is also plotted (dashed line), which clearly lacks such non-resonant enhancement. The light emission in MDM waveguides is also preferentially directed into plasmonic waveguide modes (Fig. 7b) over a large range of gap sizes. This behavior is in stark contrast to that observed for an emitter near a single metal surface, which exhibits relatively stronger coupling to lossy surface waves and free-space modes (Fig. 7c).

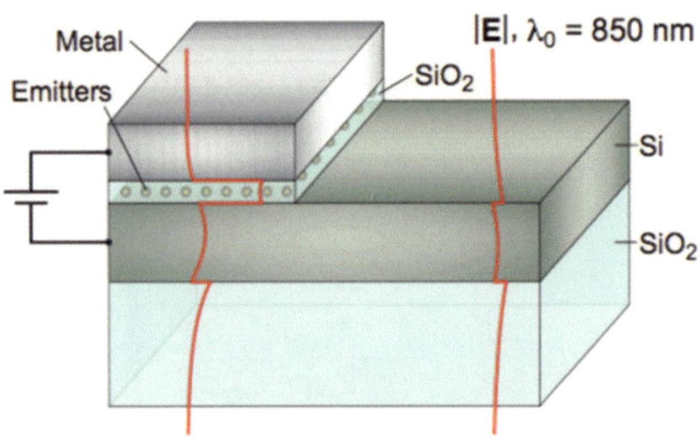

Figure 8. Schematic of a CMOS-compatible light-emitting device (Hryciw 2009). Typical electric field profiles for a TM mode are shown in red. Electrically pumped light emission in the MDS waveguide excites hybrid plasmonic modes and later can be outcoupled to SOI waveguide modes.

A high-index silicon slab supports guided dielectric modes. When this silicon slab is brought closer to a metal surface which supports SPP modes, a new hybrid plasmonic mode can form. A metal-dielectric-semiconductor (MDS) structure resembles MOS layers in conventional semiconductor devices and thus can be used as CMOS-compatible light sources. We can still achieve large field enhancement in the low-index dielectric region of such hybrid structures. Spontaneous emission enhancement in MDS waveguides is still significantly larger than those from a single metal surface. MDS waveguides exhibit broadband emission enhancement and high waveguide coupling ratio, similar to MDM waveguides (Hryciw 2009, Hryciw 2010). Furthermore, this MDS structure is more suitable for electrical carrier injection. The plasmonic modes in MDS waveguides can be efficiently converted into silicon slab waveguide modes for a longer distance transport (Fig. 8).

In addition to these plasmon-emitting diodes, there have been demonstrations on nanocavity lasers based on electrically pumped metal-coated cavities, MDM waveguides, and optically pumped hybrid plasmonic cavities (Martin 2010). In a 'spaser' (surface plasmon amplification by stimulated emission of radiation), the lasing mode may even be non-radiating, i.e. a near-field laser. Such coherent nanoscale light sources can lead to new applications in sensing and lithography.

6. Electrical control of light emission in plasmonic nanogaps

It was recently demonstrated that nanopatterned metal films such as metal slit-grooves can collimate and direct quantum dot (QD) emission in a designed fashion, working effectively as optical antennas (Jun 2011). The patterned metal films next to a slit can be used as electrical contacts to apply a field across the slit. In this section, we consider active electrical control of QD emission in such plasmonic nanogaps. We show how the metals defining the plasmonic cavity can be utilized to electrically control the emission intensity and wavelength. The unmatched, combined directional and electrical control over the emission of a large number of quantum emitters opens up a broad range of new opto-electronic applications for plasmonic antennas, facilitating the realization of a new class of active optical antennas.

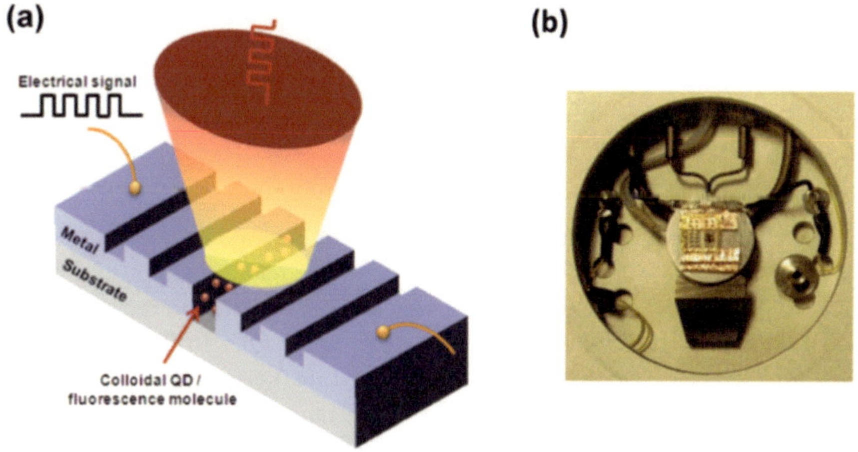

Figure 9. Electrical control of QD photoluminescence (Jun 2011). (a) Schematic of a plasmonic antenna device, in which colloidal CdSe/ZnS core/shell quantum dots are placed within the nanslit. An optical pump at 514 nm was directed from the below, and QD emission was collected from the top. (b) Top view of the mounted sample on the Linkam thermal stage.

First, two large metal contact pads were defined by standard optical lithography. Two metal pads were connected with a narrow metal stripe. A slit-groove structure was patterned later on this stripe with focused ion beam (FIB) milling (Fig. 9a). After FIB milling, the two metal pads were electrically disconnected. Then, a thin layer of colloidal CdSe/ZnS QDs mixed with PMMA (poly-methylmethacrylate) was spin-coated on top of the patterned metal structures. Electrical measurements were conducted at a low temperature using the Linkam thermal stage. A sample was mounted on the silver chuck of the Linkam thermal stage (Fig. 9b), which was cooled with liquid nitrogen flowing. Two metal contacts in the sample were wire-bonded for electric biasing. Because the chuck had a hole in the center, we could direct laser light (CW laser light at 532 nm) from the bottom and collect QD emission from the top. We applied external voltages across the metal gap to see the voltage-dependent QD emission spectrum.

Figure 10a shows the voltage-dependent QD emission spectrum measured from this slit-groove sample. As the applied DC voltage increases, the QD emission spectrum exhibited a red-shift and luminescence quenching. In most cases, the spectral changes were reversible by alternating external voltages, although sometimes QDs or the metal nanoslits were damaged under the strong field. The spectrum shift can be understood as the Stark effect of emission. The large electric field pulls apart electron-hole pairs inside QDs and induces the red-shift and broadening of the whole emission spectrum. Note that despite the intensity drop at the spectrum center, the QD emission intensity increased at longer wavelengths with applied voltages. However, the Stark effect alone does not explain the observed, large luminescence quenching. We believe this is possibly due to enhanced nonradiative recombination. A very large electric field in the gap can 'dissociate' electron-hole pairs, and those carriers can induce luminescence quenching by participating in nonradiative recombination processes such as Auger recombination.

Figure 10b shows the measured emission frequency shift as a function of applied voltage. The shift initially increases quadratically with voltage (as expected for the Stark shift) and then saturates at larger voltages (shaded region). According to quantum mechanical perturbation theory, the stark shift is given by

$$\Delta E = \mu F + \alpha F^2 + ... \qquad (6)$$

where F is the electric field magnitude, μ permanent dipole moment, and α polarizability. It is known that a CdSe QD has a Wurtzite crystal structure and a permanent dipole moment. Thus, depending on the orientation of a QD, we can observe either a red- or blue-shift with the applied external field. However, averaging over many QDs in the gap, the linear term cancels out, and we obtain only a quadratic stark shift (i.e. red-shift). Thus, for our QD ensemble measurement, we expect to see the quadratic red-shift with external field, in agreement with Fig. 10b.

We can deduce the polarizability of QDs from this Stark shift data: we first obtain the electric field magnitude inside a QD as a function of applied voltage. Assuming that a QD is a sphere dielectric, we can analytically calculate the electric field F_{in} inside the QD:

$$F_{in} = -\frac{d\varphi_{in}}{dr} = \frac{3}{\varepsilon_{QD} / \varepsilon_{matrix} + 2} F_0 = \frac{3}{\varepsilon_{QD} / \varepsilon_{matrix} + 2} \frac{V_a}{W} \quad (7)$$

where ε_{QD} and ε_{matrix} are the dielectric constants of QD and surrounding matrix medium, and F_0 is the incident field magnitude. The numbers 3 and 2 originate from the geometry (i.e. sphere).

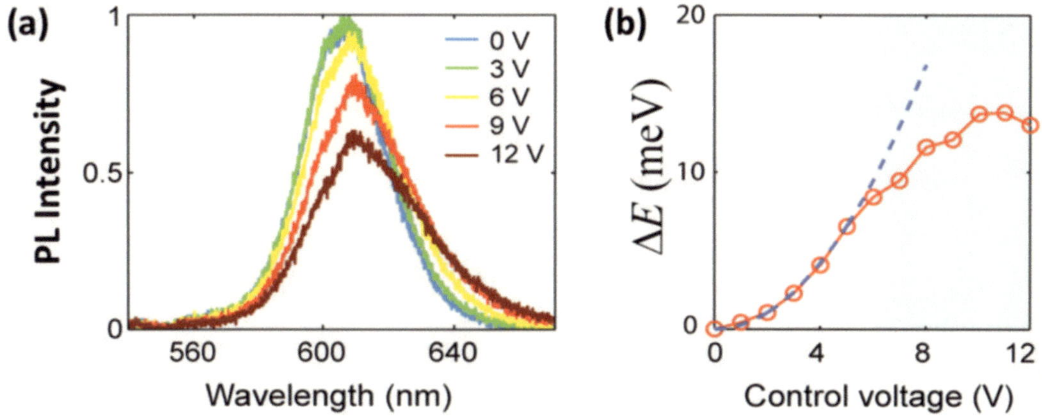

Figure 10. Bias-dependent QD emission (Jun 2011). (a) Emission spectra measured at different applied DC bias voltages across the nanogap. (b) The Stark shift of the emission as a function of applied voltage. The dashed line corresponds to a quadratic fitting curve, and the shaded region indicates a saturation effect for large voltages.

We have $F_0 = V_a/W$, considering that we have a voltage drop V_a across the gap W. In this experiment, we have W = 60 nm, $\varepsilon_{CdSe} = (2\varepsilon_{\perp} + \varepsilon_{||})/3 = 2/3 \times 9.23 + 1/3 \times 10.16 \approx 9.58$, $\varepsilon_{matrix} = \varepsilon_{PMMA} = 2.22$. By fitting data in Fig. 10b to the following equation (i.e. blue dotted curve in Fig. 10b), we obtain the QD polarizability.

$$\Delta E = \alpha F^2; \ \alpha = 4 \times 10^{-5} meV \cdot cm^2 / (kV)^2 \quad (8)$$

Furthermore, we can apply a time-varying voltage signal and modulate the QD emission dynamically (Fig. 9a). A square-wave voltage signal was applied across the gap with a function generator. A 100 kHz voltage signal modulated between 0V and 10V with a 50% duty cycle. The QD emission was detected with a single photon avalanche detector (SPAD). We used the sync signal of the function generator as a start signal and the SPAD output as a stop signal in the photon counting module (Picoharp 300, Picoquant) to obtain the accumulated histogram of the modulated QD emission intensity (Fig. 11). A clear step appears in the time-resolved histogram: this means that the QD emission intensity was modulated with the external voltage signal. We accumulated the SPAD output for a few minutes to obtain a clear modulation step. This improved the signal-to-noise ratio but did not change the actual modulation depth. A modulation depth of M ~ 24 % was achieved, where $M = (I_{max} - I_{min}) / I_{max}$.

This step appears only when we see QD emission spectrum change with voltage. For example, if incident laser light illuminated QDs on the bare quartz region (i.e. outside the metal stripe), we did not observe such a modulation step. The histogram just appears flat, i.e. no voltage dependent change. We also note that the observed modulation depth also depends on the detection wavelength. When there is a large change in QD emission spectrum, we observed larger modulation depth. This result presents new opportunities for electrically modulated optical antenna devices. In the above experiment, the modulation speed was rather moderate (100 kHz), but we can increase the speed to well above 100 MHz.

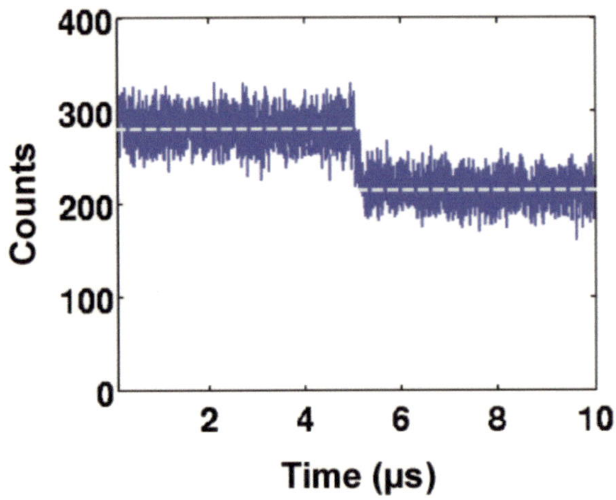

Figure 11. The time-resolved histogram of electrically modulated QD emission intensity (Jun 2011). The modulated square-wave signal (0-10 V, 100 kHz, 50% duty cycle) is applied. The step means that QD emission intensity is dynamically modulated with the external voltage signal.

Plasmonic modes in small metal nanogaps (< 100 nm) can further contribute to this electrical modulation. Spontaneous emission can be enhanced greatly in a small metal nanogap. The reduced lifetime can improve the responsivity of devices and increase the upper limit of the modulation speed. Furthermore, optimized side grooves can enable optical beaming. This electrically driven slit-groove structure acts as an active optical antenna, which can collimate and modulate QD emission in one device. The unmatched, combined directional and electrical control over the emission of a large number of quantum emitters can support a broad range of new opto-electronic applications.

More recently, electrically controlled nonlinear harmonic generation of light in plasmonic nanogaps was demonstrated (Cai 2011). The device structure is similar to the previous one (Fig. 9a). A commonly used polymer (PMMA, poly-methyl methacrylate) was inserted into the metal slit. The frequency doubled light originated from a third-order nonlinear susceptibility $\chi(2\omega; \omega, \omega, 0)$. The change in the second-harmonic signal as a function of applied DC voltage indicated a linear dependence of the frequency doubled output on the driving voltage, with the magnitude of the normalized change being over 7% per volt. Unlike the conventional second harmonic generation, which imposes a rigorous requirement on the lattice asymmetry of the nonlinear medium, third-order nonlinear responses are present in all materials. The subwavelength device size also eases the strict phase-matching condition and makes them suitable for chipscale, high density integration.

7. Conclusions

Plasmonics is emerging as a new chip-scale device technology that can bridge the size mismatch between nanoscale electronics and microscale photonics. It can combine the size of electronics and the speed/bandwidth of photonics in one device. It exploits the unique optical properties of metallic nanostructures to confine and manipulate light at the nanoscale. Strong optical field enhancement in plasmonic nanostructures can also modify light-matter interactions to unprecedented levels. Furthermore, plasmonic structures can serve as electrodes and thus perform optical and electrical functions simultaneously. This brings about many new opportunities for opto-electronic devices.

This chapter reviewed novel active plasmonic device concepts explored in the literature, including the author's own work. Electrically driven plasmonic device structures were introduced, and their device operation and physics were explained. Various physical mechanisms can be utilized for electrical control of optical properties, such as liquid crystal reorientation, thermal heating, phase transition, carrier injection/depletion, Stark shift of emitters, etc. The next step would be to develop device structures that are more efficient, compact, and scalable for large area, dense integration.

Author details

Young Chul Jun
Center for Integrated Nanotechnologies (CINT), Sandia National Laboratories, NM, USA

Acknowledgement

This work was performed, in part, at the Center for Integrated Nanotechnologies, a U.S. Department of Energy, Office of Basic Energy Sciences user facility. Sandia National Laboratories is a multi-program laboratory managed and operated by Sandia Corporation, a wholly owned subsidiary of Lockheed Martin Corporation, for the U.S. Department of Energy's National Nuclear Security Administration under contract DE-AC04-94AL85000. The author wishes to thank Dr. Igal Brener for his support in the preparation of this article.

8. References

Atwater, H.A. (2007), The promise of plasmonics, *Scientific American*, (April 2007), pp. 56-63

Brongersma, M.L. (2009), Schuller, J.A., White, J., Jun, Y.C., Bozhevolnyi, S.I., Sondergaard, T., Zia, R. Nanoplasmonics : components, devices, and circuits in *Plasmonic Nanoguides and Circuits*, (Pan Stanford Publishing, 2009), edited by Bozhevolnyi, S.I.

Cai, W. (2011), Vasudev, A.P., Brongersma, M.L. Electrically controlled nonlinear generation of light with plasmonics. *Science*, Vol.333, No.6050, (2011), pp. 1720-1723

Cai, W. (2012), Jun, Y.C., Brongersma, M.L. Electrical control of plasmonic nanodevices. SPIE Newsroom, (2012), doi: 10.1117/2.1201112.004060

Chan, W.L. (2009), Chen, H.-T., Taylor, A.J., Brener, I., Cich, M.J., Mittleman, D.M. A spatial light modulator for terahertz beams. *Appl. Phys. Lett.*, Vol.94, No.21, (2009), p. 213511

Chen, H.-T. (2006), Padilla, W.J., Zide, J.M.O., Gossard, A.C., Taylor, A.J., Averitt, R.D. Active terahertz metamaterial devices. *Nature*, Vol.444, (2006), pp. 597-600

Chu, K.C. (2006), Chao, C.Y., Chen, Y.F., Wu, Y.C., Chen, C.C. Electrically controlled surface plasmon resonance frequency of gold nanorods. Appl. Phys. Lett., Vol.89, No.10, (2006), p. 103107

Dicken, M.J. (2008), Sweatlock, L.A., Pacifici, D., Lezec, H.J., Bhattacharya, K., Atwater, H.A. Electrooptic modulation in thin film barium titanate plasmonic interferometers. *Nano Lett.*, Vol.8, No.11, (2008), pp. 4048-4052

Dickson, W. (2008), Wurtz, G.A., Evans, P.R., Pollard, R.J., Zayats, A.V. Electronically controlled surface plasmon dispersion and optical transmission through metallic hole arrays using liquid crystal. *Nano Lett.*, Vol.8, No.1, (2008), pp. 281-286

Dionne, J.A. (2009), Diest, K., Sweatlock, L.A., Atwater, H.A. PlasMOStor : a Metal-Oxide-Si field effect plasmonic modulator. *Nano Lett.*, Vol.9, No.2, (2009), pp. 897-902

Driscoll, T. (2009), Kim, H.-T., Chae, B.-G., Kim, B.-J., Lee, Y.-W., Jokerst, N.M., Palit, S., Smith, D.R., Ventra, M.D., Basov, D.N. Memory Metamaterials. *Science*, Vol.325, No.5947, (2009), pp. 1518-1521

Feigenbaum, E. (2010), Diest, K., Atwater, H.A. Unity-order index change in transparent conduction oxides at visible frequencies. *Nano Lett.*, Vol. 10, No. 6, (2010), pp. 2111-2116

Gabbay, A. (2011), Reno, J., Wendt, J.R., Gin, A., Wanke, M.C., Sinclair, M.B., Shaner, E., Brener, I. Interaction between metamaterial resonators and intersubband transitions in semiconductor quantum wells. *Appl. Phys. Lett.*, Vol. 98, No. 20, (2011), p. 203103

Gabbay, A. (2012), Brener, I. Theory and modeling of electrically tunable metamaterial devices using intersubband transitions in semiconductor quantum wells. Opt. Express, Vol.20, No.6, (2012), pp. 6584-6597

Hryciw, A. (2009), Jun, Y.C., Brongersma, M.L. Plasmon-enhanced emission from optically-doped MOS light sources. *Opt. Express*, Vol.17, No.1, (2009), pp. 185-192

Hryciw, A. (2010), Jun, Y.C., Brongersma, M.L. Electrifying plasmonics on silicon. *Nature Mater.*, Vol.9, (2010), pp. 3-4

Jun, Y.C. (2008), Kekatpure, R.D., White, J.S., Brongersma, M.L. Nonresonant enhancement of spotaneous emission in metal-dielectric-metal plasmon waveguide structures. *Phys. Rev. B*, Vol.78, No.15, (2008), p. 153111

Jun, Y.C. (2011), Huang, K.C.Y., Brongersma, M.L. Plasmonic beaming and active control over fluorescent emission. *Nature Comm.*, Vol.2, (2011), p. 283

Jun, Y.C. (2012-1), Gonzales, E., Reno, J.L., Shaner, E.A., Gabbay, A., Brener, I. Active tuning of mid-infrared metamaterials by electric control of carrier densities. *Opt. Express*, Vol.20, No.2, (2012), pp. 1903-1911

Jun, Y.C. (2012-2), Brener, I. Electrically tunable infrared metamaterials based on depletion-type semiconductor devices. *submitted* (2012)

Koller, D.M. (2008), Hohenau, A., Ditlbacher, H., Galler, N., Reil, F., Aussenegg, F.R., Leitner, A., List, E.J.W., Krenn, J.R. Organic plasmon-emitting diode. *Nature Photon.*, Vol.2, (2008), pp. 684-687

Kossyrev, P.A. (2005), Yin, A., Cloutier, S.G., Cardimona, D.A., Huang, D., Alsing, P.M., Xu, J.M. Electric field tuning of plasmonic response of nanodot array in liquid crystal matrix. Nano Lett., Vol.5, No.10, (2005), pp. 1978-1981

MacDonald, K.F. (2010), Zheludev, N.I. Active plasmonics : current status. *Laser Photonics Rev.*, Vol.4, No.4, (2010), pp. 562-567

Martin, T.H. (2010), Status and prospects for metallic and plasmonic nano-lasers. J. *Opt. Soc. Am. B*, Vol.27, No.11, pp. B36-B44.

Nikolajsen, T. (2004), Leosson, K., Bozhevolnyi, S.I. Surface plasmon polariton based modulators and switches operating at telecom wavelengths. *Appl. Phys. Lett.*, Vol.85, No.24, (2004), pp. 5833-5835

Papasimakis, N. (2010), Luo, Z., Shen, Z.X., Angelis, F.D., Fabrizio, E.D., Nikolaenko, A.E., Zheludev, N.I. Graphene in a photonic metamaterial, *Opt. Express*, Vo.18, No.8, (2010), pp. 8353-8359

Raether, H. (1988), *Surface Plasmons on Smooth and Rough Surfaces and on Gratings*, Springer-Verlag.

Reed, G.T. (2010), Mashanovich, G., Gardes, F.Y., Thomson, D.J. Silicon optical modulators. *Nature Photon.*, Vol.4, (2010), pp. 518-526

Samson, Z.L., MacDonald, K.F., Angelis, F.D, Gholipour, B., Knight, K., Huang, C.C., Fabrizio, E.D., Hewak, D.W., Zheludev, N.I. Metamaterial electro-optic switch of nanoscale thickness. *Appl. Phys. Lett.*, Vol.96, (2010), p. 143105

Schuller, J.A., Barnard, E.S., Cai, W., Jun, Y.C., White, J.S., Brongersma, M.L. Plasmonics for extreme light concentration and manipulation. *Nature Mater.*, Vol.9, (2010), pp. 193-204

Shaner, E.A. (2007), Cederberg, J.G., Wasserman, D. Electrically tunable extraordinary optical transmission gratings. *Appl. Phys. Lett.*, Vol.91, (2007), p. 181110

Soref, R.A. (1987), Bennett, B.R. Electrooptical effects in silicon. *IEEE J. Quantum Electron.*, Vol.QE-23, No.1, (1987), pp. 123-129

Walters, R.J. (2010), van Loon, R.V.A., Brunets, I., Schmitz, J., Polman, A. A silicon-based electrical source of surface plasmon polaritons. *Nature Mater.*, Vol.9, (2010), pp. 21-25

Wang, F. (2008), Zhang, Y., Tian, C.S., Girit, C., Zettl, A., Crommie, M.F., Shen, Y.R. Gate-variable optical transitions in graphene. Science, Vol.320, No.5873, (2008), pp. 206-209

Section 4

Plasmonics Applications

Chapter 16

Plasmonic Conducting Polymers for Heavy Metal Sensing

Mahnaz M. Abdi, Wan Mahmood Mat Yunus, Majid Reayi and Afarin Bahrami

Additional information is available at the end of the chapter

http://dx.doi.org/10.5772/51337

1. Introduction

Conducting polymers are a sub-group of organic and inorganic electrical conductors which can be considered as interdisciplinary science and technology. They are in the intersection of three disciplines: physics, chemistry and engineering. These macromolecular materials are unique combination of electronic and optical properties of metals and semiconductors with the processing advantages and mechanical properties of polymers [1].

Conducting polymers (CPs) such as polyaniline, polypyrrole and polythiophen have received great attention to the chemists and physicists during the last decade due to their potential applications in different fields such as mechanical and optical sensors, actuators, light-emitting diodes (LED), transistors, energy storage, supercapacitors. [2-5]. Conducting sensors based on PPy film with different dopant were developed for detecting volatile aromatic hydrocarbons [6] and volatile organic solvent sensors have been fabricated using PPy films on conducting glass substrates [7].

Different techniques including electrochemical quartz crystal microbalance [5], Fourier transform infrared spectroscopy (FTIR) [8], photoacoustic effect [9], and electromagnetic interference shielding effectiveness (EMI SE) [10-11] have been used to study the optical and electrical properties of conducting polymers. Surface plasmon resonance (SPR) is one of these methods that provided considerable information about electropolymerization, doping and dedoping processes [12].

Surface plasmon resonance (SPR) is a quantum optical-electrical phenomenon arising from the interaction of light with a metal surface. Surface plasmons are surface electromagnetic waves that propagate in a direction parallel to the metal/dielectric (or metal/vacuum) interface. The waves are on the boundary of the metal and the external medium (air or water for example); thereby these oscillations are very sensitive to any change of this boundary, such as the adsorption of molecules to the metal surface [13].

2. Basic of surface plasmon resonance

When a polarized light strikes the area between the glass and air, at a certain angle (θ), a total internal reflection (TIR) is generated. In this case the incident light is equal to the reflective light and an electric field intensity, known as evanescent wave is generated. When a metal is placed at the interface between the glass and air, at a specific angle, the plasmon on the metal surface couples with the energy from the light and cause a reduction in the intensity of the reflected light, achieve the resonance at the given wavelength and angle, where SPR is occurring. Any changes in the chemical composition of the environmental within the metal surface of the plasmon field cause a shift in the angle of the reflected light. Thus the magnitude of this change can be determined by measuring the angle of the reflected light by the metal surface using a sensogram.

SPR is an optical phenomenon related to charge density oscillation at the interface between two materials that their signs of dielectric constant are opposite [14]. The basic of SPR is the propagation of surface plasmon wave (SPW) along the interface of metal and dielectric which behaves like quasii-free electron plasma. The wave number of light should be matched with the wave number of surface plasmon given as:

$$n_p \sin\theta = \mathrm{Re}\left[\sqrt{\frac{\varepsilon_d \varepsilon_m}{\varepsilon_d + \varepsilon_m}}\right] \qquad (1)$$

where n_p , θ are refractive index of the prism and angle of resonance, while ε_m and ε_d are the dielectric constants of metal and last medium, respectively [13]. The thickness and dielectric constant of thin films deposited on the gold surface can be measured by this optical technique.

The electrochemical surface plasmon resonance (ESPR) technique is good combination of surface plasmon resonance with electrochemical measurements, which can be used for simultaneous optical characterization and electrochemical polymerization of the electrode–electrolyte interface [5]. In this technique, the real ($\Delta\varepsilon_{real}$) and imaginary ($\Delta\varepsilon_{imaginary}$) parts of the dielectric constant are parameter of the average thickness of the adsorbed layer, electrode potential modulation (ΔV), as well as changes in the surface charge density (Δδ) as follows [15-16].

$$\frac{\Delta\theta_{SPR}}{\Delta V} \approx C_1\left(\frac{\Delta\varepsilon_{real}}{\Delta V}\right) + C_2\left(\frac{\Delta\varepsilon_{imaginary}}{\Delta V}\right) + C_3\left(\frac{\Delta d}{\Delta V}\right) + C_4\left(\frac{\Delta\delta}{\Delta V}\right) \qquad (2)$$

where C_1, C_2, C_3 and C_4 are constants, Δd is the change of the average thickness of the adsorbed layer, and Δδ represents the effects of the electron density on the surface plasmon resonance angle. Above mentioned formula shows a good combination of surface plasmon resonance with electrochemical measurements, which can be used for simultaneous optical characterization and electrochemical polymerization of the electrode–electrolyte interface [5].

If the dielectric constant experessed as:

$$\varepsilon = \varepsilon_{real}(\lambda) + i\left(\frac{2\sigma\lambda}{C}\right) \tag{3}$$

It has been shown that the imaginary part of the dielectric constant can be changed by variation of conductivity of material layer and this relation is monitored as:

$$\varepsilon_{imaginary} = \frac{2\sigma\lambda}{C} \tag{4}$$

where C is the speed of light with wavelength of λ incident on the material with conductivity of σ.

3. Application of SPR in Conducting polymers

In recent years, surface plasmon resonance (SPR) has been widely demonstrated as an effective optical technique for in situ investigation of optical and electrical properties of conducting polymers such as polypyrrole films [5, 15]. Damos and his coworker showed that during electropolymerization and doping/dedoping processes of thin polypyrrole films on flat gold surfaces a significant change in the surface plasmon angle position was produced. These changes in the electrochemical and optical properties of the polypyrrole films were due to the changes in the real and imaginary parts of the complex dielectric constant during doping/dedoping processes [5].

They prepared PPy films by in situ electropolymerization using potentiostatic, galvanostatic and potential cycling process (potentiodynamic) methods and showed that the shape of the reflectivity curves changed during polymerization process, indicating that the growth processes are also efficient in producing polymer films with high conductivity. In addition, they observed that during polymerization the resonance angle shifted toward higher values which can be related to the changes in the optical properties at the electrode–solution interface when polymer grow at the surface of the metal.

On the other hand, the changes of real and imaginary part of dielectric constant for polypyrrole films polymerized by potentiodynamic method was more identical and reversible compared with those polymerized by potentiostatic or galvanostatic conditions indicating higher ability for the ions exchange of the films prepared by potentiodynamic. It was shown that the real part of the dielectric constant for polypyrrole film shows a small decrease and the imaginary part increased by switching the potential between the reduced and oxidized states of the polypyrrole films. It was rsulted that the changes in the imaginary part of the dielectric constant are associated with changes of the electronic energy states of the polypyrrole according to the polaron/bipolaron model [5].

During the oxidation and doping process, intermediary energy levels are created between the valance band (VB) and conduction band (CB); causing a decrease in the energy required for the electronic transition. As a result, the polymers with higher conductivity exhibit high values of "$\varepsilon_{imaginary}$" which results in a decrease in the real part of the complex dielectric constant. The study on the optical and electrical properties of polypyrrole at different oxidation levels showed that there is a correlation between conductivity and formation of polaron and bipolaron species [17].

The electrical properties of the PPy-CHI composite films was investigated in our previous work [18] and it was revealed the enhanced electrical properties of composite films in the presence of a certain amount of CHI. It was observed that the wavelength of electron transition in Uv-vis spectrum increased with the increase in CHI content, indicating a decrease of the band gaps between valance band (VB) and conduction band (CB).

SPR technique was used to detect ochratoxin A (OTA) which is a toxic fungal metabolite and acts as a DNA damaging agent [19]. In this research, Jorn prepared sensing area by in situ electropolymerziation of pyrrole in the presence of chloride dopant. He observed that PPy prepared in the presence higher amount of NaCl presented lower conductivity and so that the SPR angle increased. On the other hand, they believed that adsorption of Cl^- on the gold layer affected double-layer charging, resulted in a decrease in SPR angle. The linearity range of 0.1-10 μg/ml (with $R^2 = 0.9689$) was expersed for this ochratoxin A (OTA) sensor.

Guedon *et al*, [20], reported a DNA hybridization sensor based on ppy- ODN matrix coupled with SPR imaging. They prepared SPR chip by electrospotting pyrrole and pyrrole linked to an oligonucleotide (ODN). The SPR imaging configuration is closely based on the same configuration with regular SPR, in which the metal surface is imaged on a CCD camera via an imaging lens. This kind of SPR has advantages of sensing on several areas of the gold surface at the same time [21-22]. Electrospotting method is used for the construction of biochips on individually addressable microelectrode arrays. In this work [20] they optimized hybridization response of the polypyrrole/DNA probes versus spot thickness and density of immobilized probe. The optimal thickness of the spot was found to be close to 11 nm and a surface density of polypyrrole/DNA probes of 130 fmol/ mm^2 (590 pg/mm^2) was reported to optimize the hybridization signal that can be detected directly.

4. Heavy metals and SPR detection

Heavy metals such as mercury and lead ions have long been recognized as a harmful environmental pollutant. Lower levels of lead can affect on the central nervous system, kidney, and blood cells [23] and in severe cases can cause convulsions, coma, and even death. Lead interferes with the development of the nervous system and is therefore particularly toxic to children and unborn babies, causing potentially permanent learning and behavior disorders [24]. Overexposure to mercury can damage the central nervous system which can cause memory loss, loss of appetite, personality changes, and lack of coordination, reproductive problems, and possibly death [25].

There are many techniques for analysis trace metal e.g., atomic absorption, atomic emission, and fluorescence spectrometry [26], inductively coupled plasma-mass spectrometry, (ICP-MS), [27] and electrochemical techniques (such as ion-selective potentiometry and anodic stripping voltammetry) [8]. In contrast of their attractive analytical "figures of merit", each of these techniques suffers of some disadvantages. Applications of all these methods require knowledge of chemistry and instrumentation and need exactitude apparatuses (28-29)

A major disadvantage of ICP-MS is the high capital cost of the instrumentation. In addition, this instrument is bulky and not selective to different charge states of an element. The non linearity of the calibration curves is a disadvantage of the atomic absorption spectroscopy, (AAS), technique (in absorbance range higher than 0.5 to 1). Running and investment costs of inductively coupled plasma atomic emission spectroscopy, (ICP-AES), are high. Another disadvantage of ICP-AES is spectral interferences (many emission lines). Voltammetric methods are simple, inexpensive, and portable but they suffer of interferences inherent in complex sample matrix. Anodic stripping voltammetry can only measure amalgam-forming metal species. Thereby, the complementary methods have developed to overcome some of these shortcomings [29-30].

Surface plasmon resonance (SPR) is an effective optical technique to detect monolayer thicknesses of the material on the conducting surface and has been used to study the optical and electrical properties of conducting polymers such as polypyrrole films. This technique exhibited a good sensitivity, stability, and reproducibility and some times the changes in refractive index of approximately 10^{-5} can be detected by SPR technique [31]. Usually a thin layer of gold is used as the SPR excitation layer. This technique was used to detect and characterize deposited organic layers and functionalized polymer films on SPReeta sensors[32], and also in biosensor such as pseudomonas aeruginosa cells [33]. One of the important parameters used for evaluating the quality of biodiesel is copper corrosion and the standard test of copper strip is used for detection this ion. It has been shown [34] that surface Plasmon resonance based on the self-assembly of squarylium dye containing cystamine can be used as sensing area for determining copper concentration.

In our previous work [35], we prepared a sensing layer based on a composite of conducting polymer (polypyrrole) and chitosan for the detection of Cu^{2+}. Different corrosion levels classified as class 1a according to the standard copper strip test (CST) could be recognized in the sample using this optical sensor. In this work four types of biodiesel samples were used to detect different levels of corrosion and it was found that the sensor based on polypyyrole-chitosan can measure different levels of corrosion more precisely than the standard CST. In addition, the variation of reflectivity with thickness was studied in this research and it was observed that the reflectivity monotonically increased with increasing thickness of polymer film.

In addition, the refractive index and resonance angle of the composite film were determined by Feresenel theory and the results are presented in table 1. It was found that the real part of the refractive index decreased and the imaginary part increased with increasing thickness, as expected from the Kramers–Kroning relation [36-37].

Yu *et al.*, [31] developed a SPReeta sensor for Hg(II) detection using polypyrrole (PPy) and 2-mercaptobenzothiazole (2-MBT). Binding interactions of mercury ion with polypyrrole (PPy) and (2-MBT) was monitored by surface plasmon resonance (SPR) technique. They used 2-MBT as chelating agent to forms complexes with Hg(II) in order to enhance the specificity of Hg^{2+} determination. The PPy-modified SPReeta surface area was prepared by in situ pyrrole electropolymerization on the gold surface at a constant potential of +0.85V for 100–150 s. A solution of 2-MBT in ethanol/water (1:9 v/v) was employed to serve as a sensitivity enhancement agent for detecting Hg^{2+} bound on the PPy surface. They reported the detection limit of 0.01 ppm with an increase of SPR angle of 20±10 RU.

Thickness t (nm)	Resonance angle θ_R (deg)	Real part ε_{real}	Imaginary Part ($\varepsilon_{imaginary}$)	Reflectivity R
20.8	58.224	1.6654	0.153	0.260
31.8	60.726	1.6321	0.169	0.381
40.2	60.896	1.6035	0.186	0.458
49.7	62.818	1.5812	0.200	0.520
58.6	63.336	1.5661	0.230	0.579

Table 1. Refractive index and resonance angle of PPy–CHI film.The refractive index and the thickness of the gold layer were 0.235+3.31*i* and 49 nm, respectively.

5. SPR Detection of Mercury and Lead Ions

5.1. Set up configuration and Formulation

The aim of this research was to fabricate a sensor based on surface plasmon resonance (SPR) using conducting polymer to detect trace amounts of mercury and lead ions. SPR characterisation of the thin films was performed by using a set up assembled according to the Kretschmann's prism configuration (Figure 1). Data analysis was done by Matlab software using Fresnel formula .The He-Ne laser (with wavelength of 632.8nm) was used as light source and silicon detector (Newport, 818-SL) and Lock-In-Amplifier were used for detection the light. The sensitivity of amplifier was 20 mV and the chopper frequency was 125 Hz.

The values of optical constants were calculated by comparing the experimental SPR data to the theoretical values provided by the Fresnel theory applied to the SPR optical configuration used for our study [38]. The values of the real and imaginary parts of the refractive indexes of PPy and PPy-CHI films were obtained by nonlinear least square fitting using Fresnel equations to the three-layer of SPR system.

In the Fresnel equation for p-polarized light the value of reflection coefficient can be written as:

$$r = \frac{E_i}{E_r} \tag{5}$$

where E_i and E_r are the incident and reflected electrical fields, respectively [38]. At the interfaces between N layers, the reflection coefficient is as follow:

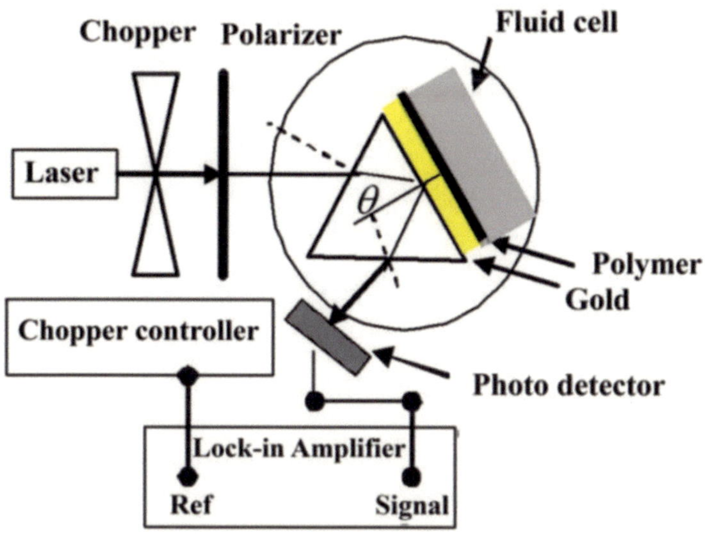

Figure 1. Set up configuration for SPR measurements.

$$r = \frac{m_{21} + m_{22}\tau_2 - m_{11}\tau_0 - m_{12}\tau_2\tau_0}{m_{21} + m_{22}\tau_2 + m_{11}\tau_0 + m_{12}\tau_2\tau_0} \tag{6}$$

where m_{ij} are the matrix layer elements and for single layer with the thickness t, the matrix element is

$$M = \begin{bmatrix} \cos\delta & -i\frac{\sin\delta}{\tau_1} \\ -i\tau_1 \sin\delta & \cos\delta \end{bmatrix} \tag{7}$$

Where the δ (the phase shift due to the beam passing through different layer) and τ_1 can be written as:

$$\delta = \frac{2\pi}{\lambda} t n_1 \cos\theta_{t_1} \tag{8}$$

$$\tau_1 = \frac{n_1}{\cos\theta_{t_N}} \sqrt{\varepsilon_0 \mu_0} \tag{9}$$

Where λ ε_0 μ_0 are the wavelength of light used during the measurements, dielectric constants, and permeability. If the refractive index of gold and the sample obey following relation, resonance takes place [13].

$$n_p \sin\theta_R = \sqrt{\frac{n_m^2 n_d^2}{n_m^2 + n_d^2}} \tag{10}$$

Where θ_R, n_p, n_m, and n_d are the resonance angle, refractive indices of the prism, metal (gold layer) and dielectric medium (sample), respectively and the refractive index of the sample can be determined from:

$$n_d = \sqrt{\frac{n_m^2 n_p^2 \sin^2 \theta_R}{n_m^2 - n_p^2 \sin^2 \theta_R}} \tag{11}$$

The angle of incidence at the interface between the prism and the gold layer is obtained from following Equation [39]:

$$\theta_2 = A - \arcsin\left(\left(n_{air} / n_p\right)\sin\theta_1\right) \tag{12}$$

Where A is the angle of the prism, θ_1 is the angle of incidence of the light beam directed to the prism, and n_{air} is the refractive index of air.The refractive index and resonance angle of the sample will be found by minimizing the sum [40]

$$\Gamma = \sum_{\theta}\left[R_{Exp}(\theta_2, n_2) - R_{Theory}(\theta_2, n_2)\right] \tag{13}$$

Where R_{Exp} and R_{Theory} are the experimental and theoretical, reflectivity.

In the Fresnel equation the reflectivity is a function of θ (angle of incidence light at the interface between the prism and surface), wavelength of light, refractive index of sample as well as refractive index and thickness of gold layer [41].

5.2. Pyrrole electropolymerization

A thin layer of gold (49 nm) was deposited onto a glass microscope slide and electrochemical deposition of PPy-CHI and PPy films on the gold substrate were performed potentiostatically imposing a constant potential of 0.85 V (vs SCE) for 150 sec. In order to form a thin film on the gold substrate with a homogeneous surface, electrodeposition should be done at low potential in short time. The sample of PPy-CHI was prepared in a solution containing 0.015 M Pyrrole, 0.005 M P-TS and 0.07% (w/v) chitosan in acetic acid. The results from FT-IR, electrical conductivity, and XRD study confirmed the incorporation of chitosan in PPy structure [42].

The PPy film was prepared using the same composition of pyrrole and P-TS without chitosan. The surface of the polymer films was washed with distilled deionized water (DDW). The thin slides were attached to the prism with high index glass (ZF52, n=1.85) by using index-matching fluid (n=1.52 new port company). The precision of rotator stage was 0.003. The refractive indexes of PPy and PPy-CHI thin films (in contact with air) by SPR technique were obtained in our previous work [43]. Since they produced a sharp peak of resonance angle, it was shown these polymers are capable for using in sensitive optical sensors.

DDW was run on the surface of the films to establish a baseline. When a stable baseline of SPR angle was obtained, the PPy and PPy-CHI film exhibited a SPR angular profiles that was deep enough for precise definition of the resonance angle. The values of the real and imaginary parts of the refractive indexes of PPy and PPy-CHI films were obtained by nonlinear least square fitting using Fresnel equations to the three-layer of SPR system.

The resonance of the Au-PPy thin film occurred at larger angles compared to the Au, which can be ascribed to a change in the actual refractive index of the gold layer when the PPy film is deposited on it. Figure 2 shows the surface plasmon resonance (SPR) reflectivity curves for the glass/Au/DDW, (Au), and glass/Au/PPy/DDW, (Au-PPy), and glass/Au/PPy-CHI/DDW, (Au-PPy-CHI), systems.

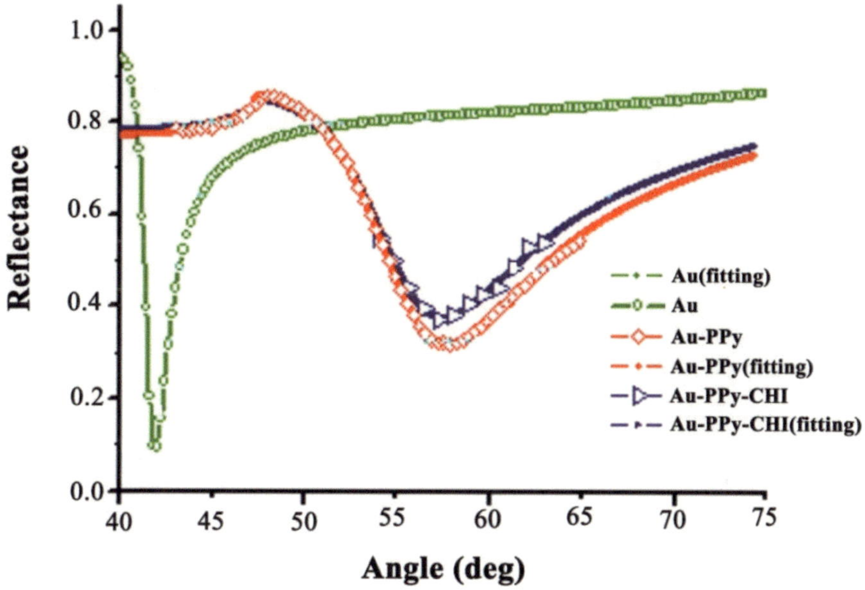

Figure 2. SPR angular profiles for DDW on the Au, Au-PPy and Au- PPy-CHI surface.

The resonance angle (θ_{spr}) and the optical constant for the PPy film were $(57.67\pm0.01)^0$ and 1.598+0.159i, respectively. SPR angle for the PPy-CHI film appeared in lower value indicating formation of a thin film of polypyrrole with chitosan on the gold surface. The resonance angle (θ_{spr}) and the optical constant of the PPy-CHI film were $(56.75\pm0.01)^0$ and 1.601+0.165i, respectively. The dip widths of the curves for the PPy-CHI and PPy films were larger than Au/DDW, which was proportional to the attenuation of the surface plasmon. The thickness of the thin films was calculated using the Fresnel equation and the values of 26 nm and 28 nm were obtained for the PPy-CHI and PPy thin films, respectively. When a stable baseline of SPR angle was obtained for PPy and PPy-CHI films, the surface of the films were ready for binding tests for Hg^{2+} and Pb^{2+} samples.

5.3. Binding of Hg^{2+} and Pb^{2+} on to PPy and PPy-CHI films

Different concentrations of Hg^{2+} and Pb^{2+} in aqueous solution were injected in to the fluid cell attached to the films, and the SPR angle (resonance units = RU) was monitored to detect any binding interactions. The increase in the SPR angle (ΔRU) for each Hg^{2+} and Pb^{2+} concentration was determined based on the resonance angle of the ions (after the ion binding event) and the DDW baselines before running ions samples over a time duration of Δt.

These ΔRU results indicated that Hg^{2+} and Pb^{2+} were binding successfully with the PPy and PPy-CHI. Each sample was run for 20 min. The concentrations of ions in the parts per million range produced the changes in the SPR angle minimum in the region of 0.03 to 0.07. Due to the high sensitivity of SPR, it also is possible to detect concentrations in the parts per billion ranges. A control experiment for the Hg^{2+} and Pb^{2+} ions with bare Au surface was done for 30 min, which produced no changes in the resonance angle indicating that the polymer was necessary to react with the ions. After exposure to each sample concentration, all of the PPy and PPy-CHI films were washed with DDW to establish a baseline, as the history of the films was irrelevant.

5.4. Sensitive optical sensor for Hg^{2+} and Pb^{2+}detection

The increase in the SPR angle for each Hg^{2+} and Pb^{2+} concentration indicated that these ions were binding successfully with PPy. The typical sensograms for 0.5–12 ppm Hg^{2+} and Pb^{2+} binding on PPy are presented in Figure 3 and 4.

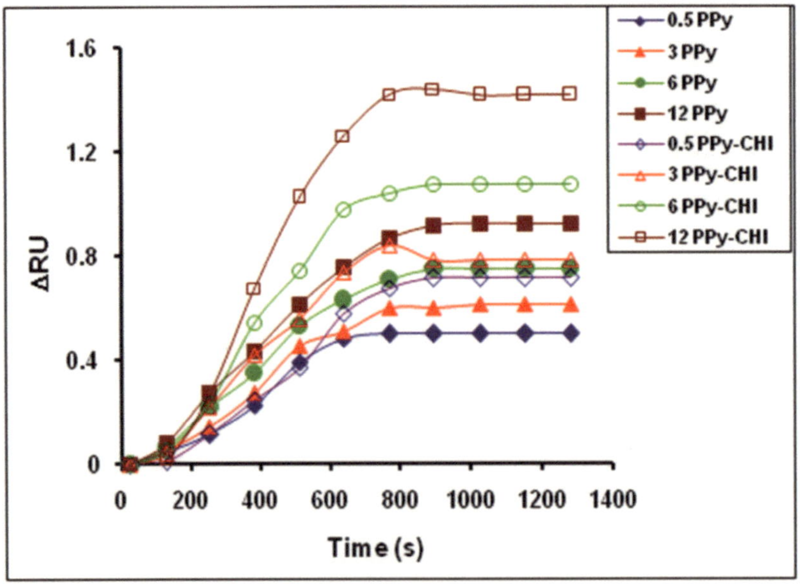

Figure 3. The sensograms for 0.5–12 ppm Hg^{2+} binding on PPy and PPy-CHI surface. Filled symbols (PPy surface), Open symbols (PPy-CHI surface).

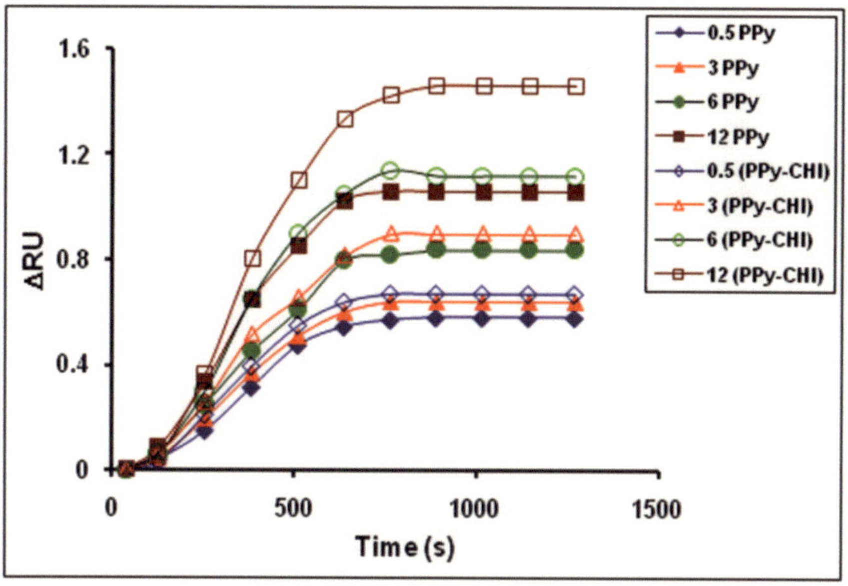

Figure 4. The sensograms for 0.5–12 ppm Pb^{2+} binding on PPy and PPy-CHI surface. Filled symbols (PPy surface), Open symbols (PPy-CHI surface).

It can be observed that, first, ΔRU increased over the time for each sample, but after almost 700 sec the graphs showed a constant value. This result can be explained by saturation of the binding sites available on the PPy surface. It seems that there is a chemical binding of ions in water with the PPy thin film that was immobilized on the gold surface. The interaction between Hg^{2+} and PPy was also reported by Yu et al. [31], as the natural preference for a combination of soft acid (Hg^{2+}) and soft base (PPy). On the other hand, there are some studies on the treatment of solutions of metal ions with redox polymers, such as polypyrroles, polyanilines and polythiophenes, to reduce the ions to a lower valence [44]. These literatures show that PPy can interact with certain heavy metals through an acid and base interaction or an oxidation reaction.

5.5. Sensitivity enhancement using CHI

Chitosan was used to prepare the PPy-CHI composite film to function as a sensitivity enhancement agent for detecting for heavy metals ions. DDW was run to establish a baseline. The decreasing SPR angle for the PPy-CHI film when compared to the PPy film indicated that chitosan incorporated inside the PPy structure. The sensograms for 0.5–12 ppm Hg^{2+} and Pb^{2+} binding on the PPy- CHI are also presented in Figure 3 and 4. The composite films of PPy- CHI showed the higher ΔRU compared to the PPy film for each time and concentration, indicating a higher sensitivity of the PPy in the presence of chitosan. In fact, the natural biopolymers have adsorption properties toward certain heavy metals. Of the many absorbents identified, chitosan has the highest sorption capacity for several metal ions [45]. The sorption proceeds by electrostatic attraction on protonated amine groups in acidic solutions resulted in an enhancement of the original ΔRU and improvement of sensitivity of sensor.

In order to show the sensitivity of the PPy and PPy-CHI films for metal ions, standard calibration curves were plotted for sensors and are shown in Figure 5 and 6. Different concentrations of ions were run for 20 min and all ΔRU were taken from the first 13 min of each binding curves. Each concentration was repeated 3 times and the standard deviation data were in the region of 0.06 to 0.18 and it is shown as error bar on the diagram of sensitivity. These figures clearly show the enhancement of sensitivity for sample determination in the presence of chitosan. In the both sensograms the SPR dip shift is directly correlated with the ions concentration ($[Hg^{2+}]$ or $[Pb^{2+}]$). Linear regression analysis of the sensors yielded the following equations with all correlation coefficients R^2 greater than 0.97, revealing good linearity for the relationship between the SPR dip shift and ions concentration:

$$\Delta\theta = 6.50 \times 10^{-1} + 6.41 \times 10^{-2} [Hg^{2+}] \text{ (PPy-CHI)}$$

$$\Delta\theta = 4.97 \times 10^{-1} + 3.69 \times 10^{-2} [Hg^{2+}] \text{ (PPy)}$$

$$\Delta\theta = 6.75 \times 10^{-1} + 6.77 \times 10^{-2} [Pb^{2+}] \text{ (PPy-CHI)}$$

$$\Delta\theta = 5.50 \times 10^{-1} + 4.24 \times 10^{-2} [Pb^{2+}] \text{ (PPy)}$$

From standard calibration curve, it can be seen that the sensitivity of PPy-CHI sensor was almost 2 times higher than that for the PPy sensor.

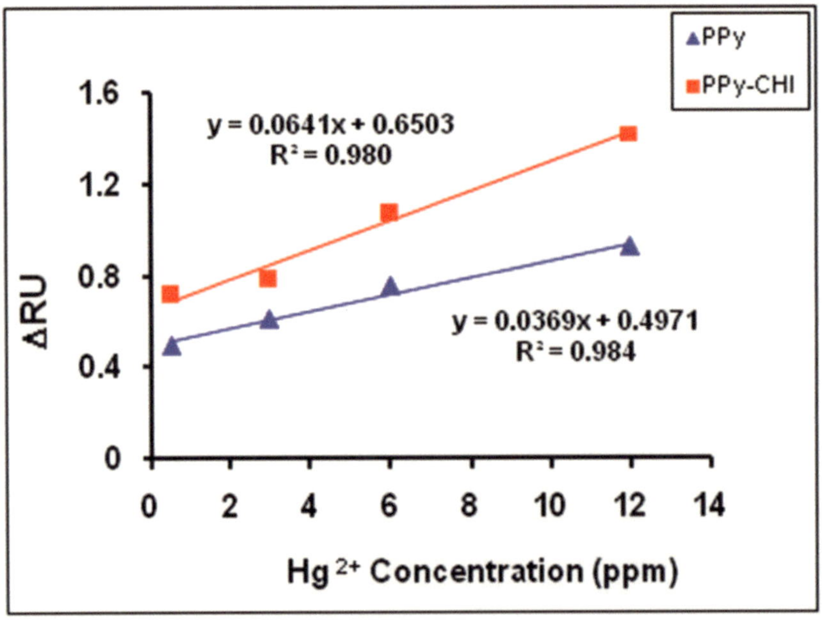

Figure 5. Standard calibration curves of ΔRU versus Hg^{2+} ion concentration for PPy and PPy-CHI films.

It can be seen that all the standard calibration curves showed a significant deviation from 0. This deviation was due to saturatation of the binding sites available on the polymer surface, a fundamental limitation of maximum uptake capacity [46].

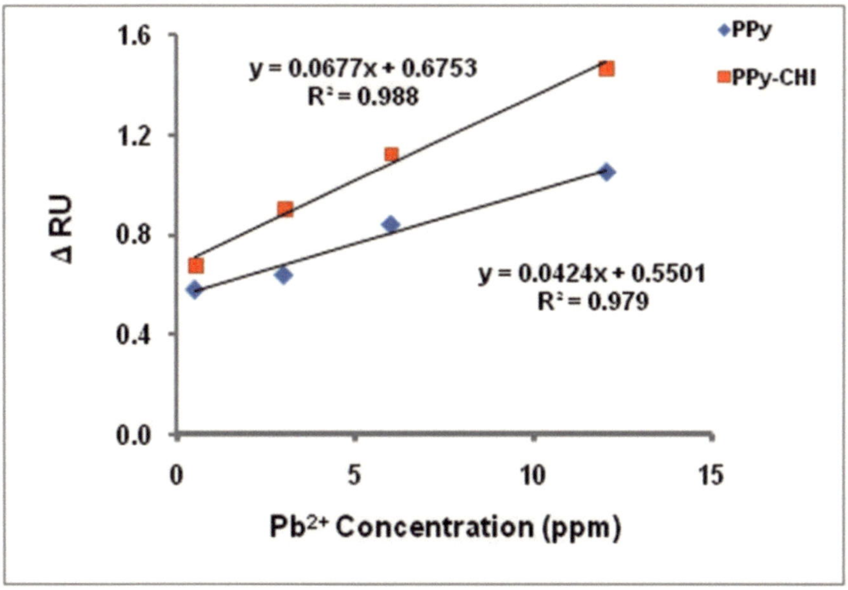

Figure 6. Standard calibration curves of ΔRU versus Pb^{2+} ion concentration for PPy and PPy-CHI films.

6. Conclusion

Application of biopolymers such as chitosan, as an electrical or optical material has rarely been reported. However, using of these materials in electrical devices are now of interest because of the current environmental threat and societal concern. Chitosan as a biodegradable, biocompatible, nontoxic and low-cost polymer was found to improve the electrical and mechanical properties of the PPy film. The changes in the electrochemical and optical properties of the polypyrrole films are due to the changes in the real and imaginary parts of the complex dielectric constant when is combined with chitosan.

Chitosan as natural biopolymers shows a high sorption capacity for several metal ions. Following the interaction of CHI and metal ions to enhance the original ΔRU, lower concentrations of ions could be readily detected with less interference. The aim of this study was to prepare a sensor area based on PPy-CHI composite films for detecting Mercury and Lead ions. The refractive indexes of PPy-CHI were successfully measured by surface plasmon resonance (SPR) technique. This technique was used to fabricate the PPy and PPy-CHI optical sensors for monitoring trace amount of Hg^{2+} and Pb^{2+}. Specific binding of chitosan with Hg^{2+} and Pb^{2+} ions were presented as an increase in resonance angles. Pb^{2+} ion showed higher value for ΔRU due to stronger biding with polymer and the sensors based on PPy and PPy-CHI were more sensitive to Pb^{2+} compared to the Hg^{2+}.

The research on the preparation and characterization of conducting polymers is still continuing in order to find conducting polymers with improved electrical and mechanical properties. Since chitosan could improve the electrical and optical properties of the PPy film, further research could be extended to prepare the composite films using other biopolymer such as Nano cellulose with other conducting polymers.

Author details

Mahnaz M. Abdi
Deprtemet of Chemistry, Faculty of Science, Universiti Putra Malaysia, Serdang, Selangor, Malaysia
Institute of Tropical Forestry and Forest Products (INTROP), Universiti Putra Malaysia, Serdang, Selangor, Malaysia

Wan Mahmood Mat Yunus
Department of Physics, Faculty of Science, Universiti Putra Malaysia, Serdang, Selangor, Malaysia

Majid Reayi
School of Chemical Sciences and Food Technology, Faculty of Science and Technology, Universiti Kebangsaan Malaysia, Bangi, Selangor D.E., Malaysia

Afarin Bahrami
Department of Physics, Faculty of Science, Universiti Putra Malaysia, Serdang, Selangor, Malaysia
Department of Physics, Faculty of Science, Islamic Azad University, Eslamshahr Branch, Iran

Acknowledgement

The authors would like to thank the Ministry of Higher Education of Malaysia for financial supporting (as Graduate Research Fellowship). We also gratefully acknowledge the Department of Chemistry and Physics, University Putra Malaysia for their continued support throughout this project.

7. References

[1] Hadziioannou G., Malliaras G.G. (2007) Semiconducting polymers, chemistry, physics and engineering; 1, ed.; WILEY-VCH.
[2] Nicolas M., Faber B., Simonet J. (2001) Electrochemical sensing of F^- and Cl^- with a boronic ester-functionalized polypyrrole. J. Electroanal. Chem 509: 73-79.
[3] Migahed M D., Fahmy T., Ishra M., Barakat A (2004) Preparation, characterization, and electrical conductivity of polypyrrole composite films. Polym Test 23: 361–365.
[4] Baba A., Lubben J., Tamada K., Knoll W. (2003) Optical properties of ultrathin poly(3,4-ethylenedioxythiophene) films at several doping levels studied by in situ electrochemical surface plasmon resonance spectroscopy. Langmuir 19: 9058-9064.
[5] Damos FS, Luz RCS, Kubota LT (2006) Investigations of ultrathin polypyrrole films: Formation and effects of doping/dedoping processes on its optical properties by electrochemical surface plasmon resonance (ESPR). Electrochim Acta 51: 304–1312.
[6] Barisci JN, Wallace GG, Andrews MK, Partridge AC, Harris PD (2002) Conducting polymer sensors for monitoring aromatic hydrocarbons using an electronic nose. Sens. Actuators B: Chem. 84: 252–257.
[7] Souza JEG, Santos FL, Neto BB, Santos CG, Santos MVB, et al. (2003) Freegrown polypyrrole thin films as aroma sensors. Sens. Actuators B: Chem. 88: 246–259.
[8] Mahmud HNME., Kassim A., Zainal Z., Mat Yunus WM (2007) Fouriertransform infrared study of polypyrrole-poly (vinyl alcohol) conducting polymer composite films: Evidence of film formation and characterization. J. Appl. Polym Sci 100 (5): 4107-4113.

[9] Costa ACR., Siqueira, AF (1996) Thermal diffusivity of conducting polypyrrole. J.Appl. Phys 80: 5579-5580
[10] Kim MS., Kim HK., Byun SW., Jeong SH., Hong YK., Joo JS., Song KT., Kim .K., Lee CJ., Lee JY (2002) PET fabric/polypyrrole composite with high electrical conductivity for EMI shielding. *Synth. Met* 126: 233–239.
[11] Kim H K., Kim M S., Chun S Y., Park YH., Jeon BS., Lee JY (2003) Characteristics of electrically conducting polymercoated textiles. Mol. Cryst. Liq. Crys 40: 161–169.
[12] Baba A., Advincula R.C., Knoll W (2002) In situ investigations on the electrochemical polymerization and properties of polyaniline thin films by surface plasmon optical techniques. J Phys Chem. B 106: 1581-1587.
[13] Homola J., (2006) Surface Plasmon Resonance Based sensors; ed.; Springer: Berlin Heidelberg.
[14] Peyghambarian N, Koch SW, Mysyrowicz, A (1993) Introduction to semiconductor optics. Cliffs E, ed. Prentice Hall: Newjersey.
[15] Garland JE, Assiongbon KA, Pettit CM, Roy D (2003) Surface Plasmon resonance transients at an electrochemical Interface: Time resolved measurements using a bicell photodiode. Anal Chim Acta 475: 47–58.
[16] Wang, S., Boussaad S., Wong S., Tao, N.J (2000) High-Sensitivity Stark Spectroscopy Obtained by Surface Plasmon Resonance Measurement. Anal. Chem 72: 4003-4008.
[17] Bredas JL., Scott JC., Yakushi K., Street GB (1984) polarons and bipolarons in polypyrrole: evolution of the band structure and optical spectrum upon doping. Phys. Rev. B 30: 1023-1025.
[18] Abdi MM., Mahmud HNME., Abdullah LC., Kassim A, Rhman MZ., Chyi JLY (2012) Optical band gap and conductivity measurement of polypyrrole-chitosan composite thin film. Chinese Journal of Polymer Science 30(1): 93-100
[19] Jorn CC Yu., Edward PC Lai (2004) Polypyrrole film on miniaturized surface Plasmon resonance sensor for ochratoxin A detection. Synthetic Metals 143: 253–258.
[20] Guedon P., Livache T., Martin F,. Lesbre F., Roget A., Bidan G., Levy Y (2000) Characterization and Optimization of a Real-Time, Parallel, Label-Free, Polypyrrole-Based DNA Sensor by Surface Plasmon Resonance Imaging. Anal. Chem. 72: 6003–6009.
[21] Thiel AJ., Frutos AG., Jordan C., Corn R., Smith L (1997) In situ surface Plasmon resonance imaging detection of DNA hybridization to oligonucleotide arrays on gold surface. Anal. Chem.69: 4948-4956.
[22] Zizlsperger M., Knoll W (1998) Multispot Parallel on-Line Monitoring of Interfacial Binding Reactions by Surface Plasmon Microscopy. Prog. Colloid Polym. Sci. 109: 244- 253.
[23] Sathawara NG, Parikh DJ, Agarwal YK (2004) Essential Heavy Metals in Environmental Samples from Western India. Bulletin of Environmental Contamination and Toxicology 73: 756–761.
[24] Pearce JM (2007) Burton's line in lead poisoning. European neurology 57(2): 118–9.
[25] Clifton JC (2007) Mercury exposure and public health. Pediatr Clin North Am 54 (2): 237–69.
[26] Skoog DA, Holler FJ, Nieman TA (1992) Principles of Instrumental Analysis. 5th ed. Philadelphia: Saunders College Publishing. pp 206–229.
[27] Taylor HE (2001) Inductively Coupled Plasma Mass Spectrometry: Practices and Techniques, Academic Press, San Diego.

[28] Tang Y, Zhai YF, Xiang JJ, Wang H, Liu B, et al. (2010) Colloidal gold probe based immunochromatographic assay for the rapid detection of lead ions in water samples. Environmental Pollution 158: 2074–2077.

[29] Bard AJ., Faulkner LR (2000) Electrochemical Methods: Fundamentals and Applications, 2nd ed, John Wiley & Sons, New York. pp. 458–464.

[30] Zhang Y, Xu M, Wang , Toledo F, Zhoua F (2007) Studies of metal ions binding by apo-metallothioneins attached onto preformed self-assembled monolayers using a highly sensitive surface plasmon resonance spectrometer Sensors and Actuators B. 123: 784–792.

[31] Yu JCC, Lai EPC, Sadeghi S (2004) Surface plasmon resonance sensor for Hg(II) detection by binding interactions with polypyrrole and mercaptobenzothiazole. Sens. Actuators B 101: 236–241.

[32] Matejka P, Hruby P, Volka K (2003) Surface plasmon resonance and Raman scattering effects studied for layers deposited on Spreeta sensors, Anal. Bioanal. Chem. 375: 1240–1245.

[33] Park JS, Lee CM, Lee KY (2007) A surface plasmon resonance biosensor for detecting Pseudomonas aeruginosa cells with self-assembled chitosan-alginate multilayers. Talanta 72: 859–862.

[34] Kim S H., Han S K., Kim J H., Lee M B., Koh K N., Kang S W (2000)A self-assembled squarylium dye monolayer for the detection of metal ions by surface plasmon resonances Dyes. Pigm. 44: 55-61.

[35] Sadrolhosseini AR., Moskin MM., Mat Yunus WM, Talib ZA., Abdi MM (2011) Surface Plasmon Resonance Detection of Copper Corrosion in Biodiesel Using Polypyrrole–Chitosan Layer Sensor. Optical review 18 (4): 1–7.

[36] Kittle C (1996) introduction to Solid State Physics. Wiley, New York. p. 390.

[37] Azzam RMA., Bashara N M (1997) Ellipsometry and Polarized Light. Elsevier, Amsterdam. p. 269.

[38] Schasfoort RBM, Tudos AJ (2008) Handbook of surface Plasmon resonance, RSC publishing UK: Cambridge. pp 15–25.

[39] [39] Blonder B. Sensing Application of Surface Plasmon Resonance. Available online: http://www.eduprograms.seas.harvard.edu/reu05_papers/Blonder_Benjamin.pdf

[40] Jaaskelainen AJ., Peiponen KE., Raty JA (2001) On reflectometric measurement of a refractive index of milk. *J. Dairy Sci. 84*: 38–43.

[41] Pedrorotti LF, Pedrotti LS, Pedrotti LM (2007) Introduction to optics. 3th ed. New York: Addison Wesley. pp 476–482.

[42] Abdi MM, Mahmud HNME, Kassim A, Mat Yunus WM, Talib ZBMohd, et al. (2010) Synthesis and Characterization of a New Conducting Polymer Composite. Polymer Science Ser B 52: Nos. 11–12, 662–669.

[43] Abdi MM, Kassim A, Mahmud HNME, Mat Yunus WM, Talib ZB, et al. (2009) Physical, optical, and electrical properties of a new conducting polymer. J Mater Sci 44: 3682–3686.

[44] Sanjay B, Krishna R, Chang W (1994) Redox polymer films for metal recovery applications US Patent 5368632.

[45] Nomanbhay SM, Palanisamy K (2005) Removal of heavy metal from industrial wastewater using chitosan coated oil palm shell charcoal. Electron j biotechn 8: 44–53.

[46] Zhang X, Bai R (2002) Adsorption behavior of humic acid onto polypyrrolecoated nylon 6,6 granules. J Mater Chem 12: 2733–2739. SPR Sensor Based on PPy-CHI

Chapter 17

Innovative Exploitation of Grating-Coupled Surface Plasmon Resonance for Sensing

G. Ruffato, G. Zacco and F. Romanato

Additional information is available at the end of the chapter

http://dx.doi.org/10.5772/51044

1. Introduction

Surface Plasmon Polaritons (SPPs) are confined solutions of Maxwell's equations which propagate at the interface between a metal and a dielectric medium and have origin from the coupling of the electromagnetic field with electron-plasma density oscillations inside the metal [1]. SPPs are localized in the direction perpendicular to the interface: field intensity decays exponentially from the surface with an extension length of the same order of the wavelength inside the dielectric and almost one order shorter in the metal [2]. These features make SPPs extremely sensitive to optical and geometrical properties of the supporting interface, such as shape, roughness and refractive indices of the facing media. Since these modes have a non-radiative nature, the excitation by means of a wave illuminating the metallic surface is possible only in the configurations providing the wavevector-matching between the incident light and SPP dispersion law (Surface Plasmon Resonance – SPR, see Figure 1). Prism-Coupling SPR (PCSPR) exploits a prism in order to properly increase incident light momentum and achieve SPP excitation, however this setup suffers from cumbersome prism alignment and it is not suitable for miniaturization and integration. A more amenable and cheaper solution consists in Grating-Coupling SPR (GCSPR), where the metal surface is modulated with a periodic corrugation. The plasmonic behaviour of these modulated metallic surfaces had been discovered since the early years of the last century by R.W. Wood [3] and the connection between Wood's anomalies and surface plasmons was finally established by J.J. Cowan and E.T. Arakawa [4]. A plane-wave illuminating the patterned area is diffracted by the periodic structure and it is possible for at least one of the diffracted orders to couple with SPP modes.

SPR has known an increasing interest in the design and realization of miniaturized label-free sensing devices based on plasmonic platforms. Surface plasmon modes in fact are extremely sensitive to changes in the refractive index of the facing dielectric medium: a thin coating film or the flowing of a liquid solution alter SPP dispersion curve and cause resonance conditions to change (Figure 2). Thus it is possible to detect refractive index variations by simply analyzing the resonance shift: it is the basis of modern SPR-sensing devices [5]. SPR reveals itself as a useful instrument for the study of surface optical properties and it is a highly suitable candidate as application with sensing purposes in a large variety of fields: environmental protection, biotechnology, medical diagnostics, drug screening, food safety and security.

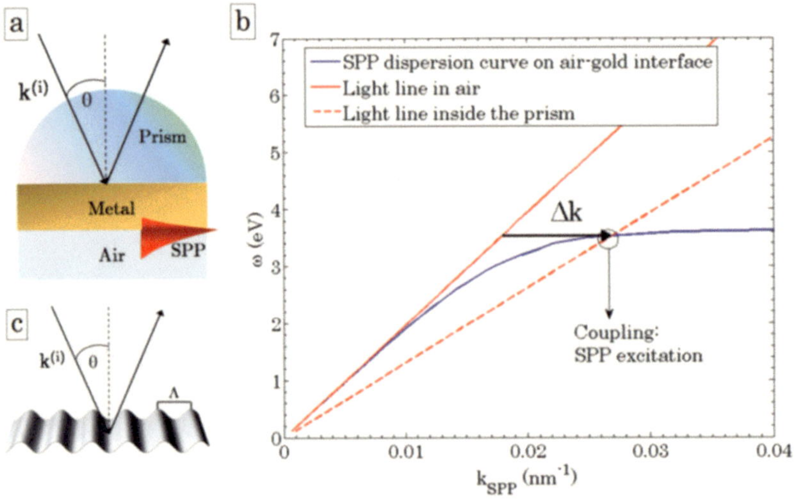

Figure 1. a) Kretschmann SPR configuration. b) SPP excitation with Prism-Coupling. The same effect of momentum-supply can be achieved by corrugating the metallic surface in the so-called Grating-Coupled SPR (c).

Currently, several groups are using different SPR approaches to detect the change of refractive index. Sensors using Prism-Coupled SPR (PCSPR) with Kretschmann configuration [6] can be readily combined with any type of interrogation: angular, wavelength, intensity or phase modulation [7]. PCSPR typically show refractive index sensitivity for angular interrogation architecture that ranges between 50-150°/RIU [8], with higher sensitivity at shorter wavelengths [9], and refractive index resolutions in the orders 10^{-6}-10^{-7} Refractive Index Units (RIU). However, PCSPR sensors suffer from cumbersome optical alignment and are not amenable to miniaturization and integration [10].

Grating-Coupled SPR (GCSPR) sensors instead, with either wavelength or angular interrogation, have been demonstrated to have sensitivity 2-3 times lower than PCSPR [11]. However, GCSPR has the intrinsic possibility to be used with different sensing architectures and interrogation systems. A parallel SPR angular detection was shown by Unfrict et al. to have the possibility for multi-detection for proteomic multi-array [12]. Homola's group demonstrated how a miniaturized GCSPR sensor implemented with a CCD allowed detection sensitivity of 50°/RIU and resolution of $5 \cdot 10^{-6}$ over 200 sensing parallel channels [13]. Alleyne has exploited the generation of an optical band gap by using prism-coupled to achieve sensitivity up to 680°/RIU by bandgap-assisted GCSPR [14]. A recent approach was reported by Telezhnikova and Homola [15] with the development of a sensor based on spectroscopy of SPPs down to $5 \cdot 10^{-7}$ RIU. The strong compatibility of gratings with mass production makes SPP couplers extremely attractive for fabrication of low-cost SPR platforms .

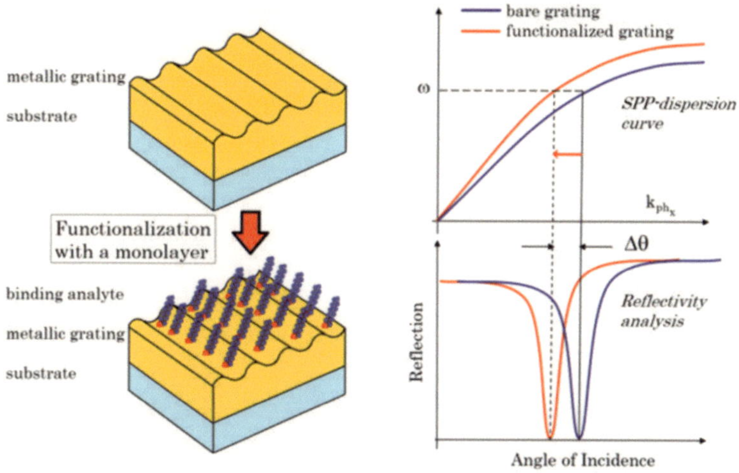

Figure 2. Surface Plasmon Resonance for sensing: the deposition of the analyte on the SPP-supporting surface results in a change of the dispersion curve and a consequent shift of the resonance dip.

Recently we have experimentally and theoretically described the effects of grating azimuthal rotation on surface plasmon excitation and propagation. More SPPs can be supported with the same illuminating wavelength [16] and sensitivity up to 1000°/RIU is achievable for the second dip in angular interrogation [17], which is one order of magnitude greater than that in a conventional configuration. On the top of that, the symmetry breaking after grating rotation makes polarization have a fundamental role on surface plasmon polaritons excitation [18] and the incident polarization must be properly tuned in order to optimize the coupling strength. This result leads to the possibility of exploiting the dependence of the optimal polarization on the resonance conditions, in order to design an innovative GCSPR configuration based on polarization interrogation in the conical mounting.

2. Nanofabrication of plasmonic gratings

2.1. Interferential lithography

Interferential Lithography (IL) is the preferred method for the fabrication of periodic patterns that must be spatially coherent over large areas [19]. It is a conceptually simple process where two coherent beams interfere in order to produce a standing wave which can be recorded over a sensitive substrate. The spatial period of the pattern can be as low as half the wavelength of the interfering light, allowing for structures down to about 100 nm using UV radiation. In Lloyd's Mirror Interferometer, one source is replaced by a mirror. Lloyd's mirror is rigidly fixed perpendicular to the surface and used to reflect a portion of incident wavefront back to the other half. The angle of interference and thus the grating periodicity are set just by rotating the mirror/substrate assembly around the point of intersection between the mirror and the substrate. Simple trigonometry guarantees that the light reflected off the mirror is always incident at the same angle as the original beam. Moreover since the mirror is in a rigid connection with the wafer chuck, vibration of the setup or wandering of the incoming beam do not affect the exposure: this results in a more stable configuration and prevents the need of phase locking systems. Furthermore, it allows to change and control the grating period without any alignment or critical adjustment between two different sources and thus the system is very convenient to calibrate and tune.

The basic principles in order to record a pattern of desired period and amplitude are the following:

- *Grating period*: the incident beam angle has to be properly adjusted in order to obtain the desired periodicity according to the following law:

$$\Lambda = \frac{\lambda}{2\sin\theta} \tag{1}$$

where Λ is the resulting spatial periodicity of the grating, λ the beam wavelength and θ the incident beam angle (see Figure 3).

- *Grating amplitude*: amplitude is controlled by varying the exposure dose. Keeping fixed the beam intensity the exposure dose can be varied by changing the exposure time. The higher the dose on the resist, the higher the amplitude that is obtained after resist development.

In our system a 50 mW Helium-Cadmium (HeCd) laser emitting TEM00 single mode at 325 nm was used as light source. After a 2 m long free-space propagation, the expanded laser beam illuminates both the sample and the perpendicular mirror. The designed sample holder offers translational and rotational degrees of freedom, while a rigid mechanical connection between the mirror and the sample-chuck prevents phase distortion. The possibility to translate the sample stage in two directions allows a fine positioning of the system in the zone where the Gaussian beam distribution reaches its maximum and the best conditions in terms of beam intensity, uniformity and spatial coherence are achieved. On the other hand the sample stage rotation around vertical axis, with 8 mrad resolution, allows a fine setting of fringes periodicity.

Metallic gratings can be processed through a sequence of steps [20]: resist spinning, IL exposure, resist devolpment, master replication, methal evaporation over the replica. Exposures can be performed over silicon samples of 2 x 2 cm^2 surface area. Silicon wafers have to be cleaned and pre-baked for 25 minutes at 120°C. A bottom anti-reflection layer can be also spun by spin coating in order to improve the exposure quality. Thus, a 120 nm thick film of photoresist S1805 (Microposit, Shipley European Limited, U.K.) and Propylene glycol monomethyl ether-1,2-acetate (PGMEA) solution (ratio 2:3) can be spun at the spinning rate of 3000 RPM for 30 seconds. After the exposure, a developing solution of MICROPOSIT MF-321 or MF-319 can be used.

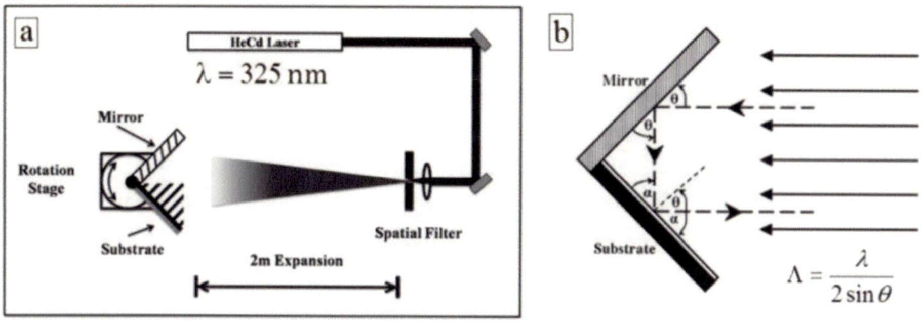

Figure 3. a) Scheme of the experimental IL setup. b) Detail of the sample holder configuration during the exposure.

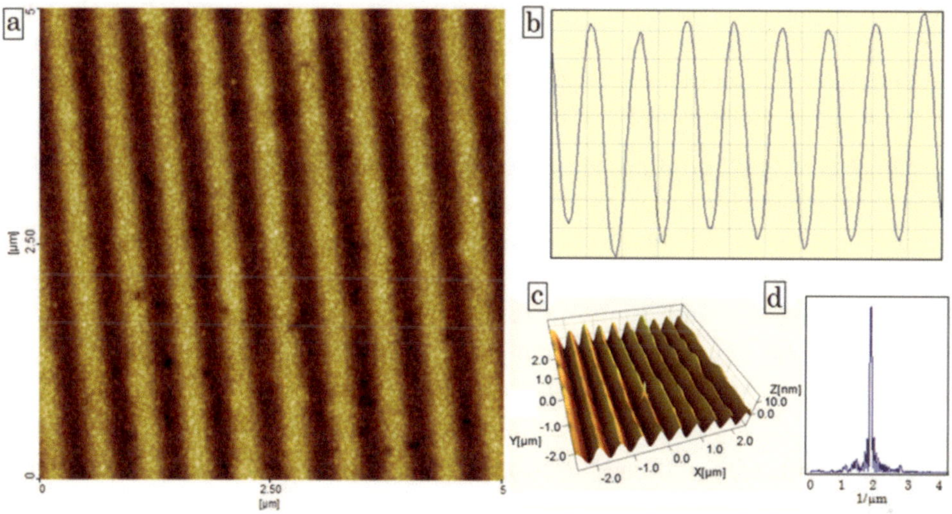

Figure 4. AFM analysis of a gold sinusoidal grating fabricated by interferential lithography of resist S1805 (period 505 nm, amplitude ~30 nm): top view (a), profile along the grating vector direction (b), 3D reconstruction (c), Fourier transform (d).

After the exposed resist has been developed, the result is a photonic crystal: a dielectric periodic surface which has no plasmonic feature yet. In order to realize a plasmonic crystal, the dielectric grating must be coated with a metallic layer of proper thickness, usually noble metals (silver or gold), for example by thermal evaporation. Common recipes consists in a silver layer with optimized thickness coated by a gold layer in order to prevent oxidation, or as an alternative an optimized gold layer over an adhesion layer of chrumium.

2.2. Soft-lithography replica

Soft-lithography is a useful technique in order to replicate pre-fabricated patterns with a nanometric resolution. It consists in making at first the negative replica of a pattern master onto a siliconic polymer and then in imprinting this pattern onto a photopolymeric substrate that cures when exposed to UV light.

In principle, a plasmonic grating could be obtained, as explained in the previous section, by simply coating a developed resist surface with specific noble metals, such as silver and gold over few nanometers of chromium or titanium as adhesion layers. However, many bio-molecular surface functionalization methods involve the use of organic solvents that may attack the photoresist pattern. To avoid this problem, a replica molding approach can be adopted in order to produce thiolene copies of the gratings [21]. Commercial thiolene resin (Norland Optical Adhesive – NOA61 [22]) can be chosen in order to exploit its relatively good resistance to organic solvents.

Resist patterns are usually replicated onto the thiolene resist film supported on microscope glass slides using polydimethylsiloxane (PDMS) molds. The latter are obtained by replicating the resist grating masters using RTV615 silicone. Base and catalyst of the two component silicone are mixed 10:1 ratio and degassed under vacuum. The PDMS is then cast against the resin masters and cured at 60°C, well below the resist post-exposure bake temperature (115° C) in order to prevent the resist pattern from distortion and after 2 hours the PDMS is peeled off from the resist master.

In order to obtain rigid and stable supports, the PDMS molds are bound to glass slides by exposing the flat backside of the PDMS mold and the glass slide to oxygen plasma before contacting them. The PDMS mold is then used to UV imprint the initial pattern onto a drop of NOA61 resin dripped on top of a glass slide by just slightly pressing the mold onto the liquid resin and exposing it to the UV light (365 nm) of a Hg vapor flood lamp at the proper distance and for enough time. After removing the PDMS mold, the replicated NOA61 grating can be coated with metal layers by thermal evaporation.

3. Grating-coupled Surface Plasmon Resonance in the conical mounting: simulation and results

3.1. Chandezon's method

A rigorous approach is necessary in order to exactly solve the problem of a monochromatic plane wave incident on a patterned surface and to simulate the optical response of such multilayered patterned structures. In the past decades, several numerical methods have been developed in order to compute the optical response of periodically modulated multilayered stacks. Among these algorithms, Chandezon's method (hereafter the C-method) revealed itself as one of the most efficient techniques for a rigorous solution of smooth grating diffraction problem. The algorithm is a curvilinear coordinate modal method by Fourier expansion that has gone through many stages of extension and improvement. The original theory was formulated by J. Chandezon et al. [23][24], for uncoated perfectly conducting gratings in classical mountings. Various author extended the method to conical diffraction gratings [25]. G. Granet et al. [26], T.W. Preist et al. [27] and L. Li et al. [28] allowed the various profile of a stack of gratings to be different from each other, although keeping the same periodicity. Solving the vertical faces case in a simple manner, J.P. Plumey et al. [29] have showed that the method can be applied to overhanging gratings.

In the numerical context, L. Li improved the numerical stability and efficiency of the C-method [30][31].

3.2. The coordinate transformation

The setup is as it follows: the grating profile is described by a differential curve $y = s(x)$, periodic in the x-direction with periodicity Λ. The basic feature of the C-method consists in the choice of a non-orthogonal coordinate system that maps the interfaces between different media to parallel planes:

$$\begin{aligned} u &= y - s(x) \\ v &= x \\ w &= z \end{aligned} \tag{2}$$

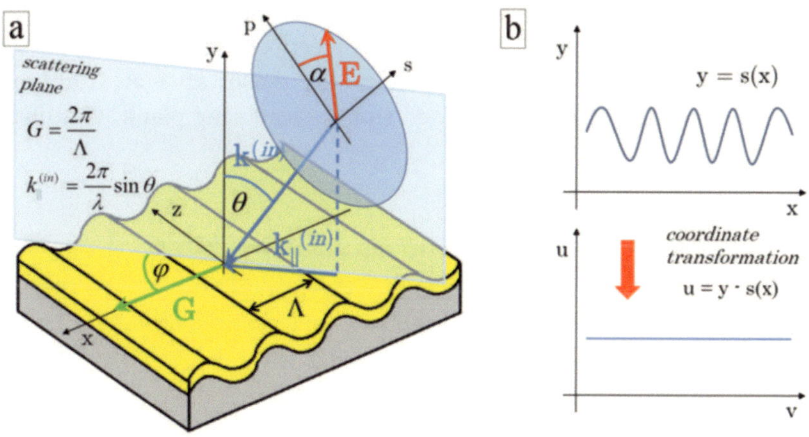

Figure 5. a) Scheme of the reference frame x-y-z and incidence angles θ (polar angle), φ (azimuth) and α (polarization). b) Pictorial description of the coordinate change effect under Chandezon's method.

Since it is a global coordinates transformation, Maxwell's equations covariant formalism is necessary. In a source-free medium the time-harmonic Maxwell equations in term of the covariant field component can be written as [32]

$$\partial_i \sqrt{g} g^{ij} E_j = 0 \tag{3}$$

$$\frac{1}{\sqrt{g}} \varepsilon^{ijk} \partial_j E_k = i\mu \frac{\omega}{c} g^{ij} H_j \tag{4}$$

$$\partial_i \sqrt{g} g^{ij} H_j = 0 \tag{5}$$

$$\frac{1}{\sqrt{g}} \varepsilon^{ijk} \partial_j H_k = -i\varepsilon \frac{\omega}{c} g^{ij} E_j \tag{6}$$

ε^{ijk} being the completely antisymmetric Levi-Civita tensor[1]. The parameter g is the determinant of the contravariant metric tensor g^{ij}, ε and μ are respectively the dielectric permittivity and the magnetic permeability of the medium. The covariant form[2] g_{ij} is the following:

$$g = \begin{pmatrix} 1+s'^2 & s' & 0 \\ s' & 1 & 0 \\ 0 & 0 & 1 \end{pmatrix} \tag{7}$$

In this case $g = 1$, since the coordinate transformation conserves the volume.

The reference frame is the following: the y-axis is perpendicular to the grating plane and the grating vector $\boldsymbol{G}$ is oriented along the x-axis positive direction. The scattering plane is perpendicular to the grating plane and forms an azimuth angle φ with the grating symmetry plane (see Figure 5.a). This reference frame has been chosen in order to simplify the coordinate transformation and the form of the metric tensor after an azimuthal rotation: under this convention, the grating is kept fixed and the scattering plane azimuthally rotates.

The incident wavevector $\boldsymbol{k}^{(in)}$ is given by:

$$\mathrm{k}^{(i)} = \frac{\omega}{c}(\sin\theta\cos\varphi, -\cos\theta, \sin\theta\sin\varphi) = (\alpha_0, -\beta_0, \gamma) \tag{8}$$

and the wavevector of the n-th diffraction order has the following form:

$$\mathrm{k}^{(n)} = (\alpha_n, \beta_n, \gamma) \tag{9}$$

where

$$\begin{aligned} \alpha_n &= \alpha_0 + n \cdot G \\ \beta_n &= \sqrt{k^{(i)2} - \alpha_n^2 - \gamma^2} \end{aligned} \tag{10}$$

With these definitions and after some algebraic manipulations, curl equations can be rearranged into the following system of differential equations in the tangential components (H_x, H_z, E_x, E_z) unknown:

$$\frac{\partial H_x}{\partial u} = i\omega\varepsilon E_z + \frac{\partial}{\partial x}\left(\frac{s'}{1+s'^2} H_x\right) + \frac{\partial}{\partial x}\left(\frac{1}{1+s'^2}\frac{i}{\omega\mu}\frac{\partial E_z}{\partial x} + \frac{1}{1+s'^2}\frac{\gamma}{\omega\mu} E_x\right) \tag{11}$$

---

[1] ε^{ijk} equals 1 if (i,j,k) is an *even* permutation of (1,2,3), -1 if it is an *odd* permutation, 0 if any index is repeated.

[2] $g_{i'j} = \Lambda^{i}_{i'} \Lambda^{j}_{j'} \delta_{ij}$ where δ_{ij} is Kronecker's delta, Λ^{i}_{j} is the Jacobi's matrix associated to the coordinate transformation: $\Lambda^{i}_{j} = \partial x^{i} / \partial x^{j}$. The metric tensor g_{ij} allows calculating the covariant components v_i of a vector $\boldsymbol{v}$ from the contravariant ones v^j and vice-versa: $v_i = g_{ij} v^j$, $v^i = g^{ij} v_j$.

$$\frac{\partial E_z}{\partial u} = \frac{s'}{1+s'^2}\frac{\partial E_z}{\partial x} + \frac{1}{1+s'^2}\frac{\gamma}{\omega\varepsilon}\frac{\partial H_z}{\partial x} + \frac{1}{1+s'^2}\left(i\omega\mu - i\frac{\gamma^2}{\omega\varepsilon}\right)H_x \tag{12}$$

$$\frac{\partial H_z}{\partial u} = \frac{s'}{1+s'^2}\frac{\partial H_z}{\partial x} - \frac{1}{1+s'^2}\frac{\gamma}{\omega\mu}\frac{\partial E_z}{\partial x} - \frac{1}{1+s'^2}\left(i\omega\varepsilon - i\frac{\gamma^2}{\omega\mu}\right)E_x \tag{13}$$

$$\frac{\partial E_x}{\partial u} = -i\omega\mu H_z + \frac{\partial}{\partial x}\left(\frac{s'}{1+s'^2}E_x\right) - \frac{\partial}{\partial x}\left(\frac{1}{1+s'^2}\frac{i}{\omega\varepsilon}\frac{\partial H_z}{\partial x} + \frac{1}{1+s'^2}\frac{\gamma}{\omega\varepsilon}H_x\right) \tag{14}$$

where we used the following relation:

$$\frac{\partial F_i}{\partial z} = i\gamma F_i \tag{15}$$

where F_i may stand for (H_x, H_z, E_x, E_z). In fact, since the grating vector has no components in the *z*-direction, the *z*-component γ of the incident momentum is conserved and is a constant parameter of the problem.

3.3. Numerical solution and boundary conditions

As suggested by Bloch-Floquet's theorem [33], thanks to the periodicity of the media in the *x*-direction, a generic field can be expanded in pseudo-Fourier series:

$$F(x,u,z) = e^{i\gamma z}\sum_{m=-\infty}^{+\infty} F_m(u)e^{i\alpha_m x} \tag{16}$$

where $F_m(u)$ is a function periodic in *x*. Likewise, also the profile functions *C(x)* and *D(x)* are periodic in *x* and can be expanded as well. Laurent's rule can been applied for the Fourier factorization, assuming the continuity of profile function derivative [34]. In the case of sharp edges instead, equations should be rearranged in order to make the inverse-rule factorization applicable [35]. Since in our case of interest the grating profile is a regular function, we can apply Laurent's factorization. After truncation to a finite order *N* [36] , the problem consists in the numerical solution of a system of 8*N*+4 first order differential equations in each medium:

$$-i\frac{\partial \mathrm{U}}{\partial u} = T\mathrm{U} \tag{17}$$

The problem is led to the diagonalization problem of the matrix *T* and the solution $\boldsymbol{U}^j$ in the *j*-th medium can be expressed as a function of eigenvectors $\boldsymbol{V}_q^j$ and eigenvalues λ_q^j of the matrix *T*:

$$\mathrm{U}^j(u) = \sum_q b_q^j \mathrm{V}_q^j e^{i\lambda_q^j u} \tag{18}$$

where b_j^q are the weights of the corresponding eigenmodes in the expansion. Thereafter boundary conditions in each medium must be imposed: continuity of the tangential components at each interfaces and outgoing-wave conditions in the first (air) and last (substrate) media [24]. This leads to a system of 8*N*+4 equations in 8*N*+4 unknowns that is now possible to be solved numerically. Once the mathematical problem has been solved, the electromagnetic fields can be computed inside each *j*-th medium:

$$F^j\left(x,u,z\right)=e^{i\gamma z}\sum_{m=-N}^{+N}\sum_{n=-N}^{+N}F_{mn}^j e^{i\lambda_n^j u}e^{i\alpha_m x} \tag{19}$$

where F^j stands for (H_x, H_z, E_x, E_z), and F^j_{mn} are the *j*-field *mn*-Fourier weight. Transversal components (H_y,E_y) are calculated from Maxwell's equations as a function of the tangential components. By applying the metric tensor to covariant components F_j, the contravariant components F^j, which represent the physical fields, can be obtained. For incident wavelength λ, polar and azimuth angles (θ,φ) and polarization α, the implemented algorithm yields the spatial dependency of the diffracted fields everywhere in space for the modelled grating stack. An estimation of reflection and transmission coefficients for the different diffraction orders, in particular transmittance and reflectivity values (0-diffraction orders) can be obtained. By analysing both real and imaginary parts of eigenvalues λ_q^j, it is possible to distinguish between propagating and evanescent modes, which respectively contribute to the far-field and the near-field parts of the solutions. Thus by selecting only the evanescent contributions in the configurations where SPPs are excited, it is possible to describe the localized plasmonic fields of the excited modes.

3.4. Numerical results and comparison with experimental data

3.4.1. Grating optimization

C-Method has been implemented in MATLAB® code in order to compute the optical response of sinusoidal metallic gratings and simulations have been performed [37] with truncation order N = 6. The stack reproduces the typical multilayer bi-metallic gratings fabricated in laboratory by laser interference lithography in the so called Lloyd's configuration (section 2.1). Simulation code provides a fundamental tool in order to select the grating geometries that exhibit the best optical features and provide to nanofabrication the optimal windows of process for the realization of optimized optical components. Silver thickness must be properly chosen in order to optimize the plasmonic response of the structure and the optimal thickness strictly depends on the amplitude of grating modulation. Simulations have been performed at the incident wavelength λ = 632 nm for several values of the profile amplitude in the range 20-60 nm and for varying silver thickness in the range 10-80 nm, in the case of a sinusoidal grating with fixed period Λ = 500 nm, gold thickness 7 nm and a fixed chromium adhesion layer of 5 nm over a NOA61 substrate.

As Figure 6.a shows, for each amplitude value a silver thickness exists that optimizes the coupling of incident light with SPP modes, i.e. that minimizes the depth of the reflectivity dip. Figure 6.b shows some examples of reflectivity curves in angular scan for optimal combinations of profile amplitude and silver thickness. For increasing amplitude of the grating profile, the optimal silver thickness decreases (Figure 6.a) and the corresponding reflectivity curve becomes broader (Figure 6.b). This result seems to suggest the choice of a shallow grating modulation with the evaporation of the corresponding optimal thickness of silver: for an amplitude A = 30 nm the optimal silver thickness is around 80 nm. On the other hand, the coating with a great quantity of metal could affect the preservation of the original pattern and cause lack in accuracy of the final profile. The evaporation of about 40 nm of silver instead, could assure the control of the grating profile and at the same time a reasonable value for the reflectivity-dip FWHM (Full Width at Half Maximum).

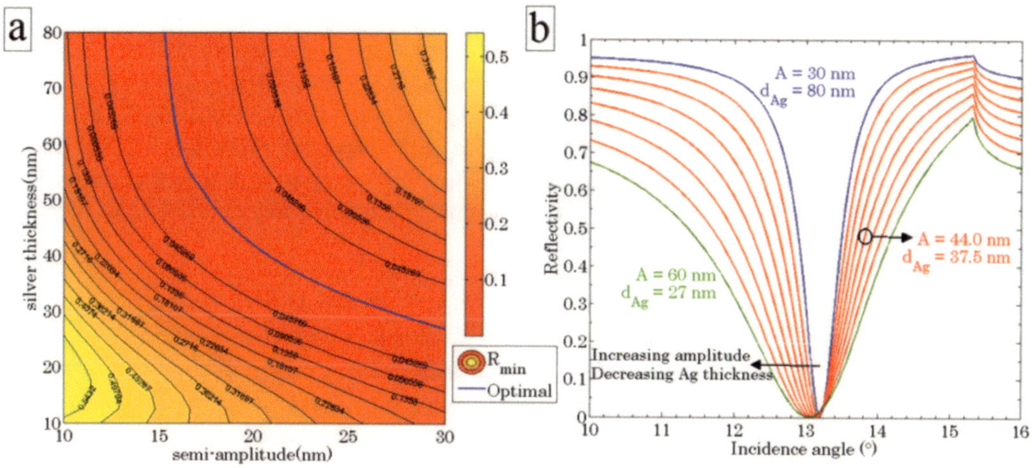

Figure 6. a) Reflectivity minimum as a function of profile semi-amplitude and silver-film thickness for fixed period 500 nm, fixed gold-film thickness 7 nm, incident wavelength 632 nm (sinusoidal profile). Superimposed blue line: optimal configurations. b) Simulated reflectivity in angle scan for values of amplitude and silver thickness along the optimal configuration line in (a): amplitude range 30-60 nm, step 4 nm.

3.4.2. Comparison with experimental data

Numerical results have been compared with experimental data from reflectivity analysis of the corresponding fabricated sample. The simulated stack reproduces the multilayer structure of the considered grating: air (upper medium), Au (8 nm), Ag (35 nm), Cr (9 nm), photoresist (70 nm), Si (substrate). For each layer the optical constants (refractive index n, extinction coefficient k) have been extrapolated from ellipsometric analysis and have been inserted into the code. From AFM analysis, the grating profile results sinusoidal with period 505 nm and peak-to-valley amplitude 26 nm.

As figure 7 shows, numerical estimation of grating reflectivity well fits experimental data within instrumental errors (~2%). Reflectivity measurements have been performed by means of the monochromatized 75 W Xe-Ne lamp of a spectroscopic ellipsometer VASE (J. A. Woollam), with angular and spectroscopic resolution respectively 0.001° and 0.3 nm

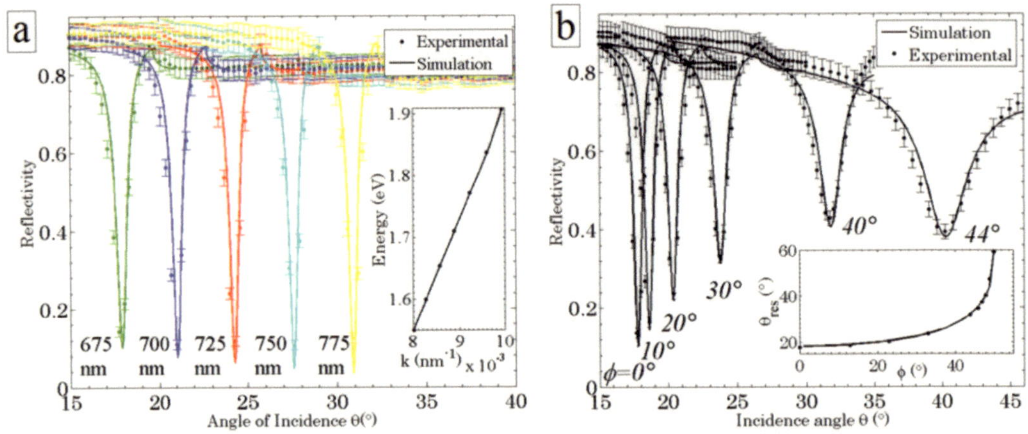

Figure 7. a) Reflectivity spectra for angular interrogation at null azimuth $\varphi = 0°$ for incident wavelengths λ in the range 675-775 nm with step 25 nm, *p*-polarization ($\alpha = 0°$). Experimental data points and simulation results (solid lines). In the inset graph: SPP dispersion curve ω–k. b) Reflectivity spectra for angular interrogation, variable azimuth φ for the $\lambda = 675$ nm, *p*-polarization. In the inset graph: resonance angle θ_{res} as a function of the azimuth angle φ: experimental points and simulation curve (solid line).

Figure 7.a shows reflectivity spectra for angular interrogation at null azimuth $\varphi = 0°$ for incident wavelengths λ in the range 675-775 nm with step size 25 nm, *p*-polarization ($\alpha = 0°$). In angular interrogation, the reflectivity dip shifts towards greater resonance angles for increasing wavelength. Numerical results perfectly reproduce experimental data trends and from dip position it is possible to reconstruct SPP dispersion (Figure 7.a, inset graph).

Figure 7.b exhibits reflectivity spectra in angular interrogation for the azimuth values $\varphi = 0°$, 10°, 20°, 30°, 40°, 44°, incident wavelength $\lambda = 675$ nm and *p*-polarization ($\alpha = 0°$).

4. Sensitivity enhancement by azimuthal control

The analysis of the wavevector components allows a description of double SPP excitation using the schematic shown in Figure 8. The excitation of SPPs on a grating is achieved when the on-plane component $\mathbf{k}_{||}^{(in)}$ of the incident light wavevector and the diffracted SPP wavevector $\mathbf{k}_{SPP}$ satisfy the momentum conservation condition:

$$\begin{aligned} \mathrm{k}_{SPP} &= \mathrm{k}_{||}^{(in)} + n \cdot \mathrm{G} \\ \mathrm{k}_{SPP} &= \frac{2\pi}{\lambda}\sin\theta_{res}\cdot(1,0) - \frac{2\pi}{\Lambda}(\cos\varphi, \sin\varphi) \end{aligned} \quad (20)$$

where θ_{res} is the resonance polar angle, φ is the azimuth angle, Λ is the grating pitch, λ is the illuminating wavelength. Only the first diffraction order ($n = -1$) is considered because in our cases of interest, grating momentum G is always greater than k_{SPP}.

All quadrants of the circle in Figure 8 can be explored for SPP excitation as long as momentum conservation is satisfied. For symmetry reason, only k_y positive half space is considered. Soving eq. (20) in the resonance angle unknown, the following expression is obtained as a function of the azimuth angle φ:

$$\theta_{\mp} = \arcsin\left(\frac{\lambda}{\Lambda}\cos\varphi \mp \sqrt{M(\lambda)^2 - \left(\frac{\lambda}{\Lambda}\sin\varphi\right)^2}\right) \tag{21}$$

where $M(\lambda) = k_{SPP}/(2\pi/\lambda)$.

The largest circle in the *k*-space represents equi-magnitude ***G*** vectors at different azimuthal orientation. The two smaller circles represent all possible $\boldsymbol{k_{SPP}}$ vectors with equal magnitude respectively before and after surface functionalization and whose modulus, for shallow gratings, can be approximated by [2]:

$$k_{SPP} = \frac{2\pi}{\lambda} M(\lambda) = \frac{2\pi}{\lambda}\sqrt{\frac{\varepsilon_m \varepsilon_{eff}}{\varepsilon_m + \varepsilon_{eff}}} \tag{22}$$

where ε_m and ε_{eff} are the dielectric permittivity of the metal and the dielectric side respectively. After functionalization, SPP modulus increases because of the small increase in ε_{eff} due to the surface coating.

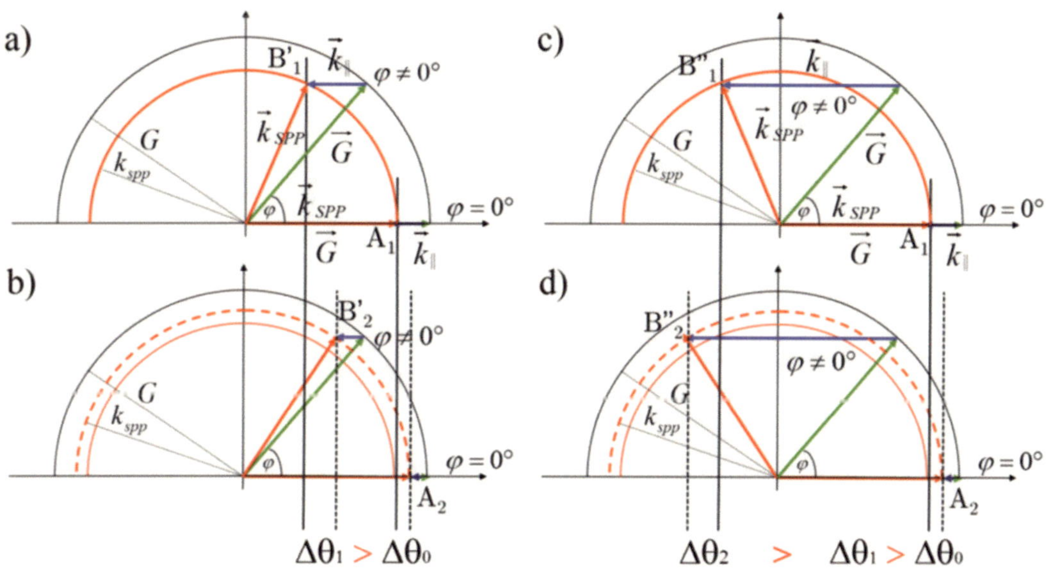

Figure 8. Schematic picture of wave-vector combination at SPP resonance (n = -1). The large circle represents equi-magnitude ***G*** vectors. The smaller circles represent equi-magnitude $\boldsymbol{k_{SPP}}$ vectors before (a, c), and after (b, d) surface functionalization for the first (a, b) and the second (c, d) SPP excitation. The blue arrows represent the photon on-plane wavevector and the red arrows represent the SPP propagation direction. The letters *A* and *B* represent the vector with azimuthal rotation $\varphi = 0°$ and $\varphi \neq 0°$ respectively.

The dashed line at the tip of the circle of radius G represents the x-component of the photon wave-vector $\mathbf{k}_{||}^{(in)}$, the only component that participates in SPP excitation. The line is scaled linearly in $\sin\theta$ so that the full length of the line at the incident angle θ of 90° corresponds to the maximum value of $k_{||}^{(in)}$. The intersections of the $\mathbf{k}_{||}^{(in)}$ dashed horizontal line with the smaller $\mathbf{k}_{SPP}$ circle determine the conditions for which eq. (20) is satisfied and allows the identification of both incident angle θ_{res} for SPP resonance excitation and SPP propagation direction β . We consider first the case of the uncoated sample - the smallest of the semicircles. For example, point B on the $\mathbf{G}$ circle is identified by the azimuthal angle φ and allows the excitation of SPP at two possible conditions: B_1' and B_1'', with β^- and β^+ respectively. Within the double SPP range (point B), a small increment in wavelength makes the points B_1' and B_1'' merge to form a very broad. On the contrary, at $\varphi = 0°$ (point A_1) it is clear that the photon wave-vector can intersect the SPP circle only in the first quadrant but not the second, thus exciting only a single SPP for each wavelength. The same argument is applicable for light exciting SPP on the functionalized sample. Due to the larger k_{SPP}, a different excitation condition is expected. The intersection points changes from B_1' and B_1'' to B_2' and B_2'' and from A_1 to A_2. The sensitivity of the GCSPP is higher at high azimuthal angles because the condition for double SPP excitation around the circumference of the k_{SPP} circle generates a shifts in k-space between points B_1' and B_2', which is much larger than that between points A_1 and A_2 provided by a single SPP excitation condition for $\varphi = 0°$. The estimated refractive index sensitivity S of this configuration can be defined as:

$$S = \frac{\partial \theta}{\partial n} = \frac{\partial \theta}{\partial k_{||}^{(in)}} \frac{\partial k_{||}^{(in)}}{\partial k_{SPP}} \frac{\partial k_{SPP}}{\partial n} \tag{23}$$

In order to calculate S, we assumed the rippling amplitude A of the grating is so shallow ($A/\Lambda \sim 0.05$ in our case of interest) that the dispersion curve of SPPs traveling at the metal-dielectric interface of a grating can be approximated by the case of a flat sample (eq. (22)). The analytical expression for the sensitivity in angular interrogation can be found as:

$$S = -\frac{1}{\cos\theta_{res}} \left(\frac{M}{n_0}\right)^3 \frac{\sqrt{\frac{1}{\Lambda^2} + \frac{\sin^2\theta_{res}}{\lambda^2} - \frac{2\cos\varphi \sin\theta_{res}}{\Lambda\lambda}}}{\frac{\cos\varphi}{\Lambda} - \frac{\sin\theta_{res}}{\lambda}} \tag{24}$$

where M is defined in eq. (22), n_0 is the refractive index of the surrounding dielectric medium, the resonance angle θ_{res} is given by eq. (21) for fixed λ.

The functional behavior of sensitivity for the first and second SPP dip is shown in Figure 9 for a typical wavelength of λ = 606 nm. Both sensitivities diverge when φ approaches its critical value φ_{MAX}, i.e. the maximum φ angle that supports SPP resonances. In this configuration, incident photon momentum is tangential to the k_{SPP} circle and its length equals the x-component of the grating momentum so that the denominator in eq. (24) becomes null. Another condition for second dip sensitivity singularity is when φ approaches the critical azimuthal angle, φ_c, necessary to excite double SPP resonances, namely when the full length of the incident photon momentum is required to intersect the edge of the k_{SPP} circle. Since the incident angle $\theta = 90°$, $\cos\theta_{res}$ in the denominator of the first term eq. (24) approaches 0 and S diverges.

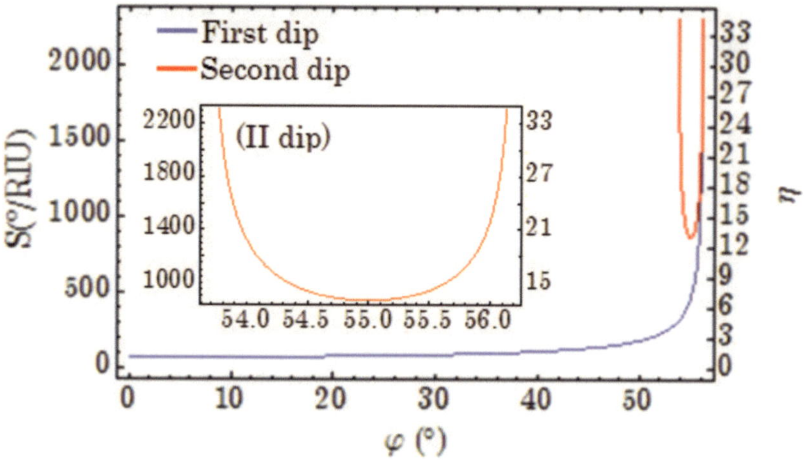

Figure 9. Sensitivity S_θ as a function of grating azimuthal angle φ for the two dips in SPR sensor with angular interrogation. The right y-scale refers to sensitivity values normalized to the first dip sensitivity at $\varphi = 0°$: $\eta = S_\theta(\varphi)/ S_\theta(\varphi=0°)$.

Although azimuth φ values close to the critical values provides a great enhancement in S (up to 2400°/RIU, 35 times higher than the case $\varphi = 0°$), these configurations should be avoided because of experimental limits. For $\varphi \cong \varphi_c$, θ_{res} becomes large (>70-80°) and broad, becoming impossible to resolve the SPP minimum. In addition, when $\varphi \cong \varphi_{MAX}$ the two resonance dips merge into a single broad dip which makes the two minima hardly distinguishable. Thus only limited parts of the azimuthal angular range are suitable to enhance sensitivity significantly. The best conditions correspond to the middle of the "U-shape" of second dip functional behavior (Figure 9) where the sensitivity ranges from 900°/RIU to 1100°/RIU, about 15 times higher than $\varphi = 0°$ whose value is 67°/RIU. The sensitivity computed for the first SPP dip is smaller all over this range but it still provides values of the order of 500°/RIU.

4.1. Application to functionalization detection

Sensitivity enhancement with azimuthal rotation has been tested with a C12-functionalization of a bimetallic (37Ag/7Au) grating, period 487 nm, amplitude 25 nm. Reflectivity analyses performed before and after C12-coating are collected in Figure 10.a-b. The figure shows the experimental evidence of the azimuth rotation, where reflectivity spectra are reported at different wavelengths. As a reference Figure 10.a reports the reflectivity spectra using the conventional GCSPR configuration with $\varphi = 0°$. In this configuration, the difference in the reflectivity minima resonance angle θ_{res} before and after

C12 is typically less than 0.05°. On the contrary after an azimuthal rotation of the grating to an angle φ of about 60°, larger angular differences can be observed between the reflectivity dips (Figure 10.b). The resonance shift can reach values up to 3.1° as for the incident wavelength λ = 618 nm. By increasing wavelength from 606 to 618 nm, the two resonance dips in the reflectivity spectra get closer while resonance shift increases from 1.8° to 3.1°, until the two resonances merge together into a single broad dip at 620 nm.

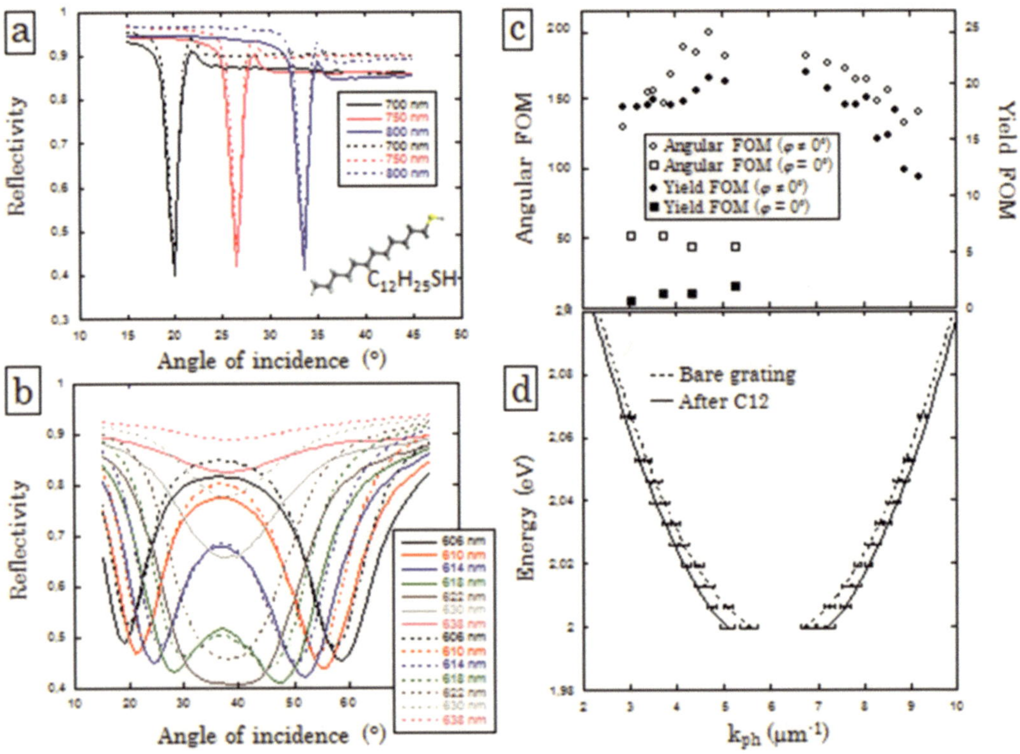

Figure 10. Comparison of SPR spectrum between uncoated (dashed line) and C12-coated (solid line) Au grating. (a) Grating with φ = 0°, for incident wavelength λ = 700-800 nm, step 50 nm, and (b) φ = 60°, (2 SPPs by single wavelength excitation condition), for incident wavelength λ = 606-638 nm, step 4 nm. c) Figure of merit for yield (• , ■) and angular (o, □) interrogation of the SPR reflectivity minima. (d) Energy dispersion curve for the $k_{\shortparallel}^{(in)}$ necessary for SPP excitation before (dashed line) and after (solid line) C12 SAM functionalization.

The experimental determinations of the wavevector $k_{\shortparallel}^{(in)}$ of the incoming light necessary for the SPP excitation have been successfully fitted with the help of eq. (20), using as fitting parameters only the effective index of refraction and azimuthal angle (see Figure 10.d). After surface functionalization we have determined a total increment in index of refraction equal to Δn = 0.00357 ± 0.00007 RIU. This is a value that agrees with the estimation of RIU performed for effective refractive index change generated by a full surface coverage of close-packed self assembled C12 molecules, considering their length 1.46 nm with dielectric constant $\varepsilon_{C12} = 2.12$. Also the theoretical sensitivity determination (Figure 9) is in good agreement with the experimental determination obtained as a ratio between Δn, the angular differences before and after C12 coating and the Δn determination. The average experimental sensitivity is of the order of 520°/RIU for the first dip and reaches maximum values of 857°/RIU for the second dip. The final error of the refractive index determination has been estimated on the basis of chi-square minimization based on *a priori* determination of the SPP angular position. The reflectivity minima of this preliminary data set have been determined with a typical uncertainty of 0.07°. However the *a posteriori* determination of the SPP angular average deviation with respect to the dispersion curve best fit is much smaller, on the order of 0.015°, as can be confirmed by a simple graphical inspection. Taking into account this more realistic value for the final evaluation of the uncertainty, the value of 10^{-5} RIU. However, because our experimental system has an instrumental resolution of 0.001°, we believe it will be possible to greatly decrease the present angular uncertainty by increasing the statistical signal-to-noise ratio and using appropriate algorithms for data analysis. We expect that experimental uncertainties of Δn on the order of $5 \cdot 10^{-7}$ RIU is achievable. In order to better describe the detection improvement given by azimuthal rotation, we have also measured the typical figure of merit for angular and yield interrogation respectively defined as:

$$\begin{aligned} FOM_\theta &= \frac{S_\theta}{\Delta\theta_{FWHM}} \\ FOM_Y &= \frac{Y_{coat} - Y_{uncoat}}{Y_{uncoat}} \end{aligned} \tag{25}$$

where $\Delta\theta_{FWHM}$ is the angular full width at half maximum of the reflectivity minima, whereas Y_{coat} and Y_{uncoat} are the minimum yield of the reflectivity spectra collected before and after C12 functionalization at SPR resonances.

Figure 10.c shows the angular and yield *FOM* at zero and after the azimuthal rotation for all the reflectivity spectra. It clearly appears an enhancement of both the figures of merit after the azimuthal rotation that amounts up to a factor 4 and 10 for the angular and yield *FOM* respectively. This means that the distance between two dips, before and after the functionalization, scales with a factor greater than the enlargement of the reflectivity dip width. Moreover, it is worth to note that the reflectivity yield is even more sensitive than the angular position. It clearly appears from the reflectivity spectra of Figure 10.b that whereas the minimum yield between coated and uncoated are almost the same for zero azimuth, it changes dramatically after azimuthal rotation. Finally, we note that both the angular and the yield *FOM* have similar functional behavior: they increase approaching the condition of two dips merging when $\beta_+ = \beta_- = 90°$.

4.2. Application to solution analysis

A microfluidic cell was fabricated in PDMS by soft-lithography technique and embodied to the metallic grating. The considered metallic gratings exhibits a sensitivity $S_\theta(0°)$ = 64.9°/RIU for λ = 840 nm (Figure 11.a-b). As demonstrated in the previous section, angular sensitivity S_θ can be improved just with an azimuthal rotation of the grating plane. For a same fixed concentration of 10g(Nacl) in 200ml(water), corresponding to a refractive index variation with respect to distilled water $\Delta n = 9.6 \cdot 10^{-3}$ [38], reflectivity spectra have been collected for increasing azimuth angle φ up to 43° at the same incident wavelength. As Figure 11.c shows, resonance angle shift increases monotonically in modulus with increasing azimuth angle from $\Delta\theta(0°) = 0.59°$ to $\Delta\theta(43°) = 5.57°$ corresponding to a sensitivity enhancement $\Delta\theta(43°)/\Delta\theta(0°) = 9.5$ of almost one order of magnitude: $S_\theta(43°)$ = 616.8°/RIU.

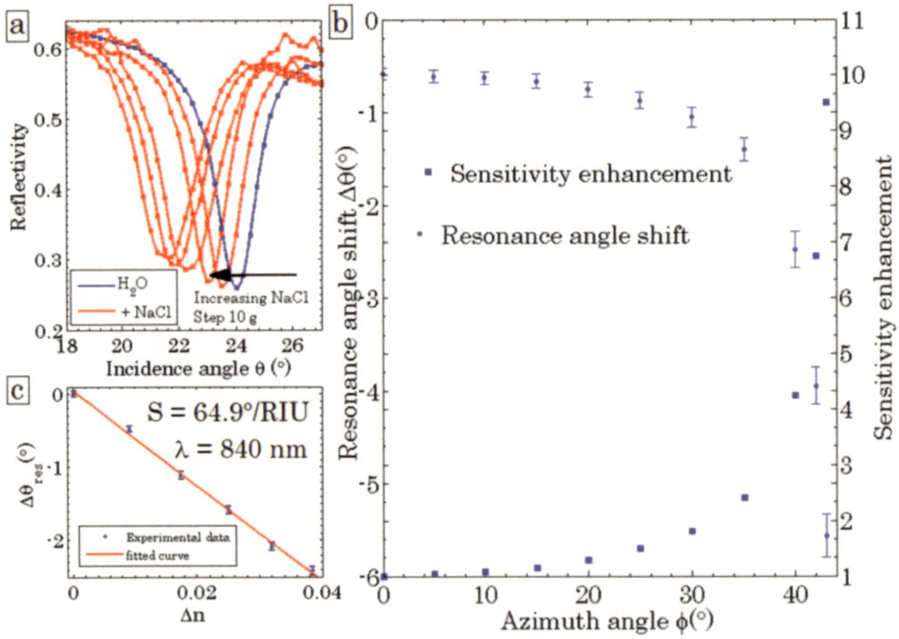

Figure 11. (a) Reflectivity in polar angular scan at incident wavelength 840 nm, null-azimuth, *p*-polarization, for increasing NaCl concentration in water solution: 0-50 g, step 10 g, in 200 ml of water. (b) Resonance angle shift as a function of refractive index variation of water solution with increasing NaCl concentration. (c) Resonance angle shift and sensitivity enhancement as a function of azimuth angle in the range 0-43°. NaCl concentration: 10g in 200ml of water ($\Delta n = 9.6 \cdot 10^{-3}$).

This sensitivity-enhancement technique with azimuthal rotation has been exploited to reveal a lower sodium-chloride concentration than the previous one. A mass of 0.56 g has been dissolved into 200 ml of water and the resulting refractive index change is $\Delta n = 5.4 \cdot 10^{-4}$. With this solution flowing through the microfluidic cell, reflectivity measurements have been performed for increasing grating rotation (Figure 12). The inset picture in Figure 12 shows the reflectivity spectra before and after salt dissolution for null azimuth and after grating azimuthal rotation up to 43°. While in the classical mounting, i.e. null azimuth, this variation is not detectable within the experimental error, in conical mounting instead grating azimuthal rotation increases SPR sensitivity and thus the system allows revealing the resonance angle shift. Resonance dips become broader when azimuthal angle increases, however the resonance angle shift $\Delta\theta$ scales with a factor greater than the enlargement of the dip full width half maximum $\Delta\theta_{FWHM}$. Thus the angular figure of merit FOM_θ increases and the detection improvement by azimuthal rotation is preserved. In this case we have $FOM_\theta(43°)/ FOM_\theta(0°)\cong 4.3$.

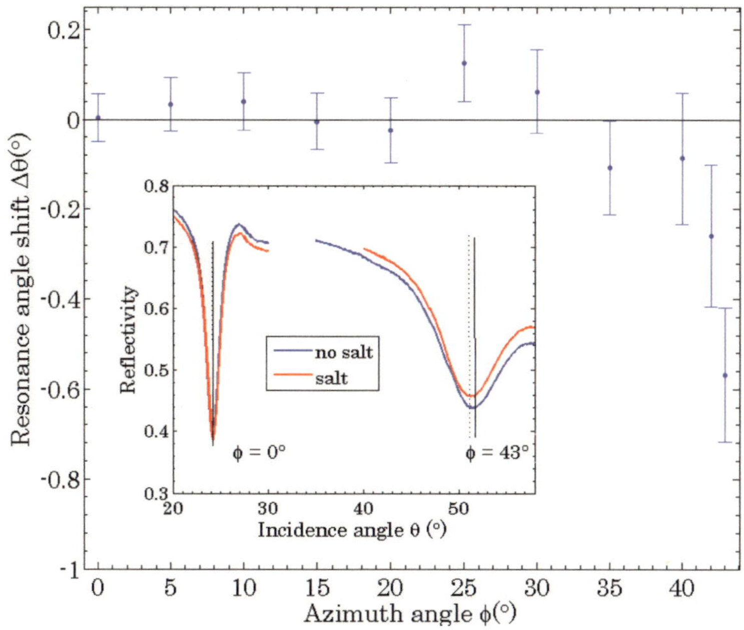

Figure 12. Resonance angle shift as a function of azimuth angle for sodium-chloride concentration: 0.56g/200 ml(water), $\Delta n = 5.4 \cdot 10^{-4}$. In the inset graph: reflectivity curve before (blue) and after (red) NaCl dissolution into water, angular scan for azimuth values $\varphi = 0°$ and 43°.

5. Innovative SPR with polarization modulation

In previous sections we experimentally and theoretically described the effects of grating azimuthal rotation on surface plasmon excitation and propagation. More SPPs can be supported with the same illuminating wavelength and a sensitivity enhancement of al least one order of magnitude is achievable than that in the conventional configurations. Here we consider how the symmetry breaking with grating rotation makes polarization have a fundamental role on surface plasmon polaritons excitation. As Figure 7.b shows, *p*-polarization becomes less and less effective for increasing azimuthal rotation and the reflectivity depth decreases. If the polar angle is fixed at resonance and a polarization scan is performed, the minimum of reflectivity R_{min} exhibits a harmonic dependency on the incidence polarization α with a periodicity of 180° (Figure 13):

$$R_{\min} = f_0 - f_1 \cos(2\alpha + \alpha_0) \tag{26}$$

where f_0, f_1 and α_0 are fitting parameters that depend on the incidence angles θ and φ, incident wavelength λ and on the optical properties of the stack (thickness and dielectric permittivity of each layer). While for null azimuth $\alpha_0 = 0°$ and *p*-polarization is the most effective for SPP excitation, after grating rotation the phase term α_0 is non null and strictly depends on the incidence angles. By assuming that only the electric field component lying on the grating symmetry plane is effective for SPP excitation, on the basis of a vectorial approach, an analytical expression can be obtained [18] for the optimal polarization α_{opt} as a function of the azimuth angle φ and the resonance angle θ_{res}:

$$\tan\alpha_{\min} = \tan\varphi \cdot \cos\theta_{res} \tag{27}$$

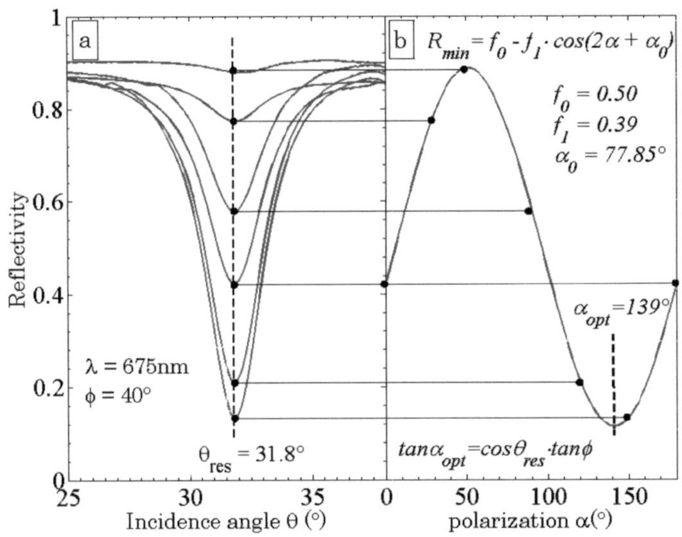

Figure 13. (a) Reflectivity for polar angle scan in the range 25°-38°, step 0.1°, at λ = 675 nm and azimuth φ = 40° for varying incident polarization α in the range 0°-180°, step 30°. (b) Reflectivity minima as a function of polarization and fit curve.

If the grating surface is functionalized, the effective refractive index n_{eff} of the dielectric medium changes and resonance conditions are different. As a consequence of the shift in the resonance angle θ for a fixed azimuth φ, there is a shift $\Delta\alpha_0$ in the phase term α_0:

$$\Delta\alpha_0 = \frac{\partial\alpha_0}{\partial n}\Delta n \tag{28}$$

This result opens the route to a new GCSPR-configuration with polarization interrogation[39]. In this setup the grating is rotated of an azimuthal angle which is kept fixed. The illuminating wavelength is fixed and the incoming light impinges on the grating at the polar resonance angle. A rotating polarizer between source and sample-holder allows changing the polarization incident on the grating. Reflectivity data collected during a polarization scan can be fitted using eq. (26) and a variation of fitting parameters, e.g. amplitude f_1 or phase α_0, can be used in order to detect grating functionalization or for solution-concentration analysis, once the system has been properly calibrated.

A metallic grating with a period of 505 nm and amplitude of 26 nm, fabricated by interferential lithography (IL) followed by thermal evaporation of a gold (40 nm) metallic layer over 5 nm of chromium adhesion layer. Optical measurements have been performed in θ/2θ symmetric reflectivity configuration, using the 75W Xe lamp of VASE Ellipsometer (J. A. Woollam). A self-assembled monolayer of dodecanethiol was deposited on the gold coated grating surfaces at room temperature [40].

First of all, reflectivity spectra have been collected in angular scan to identify the resonance angle position with a weighted centroid algorithm [41]. In order to exploit the shift enhancement, grating was azimuthally rotated and kept fixed at the value $\varphi = 57.8°$ wherein double SPP excitation is supported for the selected wavelength $\lambda = 625$ nm. In correspondence of the resonance angles, respectively $\theta_I = 30.8°$ and $\theta_{II} = 55.1°$ for the first and second dip, a polarization scan has been collected in the range 0-180°, step 10°, before and after C12 functionalization. The same analysis has been performed at the wavelength 635 nm, for the same azimuth, when the two dips merge into a single broad one centered in $\theta = 43°$.

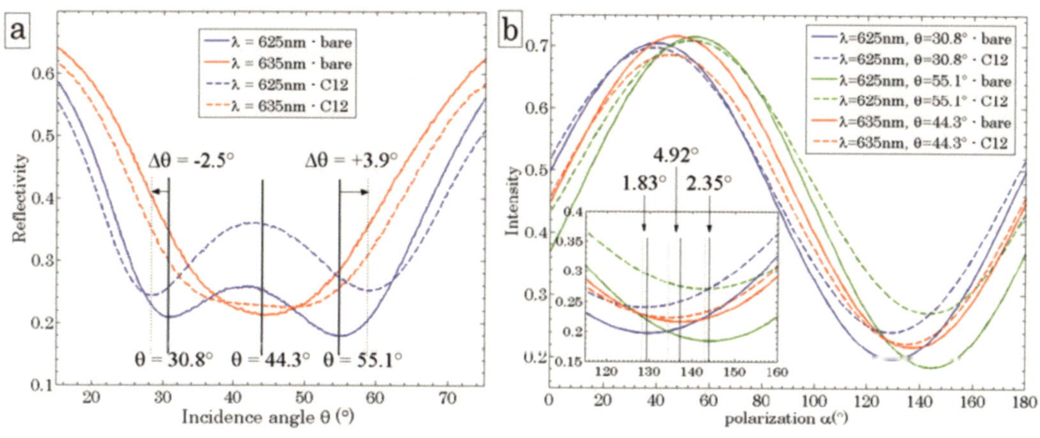

Figure 14. (a) Reflectivity for angular scan in the range 15°-75°, step 0.2°, at $\lambda = 625$ nm (blue lines) and $\lambda = 635$ nm (red lines), azimuth $\varphi = 57.8°$, incident polarization $\alpha = 140°$, before (solid lines) and after (dashed lines) functionalization with C12. (b) Polarization scan for $\lambda = 625$ nm (blue lines: I dip ($\theta_I = 30.8°$), green lines: II dip ($\theta_{II} = 55.1°$)), $\lambda = 635$ nm (red lines: merged dips ($\theta = 44.3°$)), azimuth $\varphi = 57.8°$, before (solid lines) and after (dashed lines) C12-coating. Inset picture: phase shifts $\Delta\alpha_0$.

By modelling the effective refractive index change Δn_{eff} with an effective medium approximation [42], it is possible to estimate the corresponding phase sensitivity S_α and moreover to calculate the refractive index resolution:

$$\sigma_{n,\alpha} = \frac{\sigma_\alpha}{S_\alpha} \tag{29}$$

C12 has been assumed to form a monolayer 1.46 nm-thick with refractive index n=1.458 [43] and results in Δn_{eff} (625 nm) = 50.5 · 10^{-4} and Δn_{eff} (635 nm) = 48.1 · 10^{-4}. In the case of double SPP excitation (λ=625 nm), the phase shifts result $\Delta\alpha_0(I)$=1.834 ± 0.001° (first dip) and $\Delta\alpha_0(II)$= 2.353 ± 0.001° (second dip), corresponding respectively to sensitivity values $S_\alpha(I)$=363.2°/RIU and $S_\alpha(II)$= 465.9°/RIU, refractive index resolutions $\sigma_{n,\alpha}(I)$=2.7 · 10^{-6} RIU and $\sigma_{n,\alpha}(II)$=2.1 · 10^{-6} RIU. For λ=635 nm, we get S_α= 1022.7°/RIU and 2 · 10^{-6} RIU.

Thus refractive index changes of order 10^{-6} RIU are easily detectable and the resolution can be further improved to 10^{-7}-10^{-8} by reducing output noise σ_I or by increasing the number N of the points collected during the polarization scan.

Furthermore this technique provides a resolution at least two order greater than polar angle modulation with the same setup, which results around $\sigma_{n,\theta}$~10^{-4}-10^{-5} RIU. Moreover, while angle-modulation SPR becomes difficult near the merging dip condition, since dip position is hardly detectable, in the polarization-modulation case the analysis is still valid and it assures a greater sensitivity. On the top of that, since the output trend is a well-known function of polarization, this method assures a great accuracy on fitting parameters and their dependence on grating surface conditions provides a solution to detect and quantify surface functionalization or solution concentration. This method assures a competitive resolution down to 10^{-8} and limits the mechanical degrees of freedom just to the polarization control. The option of using an electronic modulator instead of a rotating polarizer, further assures the possibility to realize very compact, fast and low-cost high-resolution plasmonic sensors.

6. Conclusions

Plasmonic gratings have been demonstrated to assure a high-sensitive optical response to surface functionalization and to represent a promising and irreplaceable component for the realization of label-free devices for sensing purposes with considerable performance in refractive index sensitivity and resolution. The problem of designing and realizing metallic gratings for sensing applications has been studied and analysed through each step of the process-chain: simulation - nanofabrication - characterization.

Chandezon's method provides an algorithm for a rigorous and complete analysis of the diffraction problem of a multi-layered patterned. The numerical code provides a precise estimation of grating reflectivity and is an essential tool to design, for given geometry and material choice, the optimal profile that optimizes the coupling strength of incident light with surface plasmon polaritons. Simulation is essential to provide to nanofabrication the proper windows of process for the production of optimized supports.

As regards the nanofabrication of these components, interferential lithography is the preferred method to fabricate periodic pattern with a spatial coherence over large areas, while grating-replica process by soft-lithography assures the possibility of a fast and cheap throughput of perfectly replicated gratings. With respect to prism-coupling, grating nanofabrication technology assures the possibility to miniaturize and integrate the sensing components without a considerable increase of the total expenditure. Moreover, our results further highlight grating-coupling advantages rather than prism-coupling thanks to the sensitivity enhancement with azimuthal rotation. An enhancement at least one order greater than the conventional mounting has been theoretically and experimentally demonstrated.

If the azimuthal rotation of the grating support increases refractive index sensitivity, on the other hand incident polarization must be tuned in order to best couple incident light and optimize the optical response. The dependence of polarization angle on the resonance conditions suggests the exploitation of a new SPR configuration based on polarization modulation in the conical mounting. The phase term of a polarization scan, which is proportional to the optimal polarization, has been experimentally demonstrated to be a sensitive parameter for surface functionalization analysis. This innovative sensing configuration hugely simplifies the mechanical complexity of the device by limiting the degrees of freedom just to the rotating polarizer and assures the possibility of realizing very compact, fast and low-cost high-resolution plasmonic sensors based on polarization modulation.

Author details

G. Ruffato, G. Zacco and F. Romanato
University of Padova, Department of Physics "G. Galilei", Padova, Italy
Laboratory for Nanofabrication of Nanodevices (LaNN - Venetonanotech), Padova, Italy
Istituto Officina dei Materiali IOM-CNR National Laboratory, Trieste, Italy

Acknowledgement

This work has been supported by a grant from "Fondazione Cariparo" - Surface PLasmonics for Enhanced Nano Detectors and Innovative Devices (SPLENDID) – Progetto Eccellenza 2008 and from University of Padova – Progetto di Eccellenza "PLATFORM".

7. References

[1] Raether H., Surface Plasmons on Smooth and Rough Surfaces and on Gratings. Springer-Verlag, 1988.
[2] Maier S. A., editor. Plasmonics - Fundamentals and Applications. Springer, 2007.
[3] Wood R. W., On a remarkable case of uneven distribution of light in a diffraction grating spectrum, *Philos. Mag.* 4, 396-402 (1902).
[4] Cowan J. J., Arakawa E. T., Dispersion of surface plasmons in dielectric-metal coatings on concave diffraction gratings, *Zeitschrift fur Physik* 235, 97 (1970).

[5] Homola J., Yee S. S., Gauglitz G., Surface plasmon resonance sensors: review, *Sens. Actuators B* 54, 3-15 (1999).

[6] Kretschmann E., *Z. Phys.* 241, 313-324 (1971).

[7] Homola J., Surface Plasmon Resonance Based Sensors. Springer, 2006.

[8] Karlsson R., Stahlberg R., Surface plasmon resonance detection and multispot sensing for direct monitoring of interactions involving low-molecular-weight analytes and for determination of low affinities, *Anal. Biochem.* 228, 274-280 (1995).

[9] Homola J., Koudela I., Yee S. S., Surface plasmon resonance sensor based on diffraction gratings and prism couplers: sensitivity comparison, *Sens. Actuators B* 54, 16-24 (1999).

[10] Hoa X. D., Kirk A. G., and Tabrizian M., Towards integrated and sensitive surface plasmon resonance biosensors: a review of recent progress, *Biosens. Bioelectron.* 23, 151-160 (2007).

[11] Yoon K. H., Shuler M. L., and Kim S. J., Design and optimization of nano-grating surface plasmon resonance sensors, *Opt. Express* 14, 4842-4249 (2006).

[12] Unfricht D. W., Colpitts S. L., Fernandez S. M. and Lynes M. A., Grating-coupled surface plasmon resonance: a cell and protein microarray platform, *Proteomics* 5, 4432-4442 (2005).

[13] Dostalek J., Homola J., Miler M., Rich information format surface plasmon resonance biosensor based on array of diffraction gratings, *Sens. Actuators B* 107, 154-161 (2005).

[14] Alleyne C. J., Kirk A. G., McPhedran R. C., Nicorovici N. A. P., and Maystre D., Enhanced SPR sensitivity using periodic metallic nanostructures, *Opt. Express* 15, 8163-8169 (2007).

[15] Telezhnikova O., Homola J., New approach to spectroscopy of surface plasmons, *Opt. Lett.* 31, 3339-3341 (2006).

[16] Romanato F., Lee K. H., Kang H. K., Wong C. C., Zong Y. and Knoll W., Azimuthal dispersion and energy mode condensation of grating-coupled surface plasmon polaritons, *Phys. Rev. B* 77, 245435-245441 (2008).

[17] Romanato F., Lee K. H., Kang H. K., Ruffato G. and Wong C. C., Sensitivity enhancement in grating coupled surface plasmon resonance by azimuthal control, *Opt. Express* 17, 12145-12154 (2009).

[18] Romanato F., Lee K. H., Ruffato G. and Wong C. C., The role of polarization on surface plasmon polariton excitation on metallic gratings in the conical mounting, *Appl. Phys. Lett.* 96, 111103 (2010).

[19] Maldovan M., Thomas E. L., Periodic Materials and Interference Lithography. Wiley-VCH Verlag GmbH & Co. KGaA, 2008.

[20] Zacco G., Romanato F., Sonato A., Sammito D., Ruffato G., Morpurgo M., Silvestri D., Carli M., Schiavuta P. and Brusatin G., Sinusoidal plasmonic crystals for bio-detection sensors, *Microelectron. Eng.* 88, 1898-1901 (2011).

[21] Cabral J. T., Hudson S. D., Harrison C., and Douglas J. F., Frontal Photopolymerization for Microfluidic Applications, *Langmuir* 20, 10020-10029 (2004).

[22] Technical data sheet for NOA optical adhesives; Norland Products, Inc.: New Brunswick, NJ.

[23] Chandezon J., Maystre D. and Raoult G., A new theoretical method for diffraction gratings and its numerical application, *J. Optics (Paris)* 11, No. 4, 235-241 (1980).

[24] Chandezon J., Dupuis M. T. and Cornet G., Multicoated gratings: a differential formalism applicable in the entire optical region, *J. Opt. Soc. Am.* 72, No. 7, 839-846 (1982).

[25] Elston S. J., Bryan-Brown G. P. and Sambles J. R., Polarization conversion from diffraction gratings, *Phys. Rev. B* 44, No.12, 6393-6400 (1991).

[26] Granet G., Plumey J. P. and Chandezon J., Scattering by a periodically corrugated dielectric layer with non identical faces, *Pure Appl. Opt.* 4, No.1, 1-5 (1995).

[27] Preist T. W., Cotter N. P. K. and Sambles J. R., Periodic multilayer gratings of arbitrary shape, *J. Opt. Soc. Am. A* 12, No. 8, 1740-1749 (1995).

[28] Li L., Granet G., Plumey J. P. and Chandezon J., Some topics in extending the C-method to multilayer-coated gratings of different profiles, *Pure Appl. Opt.* 5, No.2, 141-156 (1996).

[29] Plumey J. P., Guizal B. and Candezon J., Coordinate transformation method as applied to asymmetric gratings with vertical facets, *J. Opt. Soc. Am. A* 14, No. 3, 610-617 (1997).

[30] Li L., Multilayer-coated diffraction gratings: differential method of Chandezon et al. revisited, *J. Opt. Soc. Am. A* 11, 2816-2828 (1994).

[31] Li L., Using symmetries of grating groove profiles to reduce computation cost of the C-method, *J. Opt. Soc. Am. A* 24, 1085-1096 (2007).

[32] Post E. J., Formal Structure of Electromagnetics. North-Holland, Amsterdam, 1962.

[33] Ashcroft N. W. and Mermin N. D., Solid State Physics. Saunders College Publishing, 1976.

[34] Zygmund A., Trigonometric Series. Cambridge U. Press, Cambridge, 1977.

[35] Li L., Use of Fourier series in the analysis of discontinuous periodic structures, *J. Opt. Soc. Am. A* 13, No. 9 (1996).

[36] Li L., Justification of matrix truncation in the modal methods of diffraction gratings, *J. Opt. A: Pure Appl. Opt.* 1, 531-536 (1999).

[37] Ruffato G., Zacco G. and Romanato F., Surface Plasmon Polaritons Excitation and Propagation on Metallic Gratings: Far-Field and Near-Field Numerical Simulations, *Journal of Materials Science and Engineering A* 1, No. 6, 768-777 (2011).

[38] Dorsey E., Properties of Ordinary Water-Substance, Reinhold Publishing Corporation 1940.

[39] Ruffato G. and Romanato F., Grating-Coupled Surface Plasmon Resonance in conical mounting with Polarization Modulation, *Opt. Lett.* 37 (2012).

[40] Ang X. F., Li F. Y., Tan W. L., Chen Z. and Wong C. C., Self-assembled monolayer for reduced temperature direct metal thermocompression bonding, *Appl. Phys. Lett.* 91, 061913 (2007).

[41] Johansen K., Stalberg R., Lundstrom I., Liedberg B., Surface plasmon resonance: instrumental resolution using photo diode arrays, *Meas. Sci. Technol.* 11, 1630-1638 (2000).

[42] Jung L. S., Campbell C. T., Chinowsky T. M., Mar M. N. and Yee S. S., Quantitative interpretation of the response of surface plasmon resonance sensors to adsorbed films, *Langmuir* 14, 5636-5648 (1998).

[43] Bain C. D., Troughton E. B., Tao Y. T., Evall J., Whitesides G. M., Nuzzo R. G., Formation of monolayer films by the spontaneous assembly organic thiols from solution onto gold, *J. Am. Chem. Soc.* 111, 321-335 (1989).

Chapter 18

Plasmon Mediated Energy Transport in PV Systems with Photo-Active Surface Modified Metallically in Nano-Scale and in Metallic Nano-Chains

Witold Jacak

Additional information is available at the end of the chapter

http://dx.doi.org/10.5772/50750

1. Introduction

Experimental and theoretical investigations of plasmon excitations in metallic nano-crystals rapidly grew up mainly due to possible applications in photo-voltaics and microelectronics. A significant enhancement of absorption of incident light in photo-diode-systems with active surfaces covered with nano-size metallic particles (of Au, Ag or Cu) with planar density 10^8-$10^{10}/\mathrm{cm}^2$ was observed [24, 25, 27, 30, 32, 34]. These findings are of practical importance for enhancement of solar cell efficiency, especially for development of thin film cell technology. On the other hand, hybridized states of surface plasmons and photons result in plasmon-polaritons [20, 36], which are of high importance for applications in photonics and microelectronics [10, 20], in particular, for sub-diffraction transportation of converted light energy and information in metallically modified structures in nano-scale [21, 36].

Surface plasmons in nano-particles have been widely investigated since their classical description by Mie [23]. Many particular studies, including numerical modeling of multi-electron clusters, have been carried out [4, 5]. They were mostly developments of Kohn-Sham attitude in form of LDA (Local Density Approximation) or TDLDA (Time Dependent LDA) for small metallic clusters only [4, 5, 8, 9, 17], up to ca 200 electrons (as limited by numerical calculation constraints that grow rapidly with the number of electrons). The random phase approximation (RPA) was formulated [28] for description of volume plasmons in bulk metals and utilized also for confined geometry mainly in a numerical or semi-numerical manner [4, 5, 17]. Usually, in these analyzes the *jellium* model was assumed for description of positive ion background in the metal and the dynamics was addressed to the electron system only [5, 8, 17]. Such an attitude is preferable for clusters of simple metals, including noble metals (also transition and alkali metals).

The classical Mie frequencies [23] are not dependent on the sphere radius, in contrast to the experimental observations for both small metallic clusters and larger spheres. The emerging of the Mie response from the more general behavior was presented [4, 5, 8]. Numerical analyzes (for clusters up to 200 electrons) revealed the red-shift of Mie resonance mainly due to so-called spill-out of the electron cloud beyond the jellium rim. This effect is, however, not important for particles with radii larger than 2 nm, and for nanospheres, of size of $10 - 60$ nm, the other effects are responsible for size-dependent shifts of resonances, pronounced in large particles and opposite (with respect to the dependence versus a) to the red shift for small and ultra-small clusters.

The main factor for plasmon oscillations in large nanospheres was identified [11, 12, 15] as radiation phenomena, which for radius larger than 10 nm dominate plasmon damping as growing with the radius a as a^3. The pronounced cross-over in the vicinity of 10 nm (for Au, Ag and Cu) has been predicted [15]—for smaller nanospheres, the scattering damping scaling with a as $\frac{1}{a}$ (including inter-particle and boundary scattering and so-called Landau damping [35]) is important, while for larger spheres, the radiation losses are overwhelming and rapidly grow with a as a^3. Thus rather large nanoparticles would play a role in phenomena linked with energy transport employing plasmon radiation, which well corresponds to experimental observations [26, 27, 30, 31, 33, 34].

Metallic nanospheres (or nanoparticles of other shape) can act as light converters, collecting energy of incident photons in surface plasmon oscillations. This energy can be next transferred to the semiconductor substrate in a more efficient manner in comparison to the direct photo-effect. Experimental observations [26, 27, 30, 31, 33, 34] suggest, that the near-field coupling between plasmons in nanospheres and band electrons in the semiconductor substrate allows for significant growth of selective light energy transformation into a photo-current in the diode system. This phenomenon is not described in detail as of yet and moreover, some competitive mechanisms apparently contribute. Nevertheless, one can argue generally, that due to the nano-scale size of the metallic components, the momentum is not conserved, which leads to the allowance of all indirect optical inter-band transitions in the semiconductor layer, resulting in enhancement of the photo-current in comparison to the ordinary photo-effect, when only direct inter-band transitions are admitted.

In the present paper we review the RPA (random phase approximation) semi-classical theory of plasmons, formulated [11] for large metallic nanospheres in analogy to RPA theory of plasmons in bulk metal [28, 29] (Pines-Bohm theory). The advantage of this description [11, 15] consists in full-analytic formulation allowing for application to more complicated physical situations, as e.g., plasmons interaction with surrounding medium, in order to describe plasmonic enhancement of photo-voltaic (PV) effect in the case of metallic particles deposited on the photo-active layer of the semiconductor, or for description of transport of plasmon-polaritons, with potential applications for sub-diffraction nano-photonics [6].

The coupling of surface plasmons in the near-field zone with another systems, like a semiconductor substrate or other metallic nanosphere in the chain, is the main topic of the present paper. An assessment of the energy transfer via this channel, as made in the present paper within the Fermi golden rule scheme, reveals a high efficiency of plasmon-mediated energy transport and explains the experimentally observed PV efficiency growth of solar cell setups with surface metallic nano-modifications. In seems to be of a particular significance for thin film semiconductor solar cell technology including conjugated polymer semiconductor photo-active matrices with a great potential commercial usage, providing an increase of their efficiency by relatively not-costly modifications, e.g., by sparse coverings of photo-active surface with noble metal nanoparticles, which is feasible in various techniques.

The paper is organized as follows. In the next section a short review of the model is given including results of analyzes both for volume and surface plasmon excitations in spherical geometry. Damping phenomena with particular role of radiation losses for large nanospheres is described in the following section, in terms of the Lorentz friction. The separate discussion is addressed to giant enhancement of photo-voltaic effect by plasmon mediation in near-field zone. The comparison with experimental data related to radiation of plasmons in metallic nanospheres is also given. In the last section, the undamped mode of collective surface plasmon excitation along the metallic nanosphere chain is demonstrated, including nonlinear effects.

2. Surface and volume plasmons in large metallic nanospheres

2.1. RPA equation for collective electron oscillations in confined geometry

The Hamiltonian for the metallic nanosphere has the form:

$$\begin{aligned}\hat{H} = &-\sum_{\nu=1}^{N}\frac{\hbar^2\nabla_\nu^2}{2M} + \frac{1}{2}\sum_{\nu\neq\nu'} u(\boldsymbol{R}_\nu - \boldsymbol{R}_{\nu'}) - \sum_{j=1}^{N_e}\frac{\hbar^2\nabla_j^2}{2m} + \frac{1}{2}\sum_{j\neq j'}\frac{e^2}{|\boldsymbol{r}_j - \boldsymbol{r}_{j'}|}\\ &+\sum_{\nu,j} w(\boldsymbol{R}_\nu - \boldsymbol{r}_j),\end{aligned} \tag{1}$$

where $\boldsymbol{R}_\nu$, $\boldsymbol{r}_j$ and M, m are the positions and masses of the ions and electrons, respectively; N is the number of ions in the sphere, $N_e = ZN$ is the number of collective electrons, $u(\boldsymbol{R}_\nu - \boldsymbol{R}_{\nu'})$ is the interaction of ions (ion is treated as a nucleus with electron core of closed shells) and $w(\boldsymbol{R}_\nu - \boldsymbol{r}_j)$ is the local pseudopotential of electron-ion interaction. Assuming the jellium model [3, 5, 7] one can write for the background ion charge uniformly distributed over the sphere: $n_e(\boldsymbol{r}) = n_e\Theta(a-r)$, where $n_e = N_e/V$ and $n_e|e|$ is the averaged positive charge density, $V = \frac{4\pi a^3}{3}$ is the sphere volume and Θ is the Heaviside step-function.

A local electron density can be written as follows [28, 29]:

$$\rho(\boldsymbol{r},t) = <\Psi_e(t)|\sum_j \delta(\boldsymbol{r}-\boldsymbol{r}_j)|\Psi_e(t)>, \tag{2}$$

with the Fourier picture:

$$\tilde{\rho}(\boldsymbol{k},t) = \int \rho(\boldsymbol{r},t)e^{-i\boldsymbol{k}\cdot\boldsymbol{r}}d^3r = <\Psi_e(t)|\hat{\rho}(\boldsymbol{k})|\Psi_e(t)>, \tag{3}$$

where the operator $\hat{\rho}(\boldsymbol{k}) = \sum_j e^{-i\boldsymbol{k}\cdot\boldsymbol{r}_j}$.

Using the above notation one can rewrite the electron part of the Hamiltonian (1), $\hat{H}_e$, in the following form [11]:

$$\begin{aligned}\hat{H}_e = &\sum_{j=1}^{N_e}\left[-\frac{\hbar^2\nabla_j^2}{2m}\right] - \frac{e^2}{(2\pi)^3}\int d^3k\tilde{n}_e(\boldsymbol{k})\frac{2\pi}{k^2}\left(\hat{\rho^+}(\boldsymbol{k}) + \hat{\rho}(\boldsymbol{k})\right)\\ &+\frac{e^2}{(2\pi)^3}\int d^3k\frac{2\pi}{k^2}\left[\hat{\rho^+}(\boldsymbol{k})\hat{\rho}(\boldsymbol{k}) - N_e\right],\end{aligned} \tag{4}$$

where: $\tilde{n}_e(\boldsymbol{k}) = \int d^3 r n_e(\boldsymbol{r}) e^{-i\boldsymbol{k}\cdot\boldsymbol{r}}$, $\frac{4\pi}{k^2} = \int d^3 r \frac{1}{r} e^{-i\boldsymbol{k}\cdot\boldsymbol{r}}$.

The motion equation has the form:

$$\frac{d^2\hat{\rho}(\boldsymbol{k})}{dt^2} = \frac{1}{(i\hbar)^2}\left[\left[\hat{\rho}(\boldsymbol{k}), \hat{H}_e\right], \hat{H}_e\right]. \tag{5}$$

Within the RPA and for Thomas-Fermi averaged kinetic energy formula [28], Eq. (5) attains the form [11]:

$$\begin{array}{l}\frac{\partial^2\delta\tilde{\rho}(\boldsymbol{r},t)}{\partial t^2} = \left[\frac{2}{3}\frac{\epsilon_F}{m}\nabla^2\delta\tilde{\rho}(\boldsymbol{r},t) - \omega_p^2\delta\tilde{\rho}(\boldsymbol{r},t)\right]\Theta(a-r) \\ -\frac{2}{3m}\nabla\left\{\left[\frac{3}{5}\epsilon_F n_e + \epsilon_F\delta\tilde{\rho}(\boldsymbol{r},t)\right]\frac{\boldsymbol{r}}{r}\delta(a-r)\right\} \\ -\left[\frac{2}{3}\frac{\epsilon_F}{m}\frac{\boldsymbol{r}}{r}\nabla\delta\tilde{\rho}(\boldsymbol{r},t) + \frac{\omega_p^2}{4\pi}\frac{\boldsymbol{r}}{r}\nabla\int d^3 r_1\frac{1}{|\boldsymbol{r}-\boldsymbol{r}_1|}\delta\tilde{\rho}(\boldsymbol{r}_1,t)\right]\delta(a-r).\end{array} \tag{6}$$

In the above formula ω_p is the bulk plasmon frequency, $\omega_p^2 = \frac{4\pi n_e e^2}{m}$ (it was taken into account that $\nabla\Theta(a-r) = -\frac{\boldsymbol{r}}{r}\delta(a-r)$). The solution of Eq. (6) can be decomposed into two parts related to the distinct domains:

$$\delta\tilde{\rho}(\boldsymbol{r},t) = \begin{cases}\delta\tilde{\rho}_1(\boldsymbol{r},t), \ for\ r < a, \\ \delta\tilde{\rho}_2(\boldsymbol{r},t), \ for\ r \geq a, \ (r \to a+),\end{cases} \tag{7}$$

corresponding to the volume and surface excitations, respectively. These two parts of local electron density fluctuations satisfy the equations (according to Eq. (6)):

$$\frac{\partial^2\delta\tilde{\rho}_1(\boldsymbol{r},t)}{\partial t^2} = \frac{2}{3}\frac{\epsilon_F}{m}\nabla^2\delta\tilde{\rho}_1(\boldsymbol{r},t) - \omega_p^2\delta\tilde{\rho}_1(\boldsymbol{r},t), \tag{8}$$

and (here $\epsilon = 0+$)

$$\begin{array}{l}\frac{\partial^2\delta\tilde{\rho}_2(\boldsymbol{r},t)}{\partial t^2} = -\frac{2}{3m}\nabla\left\{\left[\frac{3}{5}\epsilon_F n_e + \epsilon_F\delta\tilde{\rho}_2(\boldsymbol{r},t)\right]\frac{\boldsymbol{r}}{r}\delta(a+\epsilon-r)\right\} \\ -\left[\frac{2}{3}\frac{\epsilon_F}{m}\frac{\boldsymbol{r}}{r}\nabla\delta\tilde{\rho}_2(\boldsymbol{r},t) + \frac{\omega_p^2}{4\pi}\frac{\boldsymbol{r}}{r}\nabla\int d^3 r_1\frac{1}{|\boldsymbol{r}-\boldsymbol{r}_1|}(\delta\tilde{\rho}_1(\boldsymbol{r}_1,t)\Theta(a-r_1)\right. \\ \left.+\delta\tilde{\rho}_2(\boldsymbol{r}_1,t)\Theta(r_1-a))\right]\delta(a+\epsilon-r).\end{array} \tag{9}$$

2.2. Solutions of plasmon RPA equations

Eqs (8) and (9) can be solved upon imposed boundary and symmetry conditions. Let us represent both parts of the electron fluctuation in the following manner:

$$\begin{array}{l}\delta\tilde{\rho}_1(\boldsymbol{r},t) = n_e\left[f_1(r) + F(\boldsymbol{r},t)\right], \ for\ r < a, \\ \delta\tilde{\rho}_2(\boldsymbol{r},t) = n_e f_2(r) + \sigma(\Omega,t)\delta(r+\epsilon-a), \ \epsilon = 0+, \ for\ r \geq a, \ (r \to a+),\end{array} \tag{10}$$

and now let us choose the convenient initial conditions, $F(\boldsymbol{r},t)|_{t=0} = 0$, $\sigma(\Omega,t)|_{t=0} = 0$, ($\Omega$ is the spherical angle), moreover $(1 + f_1(r))|_{r=a} = f_2(r)|_{r=a}$ (continuity condition), $F(\boldsymbol{r},t)|_{r\to a} = 0$, $\int\rho(\boldsymbol{r},t)d^3r = N_e$ (neutrality condition).

We arrive [11, 15] thus with the explicit form of the solutions of Eqs (8, 9):

$$\begin{array}{l}f_1(r) = -\frac{k_T a+1}{2}e^{-k_T(a-r)}\frac{1-e^{-2k_T r}}{k_T r}, \ for\ \ r < a, \\ f_2(r) = \left[k_T a - \frac{k_T a+1}{2}\left(1-e^{-2k_T a}\right)\right]\frac{e^{-k_T(r-a)}}{k_T r}, \ for\ \ r \geq a,\end{array} \tag{11}$$

where $k_T = \sqrt{\frac{6\pi n_e e^2}{\epsilon_F}} = \sqrt{\frac{3\omega_p^2}{v_F^2}}$ (Thomas-Fermi inverse radius). For the time-dependent parts of the electron fluctuations we find [11, 12]:

$$F(\boldsymbol{r},t) = \sum_{l=1}^{\infty}\sum_{m=-l}^{l}\sum_{n=1}^{\infty} A_{lmn} j_l(k_{nl}r) Y_{lm}(\Omega) sin(\omega_{nl}t), \tag{12}$$

and

$$\begin{aligned}\sigma(\Omega,t) &= \sum_{l=1}^{\infty}\sum_{m=-l}^{l} \frac{B_{lm}}{a^2} Y_{lm}(\Omega) sin(\omega_{0l}t)\\ &+ \sum_{l=1}^{\infty}\sum_{m=-l}^{l}\sum_{n=1}^{\infty} A_{lmn} \frac{(l+1)\omega_p^2}{l\omega_p^2-(2l+1)\omega_{nl}^2} Y_{lm}(\Omega) n_e \int_0^a dr_1 \frac{r_1^{l+2}}{a^{l+2}} j_l(k_{nl}r_1) sin(\omega_{nl}t),\end{aligned} \tag{13}$$

where $j_l(\xi) = \sqrt{\frac{\pi}{2\xi}} I_{l+1/2}(\xi)$ is the spherical Bessel function, $Y_{lm}(\Omega)$ is the spherical function, $\omega_{nl} = \omega_p\sqrt{1+\frac{x_{nl}^2}{k_T^2 a^2}}$ are the frequencies of electron volume self-oscillations (volume plasmon frequencies), x_{nl} are the nodes of the Bessel function $j_l(\xi)$, $k_{nl} = x_{nl}/a$, $\omega_{0l} = \omega_p\sqrt{\frac{l}{2l+1}}$ are the frequencies of electron surface self-oscillations (surface plasmon frequencies).

From the above equations it follows thus that the local electron density (within RPA attitude) has the form:

$$\rho(\boldsymbol{r},t) = \rho_0(r) + \rho_{neq}(\boldsymbol{r},t), \tag{14}$$

where the RPA equilibrium electron distribution is (correcting the uniform distribution n_e):

$$\rho_0(r) = \begin{cases} n_e\left[1+f_1(r)\right], \ for\ \ r<a,\\ n_e f_2(r),\ for\ \ r\geq a,\ r\to a+ \end{cases} \tag{15}$$

and the nonequilibrium, of plasmon oscillation type, is:

$$\rho_{neq}(\boldsymbol{r},t) = \begin{cases} n_e F(\boldsymbol{r},t),\ for\ \ r<a,\\ \sigma(\Omega,t)\delta(a+\epsilon-r),\ \epsilon=0+,\ for\ \ r\geq a,\ r\to a+. \end{cases} \tag{16}$$

The function $F(\boldsymbol{r},t)$ displays volume plasmon oscillations, while $\sigma(\Omega,t)$ describes the surface plasmon oscillations. Let us emphasize that in the formula for $\sigma(\Omega,t)$, Eq. (13), the first term corresponds to surface self-oscillations, while the second term describes the surface oscillations induced by the volume plasmons. The frequencies of the surface self-oscillations are

$$\omega_{0l} = \omega_p\sqrt{\frac{l}{2l+1}}, \tag{17}$$

which, for $l=1$, agrees with the dipole type surface oscillations described originally by Mie [23], $\omega_{01} = \omega_p/\sqrt{3}$ (for simplicity, denoted hereafter as $\omega_1 = \omega_{01}$).

3. Damping of plasmons in large nanospheres

One can phenomenologically include damping of plasmons in analogy to oscillator damping via the additional term, $-\frac{2}{\tau_0}\frac{\partial\delta\rho(\boldsymbol{r},t)}{\partial t}$, to the right-hand-side of plasmon dynamic equations. They attain the form:

$$\frac{\partial^2\delta\rho_1(\boldsymbol{r},t)}{\partial t^2} + \frac{2}{\tau_0}\frac{\partial\delta\rho_1(\boldsymbol{r},t)}{\partial t} = \frac{2}{3}\frac{\epsilon_F}{m}\nabla^2\delta\rho_1(\boldsymbol{r},t) - \omega_p^2\delta\rho_1(\boldsymbol{r},t), \tag{18}$$

and
$$\frac{\partial^2\delta\rho_2(\boldsymbol{r},t)}{\partial t^2}+\frac{2}{\tau_0}\frac{\partial\delta\rho_2(\boldsymbol{r},t)}{\partial t}=-\frac{2}{3m}\nabla\left\{\left[\frac{3}{5}\epsilon_F n_e+\epsilon_F\delta\rho_2(\boldsymbol{r},t)\right]\frac{\boldsymbol{r}}{r}\delta(a+\epsilon-r)\right\}$$
$$-\left[\frac{2}{3}\frac{\epsilon_F}{m}\frac{\boldsymbol{r}}{r}\nabla\delta\rho_2(\boldsymbol{r},t)+\frac{\omega_p^2}{4\pi}\frac{\boldsymbol{r}}{r}\nabla\int d^3r_1\frac{1}{|\boldsymbol{r}-\boldsymbol{r}_1|}\left(\delta\rho_1(\boldsymbol{r}_1,t)\Theta(a-r_1)\right.\right. \tag{19}$$
$$\left.\left.+\frac{1}{\varepsilon}\delta\rho_2(\boldsymbol{r}_1,t)\Theta(r_1-a)\right)+\frac{en_e}{m}\frac{\boldsymbol{r}}{r}\cdot\boldsymbol{E}(t)\right]\delta(a+\epsilon-r).$$

For the homogeneous forcing field $\boldsymbol{E}(t)$ (this corresponds to dipole approximation satisfied for $a \sim 10-50$ nm, when $\lambda \sim 500$ nm), only dipole surface mode can be excited and the electron dynamics resolves to the equation for a single dipole type mode, described by the function $Q_{1m}(t)$. The function $Q_{1m}(t)$ satisfies the equation:

$$\frac{\partial^2 Q_{1m}(t)}{\partial t^2}+\frac{2}{\tau_0}\frac{\partial Q_{1m}(t)}{\partial t}+\omega_1^2 Q_{1m}(t)$$
$$=\sqrt{\frac{4\pi}{3}}\frac{en_e}{m}\left[E_z(t)\delta_{m0}+\sqrt{2}\left(E_x(t)\delta_{m1}+E_y(t)\delta_{m-1}\right)\right], \tag{20}$$

where $\omega_1=\omega_{01}=\frac{\omega_p}{\sqrt{3\varepsilon}}$ (it is a dipole-type surface plasmon Mie frequency [23]). Only this function contributes to the dynamical response to the homogeneous electric field. Thus for the homogeneous forcing field, electron density fluctuations:

$$\delta\rho(\boldsymbol{r},t)=\begin{cases}0, \; for \;\; r<a,\\ \sum\limits_{m=-1}^{1} Q_{1m}(t)Y_{1m}(\Omega)\; for \;\; r\geq a,\; r\rightarrow a+.\end{cases} \tag{21}$$

For plasmon oscillations given by Eq. (21) one can calculate the corresponding dipole,

$$\boldsymbol{D}(t)=e\int d^3r\boldsymbol{r}\delta\rho(\boldsymbol{r},t)=\frac{4\pi}{3}e\boldsymbol{q}(t)a^3, \tag{22}$$

where $Q_{11}(t)=\sqrt{\frac{8\pi}{3}}q_x(t)$, $Q_{1-1}(t)=\sqrt{\frac{8\pi}{3}}q_y(t)$, $Q_{10}(t)=\sqrt{\frac{4\pi}{3}}q_x(t)$ and $\boldsymbol{q}(t)$ satisfies the equation (cf. Eq. (20)),

$$\left[\frac{\partial^2}{\partial t^2}+\frac{2}{\tau_0}\frac{\partial}{\partial t}+\omega_1^2\right]\boldsymbol{q}(t)=\frac{en_e}{m}\boldsymbol{E}(t). \tag{23}$$

There are various mechanisms of plasmon damping, which could be effectively accounted for via phenomenological oscillator type damping term. All types of scattering phenomena, including electron-electron and electron-phonon interactions, as well contribution of boundary scattering effect [6] cause significant attenuation of plasmons, in particular, in small metal clusters. All these contributions to damping time ratio scale as $\frac{1}{a}$ and are of lowering significance with the radius growth. In the following subsection we argue that damping of plasmons caused by radiation losses scales conversely, as a^3, and for large nanospheres this channel dominates plasmon attenuation.

3.1. Lorentz friction for plasmons

The nanosphere surface plasmons can be induced by a homogeneous electric field [15], while the volume mode excitations need field inhomogeneity on the radius scale (and therefore the visible light cannot excite volume modes in the nanospheres with radii of $10-50$ nm). Plasmon oscillations are themselves a source of the e-m radiation. This radiation takes away the energy of plasmons resulting in their damping, which can be described as the

Lorentz friction [18]. This damping was not included in τ_0 in Eq. (23). The latter accounted for scattering of electrons on other electrons, on defects, on phonons and on nanoparticle boundary—all they lead to damping rate expressed by the simplified formula [6]:

$$\frac{1}{\tau_0} \simeq \frac{v_F}{2\lambda_B} + \frac{Cv_F}{2a}, \tag{24}$$

where, C is the constant of unity order, a is the nanosphere radius, v_F is the Fermi velocity in metal, λ_B is the electron free path in bulk (including scattering of electrons on other electrons, on impurities and on phonons [6]); for Ag, $v_F = 1.4 \times 10^6$ m/s and $\lambda_B \simeq 57$ nm (at room temperature); the latter term in the formula (24) accounts for scattering of electrons on the boundary of the nanoparticle, while the former one corresponds to scattering processes similar as in bulk. The other effects, as the so-called Landau damping (especially important in small clusters [9, 35]), corresponding to decay of plasmon for high energy particle-hole pair, are of lowering significance for nanosphere radii larger than $2-3$ nm [35] and are completely negligible for radii larger than 10 nm. Note that the similarly lowering role with the radius growth plays also electron liquid spill-out effect [5, 8], though it was of primary importance for small clusters [5, 17].

The e-m wave emission caused electron friction can be described as the additional electric field [18],

$$\boldsymbol{E}_L = \frac{2}{3\varepsilon^{3/2}v^3}\frac{\partial^3 \boldsymbol{D}(t)}{\partial t^3}, \tag{25}$$

where $v = \frac{c}{\sqrt{\varepsilon}}$ is the light velocity in the dielectric medium, and $\boldsymbol{D}(t)$ is the dipole of the nanosphere. According to Eq. (22) we arrive at the following:

$$\boldsymbol{E}_L = \frac{2e}{3\varepsilon v^2}\frac{4\pi}{3}a^3\frac{\partial^3 \boldsymbol{q}(t)}{\partial t^3}. \tag{26}$$

Substituting this into Eq. (23), we get

$$\left[\frac{\partial^2}{\partial t^2} + \frac{2}{\tau_0}\frac{\partial}{\partial t} + \omega_1^2\right]\boldsymbol{q}(t) = \frac{en_e}{m}\boldsymbol{E}(t) + \frac{2}{3\omega_1}\left(\frac{\omega_1 a}{v}\right)^3\frac{\partial^3 \boldsymbol{q}(t)}{\partial t^3}. \tag{27}$$

If one rewrites the above equation (for $\boldsymbol{E}$=0) in the form

$$\left[\frac{\partial^2}{\partial t^2} + \omega_1^2\right]\boldsymbol{q}(t) = \frac{\partial}{\partial t}\left[-\frac{2}{\tau_0} + \frac{2}{3\omega_1}\left(\frac{\omega_1 a}{v}\right)^3\frac{\partial^2 \boldsymbol{q}(t)}{\partial t^2}\right], \tag{28}$$

thus, one notes that the zeroth order approximation (neglecting attenuation) corresponds to the equation:

$$\left[\frac{\partial^2}{\partial t^2} + \omega_1^2\right]\boldsymbol{q}(t) = 0. \tag{29}$$

In order to solve Eq. (28) in the next step of perturbation iteration, one can substitute, in the right-hand-side of this equation, $\frac{\partial^2 \boldsymbol{q}(t)}{\partial t^2}$ by $-\omega_1^2\boldsymbol{q}(t)$ (acc. to Eq. (29)).

Therefore, if one assumes the above estimation, $\frac{\partial^3 \boldsymbol{q}(t)}{\partial t^3} \simeq -\omega_1^2\frac{\partial \boldsymbol{q}(t)}{\partial t}$, one can include the Lorentz friction in a renormalized damping term:

$$\left[\frac{\partial^2}{\partial t^2} + \frac{2}{\tau}\frac{\partial}{\partial t} + \omega_1^2\right]\boldsymbol{q}(t) = \frac{en_e}{m}\boldsymbol{E}(t), \tag{30}$$

where

$$\frac{1}{\tau} = \frac{1}{\tau_0} + \frac{\omega_1}{3}\left(\frac{\omega_1 a}{v}\right)^3 \simeq \frac{v_F}{2\lambda_B} + \frac{Cv_F}{2a} + \frac{\omega_1}{3}\left(\frac{\omega_1 a}{v}\right)^3, \quad (31)$$

where we used for $\frac{1}{\tau_0} \simeq \frac{v_F}{2\lambda_B} + \frac{Cv_F}{2a}$ [6]. Renormalized damping causes a change in the shift of self-frequencies of free surface plasmons, $\omega_1' = \sqrt{\omega_1^2 - \frac{1}{\tau^2}}$.

Note also, that one can verify [15] the above calculated Lorentz friction contribution to plasmon damping by estimation of the energy transfer in the far-field zone (which can be expressed by the Poynting vector) and via comparison with the energy loss of plasmon oscillation. We have arrived [11, 15] at the same formula for damping time rate as given by Eq. (30). The radius dependent shift of the resonance resulting due to strong irradiation-induced plasmon damping was verified also experimentally [15] by measurement of light extinction in colloidal solutions of nanoparticles with different size (it is done [15] for Au, $10-80$ nm, and Ag, $10-60$ nm). These measurements clearly support the a^3 plasmon damping behavior, as described above for the far-field zone radiation losses in a dielectric surroundings.

If, however, in the vicinity of the nanosphere the another system is located, the situation would change. For the case when the nanosphere is deposited on the semiconductor surface, the near-field coupling of plasmons with semiconductor band electrons must be included.

4. Mediating of light energy transfer by surface plasmons in near-field regime

Let us calculate the probability (per time unit) of inter-band transitions of electrons in substrate semiconductor covered with metallic nanospheres (with radius a), induced by photon-excited surface plasmons in metallic components. These plasmons excited in nanospheres deposited on the semiconductor layer, couple to the semiconductor band electrons in the near-field regime. The situation is similar to the ordinary photo-effect, though the perturbation of the electron system in semiconductor is not of a plane-wave form as it was in the case of direct illumination of semiconductor by incident photons, but attains the form of dipole-type near-field electric interaction [11, 18]. This causes a change of the matrix element in the relevant Fermi golden formula. The potential for the near-field interaction of the surface Mie type plasmons with the band electrons can be written as [18]:

$$\begin{aligned} w &= e\psi(\mathbf{R},t) = \frac{e}{\varepsilon_0 R^2}\hat{\mathbf{n}}\cdot\mathbf{D}_0 sin(\omega t+\alpha) = w^+e^{i\omega t} + w^- e^{-i\omega t}, \\ w^+ &= \left(w^-\right)^* = \frac{e}{\varepsilon_0 R^2}\frac{e^{i\alpha}}{2i}\hat{\mathbf{n}}\cdot\mathbf{D}_0. \end{aligned} \quad (32)$$

The terms w^+, w^- correspond to emission and absorption, respectively ($\hat{\mathbf{n}} = \frac{\mathbf{R}}{R}$, $\mathbf{D}_0$ is the dipole plasmon amplitude). We choose the first one in order to consider emission of energy from the plasmon oscillations and transfer of it to the electron system in the semiconductor substrate. The semiconductor band system we model in the simplest single-band parabolic effective mass approximation. The inter-band transition probability is given by the Fermi golden rule,

$$w(\mathbf{k}_1,\mathbf{k}_2) = \frac{2\pi}{\hbar}\left|<\mathbf{k}_1|w^+|\mathbf{k}_2>\right|^2 \delta(E(\mathbf{k}_1) - E(\mathbf{k}_2) + \hbar\omega), \quad (33)$$

where Bloch states in the conduction and valence bands we assume as planar waves (for the sake of simplicity),

$$\begin{array}{l} \Psi_{\mathbf{k}_1} = \frac{1}{(2\pi)^{3/2}} e^{i\mathbf{k}_1 \cdot \mathbf{R} - iE(\mathbf{k}_1)t/\hbar}, \Psi_{\mathbf{k}_2} = \frac{1}{(2\pi)^{3/2}} e^{i\mathbf{k}_2 \cdot \mathbf{R} - iE(\mathbf{k}_2)t/\hbar}, \\ E(\mathbf{k}_1) = -\frac{\hbar^2 \mathbf{k}_1^2}{2m_p^*} - E_g, E(\mathbf{k}_2) = \frac{\hbar^2 \mathbf{k}_2^2}{2m_n^*}, \end{array} \tag{34}$$

(indices n, p refer to electrons from the conduction and valence bands, respectively, E_g is the forbidden gap).

The electron wave functions are normalized to the Dirac delta, which corresponds to infinite movement and the continuous energy spectrum. The wave function modulus squares do not have, in this case, probability interpretation (which must be normalized to the unity), therefore the expression, $w(\mathbf{k}_1, \mathbf{k}_2) = \frac{2\pi}{\hbar} \left|< \mathbf{k}_1 | w^+ | \mathbf{k}_2 >\right|^2 \delta(E(\mathbf{k}_1) - E(\mathbf{k}_2) + \hbar\omega)$, has to be divided by the delta Dirac square, i.e., by the factor, $\left(\frac{V}{(2\pi)^3}\right)^2$. This factor corresponds to the probability proper normalization, because,

$$\frac{1}{(2\pi)^3} \int d^3 r e^{i\mathbf{k} \cdot \mathbf{r}} = \delta(\mathbf{k}) \simeq \ (for\ \mathbf{k} = 0)\ \frac{V}{(2\pi)^3}\ (V \to \infty). \tag{35}$$

Note, that the same factor occurs also due to density of states, when one integrates over all initial and final states $\mathbf{k}_1, \mathbf{k}_2$. Two integrals will give the factor $\left(\frac{2V}{(2\pi)^3}\right)^2$, (2 is caused by spin degeneration). Thus, both renormalized factors cancel mutually themselves.

Taking into account the above described renormalization, one can find the matrix element,

$$< \mathbf{k}_1 | w^+ | \mathbf{k}_2 >= \frac{1}{(2\pi)^3} \int d^3 R \frac{e}{\varepsilon_0 2i} e^{i\alpha} \hat{\mathbf{n}} \cdot \mathbf{D}_0 \frac{1}{R^2} e^{-i(\mathbf{k}_1 - \mathbf{k}_2) \cdot \mathbf{R}}. \tag{36}$$

Let us introduce the vector $\mathbf{q} = \mathbf{k}_2 - \mathbf{k}_1$. One can choose the coordinate system in such a way, that the vector $\mathbf{q}$ is oriented along the z axis, and the vector $\mathbf{D}_0$ lies in the plane zx (as is depicted in the Fig. 1). Then, $\mathbf{q} = (0, 0, q)$, $\hat{\mathbf{n}} = \frac{\mathbf{R}}{R} = (sin\Theta_1 cos\psi_1, sin\Theta_1 sin\psi_1, cos\Theta_1)$, $\mathbf{R} = R(sin\Theta_1 cos\psi_1, sin\Theta_1 sin\psi_1, cos\Theta_1)$, $\mathbf{D}_0 = D_0(sin\Theta, 0, cos\Theta)$ and

$$\mathbf{q} \cdot \mathbf{R} = qRcos\Theta_1, \tag{37}$$

$$\hat{\mathbf{n}} \cdot \mathbf{D}_0 = D_0(sin\Theta sin\Theta_1 cos\psi_1 + cos\Theta cos\Theta_1). \tag{38}$$

Hence,

$$\begin{array}{l} < \mathbf{k}_1 | w^+ | \mathbf{k}_2 > \\ = \frac{1}{(2\pi)^3} \int d^3 R \frac{e}{\varepsilon_0 2i} e^{i\alpha} \hat{\mathbf{n}} \cdot \mathbf{D}_0 \frac{1}{R^2} e^{-i(\mathbf{k}_1 - \mathbf{k}_2) \cdot \mathbf{R}} \\ = \frac{1}{(2\pi)^3} \frac{e e^{i\alpha}}{\varepsilon_0 2i} D_0 \int_a^\infty \frac{R^2}{R^2} dR \int_0^\pi \\ \times sin\Theta_1 d\Theta_1 \int_0^{2\pi} d\psi_1 \{cos\Theta cos\Theta_1 + sin\Theta sin\Theta_1 cos\psi_1\} e^{iqRcos\Theta_1} \\ = \frac{1}{(2\pi)^3} \frac{e e^{i\alpha}}{\varepsilon_0 2i} D_0 cos\Theta 2\pi \int_a^\infty dR \int_0^\pi cos\Theta_1 sin\Theta_1 d\Theta_1 e^{iqRcos\Theta_1}, \end{array} \tag{39}$$

the integer over $d\psi_1$ vanishes the second term in the parenthesis and only the first term contributes with the factor 2π. Note that,

$$\int_0^\pi cos\Theta_1 sin\Theta_1 d\Theta_1 e^{ixcos\Theta_1} = -i\frac{d}{dx} \int_0^\pi sin\Theta_1 e^{ixcos\Theta_1} = -i\frac{d}{dx} 2\frac{sinx}{x}. \tag{40}$$

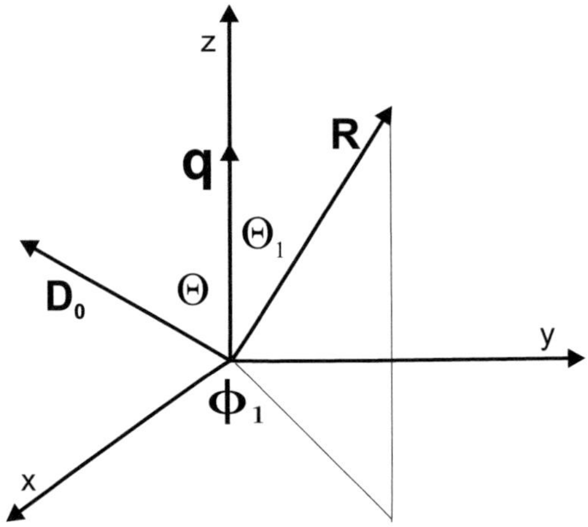

Figure 1. The reference frame is chosen in the way that the vector $\mathbf{q}$ is oriented along the axis z, while the vector $\mathbf{D}_0$ lies in the plane xz

Thus,

$$\begin{aligned} < \mathbf{k}_1|w^+|\mathbf{k}_2 >= \frac{-1}{(2\pi)^3}\frac{ee^{i\alpha}}{\varepsilon_0}D_0cos\Theta(2\pi)\int_a^\infty dR\frac{1}{q}\frac{d}{dR}\frac{sinqR}{qR} \\ = \frac{1}{(2\pi)^2}\frac{ee^{i\alpha}}{\varepsilon_0}D_0cos\Theta\frac{1}{q}\frac{sinqa}{qa} \rightarrow_{a\rightarrow 0}\frac{1}{(2\pi)^2}\frac{ee^{i\alpha}}{\varepsilon_0}D_0cos\Theta\frac{1}{q}. \end{aligned} \tag{41}$$

The lower limit of integral with respect to R is taken as a (the nanosphere radius). In this manner one can limit the accessible space for planar waves in the semiconductor, if assumes, for a model, a completely embedded nanosphere in surrounding material. This is, however, a limiting approximation as tunneling effect certainly would allow an access to the inner space, to some extent, at least. In order to account for that the nanosphere is deposited on the semiconductor surface we will confine integration with respect to $d\Theta_1$ to the segment $[0, \pi/2]$, instead of $[0, \pi]$. We obtain in this case the following matrix element,

$$\begin{aligned} < \mathbf{k}_1|w^+|\mathbf{k}_2 >= \frac{1}{(2\pi)^3}\frac{ee^{i\alpha}}{2i\varepsilon_0}D_0cos\Theta(2\pi)\int_a^\infty dR\frac{1}{q}\frac{d}{dR}\frac{1-e^{iqR}}{qR} \\ = \frac{1}{(2\pi)^2}\frac{ee^{i\alpha}}{2i\varepsilon_0}D_0cos\Theta\frac{1}{q}\frac{1-e^{iqa}}{qa} \rightarrow_{a\rightarrow 0}\frac{1}{(2\pi)^2}\frac{ee^{i\alpha}}{2i\varepsilon_0}D_0cos\Theta\frac{1}{q}, \end{aligned} \tag{42}$$

because,

$$\int_0^{\pi/2} cos\Theta_1 sin\Theta_1 d\Theta_1 e^{ixcos\Theta_1} = -i\frac{d}{dx}\int_0^{\pi/2} sin\Theta_1 e^{ixcos\Theta_1} = \frac{d}{dx}\frac{1-e^{ix}}{x}. \tag{43}$$

In the limit $a = 0$, the modulus of the matrix element will diminish twice in comparison to the case when integration over $d\Theta_1$ had been taken over the whole space.

Now we will integrate over all initial and final states of both bands. The related integration over $\mathbf{k}_1, \mathbf{k}_2$, one can substitute with integration over $\mathbf{q}, \mathbf{k}_2$. Scalar products are invariant against coordinate systems rotations, therefore the result of integration will be the same if $\mathbf{q}$ was along z axis or $\mathbf{D}_0$ is oriented now along the z axis – Θ gives, in the latter case, the deviation of $\mathbf{q}$ with respect to z direction – this choice of the reference frame is convenient for integration with respect to $d\mathbf{q}$.

Thus we arrive with the formula for transition probability in the following form,

$$\delta w = \int d^3k_1 \int d^3k_2 \left[f_1(1-f_2)w(\mathbf{k}_1,\mathbf{k}_2) - f_2(1-f_1)w(\mathbf{k}_2,\mathbf{k}_1)\right], \tag{44}$$

where, f_1, f_2 assign the temperature dependent distribution functions (Fermi-Dirac distribution functions) for initial and final states, respectively. The emission and absorption were included, but for room temperatures one can assume, $f_2 \simeq 0$ and $f_1 \simeq 1$, which leads to,

$$\delta w = \int d^3k_1 \int d^3k_2 w(\mathbf{k}_1, \mathbf{k}_2). \tag{45}$$

In the above formula we avoided the density state factors canceled by probability renormalization, as was mentioned above.

We have to calculate the following integral,

$$\delta w = \int d^3k_2 \int d^3q \frac{e^2}{(2\pi)^3\hbar\varepsilon_0} \frac{D_0^2 cos^2\Theta}{q^2} \frac{sin^2(qa)}{(qa)^2} \delta\left(\frac{\hbar^2k_1^2}{2m_n^*} + \frac{\hbar^2k_2}{2m_p^*} - (\hbar\omega - E_g)\right), \tag{46}$$

with $\delta\left(\frac{\hbar^2k_1^2}{2m_n^*} + \frac{\hbar^2k_2}{2m_p^*} - (\hbar\omega - E_g)\right) = \frac{1}{\alpha+\beta}\frac{1}{2\beta' qk_2}\delta\left(cos\Theta_2 - \frac{k_2^2+\beta' q^2-\gamma'}{2\beta' qk_2}\right)$, where, $\alpha = \frac{\hbar^2}{2m_n^*}$, $\beta = \frac{\hbar^2}{2m_p^*}$, $\gamma = \hbar\omega - E_g$, $\alpha' = \frac{\alpha}{\alpha+\beta}$, $\beta' = \frac{\beta}{\alpha+\beta}$, $\gamma' = \frac{\gamma}{\alpha+\beta}$. For each integral, with respect to $\mathbf{k}_2$ and $\mathbf{q}$ the coordinate system can be rotated independently, and for integration over $d\mathbf{k}_2$ we choose again orientation of $\mathbf{q}$ along the axis z, which leads to the spherical angle Θ_2 resulting form the product, $\mathbf{k}_2 \cdot \mathbf{q} = k_2 q cos\Theta_2$.

The expression for δw attains thus the form,

$$\begin{array}{l} \delta w = \frac{e^2D_0^2}{(2\pi)^3\hbar\varepsilon_0^2} \int d^3q \frac{sin^2(qa)}{q^2a^2} \frac{cos\Theta}{q^2} \\ \int_0^\infty dk_2k_2^2 \int_0^\pi d\Theta_2 sin\Theta_2 \int_0^{2\pi} d\psi_2 \frac{1}{\alpha+\beta}\frac{1}{2\beta' qk_2}\delta\left(cos\Theta_2 - \frac{k_2^2+\beta' q^2-\gamma'}{2\beta' qk_2}\right). \end{array} \tag{47}$$

Integration over $d\Theta_2$ employs the Dirac delta. The relevant nonzero contribution (due to integration over $dcos\Theta_2$) is conditioned by the inequality,

$$-1 < \frac{k_2^2 + \beta' q^2 - \gamma'}{2\beta' qk_2} < 1, \tag{48}$$

which resolves itself to the condition,

$$|\beta' q - \sqrt{\gamma' - (1-\beta')\beta' q^2}| < k_2 < \beta' q + \sqrt{\gamma' - (1-\beta')\beta' q^2}. \tag{49}$$

Thus,

$$\begin{array}{l} \delta w = \frac{1}{(\alpha+\beta)2\beta'} \frac{e^2D_0^2}{2\pi\hbar\varepsilon_0^2} \int_0^{\sqrt{\frac{\gamma'}{\beta'(1-\beta')}}} dq q^2 \frac{sin^2 qa}{(qa)^2} \\ \times \int_0^\pi d\Theta sin\Theta cos^2\Theta \frac{1}{q^3} \int_{|\beta' q - \sqrt{\gamma'-(1-\beta')\beta' q^2}|}^{\beta' q + \sqrt{\gamma'-(1-\beta')\beta' q^2}} dk_2 k_2. \end{array} \tag{50}$$

and hence,

$$\delta w = \frac{2}{3}\frac{\sqrt{\gamma'}}{\alpha+\beta}\frac{e^2D_0^2}{2\pi\hbar\varepsilon_0^2}\frac{1}{\zeta}\int_0^{\frac{1}{\zeta}} dx \frac{sin^2(x(a/\zeta))}{(x)^2(a/\zeta)^2}\sqrt{1-x)^2}, \tag{51}$$

where, $\zeta = \sqrt{\frac{(1-\beta')\beta'}{\gamma'}}$. Therefore,

$$\begin{array}{l} \delta w = \frac{4}{3}\frac{\mu^2(m_n^*+m_p^*)2(\hbar\omega - E_g)e^2D_0^2}{\sqrt{m_n^*m_p^*}2\pi\hbar^5\varepsilon_0^2} \int_0^1 dx \frac{sin^2(xa/\zeta)}{(xa/\zeta)^2}\sqrt{1-x^2} \\ = \frac{4}{3}\frac{\mu^2}{\sqrt{m_n^*m_p^*}}\frac{e^2D_0^2}{2\pi\hbar^3\varepsilon_0^2}\xi^2 \int_0^1 dx \frac{sin^2(xa\xi)}{(xa\xi)^2}\sqrt{1-x^2}, \end{array} \tag{52}$$

where, $\xi = 1/\zeta = \frac{\sqrt{2(\hbar\omega - E_g)(m_n^* + m_p^*)}}{\hbar}$.

In limiting cases, we obtain finally,

$$\delta w = \begin{cases} \frac{1}{3}\frac{\mu^2(m_n^*+m_p^*)2(\hbar\omega - E_g)e^2D_0^2}{\sqrt{m_n^* m_p^* 2\hbar^5 \varepsilon_0^2}}, & a\xi \ll 1, \\ \frac{4}{3}\frac{\mu^2(m_n^*+m_p^*)2(\hbar\omega - E_g)e^2D_0^2}{\sqrt{m_n^* m_p^* 4a\xi\hbar^5 \varepsilon_0^2}}, & a\xi \gg 1, \end{cases} \tag{53}$$

In the latter case the following approximation was applied,

$$\int_0^1 \frac{sin^2(xa\xi)}{(xa\xi)^2} \approx (for\ a\xi \gg 1)\frac{1}{a\xi}\int_0^\infty d(xa\xi)\frac{sin^2(xa\xi)}{(xa\xi)^2} = \frac{\pi}{2a\xi},$$

while in the former one, $\int_0^1 dx\sqrt{1-x^2} = \pi/4$.

The result must be multiplied by 4 (due to spin) and, in the case of half-space, divided by 4. In the considered limiting cases we obtain thus,

$$\delta w = \begin{cases} \frac{1(4)}{3}\frac{\mu\sqrt{m_n^* m_p^*}(\hbar\omega - E_g)e^2D_0^2}{\hbar^5\varepsilon_0^2} & for\ a\xi \ll 1, \\ \frac{1(4)}{3}\frac{\mu^{3/2}\sqrt{2}\sqrt{\hbar\omega - E_g}e^2D_0^2}{a\hbar^4\varepsilon_0^2} & for\ a\xi \gg 1, \end{cases} \tag{54}$$

where (4) is addressed to the completely embedded nanosphere.

One can notice that the above formulae are quite distinct in comparison to the ordinary photo-effect. In the latter case, the perturbation is given by the vector potential in the kinematic momentum (for gauge, $div\mathbf{A} = 0$),

$$\hat{H} = \frac{(-i\hbar\nabla - \frac{e}{c}\mathbf{A}(\mathbf{R},t))^2}{2m^*} \simeq \frac{(-i\hbar\nabla))^2}{2m^*} + \frac{i\hbar e}{m^* c}\mathbf{A}(\mathbf{R},t)\cdot\nabla. \tag{55}$$

For monochromatic plane wave, this perturbation has the form,

$$\begin{aligned} w(\mathbf{R},t) &= \tfrac{ie\hbar}{cm^*}(\mathbf{A}_0\cdot\nabla)cos(\omega t - \mathbf{k}\cdot\mathbf{R} + \alpha) \\ &= \tfrac{ie\hbar}{2cm^*}(\mathbf{A}_0\cdot\nabla)(e^{i(\omega t - \mathbf{k}\cdot\mathbf{R}+\alpha)} + e^{-i(\omega t - \mathbf{k}\cdot\mathbf{R}+\alpha)}). \end{aligned} \tag{56}$$

Because in this case, both states of band electrons and photon have the form of plane waves, the matrix element in the Fermi golden rule will be proportional to the Dirac delta with respect to the momentum sum. This expresses momentum conservation in translationally invariant system, what is a case for photon interacting with the semiconductor,

$$\begin{aligned} <\mathbf{k}_1|w(\mathbf{R},t)|\mathbf{k}_2> &= \tfrac{i\hbar e}{2cm^*(2\pi)^3}\int d^3Re^{-i(\mathbf{k}_1+\mathbf{k})\cdot\mathbf{R}}(\mathbf{A}_0\cdot\nabla)e^{i\mathbf{k}_2\cdot\mathbf{R}} \\ &= -\tfrac{e\hbar}{2cm^*}\mathbf{A}_0\cdot\mathbf{k}_2\delta(\mathbf{k}_1+\mathbf{k}-\mathbf{k}_2). \end{aligned} \tag{57}$$

Because of high value of photon velocity c, only vertical transitions, $\mathbf{k}_1 = \mathbf{k}_2$, are admitted in the ordinary photo-effect.

Nevertheless, this is no case for nanosphere plasmon interacting in near-field regime with the semiconductor substrate. This system is not translationally invariant and indirect inter-band electron excitations are admitted resulting in enhancement of total transition probability.

To compare with the ordinary photo-effect, let us recall the appropriate calculus upon the Fermi golden rule scheme,

$$w(\mathbf{k}_1, \mathbf{k}_2) = \frac{\pi e^2 \hbar}{2c^2 m^{*2}} k_2^2 A_0^2 cos^2\Theta \delta^2(\mathbf{k}_1 - \mathbf{k}_2)\delta(E_1 - E_2 + \hbar\omega), \quad (58)$$

where, Θ is the angle between $\mathbf{k}_2$ i $\mathbf{A}_0$. One can use the following approximation in order to get rid the Dirac delta square,

$$\delta^2(\mathbf{k}_1 - \mathbf{k}_2) = \delta(0)\delta(\mathbf{k}_1 - \mathbf{k}_2) \simeq \frac{V}{(2\pi)^3}\delta(\mathbf{k}_1 - \mathbf{k}_2), \quad (59)$$

Similarly as was previously discussed, the normalization of probability must be performed. Thus we get,

$$w(\mathbf{k}_1, \mathbf{k}_2) = \frac{\pi e^2 \hbar}{2c^2 m^{*2}} \frac{V}{(2\pi)^3} k_2^2 A_0^2 cos^2\Theta \delta(\mathbf{k}_1 - \mathbf{k}_2)\delta(E_1 - E_2 + \hbar\omega). \quad (60)$$

The integration over all states in both electron bands results in,

$$\delta w = \int d^3k_1 \int d^3k_2 \frac{\pi e^2 \hbar}{2c^2 m^{*2}} \frac{V}{(2\pi)^3} k_2^2 A_0^2 cos^2\Theta \delta(\mathbf{k}_1 - \mathbf{k}_2)\delta(E_1 - E_2 + \hbar\omega), \quad (61)$$

or

$$\delta w = \int d^3k_2 \frac{\pi e^2 \hbar}{2c^2 m_p^{*2}} \frac{V}{(2\pi)^3} k_2^2 A_0^2 cos^2\Theta \delta(E_1(\mathbf{k}_2) - E_2(\mathbf{k}_2) + \hbar\omega), \quad (62)$$

where the effective mass in the initial valence state is m_p^* (moreover, $A_0 = \frac{c}{\omega}E_0$). We thus obtain,

$$\delta w = \frac{1}{12}\frac{e^2\hbar}{m_p^{*2}}\frac{V}{\pi}\frac{E_0^2}{\omega^2}\frac{2^{3/2}\mu^{5/2}}{\hbar^5}(\hbar\omega - E_g)^{3/2}. \quad (63)$$

One can rewrite it in the form,

$$\delta w = \frac{4\sqrt{2}}{3}\frac{\mu^{5/2}e^2}{m^{*2}\omega\varepsilon_0\hbar^3}\left(\frac{\varepsilon_0 E_0^2 V}{8\pi\hbar\omega}\right)(\hbar\omega - E_g)^{3/2}, \quad (64)$$

this expression should be multiplied by 4 (due to spin degeneracy, though for circular polarization of photons this factor is only 2, due to angular momentum selection rules).

Taking into account, that the number of photons in the volume V equals to, $\left(\frac{\varepsilon_0 E_0^2 V}{8\pi\hbar\omega}\right)$, then the probability of single photon attenuation by the semiconductor per time unit, attains the form,

$$q = \delta w \left(\frac{\varepsilon_0 E_0^2 V}{8\pi\hbar\omega}\right)^{-1} = \frac{4(4)\sqrt{2}}{3}\frac{\mu^{5/2}e^2}{m^{*2}\omega\varepsilon_0\hbar^3}(\hbar\omega - E_g)^{3/2}, \quad (65)$$

(factor (4) corresponds here to spin degeneration of band electrons).

In order to assess efficiency of the plasmon near-field coupling channel one can estimate the ratio of the probability of energy absorption in the semiconductor via mediation of surface plasmons (per single photon incident on the metallic nanospheres, q_m) to the energy attenuation in the semiconductor directly from a planar wave illumination (also per single photon, q). The ratio $\frac{q_m}{q}$ turns out to be of order of $10^4 \frac{\beta 40}{H[nm]}$ (at a typical surface density of nanoparticles, $n_s \sim 10^8/\text{cm}^2$), which (including the phenomenological factor β [to account for proximity surface effects, not directly included], and H—the semiconductor photo-active layer depth) is sufficient to explain the scale of the experimentally observed strong enhancement of absorption and emission rates. The strong enhancement of this transition probability is linked with allowance of momentum-non-conserved transitions, which is, however, reducing with the radius a growth, according to formulae (54). For e.g., $a \sim 10 - 60$ nm, this gives reducing of the transition probability, which we have included by the effective phenomenological factor β (fitted from the experiment, $\beta \simeq 28 \times 10^{-3} \left(\frac{50}{a[nm]}\right)^2$) in the case of atomic limit, $a = 0$ [11]. The enhancement of the near-field induced inter-band transition, in the case of large nanospheres, is, however, still significant as the reducing role of size-related quenching of transitions is partly compensated by $\sim a^3$ growth of the dipole amplitude of plasmon oscillations.

High efficiency (even if decreased with growing a) of the near-field energy transfer from surface plasmons to the semiconductor substrate is caused mainly by a contribution of all inter-band transitions, not restricted here to the direct (vertical) ones as for ordinary photo-effect, due to the absence of the momentum conservation constraints for the nanosystem. The strengthening of the probability transition due to all indirect inter-band paths of excitations in the semiconductor is probably responsible for the observed experimentally strong enhancement of light absorption and emission in diode systems mediated by surface plasmons in nanoparticle surface coverings [26, 27, 30, 31, 33, 34].

In the balanced state, when the incoming energy of light is transferred to the semiconductor via plasmon near-field coupling, we deal with the stationary solution of driven and damped oscillator. The driving force is the electric field of the incident planar wave, and the damping force is the near-field energy transfer described by the $\frac{1}{\tau}$ (calculated using formulae (54) in the manner as described in [13]). The resulting red-shifted resonance with simultaneously reduced amplitude allows for the accommodation to the balance of energy transfer to the semiconductor with incident photon energy. Within this model, the amplitude of resonant plasmon oscillations $D_0(\omega)$ is thus shaped by $f(\omega) = \frac{1}{\sqrt{(\omega_1^2-\omega^2)^2+4\omega^2/\tau^2}}$. The extremum of red-shifted resonance is attained at $\omega_m = \omega_1\sqrt{1-2(\omega_1\tau)^{-2}}$ with corresponding amplitude $\sim \tau/\left(2\sqrt{\omega_1^2-\tau^{-2}}\right)$. This shift is proportional to $1/(\omega_1\tau^2)$ and scales with nanosphere radius a similarly (diminishes with decreasing a) as in the experimental observations [30] (note again that for scattering-induced $1/\tau_0$ the dependence on a is opposite [grows with decreasing a]).

In order to compare with the experiment we can estimate the photo-current in the case of a metallic modified surface. This photo-current is given by $I' = |e|N(q+q_m)A$, where N is the number of incident photons and q and q_m are the probabilities of single photon attenuation in the ordinary photo-effect [16] and in that one enhanced due to the presence of metallic nanospheres; $A = \frac{\tau_f^n}{t_n} + \frac{\tau_f^p}{t_p}$ is the amplification factor ($\tau_f^{n(p)}$ is the annihilation time of both sign carriers, $t_{n(p)}$ is the drive time for carriers [the time of traversing the distance between electrodes]). From the above formulae, it follows that (here $I = I'(q_m = 0)$, i.e., the photo-current without metallic modifications)

$$\frac{I'}{I} = 1 + 7.95 \cdot 10^5 c_0 \frac{m_p^*}{m_n^*}\left(\frac{2a}{100[1nm]}\sqrt{\frac{\hbar\omega_1[eV]}{x}}\left(\frac{m_p^*}{m}+\frac{m_n^*}{m}\right)\right)^3 \phi(x), \tag{66}$$

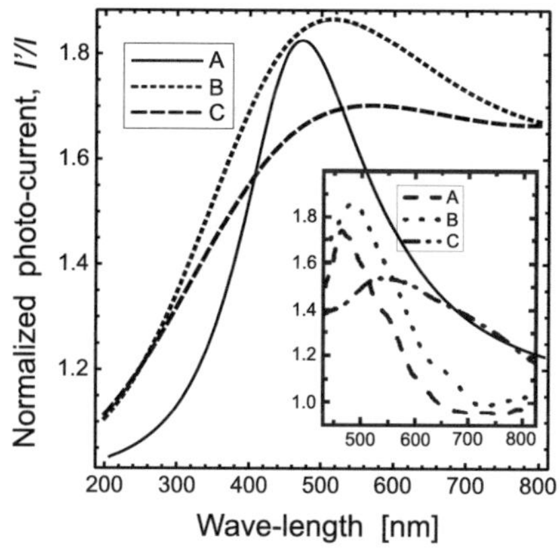

Figure 2. Dependence of the normalized photo-current $\frac{I'}{I}(\lambda)$–comparison with the experimental data [30] for A: $a = 25$ nm (better fitting for 19 nm), $n_s = 6.6 \times 10^8$ $1/\text{cm}^2$, B: $a = 40$ nm, $n_s = 1.6 \times 10^8$ $1/\text{cm}^2$, C: $a = 50$ nm, $n_s = 0.8 \times 10^8$ $1/\text{cm}^2$ ($H = 3$ μm)

where $c_0 = \frac{4\pi a^3}{3}\beta\frac{n_s}{H}$, with n_s as the surface density of metallic nanospheres, H as the semiconductor layer depth, $\phi(x) = \frac{x^2}{(x^2-1)^2+4x^2/x_1^2}\frac{1}{\sqrt{x-x_g}}$, $x = \omega/\omega_1$, $x_1 = \tau\omega_1$, $x_g = E_g/(\hbar\omega_1)$, $\hbar\omega_1 = 2.72$ eV, $m_{n(p)}$ as the effective mass of conduction band and valence band carriers (for Si, $m_n^* = 0.19(0.98)$ m and $m_p^* = 0.16(0.52)$ m, for light (heavy) carriers, band gap $E_g = 1.14$ eV, $\varepsilon = 12$), m as the bare electron mass.

The results are summarized in Fig. 2, and reproduce well the experimental behavior [30]. Both channels of photon absorption resulting in photo-current in the semiconductor sample are included, the direct ordinary photo-effect absorption with probability of transitions given by q and the plasmon mediated absorption with probability q_m, respectively. Note also that some additional effects like reflection of the incident photons or destructive interference on metallic net would contribute and it was phenomenologically accounted for in the plasmon mediated channel by the experiment fitted factor β. These corrections are, however, rather not strong for the considered low densities of metallic coverings of order of $10^8/\text{cm}^2$, and nanosphere sizes well lower than the resonant wave-length, though for larger concentrations and larger nanosphere sizes, would play a stronger reducing role [19, 32]. The resonance threshold was accounted for by the damped resonance envelope function $\phi(x)$ in Eq. (66) including also semiconductor band-gap limit.

As indicated in Fig. 2, the relatively high value of $\frac{q_m}{q} \sim 10^4\frac{\beta 40}{H[nm]}$ enables a significant growth of the efficiency of the photo-energy transfer to the semiconductor, mediated by surface plasmons in nanoparticles deposited on the active layer, by increasing β or reducing H (at constant n_s). However, because of the fact that an enhancement of β easily induces the overdamped regime, a greater perspective would be thus lowering H, the layer depth. The overall behavior of $I'/I(\omega) = 1 + q_m/q$ calculated according to the relation (66), and depicted in Fig. 2, agrees quite well with the experimental observations [30], in the position, height and shape of the photo-current curves for distinct samples (the strongest enhancement is achieved for $a = 40$ nm).

5. Nonlinear effect for undamped plasmon-polariton propagation along metallic nano-chain

In the present chapter we report the RPA description using a semiclassical approach for a large metallic nanosphere (with radius of several tens nm, and with 10^5-10^7 electrons), in an all-analytical calculus version [11]. The electron liquid oscillations of compressional and translational type result in excitations inside the sphere and on its surface, respectively. They are referred to as volume and surface plasmons. Damping of plasmons due to electron scattering and due to radiation losses (accounted for via the Lorentz friction force) is included. The shift of the resonance frequency of dipole-type surface plasmons (only such plasmons are induced by homogeneous time-dependent electric field), due to damping phenomena, well fits with the experimental data for various nanosphere radii [15].

Collective dipole-type surface plasmon oscillations in the linear chain of metallic nanospheres were then analyzed and wave-type plasmon propagation along the chain was described [13, 36]. A coupling in the near field regime between oscillating dipoles in neighboring nanospheres, together with retardation effects for energy irradiation, allowed for appearance of undamped propagation of plasmon waves (called plasmon-polaritons) along the chain in the experimentally realistic region of parameters (for separation of spheres in the chain and nanosphere radii). These effects are of a particular significance for plasmon arranged non-dissipative and sub-diffraction transport of converted light energy and information along metallic chains with promising expected applications in nano-electronics.

The undamped mode of plasmon-polaritons occurs, however, on the rim of stability of the linear approach [13]. It means that the zero damping rate separates the region with positive its value (corresponding to ordinary attenuation of plasmon-polaritons) and the region with negative damping rate (corresponding to instable plasmons-polariton modes). The latter indicates an unphysical behavior. i.e., the arte-fact of the linear approximation. In order to regularize this description the nonlinear corrections must be thus included. Inclusion of nonlinear corrections induced by the invariant form of the Lorentz friction results in such regularization. The instability region of linear approach is entirely covered by the region of undamped wave propagation with the amplitude accommodated to the nonlinearity scale and is independent of the initial condition (despite of its magnitude). This phenomenon, familiar in other nonlinear systems [2], seems to be of a particular significance for understanding of collective plasmon excitations with interesting possible applications.

5.1. Nonlinear corrections to plasmon dynamics in phenomenological approach

Even if the derivation of plasmon dynamics equation in the form of effective harmonic oscillator equation is rigid upon quantum approach of quasiclassical RPA method [11], the inclusion of plasmon attenuation of scattering type and of radiation losses type needs some phenomenological assumptions. They resolve themselves to extension of quantum RPA harmonic oscillator formulation to the damped oscillator equation form with attenuation described by heuristically assumed damping rates. It is proved [11, 14] that radiation losses in the case of the free far-field zone radiation (i.e., in the case of vacuum or dielectric surroundings of metallic nanosphere with oscillating plasmons) can be accounted for as the Lorentz friction [18]. Nevertheless, when in the near-field zone (closer than the wave length corresponding to plasmon frequency) the energy receiver (i.e., other charge system, like semiconductor with its band system or another metallic nanosphere, as in the chain) is located, the irradiation losses are dominated by energy transfer via this near-field zone coupling channel. Existence of charged system of energy receiver modifies retarded e-m potential of emitting system and this modifies the Lorentz friction formula, which was derived in the standard form for dielectric surroundings. The enhancement of radiation losses in the case when the nanoparticles with dipole Mie surface plasmons (excited by incident external light) are deposited on the semiconductor surface lies behind the observed PV efficiency growth in new generation of solar cells metallically modified [24, 25, 27, 30, 32, 34]. In this case the related attenuation rate can be also estimated by application of the Fermi golden rule to the semiconductor inter-band transitions induced by dipole near-field coupling with plasmons [11, 14], as was demonstrated above. As it was proved [11], the resulting attenuation rate scales with nano-sphere radius, a, in a different way in comparison to far-field radiation losses in dielectric medium, as is displayed by the formulae (53), additionally with a renormalization coefficient expressed in terms of the band system parameters [11]. One can expect the similar behavior in the case of the near-field coupling between nanospheres in the chain, but for the sake of effectiveness of modeling one can assume that related attenuation rate has the form as that for the Lorentz friction renormalized only by some coefficient phenomenologically assumed and accounting for a mentioned modification of e-m potential space distribution, caused by the receiver presence apart of the source.

Let us consider first a single metallic nanosphere with dipole type surface oscillations with the dipole,

$$\boldsymbol{D} = eN_e a\boldsymbol{R}, \tag{67}$$

where a is the nanosphere radius (of order of $10-50$ nm), N_e is the number of electrons, e is the electron charge. Oscillating dipole emits radiation and related energy losses can be expressed by the Lorentz friction (eventually renormalized by some phenomenological factor, as commented above). The general relativistic invariant formula for Lorentz friction force is as follows [18],

$$\begin{array}{l} g_i = \frac{2}{3}\frac{(eN_e)^2}{c}\left(\frac{d^2u_i}{ds^2} + u_i u_k \frac{d^2u_k}{ds^2}\right), \\ ds = cdt\sqrt{1-\frac{v^2}{c^2}}, \\ \boldsymbol{u} = \frac{\boldsymbol{v}}{c\sqrt{1-\frac{v^2}{c^2}}}, u_4 = \frac{i}{\sqrt{1-\frac{v^2}{c^2}}}, \\ \boldsymbol{v} = a\omega_1\dot{\boldsymbol{R}}, \dot{\boldsymbol{R}} \equiv \frac{1}{\omega_1}\frac{d\boldsymbol{R}}{dt}, \end{array} \tag{68}$$

The latter expression slightly modifies the definition of dipole derivation in (68) as it incorporates plasmon frequency, for the sake of simplicity of further formulae. As the velocity can be expressed by the dipole derivation, the lowest order of power expansion of relativistic denominator gives the nonlinear corrections to the Lorentz friction, which can be accounted for in the formula for the Lorentz friction induced electric field,

$$\boldsymbol{E}_l = \frac{2}{3}\frac{\sqrt{\epsilon}}{c^3}(eN_e)a\omega_1{}^3\left\{\dddot{\boldsymbol{R}} + \frac{\epsilon a^2\omega_1{}^2}{c^2}a\left(\frac{3}{2}\boldsymbol{R}^{(3)}\dot{\boldsymbol{R}}^2 + 3\ddot{\boldsymbol{R}}\left(\dot{\boldsymbol{R}}\cdot\ddot{\boldsymbol{R}}\right) + \dot{\boldsymbol{R}}\left(\dot{\boldsymbol{R}}\cdot\dddot{\boldsymbol{R}}\right)\right)\right\}, \tag{69}$$

with nanosphere dipole plasmon frequency, $\omega_1 = \frac{\omega_p}{\sqrt{3\epsilon}}$ (in a dielectric surroundings). In the above expression for the Lorentz friction induced electric field, the first term with the third order derivation of the dipole is the usual linear part of the Lorentz friction, while the next contribution (in parenthesis) is the lowest order nonlinear correction.

In the framework of perturbation method of solution of dynamic equation of oscillatory type for the dipole, one can notice that in the zero order of perturbation (when dissipation is neglected), $\ddot{\boldsymbol{R}}+\boldsymbol{R}=0$, from which it follows that $\dddot{\boldsymbol{R}}=-\dot{\boldsymbol{R}}$. Thus one can simplify the formula (69), which leads to dynamical equation for the plasmon dipole including, in the first order of perturbation, the damping of plasmons due to scattering $\left(\frac{1}{\tau_0}\right)$ and due to radiation losses, i.e., due to the Lorentz friction (expressed by the rest of the right-hand-side of the equation (70) with the first term of linear contribution and with the next term of nonlinear type; the overall renormalization coefficient due to near-field coupling as commented above, is assumed here 1 for the sake of simplicity, but it could be changed in the final result),

$$\ddot{\boldsymbol{R}}+\boldsymbol{R}=-\frac{2}{\tau_0\omega_1}\dot{\boldsymbol{R}}-\frac{2}{3}\left(\frac{\omega_p a}{\sqrt{3}c}\right)^3\dot{\boldsymbol{R}}+\frac{2}{3}\left(\frac{\omega_p a}{\sqrt{3}c}\right)^5\left\{-\frac{5}{2}\dot{\boldsymbol{R}}\left(\dot{\boldsymbol{R}}\cdot\dot{\boldsymbol{R}}\right)+3\boldsymbol{R}\left(\dot{\boldsymbol{R}}\cdot\boldsymbol{R}\right)\right\}. \qquad (70)$$

The above equation is complicated in mathematical sense and advanced methods of solutions must be applied, as described in [2]. According to this asymptotical methods for solution of nonlinear differential equation (70), one can find the solution in the following form,

$$\begin{aligned}
R(t) &= \frac{A_0 e^{-\frac{t}{\tau}}}{\sqrt{1+\frac{9}{8}\gamma A_0^2\left(1-e^{-\frac{2t}{\tau}}\right)}}\cos\left(\omega_1 t+\Theta_0\right), \boldsymbol{R}=R\hat{n}, \hat{n}=\frac{\boldsymbol{r}}{r},\\
\frac{1}{\tau\omega_1} &= \frac{1}{\tau_0\omega_1}+\frac{1}{3}\left(\frac{\omega_p a}{\sqrt{3}c}\right)^3\approx\frac{1}{3}\left(\frac{\omega_p a}{\sqrt{3}c}\right)^3,\\
\gamma &= \tau\omega_1\frac{1}{3}\left(\frac{\omega_p a}{\sqrt{3}c}\right)^5.
\end{aligned} \qquad (71)$$

In the formulae (71) both coefficients $\frac{1}{\tau\omega_1}$ and γ can be renormalized eventually by the mentioned above phenomenological factor (still held here 1, for simplicity). From the form of equation (71) it follows that $\frac{1}{\tau\omega_1}$ is always positive. Note that the scattering term, $\frac{1}{\tau_0}=\frac{v_F}{2a}+\frac{Cv_F}{2\lambda_B}$, where v_F is the Fermi velocity in metal material of the nanosphere, λ_B is the bulk mean free path and C is the constant of unity order [11, 14]. This scattering term is negligible (for nanosphere radius beyond 10 nm) in comparison with the linear contribution of the Lorentz friction, as it is demonstrated in Fig. (3). The nonlinear correction scale is given by $\gamma\approx10^{-4}\left(\frac{a[nm]}{10}\right)^2$. As this coefficient is small, one can neglect the related contribution in the denominator for the dipole solution (71), which results in ordinary linear solution of damped oscillations, which means that the nonlinear corrections to the Lorentz friction have no significance in the case of plasmon oscillations of a single nano-sphere.

This situation changes, however, considerably in the case of collective plasmon excitation propagating along the metallic nano-chain, as it will be described in the following paragraph.

5.2. Collective plasmon wave-type propagation along the nano-chain in the nonlinear regime

In the case of the metallic nano-chain one has to take into account the mutual affecting of nanospheres in the chain. Assuming that in the sphere located in the point $\boldsymbol{r}$ we deal with the

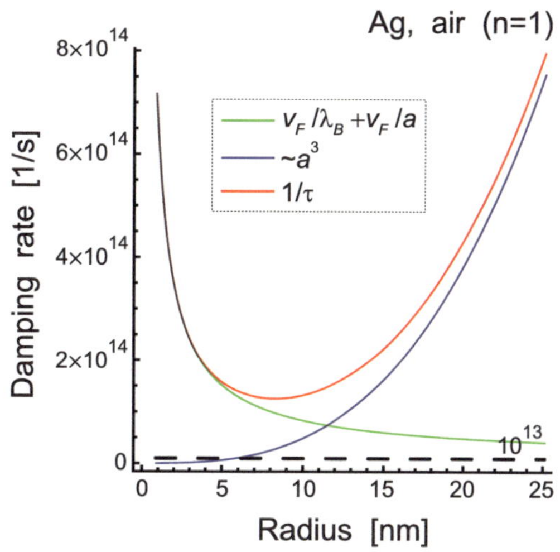

Figure 3. Contribution to the damping rate of surface plasmon oscillations in the nano-sphere versus the nano-sphere radius, including the scattering attenuation (green line) and the linear Lorentz friction damping (blue line); for radii greater than ca 10 nm the second channel dominates in the overall damping (red line)

dipole $\boldsymbol{D}$, then in the other place $\boldsymbol{r}_0$, the dipole type electric field attains the form as follows (including electro-magnetic retardation),

$$\begin{array}{l} \boldsymbol{E}\left(\boldsymbol{r},\boldsymbol{r}_0,t\right) = \frac{1}{\epsilon r_0{}^3}\left\{3\boldsymbol{n}_0\left(\boldsymbol{n}_0\cdot\boldsymbol{D}\left(\boldsymbol{r},t-\frac{r_0}{v}\right)\right)-\boldsymbol{D}\left(\boldsymbol{r},t-\frac{r_0}{v}\right)\right\}, \\ \boldsymbol{n}_0 = \frac{\boldsymbol{r}_0}{r_0}, v = \frac{c}{\sqrt{\epsilon}}. \end{array} \tag{72}$$

This allows for writing of the dynamical equation for plasmons at each nanosphere of the chain which can be numbered by integer l (d is here the separation between nanospheres in the chain). The first term of the right-hand-side in the following formula (73) describes the dipole type coupling between nanospheres [13] and the other two terms correspond to contribution due to plasmon attenuation (in the latter term the Lorentz friction caused electric field accounts also for nonlinear corrections). The index α enumerates polarizations, longitudinal and transversal ones with respect to the chain orientation.

$$\begin{array}{l} \ddot{R}_\alpha + R_\alpha\left(ld,t\right) = \sigma_\alpha\frac{a^3}{d^3}\sum\limits_{m=-\infty,m\neq 1}^{\infty}\frac{R_\alpha\left(md,t-\frac{d|l-m|}{v}\right)}{|l-m|^3} - \frac{2}{\tau_0\omega_1}\dot{R}_\alpha\left(ld,t\right) + \frac{e}{ma\omega_1{}^2}E_\alpha\left(ld,t\right), \\ \sigma_\alpha = \begin{cases} -1, \alpha = x,y \\ 2, \alpha = z \end{cases} \end{array}. \tag{73}$$

The summation in the first term of the right-hand-side of the equation (73) can be explicitly performed in the manner as presented in [13], because it is the same as for the linear theory formulation. Similarly as in the linear theory framework, one can change to the Fourier picture taking advantage of the chain periodicity (in analogy to crystals with the reciprocal lattice), i.e.,

$$\begin{array}{l} R_\alpha\left(ld,t\right) = \tilde{R}_\alpha\left(k,t\right)e^{-ikld}, \\ 0 \leq k \leq \frac{2\pi}{d}, \tilde{R}_\alpha\left(k\right)\sin\left(t\omega_1+\beta\right). \end{array} \tag{74}$$

Thus the equation (73) can be rewritten in the following form (the Lorentz friction term was represented similarly as in equation (70)),

$$
\begin{aligned}
&\ddot{\tilde{R}}_\alpha(k,t) + \tilde{\omega}_\alpha^2 \tilde{R}_\alpha(k,t)\left\{\frac{1}{\tau_\alpha\omega_1} + \frac{1}{3}\left(\frac{\omega_p}{\sqrt{3}c}\right)^5 \left(\frac{5}{2}|\dot{\tilde{R}}_\alpha(k,t)|^2 - 3|\tilde{R}_\alpha(k,t)|^2\right)\right\},\\
&\tilde{\omega}_\alpha^2 = 1 - 2\sigma_\alpha \frac{a^3}{d^3}\cos(kd)\cos\left(\frac{d\omega_1}{v}\right),\\
&\frac{1}{\tau_{x,y}\omega_1} = \frac{1}{\tau_0\omega_1} + \frac{1}{4}\left(\frac{\omega_1 d}{v}\right)\frac{a^3}{d^3}\left(\left(\frac{\omega_1 d}{v}\right)^2 - (kd-\pi)^2 + \frac{\pi^2}{3}\right),\\
&\frac{1}{\tau_z\omega_1} = \frac{1}{\tau_0\omega_1} + \frac{1}{2}\left(\frac{\omega_1 d}{v}\right)\frac{a^3}{d^3}\left(\left(\frac{\omega_1 d}{v}\right)^2 - (kd-\pi)^2 + \frac{\pi^2}{3}\right),
\end{aligned}
\tag{75}
$$

In the above formulae the remarkable property is linked with the expressions for the attenuation rate for both polarizations. Two last expressions in equation (75) give these damping rates explicitly and one can notice that they may change their signs depending on values for d,a,k. In Fig. (4) the region of negative value for damping rates is depicted (for both polarizations).

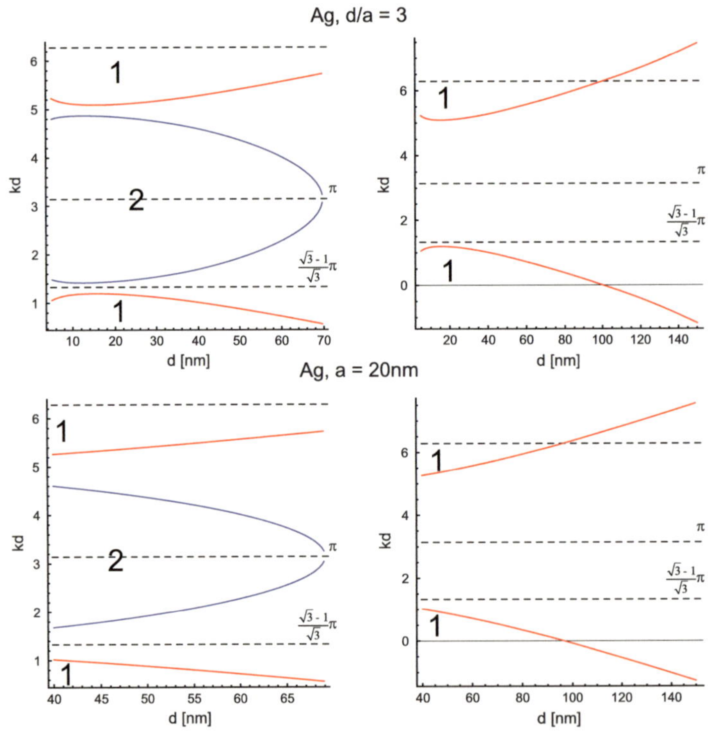

Figure 4. Regions for negative value of damping rates for plasmon-polaritons in the chain (1 for longitudinal polarization modes and 2 for transversal ones, in the nolinear formulation framework; for the linear theory (red lines gives the position of vanishing damping rate for longitudinal modes of plasmon-polaritons and blue lines the same for transversal modes)

Applying the same methods for solution of the nonlinear equation (75) as in the former paragraph using the asymptotic methods [2], one can find the corresponding solutions for both regions with positive and negative damping rate, respectively,

$$\begin{aligned}
&for\ \frac{1}{\tau_\alpha\omega_1} > 0,\\
&\tilde{R}_\alpha(k,t) = \frac{A_{\alpha 0}e^{-\frac{t}{\tau_\alpha}}}{\sqrt{1+\gamma_\alpha A_{\alpha 0}^2\left(1-e^{-\frac{2t}{\tau_\alpha}}\right)}}\cos(\omega_\alpha t+\Theta_0) \rightarrow_{t\to\infty} 0,\\
&\gamma_\alpha = |\tau_\alpha\omega_1|\left(\frac{\omega_1 a}{c}\right)^3\frac{1}{4}\left(\frac{5}{2}\tilde{\omega}_\alpha^2-1\right);\\
&for\ \frac{1}{\tau_\alpha\omega_1} < 0,\\
&\tilde{R}_\alpha(k,t) = \frac{A_{\alpha 0}e^{\frac{t}{|\tau_\alpha|}}}{\sqrt{1+\gamma_\alpha A_{\alpha 0}^2\left(e^{\frac{2t}{|\tau_\alpha|}}-1\right)}}\cos(\omega_\alpha t+\Theta_0) \rightarrow_{t\to\infty} \frac{1}{\sqrt{\gamma_\alpha}}\cos(t\omega_\alpha+\Theta_0),\\
&D_\alpha = \frac{eN_e a}{\sqrt{\gamma_\alpha}}\frac{1}{2}\left\{e^{i(\omega_\alpha t-kld)}+e^{-i(\omega_\alpha+kld)}\right\}.
\end{aligned} \tag{76}$$

From the above formulae it follows that for positive attenuation rate we deal with ordinary damped plasmon-polariton propagation, not strongly modified in comparison to linear theory (due to small value of the factor γ). Nevertheless, in the case of negative damping rate the solution behaves differently, i.e., in longer time scale this solution stabilizes on the constant amplitude value independently of initial conditions. This remarkable property characterizes undamped propagation of plasmon-polariton along the chain (if one turns back to dipole explicit form (67), then typical planar wave formula with constant amplitude describes this undamped mode, as written in the last equation of (76)). The region of negative damping corresponds thus, within the nonlinear approach, to undamped modes with the fixed amplitude. Let us note that the same region was linked with instability of the linear theory (which was, however, the arte-fact of the linear approach).

Finally, one can calculate the group velocity of the undamped plasmon-polariton mode in the following form,

$$v_\alpha = \frac{d\omega_\alpha}{dk} = \omega_1 d\frac{\sigma_{alpha}a^3\sin(kd)\cos\left(\frac{d\omega_1}{c}\right)}{d^3\sqrt{1-2\sigma_\alpha\frac{a^3}{d^3}\cos(kd)\cos\left(\frac{d\omega_1}{c}\right)}}. \tag{77}$$

From this formula it follows that the group velocity of the undamped wave type collective plasmon excitation (called plasmon-polariton) may attain different values depending on a,d,k.

Indicated above undamped mode of propagation of collective surface plasmons seems to match with experimentally observed long range propagation of plasmon excitations along the metallic nano-chain [1, 6, 21, 22, 36]. The constant and fixed value of the amplitude for these oscillations (76) are independent of initial conditions, which means that these excitations will be present in the system even if are excited by arbitrarily small fluctuations. Thus one can conclude that these are self-exciting modes which are always present in the system provided that radii of spheres and their separation in the chain have values for which attenuation rates (75) are negative.

6. Conclusions

We have demonstrated the practical utilization of RPA semiclassical description of plasmon oscillations in metallic nanospheres. The oscillatory form of dynamics both for volume and surface plasmons, rigorously described upon the RPA semiclassical limit fits well with the large nanosphere case, when the nanosphere radius is greater than 10 nm and lower than 60 nm, (for Au, Ag or Cu material), what is confirmed by experimental observations, on the other hand. The most important property of plasmons on such large nanospheres is the very strong e-m irradiation caused by these excitations, which results in quick damping of oscillations. The attenuation effects for plasmons were not, however, included into the quantum RPA model. Nevertheless, they could be included in a phanomenological manner, taking advantage of the oscillatory form of dynamic equations. Some information on plasmon damping can be taken from microscopic analyzes of small metallic clusters (especially made by LDA and TDLDA methods of numerical simulations employing Kohn-Sham equation). For larger nanospheres, these effects, mainly of scattering type (also Landau damping), are, however, not specially important as diminishing with radius growth as $\frac{1}{a}$. The irradiation effects overwhelming the energy losses in the case of large nanospheres can be grasped in terms of the Lorentz friction, which reduces the charge movement. This approach has been analyzed in the present paper. Two distinct situations were indicated, the first one–of the free radiation to far-field zone in dielectric/vacuum surroundings of single nanoparticle and the second one, when in the near-field zone of plasmons on the nanosphere, an additional charged system is located. This additional charge system of the e-m energy receiver strongly modifies the e-m potentials of the source and in this way modifies energy emission. In particular, the Lorentz friction is modified in this case in comparison to simple free emission to far-field zone. The e-m energy receiver located close to emitting nanosphere, could be semiconductor (as in the case of metallically modified solar cells) or other metallic nanospheres (as in the case of metallic nano-chain). Both these situations have been analyzed in this chapter.

The energy transfer from surface plasmon oscillations in metallic nanospheres deposited on photo-active surface of semiconductor to the substrate electron band system has been described upon the scheme of Fermi golden rule. The dependence of the effect with respect to the nanosphere dimension has been analyzed in details and the comparison with experimental data is given. We have elucidated the reason of the experimentally observed giant enhancement of PV efficiency caused by plasmon effect. Surface plasmons in metallic nanocomponents deposited on the semiconductor surface act as solar energy converters. The surface plasmons excited in the metallic nanospheres by incident photos couple with band electrons in the semiconductor substrate in the near-field regime. This dipole type coupling is more effective in comparison to the ordinary photo-effect coupling of electrons with incident photons, due to allowance of indirect inter-band transitions in the metallically locally modified systems. The metallic nanosphere coupled with substrate electrons in near-field zone of plasmon oscillations is not translational invariant system and thus the momentum conservation does not constrain quantum transitions in this case. This leads to strong enhancement of the inter-band transition probability, quenched, however, with the nanosphere radius growth, in the manner as presented in this paper.

With regard to collective plasmon fluctuations in ordered metallic systems, we have shown previously that along the infinite nano-chain the collective plasmon-polaritons can propagate (being collective surface plasmons coupled by e-m field in near-field zone), which at certain values of nanosphere radii and separation in the chain, appear as undamped modes. Simultaneously, the instability regions of linear theory of plasmon-polariton dynamics occur, which shows that the nonlinear corrections must be included. In this paper we have developed the nonlinear theory of collective plasmon-polariton dynamics along the chain, including nonlinear corrections to Lorentz friction force. Even though the related nonlinearity is small, it suffices to regularize the instable linear approach. As the most important observation, we noted the presence of undamped excitations (instead of those instable within the linear approach), which have fixed amplitude independently how small or large the initial conditions were. This excitations, typical for nonlinear systems, would have some practical significance, e.g., to enhance sensitivity of antennas with coverings by plasmon systems offering self-induced collective plasmon-polaritons in wide range of frequencies, which would be excited by even very small signal (the energy to attain the stable level of plasmon-polariton amplitude would be supplied, in this case, by an external auxiliary source).

Acknowledgments

Supported by NCN project No: 166719 Sonata 2

Author details

Jacak Witold
Institute of Physics, Wrocław University of Technology, Wyb. Wyspiańskiego 27, 50-370 Wrocław, Poland

7. References

[1] Barnes, W. L., Dereux, A. & Ebbesen, T. W. [2003]. *Nature* 424: 824.
[2] Bogolubov, N. N. & Mitropolskyj, J. A. [2005]. *Asymptotical methods in theory of nonlinear oscillations*, Nauka, Moscow.
[3] Bohren, C. F. & Huffman, D. R. [1983]. *Absorption and Scattering of Light by Small Particles*, Wiley, New York.
[4] Brack, M. [1989]. *Phys. Rev. B* 39: 3533.
[5] Brack, M. [1993]. *Rev. of Mod. Phys.* 65: 667.
[6] Brongersma, M. L., Hartman, J. W. & Atwater, H. A. [2000]. *Phys. Rev. B* 62: R16356.
[7] Ekardt, W. [1984]. *Phys. Rev. Lett.* 52: 1925.
[8] Ekardt, W. [1985]. *Phys. Rev. B* 31: 6360.
[9] Ekardt, W. [1986]. *Phys. Rev. B* 33: 8803.
[10] Garcia de Abajo, F. J. [2010]. *Rev. Mod. Phys.* 82: 209.
[11] Jacak, J., Krasnyj, J., Jacak, W., Gonczarek, R., A.Chepok & Jacak, L. [2010]. *Phys. Rev. B* 82: 035418.
[12] Jacak, L., Krasnyj, J. & Chepok, A. [2009]. *Fiz. Nisk. Temp.* 35: 491.
[13] Jacak, W., Krasnyj, J., Jacak, J., Chepok, A., Jacak, L., Donderowicz, W., Hu, D. & Schaadt, D. [2010]. *J. Appl. Phys.* 108: 084304.
[14] Jacak, W., Krasnyj, J., Jacak, J., Donderowicz, W. & Jacak, L. [2011]. *J. Phys. D: Appl. Phys.* 44: 055301.
[15] Jacak, W., Krasnyj, J., Jacak, J., Gonczarek, R., Chepok, A., Jacak, L., Hu, D. & Schaadt, D. [2010]. *J. Appl. Phys.* 107: 124317.
[16] Kiriejew, P. S. [1969]. *Physics of Semiconductors*, PWN, Warsaw.
[17] Kresin, V. V. [1992]. *Phys. Rep.* 220: 1.

[18] Landau, L. D. & Lifshitz, E. M. [1973]. *Field Theory*, Nauka, Moscow.

[19] Losurdo, M., Giangregorio, M. M., Bianco, G. V., Sacchetti, A., Capezzuto, P. & Bruno, G. [2009]. *Solar Energy Mat. & Solar Cells* 93: 1749.

[20] Maier, S. A. [2007]. *Plasmonics: Fundamentals and Applications*, Springer, Berlin.

[21] Maier, S. A., Kik, P. G. & Atwater, H. A. [2003]. *Phys. Rev. B* 67: 205402.

[22] Markel, V. A. & Sarychev, A. K. [2007]. *Phys. Rev. B* 75: 085426.

[23] Mie, G. [1908]. *Ann. Phys.* 25: 337.

[24] Morfa, A. J., Rowlen, K. L., Reilly, T. H., Romero, M. J. & Lagemaat, J. [2008]. *Appl. Phys. Lett.* 92: 013504.

[25] Okamoto, K., Niki, I., Scherer, A., Narukawa, Y. & Kawakami, Y. [2005]. *Appl. Phys. Lett.* 87: 071102.

[26] Okamoto, K., Niki, I., Shvartser, A., Narukawa, Y., Mukai, T. & Scherer, A. [2004]. *Nature Mat.* 3: 601.

[27] Pillai, S., Catchpole, K. R., Trupke, T., Zhang, G., Zhao, J. & Green, M. A., [2006]. *Appl. Phys. Lett.* 88: 161102.

[28] Pines, D. [1999]. *Elementary Excitations in Solids*, ABP Perseus Books, Massachusetts.

[29] Pines, D. & Bohm, D. [1952]. *Phys. Rev.* 92: 609.

[30] Schaadt, D. M., Feng, B. & Yu, E. T. [2005]. *Appl. Phys. Lett.* 86: 063106.

[31] Stuart, H. R. & Hall, D. G. [1998]. *Appl. Phys. Lett.* 73: 3815.

[32] Sundararajan, S. P., Grandy, N. K., Mirin, N. & Halas, N. J. [2008]. *Nano Lett.* 8: 624.

[33] Wen, C., Ishikawa, K., Kishima, M. & Yamada, K. [2000]. *Sol. Cells* 61: 339.

[34] Westphalen, M., Kreibig, U., Rostalski, J., Lüth, H. & Meissner, D. [2003]. *Sol. Energy Mater. Sol. Cells* 61: 97.

[35] Yannouleas, C., Broglia, R. A., Brack, M. & Bortignon, P. F. [1989]. *Phys. Rev. Lett.* 63: 255.

[36] Zayats, A. V., Smolyaninov, I. I. & Maradudin, A. A. [2005]. *Phys. Rep.* 408: 131.

Chapter 19

Plasmonic Rectenna for Efficient Conversion of Light into Electricity

Fuyi Chen, Jian Liu and Negash Alemu

Additional information is available at the end of the chapter

http://dx.doi.org/10.5772/51045

1. Introduction

Solar energy has been widely studied as a green energy source. Present photovoltaic techniques are mainly semiconductor solar cells. The silicon solar cell has a good power conversion efficiency of 24%, and the semiconductor solar cell based on GaAs multijunction has the highest conversion efficiency of 32% at present [1]. Though these semiconductor solar cells have relatively high conversion efficiency, their high cost and some poisonous byproduct produced during the manufacturing process limit their commercial use. As an alternative for large scale application, the dye-sensitized solar cell (DSSC) has obtained conversion efficiency of 12% at maximum in these days [2], only a few percentages has been increased since the Grazel's research, in which the conversion efficiency of DSSC is improved using nanoporous photoanodes [3].

In 1998, Stuart et al. [4] reported the first application of surface plasmons in the photovoltaic technology. It was found that the photocurrent of the thin-film Silicon-on-insulator (SOI) device was 18 times improved at the wavelength of 800 nm by metallic nanoparticles. So far, the plasmonic effect of the metal nanostructure is one of the main directions [5, 6] to improve the conversion efficiency for DSSC. The United States of America (USA), the European Union and many other countries undertake some related projects, such as the SOLAMON project, and some typical research achievements are obtained in the plasmonic DSSC field, which are described as follows according to the role of metal surface plasmons:

Scattering effect of localized surface plasmon resonance (LSPR). Brown et al. [7] reported that adding Au-Si nanoparticles with the shell-core structure with size of 20 nm into the typical solid electrolyte DSSC led to an improvement of the conversion efficiency; Hagglund et al. [8] studied the influence of Au nanoplate arrays on the photoelectric conduction of the dye-sensitized TiO_2 thin-film.

Concentration effect of light using localized surface plasmon resonance (LSPR). Du et al. [9] theoretically demonstrated that the Au nanodimer improved the ultraviolet absorption of the TiO_2 nanoparticles; Guilatt et al. [10] studied the light absorption property of the Au nanoshell in the Si thin-film.

Trapping effect of surface plasmon polariton (SPP). Fu et al. [11] reported the conversion efficiency of the typical DSSC with the Pt fishnet arrays as the counter electrode, Durr et al. [12] studied the effect of the Ag thin-film on the short circuit current in DSSC with a planar waveguide structure.

It is clear that metallic nanostructures possess the excellent light-harvesting capability in the visible spectrum, and they can collect almost all the photon energy which is concentrated on the nanoscale plasmonic hotspots from an incident light wavelength region of several hundred nanometers. Therefore, the electromagnetic fields in these regions are enhanced exponentially.

However, the total energy conversion efficiency of DSSC with the plasmonic optical effect is not increased simultaneously as we expect. As shown in the experiments, the power conversion efficiencies were only 1.95 % [7] and 5.77 % [11] with the plasmonic effect considered. We show in next section that the parasitic absorption of small metal nanoparticles is the main reason, and the power conversion efficiency enhanced by metallic nanoparticles is much less than their enhanced efficiency of light collection because the metallic nanoparticles absorb one part of the incident light during surface plasmon resonance (SPR). The parasitic absorption had been mentioned in previous thin film devices [13, 14] which demonstrated that the metallic absorption had an adverse effect on the solar cell performance.

2. Parasitic absorption of metal nanoparticles in the plasmonic DSSC

Our numerical calculations were performed using the finite difference time domain technique (FDTD) to solve the three-dimensional vector Maxwell equations in the optical structure of the plasmonic DSSC, as showed in Figure 1. The distance of the two transparent conductive oxides (TCO) electrodes was 1600 nm, the dye-sensitized layer was 500 nm in thickness. The FCC-arranged Ag spherical nanoparticle array was on top of the dye-sensitized layer with a 2 nm space. The plasmonic DSSC was illuminated with plane wave from the top TCO cathode with the electric field in the Ag spherical nanoparticle array. A perfectly matched layer and a conformal mesh region self-adapted to the structure were used.

The transmission parameters were calculated as the ratio of the power transmitted through the structure to the power incident, and the reflection parameters were calculated as the ratio of the power reflected from the structure to the power incident. The particle swarm optimization (PSO) was used to evaluate different design geometries to locate an optimum solution. PSO, based on the movement and intelligence of swarms, was a robust stochastic evolutionary computation technique inspired in the behavior of bee flocks [15]. It has been shown in certain instances to outperform other methods of optimization like genetic algorithms (GA) in the electromagnetic community [16].

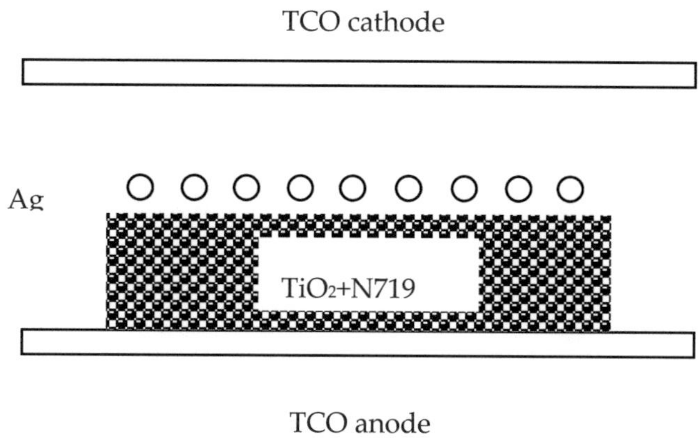

Figure 1. Schematic of the plasmonic DSSC structure. Dye-sensitized layer is 500 nm in thickness, Ag nanoparticle is 210 nm in diameter, the spacing between sensitized layer and Ag nanoparticles is 2 nm, the distance of the two transparent conductive oxides (TCO) electrodes is 1600 nm.

Figure 2 shows the monochromatic absorption enhancement and the integral quantum efficiency (IQE) enhancement as a function of Ag nanoparticles diameter and period of the Ag nanoparticles arrays in the wavelength range of 400~1100 nm. The optimal optical structure designed based on PSO for the present study is the Ag nanoparticles array of 210 nm in diameter and 500 nm in period, which has an improved integral quantum efficiency (IQE) enhancement of 11.3%. The monochromatic absorption enhancement of the Ag nanoparticles of diameter = 210 nm and period = 500 nm is improved by 20-30 % in the wavelength range 630 nm to 1100 nm, it has a maximum enhancement of absorption at the wavelength of 796 nm, and it is far higher than that of the Ag nanoparticle arrays of diameter = 70 nm and period = 250 nm which does not improve the conversion efficiency significantly.

Figure 3 shows the optical properties, the electric near-field distribution and the visible light absorption distribution at wavelength of 796 nm in the optimized plasmonic DSSC. It can be seen from Figure 3 (left) that even under optimization condition the metal absorption takes up 20-30 % of total absorption in the visible bands of 400-700 nm, which is consistent with the monochromatic absorption enhancement spectrum. This may be because of the electric field is mainly distributed at the outside of the Ag nanoparticles, and the metal nanoparticle absorption intensity is higher than the dyes, which can be seen from Figure 3 (right). It is clear from Figure 3 (middle) that the silver particle increases absorption of the dye by the enhanced forward scattering due to the surface plasmon resonance.

Although the conversion efficiency of photoactiviated thin film can be improved by about 15-20% theoretically by optimized metal nanoparticle arrays, the absorption of metal nanoparticles (210 nm in diameter) take up about 20-30% of total absorption of the DSSC in visible light, *i.e.*, lights absorbed by DSSC does not completely convert into photocurrent. A part of the energy is lost in the form of Joule heat (ohmic loss), which led to the low external quantum efficiency of photoelectric conversion.

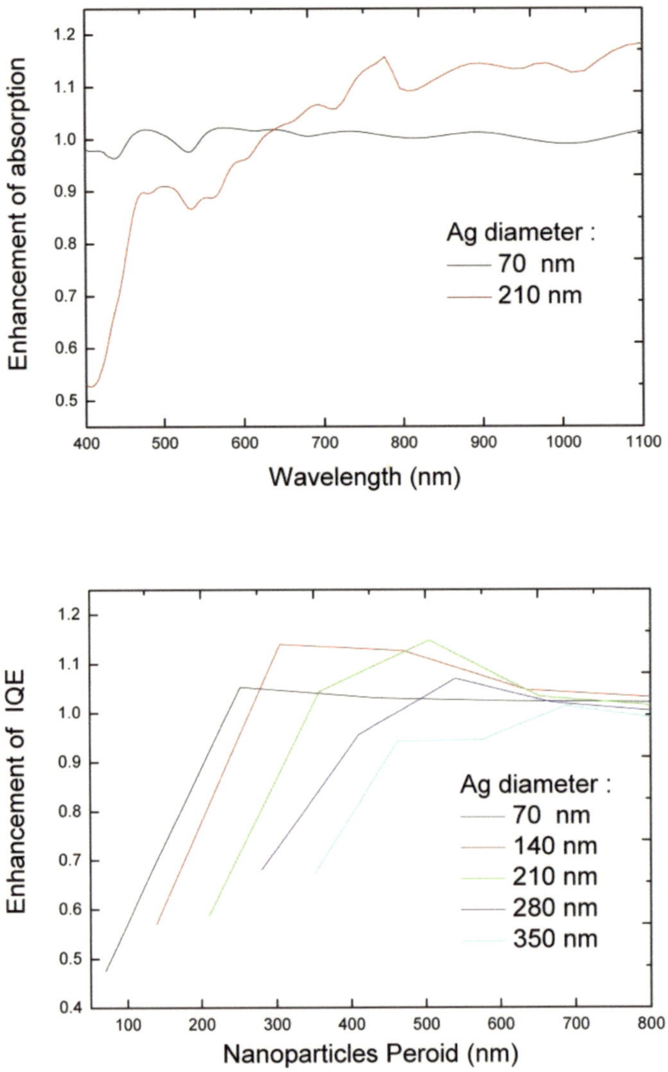

Figure 2. (a) Monochromatic light absorption enhancement of the Ag nanoparticle arrays of 70 nm and 210 nm in diameter and (b) the integral quantum efficiency enhancement of Ag nanoparticle arrays with different period (70-800 nm) and diameter (70-210 nm).

In typical DSSC device, the dye sensitized TiO_2 is 10-80 nm in diameter [17], in order to build the electrostatic localized surface plasmon resonances of metallic nanoparticles inside the dye sensitized TiO_2, the smaller Ag nanoparticles were buried to improve the absorption of dye sensitized TiO_2 nanoparticles, the intrinsic heat loss of small metal nanoparticles (such as 70 nm) at visible light bands is far greater than the electromagnetic energy scattered into the optical activity medium by these metal nanoparticles. The problem has to be considered in the plasmonic DSSC device, which has been highlighted in a recent review [18].

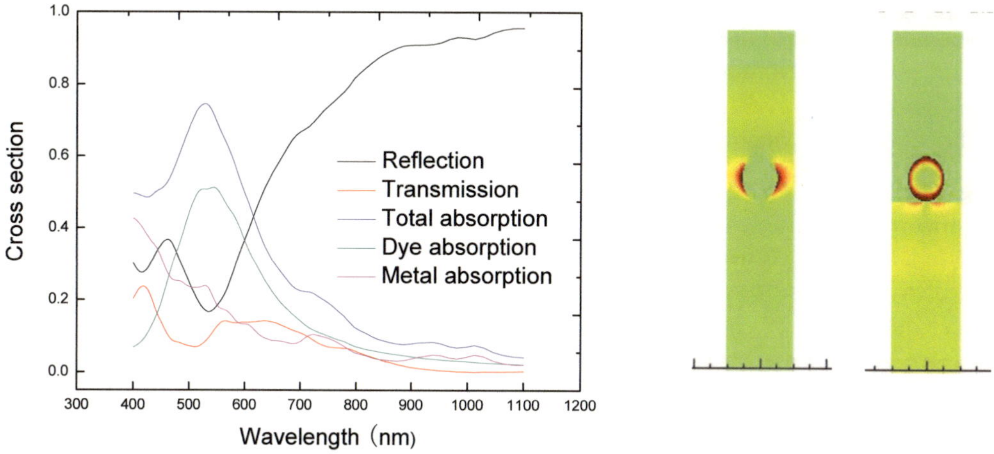

Figure 3. Optical properties of the plasmonic DSSC (left) under optimized conditions (Ag diameter of 210 nm, period 500 nm), the electric near-field distribution (middle) and the visible light absorption distribution (right) in DSSC at wavelength of 796 nm where the enhancement of monochromatic absorption is maximum.

To solve the problem of metal parasitic absorption in DSSC caused by surface plasmon optical effect, Guilatt et al. [10] theoretically studied the method of using metal shell-core nanostructure to substitute pure metal nanoparticles. But the potential development of this method was limited by the metallic skin absorption property. Designing and fabricating the metal-insulator-metal (MIM) sandwich structure optical rectenna as photon acceptor and electron donor may be a research direction for the further development of the plasmonic photovoltaic techniques. In this chapter, we further develop the idea of optical rectification (OR) that can convert visible/near-infrared (VIS/NIR) wavelengths to direct current. Section 3 includes the antenna properties of the noble metal (Ag) and transition metal (Ti), and metal-insulator-metal (MIM) nanostructure and Section 4 includes the optical rectification properties of Ag-TiO_2-Ti and AgCu-TiO_2/Ti metal-insulator-metal (MIM) diode devices. We exploit these contents to address the effect of plasmonic properties on the electronic transport behaviors in noble-transition metal heterodimers in order to put forth a new theory on the plasmonic optical rectification.

3. Antenna properties of the noble, transition metal nanowires, metal-insulator-metal (MIM) nanostructures

3.1. Review of optical rectification (OR) effect

Metallic nanostructures provide a means to transduce free-propagating electromagnetic waves into localized surface plasmon resonance modes [19] and can function as optical antennas [20]. In the traditional radiofrequency and microwave regime, antennas (such as antennas in mobile phones of our everyday life) are usually used to convert electromagnetic radiations into electric currents, however, most of the optical antennas studied so far operate on a "light-in and light-out" basis [21], the utilizing of metallic optical antennas to convert the optical radiations into photocurrents is one of the most imperative tasks for the state-of-the-art plasmonic technology. Conversion of visible/near-infrared (VIS/NIR) wavelengths to effective direct currents (DC) is a second-order nonlinear optical phenomena, e.g., optical rectification (OR) effect.

The search for OR effect in metallic optical antennas was started in 1964, when Brown of the Raytheon Company demonstrated the first flight of a microwave-powered helicopter using rectenna [22], a high up to 90% efficiency was obtained for a rectenna operating at single microwave frequency in 1976 [23]. Generally speaking, the OR effect could achieve up to 80 percent efficiency of energy conversion and power transmission in microwave range 300 MHz to 300 GHz, and typical antenna structure size was around few millimeters.

The endeavor for conversion of light radiation to DC power was started in 1968 [24], following the example of the microwave point-contact diode rectification structure, much research has been performed to extend the concept and application of rectenna into ultrahigh (visible) frequencies (1000THz), and progress has been made in the fabrication and characterization of the metal-insulator-metal (MIM) diodes for use in rectification device. It has been demonstrated that optical antennas can couple electromagnetic radiation in the visible in the same way as radio antennas do at their corresponding wavelength, three kinds of the metallic nanostructures with OR effect in the visible/near-infrared wavelength range had been located: whisker diodes [24,25], Schottky diodes [26,27] and metal-insulator-metal tunneling diodes [28,29]. Meanwhile, for the application in the solar cell technology, Bailey originally proposed that broadband rectifying antennas can be used for direct conversion of solar energy to DC power [30], the first patent on solar rectification was issued to Marks [31].

It is clear that metallic optical antennas possess as good capability of receiving electromagnetic energy in the visible/near-infrared spectrum as the radiofrequency antennas do, and that they can collect almost all the photon energy in a wavelength range of several hundreds of nanometers. However, in comparison with the high conversion efficiency of the radio frequency antennas, the theory and application of OR effect of visible and infrared radiation is still in its infancy, a major technological difficulties that impede the performance of OR device at high frequency is that in the infrared and visible wavelength, metals are no longer perfect conductor.

The main difference between low frequency and ultrahigh frequencies electromagnetic waves when it comes to interaction with the free electrons in a finite piece of metal is the fact that electrons have an effective mass which causes them to react with increasing phase lag to an oscillating electromagnetic field as the frequency increases. For increasing frequency of the excitation, they exhibit an increasing oscillation amplitude as well as an increasing phase lag. As soon as the phase lag approaches 90° the amplitude of the charge oscillation goes through a maximum and is only limited by the internal (ohmic and radiation) damping of the system.

This resonance corresponds to the surface plasmon resonance, for certain metals (such as gold, silver and copper) which consist of the anode and cathode in the metal-insulator-metal (MIM) structure optical rectennas, the surface plasmon resonance happens to appear in or close to the visible spectral range. Plasmon resonances do not appear in perfect conductors (metals at low enough frequencies) since in those materials by definition no phase lag exists between excitation and charge response. The presence of plasmon resonance is therefore characteristic for optical frequencies and give rise to drawbacks of antenna systems in this frequency range, such as enhance Ohmic losses compared to the radiofrequency regime.

3.2. Modeling of the nanowire antenna

Nanowires which are made up of either noble or transition metals exhibit distinctive electromagnetic properties that make them appropriate candidate for nanoantenna at higher frequencies. This nanoantenna is put to practical application for efficient conversion of visible-near infrared spectrum energy into direct current electricity since its operation is on the basis of the wave nature of optical radiation.

In this section, long vertical nanowires were theoretically investigated as a model system to understand the behavior of dipole optical antennas. Figure 4(a) shows the geometry of the nanowire structure. The nanowire consists of the perfectly conductor (PEC), silver (Ag) or titanium (Ti) with the diameter of 100 nm and the length of 2000 nm. The plane wave source is located to the left of the structures in vacuum. The results of field enhancement and vertical components of the Poynting vector of antenna radiation simulations are depicted in Figures 5-11.

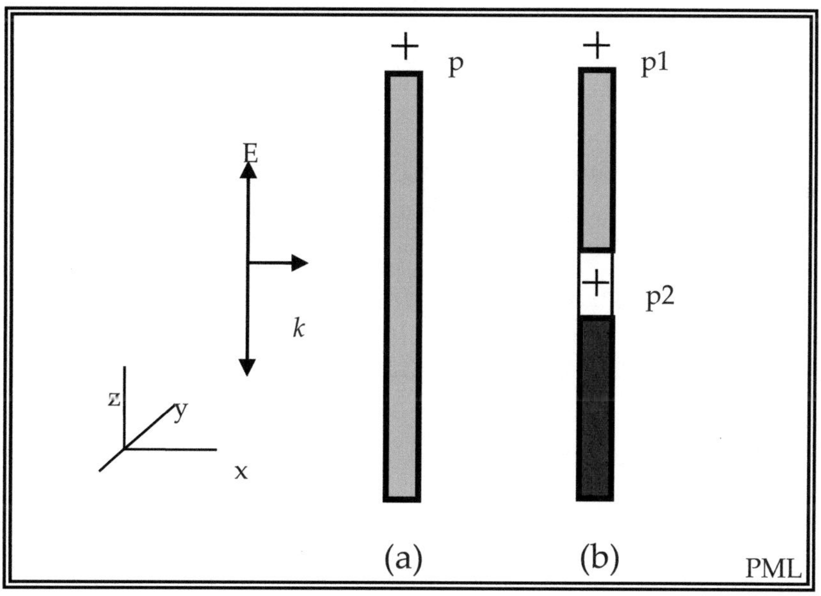

Figure 4. Schematic of the nanowire (a) and its heterodimer (b) as the calculation model for the antenna properties using the finite difference time domain (FDTD) method. The polarization direction (E) is along the long symmetric axis of the nanowires. The position of p and p1 are 5 nm from the nanowire apex, and p2 is at the middle of the dimer. The mesh is truncated using PML absorbing boundary conditions.

3.3. Perfectly conducting nanowire

According to classical antenna theory, metal antennas resonate at a wavelength λ when their length equals approximately (2n+1)λ/2, where n=0, 1, 2, etc. At resonance the charge distribution inside the antenna intensifies the incident electric field (E-field) locally at its end. In Figure 5 and 6, we plot the calculated values for the field enhancement |E|/|E₀|, normalized to the incident electric field, at the apex of the nanorod, for a 2000 nm perfectly conducting nanowires.

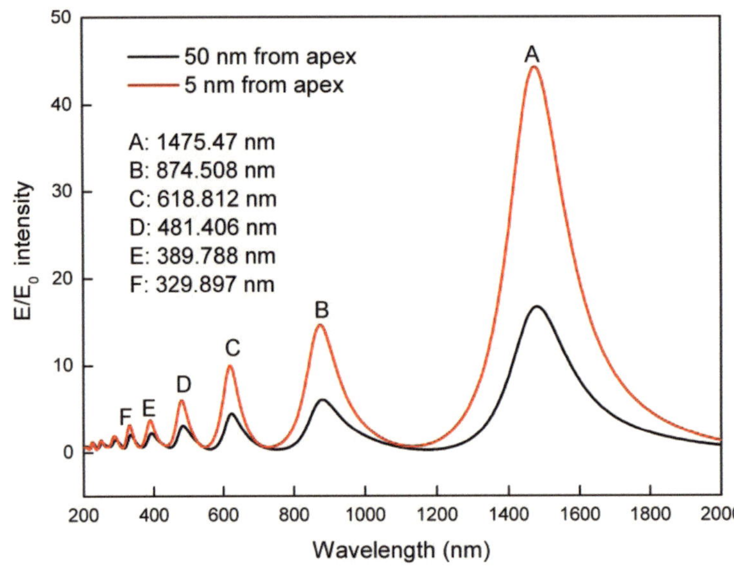

Figure 5. Multiple resonance spectrum around a perfectly conducting nanowire. The resonance is shown as a value of $|E|/|E_0|$ at a point of 5 or 50 nm above the nanowire as a function of wavelength of incident light (200 nm to 2000 nm).

Figure 5 and 6 show the simulated local field intensity enhancement spectra for perfectly conducting nanowire, its intensity distribution and Poynting vector corresponding to the higher harmonic mode resonances. Multiple resonances have seen in the field enhancement spectrum (Figure 5) at the given wavelength range. Each resonance corresponds to specific harmonic modes in the near field.

3.4. Ag nanowire

Any specific structure to consider as an optical antenna is supposed to be able to localize and enhance the propagating electromagnetic wave within a certain bandwidth. The results for silver nanowire with the parameters mentioned above are presented in Figure 7, 8 and 11. The results show that silver nanowire has enhancement spectra with very distinct peaks. The peak at shorter wavelength (350.477nm) attributed to surface plasmon resonance and for the wavelengths in the remaining band strong multiple resonances are observed.

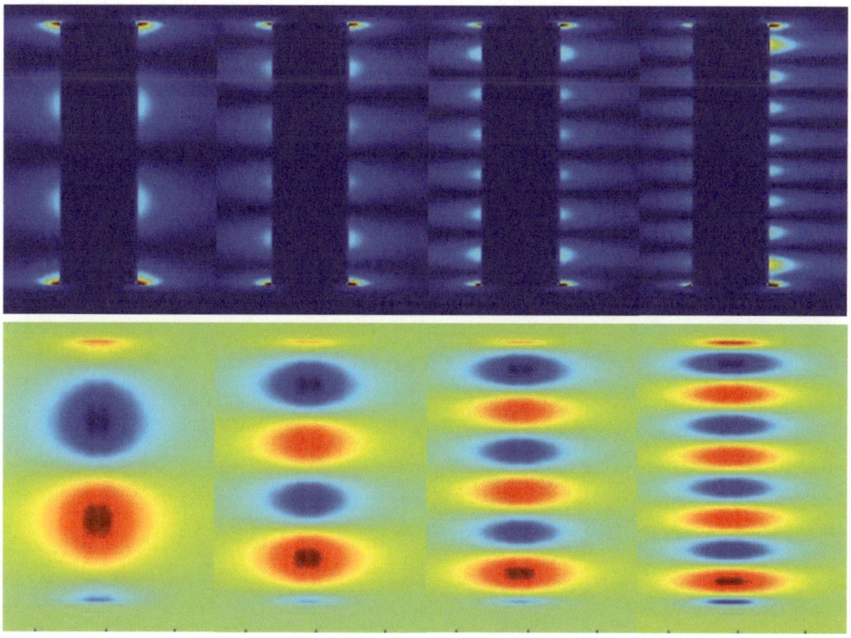

Figure 6. Simulated results for electric field profiles (top four panels) and vertical components of the Poynting vector of antenna radiations (bottom four panels) around a perfectly conducting nanowire, corresponding to the antenna mode (3λ/2, 5λ/2, 7λ/2 and 9λ/2 resonances). The electric field profile running through the center of the nanowires and the Poynting vector calculated at 55 nm above the nanowire axis.

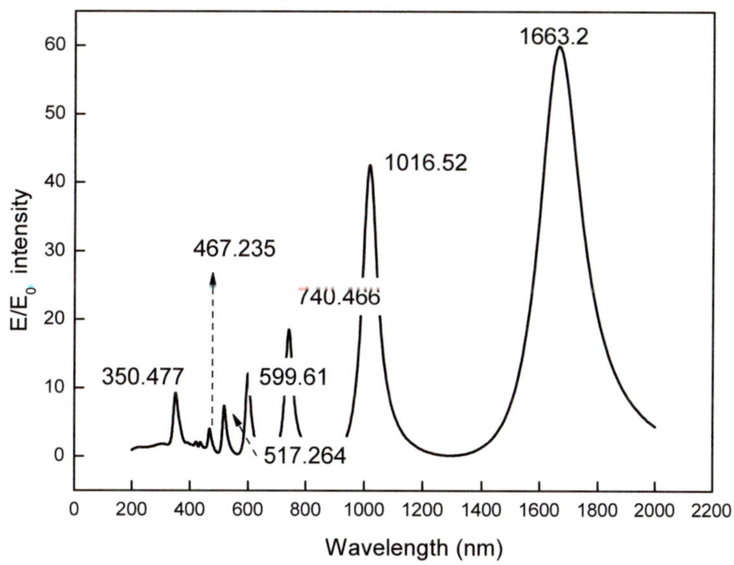

Figure 7. Multiple resonance spectrum around a silver nanowire. The resonance is shown as a value of $|E|/|E_0|$ at a point of 5 nm above the nanowire as a function of wavelength of incident light (200 nm to 2000 nm).

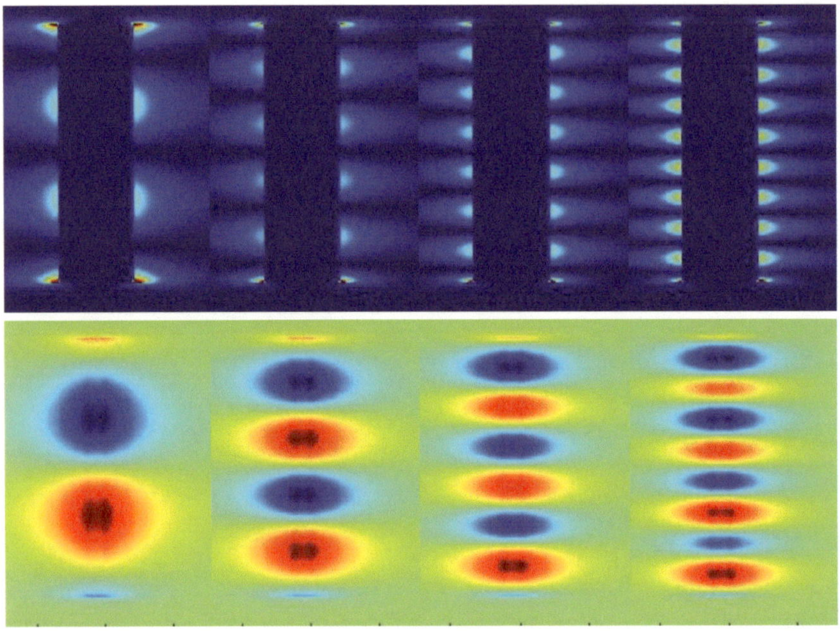

Figure 8. Simulated results for electric field profiles (top four panels) and vertical components of the Poynting vector of antenna radiations (bottom four panels) around a silver nanowire, corresponding to the antenna mode (3λ/2, 5λ/2, 7λ/2 and 9λ/2 resonances). The electric field profile running through the center of the nanowires and the Poynting vector calculated at 55 nm above the nanowire axis.

3.5. Ti nanowire

Figure 9 and 10 show the calculated result of amplitude enhancement, electric field profile and poynting vector as a function of wavelength for Ti nanowire. In the spectral range of calculation a feature of the Ti nanowires is the broaden peaks and weak field distribution as compared to perfectly conducting and Ag nanowire structures.

When comparing the field enhancement vs. wavelength for the silver and titanium nanowires to the perfect conducting nanowire, we can understand the optical antennas based on the background of both well-developed radiowave antenna engineering and the plasmonic behavior. According to the classical antenna theory that a fundamental mode occurs when an antenna length is half of the wavelength (λ/2), the fundamental antenna mode (λ/2 resonance) for a 2000 nm long perfect conducting nanowire is around 4000 nm, the observed modes (A-D) shown in Figures 5 and 6 should be assigned as higher harmonic modes (3λ/2, 5λ/2, 7λ/2, and 9λ/2 resonances). Comparing the multiple resonances spectrum between the perfect conducting and silver nanowires (Figures 7 and 8), it can be pointed out that a silver nanowire is a superb candidate for broadband optical antenna, which support four harmonic resonance modes in spectral range 517 to 1663 nm and one surface plasmon resonance (SPR) at 350.477 nm, as shown in Figure 11. The silver nanowire antenna is a one-dimensional SPR cavity, support the plasmon mode by reflecting the SPR currents at both ends of the nanowires.

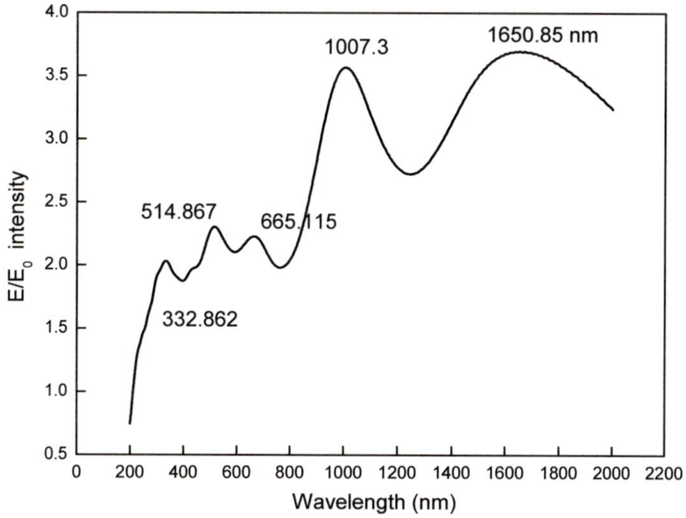

Figure 9. Multiple resonance spectrum around a titanium nanowire. The resonance is shown as a value of $|E|/|E_0|$ at a point of 5 nm above the nanowire as a function of wavelength of incident light (200 nm to 2000 nm).

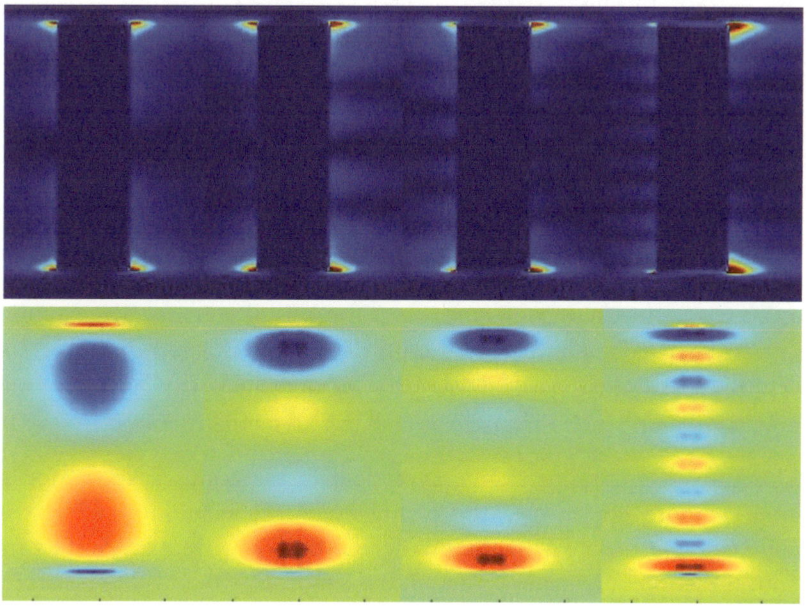

Figure 10. Simulated results for electric field profiles (top four panels) and vertical components of the Poynting vector of antenna radiations (bottom four panels) around a 2000 nm long titanium nanowire, corresponding to the antenna mode (3λ/2, 5λ/2, 7λ/2 and 9λ/2 resonances). The electric field profile running through the center of the nanowires and the Poynting vector calculated at 55 nm above the nanowire axis.

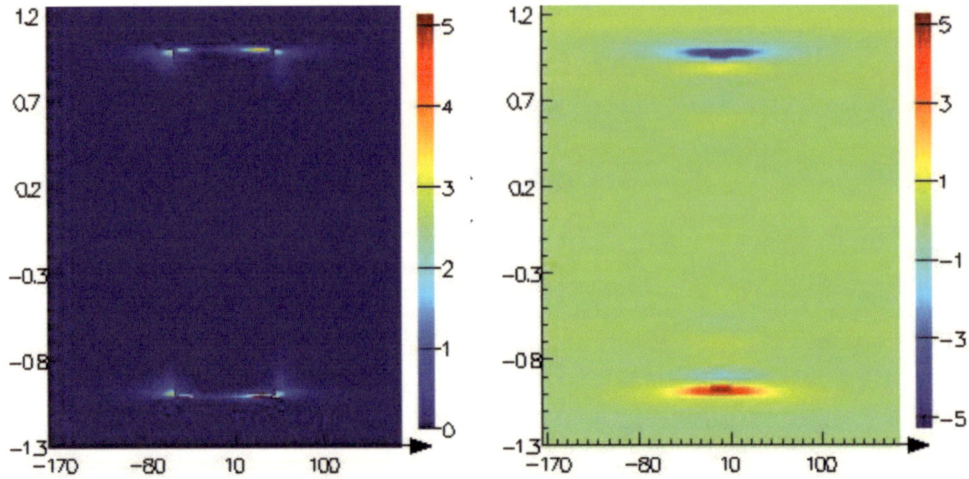

Figure 11. The electric field profiles (left) and vertical components of the Poynting vector of antenna radiations (right) around a silver nanowire at 350.477 nm.

For the titanium nanowire (Figures 9 and 10), their |E|/|E$_0$| profiles and Poynting vector components are much weaker than the perfect conducting and silver nanowires, the possible research is that titanium has interband electronic transition at visible range [32]. The noble metals silver has higher energy of interband transitions 3.8ev (326nm) with respect to its surface plasmon resonance energy 3.38 eV (350nm) which reduces the damping and enhances the appearance of the plasmon mode, however, the transition metal titanium has lower energy of interband transition 2.0ev (610nm) which is liable for strong damping.

3.6. Ag-TiO_2-Ag MIM nanostructures

The end to end coupling of two nanowires with a small distance between them can create highly localized and strongly enhanced optical fields in the gap and due to this phenomenon such structures are well competent for optical antenna. The coupling between two nanowire antennas robustly increases with decreasing the gap distance. To study this effect in the spectral response and its intensity enhancement we therefore calculated the symmetric and asymmetric metal-insulator-metal (MIM) nanostructures

MIM nanostructures were theoretically investigated as model systems for split-dipole optical antennas. Figure 4(b) shows the geometry of the heteronanowire structure. The heteronanowire is a split-dipole nanoantennas consisting of a silver nanowire (diameter, 100 nm) and identical titanium (Ti) nanowire with the length of 950 nm, aligned on the long axis along the dimer symmetric axes. The gap in the dimer is bridged by the titania (TiO_2) nanowire with the same diameter and 100 nm in length. The position p1 are 5 nm from the heterodimer apex, and the point p2 is at the middle of the titania nanowire. The nanoantenna-insulator-metal substrate structure imitated the nanorectenna, where an ac voltage could generate in the insulator and stimulate a tunnel current in an asymmetric

MIM tunneling diode. To observe the antenna electric-field profiles, full-field finite-difference-time-domain (FDTD) electromagnetic simulations were carried out for Ag-TiO_2-Ag and an Ag-TiO_2-Ti asymmetric MIM nanostructure under plane wave excitation.

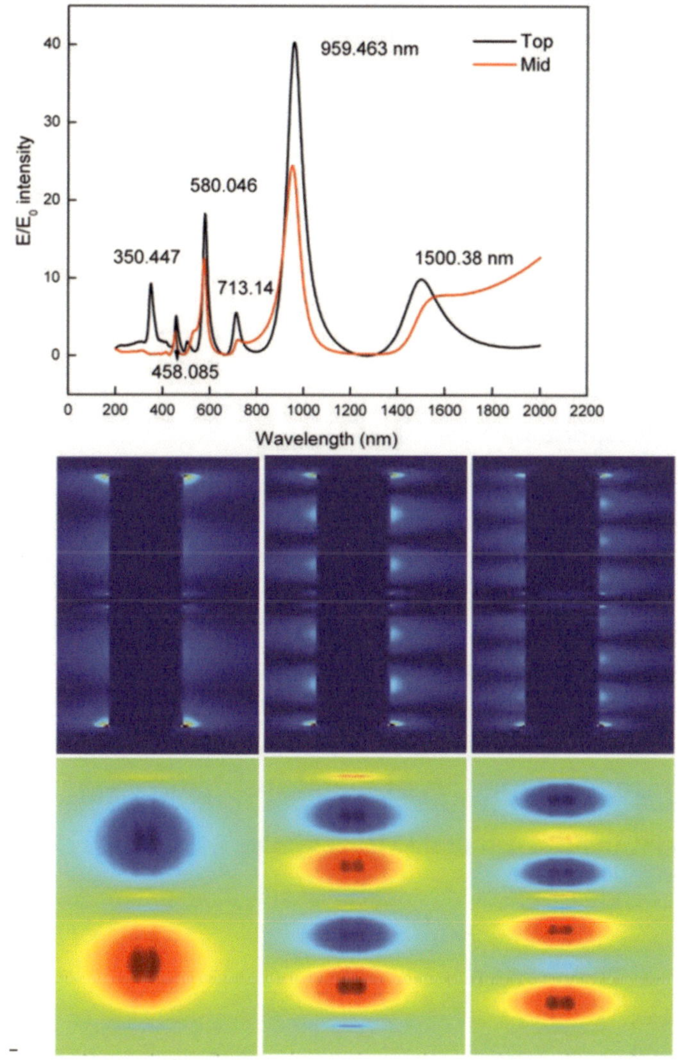

Figure 12. Resonance spectrum around a Ag-TiO_2-Ag MIM nanostructure shown as a value of $|E|/|E_0|$ at a point of 5 nm on top and in the middle of the homodimer as a function of wavelength (upper panel), the electric field profiles (middle panel) and vertical components of the Poynting vector of antenna radiations (bottom panel) corresponding to the antenna mode ($\lambda/2$, $3\lambda/2$, and $5\lambda/2$ resonances)

Figure 12 shows the local field enhancement spectra along with field profile and poynting vector component for the Ag-TiO_2-Ag symmetric. The Ag-TiO_2-Ag nanostructure is resonant at wavelength of 1500.38 nm, 959.463 nm and 713.14 nm and etc, corresponding to the first three order of the antenna resonances, and the near-field profiles at these $\lambda/2$, $3\lambda/2$ and $5\lambda/2$

resonances antenna mode are viewed from the electric field profiles and vertical components of the Poynting vector of antenna radiations. The near-field profiles clearly show that the electric field of the 2000 nm long Ag-TiO_2-Ag MIM antenna has a standing wave pattern, which is quite similar to that of single Ag nanowire antenna with 2000 nm length, their antenna radiations along the nanowires have same number of lobes and the lobe spacing of the fundamental mode is a little enlarged in the MIM antenna due to the gap between nanowires.

3.7. Ag-TiO_2-Ti MIM nanostructures

For asymmetric Ag-TiO_2-Ti MIM nanostructure, we also calculated the electric field, filed amplitude enhancement and vertical component of the Poynting vector antenna radiation corresponding to resonance modes, and the results are presented in Figure 13. It can be seen that the Ag-TiO_2-Ti nanostructure is resonant at wavelength of 1401.05 nm, 914.077 nm and 683.527 nm and etc, corresponding to the $\lambda/2$, $3\lambda/2$ and $5\lambda/2$ resonances antenna mode and the near-field profiles at first three order of the antenna resonances shows that the main contribution of the dimer antenna intensity field enhancement and field distribution are from silver in the asymmetric dimer.

For the Ag-TiO_2-Ti MIM antenna, the electric field is dominated by the Ag nanowires, the antenna radiations along the Ag nanowire of the MIM structure have same lobe patterns to the single Ag nanowire antenna with 950 nm lengths, and a fundamental antenna mode $\lambda/2$ can be identified unambiguously at the wavelength of 1401.25 nm. The electric field enhancement is bigger at the second harmonic mode ($3\lambda/2$) and reaches up to 22 and 7.5 in the middle of 100 nm thick TiO_2 insulators for the Ag-TiO_2-Ag and Ag-TiO_2-Ti MIM structures.

4. Optical rectification properties of metal-insulator-metal (MIM) devices

4.1. Review of metal-insulator-metal (MIM) diodes

The characterization and design of the optical rectification device focus mainly on the following areas:

1. The antenna;
2. The rectifier;
3. System integration.

The rectifier can transform an ac voltage to a dc voltage by means of a non-linear device, such as diode. To convert electromagnetic energy efficiently, a diode should be coplanar and couple to an optical antenna in order to take full advantage of the enhanced electric-field at the top of the metal electrodes or in the center gap of the antenna. Schottky diodes are routinely used in high frequency rectifiers because of their fast response time. These diodes have an operating frequency upper limit of approximately 3 THz, which is far below visible light frequencies. Antenna-couple metal-insulator-metal (MIM) diodes have been the subject of increasing interest due to their small size, CMOS compatibility, and ability to offer full functionality without cooling and applied bias [33-36]. The diode of choice for the optical frequency rectifier is a metal-insulator-metal (MIM) diode, it is previously considered that these diodes are the fastest available diodes for detection in the optical region.

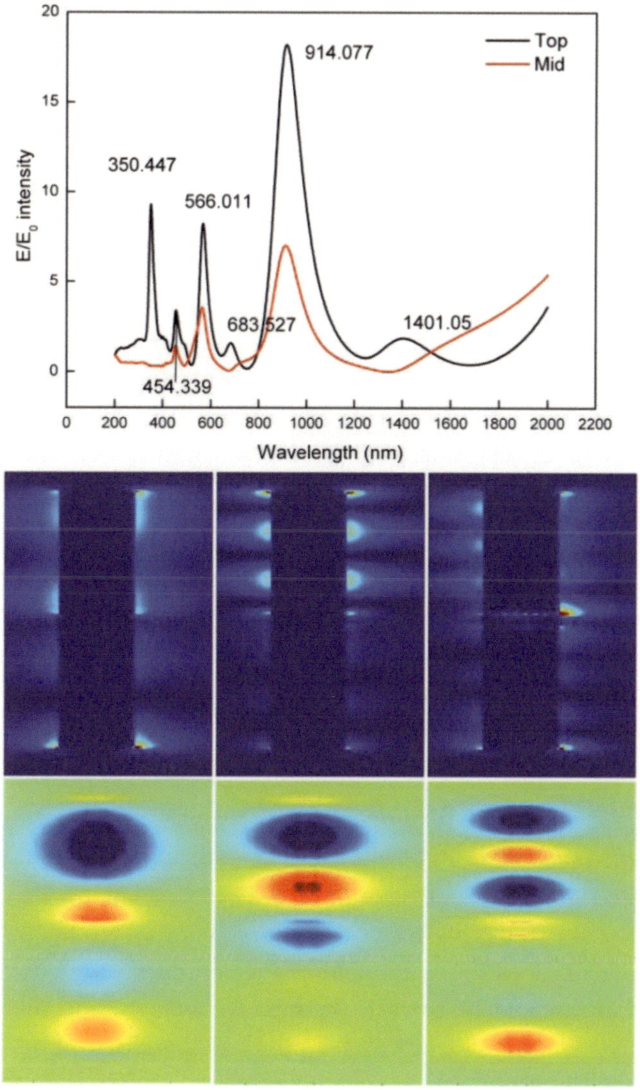

Figure 13. Resonance spectrum around a Ag-TiO_2-Ti MIM nanostructure shown as a value of $|E|/|E_0|$ at a point of 5 nm on top and in the middle of the heterodimer as a function of wavelength (upper panel), the electric field profiles (middle panel) and vertical components of the Poynting vector of antenna radiations (bottom panel) corresponding to the antenna mode ($\lambda/2$, $3\lambda/2$, and $5\lambda/2$ resonances)

However, plasmon absorption in metal-TiO_2 Schottky diode structures have recently been applied to photovoltaic devices [37, 38] and photocatalysts [39, 40] in the UV-visible and wavelength range, and an enhanced light harvesting property and a visible-light-induced charge separation are obtained for the benefit of the metal nanoparticles. Furthermore, different explanations have been presented about the role of metal nanoparticles in the observed improvement in light conversion efficiency. These include (i) metal nanoparticles increased absorption due to surface plasmons and light trapping effects [41], (ii) metal nanoparticles functioned as electron donor promoting electron transfer from metal to semiconductor [37, 38, 42] and (iii) metal nanoparticles served as electron trapping media that can minimize the surface charge recombination in semiconductor [43 , 44].

Meanwhile, two main mechanisms have been mentioned for the electron transfer between the metal nanoparticle and the semiconductor during the energy conversion process. First, Tatsuma et al. [37, 38, 45] and other workers [42, 46] proposed that the photoexcited electrons in the metal nanoparticles transferred from the metal particle to the TiO_2 conduction band since the photoresponse of these metal-TiO_2 diode structures was consistent with the absorption spectra of Au or Ag nanoparticles. Second, Kamat et al [43, 44] and Li et al. [47] have suggested that the noble metal nanoparticles act as electron sinks or traps in the metal-TiO_2 diode structures to accumulate the photogenerated electrons, which could minimize charge recombination in the semiconductor films. Obviously, a better understanding of these effects is crucial in exploiting the beneficial aspects of metal nanoparticles in photovoltaics.

In this section, we modeled and fabricated a MIM diode located at the mid-point of an Ag-TiO_2-Ti MIM antenna. The photodeposition was used to synthesis Ag-Cu nanoparticles in the TiO_2 nanotube grown from Ti foils to obtain the AgCu-TiO_2/Ti MIM diode structures, and their photoelectronic properties were measured under simulated sunlight and visible light. The Schottky barrier analyses indicated that the electron transport direction was the electrons transferred from metal nanoparticles to TiO_2.

4.2. Theoretical modeling of Ag-TiO_2-Ti MIM diodes

As shown in Figure 14, a MIM diode consists of three sections: a polished metal on a substrate as the base (anode), a natural oxide layer as the insulator (barrier), and another metal layer (cathode). When the barrier layer is extremely thin (10 to 50 Å), some quantum confinement effects may result in the tunnelling of electrons through the insulator layer. In this device, metals are treated as perfect electrical conductors, and are assumed to be in thermal equilibrium. Typically, the electrons in the metal are driven out of equilibrium by the absorbed power. The absorbed photons energize the electrons in the metal, allowing them to overcome the Schottky barrier and be collected in the cathode, eventually leaving through the back contact as measured current.

However, in these recent metal-TiO_2 Schottky diode structures [37-47], it would appear that the barrier layer was actually quite a bit thicker than 10 nm (probably in excess of 1 um) and further details was unable to be found in these papers. Semi-classical models did not account for non-equilibrium energy distributions of carriers, or do so through a localize lattice temperature. This problem in general is difficult for a traditional drift-diffusion model to capture the plasmonic effect in the illuminated metal layer. The logical approach to modeling the MIM rectifier would be to start with a customary potential-energy diagram.

Figure 14 shows the energy-band diagram of an Ag-TiO_2-Ti MIM diode. For smooth electron transportation through the interface between the semiconductor TiO_2 and the metal electrodes, an intervening ohmic contact is necessary. Since TiO_2 is an n-type semiconductor and its conduction band level is close to 4.2 eV, the electron affinity of TiO_2 is little greater than 4.2 eV, the work function of Ti is 4.33 eV, and the Ti-TiO_2 interface therefore becomes Schottky contact. Ohmic contact can be formed using Ag with the work function of 4.0 eV.

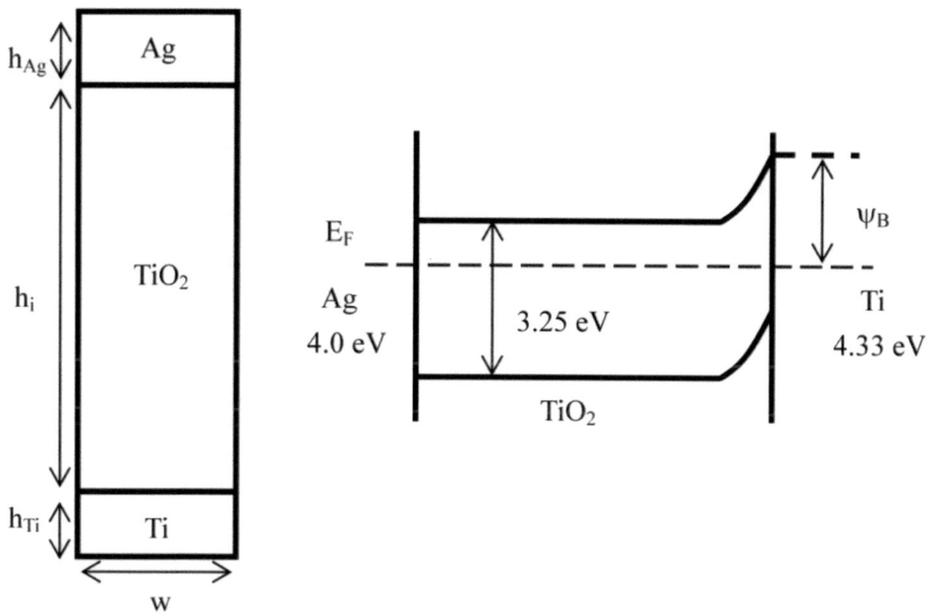

Figure 14. Energy-band diagram of a Ag-TiO_2-Ti MIM diode. Two MIM diodes are considered here, one is conventional diode with the diameter w= 2 μm, the TiO_2 layer thickness h_i = 60 μm, and the metal electrode thickness $h_{Ag} = h_{Ti}$ = 2 μm, the other is MIM nanodiode with the diameter w= 100 nm, the TiO_2 layer thickness h_i = 100 nm, and metal electrode thickness $h_{Ag} = h_{Ti}$ = 100 nm.

In order to understand the electron transport properties in the Ag-TiO_2-Ti MIM nanodiode, we performed the two-dimensional finite element (FEM) calculation based on the drift and diffusion equation for electrons and holes. In Figure 15, the calculated I-V curves are depicted for a conventional diode (a) and a MIM nanodiode (b) under various n-doping concentrations. It is show that the I-V curves show clear nonlinear and asymmetric current characteristic. The nonlinearity of current is basic mechanism for the rectification of an incident wave to a DC output.

As shown in Figure 15(a), for a conventional Ag-TiO_2-Ti MIM diode, the high doping leads to the high current density at zero bias. When doping is added, it will shift the Fermi level in the TiO_2, which will then define the Schottky barrier height (SBH) at the Ti-TiO_2 interface, the greater SBH results in larger charge transfer across metal-semiconductors interface, creating a large potential drop across depletion width and allowing a more efficient collection of electrons and holes.

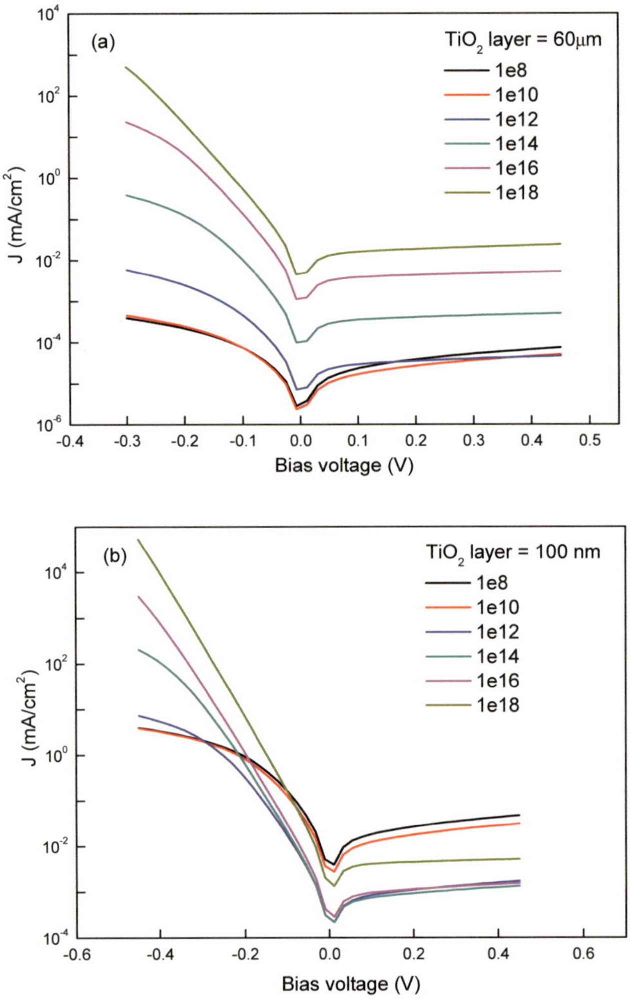

Figure 15. Simulated I-V characteristic for the Ag-TiO_2-Ti MIM diode. (a) is conventional diode with the diameter w= 2 μm, the TiO_2 layer thickness h_i = 60 μm, and the metal electrode thickness $h_{Ag} = h_{Ti}$ = 2 μm; (b) is MIM nanodiode with the diameter w= 100 nm, the TiO_2 layer thickness h_i = 100 nm, and the metal electrode thickness $h_{Ag} = h_{Ti}$ = 100 nm.

4.3. Experimental work of AgCu-TiO_2/Ti MIM nanostructures

The AgCu-TiO_2/Ti MIM nanoantenna heterostructures were fabricated using an electrochemical process and several measurements were carried out to determine if the MIM diode was producing direct current through plasmonic rectenna action. The TiO_2 nanotube layer was grown by anodizing 300-600 nm of titanium [48] on which the bimetallic Ag-Cu nanoparticle film were deposited using a photodeposition method. We chose the bimetallic Ag-Cu nanoparticles as the plasmonic medium where Cu alloying was used to prevent the natural oxidation of the silver nanoparticles and keep its good plasmonic property [49-51]. The surface morphologies of samples were characterized by scanning electron microscopy (JSM, 6390A) with energy dispersive X-ray spectroscopy (SEM-EDS). The photocurrent density - voltage curves (J-V) were measured at a potential sweep rate of 10 mV/s, with the Pt net as counter electrode, and a saturated calomel electrode (SCE) as reference electrode. Electrochemical Impedance Spectroscopy (EIS) was used to evaluate the properties of the different electrodes under AC polarization. The frequency range was 0.1 Hz to 100 kHz for amplitude of 5 mV in a DC potential of $-0.2\ V_{SCE}$.

a

b

c

Figure 16. SEM images of Cu (a), Ag (b) and Ag-Cu (c) nanoparticle layer deposited on TiO_2/Ti substrate.

Figure 16 (a, b and c) shows the surface morphology of the Cu, Ag and Ag-Cu nanoparticles deposited on TiO_2 nanotube, respectively. The Cu nanoparticles were evenly and densely distributed on the surface of TiO_2 nanotube layer, the measured nanoparticle size ranged 46 nm to 120 nm. The Ag nanoparticles are bigger than Cu nanoparticles and are not in uniform size with the big ones over 500 nm and the small ones below 100 nm. The possible reason is that the standard electrode potential of Ag^+/Ag^0 (0.78 eV) is higher than that of Cu^{2+}/Cu^0 (0.34 eV), so Ag^+ is reduced more rapidly and grows faster than Cu^{2+}. For the surface morphology of the Ag-Cu nanoparticles on TiO_2 nanotube layers, it is obvious that the Ag-Cu nanoparticles were denser than Ag nanoparticles. The energy dispersive X-ray spectroscopy (EDS) show that the atom ratios of Ag to Cu was 1.68:3.72 and the Cu content was less than that in the electrolyte because the reduced Cu can be further oxidized by Ag^+. Actually, this galvanic reaction has a dominated effect on the formation of Ag-Cu nanoparticles. Once the Cu nanoparticles formed firstly, the Ag^+ will be reduced at the surface of Cu nanoparticles and form the core-shell structure, an alloyed Ag-Cu nanoparticles formed after the reductive reaction and the atomic mutual diffusion at the interface of Cu and Ag atoms.

Figure 17 shows the measured J-V curves under simulated sunlight and visible light for the AgCu-TiO_2/Ti MIM structure, which had a short current density of -1.201 mA/cm^2 under simulated sunlight, and decreased to -0.734 mA/cm^2 under visible light, indicating that Ag-Cu nanoparticles were very photosensitive to the UV light.

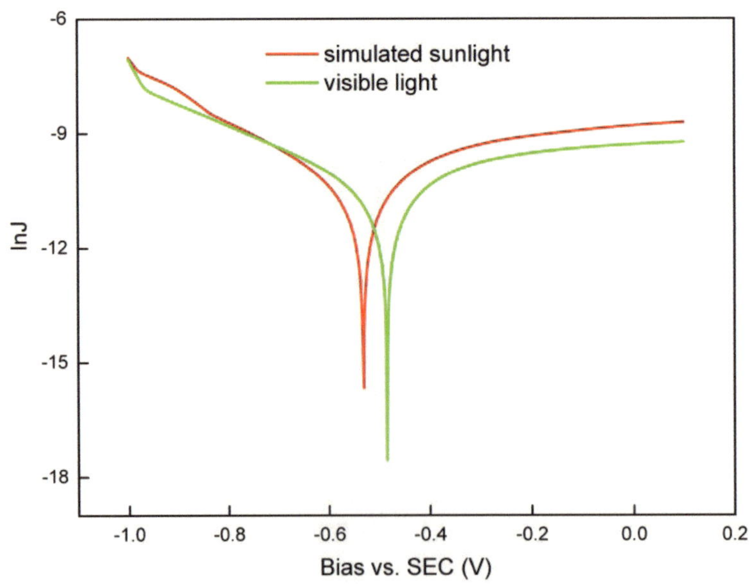

Figure 17. Measured current density and voltage (J-V) curves for the AgCu-TiO_2/Ti MIM structure under visible light and under simulated sunlight. The applied bias is versus the saturated calomel electrode (SCE).

The work function difference between the metal and the n-TiO_2 results in electrons transferred from TiO_2 to the metal nanoparticles yielding a Schottky junction. We can determine the electron transport direction by comparing the SBH changes under different irradiation conditions. The SBH values were calculated in lnJ-V diagram shown in Figure 17 by the following Equation:

$$J_s = A^* T^2 \exp(-q\phi_{SBH} / k_B T) \tag{1}$$

where ϕ_{SBH} is SBH at the zero bias, A^* is the Richardson constant, k_B is the Boltzmann constant, and J_s is the zero bias saturation current density. The calculated SBH values were 1.021eV and 1.006 eV for the simulated sunlight and under visible light irradiations, *i.e.*, the ϕ_{SBH} values calculated under the simulated sunlight are higher than that measured under visible light and less electrons were produced and transferred to TiO_2 under visible light. This result clearly indicates that the metal nanoparticles can be photoexcited as electron donors, and that the electrons transferred from metal nanoparticles to TiO_2.

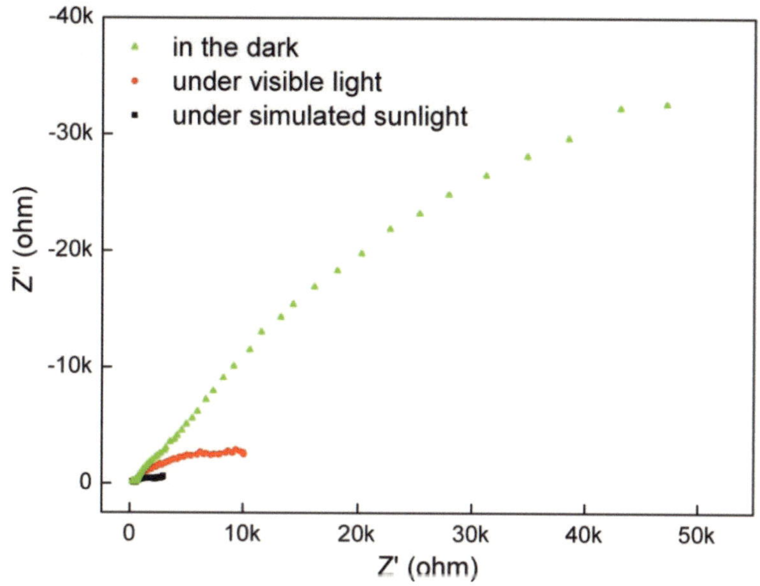

Figure 18. Nyquist diagram measured for the AgCu-TiO_2/Ti MIM structure under simulated sunlight, visible light, and in the dark.

Figure 18 exhibits that impedance measured under simulated sunlight and visible light were sharply decreased in comparison with that measured in the dark at the low frequency region (< 100Hz), indicating a decrease of charge transfer resistance. This can be justified by the bending of the impedance arc in the low frequency region (the second arc) in the Nyquist diagram, because the bending of the arc in this region indicated a process of charge transfer while the linear relationship between the imaginary and real component of the impedance mean a diffusion process controlled step. Among the three photoelectrodes, the low frequency region arc of the AgCu-TiO_2/Ti MIM structure was largely bended under both light irradiations, which indicated that faster charge transfer was obtained in the AgCu-TiO_2/Ti diode.

5. Summary

The plasmonic effect of the metal nanostructure has been explored to improve the conversion efficiency for DSSCs. We found that metal nanoparticles in the plasmonic DSSC can lead to serious parasitic absorption even under the optimized conditions, the absorption of metal nanoparticles takes up about 20–30% of the total absorption of the DSSC at visible wavelengths. To solve the problem of the metal parasitic absorption in the DSSC caused by the surface plasmon optical effect, designing and fabricating metal-insulator-metal sandwich structure as photon acceptor and electron donor may be a research direction for the further development of the plasmonic photovoltaic techniques.

The metal-insulator-metal (MIM) nanostructures were theoretically investigated as model systems for optical antennas. The field enhancements vs. wavelength for the silver and titanium nanowires with 2000 nm length have shown the higher harmonic modes ($3\lambda/2$, $5\lambda/2$, $7\lambda/2$, and $9\lambda/2$ resonances). A silver nanowire is a superb candidate for broadband optical antenna, which support four harmonic resonance modes in spectral range 517 to 1663 nm and one surface plasmon resonance (SPR) at 350.477 nm. The Ag nanowires dominated the electric field for the Ag-TiO_2-Ti MIM antenna.

The current density and voltage curves of Ag-TiO_2-Ti (MIM) nanostructure have been modeled at the micrometer and nanometer scale. We are carrying out experiments on the optical rectification properties of AgCu-TiO_2/Ti metal-insulator-metal (MIM) diode devices. Cu, Ag, and Ag-Cu nanoparticle films exhibits different morphological characters. The measured J-V curves under simulated sunlight and visible light for the AgCu-TiO_2/Ti MIM structure exhibit a short current density of -1.201 mA/cm^2 under simulated sunlight, and decreased to -0.734 mA/cm^2 under visible light, indicating that the electrons transferred from metal nanoparticles to TiO_2 layer.

Author details

Fuyi Chen*, Jian Liu and Negash Alemu
State Key Laboratory of Solidification Processing,
Northwestern Polytechnical University, Xian, China

Acknowledgement

This study was supported by the National Natural Science Foundation of China (Grant Nos. 50971100 and 50671082), the Research Fund of State Key Laboratory of Solidification Processing in China (Grant No. 30-TP-2009), and the NPU Foundation for Fundamental Research (Grant No. NPU-FFR-ZC200931).

* Corresponding Author

6. References

[1] Takamoto T, Kaneiwa M, Imaizumi M, et al. InGaP/GaAs-based Multijunction Solar Cells. Progress in Photovoltaics 2005; 13(6) 495-511

[2] Nazeeruddin M K, DeAngelis F, Fantacci S, Selloni A, Viscardi G, Liska P, Ito S, Takeru B, Gratzel B M. Combined Experimental and DFT-TDDFT Computational Study of Photoelectrochemical Cell Ruthenium Sensitizers. Journal of American Chemistry Society 2005; 127(48)16835-16847

[3] O'Regan B, Gratzel M, A Low-cost, High-efficiency Solar Cell Based on Dye-sensitized Colloidal TiO_2 Films. Nature 1991; 353(24)737-739

[4] Stuart H R, Hall D G. Island Size Effects in Nanoparticle-enhanced Photodetectors. Applied Physics Letters 1998; 73, 3815,

[5] Atwater H A, Polman A. Plasmonics for Improved Photovoltaic Device. Nature Materials 2010; 9,205-213

[6] Ferry V E, Munday J N, Atwater H A. Design Consideration for Plasmonic Photovoltaics. Advanced Materials 2010; 22, 4794-4808

[7] Brown M D, Suteewong T, Kumar R S S, Innocenzo V D, Petrozza A, Lee M M, Wiesner U, Snaith H J. Plasmonic Dye-sensitized Solar Cells Using Core-shell Metal-insulator Nanoparticles. Nano Letters 2011; 11(2)438-445

[8] Hagglund C, Zach M, and Kasemo B. Enhanced Charge Carrier Generation in Dye sensitized Solar Cells by Nanoparticle Plasmons. Applied Physics Letters 2008; 92, 013113

[9] Du S, Li Z. Enhanced Light Absorption of TiO_2 in the Near-ultraviolet Band by Au Nanoparticles. Optics Letters 2010; 35(20)3402-3404

[10] Guilatt O, Apter B, Efron U, Light Absorption Enhancement in Thin Silicon Film by Embedded Metallic Nanoshells. Optics Letters 2010; 35(8)1139-1141

[11] Fu D, Zhang X, Barber R L and Bach U. Dye-sensitized Back-contact Solar Cells. Advanced Materials 2010; 22, 4270–4274

[12] Durr M, Menges B, Knoll W, Yasuda A, and Nelles G. Direct Measurement of Increased Light Intensity in Optical Waveguides Coupled to a Surface Plasmon Spectroscopy setup. Applied Physics Letters 2007; 91, 021113

[13] Atwater H A, Polman A. Plasmonics for Improved Photovoltaic Device. Nature Materials 2010; 9,205-213

[14] Yuan L, Chen F Y, Zheng C F, Liu J, Alemu N. Parasitic Absorption Effect of Metal Nanoparticles in the Dye-sensitized Solar Cells. Physica Status Solidi A 2012; 209,1376-1379.

[15] Robinson J and Rahmat-Samii Y. Particle Swarm Optimization in Electromagnetics. IEEE Transaction on Antennas and Propagation 2004; 52, 397.

[16] Robinson J, Sinton S and Rahmat-Samii Y. Particle Swarm, Genetic Algorithm, and their Hybrids: Optimization of a Profiled Corrugated Horn Antenna. In: Proc. IEEE Int. Symp. Antennas Propagation, San Antonio, TX, 1(314-317), 2002

[17] O'Regan B, Gratzel M. A Low-cost, High-efficiency Solar Cell Based on Dye-sensitized Colloidal TiO_2 Films. Nature 1991; 353,737
[18] Green M A and Pillai S. Harnessing Plasmonics for Solar Cells. Nature Photonics 2012; 6, 130
[19] Schuck P J, Fromm D P, Sundaramurthy A, Kino G S, Moerner W E. Improving the Mismatch between Light and Nanoscale Objects with Gold Bowtie Nanoantennas. Physics Review Letters 2005; 94, 017402
[20] Muhlschlege P, Eisler H, Martin O, Hecht B, Pohl D. Resonant Optical Antennas. Science 2005; 308, 1607-1609
[21] Novotny L, van Hulst N. Antennas for Light. Nature Photonics 2011; 5, 83-90,
[22] Brown W C. The history of Power Transmission by Radio Waves. IEEE Transactions on Microwave Theory and Techniques 1984; 32(9)1230-1242
[23] Brown W C. Optimization of the Efficiency and other Properties of the rectenna element. IN: 1976 IEEE-MTT-S International Microwave Symposium, 142-144, 1976
[24] Hocker L O, Sokoloff D R, Daneu V, Szoke A and Javan A. Frequency Mixing in the Infrared and Far-infrared using a Metal-to-metal Point Contact Diode. Applied Physics Letters 1968; 12, 401-402
[25] Matarrese L M and Evenson K M. Improved Coupling to Infrared Whisker Diodes by Use of Antenna Theory. Applied Physics Letters 1970; 17, 8-10
[26] Fetterman H R, Clifton B J, Tannenwald P E and Parker C D. Submillimeter detection and mixing using Schottky diodes. Applied Physics Letters 1974; 24, 70
[27] Fetterman H R, Tannenwald P E, Clifton B J, Parker C D, Fitzgerald W D and Erickson N R. Far-ir Heterodyne Radiometric Measurements with Quasioptical Schottky Diode Mixers. Applied Physics Letters 1978; 33, 151
[28] Small J G, Elchinger G M, Javan A, Sanchez A, Bachner F J and Smythe D L, An Electron Tunnelling at Infrared Frequencies: Thin-film M-O-M Diode Structure with Broad-band Characteristics. Applied Physics Letters 1974; 24,275-279
[29] Gustafson T K, Schmidt R V, and Perucca J R. Optical Detection in Thin-film metal - oxide - metal diodes. Applied Physics Letters 1974; 24, 620
[30] Bailey R L. A Proposed New for a Solar-energy Converter. Journal of Engineering for Power 1972; 94, 73-77
[31] Marks A M. Device for Light Power to Electric Power. U.S. Patent 4 445 050, 1984
[32] Chen F Y, Alemu N, Johnston R L. Collective Plasmon Modes in a Compositionally Asymmetric Nanoparticle Dimer. AIP Advances 2011; 1,302134
[33] Fumeaux C, Herrmann W, Rothuizen H, De Natale P and Kneubühl F K. Mixing of 30 THz Laser Radiation with Nanometer Thin Film Ni-NiO-Ni Diodes and Integrated Bow-tie antennas. Applied Physics B Laser and Optics 1996; 63(2)135–140.
[34] Codreanu I, González F and Boreman G. Detection Mechanisms in Microstrip Dipole Antenna-coupled Infrared Detectors. Infrared Physics and Technology 2003; 44(3)155–163.

[35] Hobbs P C, Laibowitz R B and Libsch F R. Ni-NiO-Ni Tunnel Junctions for Terahertz and Infrared Detection. Applied Optics 2005; 44(32)6813–6822

[36] Chen F Y, Yuan L, Johnston R. L. Low-loss Optical Magnetic Metamaterials on Ag-Au Bimetallic Fishnets. Journal of Magnetism and Magnetic Materials 2012; 324, 2625-2630.

[37] Tian Y, Tatsuma. T. Plasmon-induced Photoelectrochemistry at Metal Nanoparticles Supported on Nanoporous TiO_2. Chemistry Communication 2004; 16, 1810-1811.

[38] Tian Y, Tatsuma T. Mechanisms and Applications of Plasmon-Induced Charge Separation at TiO_2 Films Loaded with Gold Nanoparticles. Journal of American Chemistry Society 2005; 127, 7632-7637.

[39] Awazu K, Fujimaki M, Rockstuhl C, Tominaga J, Murakami H, Ohki Y, Yoshida N, Watanabe T A. Plasmonic Photocatalyst Consisting of Silver Nanoparticles Embedded in Titanium Dioxide. Journal of American Chemistry Society 2008; 130, 1676-1680

[40] Irie H, Kamiya K, Shibanuma T, Miura S, Tryk D A, Yokoyama T, Hashimoto K. Visible Light-Sensitive Cu(II)-Grafted TiO_2 Photocatalysts: Activities and X-ray Absorption Fine Structure Analyses. Journal of Physical and Chemistry C 2009; 113,10761–10766.

[41] Stuart H R and Hall D G. Island Size Effects in Nanoparticle-enhanced Photo Detectors. Applied Physics Letters 1998; 73, 3815-3817.

[42] Mubeen S, Hernandez-Sosa G, Moses D, Lee J, Moskovits M. Plasmonic Photosensitization of a Wide Band Gap Semiconductor: Converting Plasmons to Charge Carriers. Nano Letters 2011; 11, 5548–5552.

[43] Chandrasekharan N, Kamat P V. Improving the Photoelectrochemical Performance of Nanostructured TiO_2 Films by Adsorption of Gold Nanoparticles. Journal of Physical and Chemistry B 2000; 104, 10851–10857.

[44] Takai A, Kamat P V. Capture, Store, and Discharge. Shuttling Photogenerated Electrons across TiO_2 Silver Interface. ACS Nano 2011; 5, 7369-7376.

[45] Sakai N, Fujiwara Y, Takahashi Y, Tatsuma T. Plasmon-Resonance-Based Generation of Cathodic Photocurrent at Electrodeposited Gold Nanoparticles Coated with TiO_2 Films. ChemPhysChem 2009; 10, 766 – 769.

[46] Furube A, Du L, Hara K, Katoh R, Tachiya M. Ultrafast Plasmon-Induced Electron Transfer from Gold Nanodots into TiO_2 Nanoparticles. Journal of American Chemistry Society 2007; 129, 14852-14853.

[47] Liu L, Wang G, Li Y, Li Y, Zhang J. Z. CdSe Quantum Dot-Sensitized Au/TiO2 Hybrid Mesoporous Films and Their Enhanced Photoelectrochemical Performance. Nano Research 2011; 4, 249–258.

[48] Chen F Y, Liu J. A Plasmonic Rectenna and its Fabrication Method. Chinese Patent, 201210002179.5, 2012.

[49] Chen F Y, Zheng C F, A Silver-copper Nanoalloy and its Electrical Synthesis Method. Chinese Patent, 201110310987.3, 2011

[50] Chen F Y, Johnston R L. Charge Transfer Driven Surface Segregation of 13-atom Au-Ag Nanoalloy and its Relevance to Structural, Optical and Electronic properties. Acta Materialia 2008; 56, 2374-2380

[51] Chen F Y, Johnston R L. Energetic, Electronic and Thermal effects on Structural Properties of Ag-Au Nanoalloys. ACS Nano 2008; 2, 165-175

Chapter 20

Application of Surface Plasmon Polaritons in CMOS Digital Imaging

Qin Chen, Xiaohua Shi, Yong Ma and Jin He

Additional information is available at the end of the chapter

http://dx.doi.org/10.5772/50757

1. Introduction

Recent years there has been a rapid expansion of research into nanophotonics based on surface plasmon polariton (SPP), which is a collective electron oscillation propagating along a metal-dielectric interface together with an electromagnetic wave [1]. What distinguishes SPPs from photons is that they have a much smaller wavelength at the same frequency. Therefore SPPs possess remarkable capabilities of concentrating light in a nanoscale and the resulted significant enhancement of localized field [2]. SPPs can be excited by an incident electromagnetic wave if their wavelength vectors match. This is usually achieved by nanopatterning the metal film. The resonant frequency of SPP is determined by the metal materials, dielectric materials, profiles and dimensions of the patterns, etc. As a result, the tunability of SPP enables its application as a colour filter in the visible range. Actually this was used in the stained glass manufacture hundreds years ago. Since the extraordinary optical transmission through a nanohole array in a thin metal film was reported by Ebbesen [3], the plasmonic photon sorting has been explored for the potential applications in digital imaging and light display [4,5]. In addition, SPP based light manipulating elements like planar metallic lenses, beam splitters, polarizers have been investigated both theoretically and experimentally [6-7].

In solid state digital imaging, complementary metal oxide semiconductor (CMOS) image sensors (CISs) are the leading mass-market technology. CMOS image sensors with smaller pixels are expected to enable digital imaging systems with better resolution and possibly high photosensitive area in the pixel (fill-factor). In present colour CMOS digital imaging systems, dye-doped polymer filters and curved dielectric microlenses are used to disperse light of different wavelengths and manipulate the light beams, respectively . With the down scaling of pixel size to the sub-2 μm range, these optical elements suffer from performance degradation, such as colour cross-talk, due to the large distance between the optical elements and the photodiodes underneath [8]. Therefore most high-resolution digital cameras have to use the backside illuminated CISs that are fabricated with complicated processes [9]. Furthermore different colour polymer filters have to be fabricated successively in several process steps using the back-end-of-line process. Continuous development of new applications for CISs requires that they are able to be manufactured at low cost.

The idea of introducing plasmonics into a CIS was first proposed by Catrysse *et al* [10]. They were able to show the potential of metallic nanodevices in a CIS by detailed numerical simulations, and provided preliminary experimental results [10,11]. There are several advantages: (1) this technique has the potential to produce all the required colour filters in a single metal layer by one lithography step at a low cost; (2) colour cross-talk can be reduced by integrating plasmonic colour filters in Metal 1 layer which is very close to the CMOS photodiodes; (3) the thickness of the plasmonic device is one or two orders of magnitude thinner than that of the colourant one used in a CIS or a LCD; (4) localized field enhancement effect of metallic nanostructures may increase the photodetection sensitivity; (5) the light filtering can be readily obtained by varying the nanostructures that enables tuning of the resonant wavelengths and thus achieves a large spectral design freedom; (6) other optical elements such as lenses and polarizers can be achieved based on SPP in metal layers in CMOS technology. Planar metallic lenses integrated in the metal layers in standard CMOS technology can readily tune the phase of incident light on a pixel level to achieve a great control on the light beam divergence therefore the cross-talk especially for the pixels at the edge of the whole sensor; (7) plasmonic metal like aluminium is a CMOS compatible metal and it is more stable compared to polymer.

In this chapter, we briefly review the research progress on various plasmonic optical elements for the application in digital imaging and describe our work on a plasmonic CIS (pCIS). Section 2, 3 and 4 focus on plasmonic colour filters, PLs and wire-grid polarizers, respectively. Section 5 presents our work on the integration of plasmonic colour filters on CISs. Finally, we conclude this chapter and discuss the outlook of this technique.

2. Plasmonic colour filters

2.1. Background

A colour filter that selectively transmits or reflects input light is an important element in a CIS. Established colour filtering technologies for CISs use dye-doped polymers. Each colour filter for red (R), green (G) and blue (B) must be fabricated successively in several process steps. Because of the encroaching difficulties of cross-talk and the cost of manufacture, it is desirable to find new methods for building colour filters into CISs. Other colour filtering techniques for imaging arrays have been investigated. Guided-mode resonance filters based on subwavelength dielectric gratings were shown to work as a bandpass filter [12]. They can be designed to work in both reflection and transmission modes. Hybrid metal-dielectric gratings were found to offer excellent optical transmission (87%) property in the midinfrared range due to the Fano transmission resonance [13]. One-dimensional (1D) periodic metal-insulator-metal (MIM) waveguide array supports a surface plasmon antisymmetric mode which showed colour filtering effect [14]. However, devices based on a 1D structure have an intrinsic polarization dependency. Photonic crystal colour filter was proposed for the application in a MOS image sensor, where two multilayer stack mirrors separated by a defect layer were used to form a cavity resulting a passband filtering [15]. But the dielectric mirrors have a relatively narrow bandwidth limited by the index contrast in the stack and the multilayer deposition process is complex. Silver mirrors were proposed to replace the dielectric mirrors and a CIS based on this technique was demonstrated where the different colours were achieved by tuning the cavity length [16]. The device requires multi-lithography steps to make R, G, B pixels. A bull's eye structure consisting of concentric grooves with a central hole was proposed as another candidate for colour filters [5]. This method gave very good narrow band wavelength filtering, but the low fill ratio resulted in poor transmission efficiency.

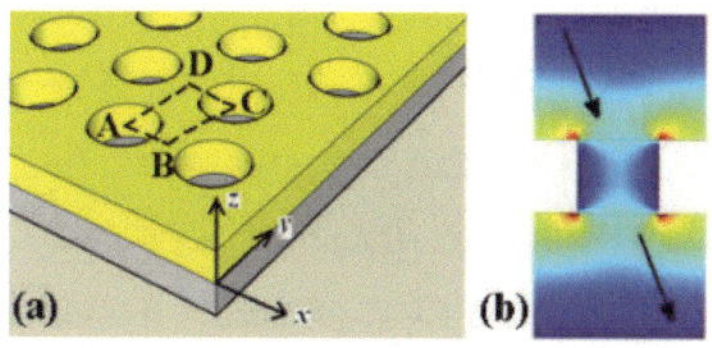

Figure 1. (a) A square-lattice hole array in a metal film. (b) Cross-section field distribution around a hole where light couples to SPPs.

Ebbeson *et al.* investigated the subwavelength holes in silver films as shown in Fig. 1(a) and observed extraordinary optical transmission (EOT) of light through the holes and wavelength filtering due to the excitation of surface plasmon resonance (SPR) [3]. The incident light first couples to the SPR at the top surface of the metal film in the presence of the 2D hole array. The SPR at the top surface then couples to the one at the bottom surface of the metal film. Finally, light reemits from the bottom surface of the patterned metal film as shown in Fig. 1(b). The peak position, λ_{max}, of the transmission spectrum at normal incidence can be approximated by

$$\lambda_{max} = \frac{a}{\sqrt{i^2 + j^2}} \sqrt{\frac{\varepsilon_m \varepsilon_d}{\varepsilon_m + \varepsilon_d}} \quad (square\ \ lattice) \tag{1}$$

$$\lambda_{max} = \frac{a}{\sqrt{\frac{4}{3}(i^2 + ij + j^2)}} \sqrt{\frac{\varepsilon_m \varepsilon_d}{\varepsilon_m + \varepsilon_d}} \quad (triangular\ \ lattice) \tag{2}$$

where a is the period of the array, ε_m and ε_d are the dielectric constants of the metal and the dielectric material in contact with the metal respectively, and i and j are the scattering orders of the array [3,4]. As shown in Eq. (1) and (2), the period determines the transmission peak positions of SPRs for a given material configuration. Following this discovery, researchers started to use 2D hole arrays in metal films as transmissive colour filters. A square-lattice hole array surrounded by a square-lattice dimples at the same period in a silver film was reported to show colour filtering function [4]. Without surrounding dimples, a square array of circular holes were designed and fabricated in a thin aluminium film as red and green colour filters [17], where unfortunately the green filter devices exhibited a yellow colour due to the high colour cross-talk.

In this section, we focus on transmitted colour filters consisting of triangular-lattice hole arrays in aluminium films that are compatible with standard CMOS technology. Both numerical simulation and experimental results are presented.

2.2. FDTD simulation

Eq. (1) and (2) theoretically predict the resonant wavelengths of SPRs. But the thickness of the metal film and the coupling between the SPRs at two interfaces has not been considered in these equations. To accurately optimize the design, a finite-difference time-domain (FDTD) algorithm based commercial software, Lumerical FDTD Solutions [18], was used to investigate the colour filters and identify the transmission peaks in the measured spectra. As shown in Eq. (1) and (2), the wavelength interval between the first two SPR peaks in a triangular array is larger than that for a square array of a same period. Therefore we focus on the triangular lattice in the following part. In our work, we focus on a triangular-lattice hole array in an aluminium film on glass. Compared to silver and gold, aluminium is compatible with standard CMOS technology and cheap although it has a relatively higher absorption loss. Aluminium also has good adhesion to many substrates making fabrication easier.

The effects of the structure dimensions were investigated numerically. As shown in Fig. 2(a), the transmission efficiency was strongly affected by the size of the circular hole. As the hole size increases, the magnitude of the main transmission peak increases from 24% to 75%. There is also a red shift and an increase in the full-width at half-maximum (FWHM) from 60 nm to 300 nm. The reflections at the transmission peaks are almost zero in all three cases. Therefore the loss at the transmission peak is larger at a smaller radius, where the absorption within the metal due to the non-zero imaginary component of the permittivity increases [4]. A tradeoff between transmission and FWHM must be considered to optimize the RGB filters for the application in a CIS. The coupling between the SPRs at both sides of the metal film has an important effect on the transmission spectra. As shown in Fig. 2(b), the structure made with a thinner aluminium film has higher transmission but also a much larger bandwidth. The enhanced coupling of SPR in the case of a thin metal film increases the splitting of the two transmission peaks, i.e. the long wavelength side peak has a red shift but the short wavelength side peak has a blue shift. In addition, the transmittance of the very low level at a thickness of 30 nm. The EOT phenomenon is more prominent in the case of a thin metal film where the SPR coupling is strong. However, the FWHM is large for a thick metal film and may cause a spectral cross-talk for the RGB filters. Predicting in Eq. (2), the period has a dominant effect on the transmission peak, i.e. the transmitted colour of the filter. The filters shown in Fig. 2(c) have the peak transmissions around 50% filters in the visible range can be readily obtained by tuning the period of the hole array. Finally, the different hole shapes with a similar area in a same triangular lattice were investigated as shown in Fig. 2(b). The circular hole array has the highest transmittance at the transmission peak and the smallest FWHM. Therefore, we focused on the circular hole array in our experiments. Plasmonic colour filters with spectral responses matching the International Commission on Illumination colour matching functions are useful for effectively communicating color between colour detection and output devices. A fully automated genetic algorithm that incorporated on-demand 3D FDTD simulations were used to determine the structure dimensions [19].

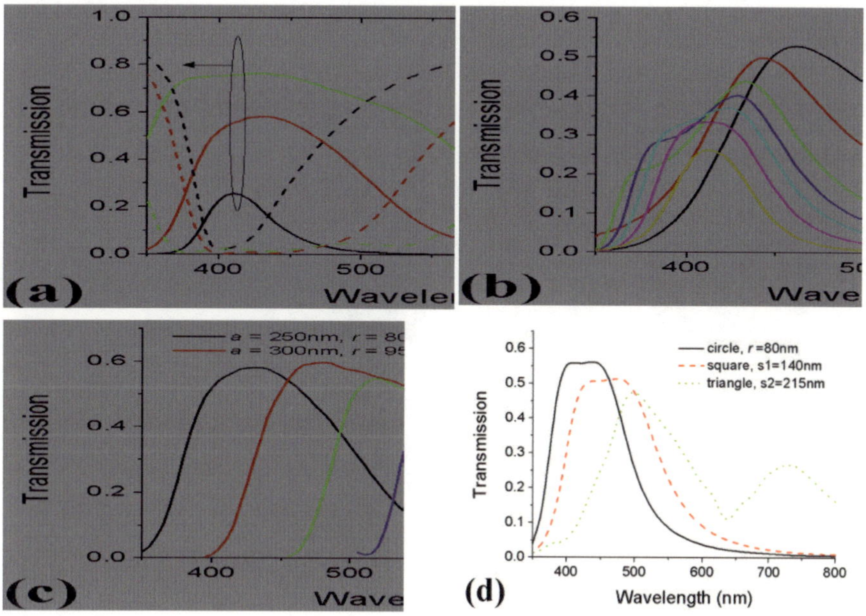

Figure 2. Simulated transmission/reflection spectra for hole arrays in a triangular lattice in an aluminium film on glass with a 200 nm SiO_2 cap layer. (a) Different circular hole radius r with a period $a = 250$ nm and a thickness $t = 150$ nm. (b) Different t at $a = 250$ nm and $r = 65$ nm. (c) Different a and r at $t = 150$ nm. (d) Different hole shapes.

2.3. Fabrication of colour filters on glass

Fig. 3(a) is the process flow for fabricating plasmonic colour filters. The first step is to evaporate a 150 nm aluminium film on a clean glass substrate at a rate of 0.3 nm/s by electron beam evaporation. Then ZEP520A electron beam resist was spin-coated on to the sample and exposed. After development, aluminium was etched using $SiCl_4$ in a Plasmalab System 100. Finally, a 200 nm SiO_2 layer is deposited on top of the patterned aluminium film after removal of the residual resist to enhance the transmission due to the symmetric SPR coupling (the refractive index of SiO_2 is close to that of glass). Fig. 3(b) shows a scanning electron microscope (SEM) image of patterned aluminium film after removing the residual resist. Vertical and smooth sidewalls can be seen from the inset SEM image for which the sample was tilted at 30º [20].

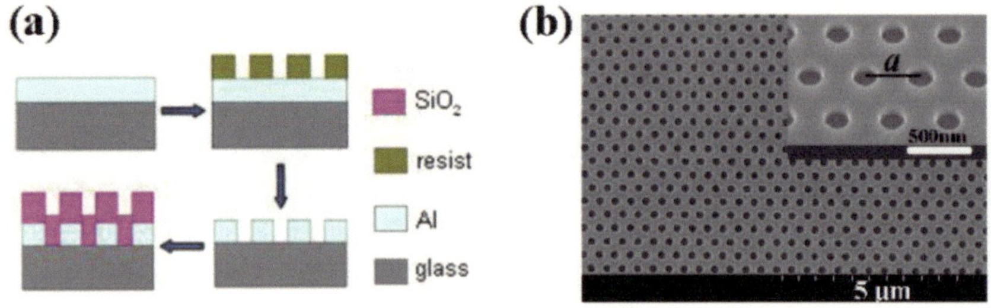

Figure 3. (a) Process flow for fabricating plasmonic colour filters. (b) A SEM image of etched holes in a triangular array with $a = 430$ nm in an aluminium film on glass. The inset is a SEM image for a sample tilted at 30º.

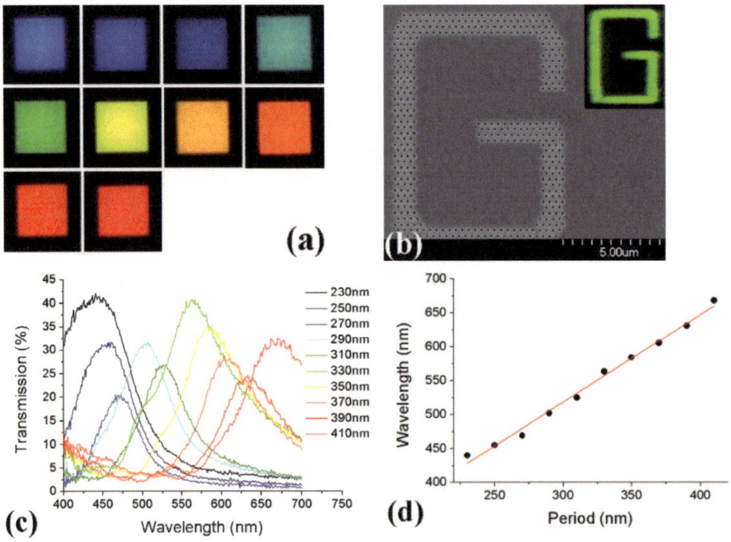

Figure 4. Performance of various plasmonic colour filters fabricated on glass. (a) Images of various plasmonic colour filters taken in microscope transmission mode under a white light illumination. (b) A SEM image of holes composing the letter 'G' with $a = 330$ nm. The transmitted light image of the structure with a 200 nm SiO_2 cap layer is shown in the inset. (c) Transmissive spectra of the filters shown in (a). (d) Wavelengths at which transmission through each peak versus the period of the plasmonic nanostructures. Reprint from [21].

The colour images of plasmonic filters were examined using an Olympus BX51 microscope fitted with a broadband halogen lamp. Fig. 4(a) shows the well-defined colour squares with a size of 50 µm × 50 µm consisting of holes in a triangular lattice with different periods. A letter 'G' consisting of holes with $a = 330$ nm was fabricated as shown in Fig. 4(b), where the period number of hole array is as small as three. The clear green letter 'G' appeared under the white light illumination as shown in the inset. Unlike the structures in earlier work [4], there is no dimple around the etched holes in our structure. This means that the pixel size could be as small as 1 µm on a high-resolution image sensor. Spectral measurement was carried out on a microscope spectrophotometer TFProbe MSP300 that can sample the signal from a minimum area of 10 µm × 10 µm. Different colour filters were written on to the same sample to ensured consistent measurement of the different structures. Due to the symmetrical properties of the triangular-lattice hole array and the circular holes, the transmittance shows no difference for illumination at different polarization angles. An unpolarized light beam from a halogen lamp was launched normally on to the back-side of the sample. The transmitted light was collected using a 40× lens with a *NA*=0.9 and guided to the spectrometer and image camera. The transmission spectra from the sample in the visible range from 400 nm to 700 nm are shown in Fig. 4(c), where transmittances are between 20% and 40%, with full width at half maximum between 70 nm and 110 nm. As predicted in Eq. (2), Fig. 4(d) shows that the wavelength at which transmission through each filter peaks was found to linearly increase with the period of the holes. This means that a complete colour filter set can be reliably made using a single lithographic step.

3. Plasmonic polarizers

3.1. Background

Polarization is a general property of light and contains information about reflecting objects that traditional intensity-based sensors like human being's eyes ignore. However, polarization offers a number of advantages for imaging [22-24]. Polarization filtering has long been used in photography through haze. Difficult computer vision tasks such as image segmentation and object orientation are made tractable with polarization vision techniques. Investigation of the polarized light backscattering enables noninvasive surface and beneath-the-surface imaging of biological systems. Traditionally, the polarization image was obtained by taking an objective at the same place twice at different polarization angle with an external polarizer mounted in front of the camera. Mathematic analysis on these data finally generates a polarization image. Obviously, this method has low efficiency, low accuracy and high cost. Micropolarizer array was proposed to implement the polarization imaging like a Baye array in colour filter array [25,26]. In each 2×2 cell, there are four polarizers at different polarization angles to collect the required polarization information. Similar to a colour image, polarization image can be obtained from a polarization image sensor with micropolarizer arrays by just one shot of the scene. In the past, micropolarizer array was fabricated using polymer which is usually above 10 μm [25]. The large thickness causes large cross-talk between neighboring pixels. Furthermore, polymer polarizer faces a stability issue. Alternatively, metal wire grids are an ideal polarizer, which has been applied for IR polarization imaging [26]. Optimization of metal wire grids for a large extinction ratio (ER), defined as the transmission ratio of TM modes (the polarization of light perpendicular to the metal wire grid) to TE modes (the polarization of light parallel to the metal wire grid), is important to a high-contrast polarization image.

In this section, we numerically investigate the plasmonic polarzers and demonstrate the experimental results of aluminium grating polarizers.

3.2. Simulation

2D FDTD simulation is used to investigate the performance of metal wire grids. Wire grids with different periods were simulated and the results are shown in Fig. 5. In all cases, the effect. With the decreasing of the period, the transmission of TM polarization keeps at a high value but that of TE polarization decreases significantly. As a result, the ER increases with the decrease of the period. The state-of-the-art CMOS technology enables a feature size below 50 nm. However, a thick metal layer increases the aspect ratio of the gratings and therefore increases the difficulty in fabrication. The aspect ratio also affects the ER of the wire grids. As shown in Fig. 6, with a period of 100 nm, the transmission of TE polarization decreases more than one order of magnitude and the ER of Al wire grids increases approximately 15 dB when t increases 50 nm. All the above results are for wire grids with a duty cycle (metal filling ratio) of 50%. In Fig. 7, we show the effect of the duty cycle.

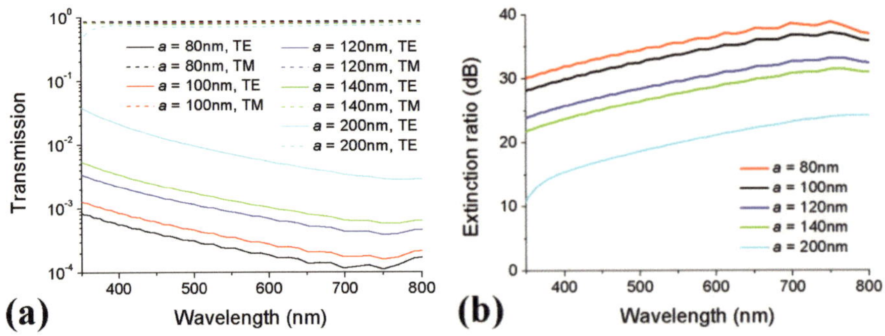

Figure 5. Transmission spectra (a) and ER (b) of Al wire grids on glass with different periods at $t = 100$ nm and $f = 50\%$.

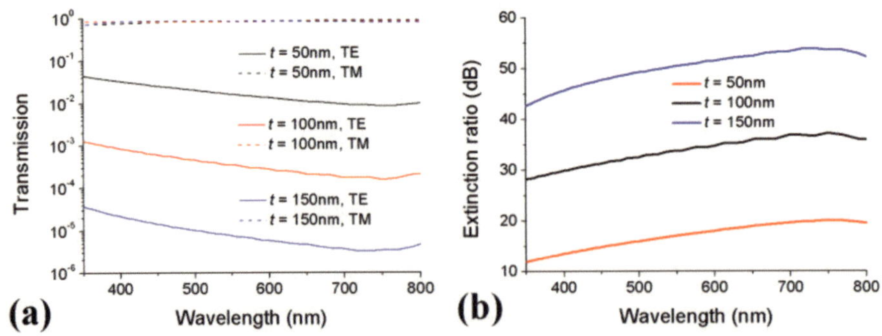

Figure 6. Transmission spectra (a) and ER (b) of Al wire grids on glass with different metal thicknesss at $a = 100$ nm and $f = 50\%$.

Transmissions of both TE and TM polarizations reduce when the duty cycle increases. Furthermore, ER is higher at a higher duty cycle.

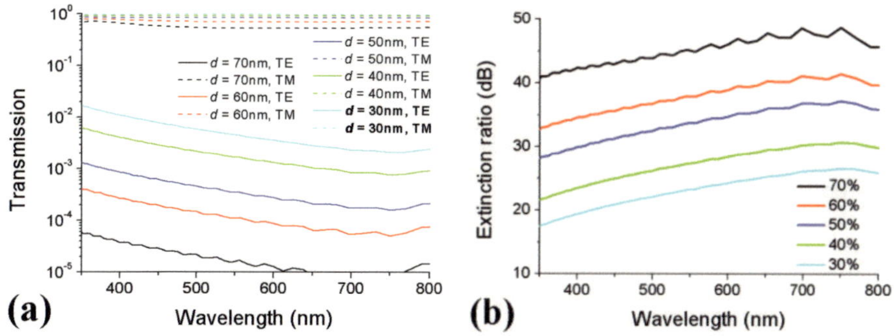

Figure 7. Transmission spectra (a) and ER (b) of Al wire grids on glass with different duty cycle at a = 100 nm and t = 100 nm.

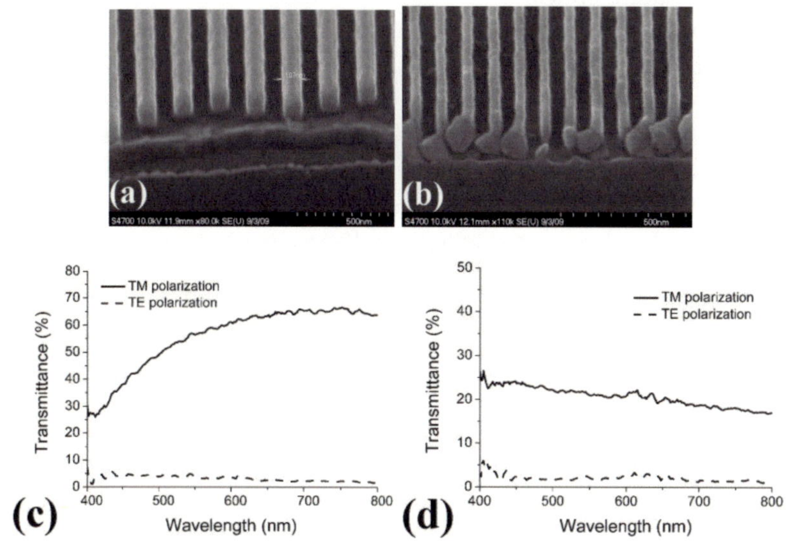

Figure 8. SEM images of fabricated Al wire grids with a period of 200 nm (a) and 100 nm (b). (c) and (d) Measured transmission spectra of polarizers in (a) and (b).

3.3. Experiments

The process flow is similar to that of plasmonic colour filters. The SEM images of devices with periods of 200 nm and 100 nm are shown in Fig. 8(a) and (b), respectively. Spectral measurement was carried out on a same microscope spectrophotometer TFProbe MSP300 as the plasmonic colour filter experiments. The incident light was polarized using a linear polarizer amounted before the sample in the light path. The transmitted light was collected using a 40× lens with a NA=0.9 and guided to the spectrometer. As the simulation results, the TM polarization has higher transmission than TE polarization as shown in Fig. 8(c) and (d). The measured ER of the device with a period of 100 nm is approximately 16, enabling a 4-bit polarization imaging. Actually, our polarizers may have better performance because the transmittance of TE polarization was closed to the background noise limited by our detector. In addition, dual grating structures may be considered to obtain a higher ER [27].

4. Plasmonic Lenses (PLs)

4.1. Background

Traditionally light is manipulated using dielectric optical elements such as refractive lens, diffractive gratings, mirrors and prisms. For example, dielectric microlens arrays are used in current CMOS image sensor to increase the light collection efficiency. However, diffraction may put the usefulness of the microlens in question in sub-2μm pixels [28]. Recent progress in nanotechnology and the theory of plasmonics has led to rapidly growing interest in the implementation of metallic optical elements on a nano-scale for light beam manipulation [29-39]. Especially, lots of theoretical predictions [30-35] and experimental demonstration [28,36-39] have been reported on both one-dimensional (1D) and 2D PLs. However, most experimental results showed a large deviation of focal length from the design [28,36-39]. The authors attributed this phenomenon to the finite size of the lenses and the resulting diffraction. The effect of the lens size on the focal length was theoretically investigated in a plano-convex refractive microlens [40]. It turned out that the upper limit of the focal length was determined by the lens aperture due to the diffraction effect.

In this section, we discuss the diffraction effect in PLs and experimentally demonstrate PLs with accurate control of the focal lengths that are important for the application in a CIS.

4.2. Theory

A 1D PL, as shown in Fig. 9(a), is a group of nano-slits in metal that form zones to modulate the phase delay distribution across the device surface. It focuses light only in the x-direction. The distribution of the phase delay at the exit of each slit is designed to provide constructive interference at the focus. Incoming laser light excites SP modes at the slit entrances that propagate through the slits before emitting into light at the exits, forming a convergent focus. In general, the SP modes can be excited in a metal/dielectric/metal waveguide, as shown in the inset of Fig. 9(b), under TM-polarized illumination with the electric field perpendicular to the slits. In a narrow slit, only the fundamental SP mode exists with a complex propagation constant, β, that can be calculated from the equation

$$\tanh\left(\sqrt{\beta^2 - k_0^2\varepsilon_d}\,\frac{w}{2}\right) = \frac{-\varepsilon_d\sqrt{\beta^2 - k_0^2\varepsilon_m}}{\varepsilon_m\sqrt{\beta^2 - k_0^2\varepsilon_d}} \tag{3}$$

where ε_d and ε_m are the permittivity of the dielectric inside the slit and the metal, w is the slit width, and k_0 is the wave vector of light in free space [41]. The real and imaginary part of the effective refractive index n_{eff}, defined to be β/k_0, determine the phase velocity and the propagation loss of the SP modes, respectively. The real part of n_{eff} and the corresponding loss are shown in Fig. 9(b) for the example of Al/air/Al structure as a function of the slit widths. We use ε_m = -56.12 + *i*21.01 for aluminum at a wavelength of 633 nm [19]. A phase delay modulation on the surface of a patterned metal film can be realized by simply adjusting the widths of patterned nano-slits. Light focusing can therefore be obtained using a planar structure. The phase delay β*d* (*d* is the thickness of the metal film) of the SP modes inside the slits dominates the phase delay distribution across the device surface. For a PL with a focal length *f* in air, the various slits and their positions can be readily obtained according to the constructive interference principle

$$\beta(x)d + 2\pi\sqrt{f2 + x2}/\lambda = \beta(0)d + 2\pi f/\lambda + 2(m-1)\pi \tag{4}$$

where λ is the wavelength of the illumination, β is a function of the position, and *m* is the zone number. β(0) is the propagation constant of the SP mode in the central slit, which is usually the narrowest slit with the largest value of β. Light coming from slits in a same zone, i.e. a same *m* (including the case of two identical slits at *x* and –*x*), has a same phase at the focus. But light from slits in different zones has a phase shift of an integer times of 2π at the focus.

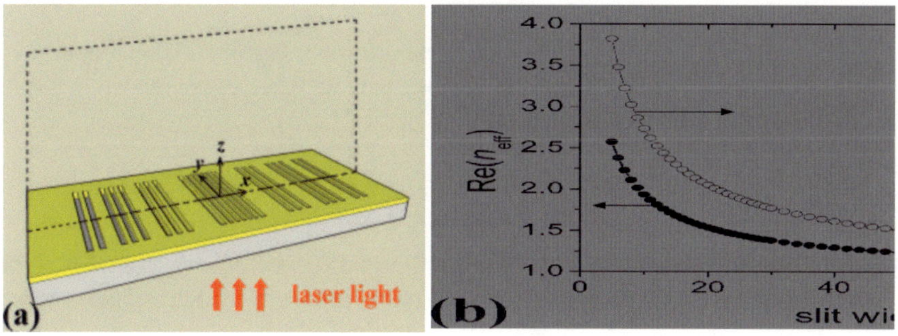

Figure 9. (a) Schematic of a PL. (b) Dependence of the real part of neff and the loss of a SP mode on the slit width. Reprint from [42].

Using Eq. (4), PLs with focal lengths of 0.5 μm, 1 μm, 2 μm, 3 μm, 6 μm and 12 μm at 633 nm were designed in a 200 nm thick aluminium film on a glass substrate. All PLs have an aperture size *D* of approximately 10 μm. The minimum width of the slit was 10 nm and the minimum gap between two neighboring slits was 50 nm, which was thicker than the skin depth. The central slit, with a width of 10 nm, was at *x* = 0 and the whole structure was symmetrical in *x* on the the *yz* plane. For an ideal planar lens with a focal length *f* at λ, the phase delay caused by the light path difference is

$$\phi = 2\pi\sqrt{f2 + x2}/\lambda - 2\pi f/\lambda \quad (x \geq 0) \tag{5}$$

ϕ for each lens calculated from Eq. (5) is shown in Fig. 14 as a solid line. Light emitting from a position at a larger *x* to the focus has a larger phase delay. By counting from the center we divide the phase delay into zones with a range of 2π. As a result, the slits in a same zone have a same *m* in Eq. (4). Optimization of the positions and widths of nano-slits gives a phase delay in the slits of the SP mode, ($\beta(x)d$ - $\beta(0)d$), obeying the constructive interference principle as shown in Eq. (4). The symbols in Fig. 10 show the phase delay, ϕ', for each slit in different zones. We can see that the symbols and the lines agree very well, i.e. $\phi=\phi'$. For a fixed aperture size, the PL with the longest (shortest) f has the smallest (largest) ϕ'_{max}. A PL with a longer focal length has more slits in each zone but has fewer zones. We note that the PL with f = 12 μm has almost the same number (21) of slits as the PL with f = 0.5 μm that has 22 slits, but the former has 2 zones and the later has 7 zones.

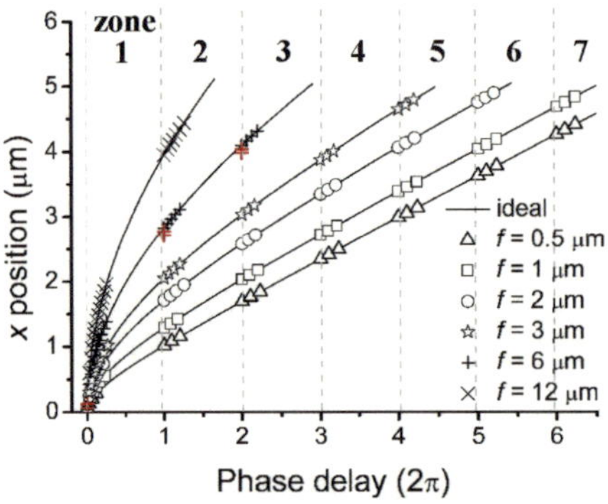

Figure 10. The phase delay ϕ and ϕ' for each slit in different zones ($x \geq 0$) for similar aperture PZPLs with f = 0.5 μm, 1 μm, 2 μm, 3 μm, 6 μm and 12 μm at 633 nm. Red crosses are related to the structure shown in Fig. 18(b). Reprint from [42].

4.3. FDTD simulation

A 2D FDTD simulation was chosen to be an adequate approximation since the length of the uniform 10 μm slits in the y-direction in our experiment is much larger than the light wavelength. In the simulation, a uniform cell of $\Delta x = \Delta z = 1$ nm was used in the metal slab and a nonuniform cell was used elsewhere. The simulation domain was bounded by perfectly matched layers. A TM-polarized plane wave source at 633 nm was launched with normal incidence to the PL surface from the glass substrate side of the metal film.

4.3.1. PLs with variant focal lengths

The Poynting vector, P_z, distribution of the PL with variant focal lengths is plotted in Fig. 11. As can be seen, P_z for each PL has a light spot at the expected focal point. The constructive interference of the light from each slit at the focus is clearly demonstrated. For the PLs with f = 0.5 μm, 1 μm, 2 μm, 3 μm and 6 μm, P_z has its maximum value at the design focus. However, although there is a local maximum in P_z around the design focus for the PL with f = 12 μm, the maximum P_z occurs at a distance of 5.4 μm from the PL surface. In [36-39], the shorter focal lengths compared to the designs were explained as a similar phenomenon found in conventional refractive microlens [49]. The effect of diffraction at the lens aperture limits the maximum of the focal length. Using FDTD method, an aperture of 10 μm in the same aluminium film was simulated and showed a peak radiation at a distance of approximately

40 μm. Although the designed focal length of 12 μm is well below this value, the decrease of the focal length still occurs. Furthermore, close examination of the field pattern for the PL with $f = 6$ μm shows an unwanted local maximum at a distance of 2.1 μm from the PL surface that has an amplitude slightly smaller than that at the focus. Although the PL with $f = 12$ μm has almost the same number of slits as the PL with $f = 0.5$ μm in a similar size aperture, the focusing quality of the former is much poorer. The much smaller $NA = 0.35$ for the PL with $f = 12$ μm compared to 0.99 for the PL with $f = 0.5$ μm may be one reason. Furthermore, ϕ'_{max} is 12.5 π for the PL with $f = 12$ μm and 2.5 π for the PL with $f = 0.5$ μm.

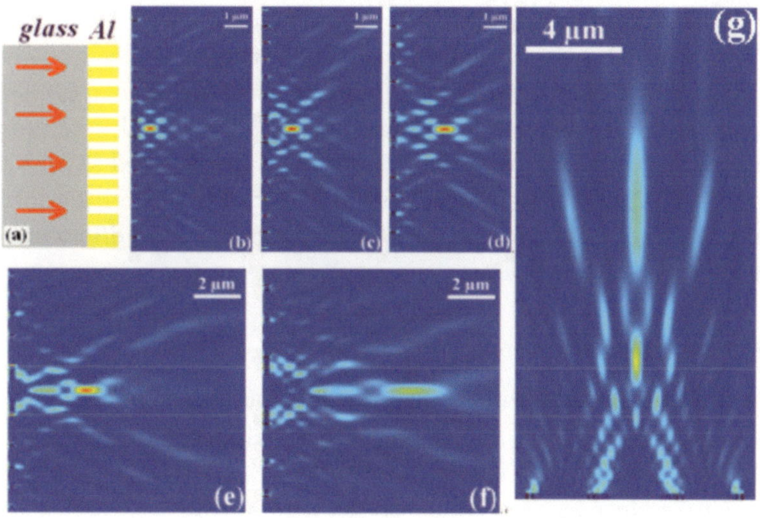

Figure 11. (a) Schematic of the PL. (b)-(g) Simulation results for the Poynting vector Pz distributions of the PZPLs with $f = 0.5$ μm, 1 μm, 2 μm, 3 μm, 6 μm and 12 μm respectively.

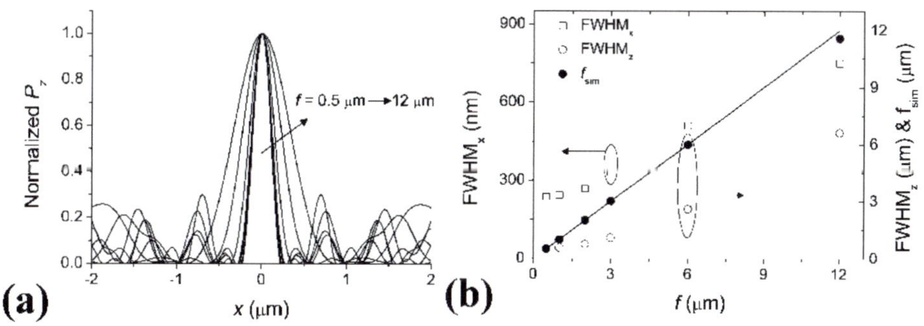

Figure 12. (a) Normalized simulation results for the Poynting vector Pz distributions along the x axis through the foci of PLs with $f = 0.5$ μm, 1 μm, 2 μm, 3 μm, 6 μm and 12 μm respectively. The optical axis is at x = 0. (b) The simulation results of the focal length, FWHMx and FWHMz for PLs with the design $f = 0.5$ μm, 1 μm, 2 μm, 3 μm, 6 μm and 12 μm respectively. fdesign is plotted as a solid line. Reprint from [42].

In Fig. 12(a) the normalized P_z distribution through the centre of the focus spot along the x axis is shown for each PL in Fig. 11. As can be seen, P_z is tightly confined in the x-direction although there are some side lobes. The simulation results of the focal length and FWHM of the beam extension in both *x*- and *z*- directions are shown in Fig. 12(b). We can see that the focal length obtained using the FDTD method is in an excellent agreement with the design. Moreover, the FWHM of the beam extension on the *z*-axis ($FWHM_z$) increases with the increasing *f*, starting from 0.49 μm for $f = 0.5$ μm to 6.6 μm for $f = 12$ μm. The FWHM of the beam in the *x*-axis at the focal point ($FWHM_x$) also increases from 236 nm (0.37λ) for $f = 0.5$ μm to 750 nm (1.18λ) for $f = 12$ μm. As a further test of the PL quality, we compared $FWHM_x$ with what we would expect for a diffraction limited beam. The Rayleigh limit for the resolving power of a lens is given by $d_s = \lambda f/D$ for a line, where d_s is the line-width, *D* is the lens aperture and λ is the wavelength. For $f = 6$ μm, the PL has $FWHM_x = 510$ nm while a conventional lens with the same aperture has $d_s = 440$ nm. The PL therefore achieves focus resolution close to the diffraction limit. One advantage of the planar PL is the ability to achieve a large aperture for a small focal length and therefore a high resolution. We would like to mention that the above formula for d_s can not be approximated as $\lambda/2NA$ because *NA* is not even close to $D/2f$ in the case of near-field application. The performance of the proposed plasmonic Fresnel zone plate (FZP), working beyond the diffraction limit in [32], was therefore overestimated.

4.3.2. *The effect of the number of zones on the PL performance*

Varying the number of zones in the structures, PLs with $f = 0.5$ μm and $f = 6$ μm were simulated. As shown in Fig. 13(a), the simulated focal length has a large deviation from the design for a PL with a small zone number. In the case of a one-zone PL with $f = 6$ μm, *NA* is 0.23 and ϕ'_{max} is 0.5 π. Increasing to three zones, the focal lengths become stable and agree well with the design for both PLs with a focal length approximately 0.8λ and 10λ, respectively. In the case of three zones in the PL with $f = 6$ μm, *NA* is 0.59 and ϕ'_{max} is 4.4 π. The increasing *NA* and ϕ'_{max} provide an accurate focal length as predicted in the design. As shown in Fig. 13(b), the lens aperture ($f = 6$ μm) was found to almost linearly increases with the increasing zone number but $FWHM_x$ reduces. $FWHM_x$ decreases from 510 nm for 3 zones to 258 nm for 16 zones. The resolution approximately doubles at a cost of a 3.3 times increase of the aperture. Although $d_s = \lambda/2NA$ is not valid in the near-field case, the $FWHM_x$ is found to be nearly proportional to $1/NA$, where the trend predicts a resolution of approximately 210 nm at $NA = 1$. So the focal length of a PL agrees with the design very well if the zone number is larger than 3. However, the extension of the focus in the focal plane, i.e. $FWHM_x$, is sensitive to the number of zones. Other authors also found the dependence of the focal lengths of a FZP on the zone numbers [34]. A straightforward understanding is that more zones integrated in the PL aperture, i.e. a larger *NA*, give a focus with a higher contrast because of more constructive light interference from each slit dominates the interference between the diffraction wave from the zones. We should notice that a 3-zone PL with $f = 6$ μm has a same $NA = 0.59$ as a one-zone PL with $f = 0.5$ μm. As can be seen from Fig. 13(a), the accuracy of the focal length of the former is much better than the later. So the number of zones or the total phase shift range of light involving in the interference at the focus is the real factor determining the device performance.

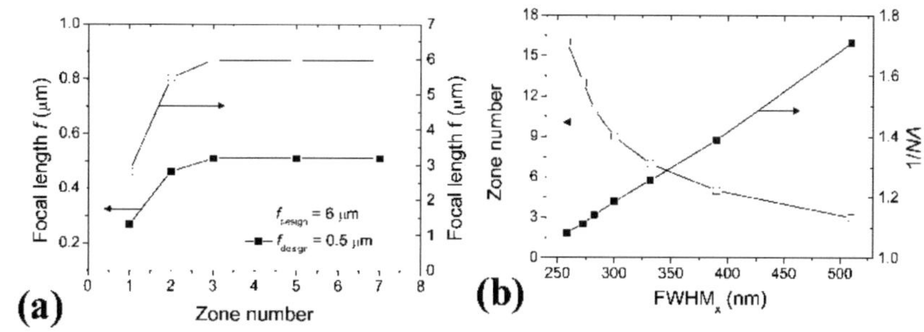

Figure 13. (a) Simulation results for the focal lengths in the PL s with variant zone numbers. The two PLs were designed to have a f of 0.5 μm and 6 μm respectively. (b) Calculated FWHM$_x$ and NA for the PL (f = 6 μm) with variant zone numbers. Reprint from [42].

Simulation results for the Poynting vector P_z distributions of the PL with f = 6 μm and in total 7 zones are shown in Fig. 14(a). Compared to the 3-zone PL shown in Fig. 11(f), the aperture size increases to 15.8 μm from 8.6μm. Accordingly, NA increases from 0.59 to 0.80 and ϕ'_{max} increases from 4.4 π to 12.3 π. The 7-zone PL has a more confined focus both in the focal plane and the light propagating direction. The amplitude ratio to the high-order foci increase as well because the additional zones contribute substantially to the main focus but negligible amount to the high-order foci due to the increasing incline. To achieve a good focusing performance, increasing the lens aperture is a straightforward way. But the footprint increases at the same time. In order to get complete constructive interference at the focus, the amplitude of the diffracted light waves from each zone should be the same. The SP modes in slits with different widths have different loss as shown in Fig. 9(b). Although the phase match at the focus in the design is nearly perfect, the amplitudes of light emitting from each zone are different. Amplitude compensation for the narrow slit is able to balance the constructive interference at the focus and therefore improve the contrast of the focus without increase of the device size. A simple method is proposed to demonstrate this effect as shown in Fig. 14(b), where some additional slits shown as red crosses in Fig. 12 were introduced to the original structure. Two more slits with a width of 10 nm were added neighbouring to each 10 nm slit in the original PL with a gap of 50 nm to enhance the transmitted light contribution at the focus. In this case, both NA and the number of zones are unchanged but very obvious improvement of the contrast around the focus is seen. In addition, the accuracy of the focal length, FWHM$_x$ and FWHM$_z$ are almost the same as the original device.

Although a well-defined focus has been demonstrated above, the transmission efficiency of a PL is still low and the side lobes are obvious. Integrating more slits in a limited lens aperture is a straightforward way to solve this problem. Etching narrower slits in a thicker metal film is one way to integrate more slits due to the increased phase delay across the lens surface but the fabrication is very difficult at a large aspect ratio. The current design has very limited modulation of the phase delay due to the quickly decreasing dependence of the mode propagation constant on the slit width (>30 nm) as shown in Fig. 9(b). Nano-slits narrower than 30 nm are very difficult to fabricate in a thick metal film (>200 nm). The dependence of Re(n_{eff}) on the refractive index of material inside the slits apparently implies a method of phase delay modulation by selectively filled the slits with different materials

[43]. Combining both the slit width and the refractive index tuning, more slits, i.e. a wider phase delay range, can be obtained. The selective refractive index tuning is feasible by applying the masked planarization and the etch-back process.

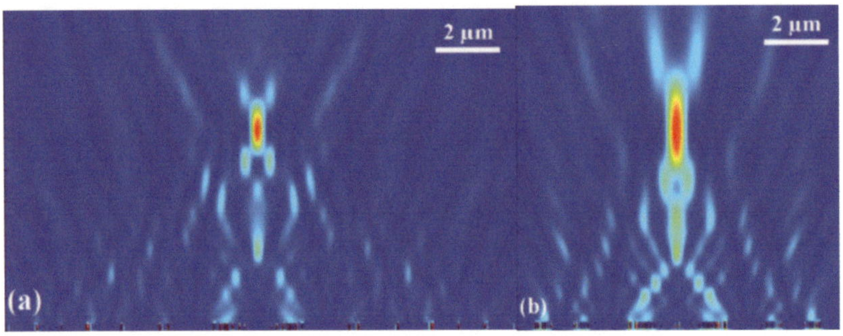

Figure 14. Simulation results for the Poynting vector Pz distributions of the modified structures of the PL shown in Fig. 11(f). (a) A PL with an extended aperture including seven zones compared to three zones in the reference PL. (b) A PL with additional slits shown as red crosses in Fig. 10 compared to the reference PL. Reprint from [42].

4.3.3. Experiments on PLs

The PLs in this work were designed to operate at 633 nm. To simplify the fabrication process, the minimum slit width was 50 nm and the minimum gap was 100 nm in the experiment. Thus the contrast around the focus will degenerate due to the less slits but the accuracy of the focal length keeps in a proper design. Electron beam lithography and dry etch were used to fabricate the nano-slits in a 200 nm thick aluminium film on glass. Fig. 15 shows SEM images of a typical PL structure that was patterned, where the focal length f = 6 μm. The nanoslits were 10 μm in length and the PL width, or aperture, was 10.84 μm. The device has in total 25 slits. From the central slit to the last one on the right hand side, the first four slits have widths of 50 nm, 54 nm, 68 nm, 125 nm and all the others are 100 nm. An enlarged image of the region inside the dash line rectangle of Fig. 15(b) is shown in Fig. 15(c). The roughness on the sidewalls of the gratings is predominantly due to the large grain size of the evaporated aluminium. The error of the slit widths of the fabricated slits is within 10%.

The far-field focusing pattern produced by the lenses was measured using a WITec alpha300S confocal scanning optical microscope (CSOM). A pure confocal mode was used for the experiments because the probe for near-field scanning optical microscopy may have caused a perturbation of the local fields. Sample illumination was with a collimated laser beam operating at 633 nm. The laser source was polarized in the TM mode with its electric field perpendicular to the slits. The light that was transmitted through the sample was collected using a 100×, NA = 0.9 objective. A multi-mode fiber with a core diameter of 25 μm was used to couple the transmitted light into a photomultiplier tube that had a sample integration time is 0.5 ms. The core of the fiber acted as the CSOM pinhole. The sample was scanned in the x and y directions using a piezoelectric scan table, and the microscope working distance was scanned to obtain the z-axis data. The step size in any direction was 200 nm.

Fig. 16(a) and (b) show the focusing light pattern in the *xz* plane measured by the CSOM for lenses designed to have $f = 3$ μm and $f = 6$ μm respectively. The diffracted light distribution clearly shows focusing for both lenses. The constructive interference at the focal point can be clearly seen. The positions of the foci and the side lobes agree extremely well with the simulated electric field intensity distribution shown in Fig. 16(c) and (d). The simulation results showed that the field intensity at the focus of the lens with $f = 6$ μm was 1.9 times that of the incident light. The overall transmission through the lens at 633 nm is 27% that is much larger than the slit filling ratio of 10%. The significant enhancement of the transmission is mainly due to the excitation of SPs [3]. The relatively stronger intensity of the side lobes observed in the experiment was caused by the tolerance limits of fabrication that most strongly affected the narrower slits towards the lens centre. It can be seen that the transmitted light from the central slits is weaker. Theoretical work has also shown that multilayer metal/dielectric films may give a much stronger light intensity at the focus [45].

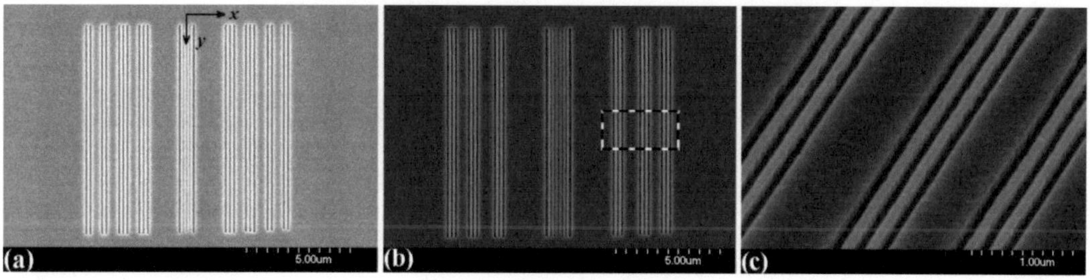

Figure 15. Scanning electron micrographs of PLs in an aluminium film. (a) and (b) are lenses designed to have $f = 3$ μm and $f = 6$ μm, respectively. (c) An enlarged image of the region inside the dash rectangle in the tilted PL in (b).

The normalized light intensity distributions through the centre of the focus spot along the *x* and *z* axis, is shown for both lenses in Fig. 17. The experimental focal lengths for the lenses (Fig. 17(a) and (b)) were 3.1 μm and 6.1 μm, which compare favorably with the simulated values of 3 μm and 6 μm, respectively. The deviation from the intended focal length is less than 3.5% for both lenses. The FWHM of the beam extension in the *z*-direction for the lens with a focal length of 3.1 μm is 1.3 μm, and the light intensity at the focus centre is approximately seven times that of the nearest side lobe. The extension of the focus of the lens with the focal length of 6.1 μm is larger as shown in Fig. 17(b). In the *x*-direction, a line plot across the focal plane of each lens shows that the light intensity drops quickly with the distance from the optical axis (at $x = 0$) as shown in both Fig. 17(c) and (d). The FWHM line-width for the $f = 3.1$ μm lens is 470 nm, and for the $f = 6.1$ μm lens it is 490 nm. The resolution of the measured data is potentially limited by the CSOM scan step size of 200 nm and the resolving power of the objective, but as can be clearly seen from Fig. 17, the measured profiles compare very well with the simulations. As a further test of lens quality, we have compared the line-width at the focal point with what we would expect for a Gaussian limited beam. The Rayleigh limit for the resolving power of a lens is given by $d_s = \lambda f/D$, where d_s is a half of the line width, λ is the wavelength, f is the focal length and D is the lens aperture. For the lens with $f = 3.1$ μm, $d_s = 200$ nm, whereas for the lens with $f = 6.1$ μm, $d_s = 356$ nm. The results therefore suggest that the $f = 6.1$ μm lens works close to its theoretical limit. Subdiffraction focusing elements based on near-field effects were reported to be able to focus electromagnetic waves to a spot of size less than $\lambda/10$ [46-48]. However, the focal length is limited to dimensions much less than the wavelength, which is about 100 times less than our devices. Alternatively, we focused on the far-field lenses.

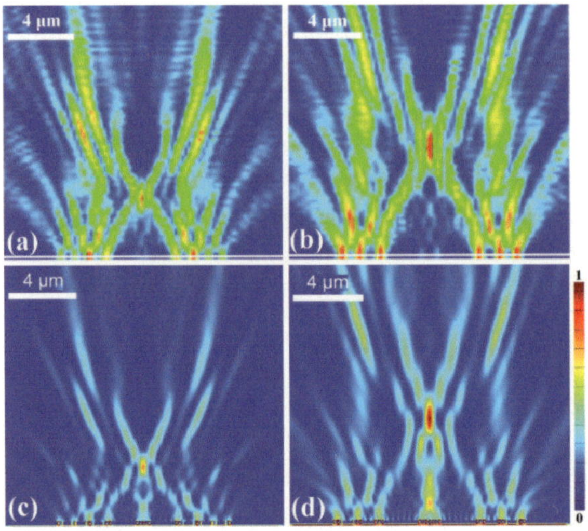

Figure 16. (a) Focusing light pattern in the *xz* plane obtained by the CSOM for a lens designed to have $f = 3$ µm and (b) for a lens designed to have $f = 6$ µm. The horizontal white line in (a) and (b) shows the position of the sample surface. (c) Simulation results for the $f = 3$ µm lens and (d) simulation results for the $f = 6$ µm lens. Reprint from [44].

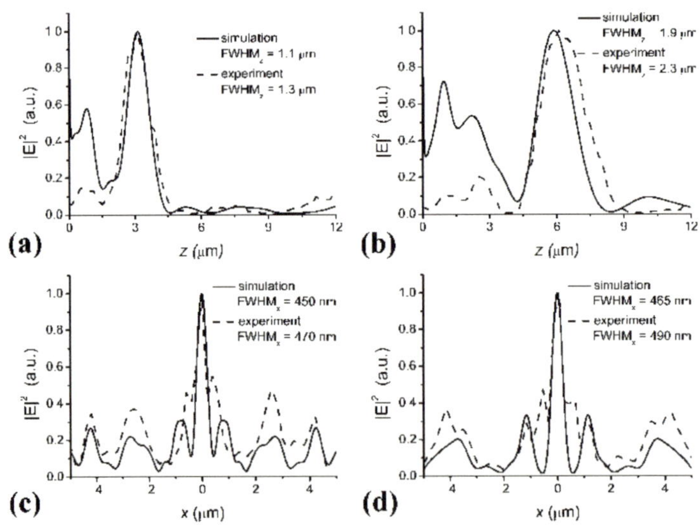

Figure 17. (a) Normalized simulation and experimental results for the light intensity distributions of the lenses. The optical axis is at $x = 0$ and the lens plane is at $z = 0$. (a) and (b) show the distribution along the z direction through the foci of the lenses with $f = 3.1$ µm and $f = 6.1$ µm respectively. (c) and (d) show the distribution along the *x*-direction through the foci of the lenses with $f = 3.1$ µm and $f = 6.1$ µm respectively. Reprint from [44].

To investigate the diffraction effect discussed in Section 4.3, a PL consisting of the central seven slits in the one shown in Fig. 15(b) was fabricated and characterized. As shown in Fig. 18(a), the PL with a single zone shows a light spot far away from the design focus, with an error of 90%. The pattern was also well predicted by the FDTD simulation result as shown in Fig. 18(b). We found that *NA*s are 0.67 and 0.1 and ϕ'_{max} s are 8π and 0.08π for the PLs in Fig. 16(b) and Fig. 18(a), respectively. For a light focusing optical element with the focal length and the aperture size comparable to the wavelength, the number of the zones or ϕ'_{max} play an important role. These experimental results confirm our conclusion that the focal length of a PL is sensitive to the number of zones. More zones give a higher accuracy of the focal length.

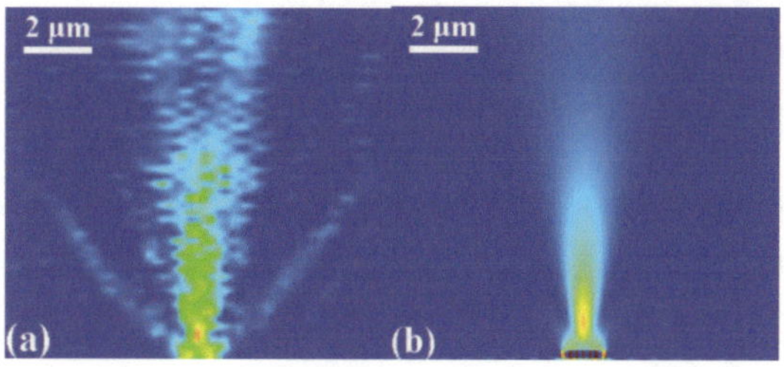

Figure 18. (a) Focusing light pattern in the *xz* plane obtained by the CSOM for a PL consisting of the central seven slits in the one shown in Fig. 15(b). (b) Simulation results for the Poynting vector P_z. Reprint from [44].

5. Plasmonic CMOS image sensors

The above plasmonic optical elements were finally integrated on to CISs. One CIS used in this work has a single-pixel photodiode manufactured with a United Microelectronics Corporation (UMC) 0.18 μm process. The fabrication procedure for pCIS must be modified from that used to make plasmonic filters on glass. To increase the transmission of the colour filters integrated on the CIS, we deposited a layer of SiO_2 on top of the SiN_x surface passivation layer of the CIS before depositing a 150 nm film of aluminium by evaporation. This SiO_2 layer also protected the bond pads of the chip during processing. Before spin-coating with ZEP520A electron beam resist, a thin layer of SiO_2 was added on to the aluminium film to improve adhesion. The sample was exposed using a Vistec VB6 UHR EWF electron beam lithography tool. After development in o-xylene, the sample was etched using CHF_3 and Ar in a Plasmalab 80 plus and then etched using $SiCl_4$ in a Plasmalab System 100. After deposition of a 200 nm thick SiO_2 cap layer, a further mask and etch step was required to reopen windows over the bond-pads of the integrated circuits. Throughout processing the chip was bonded to a Si carrier to aid handling.

As shown in Fig. 19(a)-(c), reflection microscope images of the processed CIS showed different colours of the pixels. The three primary colour filters integrated on the CIS in Fig. 19(a)-(c) were designed to transmit blue (sample S1), green (sample S2) and red (sample S3) light through to the photodiodes. In accordance with this, the reflection spectra showed a complementary minimum in the reflection coefficient for each colour filter. In Fig. 19(c) we can also see a colour variation across the whole photodiode area. This is caused by non-uniformity in the fabrication of nanohole array. The non-uniformity arose because the CIS was bonded to a carrier using photoresist and there was a non-negligible tilt error during electron beam lithography. As we can see in Fig. 19(c), there are four small slightly different coloured sections in the large pixel area. The boundaries between different sections are the boundaries of the electron beam writing fields between which there is the largest non-uniformity. These non-uniformities would be eliminated in wafer-scale manufacturing.

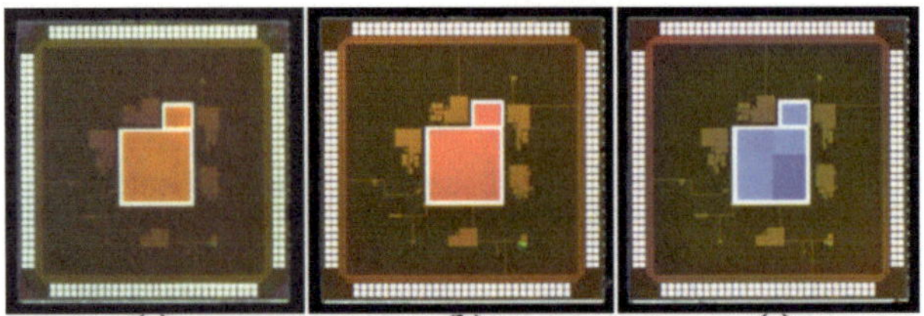

Figure 19. (a)-(c) are microscope images of CIS with integrated blue (S1), green (S2) and red (S3) plasmonic colour filters. Reprint from [49].

Simulated transmission spectra of all three samples are shown in Fig. 20(a). A complete layer stack was modelled using FDTD method to replicate the CIS structure. The stack parameters were derived from the design data of UMC 0.18 μm process. The periods and radiuses are 250nm/80nm, 340nm/90nm and 420nm/110nm for S1, S2 and S3, respectively. Since we could not measure the transmission spectra on the CIS directly, the reflection spectrum was measured instead. The result for sample S2 is shown in Fig. 20(a). As we can see, the experiment and simulation results match very well. The minimum in the reflection coefficient observed in the experimental results is slightly wider than that in simulation. This is mainly caused by the non-uniformity of the nanostructures across the 1 mm^2 photodiode area, as discussed. Photocurrent measurements for the CMOS photodiodes were conducted using a tungsten bulb, a monochromator and an Agilent/HP 4155B. The experimental wavelength resolution of 5 nm was determined by the grating and slit of the monochromator. To test the electrical variations of the photocurrent measurement and themechanical variations of the monochromator, 15 repetitive scans were conducted for one sample; the results showed negligible change. The red dash line in Fig. 20 (b) shows that the photocurrent spectrum of an unprocessed reference CIS has significant fluctuations; most notably two main dips labelled C (520 nm) and D (695 nm). A simple simulation for the whole dielectrics stack using the transfer matrix method showed that the experimental dips approximately match the dips A and B in the simulated spectrum in Fig. 20(b). The dips can be attributed to FP resonances in the CIS dielectric stack. This result is not unexpected since the CMOS process we used has not been optimized, as is usual for commercial CIS. Because we cannot directly measure the transmission spectrum, we have determined the relative transmission using the ratio of the photocurrent for sample S2 to that of an unprocessed reference CIS. The relative transmission spectrum of S2 in Fig. 20(b) has an obvious transmission band for the green colour with an average transmission of approximately 30% and a full-width at half-maximum of 130 nm. There is a sharp transmission peak near 700 nm that is caused by the shift of the labelled dip D in the photocurrent of S2 due to the change in the stack dielectric structure that arises from our processing as compared to the reference CIS. Optimisation of the CIS dielectric stack would remove this unwanted dip. The relative transmission of CIS S1, S2 and S3 is shown in Fig. 20(c). We can see the transmission bands for blue, green and red respectively. These bands are wider than the simulation results shown in Fig. 20(a). We attribute the poor performance to the non-uniformity of the nanostructures across the whole photodiode area and variation in the fabrication tolerance for the hole sizes required. Note that the unwanted transmission peaks labelled F and G are also a consequence of FP resonances in the unoptimised the unwanted transmission peaks labelled F and G are also a consequence of FP resonances in the unoptimised layer stack. The colour cross talk was evaluated in the same way as a conventional CIS in [50]. Our plasmonic CMOS photodetectors have higher cross talk due to the wider passbands of the fabricated colour filters. But these can be reduced by optimizing the plasmonic filters and the fabrication process. Note that the integration of plasmonic colour filters in a CIS would reduce the colour cross talk between the neighbouring pixels, which is not included in the calculation because the devices discussed here are single-pixel photodetectors.

A CIS with a 100×100-pixel photodiode array was investigated after the initial work on a single-pixel photodetector. As shown in Fig. 21(a), the pixel size is 10 μm × 10 μm and each pixel contains a 4.5 μm × 9 μm photodiode with an inter-photodiode gap of 0.7 μm between some pixels. The topography of the pixels has a 1.1 μm vertical step between the photodiode and the circuit regions within each pixel due to the top metal layer in the AMS 0.35 μm process as shown in Fig. 21(b). Because the CMOS chips use aluminium, it is difficult to register accurately to the pixels using our 100 kV EBL system that has a back-scatter detector. We therefore deposit approximately positioned gold electron beam markers around the CIS using a standard lift-off process. A registration EBL step is then used to write a dummy pattern on to the pixel array. With this pattern we determine the positional error between our gold markers and the pixel array using a high resolution SEM Hitachi S4700 SEM. This data is then used in a third EBL step to write the final pattern on a 150 nm aluminium film evaporated on the photodiode pixel array. Of course, only one lithography step would be needed if this process was implemented in the manufacturing flow of the foundry. A final mask and etch step was required to reopen windows over the bond-pads of the integrated circuits. A microscope image of the processed pixels is shown in Fig. 21(c), where a series of varying plasmonic components were repeated across the photodiode pixel array. In this reflection image it is possible to see the various colours of the pixels due to the SPR generated by differently patterned nanostructures. It is also possible to see black rectangles that correspond to reference pixels, from which the aluminium film above the photodiodes has been totally removed. A SEM image of the pixels with plasmonic filters is shown in Fig. 21(d), where we can see a good alignment of the plasmonic structure to each pixel. The inset in Fig. 21(d) shows a hole array with a period of 230 nm. This filter transmits a dark blue colour.

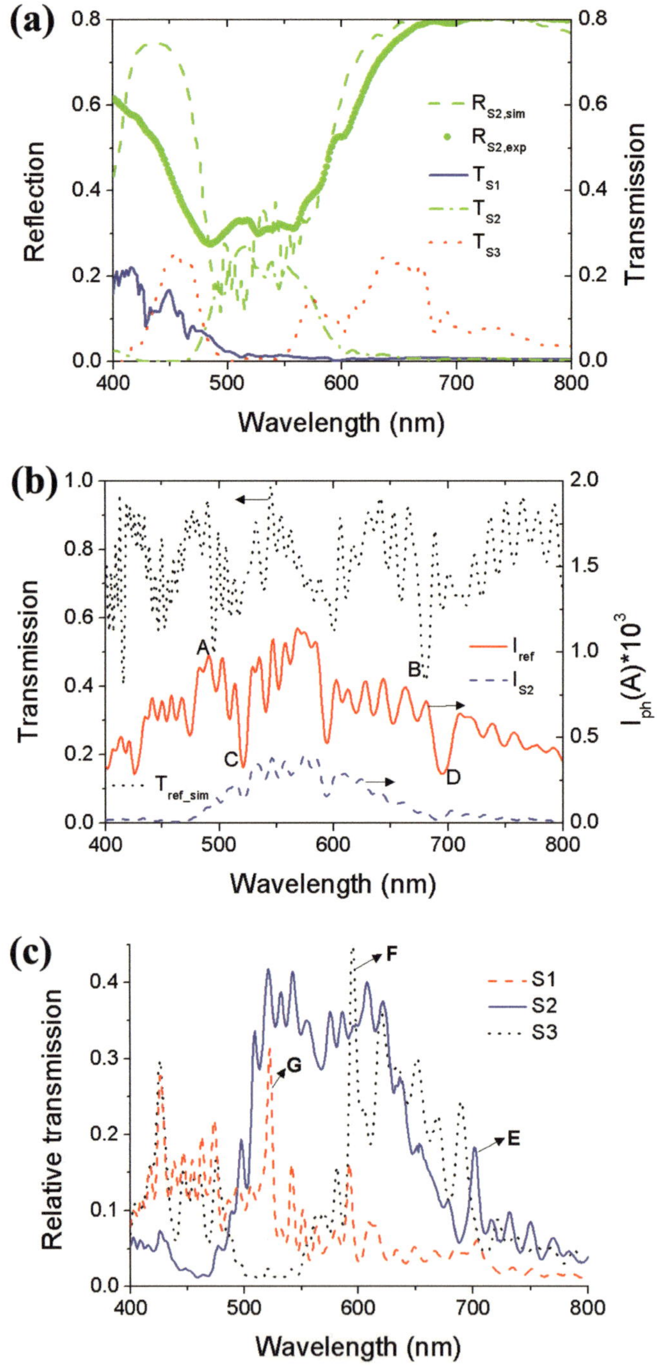

Figure 20. (a) Simulated transmission spectra of CIS S1, S2 and S2. Simulated and measured reflection spectra of S2 are shown as well. (b) Measured photocurrent of the reference sensor and S2. The simulated transmission spectrum of the reference sensor is shown for comparison. (c) Relative transmission of sample S1, S2 and S3. Reprint from [49].

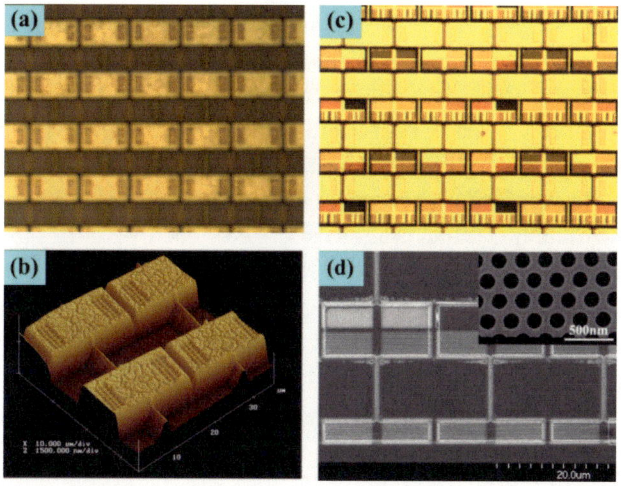

Figure 21. (a) A microscope image of the photodiode pixel array on an unprocessed chip. (b) An AFM image of the topography of pixels before processing. (c) A microscope image of the photodiode pixel array with plasmonic colour filters fabricated on top. (d) A SEM image of the plasmonic colour filters on top of a CIS. The inset is an enlarged image of a filter with the period of 230 nm. Reprint from [21].

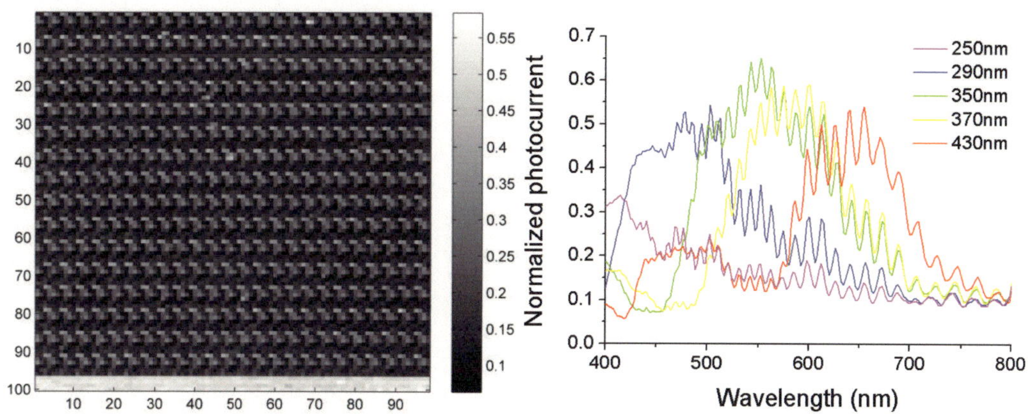

Figure 22. (a) Photocurrent distribution across the 100×100-pixel photodiode array. (b) Normalized photocurrents of photodiodes with various plasmonic colour filters. Reprint from [21].

To test the pixel array, a tungsten bulb and a monochromator were used to illuminate the array with wavelengths in the range between 400 nm and 800 nm. The slits used within the monochromator were chosen to limit the range of wavelengths during each experiment to 5 nm, whilst the centre wavelength was changed in 2 nm increments. At each centre wavelength the response of each pixel, was measured using a data-acquisition system, specifically a NI USB-6218. In order to correct for fixed pattern noise the response of each pixel was determined from the change in the output voltage after an integration time of 125 ms. In addition, to compensate for the effects of the currents that flow within each pixel even in the absence of light, the response of each pixel after an integration time of 125 ms in the absence of light was subtracted from the response at each centre wavelength. The response distribution across the pixel array when it is illuminated by light with a wavelength of 550 nm is shown in Fig. 22(a) to demonstrate that, with the exception of a very few defective pixels, there is a good uniformity to the plasmonic components. The spectral responses we present have been determined by averaging the responses of all the pixels with the same plasmonic filter, across the whole pixel array. Since it is impossible to directly measure the transmission spectrum of the filters, the performance of the filters has been assessed by normalizing the average response of each pixel design using the average response of the unpatterned reference pixels. Results for pixels with five different plasmonic filters (Fig. 22(b)), show that the filters act as bandpass filters with different centre wavelengths. The normalized responses of these filters are above 50% of the signal measured from an unprocessed pixel and the FWHMs are between 110 nm and 150 nm. One obvious feature of these results that was not seen in the results in Fig. 4(c) is the oscillations in all the responses. The fact that these oscillations were previously observed in the single-pixel plasmonic CMOS photodetector and that they are also observed in the responses of pixels before any back end of line processing has occurred leads to the conclusion that they are attributable to FP resonances in the CIS dielectric stack. The resonances occur because the layers in the dielectric stack have different permittivities, hence there is a reflection at each interface. This result is not unexpected since the CMOS process we used has not been optimised to reduce this effect, as is usual for commercial CIS.

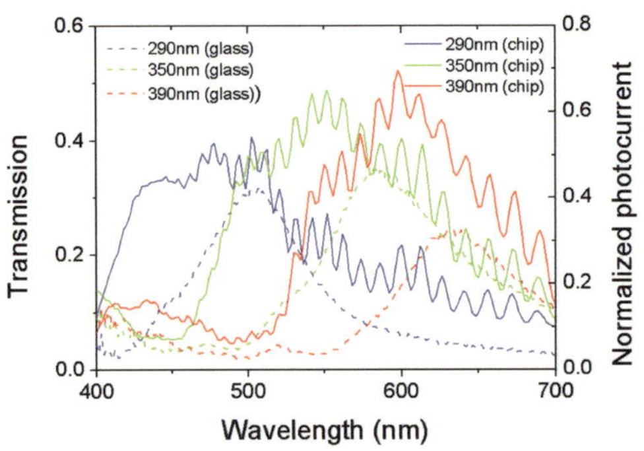

Figure 23. Normalized photocurrents of three sets of photodiodes with peak responses close to 445 nm, 555 nm and 600 nm. Transmission spectra of the same colour filters on glass are shown for comparison. Reprint from [21].

Average results of three groups of pixels with the peak response wavelengths close to Red (600 nm), Green (555 nm), Blue (445 nm), according to the 1931 International Commission on Illumination 2° standard observer colour matching functions [51], are shown in Fig. 23. The transmission spectra of the same filters on glass are shown for comparison. All photodiode pixels show a blue-shift of their peak responses compared to the peak transmission wavelengths of the plasmonic colour filters on glass. Two factors contribute to this phenomenon; one is the smaller hole sizes of the filters integrated on chip than those on glass, the other is the high-index substrate loading effect in the case of CMOS chips. This effect can be eliminated by calibration of the process for colour matching. In addition the transmission bands of the filters on the CIS seem to be wider than the equivalent filters when they are made on glass. The phenomena that could be contributed to this broadening include the possibility that a wider range of angles of incidence of light occur when the pixels are tested and cross-talk caused by the large vertical separation, 8 μm, between the filters and the photodiodes underneath. These effects mean that the results obtained are not necessarily representative of those that will be obtained in an optimized manufacturing process, especially if the plasmonic colour filters are integrated in lower metal layers in a standard CMOS process. In addition, it is anticipated that the broadening will be reduced by using narrow band filters such as low loss silver filters. The optoelectronic efficiency can be further improved if these plasmonic nanostructures are fabricated close enough to the photodiodes, where the localized field can be greatly enhanced due to the SPR effects [52]. It is anticipated that in future state-of-the-art CMOS technology (ITRS roadmap 2010) the half-pitch in Metal 1 will scale down to 32 nm in 2012, enabling mass manufacture of suitable SPR structures.

6. Conclusion

We demonstrated a detailed study of plasmonic optical elements for the application in CISs and the first plasmonic CIS with plasmonic colour filters replacing conventional polymer colour filter array. The plasmonic optical elements such as colour filters, polarizers and lenses showed promising performance where the complete control on wavelength filtering, polarization and the phase distribution were achieved by carefully optimizing the metallic nanostructures. The complete compatibility with the CMOS technology of these metallic optical devices facilitates the plasmonic CIS integration. Replacing the conventional optical elements, plasmonic devices offer various advantages such as less cross-talk, low cost and multifunction, etc. It would be an important step forward to apply nanophotonics in the CMOS imaging. It could be a new way to bring plasmonics research from the lab to the foundry.

Author details

Qin Chen
Corresponding Author
Suzhou Institute of Nano-Tech and Nano-Bionics, Chinese Academy of Sciences, China
University of Glasgow, United Kingdom
SOC Key Laboratory, Peking University Shenzhen Research Institute, China

Xiaohua Shi
Suzhou Institute of Nano-Tech and Nano-Bionics, Chinese Academy of Sciences, China

Yong Ma
University of Glasgow, United Kingdom

Jin He
Peking University Shenzhen SOC Key Laboratory, PKU-HKUST Shenzhen-Hongkong Institution, Shenzhen, P.R. China

7. References

[1] Raether H. Surface Plasmons. Berlin: Springer; 1988.

[2] Gramotnev DK, Bozhevolnyi SI. Plasmonics beyond the diffraction limit. Nature Photonics 2010;4 83-91.

[3] Ghaemi HF, Thio T, Grupp DE, Ebbesen TW, Lezec HJ. Surface plasmons enhance optical transmission through subwavelength holes. Physical Review B 1998;58 6779-6782.

[4] Genet C, Ebbesen TW. Light in tiny holes. Nature 2007;445 39-46.

[5] Laux E, Genet C, Skauli T, Ebbesen TW. Plasmonic photon sorters for spectral and polarimetric imaging. Nature Photonics 2008;2 161-164.

[6] Xu T, Du C, Wang C, Luo X. Subwavelength imaging by metallic slab lens with nanoslits. Applied Physics Letters 2007;91 201501.

[7] Verslegers L, Catrysse PB, Yu Z, Fan S. Planar metallic nanoscale slit lenses for angle compensation. Applied Physics Letters 2009;95 071112.

[8] Koo C, Kim H, Paik K, Park D, Lee K, Park Y, Moon C, Lee S, Hwang S, Lee D, Kong J. Improvement of crosstalk on 5M CMOS image sensor with 1.7x1.7μm^2 pixels. In Proc. Of SPIE 2007;6471 647115.

[9] Prima J, Roy F, Leininger H, Cowache C, Vaillant J, Pinzelli L, Benoit D, Moussy N, Giffard B. Improved colour separation for a backside illuminated image sensor with 1.4 μm pixel pitch. Proc. International Image Sensor Workshop, Bergen, Norway; 2009.

[10] Catrysse PB, Wandell BA. Integrated colour pixels in 0.18um complementary metal oxide semiconductor technology. Journal of the Optical Society of America A 2003;20 2293-2306.

[11] Catrysse PB. Monolithic integration of electronics and sub-wavelength metal optics in deep submicron CMOS technology. Materials Research Society Symposium Proceedings 2005;869 D1.5.1-D1.5.12.

[12] Kanamori Y, Shimono M, Hane K. Fabrication of transmission color filters using silicon subwavelength gratings on quartz substrates. IEEE Photonic Technology Letters 2006;18 2126-2128.

[13] Collin S, Vincent G, Haidar R, Bardou N, Rommeluere S, Pelouard J. Nearly perfect Fano transmission resonances through nanoslits drilled in a metallic membrane. Physical Review Letters 2010;104 027401.

[14] Xu T, Wu Y, Luo X, Guo LJ. Plasmonic nanoresonators for high-resolution colour filtering and spectral imaging. Nature Communications 2010;1 59.

[15] Inaba Y, Kasano M, Tanaka K, Yamaguchi T. Degradation-free MOS image sensor with photonic crystal colour filter. IEEE Electron Device Letters 2006;27 457-459.

[16] Frey L, Parrein P, Raby J, Pellé C, Hérault D, Marty M, Michailos J. Color filters including infrared cut-off integrated on CMOS image sensor. Optics Express 2011;19 13073-13080.

[17] Lee HS, Yoon YT, Lee SS, Kim SH, Lee KD. Colour filter based on a subwavelength patterned metal grating. Optics Express 2007;15 15457-15463.

[18] Lumerical FDTD Solution, http://www.lumerical.com/

[19] Walls K, Chen Q, Collins S, Cumming DRS, Drysdale TD. Automated design, fabrication and characterization of colour matching plasmonic filters. IEEE Photonics Technology Letters 2012;24 602-604.
[20] Chen Q, Cumming DRS. High transmission and low colour cross-talk plasmonic colour filters using triangular-lattice hole arrays in aluminium films. Optics Express 2010;18 14056-14060.
[21] Chen Q, Das D, Chitnis D, Walls K, Drysdale TD, Collins S, Cumming DRS. A CMOS Image Sensor Integrated with Plasmonic Colour Filters. Plasmonics, DOI: 10.1007/s11468-012-9360-6.
[22] Andreou AG, Kalayjian ZK. Polarization imaging: principles and integrated polarimeters. IEEE Sensors Journal 2002;2 566-576.
[23] Schechner YY, Narasimhan SG, Nayar SK. Polarization-based vision through haze. Applied Optics 2003;42 511-525.
[24] Tokuda T, Sato S, Yamada H, Ohta J. Polarization analyzing CMOS sensor for microchamber/microfluidic system based on image sensor technology. IEEE International Symposium on Circuit and Systems, May 18-21, 2008, Seattle, USA.
[25] Zhao X, Boussaid F, Bermak A, Chigrinov VG. Thin Photo-Patterned Micropolarizer Array for CMOS Image Sensors. IEEE Photonics Technology Letters 2009;21 805-807.
[26] Nordin GP, Meier JT, Deguzman PC, Jones MW. Micropolarizer array for infrared imaging polarimetry. Journal of the Optical Society of America A 1999;16 1168-1174.
[27] Ekinci Y, Solak HH, David C, Sigg H. Bilayer Al wire-grids as broadband and high-performance polarizers. Optics Express 2006;14 2323-2334.
[28] Huo Y, Fesenmaier CC, Catrysse PB. Microlens performance limits in sub-2μm pixel CMOS image sensors. Optics Express 2010;18 5861-5872.
[29] Lezec HJ, Degiron A, Devaux E, Linke RA, Martin-Moreno L, Garcia-Vidal FJ, Ebbesen TW. Beaming Light from a Subwavelength Aperture. Science 2002;297 820-822.
[30] Sun Z, Kim HK, Refractive transmission of light and beam shaping with metallic nano-optic lenses Applied Physics Letters 2004;85 642-644.
[31] Shi H, Wang C, Du C, Luo X, Dong X, Gao H. Beam manipulating by metallic nano-slits with variant widths. Optics Express 2005;13 6815-6820.
[32] Fu Y, Zhou W, Lim L, Du CL, Luo XG. Plasmonic microzone plate: Superfocusing at visible regime. Applied Physics Letters 2007;91 061124.
[33] Chen Y, Zhou C, Luo XG, Du C. Structured lens formed by a 2D square hole array in a metallic film. Optics Letters 2008;33 753-755.
[34] Mote RG, Yu SF, Ng BK, Zhou W, Lau SP. Near-field focusing properties of zone plates in visible regime - New insights. Optics Express 2008;16 9554-9564.
[35] Verslegers L, Catrysse PB, Yu Z, Shin W, Ruan Z, Fan S. Phase front design with metallic pillar arrays. Optics Letters 2010;35 844-846.
[36] Verslegers L, Catrysse PB, Yu Z, White JS, Barnard ES, Brongersma ML, Fan S. Planar Lenses Based on Nanoscale Slit Arrays in a Metallic Film. Nano Letters 2009;9 235-238.
[37] Lin L, Goh XM, McGuinness LP, Roberts A. Plasmonic Lenses Formed by Two-Dimensional Nanometric Cross-Shaped Aperture Arrays for Fresnel-Region Focusing. Nano Letters 2010;10 1936-1940.

[38] Fu Y, Liu Y, Zhou X, Xu Z, Fang F. Experimental investigation of superfocusing of plasmonic lens with chirped circular nanoslits. Optics Express 2010;18 3438-3443.

[39] Goh XM, Lin L, Roberts A. Planar focusing elements using spatially varying near-resonant aperture arrays. Optics Express 2010;18 11683-11688.

[40] Ruffieux P, Scharf T, Herzig HP, Völkel R, Weoble KJ. On the chromatic aberration of microlenses. Optics Express 2006;14 4687-4694.

[41] Gordon R, Brolo AG. Increased cut-off wavelength for a subwavelength hole in a real metal. Optics Express 2005;13 1933-1938.

[42] Chen Q. Effect of the Number of Zones in a One-Dimensional Plasmonic Zone Plate Lens: Simulation and Experiment. Plasmonics 2011;6 75-82.

[43] Chen Q. A novel plasmonic zone plate lens based on nano-slits with refractive index modulation. Plasmonics 2011; 6 381-385.

[44] Chen Q, Cumming DRS. Visible light focusing demonstrated by plasmonic lenses based on nano-slits in an aluminum film. Optics Express 2010;18 14788-14793.

[45] Kim HC, Ko H, Cheng M. High efficient optical focusing of a zone plate composed of metal/dielectric multilayer. Optics Express 2009;17 3078-3083.

[46] Grbic A, Jiang L, Merlin R. Near-Field Plates: Subdiffraction Focusing with Patterned Surfaces. Science 2008;320 511-513.

[47] Eleftheriades GV, Wong AMH. Holography-Inspired Screens for Sub-Wavelength Focusing in the Near Field. IEEE Microwave and Wireless Components Letters 2008;18 236-238.

[48] Gordon R. Proposal for Superfocusing at Visible Wavelengths Using Radiationless Interference of a Plasmonic Array. Physical Review Letters 2009;102 207402.

[49] Chen Q, Chitnis D, Walls K, Drysdale TD, Collins S, Cumming DRS. CMOS Photo Detectors Integrated with Plasmonic Colour Filters. IEEE Photonic Technology Letters 2012;24 197-199.

[50] Agranov G, Berezin V, Tsai RH. Crosstalk and microlens study in a color CMOS image sensor. IEEE Transactions on Electron Devices 2003;50 4-11.

[51] CIE Free Documents for Download: CIE 1931 Standard Colorimetric Observer Data. http://www.cie.co.at/main/freepubs.html.

[52] Schaadt DM, Feng B, Yu ET. Enhanced semiconductor optical absorption via surface plasmon excitation in metal naonparticles. Applied Physics Letters 2005;86 063106.

Chapter 21

Merging Plasmonics and Silicon Photonics Towards Greener and Faster "Network-on-Chip" Solutions for Data Centers and High-Performance Computing Systems

Sotirios Papaioannou, Konstantinos Vyrsokinos, Dimitrios Kalavrouziotis, Giannis Giannoulis, Dimitrios Apostolopoulos, Hercules Avramopoulos, Filimon Zacharatos, Karim Hassan, Jean-Claude Weeber, Laurent Markey, Alain Dereux, Ashwani Kumar, Sergey I. Bozhevolnyi, Alpaslan Suna, Oriol Gili de Villasante, Tolga Tekin, Michael Waldow, Odysseas Tsilipakos, Alexandros Pitilakis, Emmanouil E. Kriezis and Nikos Pleros

Additional information is available at the end of the chapter

http://dx.doi.org/10.5772/51853

1. Introduction

In recent years it has become evident that the increasing need for huge bandwidth and throughput capabilities in Data Center and High Performance Computing (HPC) environments can no longer be met by bandwidth-limited electrical interconnects. Besides their limited capacity, electrical wiring technologies impose great energy and size limitations originating from the requirements to handle the vast amount of information that needs to be exchanged across all hierarchical communication levels within Data Centers and HPCs, i.e. rack-to-rack, backplane, chip-to-chip and on-chip interconnections. This reality has inevitably led to a clearly shaped roadmap for bringing optics into the spotlight and replacing electrical with optical interconnects, thereby overtaking the bottleneck in traffic exchange imposed by electrical wires [1]. Hence, the optical technology should penetrate into intra-rack and board-to-board transmission links, considering that the optical fiber as a large-bandwidth transmission medium [2] is used only in commercial systems for rack-to-rack communication [3].

Nevertheless, data communication and power consumption are still daunting issues in Data Centers and HPCs. According to recent predictions made in [4], the barrier of 10PFlops computing performance should have been overcome in 2012 by a supercomputer that consumes 5MW of power [5]. In addition, [4] predicted that exascale supercomputing machines would consume 20MW having a power efficiency of 1mW/Gb/s [5]. Nonetheless, power consumption in such environments has been proven to be even higher than expected: Today's top-ranked supercomputer, the "K computer", has already reached the 10PFlops performance benchmark but at the expense of excessive consumed power that is more than twice [6] the value that was predicted in 2008. All the above imply that the use of optics at inter-rack communication level is not enough for delivering the necessary performance enhancements. Therefore, the optical technology should now be exploited at shrinked networking environments: The penetration of low-energy photonic solutions at board-to-board, chip-to-chip and eventually intra-chip interconnects would yield remarkable savings in energy consumption [7]. The current mainstream photonic route with high integration and low-cost perspectives relies on the Silicon-on-Insulator (SOI) photonics platform, whose growing maturity is soon expected to release Tb/s-scale data transmission and switching capabilities in datacom and computercom units ensuring low latency, low power consumption and chip-scale integration credentials [8].

Even so, photonic devices cannot reach the compactness of their electronic counterparts: the dimensions of traditional optical structures are limited by the fundamental law of diffraction, preventing the way towards high density integration for interfacing with electronics at the nanoscale. This gap in size between photonic and electronic components is called to be bridged by a promising disruptive technology named plasmonics [9]-[11]. The emerging discipline of plasmonics has started to gain ground as the "beyond photonics" chip-scale platform that can enter the interconnect area [12]-[14], holding a great promise for additional reductions in circuit size and increase in energy efficiency. Plasmonics relies on the excitation of surface plasmon polaritons (SPPs) that are electromagnetic waves coupled to oscillations of free electrons in a metal and propagate along a metal-dielectric interface at near the speed of light. These "hybrid" surface waves have transverse magnetic (TM) polarization in nature and their exhibited electromagnetic field intensity reaches its maximum value at the metal surface whereas it decays exponentially while moving away from the metal-dielectric interface [15]. In this way strong intrinsic confinement is feasible even at sub-wavelength scale [16], breaking the size barriers of diffraction-limited optics and enabling the development of compact integrated nanophotonic circuits [17]. Plasmonic technology does not only succeed in providing light manipulation at sub-wavelength dimensions but at the same time allows for the injection of electrical pulses via the metallic layer, offering thereby a seamless energy-efficient platform for merging light beams with electrical control signals towards "active" operations [18]-[25].

Among the various plasmonic waveguide structures proposed so far (i.e. band-gap structures [26], metallic nanowires [27], V-groove waveguides [28]), the low-energy credentials of plasmonics has been mainly highlighted in the case of dielectric-loaded SPP (DLSPP) waveguides, where a dielectric (e.g. polymer) ridge is deposited on top of a smooth metallic film. As a result, strong sub-wavelength (DLSPP) mode confinement is achieved at the metal-dielectric interface. Nonetheless, this performance comes at the expense of excessively increased propagation losses compared to conventional dielectric waveguides due to radiation absorption in the metallic stripe, yielding quite short propagation lengths lying in the order of few tens of micrometers [29],[30]. Apart from the strong mode confinement, the main advantage of the DLSPP waveguide technology stems from the nature of the chosen dielectric material which, depending on its thermo-optic (TO) [25],[31]-[35] or electro-optic [24],[36] properties, allows for the exploitation of the corresponding effects, enabling in this way the deployment of highly functional active plasmonic elements. Regarding the TO effect, the underlying metallic layer, which is in direct contact with the polymer and therefore with the largest part of the DLSPP mode field, serves as an energy-efficient electrode that yields immediate change of the effective index of the propagating mode at the presence of electric current, leading to fast TO responses. These advantages render the DLSPP waveguides ideal for TO manipulation of the plasmonic waves in various functional circuitry implementations like modulation [32],[33], ON/OFF gating [25] and switching [34],[35]. The progress made so far on active DLSPP-based circuits renders the TO effect as the most mature mechanism for bringing low-energy active plasmonics in true data traffic environments [25],[31].

To this end, the roadmap towards practical employment of active plasmonic circuits in the development of "greener" and faster "Network-on-Chip" (NoC) solutions for Data Centers and HPCs seems to be the following: the synergy between plasmonic, electronic and photonic components for the realization of NoC deployments with minimized power consumption, size and enhanced throughput capabilities, simply by taking advantage of the virtues of each technology. In this perspective, silicon photonics can be used for low-loss optical transmission in passive interconnection components, plasmonics can provide small footprint and low power consumption in active switching modules and electronics can be employed for intelligent decision-making mechanisms. However, considering that plasmonic technology is a premature technology with only a few years of development compared to silicon photonics and with limited functionality so far, the way towards the implementation of such hybrid NoC environments requires priorly: a) the interconnection between silicon and plasmonic waveguide structures in a SOI platform in order to avoid the employment of high-loss plasmonic waveguides for passive circuitry [30],[37], b) the proof of the credentials of plasmonics in wavelength division multiplexing (WDM) applications so as to support the complete portfolio of optics [38], c) the demonstration of functional active plasmonic circuitries in realistic data traffic environments [25],[31].

This chapter aims to cover all these issues starting from the 4×4 silicon-plasmonic router architecture, where its main building blocks and their principle of operation are described briefly. The chapter continues with the geometrical specifications of the silicon and plasmonic waveguides of the hybrid routing platform as well as with the silicon components and sub-systems, namely the photodiodes (PDs) and the multiplexers (MUXs), respectively. The silicon-plasmonic asymmetric Mach-Zehnder interferometer (A-MZI) is then presented and analyzed thoroughly since it is the constituent component of the 4×4 switching matrix of the router. Subsequently, the WDM credentials of plasmonics are experimentally confirmed by demonstrating WDM data transmission over a DLSPP waveguide as well as WDM data switching operation via a hybrid Si-DLSPP A-MZI, along with its switching performance metrics. Various optimization procedures are also listed. The chapter concludes with an estimation of the router's total power consumption and losses.

2. Tb/s-scale hybrid Router-on-Chip platform

Data Center and HPC communication systems require the development of NoC environments with compact size, high-throughput and enhanced power saving capabilities for efficient traffic management. In this perspective, the benefits of silicon photonics and plasmonics can be combined towards realizing a 4×4 "Router-on-Chip" platform [13] with Tb/s-scale data capacity and ultra-low energy requirements for chip-to-chip and on-chip optical interconnects. This hybrid routing platform employs a silicon motherboard hosting all various optical and electrical circuitry, where the "heart" of this router chip is a hybrid 4×4 switching matrix comprising 2×2 TO Si-DLSPP switches, and exploits the best out of all technology worlds: the low optical loss of Si-waveguide motherboard for its passive interconnection parts, the high energy efficiency of plasmonics in the router's "active" parts, and the intelligence provided by electronics for its decision-making mechanisms. The following sections describe in detail the Tb/s-scale Si-DLSPP router architecture and its basic building blocks, the experimental evidences of the WDM data transmission and switching capabilities of the DLSPP waveguides, various optimization approaches and finally the performance of the 4×4 router with estimations in power consumption and optical insertion losses.

2.1. 4×4 silicon-plasmonic router architecture

The 4×4 silicon-plasmonic router architecture aims at achieving 1.12Tb/s aggregate throughput with small footprint and low power consumption. Figure 1 illustrates a schematic layout of this hybrid router architecture. The router consists of SOI MUXs, Si PDs, a hybrid Si-DLSPP 4×4 switching matrix and an integrated circuit (IC) microcontroller. Size minimization and power savings are promoted by the employment of four SOI MUXs composed of cascaded Si ring resonators (RRs) placed in parallel at the frontend of the router and four identical thermo-optically addressed Si-DLSPP A-MZIs interconnected in a Beneš topology. The plasmonic components are used only where switching functionality is required in order to achieve both low optical losses and low power consumption. The Si technology is also responsible for coupling light in and out of the platform, the interconnection of all subsystems and the optoelectronic (OE) signal conversion of the incoming packets' headers. Four 7×1 MUX circuits based on SOI RRs are used to multiplex 28 time-overlapped wavelengths, spaced by 100GHz and modulated with 40Gb/s non-return-to-zero (NRZ) packet traffic, into four optical data streams. Consequently, each one of these multi-wavelength data sequences carries an aggregate traffic of 280Gb/s that will enter one of the four inputs of the plasmonic switching device and will then follow the same route through the entire platform. Apart from the 28 input ports that are utilized for the insertion of the data carrying packets, the router also incorporates 4 additional input ports where they are inserted four discrete low-rate wavelength beams containing the header information of the four incoming data streams. The header for each stream is a two pulse combination within 1μs time interval and specifies the desired output port, while the duration of the payload of the data packets can be variable. These header pulses are optoelectronically converted via low-rate all-silicon PDs into respective electrical signals that subsequently feed the IC control unit. Here the low speed electronics operating at the data packets rate carry out the intelligent decision-making process that is essential for the routing functionality: The IC microcontroller performs header processing operations and generates differential control electronic signals for driving appropriately the switching matrix. In this way, the incoming optical data streams are routed to the desired output ports as a result of fruitful cooperation between the silicon photonic, plasmonic and electronic technologies.

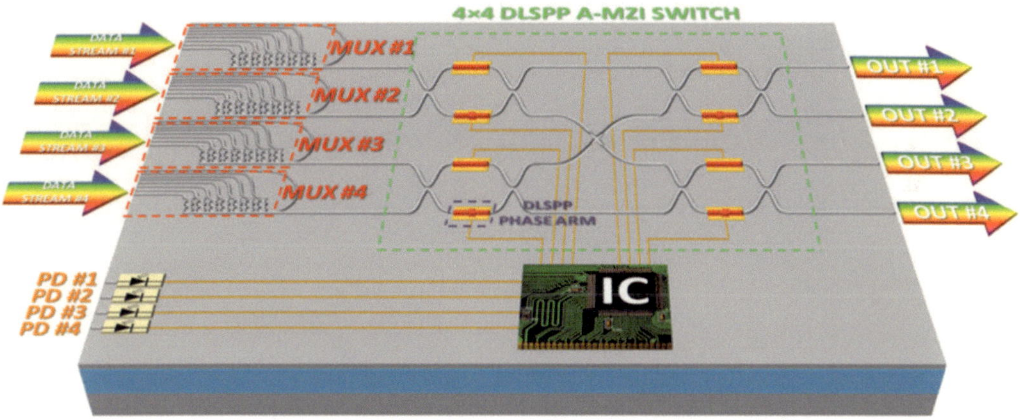

Figure 1. Schematic layout of the 4×4 silicon-plasmonic router architecture.

2.2. Silicon and DLSPP waveguides

The different technologies and building blocks involved in the hybrid Tb/s-scale router require the employment of the appropriate waveguide technology in order to merge all of them successfully on the same motherboard. The existence of a slab layer in rib waveguides compared to the strip counterparts facilitates the fabrication process for the hetero-integration of all processing components, namely the PDs, the SOI MUXs and the DLSPP waveguide structures, rendering this waveguide technology suitable for the low-loss Si-based communication links needed in the SOI routing platform. Taking into consideration the intrinsic TM nature of plasmonic waves, the geometrical dimensions of the Si rib waveguides need to be carefully chosen for compatibility with plasmonic waveguides. In this perspective, rib waveguides featuring a cross section of 400nm width by 340nm height with a 50nm-thick remaining slab are the Si waveguide technology of choice for the routing platform in order to support low-loss TM light propagation [37].

Nevertheless, the coexistence of silicon photonics and plasmonics on the same platform imposes proper formation of the SOI motherboard for efficient hetero-integration of the

DLSPP waveguides [31]: By etching the SOI motherboard down to 200nm (Figure 2(a)), the formed recess in the 2μm-thick buried oxide (BOX) of the SOI substrate serves as the hosting region of a DLSPP waveguide. Towards selecting the dimensions of the DLSPP waveguides, it should be taken into account that the DLSPP waveguide characteristics, i.e., the mode field confinement, effective index and propagation length, are strongly influenced by the width and height of the employed dielectric ridge [29] as a result of the SPP field confinement that occurs when this ridge is placed on top of a metal film. Considering methyl-methacrylate (PMMA) ridges placed on gold stripes to be operated at telecom wavelengths (~1550nm), the optimum ridge dimensions, ensuring tight mode confinement (<1μm) and relatively long propagation (~50μm) of DLSPP modes, are about 500nm and 600nm for the ridge width and height, respectively [29]. Moreover, the underlying gold stripes are about 3μm wide and 60nm thick. A microscope image taken from a Si-DLSPP waveguide is shown in Figure 2(b) where the recessed area of the BOX and the underlying gold film are evident.

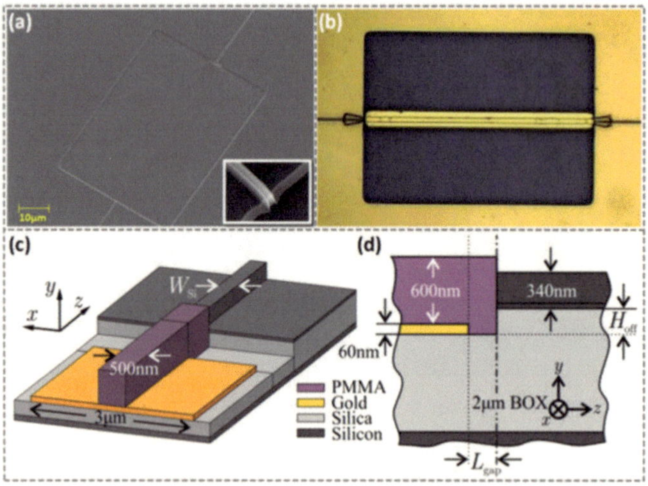

Figure 2. (a) Scanning electron microscope (SEM) image in bird's eye view of the silicon cavity after etching, including as inset the Si waveguide tip after the cavity patterning process, (b) microscope image of a straight DLSPP waveguide coupled to Si waveguides. (c) Bird's eye and (d) side schematic views of a Si-to-DLSPP waveguide transition.

Apart from the geometrical specifications of these two different waveguide technologies, it is necessary to implement an efficient transition from one type of waveguide to the other. In our case, the interface between the silicon and plasmonic waveguides is implemented by a butt-coupling approach, as depicted in bird's eye and side views in Figure 2(c)-(d), respectively. The design of this approach relies on numerical modeling conducted by means of full vectorial three-dimensional finite element method (3D-FEM) simulations. Towards finalizing the design specifications, various parameters should be examined, i.e., the Si waveguide's width, the vertical offset between the two types of waveguides, the existence of a longitudinal gap in the metallic stripe as usually exists in fabricated structures, so as to optimize the matching of the optical modes during transition [37]. The validity of the simulation outcomes is achieved by their comparison with cut-back measurements for a variety of fabricated all Si and hybrid Si-DLSPP waveguide samples [31],[37]. As a result, a coupling loss per interface variance with a mean value of -2.5dB is feasible for 175nm-wide Si waveguides with 200nm vertical offset and DLSPP waveguides with 0.5μm metal gap. The same cut-back evaluation procedure leads to silicon and plasmonic propagation losses of approximately 4.5dB/cm and 0.1dB/μm, respectively, being in good agreement with typical results reported in the literature [30].

2.3. Silicon photodiodes

Since the silicon-plasmonic routing platform comprises also microelectronics, it is necessary to develop integrated C-band (1530-1565 nm) photodiodes that enable the OE conversion of the optical header pulses into electrical signals. However, a complementary metal-oxide-semiconductor (CMOS) -compatible SOI-based integration of a material that absorbs inside this wavelength window is really challenging. One detection scheme within the C-band on a SOI platform is the hybridization of integrated Si structures with other materials (e.g. Ge [39], InGaAs [40]) for direct (linear) absorption. This hybrid approach can lead to high-performance but complex devices, requiring complicated and cost-intensive fabrication procedures and modifications compared to the standard CMOS process. For this purpose, monolithical and fully CMOS-compatible detector concepts have been proposed for the development of comparably simple devices with less fabrication costs. In this perspective, two photon absorption detection can be exploited inside a non-resonant or resonant Si device [41], revealing however non-linear effect and requiring high light intensity. Alternatively, the generation of direct (linear) absorption inside the telecom window can be achieved by the introduction of midgap energy states via Si^+-ion implantation [42]. Despite the lower (direct) absorption compared to the hybrid detectors, this type of monolithic Si detector approach, the defect state Si^+-ion implanted detector, offers the best compromise for simple integration and low-rate detection.

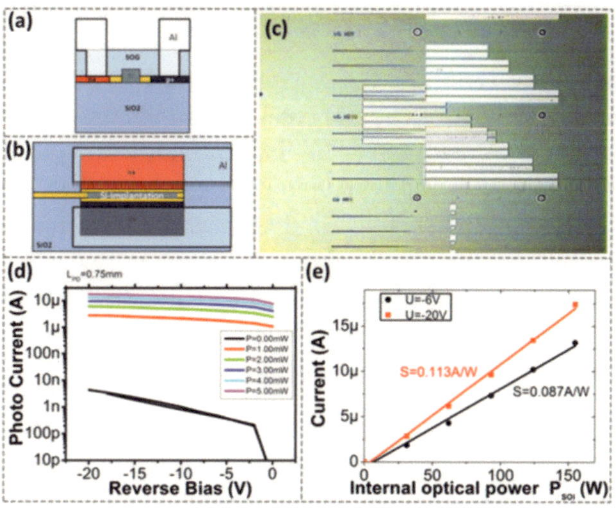

Figure 3. Integrated Si-implanted PDs: (a) Schematic cross sectional view, (b) schematic top view, (c) test chip, (d) I-V curves at various input optical power levels, (e) sensitivity at different reverse bias.

Taking into account that the 4×4 silicon-plasmonic router architecture employs low-rate optical header pulses, the defect state all-silicon implanted PDs are suitable for the procedure of the OE conversion, pursuing a trade-off between minimum signal distortion during header detection and integration complexity. Schematic illustrations of this type of PD interconnected with a Si waveguide are shown in cross sectional and top views in Figure 3(a)-(b), respectively, whereas the top view of a test chip including fabricated all-Si implanted PDs with different absorption lengths is depicted in Figure 3(c). Figure 3(d)-(e) illustrate the performance of a 0.75mm-long Si^+-ion implanted PD, exhibiting linear absorption and sensitivity values of 0.09A/W and 0.11A/W for -6V and -20V bias voltage, respectively. By increasing the device length, the sensitivity is getting higher. However, even for a three times longer device, the dark current is well below 100nA.

2.4. Silicon multiplexers

In order to support Tb/s-scale bandwidth, the hybrid router takes advantage of SOI multiplexing devices. Till now various multiplexing/demultiplexing configurations based on SOI technology have been proposed in the scientific community, such as echelle diffraction gratings (EDGs) and arrayed waveguide gratings (AWGs) [43], first or higher order RRs [44]-[49] and A-MZIs [50]. In many cases, due to fabrication limits and discrepancies, thermal tuning is performed to attain ideal spectral positioning of the MUX resonances. This controlling technique is adopted for tunability of the second-order SOI RRs employed in the silicon-plasmonic router, aiming at optimized multiplexing performance. Considering that the router is planned to handle 40Gb/s line-rates, the 3-dB passband bandwidth for each data channel should be at least 40GHz and the adjacent channel crosstalk should be kept lower than -15dB. These issues can be met by the 2nd order RRs employed in the router, since such structures enable increased flexibility in the formation of transmission peaks' spectral shape and bandwidth within the same channel spacing constraints due to the additional coupling stage employed between the two RRs [51].

A fabricated 2nd order RR featuring the Si waveguide specifications of Section 2.2 is illustrated in Figure 4(a). Indicative experimental spectra obtained at the Drop port of 2nd order ring structures are presented with blue lines in Figure 4(b)-(c). In particular, Figure 4(b) refers to a fabricated 2nd order RR with 9μm radius, 190nm bus-to-ring gaps and 380nm ring-to-ring gap, exhibiting 42.5GHz 3-dB bandwidth and 9.2nm free spectral range (FSR). In the case of a ring structure with 12μm radius, 170nm bus-to-ring gaps and 300nm ring-to-ring gap, the corresponding 3-dB bandwidth and FSR values are 48.5GHz and 6.9nm (Figure 4(c)). Replicas of these measured transfer functions are also illustrated in Figure 4(b)-(c) with red dashed line by using the circuit-level modeling tool described extensively in [13].

The aforementioned 2nd order RRs are employed as the building blocks for the design of the 7:1 SOI MUX devices required in the router. Figure 5 depicts a more generic 8:1 multiplexing layout. The MUX design comprises eight cascaded thermo-optically tunable 2nd order RRs clustered in two groups with two different radii (*R1* and *R2*). Based on this clustering approach, where the ring structures of each group have the same radius, it is feasible to comply with the limits dictated by the fabrication resolution and the maximum possible thermo-optically induced wavelength tuning of the rings. The 100GHz spacing between the MUX stages as well as the wavelength resonance matching for each pair of RRs forming the MUX's stages are achieved by considering micro-heaters on top of each ring. However, the Si-DLSPP router requires also four different SOI MUXs operating efficiently at different spectral regions within the C telecom band. This requirement can be addressed by choosing appropriate ring radii combinations and values for the bus-to-ring (*gap1*) and ring-to-ring (*gap2*) gaps and for the lengths (*L1* to *L7*) of the inter-ring waveguide sections during the design procedure.

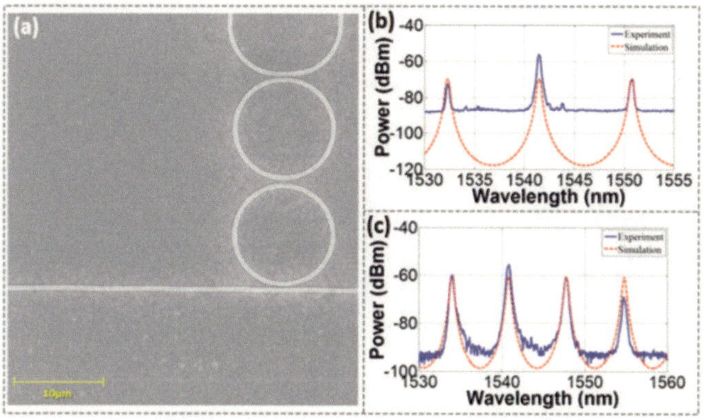

Figure 4. (a) SEM image of a fabricated 2nd order RR. Experimental spectra (blue solid line) at the Drop port for RRs with (b) 9μm and (c) 12μm radii. Red dashed lines illustrate the corresponding simulated spectral responses.

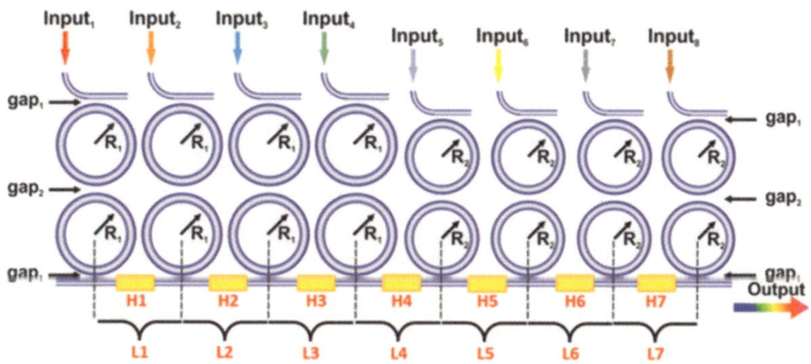

Figure 5. Generic layout of a SOI MUX based on 8 cascaded 2nd order thermo-optically tunable RRs that are grouped in two different radii clusters (*R1* and *R2*). The sections with *H* illustrate the heaters controlling the straight inter-ring waveguide lengths.

Following this rationale, the desired spectral responses of the four MUXs are generated within the 1530-1565 nm wavelength window, revealing 100GHz channel spacing, at least -20dB optical crosstalk and 40GHz 3-dB bandwidth, as illustrated in Figure 6. The first three designs rely on rings with 12μm and 11.7μm radii for the first and second cluster, respectively, whereas the corresponding ring radii values for the fourth design are 9μm and

9.2μm. Moreover, the power coupling coefficients between ring-to-ring and bus-to-ring waveguides for all MUX designs are in the range of 0.33-0.007, corresponding to gap dimensions between 170-380 nm. It should be mentioned that the low crosstalk values in adjacent channels are the result of uneven straight waveguide sections between the successive MUX stages that are long enough (~100μm) also to prevent thermal crosstalk. Since the slightest variation in their length seems to greatly affect the optical crosstalk, micro-heaters are considered to be placed on the middle of every straight waveguide section so as to restore length deviations.

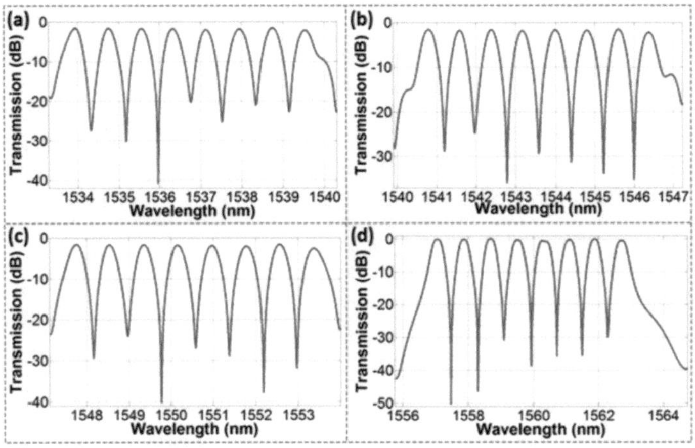

Figure 6. Spectral responses of the (a) first (1533-1540 nm band), (b) second (1540-1547 nm band), (c) third (1547-1554 nm band) and (d) fourth (1557-1563 nm band) 8:1 MUX designs.

2.5. Silicon-plasmonic A-MZIs

Targeting the best compromise between losses, footprint and power consumption, the A-MZI modules of the router rely on a hybrid implementation, where the coupling stages are based on low-loss Si waveguides and only the active phase branches exploit the DLSPP structures. The silicon-plasmonic A-MZI structures play a key role in the accomplishment of routing operation. These are the elements of the router that are capable of performing optical switching by taking advantage of the TO effect at the plasmonic parts. The principle of operation regarding the TO switching is the following: In the 500×600 nm^2 PMMA ridge, the DLSPP mode fills practically the whole ridge, while the largest fraction of the mode is concentrated at the metal-dielectric interface. The strong confinement at the metal-dielectric interface renders the mode very sensitive to any temperature variation induced through current injection in the metal; in this way the underlying gold film acts as a heating electrode for the dielectric ridge, modifying in a very efficient way the DLSPP mode effective index via changing the PMMA refractive index [32]-[34],[52] and requiring only a small amount of consumed power. At the same time, the inherent instantaneous heating of the metal is immediately transferred to the propagating SPP mode, again due to its strong confinement at the metal-dielectric interface, leading to small response times of the TO effect in the DLSPP waveguide. Such low driving powers (~1mW) and response times (~1μs) in

TO modulation and switching have already been shown in theoretical estimations [32] and calculations [52] that make use of these ridge dimensions in DLSPP waveguide structures.

Considering operation at telecom wavelengths (~1550nm) and an increase in temperature of about 61K in PMMA loadings (where the thermo-optic coefficient *TOC* is $-1.05\cdot10^{-4}K^{-1}$), a full π phase shift can be achieved for a waveguide length of 120μm, according to the formula $L_\pi=\lambda/(2\cdot\Delta T\cdot TOC)$. This length value is almost twice the maximum propagation length of the DLSPP mode and results in high propagation losses [32]. However, this length can be reduced by a factor of 2 by adopting the asymmetric MZI configuration shown in Figure 7(a). In this configuration, a permanent $\pi/2$ phase shift is induced in one of the two branches by slightly widening the PMMA ridge over a short distance, resulting in this way to a higher effective refractive index value in the wider ridge section and to a natural biasing of the MZI at its quadrature point of operation. To this end, the necessary thermo-optically imposed phase shift is reduced down to $\pi/2$. Following this rationale, the length of the phase arms in the employed hybrid A-MZIs has been set to 60μm (*L1*), while the $\pi/2$ phase asymmetry was achieved by widening the lower 600-nm-thick DLSPP waveguide from 500nm (*W1*) to 700nm (*W2*) for a length of 6μm (*L2*), as a result of an increase of the DLSPP mode effective index by ~0.06 [29],[52]. These dimensions ensure the existence of a single TM mode inside the polymer at telecom wavelengths, both for the nominal and the widened DLSPP waveguides, as depicted in Figure 7(b)-(c).

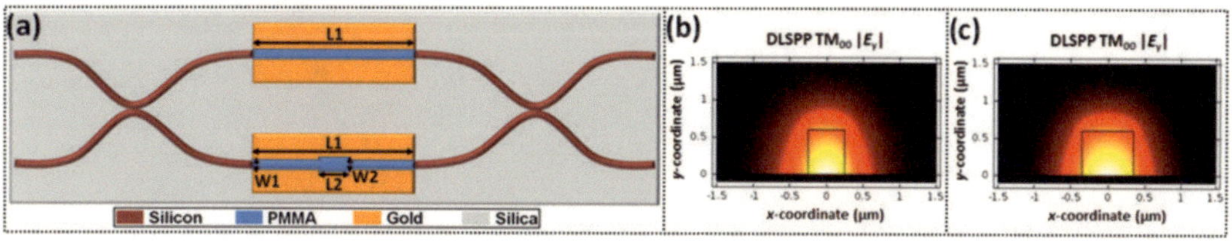

Figure 7. (a) Schematic layout of a Si-DLSPP A-MZI. The lower plasmonic branch is widened in order to introduce a default asymmetry, (b) fundamental quasi-TM mode of the 500×600 nm² PMMA-loaded SPP waveguide, (c) fundamental quasi-TM mode of the 700×600 nm² PMMA-loaded SPP waveguide.

Taking advantage of the default $\pi/2$ phase asymmetry introduced at the lower plasmonic branch, this asymmetric interferometric configuration is capable of performing switching with only a $\pi/2$ phase shift compared to all-plasmonic symmetric MZI switches [52]-[54]. The injection of electric current to a MZI arm yields a temperature rise that alters the effective index of the mode propagating on the heated arm, leading to a phase shift via this TO effect [32],[34],[52]. By electrically controlling the upper A-MZI arm, a negative phase shift is experienced by the propagating DLSPP waveguide mode as a result of the PMMA's TOC. When the induced phase shift equals $\pi/2$, the phase difference between the modes travelling through the two MZI branches equals π and therefore the whole mode power is exported to the BAR output of the device. On the contrary, when the same current level applies only to the lower MZI plasmonic branch, the default MZI phase asymmetry is cancelled out due to the $-\pi/2$ thermo-optically induced phase shift and, thus, the whole mode power emerges at the CROSS port of the MZI. To this end, the ON switching operation is achieved by electrically driving the upper MZI plasmonic arm, whereas the lower branch has to be driven by the same amount of electric current for the OFF switching operation. Nevertheless, due to the default $\pi/2$ phase asymmetry of the MZI, high-performance switching can be reached even if one of the two plasmonic branches is thermo-optically addressed, since the MZI is initially biased at the quadrature point that lies in the linear domain of its output transfer function. Consequently, this asymmetric formation constitutes a simple and passive mechanism for reducing the required energy level and the active plasmonic arm length for a given maximum service temperature.

2.6. WDM data transmission through a DLSPP waveguide

The evaluation of the WDM data transmission capabilities of plasmonics is initially carried out via a 60μm-long PMMA-loaded straight waveguide that is included in a Si-plasmonic chip [31],[38]. This chip comprises Si rib waveguides, DLSPP waveguides and Si-to-DLSPP interfaces that follow the specifications presented in Section 2.2 as well as Si TM grating couplers so as to enable optical communication with the outside world. At 1545nm TM-polarized light the cut-back measurements reveal 0.1dB/μm and 4.6dB/cm plasmonic and silicon propagation losses as well as 2.5dB and 12dB coupling losses for each Si-to-DLSPP and TM grating coupling interface, respectively.

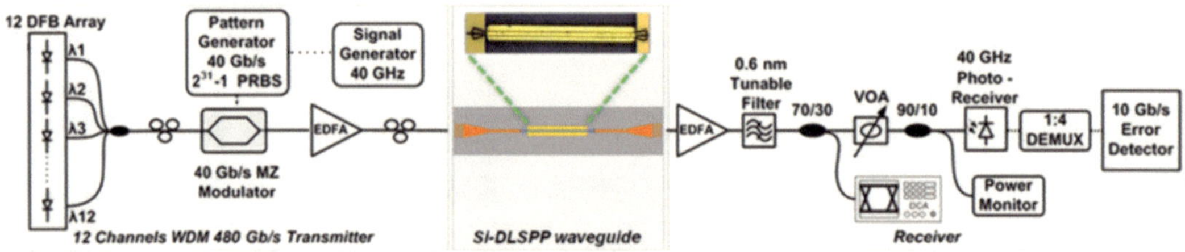

Figure 8. Experimental setup consisting of the WDM transmitter, the Si-DLSPP waveguide and the receiver.

The experimental setup used to transmit a 480Gb/s WDM optical stream through the DLSPP waveguide is presented in Figure 8. The transmitter involves twelve continuous-wave (CW) optical signals stemming from distributed feedback (DFB) lasers and spaced by 200GHz within the 1542-1560 nm spectral range. The multiplexing of these individual light beams into a single optical fiber is realized by using AWG and 3-dB couplers. The multiplexed signal is then encoded by a 2^{31}-1 pseudo-random bit sequence (PRBS) at 40Gb/s NRZ line rate in a Ti:LiNbO$_3$ Mach-Zehnder modulator (MZM). Subsequently, the 480Gb/s WDM signal is amplified by a high-power erbium-doped fiber amplifier (EDFA), providing 24dBm output power towards tackling the high losses induced by the hybrid chip. After ensuring TM polarization conditions for compliance with the inherent TM nature of plasmonics, via a polarization controller, the incoming optical data signal is inserted into the DLSPP waveguide. The multi-wavelength data stream that transmitted through the whole Si-DLSPP waveguide is amplified at the chip's output by a low-noise EDFA and demultiplexed into individual data channels by an optical bandpass tunable filter (OBPF). Each optical data channel is then converted into an electrical signal by a 40GHz 3-dB bandwidth photoreceiver that is connected to a 1:4 electrical demultiplexer, where each electrical data signal is received by a 10Gb/s error detector for bit error rate (BER) measurements.

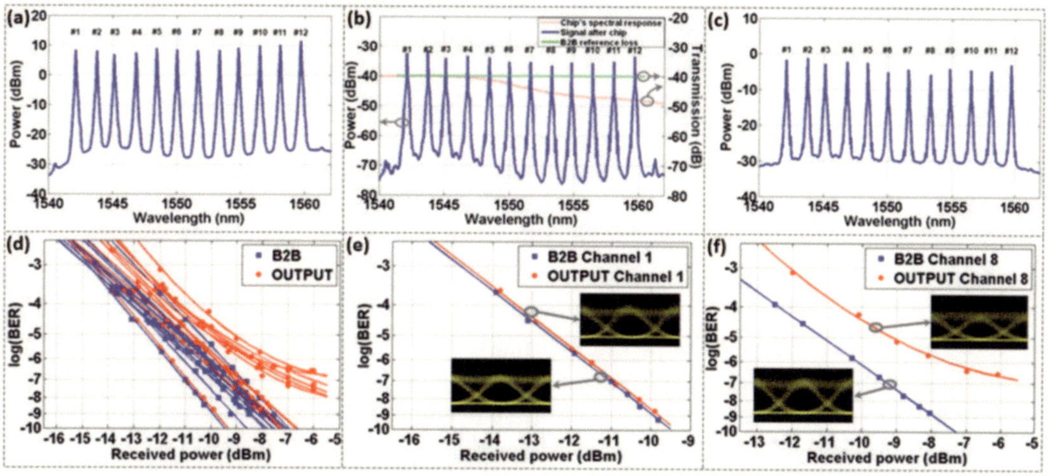

Figure 9. Spectra of the 12×40Gb/s WDM signal (a) at chip's input, (b) at chip's output, in comparison with chip's spectral response and B2B flat losses, (c) after post-chip amplification, (d) BER curves for all 40Gb/s B2B and transmitted channels, (e) BER curves for B2B and transmitted channel 1, (f) BER curves for B2B and transmitted channel 8.

The 12-channel data signal is examined, in terms of its spectrum, in three crucial places along the transmission line: just before the Si-DLSPP chip's input, after the chip's output and after the receiver's EDFA, as depicted in Figure 9(a)-(c). The TM grating couplers of the chip reveal a non-flattened spectral response (red dashed line in Figure 9(b)), dictating thereby the chip's spectral response and resulting in wavelength-dependent fiber-to-fiber transmission losses which range between 40dB and 48dB within the wavelength window of interest. As a consequence, the spectrum of the WDM signal constantly alters, even after the final amplification stage due to the absence of gain flattened filters (GFFs) at the EDFAs. To this end, a wavelength-dependent performance is observed in the receiver with respect to the optical noise-to-signal ratio (OSNR). The performance of the 480Gb/s WDM transmission through the 60μm-long DLSPP waveguide is evaluated via BER measurements that are shown in Figure 9(d)-(f). In particular, Figure 9(d) illustrates an overview of the BER curves obtained for all 40Gb/s discrete channels. Six out of the twelve channels (channels #1-5, 12) perform error-free, exhibiting a power penalty in a range between 0.2dB and 1dB for a 10^{-9} BER value against back-to-back (B2B) measurements. However, the remaining six channels (channels #6-11) reveal an error-floor at ~10^{-7} BER. Figure 9(e) corresponds to the best performing channel #1 that yields 0.2dB power penalty, whilst Figure 9(f) refers to the worst performing channel #8 that reveals an error-floor at 10^{-7} BER value. The B2B measurements are generated by replacing the hybrid chip by a variable attenuator that induces constant, wavelength-independent losses equal to the losses experienced by channels 1-4 that reside in the 1542-1548 nm spectral region (flat reference loss with green line in Figure 9(b)). The different performances between the channels transmitted via the hybrid chip originate from the unequal OSNR distribution due to the wavelength-dependent spectral responses of the grating couplers and the EDFAs. It should be noted that the error-floor in channels #1-5, 12 can be eliminated by selecting their spectral position to reside within the almost flat low-loss chip's response and by the employment of EDFAs with GFFs.

2.7. WDM data traffic switching with a hybrid Si-plasmonic A-MZI

After verifying the WDM data transmission capabilities of plasmonics through the DLSPP waveguide, the next step towards enriching the WDM portfolio of plasmonics is the demonstration of data switching [25]. In this perspective, the TO electrically controlled PMMA-based A-MZI that analyzed in Section 2.5 is employed towards performing switching of a 4×10Gb/s WDM traffic according to the rationale described in the same section. This A-MZI, along with other structures, is included in a Si-plasmonic chip [25] that is equipped with TM grating couplers, similar to the ones presented in the previous section, in order to couple light in and out of the hybrid chip. At 1542nm TM-polarized light the cut-back measurements reveal 0.1dB/μm and 4.4dB/cm plasmonic and silicon propagation losses as well as 2.5dB and 13dB coupling losses for each Si-to-DLSPP and TM grating coupling interface, respectively.

Before proceeding to the WDM approach, the Si-DLSPP A-MZI is tested in static and single-channel conditions regarding its electrical requirements and switching performance. Figure 10(a) shows the static TO transfer functions for the CROSS and BAR output ports of the A-MZI. The input signal is a 1542nm CW light beam with 6dBm optical power and the control medium is a direct current (DC) that drives the upper MZI arm and takes values up to 40mA. As it is obvious from this figure, the extinction ratio (ER) values for the CROSS and BAR ports are 14dB and 0.9dB when the DC reaches its maximum value. This unbalanced behavior between the two output ports stems from the 95:5, instead of 50:50, Si input/output couplers of the MZI as the result of unfortunate design error. Moreover, by comparing the obtained TO transfer functions with the theoretical transfer functions of a symmetric MZI, the default biasing point of the A-MZI is estimated at ~70°. Following the same procedure, the phase induced by the 40mA DC is ~-90° that corresponds to an increase in temperature by ~60K. These phase conditions in both MZI arms imply a total phase difference of ~160° between the two signal components that travel through the two branches. In terms of power requirements, the hybrid A-MZI consumes ~13.1mW when it is driven by 40mA DC, considering that the resistance of the 60μm-long DLSPP phase arm is found to be 8.2Ω.

After the initial static and single-channel switching characterization, the Si-DLSPP A-MZI is evaluated in terms of switching under dynamic and WDM data conditions. This evaluation is realized by using the experimental setup of Figure 10(b). Four CW optical signals emitted by DFB sources at 1545.1nm, 1546.7nm, 1547.7nm and 1549.1nm wavelengths are combined in pairs by 3-dB couplers in a way that channels 1 and 3 constitute the first pair and channels 2 and 4 form the other one. Subsequently, each channel pair is modulated by a 2^{31}-1 PRBS at 10Gb/s NRZ line rate in a corresponding Ti:$LiNbO_3$ MZM. The final 4×10Gb/s WDM signal is formed by using a 3-dB coupler to multiplex the two pairs of data sequences and is amplified by a high-power EDFA that provides 31dBm power at the input of the electrically controlled hybrid A-MZI. A polarization controller is used again towards establishing the required TM polarization conditions of the incoming WDM signal. Exploiting the TO effect, the switching state of the A-MZI is controlled dynamically by a pulse generator operating at 20KHz. After its transmission through the A-MZI, the multi-wavelength data signal is amplified by a two stage EDFA that comprises a 5nm midstage OBPF for out-of-band amplified spontaneous emission (ASE) noise rejection. The amplified WDM stream is then demultiplexed into its constituent channels by a 0.8nm OBPF. Each optical data channel is launched into a photoreceiver with 10GHz 3-dB bandwidth for OE conversion. Subsequently, the electrical signal exiting the photoreceiver is then received by a 10Gb/s error-detector for evaluation via BER measurements.

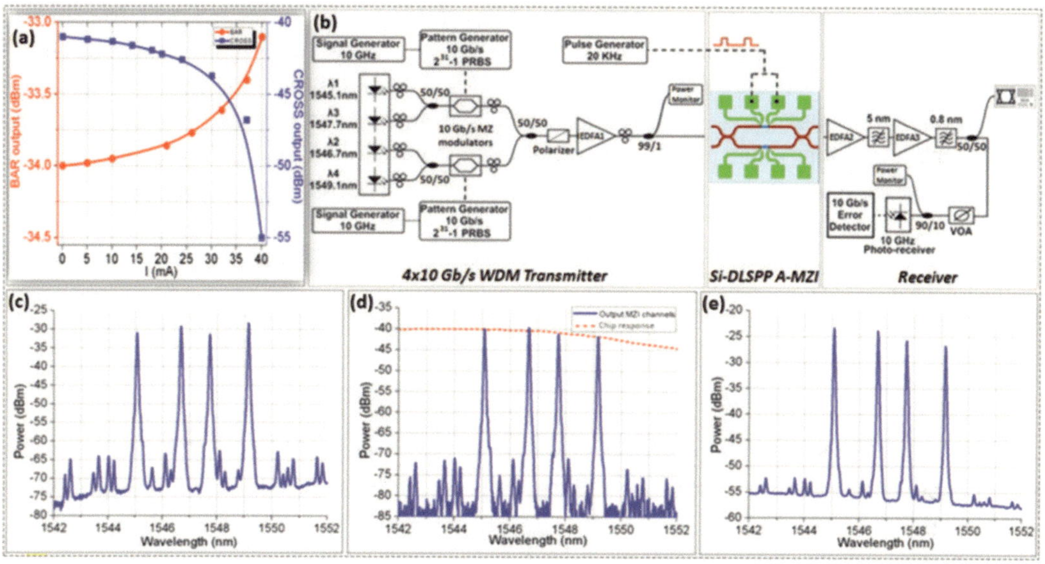

Figure 10. (a) Static TO transfer functions for the CROSS and BAR output ports of the A-MZI, (b) Experimental setup and the 4-channel spectrum at (c) MZI input, (d) directly at the MZI output before entering EDFA2, (e) after the receiver's pre-amplification stage. The spectral response of the chip including the A-MZI and the TM grating couplers is shown with the red dashed line in (d).

As depicted in Figure 10(c)-(e), the spectrum of the 4-channel data signal is observed at three places along the transmission line, that is just before the MZI's input, after the MZI's output and after the pre-amplifier (EDFA2) in the receiver. The high-power signal propagating in the fiber link between EDFA1 and the A-MZI causes the generation of four-wave mixing (FWM) terms, as it is easily noticeable in these figures. Nevertheless, by selecting unequal channel spacing ranging between 125GHz and 200GHz in the source signals, these terms are out-of-band compared to the data channels, giving the opportunity to be removed by subsequent filters. The energy transfer to the FWM components as well as the absence of GFFs after the transmitter's EDFA form the unequal power profile of the WDM signal at the

MZI's input, as shown in Figure 10(c). Due to the spectral response of the chip's TM grating couplers, the WDM spectral power profile at the MZI's output is now different (Figure 10(d)). The same figure also depicts with red dashed line the spectral response of the chip including the A-MZI and the TM grating couplers, revealing a 3dB loss variation within the 1545-1549 nm spectral band. After the receiver's pre-amplification stage, different power levels and also different OSNR values between the four channels are observed, as a result of the gain profile of EDFA2 (Figure 10(e)).

The dynamic control of the Si-DLSPP A-MZI is achieved by driving its upper arm with electrical rectangular pulses of 15μs duration, 20KHz repetition rate and 40mA peak value. The obtained data traces and eye diagrams for channel 1 (λ1) and channel 2 (λ2) signals at the CROSS and BAR output ports of the A-MZI are illustrated in Figure 11(a)-(h), where the electrical control signal is also depicted with red dashed line. According to these figures, inverted mode operation with an ER close to 14dB is attained at the CROSS port. In contrast, only 0.9dB ER performance is recorder at the BAR port as a result of the 95:5 Si couplers of the A-MZI. In terms of response time, the Si-DLSPP A-MZI exhibits fast rise and fall times lying in the 3-5 μs range, as it has been also demonstrated in a 90μm-long PMMA-based A-MZI [25]. It should be noted that similar results for both CROSS and BAR output ports are also obtained for data channels 3 and 4 at λ3 and λ4 wavelengths, respectively.

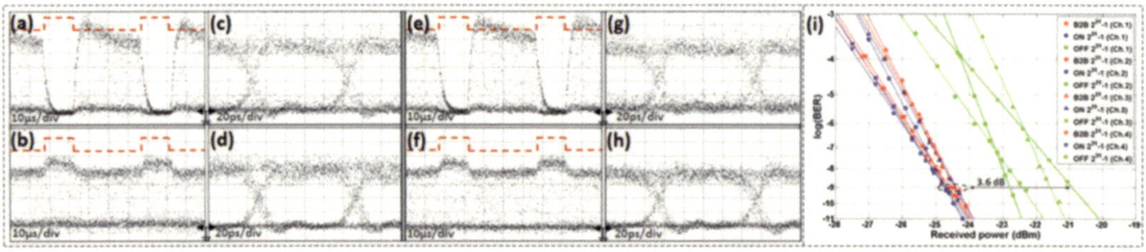

Figure 11. Modulation with 15μs electrical rectangular pulses at 20KHz repetition rate for 10Gb/s: (a) data trace at the CROSS port (channel 1), (b) data trace at the BAR port (channel 1), (c) eye diagram at the CROSS port (channel 1), (d) eye diagram at the BAR port (channel 1), (e) data trace at the CROSS port (channel 2), (f) data trace at the BAR port (channel 2), (g) eye diagram at the CROSS port (channel 2), (h) eye diagram at the BAR port (channel 2), (i) BER curves for ON/OFF operation of the A-MZI.

The performance of the 4×10Gb/s WDM switching via the 60μm-long PMMA-based A-MZI in ON and OFF switching states is evaluated via BER measurements that are depicted in Figure 11(i). In this experiment, the B2B measurements are obtained by using a straight Si waveguide, which is also included in the hybrid chip, as a reference point. In this way, only the signal degradation through the A-MZI is taken into consideration, excluding the lossy input/output TM grating couplers of the chip. Regarding the ON state, a DC value of 40mA is applied to the upper MZI branch, recording at the same time the BER measurements for the BAR output port. In case of no driving current, namely the OFF state, the BER curve is obtained at the CROSS output port of the MZI. As it is shown in Figure 11(i), the ON state reveals a negative power penalty close to the statistical error of 0.2dB for all four channels compared to the B2B curve. However, when operating at OFF state, the four channels exhibit power penalties in a range between 1.7dB and 3.6dB at a 10^{-9} BER value. The enhanced power penalty values compared to the corresponding ON state performance originate mainly from the 8dB lower power level received at MZI's CROSS output (-26dBm) against its BAR port during ON state (-18dBm). Additionally, the power penalty shows a clear tendency to increase with the channel wavelength, due to the wavelength-dependent gain and OSNR profile experienced by the four channels during amplification in EDFA2. Finally, the different slopes observed between the BER graphs of channels 1, 3 and the BER curves of channels 2 and 4 are caused due to the different modulators per channel pair at the 4×10Gb/s transmitter end.

2.8. Switching performance metrics

A clear view about the switching credentials and advantages of plasmonics over the SOI and the polymer-based TO waveguide technology platforms can be obtained by comparing performance metrics regarding the power consumption, switching time and active phase arm length. Table 1 provides an overview of the respective performance metrics reported for different SOI and polymer-based TO MZI switches. The power consumption × switching time metric is also included in this table, since it has been commonly used for the comparison of TO switching elements in applications where both low energy and fast operation are a prerequisite for high performance [55]. This table verifies the advantages of plasmonic technology: The Si-DLSPP A-MZI switch attains the shorter active region and the smallest power-time product over every type of undoped thermo-optically addressed MZI switch. These findings come as a natural consequence of the strong confinement of the DLSPP mode at the metal-dielectric interface, the inherent instantaneous heating of the underlying metallic layer and the asymmetric configuration of the MZI. On the contrary, in SOI TO structures, where the electrode is located at certain distance from the Si waveguide, it is very difficult to achieve simultaneously ultra-low power consumption and response time values. The required power can be greatly reduced by using sophisticated waveguide

Active waveguide technology [reference]	Phase arm length (active region in μm)	Power consumption (P in mW)	Switching time (τ in μs)	Power-time product ($P \times \tau$ in mW·μs)
SOI [58]	200	20	2.8	56
SOI [56]	1000	0.49	144	70.56
SOI [57]	100	0.54	141	76.14
SOI [61]	700	50	3.5	175
SOI [62]	6300	6.5	14	91
Polymer [55]	300	1.85	700	1295
Polymer [59]	100	4	200	800
PMMA-loaded SPP [Current work]	60	13.1	3.8	49.78

Table 1. Comparison with other SOI and polymer-based TO MZI switches.

engineering (e.g. suspended waveguides), thereby leading to sub-mW switching but at the expense of increased switching times and poor mechanical stability [56],[57]. A very good compromise between power consumption and response time is offered by using integrated NiSi waveguide heaters in single driving SOI-based TO switches [58]. Compared to the latter switches, lower power requirements but higher switching times are observed in polymer-based switches [55],[59]. It should be mentioned that TO switches that use the waveguide itself as the resistive heater and exploit differential driving schemes [60] have been excluded from this study to keep the same conditions for comparison.

2.9. Optimization procedures

The performance of the Si-DLSPP A-MZI switch regarding its BAR and CROSS output ports is determined by the splitting ratio of its Si coupling stages. To this end, the incorporation of 50:50 couplers, instead of the considered 95:5, can greatly improve the switching performance, yielding more than 20dB ER for both output ports. Another aspect that can be optimized is the switching time of this hybrid interferometric element. As it is already mentioned, the adoption of differential driving schemes for actively controlling both of the DLSPP MZI branches has the potential for lower time responses, even in the order of sub-μs [60]. Moreover, the power consumption in switching can be also reduced. One way is to form a kind of "suspended" plasmonic waveguides by etching the BOX layer under the gold film in order to achieve better confinement of the heat in the DLSPP waveguides and therefore enhanced power efficiency. This technique could be also exploited in the SOI MUX devices for energy-efficient thermal tuning of the RRs [63]. Another approach for bringing down the consumed power is the utilization of polymer materials with higher TOC value instead of PMMA loadings. For example, Cycloaliphatic acrylate (Cyclomer) exhibits almost three times higher TOC than PMMA and has been already successfully applied as the polymer loading in DLSPP switching structures [34],[38]. Though yet not optimized for low-energy switching operation, this route is expected to significantly decrease the required energy levels without compromising the switching performance. Alternatively, this scheme can be used to reduce the footprint and as such the losses of the A-MZI device. Considering the same amount of electric current to that used in the aforementioned WDM switching experiment, the length of the active plasmonic regions can be decreased from 60μm to 20μm, resulting in 4dB lower plasmonic propagation losses. Besides, the Cyclomer loadings can boost the TO tuning performance of plasmonic resonant devices compared to respective PMMA-based structures [31],[33] when similar temperature changes occur [38].

Apart from the hybrid A-MZI switch, optimization acts should be carried out for the TM grating couplers that are responsible for coupling light in and out of the chips. The grating couplers employed in the WDM data transmission and switching experiments exhibit 24dB overall insertion losses at best that are too high for practical datacom applications where the employment of amplifiers is prohibited. Towards tackling this issue, new TM grating couplers are considered for a far better efficiency. These coupling structures, filled with Spin-on-Glass (SOG) of 800nm height, rely on a fully etched approach with 0.8 filling factor,

0.71μm grating period, 0.13μm groove width and 10 degrees incident angle of the light. Figure 12(a) presents a layout of the new grating coupler's schematic cross section and Figure 12(b) depicts the SEM image of the fabricated one. These optimized TM grating couplers reveal 3.25dB minimum coupling loss at 1557nm and about 32nm 3-dB bandwidth. The blue line in Figure 12(c) presents also the results obtained from simulation that indicate very good agreement between theory and experiment. These results ensure that the new TM grating coupler is a major achievement since it decreases the router's overall optical losses by 17.5dB.

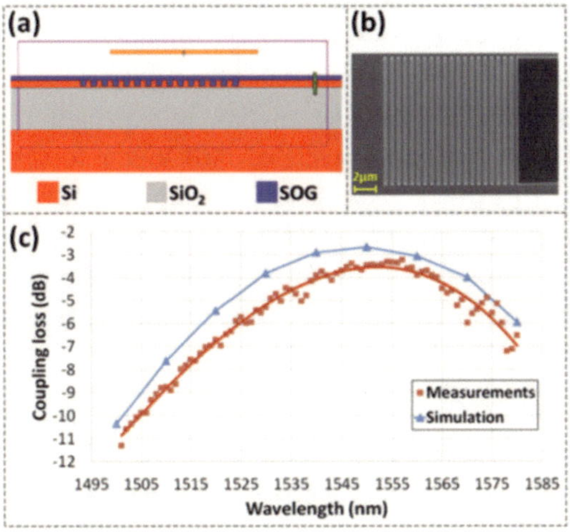

Figure 12. TM grating coupler: (a) Schematic cross sectional view, (b) SEM image, (c) Spectral responses of the simulation and experimental results.

2.10. Router performance in terms of optical insertion losses and power consumption

As a NoC solution for Data Centers and HPC systems, the hybrid Router-on-Chip platform should consume low power towards "greener" computing environments and induce low losses towards intrachip point-to-point connection without using on-chip amplifiers.

Within this framework, an estimation of the total power consumption of the entire 4×4 silicon-plasmonic router is provided here based on the characterization and experimental results obtained from the router's components and subsystems. Considering 60μm-long PMMA-phase arms in the branches of the hybrid A-MZIs that require 40mA driving current, the hybrid router is expected to consume at maximum ~2W. This power consumption stems from 1mW in the four PDs, 335mW in the four frontend SOI MUX devices due to the thermal tuning of their RRs and ~1.67W in the IC that is capable of providing at the same time two differential outputs with 48mA current. Therefore, the electronic IC is the router's component that consumes the largest part of power whereas the optical components reveal a power consumption of only ~400mW, considering the 13mW power requirement of each hybrid A-MZI. In comparison with current multicore central processing units (CPUs) for servers and HPCs that consume 150W [64] and 55W [65] respectively, the hybrid router requires only 2W, rendering this technology ideal for such shrinked networking environments in terms of energy savings. Keeping in mind that the Si-DLSPP router has an aggregate throughput capability of 1.12Tb/s, the power efficiency of this hybrid routing platform is about 1.8mW/Gb/s that is a very low value compared to other hybrid OE routers [66].

The optimization procedures of the previous section could lead to significant reductions in the total power consumption of the silicon-plasmonic router. The usage of Cyclomer, instead of PMMA, loadings in the phase arms of the A-MZIs can bring the IC's power consumption down to ~600mW. Consequently, the total power consumption is limited to ~1W. Keeping the power consumption in the original level of 2W, the same switching performance in the 4×4 DLSPP switching matrix is expected by employing 20μm-long Cyclomer-based A-MZIs but with the profit of 8dB lower losses in the router. Towards minimizing the aggregate power consumption, the approach of suspended arms for the fabrication of the RRs composing the SOI MUXs can lead to increased tuning efficiency that results in ~30mW for the whole multiplexing circuitry and therefore in ~630mW for the entire routing platform. As a result, an outstanding performance with lower than 0.6mW/Gb/s power efficiency can be achieved.

The total insertion losses of the silicon-plasmonic router can be also estimated. The aggregate losses across an input-to-output route are expected to be ~31dB that are analyzed in 6.5dB for the input and output grating couplers, 2dB for the SOI MUX device, 0.5dB for a 90:10 monitor coupler, ~8dB for the silicon waveguides, 10dB for the four Si-to-DLSPP interfaces and 4dB for the two 20μm-long Cyclomer-loaded SPP branches at the A-MZIs. However, the losses concerning the Si waveguide interconnections can be quite lower after finalizing the router prototype, since the value of 8dB relies on first layouts, where the building blocks and their interconnections are not placed in an optimal -in terms of spatial density- way. Nevertheless, even the 31dB losses of the original design can be manageable with high-power vertical-cavity surface-emitting lasers (VCSELs) [67] and high sensitivity Si-PDs [68].

3. Conclusion

Power consumption and size appear as the main set of barriers in next-generation Data Center and HPC environments. Within this framework, the penetration of optics into shrinked networking environments for chip-scale interconnects is now more vital than ever. SOI technology has already demonstrated its low-cost and high-integration credentials towards supporting fundamental operations required for optical interconnect applications. At the same time, the emerging discipline of plasmonics appears as a promising candidate for further size minimization and power savings. To this end, the idea of merging plasmonics and silicon photonics into the same platform seems to be a great solution for the implementation of faster and "greener" NoC environments. Deriving the strengths of each technology, a high-throughput, energy efficient and compact hybrid "Router-on-chip" is feasible. In this perspective, we present a 4×4 silicon-plasmonic router architecture for chip-scale applications with 1.12Tb/s aggregate bandwidth. The hybrid router relies on a SOI motherboard that incorporates Si waveguides for low-loss interconnection of SOI MUXs, PDs and the remaining subsystems, low-energy plasmonic waveguides in active A-MZI switches and an IC control unit for intelligent decision-making operations. Towards the demonstration of the router's functionality, we proceed to experimental evidences that turn the promises of plasmonics into real system-level application benefits: The transmission of a 12×40Gb/s WDM stream through a Si-DLSPP waveguide and the switching of a 4×10Gb/s WDM signal via a Si-DLSPP A-MZI. With the lowest reported power consumption × time response product among undoped SOI and polymer-based TO MZI switches, plasmonic technology becomes suitable for on-chip optical interconnects. According to experimental results, the hybrid router is estimated to consume ~2W with only ~400mW corresponding to the optical parts, yielding a power efficiency of 1.8mW/Gb/s. Moreover, the router is expected to induce about 31dB optical losses. With these power and optical loss characteristics, the Si-DLSPP routing platform seems to be appropriate for NoC environments where fast path establishment and route reconfiguration is necessary for efficient traffic management in Data Centers and HPCs.

Author details

Sotirios Papaioannou
Corresponding Author
Informatics & Telematics Institute, Center for Research & Technology Hellas, Thessaloniki, Greece
Department of Informatics, Aristotle University of Thessaloniki, Thessaloniki, Greece

Konstantinos Vyrsokinos
Informatics & Telematics Institute, Center for Research & Technology Hellas, Thessaloniki, Greece

Dimitrios Kalavrouziotis, Giannis Giannoulis,
Dimitrios Apostolopoulos and Hercules Avramopoulos
School of Electrical & Computer Engineering, National Technical University of Athens, Athens, Greece

Filimon Zacharatos, Karim Hassan, Jean-Claude Weeber,
Laurent Markey and Alain Dereux
Institut Carnot de Bourgogne, University of Burgundy, Dijon, France

Ashwani Kumar and Sergey I. Bozhevolnyi
Institute of Sensors, Signals,and Electrotechnics, University of Southern Denmark, Odense M, Denmark

Alpaslan Suna, Oriol Gili de Villasante and Tolga Tekin
Fraunhofer IZM, Berlin, Germany

Michael Waldow
AMO Gesellschaft für Angewandte Mikro- und Optoelektronik GmbH, Aachen, Germany

Odysseas Tsilipakos, Alexandros Pitilakis and Emmanouil E. Kriezis
Department of Electrical & Computer Engineering, Aristotle University of Thessaloniki, Thessaloniki, Greece

Nikos Pleros
Informatics & Telematics Institute, Center for Research & Technology Hellas, Thessaloniki, Greece
Department of Informatics, Aristotle University of Thessaloniki, Thessaloniki, Greece

Acknowledgement

This work was partially supported by the European FP7 ICT-PLATON (ICT- STREP no. 249135) project.

4. References

[1] Miller D. A. B. Device Requirements for Optical Interconnects to Silicon Chips. Proceedings of the IEEE 2009; 97(7) 1166-1185.
[2] Qian D. et al. 101.7-Tb/s (370×294-Gb/s) PDM-128QAM-OFDM Transmission over 3×55-km SSMF using Pilot-based Phase Noise Mitigation. Proceedings of OFC/NFOEC 2011, Paper PDPB5, 6-10 March 2011, Los Angeles, CA, USA.
[3] Taira Y. et al. High Channel-Count Optical Interconnection for Servers. Proceedings of ECTC 2010; 5490959, 1-4 June 2010, Las Vegas, Nevada, USA.
[4] Kash J. A. et al. Optical Interconnects in Exascale Supercomputers. Proceedings of IEEE Photonics Society 2010, 483-484, 7-11 November 2010, Denver, CO, USA.
[5] http://www.igigroup.net/download/FFS09/0930_Benner_IBM.pdf.
[6] http://www.datacenterknowledge.com/archives/2011/06/20/new-top-500-champ-the-k-supercomputer/.
[7] Taubenblatt M. A. Optical Interconnects for High-Performance Computing. IEEE/OSA Journal of Lightwave Technology 2012; 30(4) 448-458.
[8] Paniccia M. Integrating Silicon Photonics. Nature Photonics 2010; 4(8) 498-499.
[9] Barnes W. L., Dereux A., Ebbesen T. W. Surface plasmon subwavelength optics. Nature 2003; 424(6950) 824-830.
[10] Atwater H. A. The promise of plasmonics. Scientific American 2007; 296(4) 38-45.
[11] Brongersma M. L., Shalaev V. M. The case for plasmonics. Science 2010; 328(5977) 440-441.
[12] Zia R., Schuller J. A., Chandran A., Brongersma M. L. Plasmonics: the next chip-scale technology. Materials Today 2006; 9(7-8) 20-27.
[13] Papaioannou S. et al. A 320Gb/s-throughput capable 2×2 silicon-plasmonic router architecture for optical interconnects. IEEE/OSA Journal of Lightwave Technology 2011; 29(21) 3185-3195.
[14] Kim J. T. et al. Chip-to-chip optical interconnect using gold long-range surface plasmon polariton waveguides. Optics Express 2008; 16(17) 13133-13138.

[15] Raether H. Surface Plasmons on Smooth and Rough Surfaces and on Gratings. Berlin: Springer-Verlag; 1988.

[16] Gramotnev D. K., Bozhevolnyi S. I. Plasmonics beyond the diffraction limit. Nature Photonics 2010; 4(2) 83-91.

[17] Grandidier J. et al. Surface plasmon routing in dielectric-loaded surface plasmon polariton waveguides. Proceedings of SPIE 2008, 7033, 70330S-70330S-8, 10-14 August 2008, San Diego, CA, USA.

[18] Krasavin A. V., Zheludev N. I. Active plasmonics: Controlling signals in Au/Ga waveguide using nanoscale structural transformations. Applied Physics Letters 2004; 84(8) 1416-1418.

[19] Nikolajsen T., Leosson K., Bozhevolnyi S. I. Surface plasmon polariton based modulators and switches operating at telecom wavelengths. Applied Physics Letters 2004; 85(24) 5833-5835.

[20] Lereu A. L., Passian A., Goudonnet J.-P., Thundat T., Ferrell T. L. Optical modulation processes in thin films based on thermal effects of surface plasmons. Applied Physics Letters 2005; 86(15) 154101.

[21] Pacifici D., Lezec H. J., Atwater H. A. All-optical modulation by plasmonic excitation of CdSe quantum dots. Nature Photonics 2007; 1(7) 402-406.

[22] Pala R. A., Shimizu K. T., Melosh N. A., Brongersma M. L. A nonvolatile plasmonic switch employing photochromic molecules. Nano Letters 2008; 8(5) 1506-1510.

[23] MacDonald K. F., Sámson Z. L., Stockman M. I., Zheludev N. I. Ultrafast active plasmonics. Nature Photonics 2009; 3(1) 55-58.

[24] Krasavin A. V., Zayats A. V. Electro-optic switching element for dielectric-loaded surface plasmon polariton waveguides. Applied Physics Letters 2010; 97(4) 041107.

[25] Kalavrouziotis D. et al. First demonstration of active plasmonic device in true data traffic conditions: ON/OFF thermo-optic modulation using a hybrid silicon-plasmonic asymmetric MZI. Proceedings of OFC/NFOEC 2012, OW3E.3, 4-8 March 2012, Los Angeles, CA, USA.

[26] Bozhevolnyi S. I., Erland J., Leosson K., Skovgaard P. M. W., Hvam J. M. Waveguiding in Surface Plasmon Polariton Band Gap Structures. Physical Review Letters 2001; 86(14) 3008-3011.

[27] Knight M. W., Grady N. K., Bardhan R., Hao F., Nordlander P., Halas N. J. Nanoparticle-mediated coupling of light into a nanowire. Nano Letters 2007; 7(8) 2346-2350.

[28] Bozhevolnyi S. I., Volkov V. S., Devaux E., Ebbesen T. W. Channel plasmon-polariton guiding by subwavelength metal grooves. Physical Review Letters 2005; 95(4) 046802.

[29] Holmgaard T., Bozhevolnyi S. I. Theoretical analysis of dielectric-loaded surface plasmon-polariton waveguides. Physical Review B 2007; 75(24) 245405.

[30] Briggs R. M., Grandidier J., Burgos S. P., Feigenbaum E., Atwater H. A. Efficient coupling between dielectric loaded plasmonic and silicon photonic waveguides. Nano Letters 2010; 10(12) 4851-4857.

[31] Giannoulis G. et al. Data transmission and thermo-optic tuning performance of dielectric-loaded plasmonic structures hetero-integrated on a silicon chip. IEEE Photonics Technology Letters 2012; 24(5) 374-376.
[32] Gosciniak J. et al. Thermo-optic control of dielectric-loaded plasmonic waveguide components. Optics Express 2010; 18(2) 1207-1216.
[33] Hassan K., Weeber J.-C., Markey L., Dereux A. Thermo-optical control of dielectric loaded plasmonic racetrack resonators. Journal of Applied Physics 2011; 110(2) 023106.
[34] Hassan K. et al. Thermo-optic plasmo-photonic mode interference switches based on dielectric loaded waveguides. Applied Physics Letters 2011; 99(24) 241110.
[35] Hassan K. et al. Characterization of thermo-optical 2×2 switch configurations made of dielectric loaded surface plasmon polariton waveguides for telecom routing architecture. Proceedings of OFC/NFOEC 2012, OW3E.5, 4-8 March 2012, Los Angeles, CA, USA.
[36] Randhawa S. et al. Performance of electro-optical plasmonic ring resonators at telecom wavelengths. Optics Express 2012; 20(3) 2354-2362.
[37] Tsilipakos O. et al. Interfacing Dielectric-Loaded Plasmonic and Silicon Photonic Waveguides: Theoretical Analysis and Experimental Demonstration. IEEE Journal of Quantum Electronics 2012; 48(5) 678-687.
[38] Kalavrouziotis D. et al. 0.48Tb/s (12×40Gb/s) WDM transmission and high-quality thermo-optic switching in dielectric loaded plasmonics. Optics Express 2012; 20(7) 7655-7662.
[39] Vivien L., Osmond J., Fédéli J., Marris-Morini D., Crozat P., Damlencourt J., Cassan E., Lecunff Y., Laval S. 42 GHz p.i.n Germanium photodetector integrated in a silicon-on-insulator waveguide. Optics Express 2009; 17(8) 6252-6257.
[40] Sheng Z., Liu L., Brouckaert J., He S., Van Thourhout D. InGaAs PIN photodetectors integrated on silicon-on-insulator waveguides. Optics Express 2010; 18(2) 1756-1761.
[41] Bravo-Abad J., Ippen E. P., Soljačić M. Ultrafast photodetection in an all-silicon chip enabled by two-photon absorption. Applied Physics Letters 2009; 94(24) 241103.
[42] Geis M. W. et al. Silicon waveguide infrared photodiodes with >35 GHz bandwidth and phototransistors with 50 AW-1 response. Optics Express 2009; 17(7) 5193-5204.
[43] Bogaerts W. et al. Silicon-on-Insulator Spectral Filters Fabricated With CMOS Technology. IEEE Journal of Selected Topics in Quantum Electronics 2010; 16(1) 33-44.
[44] Geng M. et al. Compact four-channel reconfigurable optical add-drop multiplexer using silicon photonic wire. Optics Communications 2009; 282(17) 3477-3480.
[45] Barwicz T. et al. Reconfigurable silicon photonic circuits for telecommunication applications. Proceedings of SPIE 2008; 6872, 68720Z-1-12.
[46] Klein E. J. et al. Reconfigurable optical add-drop multiplexer using microring resonators. IEEE Photonics Technology Letters 2005; 17(11) 2358-2360.
[47] Dong P. et al. Low power and compact reconfigurable multiplexing devices based on silicon microring resonators. Optics Express 2010; 18(10) 9852-9858.
[48] Xiao S., Khan M. H., Shen H., Qi M. Multiple-channel silicon micro-resonator based filters for WDM applications. Optics Express 2007; 15(12) 7489-7498.

[49] Dahlem M. S. et al. Reconfigurable multi-channel second-order silicon microring-resonator filterbanks for on-chip WDM systems. Optics Express 2011; 19(1) 306-316.

[50] Liu A. et al. Wavelength Division Multiplexing Based Photonic Integrated Circuits on Silicon-on-Insulator Platform. Journal of Selected Topics in Quantum Electronics 2010; 16(1) 23-32.

[51] Saeung P., Yupapin P. P. Generalized analysis of multiple ring resonator filters: Modeling by using graphical approach. Optik-Int Journal for Light and Electron Optics 2008; 119(10) 465-472.

[52] Pitilakis A., Kriezis E. E. Longitudinal 2×2 switching configurations based on thermo-optically addressed dielectric-loaded plasmonic waveguides. IEEE/OSA Journal of Lightwave Technology 2011; 29(17) 2636-2646.

[53] Krasavin A. V., Zayats A. V. Three-dimensional numerical modeling of photonic integration with dielectric-loaded SPP waveguides. Physical Review B 2008; 78(4) 045425.

[54] Tsilipakos O., Pitilakis A., Tasolamprou A. C., Yioultsis T. V., Kriezis E. E. Computational techniques for the analysis and design of dielectric-loaded plasmonic circuitry. Optical and Quantum Electronics 2011; 42(8) 541-555.

[55] Al-Hetar A. M., Mohammad A. B., Supa'at A. S. M., Shamsan Z. A. MMI-MZI polymer thermo-optic switch with a high refractive index contrast. Journal of Lightwave Technology 2011; 29(2) 171-178.

[56] Fang Q. et al. Ultralow power silicon photonics thermo-optic switch with suspended phase arms. IEEE Photonics Technology Letters 2011; 23(8) 525-527.

[57] Sun P., Reano R. M. Submilliwatt thermo-optic switches using free-standing silicon-on-insulator strip waveguides. Optics Express 2010; 18(8) 8406-8411.

[58] Van Campenhout J., Green W. M., Assefa S., Vlasov Y. A. Integrated NiSi waveguide heaters for CMOS-compatible silicon thermo-optic devices. Optics Letters 2010; 35(7) 1013-1015.

[59] Xie N., Hashimoto T., Utaka K. Very low-power, polarization-independent, and high-speed polymer thermooptic switch. IEEE Photonics Technology Letters 2009; 21(24) 1861-1863.

[60] Geis M. W., Spector S. J., Williamson R. C., Lyszczarz T. M. Submicrosecond Submilliwatt Silicon-on-Insulator Thermooptic Switch. IEEE Photonics Technology Letters 2004; 16(11) 2514-2516.

[61] Espinola R. L., Tsai M.-C., Yardley J. T., Osgood R. M. Fast and low-power thermooptic switch on thin silicon-on-insulator. IEEE Photonics Technology Letters 2003; 15(10) 1366-1368.

[62] Densmore A. et al. Compact and low power thermo-optic switch using folded silicon waveguides. Optics Express 2009; 17(13) 10457-10465.

[63] Dong P. et al. Thermally tunable silicon racetrack resonators with ultralow tuning power. Optics Express 2010; 18(19) 20298-20304.

[64] http://ark.intel.com/products/64582/Intel-Xeon-Processor-E5-2687W-(20M-Cache-3_10-GHz-8_00-GTs-Intel-QPI).

[65] https://www.power.org/home/Blue_Gene_Q_Super_Efficient_Linley.pdf.

[66] Takahashi R. et al. Hybrid Optoelectronic Router for Future Energy-efficient, Large-capacity, and Flexible OPS Networks. Proceedings of IEEE Photonics Society Summer Topical Meeting Series 2011; 153-154, 18-20 July 2011, Montreal, QC, Canada.

[67] Hofmann W. et al. 44 Gb/s VCSEL for optical interconnects. Proceedings of OFC/NFOEC 2011, Paper PDPB5, 6-10 March 2011, Los Angeles, CA, USA.

[68] Geis M. W. et al. CMOS-Compatible All-Si High-Speed Waveguide Photodiodes With High Responsivity in Near-Infrared Communication Band. IEEE Photonics Technology Letters 2007; 19(3) 152-154.